10-10-12
#144

P9-DDI-647

JOHNNIE MAE BERRY LIBRARY
CINCINNATI STATE
3520 CENTRAL PARKWAY
CINCINNATI, OH 45223-2690

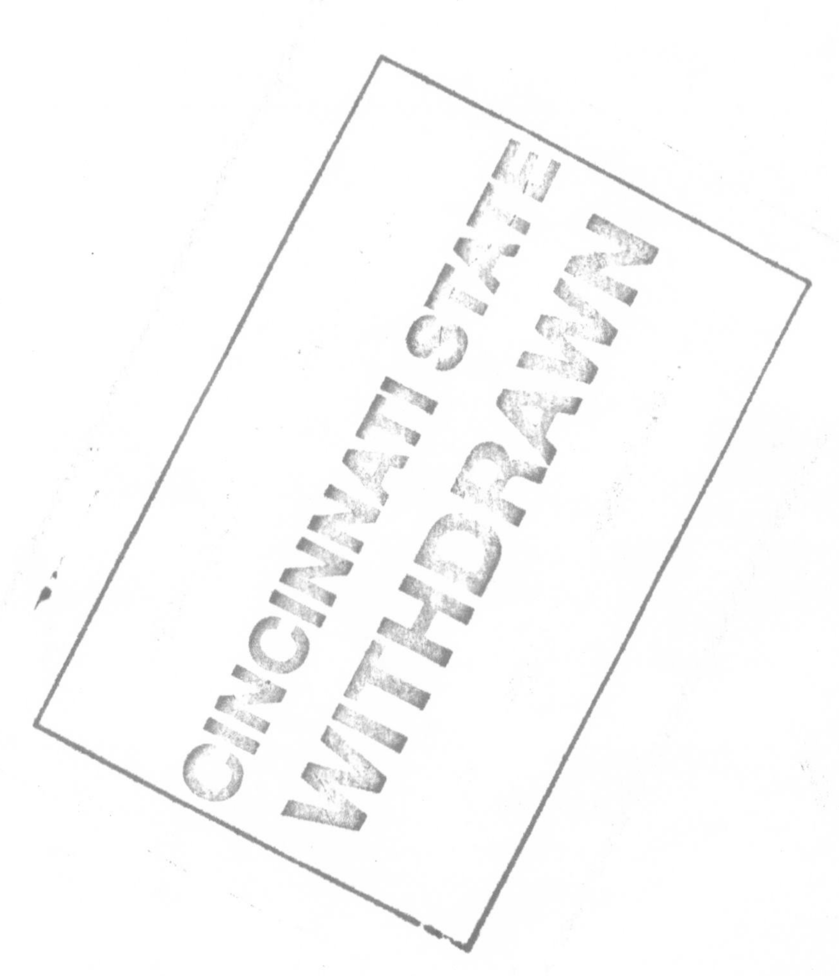

Food and Culture

394.120973
K62
2012

6TH

Food and Culture

Pamela Goyan Kittler, M.S.
Food, Culture, and Nutrition Consultant

Kathryn P. Sucher, Sc.D., R.D.
Department of Nutrition and Food Science San Jose State University

Marcia Nahikian-Nelms, Ph.D., R.D., L.D., CNSC
Medical Dietetics, College of Medicine, The Ohio State University

WADSWORTH
CENGAGE Learning™

Australia • Brazil • Japan • Korea • Mexico • Singapore • Spain • United Kingdom • United States

Food and Culture, Sixth edition
Pamela Goyan Kittler, Kathryn P. Sucher, Marcia Nahikian Nelms

Executive Editor: Yolanda Cossio

Senior Acquisitions Editor: Peggy Williams

Senior Development Editor: Shelley Parlante

Assistant Editor: Elesha Feldman

Senior Marketing Manager: Laura McGinn

Senior Marketing Communications Manager: Linda Yip

Content Project Management: PreMediaGlobal

Art Director: John Walker

Print Buyer: Rebecca Cross

Senior Rights Acquisition Specialist (Image, Text): Dean Dauphinais

Production Service: PreMediaGlobal

Cover Designer: Riezebos Holzbaur / Tae Hatayama

Cover Image: Assorted Spices in Pots (Getty Image, Photographer: Gentl & Hyers); First row—Family Having Dinner (Getty Image, Photographer: Fuse); Second row—Family of four having a meal at home (Getty Image, Photographer: Asia Images Group); Third row—Family Sharing a Meal (Getty Image, Photographer: Fuse)

Compositor: PreMediaGlobal

© 2012, 2008, 2004 Wadsworth, Cengage Learning

ALL RIGHTS RESERVED. No part of this work covered by the copyright herein may be reproduced, transmitted, stored, or used in any form or by any means graphic, electronic, or mechanical, including but not limited to photocopying, recording, scanning, digitizing, taping, Web distribution, information networks, or information storage and retrieval systems, except as permitted under Section 107 or 108 of the 1976 United States Copyright Act, without the prior written permission of the publisher.

For product information and technology assistance, contact us at
Cengage Learning Customer & Sales Support, 1-800-354-9706
For permission to use material from this text or product,
submit all requests online at **www.cengage.com/permissions**
Further permissions questions can be emailed to
permissionrequest@cengage.com

Library of Congress Control Number: 2011933853

ISBN-13: 978-0-538-73497-4

ISBN-10: 0-538-73497-3

Wadsworth
20 Davis Drive
Belmont, CA 94002-3098
USA

Cengage Learning is a leading provider of customized learning solutions with office locations around the globe, including Singapore, the United Kingdom, Australia, Mexico, Brazil and Japan. Locate your local office at **international.cengage.com/region**

Cengage Learning products are represented in Canada by Nelson Education, Ltd.

For your course and learning solutions, visit **www.cengage.com.**

Purchase any of our products at your local college storeor at our preferred online store **www.cengagebrain.com.**

Instructors: Please visit **login.cengage.com** and log in to access instructor-specific resources.

Printed in the United States of America
1 2 3 4 5 6 7 15 14 13 12 11

CONTENTS

PREFACE

The population of the United States is increasingly heterogeneous, moving toward a plurality of ethnic, religious, and regional groups. Each of these groups has traditional food habits that differ—slightly or significantly—from the so-called typical American majority diet. Effective nutrition counseling, education, and food service require that these variations be acknowledged and understood within the context of culture. It is our goal to provide dietitians, nutritionists, and food service professionals with the broad overview needed to avoid ethnocentric assumptions and the nutrition specifics helpful in working with each group discussed. We have attempted to combine the conceptual with the technical in a way that is useful to other health professionals as well.

We would like to draw attention specifically to the area of nutrition counseling: "In nutrition counseling, where many therapeutic interventions are on a personal level, sensitivity to the strong influence of culture on an individual's food intake, attitudes, and behaviors is especially imperative. . . . Multicultural competence is not a luxury or a specialty but a requirement for every registered dietitian" (Curry, 2000, p. 1,142). A model recommended for multicultural nutrition competencies specifically lists (Harris-Davis & Haughton, 2000):

1. Understand food selection, preparation, and storage with a cultural context.
2. Have knowledge of cultural eating patterns and family traditions such as core foods, traditional celebrations, and fasting.
3. Familiarize self with relevant research and latest findings regarding food practices and nutrition-related health problems of various ethnic and racial groups.
4. Possess specific knowledge of cultural values, health beliefs, and nutrition practices of particular groups served, including culturally different clients.

This book offers information fundamental to these competencies.

How the Book Is Organized

The first four chapters form an introduction to the study of food and culture. Chapter 1 discusses methods for understanding food habits within the context of culture, changing demographics, and the ways in which ethnicity may affect nutrition and health status. Chapter 2 focuses on the role of diet in traditional health beliefs. Some intercultural communication strategies are suggested in Chapter 3, and Chapter 4 outlines the major eastern and western religions and reviews their dietary practices in detail.

Chapters 5 through 14 profile North American ethnic groups and their cuisines. We have chosen breadth over depth, discussing groups with significant populations in the United States, as well as smaller, more recent immigrant groups who have had an impact on the health care system. Other groups with low numbers of immigrants but notable influences on American cooking are briefly mentioned.

Groups are considered in the approximate order of their arrival in North America. Each chapter begins with a history of the group in the United States and current demographics. Worldview (outlook on life) is then examined, including religion, family structure, and traditional health practices. This background information illuminates the cultural context from which ethnic foods and food habits emerge and evolve. The next section of each chapter outlines the traditional diet, including ingredients, some common dishes, meal patterns, special occasions, the role of food in the society, and therapeutic uses of food. The final section explains the contemporary diet of the group, such as adaptations made by the group after arrival in the United States and influences of the group on the American diet. Reported nutritional status is reviewed, and general counseling guidelines are provided.

One or more cultural food group tables are found in each of the ethnic group chapters. The emphasis is on ingredients common to the populations of the region. Important variations within regions and unique food habits are listed in the "Comments" column of the table. Known adaptations in the United States are also noted. The tables are intended as references for the reader; they do not replace either the chapter content or an in-depth interview with a client.

Chapter 15 considers the regional American fare of the Northeast, the Midwest, the South, and the West. Each section includes an examination of the foods common in the region and general nutritional status. Canadian regional fare is also briefly considered. This chapter brings the study of cultural nutrition full circle, discussing the significant influences of different ethnic and religious groups on North American fare.

New to This Edition

- New Discussion Starters in every chapter prompt student thinking about the chapter's subject matter at all levels of Bloom's Taxonomy—that is, moving from recall and comprehension toward application, analysis, synthesis, and evaluation. Many of the Discussion Starters begin by asking for individual reflection on the chapter topic and end with small group work, intended to help students communicate their ideas with others, experience a diversity of ideas about diet and culture, and collaborate with others to find consensus and to improve critical thinking. Several chapters in the latter part of Food and Culture use very similar Discussion Starter prompts in order to allow students to compare their responses to the diets and culture of different parts of the world across chapters.
- Each chapter's Discussion Starter can be assigned with the reading of that chapter and completed in class or may simply be assigned in class. Some are broken up into separate parts, such that an instructor might use only part of some Discussion Starters.

Chapter Specific Changes

- Chapter 1. Food and Culture—Population data updated; Discussion Starter on intercultural communications.
- Chapter 2. Traditional Health Beliefs and Practices—Updated data on the use of complementary and alternative medicine (CAM); Discussion Starter on the role of health care in your worldview.
- Chapter 3. Intercultural Communications—Discussion Starter on how a health practitioner might improve relations with clients/patients.
- Chapter 4. Food and Religion—Updated demographics data on religious affiliation in the U.S.; Discussion Starter on the effect of religion on food habits.
- Chapter 5. Native Americans—Updated U.S. Census data on Native American population and other demographics; updated information on current diets, nutritional status, and medical disorders related to diet and nutrition; Discussion Starter on the traditional diets of Native Americans and the possible obstacles for modifying their current diet.
- Chapter 6. Northern and Southern Europeans—Updated U.S. Census population and other demographics on European groups. Updated information on current diets, nutritional status, and medical disorders related to diet and nutrition; Discussion Starter on the adaptation of traditional European foods to accommodate vegetarianism.
- Chapter 7. Central Europeans. Updated U.S. Census population and other demographics on Central/Eastern European groups. Updated information on current diets, nutritional status, and medical disorders related to diet and nutrition; Discussion Starter on traditional ethnic foods in the U.S.
- Chapter 8. Africans. Updated U.S. Census population and other demographics on African Americans and more recent immigrant groups from Africa. Updated information on current diets, nutritional status, and medical disorders related to diet and nutrition; Discussion Starter on health concerns of African Americans.
- Chapter 9. Mexican and Central Americans—Updated U.S. Census population and other

demographics on Mexicans and Central American groups. Updated information on current diets, nutritional status, and medical disorders related to diet and nutrition; Discussion Starter on health education interventions for Central American immigrants.

- Chapter 10. Caribbean Islanders and South Americans—Updated U.S. Census population and other demographics on Caribbean and South American groups. Updated information on current diets, nutritional status, and medical disorders related to diet and nutrition; Discussion Starter on similarities and differences in health concerns of Caribbean and South American immigrant groups.
- Chapter 11. East Asians—Updated U.S. Census population and other demographics on East Asians groups. Updated information on current diets, nutritional status, and medical disorders related to diet and nutrition; Discussion Starter on steps to compare and contrast different aspects of the diet and culture of East Asian groups.
- Chapter 12. Southeast Asians and Pacific Islanders—Updated U.S. Census population and other demographics on East Asian groups. Updated information on current diets, nutritional status, and medical disorders related to diet and nutrition; Discussion Starter on activities to compare and contrast different aspects of the diet and culture of East Asian groups.
- Chapter 13. People of the Balkans and Middle East—Updated U.S. Census population and other demographics on Balkan and Middle Eastern groups. Updated information on current diets, nutritional status, and medical disorders related to diet and nutrition; Discussion Starter on activities to compare and contrast different aspects of the diet and culture of population groups from the Balkans and Middle East countries.
- Chapter 14. South Asians—Updated U.S. Census population and other demographics on South Asian groups. Updated information on current diets, nutritional status and medical disorders related to diet and nutrition; Discussion Starter focuses on development of hypotheses about the diet and food habits of South Asian groups with data that was collected from previous Discussion Starters.
- Chapter 15. Regional Americans—Updated U.S. Census regional population and other demographics. Updated information on current diets, nutritional status, and medical disorders related to diet and nutrition; Discussion Starter that guides students in identifying characteristics of their own regional identities and how this may affect their own health risks.

Before You Begin

Food is so essential to ethnic, religious, and regional identity that dietary descriptions must be as objective as possible to prevent inadvertent criticism of the underlying culture. Yet as members of two Western ethnic and religious groups, we recognize that our own cultural assumptions are unavoidable and, in fact, serve as a starting point for our work. One would be lost without such a cultural footing. Any instances of bias are unintentional.

Any definition of a group's food habits implies homogeneity in the described group. In daily life, however, each member of a group has a distinctive diet, combining traditional practices with new influences. We do not want to stereotype the fare of any cultural group. Rather, we strive to generalize common U.S. food and culture trends as a basis for understanding the personal preferences of individual clients.

We have tried to be sensitive to the designations used by each cultural group, though sometimes there is no consensus among members regarding the preferred name for the group. Also, there may be some confusion about dates in the book. Nearly all religious traditions adhere to their own calendar of events based on solar or lunar months. These calendars frequently differ from the Gregorian calendar used throughout most of the world in business and government. Religious ceremonies often move around according to Gregorian dates, yet usually they are calculated to occur in the correct season each year. Historical events in the text are listed according to the Gregorian calendar, using the abbreviations for before common era (BCE) and common era (CE).

We believe this book will do more than introduce the concepts of food and culture. It should also encourage self-examination and individual cultural identification by the reader. We hope that it will help dietitians, nutritionists, other health care providers, and food service professionals work effectively with members of different ethnic, religious, and regional groups. If it sparks a gustatory interest in the foods of the world, we will be personally pleased. De gustibus non est disputatum!

Acknowledgments

We are forever indebted to the many researchers, especially from the fields of anthropology and sociology, who did the seminal work on food habits that provided the groundwork for this book, and to the many nutrition professionals who have shared their expertise with us over the years. We especially want to thank the many colleagues who have graciously given support and advice in the development of the fifth edition: Carmen Boyd, MS, LPC, RD, Missouri State University; Arlene Grant-Holcomb, RD, MAE, California State Polytechnic University, San Luis Obispo; Carolyn Hollingshead, PhD, RD, University of Utah; Tawni Holmes, PhD, RD, University of Central Oklahoma; Claire G. Kratz, MS, RD, LDN, Montgomery County Community College; Yvonne Moody, EdD, Chadron State College; Sudha Raj, PhD, Syracuse University; Stacey A. Roush, MS, Montgomery County Community College; Dana Wassmer, MS, RD, Cosumnes River College; Bonny Burns-Whitmore, DrPH, RD, California State Polytechnic University, Pomona; and Donna M. Winham, DrPH, Arizona State University. We are grateful for the expertise of Gerald Nelms, PhD as his development of the discussion starters in this sixth edition of the text adds an important contribution to the pedagogy for this text.

CHAPTER 1

Food and Culture

What do Americans eat? Meat and potatoes, according to popular myth. There's no denying that for every person in the United States, an average of over half a pound of beef, pork, lamb, or veal is eaten daily, and more than one hundred pounds of potatoes (mostly as chips and fries) are consumed annually. Yet the American diet cannot be so simply described. Just as the population of the United States includes numerous cultural groups, the food habits of Americans are equally diverse. It can no more be said that the typical U.S. citizen is white, Anglo-Saxon, and Protestant than it can be stated that meat and potatoes are what this citizen eats.

U.S. Census and other demographic data show that one in every four Americans is of non-European heritage, and one in every ten residents is foreign born. Over 75 different ancestry groups were reported in 2007.[1] Even these figures underestimate the number and diversity of North American cultural groups. The data do not list members of some white ethnic populations nor those of religious or regional groups. Census terminology can be ambiguous and confusing for some respondents. For instance, the category Hispanic or Latino (defined as persons born in Latin America, whose parents were born in Latin America, who have a Spanish surname, or who speak Spanish) is considered an ethnicity, and counts Puerto Ricans (who are U.S. citizens) if they reside on the mainland but not if they live on the island. Furthermore, unauthorized immigrants, estimated to be more than 11 million residents, may not be included in the census. Thus, the proportion of American ethnic group members is larger than statistics indicate, and more important, it is rapidly increasing.

The fastest and largest growing ethnic groups in America are from Latin America. Data from 2007 indicate that more than 50 percent of all immigrants to the U.S. are from Latin America. Asians make up the second largest group at 17 percent.[2]

Each American ethnic, religious, or regional group has its own culturally based food habits. Many of these customs have been influenced and modified through contact with the majority culture and, in turn, have changed and shaped American majority food habits. Today, a fast-food restaurant or street stand is as likely to offer pizza, tacos, egg rolls, or falafel as it is hamburgers. It is the intricate interplay between food habits of the past and the present, the old and the new, and the traditional and the innovative that is the hallmark of the American diet.

What Is Food?

Food, as defined in the dictionary, is any substance that provides the nutrients necessary to maintain life and growth when ingested. When most animals feed, they repeatedly consume those foods necessary for their well-being, and they do so in a similar manner at each feeding. Humans, however, do not feed. They eat.

As suggested by their names, not even hamburgers and French fries are American in origin. Chopped beef steaks were introduced to the United States from the German city of Hamburg in the late nineteenth century. The American term *French-fried potatoes* first appeared in the 1860s and was probably coined to describe the method used in France for deep-frying potato pieces until crisp. Other foods considered typically American also have foreign origins, for example, hot dogs, apple pie, and ice cream.

Data from the 2006 Canadian census indicate more than 200 different ethnic origins were documented. The most common ethnic groups noted included English, French, Scottish, Irish, German, Italian, Chinese, North American Indian, Ukrainian, and Dutch. Newer groups include individuals from Montserrat in the Carribean and African countries such as Chad, Gabon, Gambia, and Zambia.[90]

© Tom McCarthy/PhotoEdit, Inc.

▲ *Humans create complex rules, commonly called manners, about how food is to be eaten.*

Eating is distinguished from feeding by the ways humans use food. Humans not only gather or hunt food, but they also cultivate plants and raise livestock. Agriculture means that some foods are regularly available, alleviating hand-to-mouth sustenance. This permits the development of specific customs associated with foods that are the foundation of the diet, such as wheat or rice. Humans also cook, softening tough foods, including raw grains and meats, and reducing toxic substances in other items, such as certain root vegetables. This greatly expands the number and variety of edible substances available. Choosing foods to combine with other foods follows, and prompts rules regarding what can be eaten with what and creating the meal. Humans use utensils to eat meals and institute complex rules, commonly called manners, about how meals are consumed. And, significantly, humans share food. Standards for who may dine with whom in each eating situation are well-defined.

The term *food habits* (also called food culture or foodways) refers to the ways in which humans use food, including everything from how it is selected, obtained, and distributed to who prepares it, serves it, and eats it. The significance of this process is unique to humankind. Why don't people simply feed on the diet of our primitive ancestors, surviving on foraged fruits, vegetables, grains, and the occasional insect or small mammal thrown in for protein? Why do people choose to spend their time, energy, money, and creativity on eating? The answers to these questions, according to some researchers, can be found in the basic biological and psychological constitution of humans.

It is thought that children are less likely than adults to try new foods, in part because they have not yet learned cultural rules regarding what is safe and edible. A child who is exposed repeatedly to new items loses the fear of new foods faster than one who experiences a limited diet.[11]

The Omnivore's Paradox

Humans are omnivorous, meaning that they can consume and digest a wide selection of plants and animals found in their surroundings. The primary advantage to this is that they can adapt to nearly all earthly environments. The disadvantage is that no single food provides the nutrition necessary for survival. Humans must be flexible enough to eat a variety of items sufficient for physical growth and maintenance, yet cautious enough not to randomly ingest foods that are physiologically harmful and, possibly, fatal. This dilemma, the need to experiment combined with the need for conservatism, is known as the omnivore's paradox.[3,4] It results in two contradictory psychological impulses regarding diet. The first is an attraction to new foods; the second is a preference for familiar foods. The food habits developed by a group provide a framework that reduces the anxiety produced by these opposing desires. Rules about which foods are edible, how they are procured and cooked safely, how they should taste, and when they should be consumed provide guidelines for both experimentation (based on previous experience with similar plants and animals or flavors and textures) and conservatism through ritual and repetition.

Self-Identity

The choice of which foods to ingest is further complicated, however, by another psychological concept regarding eating—the incorporation of food. Consumption is understood as equaling conversion of a food and its nutrients into a human body. For many people, incorporation is not only physical but associative as well. It is the fundamental nature of the food absorbed by a person, conveyed by the proverbial saying, "You are what you eat." In its most direct interpretation, it is the physical properties of a food expressed through incorporation. Some Asian Indians eat walnuts to improve their brain, and weight lifters may dine on rare meat to build muscle. In other cases, the character of the food is incorporated. Some Native Americans believe that because milk is a food for infants, it will weaken adults. The French say a person who eats too many turnips becomes gutless, and some Vietnamese consume gelatinized tiger bones to improve their strength.

It is a small step from incorporating the traits associated with a specific food to making assumptions about a total diet. The correlation between what people eat, how others perceive them, and how they characterize themselves is striking. In one study researchers listed foods typical of five diets: vegetarian (broccoli quiche, brown rice, avocado, and bean sprout sandwich), gourmet (oysters, caviar, French roast coffee), health food (protein shake, wheat germ, yogurt), fast food (Kentucky Fried Chicken, Big Mac, pizza), and synthetic food (Carnation Instant Breakfast, Cheez Whiz). It was found that each category was associated with a certain personality type. Vegetarians were considered to be pacifists and likely to drive foreign cars. Gourmets were believed to be liberal and sophisticated. Health food fans were described as antinuclear activists and Democrats. Fast food and synthetic food eaters were believed to be religious, conservative, and fond of polyester clothing. These stereotypes were confirmed by self-description and personality tests completed by people whose diets fell into the five categories.[5]

Another study asked college students to rate profiles of people based on their diets. The persons who ate "good" foods were judged thinner, more fit, and more active than persons with the identical physical characteristics and exercise habits who ate "bad" foods. Furthermore, the people who ate "good" foods were perceived by some students as being more attractive, likable, practical, methodical, quiet, and analytical than people who ate "bad" foods. The researchers attribute the strong morality–food effect to several factors, including the concept of incorporation and a prevailing Puritan ethic that espouses self-discipline.[6]

Food choice is, in fact, influenced by self-identity, a process whereby the food likes or dislikes of someone else are accepted and internalized as personal preferences. Research suggests that children choose foods eaten by admired adults (e.g., teachers), fictional characters, peers, and especially older siblings. Parents have little long-lasting influence. Group approval or disapproval of a food can also condition a person's acceptance or rejection. This may explain why certain relatively unpalatable items, such as chili peppers or unsweetened coffee, are enjoyed if introduced through socially mediated events, such as family meals or workplace snack breaks. Although the mechanism for the internalization of food preference and self-identity is not well understood, it is considered a significant factor in the development of food habits.[7,8] A study on the consumption of organic vegetables, for example, found that those who identified themselves as green (people who are concerned with ecology and make consumer decisions based on this concern) predicted an intention to eat organic items independent of other attitudes, such as perceived flavor and health benefits.[9]

Food as self-identity is especially evident in the experience of dining out. Researchers suggest that restaurants often serve more than food, satisfying both emotional and physical needs. A diner may consider the menu, atmosphere, service, and cost or value when selecting a restaurant; and most establishments cater to a specific clientele. Some offer quick, inexpensive meals and play equipment to attract families. Business clubs feature a conservative setting suitable for financial transactions, and the candlelit ambiance of a bistro is conducive to romance. The same diner may choose the first in her role as a mother, the second while at work, and the last when meeting a date. In Japan, restaurants serve as surrogate homes where company is entertained, preserving the privacy of family life. The host chooses and pays for the meal ahead of time, all guests are provided the same dishes, and the servers are expected to partake in the conversation. Ethnic restaurants appeal to those individuals seeking familiarity and authenticity in the foods of their homeland or those interested in novelty and culinary adventure. Conversely, exposure to different foods in restaurants is sometimes the first step in adopting new food items at home.[10]

Symbolic Use of Food

The development of food habits clearly indicates that for humans, food is more than just nutrients. Incorporation has meaning specifically because people are omnivores and have a choice regarding what is consumed. Humans use foods symbolically, due to relationship, association, or convention. Bread is an excellent example—it is called the staff of life; one breaks bread with friends, and bread represents the body of Christ in the Christian sacrament of communion. White bread was traditionally eaten by the upper classes, dark bread by the poor, but whole wheat bread

The inability to express self-identity through food habits can be devastating. A study of persons with permanent feeding tubes living at home or in nursing facilities found they frequently avoided meals with families and friends. They missed their favorite foods, but more important, they mourned the loss of their self-identities reinforced by these daily social interactions.[116]

Reprinted with permission from *All Over The Map* by David Jouris. Copyright 1994 by David Jouris, Ten Speed Press, Berkeley, CA. www.tenspeed.com.

Figure 1.1
An edible map—food-related names of cities and towns in the United States. Food often means more than simply nutrients.
Source: From *All Over the Map: An Extraordinary Atlas of the United States: Featuring Towns That Actually Exist!* by David Jouris, copyright © 1994 by David Jouris. Used by permission of Ten Speed Press, an imprint of Crown Publishing Group, a division of Random House, Inc.

is consumed today by persons concerned more with health than status. A person with money has "a lot of bread." In many cultures, bread is shared by couples as part of the wedding ceremony or left for the soul of the dead. Superstitions about bread also demonstrate its importance beyond sustenance. Greek soldiers took a piece from home to ensure their safe, victorious return; English midwives placed a loaf at the foot of the mother's bed to prevent the woman and her baby from being stolen by evil spirits; and sailors traditionally brought a bun to sea to prevent shipwreck. It is the symbolic use of a food that is valued most by people, not its nutritional composition.

Cultural Identity

An essential symbolic function of food is cultural identity. Beyond self-identification, incorporation can signify collective association. What one eats defines who one is, culturally speaking, and, conversely, who one is not. In the Middle East, for example, a person who eats pork is probably Roman Catholic or Orthodox Christian, not Jewish or Muslim (pork is prohibited in Judaism and Islam). Ravioli served with roast turkey suggest an Italian-American family celebrating Thanksgiving, not a Mexican-American family, who would be more likely to dine on tamales and turkey. The food habits of each cultural group are often linked to religious beliefs or ethnic behaviors. Eating is a daily reaffirmation of cultural identity (Figure 1.1).

Foods that demonstrate affiliation with a culture are usually introduced during childhood and are associated with security or good memories. Such foods hold special worth to a person, even if other diets have been adopted due to changes in residence, religious membership, health status, or daily personal preference. They may be eaten during ethnic holidays and for personal events, such as birthdays or weddings, or during times of stress. These items are sometimes called comfort foods because they satisfy the basic psychological need for food familiarity. For example, in the United States one study found comfort foods for women required little preparation and tended to be snacks, such as potato chips, ice cream, chocolate, and cookies; men preferred foods served by their mothers, such as soup, pizza or pasta, steak, and mashed potatoes.[11] Occasionally, a person embraces a certain diet as an adult to establish association with a group. A convert to Judaism, for instance, may adhere to the kosher dietary laws. African Americans who live outside the South may occasionally choose to eat soul food (typically Southern black cuisine, such as pork ribs and greens) as an expression of ethnic solidarity.

The reverse is also true. One way to establish that a person is not a member of a certain cultural group is through diet. Researchers suggest that when one first eats the food of another cultural group, a chain of reasoning occurs, beginning with the recognition that one is experiencing a new flavor and ending with the assumption that this new flavor is an authentic marker of other group members.[12] These other individuals may be denigrated by food stereotyping, and such slurs are found in nearly all cultures. In the United States, Germans are sometimes pejoratively called "krauts," Chinese "cookies" or "dim sums," Italians "spaghetti benders," Mexicans "beaners," Irish "potatoheads," Koreans "kimchis," and poor white Southerners "crackers" (possibly from "corncracker," someone who cracks corn to distill whiskey or from early immigrants to Georgia who survived on biscuits).

Foods that come from other cultures may also be distinguished as foreign to maintain group separation. Kafir, a derogatory Arabic term for "infidel," was used to label some items found in areas they colonized, including the knobby kaffir lime of Malaysia, and kaffir corn (millet) in Africa. Similarly, when some non-Asian foods were introduced to China, they were labeled barbarian or Wester and named after items already familiar in the diet. Thus, sweet potatoes were called barbarian yams, and tomatoes became barbarian eggplants.[13] Less provocative place names are used, too, though the origins of the food are often incorrect, such as Turkey wheat (the Dutch term for native American corn, which was thought to come from Turkey) and Irish potatoes (which are indigenous to Peru but were brought to the United States by immigrants from Ireland). The powerful symbolic significance of food terms leads occasionally to renaming foreign items in an attempt to assert a new cultural identity. Turkish coffee (it was the Ottomans of Turkey who popularized this thick, dark brew from Africa and spread it through their empire) became Greek coffee in Greece after tensions between the two nations escalated in the 1920s. Examples in the United States include renaming sauerkraut liberty cabbage during World War I, and more recently, calling French fries freedom fries, when France opposed the United States in the invasion of Iraq.

Specific foods are not the only way food can symbolize cultural identity. The appropriate use of food and the behaviors associated with eating, also known as etiquette, is another expression of group membership. In the United States, entirely different manners are required when lunching with business associates at an expensive restaurant, when attending a tea, when eating in the school cafeteria, when drinking with friends at a bar, or when dining with a date. Discomfort can occur if a person is unfamiliar with the rules, and if a person deliberately breaks the rules, he or she may be ostracized or shunned.

Another function of food symbolism is to define status—a person's position or ranking within a particular cultural group. Food can be used to signify economic social standing: champagne and caviar imply wealth, mesquite-grilled foods and goat cheese suggest upward mobility, and beans or potatoes are traditionally associated with the poor. Status foods are characteristically used for social interaction. In the United States, a girlfriend appreciates a box of chocolates from her boyfriend—but not a bundle of broccoli. Wine is considered an appropriate gift to a hostess—a gallon of milk is not. In general, eating with someone connotes social equality with that person. Many societies regulate commensalism (who can dine together) as a means of establishing class relationships. Men may eat separately from women and children, or servants may eat in the kitchen, away from their employers. In India, the separate social castes did not traditionally dine together, nor were people of higher castes permitted to eat food prepared by someone of a lower caste. This class segregation was also seen in some U.S. restaurants that excluded blacks before civil rights legislation of the 1960s.

Children younger than age two will eat anything and everything. Children between three and six years of age begin to reject culturally unacceptable food items. By age seven, children are completely repulsed by foods that their culture categorizes as repugnant.[117]

© Peter Menzel/Stock, Boston Inc.

▲ *Typically, first-generation immigrants remain emotionally connected to their ethnicity, surrounding themselves with a reference group of family and friends who share their cultural background.*

What Is Culture?

Culture is broadly defined as the values, beliefs, attitudes, and practices accepted by members of a group or community. Culture is learned, not inherited; it is passed from generation to generation through language acquisition and socialization in a process called enculturation.[14] It is a collective adaptation to a specific set of environmental conditions, and cultural behavior patterns are reinforced when a group is isolated by geography or segregated by socioeconomic status. Yet culture is not a static condition. It changes over time, from place to place, and in response to social dynamics.[15]

Cultural membership is defined by ethnicity. Unlike national origin (which may include numerous ethnic groups), ethnicity is a social identity associated with shared behavior patterns, including food habits, dress, language, family structure, and often religious affiliation.[15] Members of the same ethnic group usually have a common heritage through locality or history and participate together with other cultural groups in a larger social system. As part of this greater community, each ethnic group may have different status or positions of power. Diversity within each cultural group is also common due to racial, regional, or economic divisions as well as differing rates of acculturation to the majority culture.[16]

Ethnocentric **is the term applied to a person who uses his or her own values to evaluate the behaviors of others. It may be done unconsciously or in the conscious belief that their own habits are superior to those of another culture. Ethnorelativism occurs when a person assumes that all cultural values have equal validity, resulting in moral paralysis and an inability to advocate for a belief. Prejudice is hostility directed toward persons of different cultural groups because they are members of such groups; it does not account for individual differences.[118]**

The Acculturation Process

When people from one ethnicity move to an area with different cultural norms, adaptation to the new majority society begins. This process is known as acculturation, and it takes place along a continuum of behavior patterns that can be very fluid, moving back and forth between traditional practices and adopted customs. It occurs at the micro level, reflecting an individual's change in attitudes, beliefs, and behaviors, and at the macro level, resulting in group changes that may be physical, economic, social, or political in nature.[17,18,19] Typically, first-generation immigrants remain emotionally connected to their culture of origin. They integrate into their new society by adopting some majority culture values and practices but generally surround themselves with a reference group of family and friends from their ethnic background. For example, Asian Indians living in the United States who consider themselves to be "mostly or very Asian Indian" may encourage their children to speak English and allow them to celebrate American holidays, but they do not permit them to date non-Asian Indian peers.[20] Other immigrants become bicultural, which happens when the new majority culture is seen as complementing, rather than competing with, an individual's ethnicity. The positive aspects of both societies are embraced, and the individual develops the skills needed to operate within either culture.[21] Asian Indians who call themselves Indo-Americans or Asian Indian Americans fall into this category, eating equal amounts of Indian and American foods, thinking and reading equally in an Indian language and in English. Assimilation occurs when people from one cultural group shed their ethnic identity and fully merge into the majority culture. Although some first-generation immigrants strive toward assimilation, due perhaps to personal determination to survive in a foreign country or to take advantage of opportunities, most often assimilation takes place in subsequent generations. Asian Indians who identify themselves as being "mostly American" do not consider Asian Indian culture superior to American culture, and they are willing to let their children date non-Indians. It is believed that ethnic pride is reawakened in some immigrants if they become disillusioned with life in America, particularly if the disappointment is attributed to prejudice from the majority society. A few immigrants exist at the edges of the acculturation process, either maintaining total ethnic identity or rejecting both their culture of origin and that of the majority culture.[22]

Acculturation of Food Habits

Culturally based food habits are often among the last practices people change through acculturation. Unlike speaking a foreign language or wearing traditional clothing, eating is usually done in the privacy of the home, hidden from observation by majority culture members. Adoption of new food items does not generally develop as a steady progression from traditional diet to the diet of the majority culture. Instead, research indicates that the consumption of new items is often independent of traditional food habits.[17,18] The lack of available native ingredients may force immediate acculturation, or convenience or cost factors may speed change. Samoans may be unable to find the fresh coconut cream needed to prepare favorite dishes, for instance, or an Iranian may find the expense of purchasing saffron for certain rice recipes prohibitive. Some immigrants, however, adapt the foods of the new culture to the preparation of traditional dishes.[17] Tasty foods are easily accepted—fast food, pastries, candies, and soft drinks; conversely, unpopular traditional foods may be the first to go. Mexican children living in the United States quickly reject the variety cuts of meat, such as tripe, that their parents still enjoy. It is the foods most associated with ethnic identity that are most resistant to acculturation. Muslims will probably never eat pork, regardless of where they live. Rice was considered to be the most consistent food associated with Hmong culture in a study of acculturation and food habits.[18] People from China may insist on eating rice with every meal, even if it is the only Asian food on the table.

Cultural Food Habits

Food functions vary culturally, and each group creates categorizations reflective of their priorities. In the United States, food has been typically classified by food group (as in the four basic food groups: protein, dairy, cereal and grain, fruits and vegetables), by percentage of important nutrients (as identified in the Recommended Dietary Allowances [RDA] for energy, protein, vitamins, and minerals), or according to recommendations for health. American models, especially the Dietary Guidelines for Americans 2010 and the new model—"Choose My Plate" outline current dietary recommendations to support health guidelines. These categories also suggest that Americans value food more for nutritional content and impact on health than for any symbolic use. But only limited information is provided about U.S. food habits; although these schemes list what foods people eat, they reveal nothing about how, when, or why foods are consumed.

Other less explicit categorizations are commonly used by members of each culture and are associated with the meaning of food. Examples of classifications found in both developing and industrialized societies include cultural superfoods, usually staples that have a dominant role in the diet; prestige foods, often protein items or expensive or rare foods; body image foods, believed to influence health, beauty, and well-being; sympathetic magic foods, whose traits, through association of color or form, are incorporated; and physiologic group foods, reserved for, or forbidden to, groups with certain physiologic status, such as gender, age, or health condition.[23]

Researchers have proposed numerous models to identify and understand the food habits of different cultures. Some of these categorizations are helpful in understanding the role of food within a culture, including (1) the frequency of food consumption, as described through the core and complementary foods model; (2) the ways in which a culture traditionally prepares and seasons its foods, as examined by flavor principles; (3) the daily, weekly, and yearly use of food, as found in meal patterns and meal cycles; and (4) changes in food functions that emerge during structural growth in a culture, as predicted by the developmental perspective of food culture.

Core and Complementary Foods Model

Foods selected by a culture can be grouped according to frequency of consumption. The expanded concept of core foods states that the staples regularly included in a person's diet, usually on a daily basis, are at the core of food habits.[24] These typically include complex carbohydrates, such as rice, wheat, corn, yams, cassava, taro, or plantains. Foods widely but less frequently eaten are termed secondary foods. These items, such as chicken or lettuce or apples, are consumed once a week or more, but not daily. Foods eaten only sporadically are called peripheral foods. These foods are characteristic of individual food preference, not cultural group habit.

Another version of the model suggests that in many cultures, especially agrarian societies, the core food is always served with fringe, or complementary, items to improve palatability (Figure 1.2).[25] Because most starchy staples are bland and uniform in texture, these flavorful substances, eaten in small quantities, encourage consumption of the core food as the bulk of the diet. An added component in this model, sometimes as a complementary food and sometimes as a secondary food, is legumes. It has been hypothesized that these core and complementary food pairings often

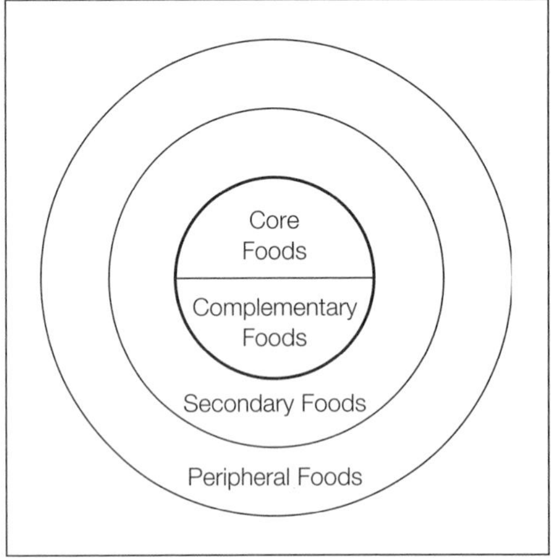

Figure 1.2
The core and complementary foods model.

combine to provide nutritionally adequate meals, especially when legumes are included. For example, in cultures where grain is a core food, sources of vitamins A and C are needed to approach adequacy. Rice, breads and pastas, and corn are frequently prepared with leafy green vegetables, abundant herbs, or tomatoes, which are high in these needed nutrients. Chinese rice with pickled vegetables, Italian noodles with tomato sauce, Mexican corn tortillas with salsa, and Middle Eastern pilaf with parsley and dried fruit are examples. When the core diet is almost adequate nutritionally, the addition of secondary foods—including legumes (soybean products in China; beans or lentils in Italy; red or pinto beans in Mexico; and chickpeas, fava beans, and lentils in the Middle East), small amounts of meats, poultry, fish, and cheeses or yogurt—can provide the necessary balance.

Changes in food behaviors are believed to happen most often with peripheral foods and least often with core foods. A person who is willing to omit items that she or he rarely eats is typically much more reluctant to change foods eaten daily and associated with her or his cultural identity. Although little has been reported on the significance of complementary foods in diet modification, presumably, if complementary items were altered or omitted, the core would no longer be palatable. The complementary foods provide the flavor familiarity associated with the core.

Flavor Principles

The significance of food flavor in a culture cannot be overestimated. The ways foods are prepared and seasoned is second in importance only to the initial selection of ingredients in the development of food habits. It is no less than the transformation of feeding into eating.

Foods demonstrate variability according to location. Much is made, for example, of wine *terroir*—the soil texture, natural minerals, drainage, source of water, sun exposure, average temperature, and other environmental factors in which grapes are grown. Each region and every vineyard are distinctive, often producing appreciable differences in the resulting product. Yet this variation is insignificant when compared to how foods in general are processed for consumption. Every technique, from preparation for cooking (e.g., washing, hulling or peeling, chopping, pounding, squeezing, soaking, leaching, and marinating) to cooking (e.g., baking, roasting, grilling, stewing, toasting, steaming, boiling, and frying) and preserving (e.g., drying, curing, canning, pickling, fermenting, and freezing), alters the original flavor of the ingredient. Nevertheless, location and manipulation practices alone do not equal cuisine. For that, foods must be seasoned.

Historians and scientists speculate there are several reasons why herbs and spices have assumed such an essential role in food habits. Foremost is palatability. Salt, one of the most widely used seasonings, prompts an innate human taste response. It is enjoyed by most people and physiologically craved by some. Researchers also suggest that the burn of chili peppers (and perhaps other spices) may trigger the release of pleasurable endorphins. Another recurrent theory on the popularity of seasoning early on was to disguise the taste of spoiled meats, though evidence for this is limited. A more plausible assertion is that spices were found effective in preserving meats. A survey of recipes worldwide suggested that the antimicrobial activity of spices accounts for their widespread use, especially in hot climates.[26] Other researchers speculate that eating chili peppers (and, by extension, other hot seasonings such as mustard, horseradish, and wasabi) is a benign form of risk taking that provides a safe thrill.[27] Additionally, the recurrent use of seasonings may provide the familiarity sought in the omnivore's dilemma.[28]

Theories aside, seasonings can be used to classify cuisines culturally.[28,29] Unique seasoning combinations, termed flavor principles, typify the foods of ethnic groups worldwide. They are so distinctive that few people mistake their use.

For example, a dish flavored with fermented fish sauce shouts Southeast Asian, not Chinese, Norwegian, or Brazilian. These seasoning combinations are often found in the complementary foods of the core and complementary foods model, providing the flavors associated with the starchy carbohydrates that are the staples of a culture. They usually include herbs, spices, vegetables, and a fat or oil, although many variations exist. A principle flavor combination in West Africa is tomatoes, onion, and chili peppers that have been sautéed in palm oil. In the Pacific Islands, a flavor principle is coconut milk or cream with a little lime juice and salt. Yams taste like West African food when topped with the tomato mixture and like Pacific Islander food when served with the coconut sauce. Some widely recognizable flavor principles include:

- Asian Indian: garam masala (curry blend of coriander, cumin, fenugreek, turmeric, black pepper, cayenne, cloves, cardamom, and chili peppers)
- Brazilian (Bahia): chili peppers, dried shrimp, ginger root, and palm oil
- Chinese: soy sauce, rice wine, and ginger root
- French: butter, cream, wine, boquet garni (selected herbs, such as tarragon, thyme, and bay leaf)
- German: sour cream, vinegar, dill, mustard, and black pepper
- Greek: lemon, onions, garlic, oregano, and olive oil
- Italian: tomato, garlic, basil, oregano, and olive oil
- Japanese: soy sauce, sugar, and rice wine vinegar
- Korean: soy sauce, garlic, ginger root, black pepper, scallions, chili peppers, and sesame seeds or oil
- Mexican: tomatoes, onions, chili peppers, and cumin
- Puerto Rican: sofrito (seasoning sauce of tomatoes, onions, garlic, bell peppers, cilantro, capers, pimento, annatto seeds, and lard)
- Russian: sour cream, onion, dill, and parsley
- Scandinavian: sour cream, onion, mustard, dill, and caraway
- Thai: fermented fish sauce, coconut milk, chili peppers, garlic, ginger root, lemon grass, and tamarind

It would be incorrect, however, to assume that every dish from each culture is flavored with its characteristic seasoning combinations. Nor is the seasoning in each culture limited to just those listed. Regional variations are especially prevalent. In China, northern cuisine often includes the flavor principle seasonings enhanced with soybean paste, garlic, and sesame oil. In the south, fermented black beans are frequently added, although in the Szechwan region hot bean paste, chili peppers, or Szechwan (fagara) pepper is more common. In the specialty cuisine of the Hakka, the addition of red rice wine is distinctive. Further, in any culture where the traditional seasoning combinations are prepared at home, not purchased, modifications to suit each family are customary.[30] Flavor principles are therefore more of a marker for each culture's cuisine than a doctrine.

Meal Patterns and Meal Cycles

People in every culture dine on at least one meal each day. The structural analysis of meal patterns and meal cycles reveals clues about complex social relations and the significance of certain events in a society.[31] The first step in decoding these patterns and cycles is to determine what elements constitute a meal within a culture.

In the United States, for instance, cocktails and appetizers or coffee and dessert are not considered meals (Figure 1.3). A meal contains a main course and side dishes; typically a meat, vegetable, and starch. In the western African nation of Cameroon, a meal is a snack unless cassava paste is served. In many Asian cultures, a meal is not considered a meal unless rice is included, no matter how much other food is consumed. A one-pot dish is considered a meal if it contains all the elements of a full meal. For example, American casserole dishes often feature meat, vegetables, and starch, such as shepherd's pie (ground beef, green beans, and tomato sauce topped with mashed potatoes) or tuna casserole (tuna, peas, and noodles).

The elements that define a meal must also be served in their proper order. In the United States, appetizers come before soup or salad, followed by the entrée, and then by dessert. In France, the salad is served after the entrée. All foods are served simultaneously in Vietnam so that each person may combine flavors and textures

A few cuisines have extremely limited seasonings, including the fare of the Inuits. Broadly speaking, cuisines offering large portions of meat and other protein foods tend to be less seasoned than those with a higher proportion of grains, fruits and vegetables, and legumes.

The sprig of parsley added to a plate of food may have originated as a way to safeguard the meal from evil.

BLONDIE • Young & Lebrun

Blondie © King Features Syndicate. Reprinted by permission

Figure 1.3
Each culture defines which foods are needed to constitute a meal.
Source: Blondie. Reprinted with special permission of King Features Syndicate.

according to taste. In addition to considering the proper serving order, foods must also be appropriate for the meal or situation. Some cultures do not distinguish which foods can be served at different meals, but in the United States eggs and bacon are considered breakfast foods, while cheese and olives are popular in the Middle East for the morning meal. Soup is commonly served at breakfast in Southeast Asia, but in the United States soup is a lunch or dinner food, and in parts of Europe fruit soup is sometimes served as dessert. Cake and ice cream are appropriate for a child's birthday party in the United States; wine and cheese are not.

Other aspects of the meal message include who prepares the meal and what culturally specific preparation rules are used. In the United States, ketchup goes with French fries; in Great Britain, vinegar is sprinkled on chips (fried potatoes). Orthodox Jews consume meat only if it has been slaughtered by an approved butcher in an approved manner and has been prepared in a particular way. (See Chapter 4, "Food and Religion," for more information on Judaism.)

Who eats the meal is also important. A meal is frequently used to define the boundaries of interpersonal relationships. Americans are comfortable inviting friends for dinner, but they usually invite acquaintances for just drinks and hors d'oeuvres. For a family dinner, people may include only some of the elements that constitute a meal, but serving a meal to guests requires that all elements be included in their proper order.

In many homes, few meals are eaten as a family. The term *grazing* refers to grabbing small amounts of food throughout the day to consume. There are an estimated 7 million vending machines in the United States, with over 100 million customers daily.

The final element of what constitutes a meal is portion size. In many cultures, one meal a day is designated the main meal and usually contains the largest portions. The amount of food considered appropriate varies, however. A traditional serving of beef in China may be limited to one ounce added to a dish of rice. In France, a three- or four-ounce filet is more typical. In the United States, a six- or even eight-ounce steak is not unusual, and some restaurants specialize in twelve-ounce cuts of prime rib. American tradition is to clean one's plate regardless of how much is served, while in other cultures, such as those in the Middle East, it is considered polite to leave some food on the plate to indicate satiety.

Beyond the individual meal is the cycle in which meals occur. These include the everyday routine, such as how many meals are usually eaten and when. In much of Europe a large main meal is customarily consumed at noontime, for example, while in most of the United States today the main meal is eaten in the evening. In poor societies only one meal per day may be eaten, whereas in wealthy cultures three or four meals are standard.

The meal cycle in most cultures also includes feasting or fasting, and often both. Feasting celebrates special events, occurring in nearly every society where a surplus of food can be accumulated. Religious holidays such as Christmas and Passover; secular holidays such as Thanksgiving and the Vietnamese New Year's Day, known as Tet; and even personal events such as births, marriages, and deaths are observed with appropriate foods. In many cultures, feasting means simply more of the foods consumed daily and is considered a time of plenty when even the poor have enough to eat. Special dishes that include costly ingredients or are time-consuming to prepare also are characteristic of feasting. The elements of a feast rarely differ from those of an everyday meal. There may be more of an everyday food or several main courses with additional side dishes and a selection of desserts, but the meal structure does not change. For example, Thanksgiving typically includes turkey and often another entrée, such as ham or a casserole (meat); several vegetables; bread or rolls, potatoes, sweet potatoes, and stuffing (starch); as well as pumpkin, mincemeat, and pecan pies or other dessert selections. Appetizers, soups, and salads may also be included.

Fasting may be partial or total. Often it is just the elimination of some items from the diet, such as the traditional Roman Catholic omission of meat on Fridays or a Hindu personal fast day, when only foods cooked in milk are eaten. Complete fasts are less common. During the holy month of Ramadan, Muslims are prohibited from taking food or drink from dawn to sunset, but they may eat in the evening. Yom Kippur, the day of atonement observed by many Jews, is a total fast from sunset to sunset. (See Chapter 4 for more details on fasting.)

© Robert Brenner/PhotoEdit, Inc.

▲ *Special dishes that include costly ingredients or are time-consuming to prepare are characteristic of feasting in many cultures.*

Developmental Perspective of Food Culture

The developmental perspective of food culture (Table 1.1) suggests how social dynamics are paralleled by trends in food, eating, and nutrition.[32] It is useful in conceptualizing broad trends in cultural food habits that emerge during structural changes in a society.

Globalization is defined as the integration of local, regional, and national phenomena into an unrestricted worldwide organization. The parallel change in cultural food habits is consumerization, the transition of a society from producers of indigenous foods to consumers of mass-produced foods. Traditionally seasonal ingredients, such as strawberries, become available any time of year from a worldwide network of growers and suppliers. Specialty products, such as ham and other deli meats, which were at one time prepared annually or only for festive occasions, can now be purchased presliced, precooked, and prepackaged for immediate consumption.

The social dynamic of modernization encompasses new technologies and the socioeconomic shifts that result, such as during the industrial revolution when muscle power was replaced by fuel-generated engine power or during the 1990s with the rise of the information age. Cultural beliefs, values, and behaviors are modified in response to the dramatic structural changes that take place. Commoditization typifies associated food habits, with foods becoming processed, marketed commodities instead of home-prepared sustenance. The fresh milk from the cow in the barn becomes the plastic gallon container of pasteurized milk shipped to another part of the country and sold online over the Internet to a consumer who has

Feasting functions to redistribute food from rich to poor, to demonstrate status, to motivate people toward a common goal (e.g., a political fundraising dinner), to mark the seasons and life-cycle events, and to symbolize devotion and faith (e.g., Passover, Eid al-Fitr, and communion).

Table 1.1 Developmental Perspective of Food Culture

Structural Change	Food Culture Change
Globalization: Local to worldwide organizations	Consumerization: Indigenous to mass-produced foods
Modernization: Muscle to fueled power	Commoditization: Homemade to manufactured foods
Urbanization: Rural to urban residence	Delocalization: Producers to consumers only
Migration: Original to new settings	Acculturation: Traditional to adopted foods

Source: Adapted from Sobal, J. 1999. Social change and foodways. In Proceedings of the Cultural and Historical Aspects of Food Symposium. Corvallis: Oregon State University.

limited time (and limited access to dairy cows) but not limited money.

Urbanization occurs when a large percentage of the population abandons the low density of rural residence in favor of higher density suburban and urban residence. Often income levels do not change in the move, but families who previously survived on subsistence farming become dependent on others for food. Delocalization occurs when the connections among growing, harvesting, cooking, and eating food are lost, as meals prepared by anonymous workers are purchased from convenience markets and fast-food restaurants.

Finally, migration of populations from their original homes to new regions or nations creates a significant shift from a home-bound, culture-bound society to one in which global travel is prevalent and immigration common. Traditional food habits are in flux during acculturation to the diet of a new culture and as novel foods are introduced and accepted into a majority cuisine. Often wholly new traditions emerge from the contact between diverse cultural food habits.

The developmental perspective of food culture assumes that cultures progress from underdeveloped to developed through the structural changes listed. Deliberate efforts to reverse that trend can be seen in the renewed popularity of farmers' markets in the United States and attacks on fast-food franchises in Europe. Other evidence of resistance includes the work of the Slow Food movement—mobilizing against the negative effects of industrialization—and the seed banks that have opened throughout the nation to promote genetic diversity and save indigenous plant populations.[32,33,34]

Individual Food Habits

Cultural values, beliefs, and practices are usually so ingrained that they are invisible in the day-to-day life of the individual. Each person lives within his or her culture, unaware of the influences exerted by that culture on food habits. Eating choices are typically made according to what is obtainable, what is acceptable, and what is preferred: a dietary domain determined by availability and by what each person considers edible or inedible. Within the limitations of this dietary domain, personal preference is most often concerned with more immediate considerations, such as taste, cost, convenience, self-expression, well-being, and variety, as explained in the consumer food choice model discussed later in this chapter.

Food Availability

A person can only select a diet from foods that are available. Local ecological considerations, such as weather, soil, and water conditions; geographical features; indigenous vegetation; the native animal population; and human manipulation of these resources through cultivation of plants and domestication of livestock, determine the food supply at a fundamental level. A society living in the cool climate of northern Europe is not going to establish rice as a core food, just as a society in the hot wet regions of southern India is not going to rely on oats or rye. Seasonal variations are a factor, as are unusual climactic events, such as droughts, that disrupt the food supply.

The political, economic, and social management of food at the local level is typically directed toward assuring a reliable and affordable source of nourishment. Advances in food production, storage, and distribution are examples. However, the development of national and international food networks has often been motivated by other needs, including profit and power. The complexity of the food supply system has been examined by many disciplinary approaches. Historians trace the introduction and replacement of foods as they spread regionally and globally. Economists describe the role of supply and demand, the commodity market, price controls, trade deficits, and farm subsidies (as well as other entitlements) on access to food. Psychologists investigate how individual experience impacts diet; political scientists detail how fear of biotechnology, bioterrorism, and disease (such as the mad-cow or bovine spongiform encephalopathy scare in Europe) can alter acceptability. Sociologists document how social structures and relations affect the obtainment of food; legal experts debate the ethics of food policies for the poor, the incarcerated, and the terminally ill. This is only a small sampling of the factors influencing food availability. However, except in regions where serious food shortages are the norm, availability issues are usually not at the forefront of individual food choice.

Edible or Inedible?

This approach: was one of the earliest food habits models, describing the individual process that establishes the available, appropriate, and personal food sphere. Each person's choice of what to eat is generally limited to the foods found in this dietary domain.[35]

1. ***Inedible foods:*** These foods are poisonous or are not eaten because of strong beliefs or taboos. Foods defined as inedible vary culturally. Examples of frequently prohibited foods (or taboo foods, from the Tongan word tabu, meaning "marked as holy") include animals useful to the cultural group, such as cattle in India; animals dangerous to catch; animals that have died of unknown reasons or of disease; animals that consume garbage or excrement; and plants or animals that resemble a human ailment (e.g., strawberries or beef during pregnancy to protect the infant, as described later).
2. ***Edible by animals, but not by me:*** These foods are items such as rodents in the United States or corn in France (where it is used primarily as a feed grain). Again, the foods in this category vary widely by culture.
3. ***Edible by humans, but not by my kind:*** These foods are recognized as acceptable in some societies, but not in one's own culture. Some East Africans are disgusted by eggs, for instance, which are associated with excrement.[36] Some rural South Africans who consider termites a delicacy are repulsed by the idea of eating scorpions, a specialty enjoyed by some Chinese.[37] Examples of foods unacceptable in the United States but acceptable elsewhere include giant snails (Africa), dog meat (Asia), iguana (the Caribbean), horse meat or blood sausage (Europe), and bear paw (Mongolia).
4. ***Edible by humans, but not by me:*** These foods include all those accepted by a person's cultural group but not by the individual, due to factors such as preference (e.g., tripe, liver, raw oysters), expense, or health reasons (a low-sodium or low-cholesterol diet may eliminate many traditional American foods). Other factors, such as religious restrictions (as in kosher law or halal practices) or ethical considerations (vegetarianism), may also influence food choices.
5. ***Edible by me:*** These are all foods accepted as part of an individual's dietary domain.

There are always exceptions to the ways in which foods are categorized. It is generally assumed, for instance, that poisonous plants and animals will always be avoided. In Japan, however, fugu (blowfish or globefish) is considered a delicacy despite the deadly toxin contained in the liver, intestines, testes, and ovaries. These organs must be deftly removed by a certified chef as the last step of cleaning (if they are accidentally damaged, the poison spreads rapidly through the flesh). Eating the fish supposedly provides a tingle in the mouth prized by the Japanese. Several people die each year from fugu poisoning.

Among the most universal of food taboos is cannibalism, although anthropologists have discovered numerous examples of prehistoric human consumption in European and New World excavations.

Insects, such as termites and ants, provide 10 percent of the protein consumed worldwide.

Consumer Food Choice Model

An individual's dietary domain is established before he or she sets foot in a restaurant, deli, or supermarket. The consumer food choice model (Figure 1.4) explains the factors influencing individual decisions within that predetermined food sphere.[38]

Food selection is primarily motivated by taste. Taste is defined broadly by the sensory properties detectable in foods: color, aroma, flavor, and texture. Humans anticipate a specific food will have certain sensory characteristics; deviations can signal that the item is poisonous or spoiled. Many of these expectations are developed through early exposure to culturally determined foods. For

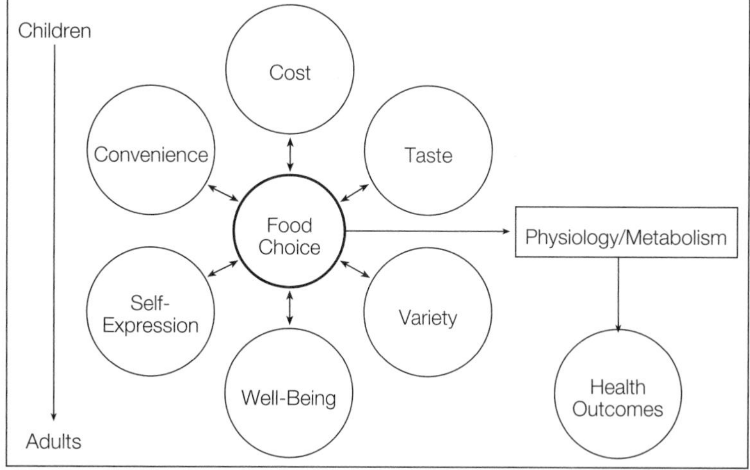

Figure 1.4
The consumer food choice model.
Source: Adapted from A. Drewnowski, *Taste, Genetics, and Food Choice.* Copyright © 2002. Used by permission of Adam Drewnowski, PhD.

Humans can detect approximately 10,000 different odors, though genetics may determine which odors can be detected. For example, nearly 50 percent of people cannot smell androstenone (also called boar pheromone), which is found in bacon, truffles, celery, parsnips, boar saliva, and many human secretions; however, researchers have found people can be taught to perceive it through daily sniffing.[119]

Though the physiological response to disgust, nose wrinkling, retraction of the lips, gaping, gagging, and even nausea, seems instinctual, it is actually a cognitively sophisticated feeling that does not develop in children until between the ages of four and seven years old. Which items are disgusting in a culture is learned from parents and peers.[120]

Some researchers contend that there is a fifth type of receptor on the tongue for umami (from the Japanese for yummy). It is the taste associated with meats, mushrooms, cheese, and the flavor-enhancer monosodium glutamate (MSG).

example, most core foods are pale white, cream, or brown in color; however, some West Africans prefer the bright orange of sweet potatoes, and Pacific Islanders consider lavender appropriate for the taro root preparation called poi. Should the core item be an unanticipated color, such as green or blue, it may be rejected. Similarly, each food has a predictable smell. Aromas that are pleasurable may trigger salivation, while those considered disgusting, such as the odor of rotting meat, can trigger an immediate gag reflex in some people. Again, which odors are agreeable and which are disagreeable are due, in part, to which foods are culturally acceptable: Strong-smelling fermented meat products are esteemed by some Inuit (*muktuk*) and some rural Filipinos (*itog*). Strong-smelling cheese (which is nothing more than the controlled rot of milk) appeals to many Europeans, but even mild cheddar may evoke distaste by many Asians and Latinos. Appropriate texture is likewise predictable. Ranging from soft and smooth to tough and coarse, each food has its expected consistency. Novel textures may be disliked: Some Americans object to gelatinous bits in liquid, as experienced in tapioca pudding or bubble tea, yet these foods are popular in China. Conversely, some Asians find the thick, sticky consistency of mashed potatoes unappetizing. Okra, which has a mucilaginous texture, is well-liked in the U.S. South but is considered too slimy by many residents of other regions.

The human tongue has receptors for the perception of sweet, sour, salty, and bitter. It is hypothesized that food choice in all societies is driven, in part, by an inborn preference for the taste of sugars and fats. These nutrients are indicative of foods that are energy dense; a predisposition for sweets and foods high in fat ensures adequate calorie intake, an evolutionary necessity for omnivores with a wide selection of available foods. Sugars and fats are especially pleasurable flavor elements, associated with palatability and satiety (including the texture factor provided by fats, called mouthfeel).

Preferences for sweets (especially when combined with fats) are found during infancy and childhood and peak in early adolescence. One study found 45 percent of kcalories eaten by young people came from discretionary sugar and fat.[39] This preference declines in later years, reducing their significance in food choice; however, it has been suggested that there are familial differences in preference for high-energy foods associated with the pleasant taste of sugar and fat lasting into adulthood.[40]

The opposite is true for bitterness, which is associated with toxic compounds found in some foods and is strongly disliked by most children. The ability to detect bitterness decreases with age, however, and many adults consume foods with otherwise unpleasant sulfides and tannins, including broccoli and coffee. There are some who remain especially sensitive to certain bitter compounds, affecting their other preferences as well; they tend to dislike sweet foods and opt for bland over spicy items. Sour alone is rarely well-liked, but is enjoyed when combined with other flavors, especially sweet. It has been suggested that a preference for the sweet-sour taste prompted human ancestors to seek fruit, an excellent source of vitamins and minerals.[41,42,43]

Unlike the tastes of sweet, bitter, and sour, which are actively liked or disliked by infants, babies generally are indifferent to salt until about four months of age. Similar to sugar, children prefer higher concentrations of salt than do adults. Their preference for salt is shaped by the frequency of exposure to it after birth, and perhaps perinatally. One study found that children born to women who experienced moderate to severe morning sickness with dehydration had a stronger preference for salt than children born to mothers with no morning sickness; other research suggests smaller babies (5.5 pounds or less) may be linked to salt preference.[44]

Finally, taste is influenced by flavor principles, the characteristic combinations of core and complementary foods, as well as traditional grouping of meal elements. These traditions are important in providing an expected taste experience and satisfying a need for familiarity in food habits.

Cost is often the second most important influence on food choice, and income level is the most significant sociodemographic factor in predicting selection. In cultures with a limited food supply due to environmental conditions or in societies where a large segment of the population is disadvantaged, food price is more imperative than taste, dictating nutritional sufficiency and well-being. The wealthier the society, the less disposable income is spent on food, and, as income increases, food choices change. Typically, the people of poorer cultures survive on a diet dependent on grains or tubers and limited amounts of protein, including meat, poultry, fish, or dairy foods. Only a small variety of fresh fruits or vegetables may be available. People with ample income consistently include more meats, sweets, and fats in their diet (a trend seen in the global popularity of inexpensive American fast foods), plus a wider assortment of fruits and vegetables.[45,46] In the United States, the affordability of foods has been found to limit the purchasing of health-promoting foods, and in some cases even families with government subsidies find it difficult to meet nutritional needs.[47] It is estimated that in 2007 over 13 million households are considered to be food insecure.[48] One analysis found that making even small dietary improvements, such as switching from white bread to whole wheat, regular ground beef to lean ground beef, and whole-fat cheese to low-fat cheese, can represent a 25 percent increase in the cost of food for a low-income family.[49,50]

The local dietary domain is also a factor in obtaining affordable food. A subsistence farmer may have greater access to fresh foods than a person with the same limited income living in a city. In urban areas supermarkets with a cheaper selection of foods often choose to locate outside low-wealth neighborhoods and residents may only have access to high-priced convenience stores or small, independently run groceries with limited selection.[47,50,51] One study identified that when seeking a list of foods recommended for patients with diabetes it was found that only 18 percent of stores located in a largely Hispanic lower-income neighborhood stocked the foods, compared to 58 percent of stores located in an adjacent mostly white affluent neighborhood.[52] Further, access to healthful restaurant dining varies. Studies suggest that predominantly African American and low-income neighborhoods have more fast-food restaurants per square mile than in white neighborhoods, with fewer healthy options.[53,54,55,56] When nutritious food is available and affordable, the prestige of certain food items, such as lobster or prime beef, is often linked to cost. Protein foods are most associated with status, although difficult-to-obtain items, such as truffles, can also be pricey.

Convenience is a major concern in food purchases, particularly by members of urbanized societies. In some cultures, everyone's jobs are near home, and the whole family joins in a leisurely midday lunch. In urbanized societies, people often work far from home; therefore, lunch is eaten with fellow employees. Instead of a large, home-cooked meal, employees may eat a quick fast-food meal. Furthermore, family structure can necessitate convenience. In the United States, the decreasing number of extended families (with help available from elder members) and increasing number of households with single parents, along with couples who both work outside the home and unassociated adults living together all reduce the possibility that any adult in the household has the time or energy to prepare meals. Studies show, for example, that the greater the number of hours a woman works outside the home each day, the fewer the hours she spends in cooking. Only 40 percent of families report cooking at least once a day, and in more than one-quarter of all homes cooking is done less than once a day.[57] Recent research indicates that a higher amount of family meals is correlated with more positive health indicators.[58] Furthermore, the quality of dietary intake improves when there is a reduction in spending for food away from home.[59] Convenience generally spurs the increasing number of takeout foods and meals purchased at restaurants. It has been the trend for several years that U.S. citizens spent more for food away from home than for that prepared in the home, but this trend may be changing due to recent economic changes. In 2010, the number of restaurants in the U.S. declined for the first time in this decade.[60] Takeout foods include not only items purchased from restaurants and fast-food franchises, but also ready-made hot or cold dishes from supermarkets and warehouse stores. For families short on time but interested in fare considered more nutritious and less expensive than what is typically found in restaurants or fast-food franchises, "easy-meal-prep" is available. Centers offering prepared ingredients ready for assembly into dishes, such as lasagna, chicken casserole, or Salisbury steak with gravy, allow meal preparation without shopping or cleanup.

Self-expression, the way in which we indicate who we are by behavior or activities, is important for some individuals in food selection, particularly as a marker of cultural identity. Although the foods associated with ethnicity, religious affiliation, or regional association are predetermined through the dietary domain, it is worth noting that every time a person makes a food choice he or she may choose to follow or ignore convention.

In addition to salt, other flavor preferences may be passed on perinatally. A study of women who ate garlic or a placebo before amniocentesis found the odor of garlic in the amniotic fluid evident from the garlic-ingesting women.[74]

In 1901, the average American family spent nearly half (45 percent) of their income on food. A century later, that figure had decreased to just 13 percent of total income.[121]

Even when supermarkets with a greater selection of healthful foods are available, less-acculturated immigrants may feel more comfortable shopping at stores where their language is spoken and ethnic ingredients are stocked.[122]

The status of food can change over time. In the early years of the United States, lobster was so plentiful it piled up on beaches after storms, but colonists considered it fit only for Indians or starving settlers.

© Kostenko Maxim/Shutterstock

▲ *Regional fare differs throughout the United States and can be consumed for self-expression. The southwestern foods shown here represent one of many distinct regional cuisines.*

Meals and snacks prepared at home are lower in calories per eating occasion, and lower in total fat, saturated fat, cholesterol, and salt per calorie than foods prepared away from home. IHOP's Big Country Breakfast contains 1,790 kcalories, more than a full day's requirement (from www.calorieking.com).

Ethnic identity may be immediate, as in persons who have recently arrived in the United States; or it may be remote, a distant heritage modified or lost over the generations through acculturation. An individual who has just immigrated to the United States from Japan, for instance, is more likely to prefer traditional Japanese cuisine than is a third- or fourth-generation Japanese American.

Religious beliefs are similar to ethnic identity in that they may have a great impact on individual food habits or an insignificant influence depending on religious affiliation and degree of adherence. Many Christian denominations have no food restrictions, but some, such as the Seventh-Day Adventists, have strict guidelines about what church members may eat. Judaism requires that only certain foods be consumed in certain combinations, yet most Jews in the United States do not follow these rules strictly (see Chapter 4).

A person may also choose foods associated with a specific region. In the United States, the food habits of New England differ from those of the Midwest, the South, and the West, and local specialties such as Pennsylvania Dutch, Cajun, and Tex-Mex may influence the cooking of all residents in those areas. Generally speaking, people in the Northeast purchase more lunch meats, breads, cakes, and butter, and drink more tea than the national average. In the South, people favor sausages, bacon, biscuits, and beans and peas; they also use more cornmeal and shortening. Midwesterners buy more roasts, salad dressing, margarine, and almost 50 percent purchase more potato chips than average. Carbonated drinks are also common in the Midwest. In the West, fresh fruits and vegetables, condiments and seasonings, and cheese are popular, and people consume more than twice the national average of whole wheat bread. Coffee is especially favored.[61,62]

Self-identity can be another factor in food selection, as discussed previously. An environmentalist may be a vegetarian who prefers organic, locally grown produce, while a gourmet or foodie may patronize small markets in ethnic neighborhoods throughout a city searching for unusual ingredients. Advertising has been directly related to self-expression, especially self-identity. Research indicates that in blind taste tests people often have difficulty discriminating between different brands of the same food item. Consumer loyalty to a particular brand is believed more related to the sensual and emotional appeal of the name and packaging.[63] For example, similar-tasting flake cereals such as Wheaties (which touts itself as the "breakfast of champions"), Special K, and Total target sports enthusiasts, dieters, and the health-conscious, respectively.

Advertising also promises food-provided pleasure, appealing to the desire of consumers to be seen as popular, fun-loving, and trendy. Exploitation of sex to sell hamburgers and beer is common, as are suggestions that eating a chocolate or drinking a soft drink will add zest to living. A study of television food ads targeting children found that 75 percent were associated with "good times," 43 percent with being "cool and hip," and 43 percent with feelings of happiness.[64,65,66] Such advertising is a reflection of a larger trend: food as entertainment, the vicarious enjoyment of eating through reading about it or watching food-related programs on television, also called food porn.[56,67] In the United States, nearly 150 food/wine magazines are published monthly, almost 500 million food/wine books are sold annually, and numerous network cooking/dining shows air daily. The impact of this media on food choice is as yet unknown. Food entertainment may popularize certain ingredients, such as arugula or mangoes, or cuisines, such as Spanish fare or updated traditional American dishes like spicy meatloaf and garlic mashed potatoes. They may also set such a high standard of preparation and presentation that some home cooks feel inadequate, choosing to dine out or select prepackaged items instead of making meals from scratch.

Physical and spiritual well-being is another food choice consideration for some individuals. Physiological characteristics, including age, gender, body image, and state of health, often impact food habits. Preferences and the ability to eat and digest foods vary throughout the life cycle. Pregnant and lactating women commonly eat differently than other adults. In the United States, women are urged to consume more food when they are pregnant, especially dairy products. They are also believed to crave unusual food combinations, such as pickles and ice cream. They may avoid certain foods, such as strawberries, because they are believed to cause red birthmarks.

In some societies with subsistence economies, pregnant women may be allowed to eat more meat than other family members; in others, pregnant women avoid beef because it is feared that the cow's cloven hoof may cause a cleft palate in the infant. Most cultures also have rules regarding which foods are appropriate for infants; milk is generally considered wholesome, and sometimes any liquid resembling milk, such as nut milk, is also believed to be nourishing.

Puberty is a time for special food rites in many cultures. In the United States, adolescents are particularly susceptible to advertising and peer pressure. They tend to eat quite differently from children and adults, rejecting those foods typically served at home and consuming more fast foods and soft drinks. A rapid rate of growth at this time also affects the amount of food that teenagers consume. One survey found teenage boys down an average of five meals per day, and teenage girls eat four meals.[68,69]

The opposite is true of older adults. As metabolism slows, caloric needs decrease. In addition, they may develop a reduced tolerance for fatty foods or highly spiced items. Elders often face other eating problems related to age, such as the inability to chew certain foods or a disinterest in cooking and dining alone. It is predicted that the shift toward an older population in the next two decades will result in both the types of foods purchased (an increase in fruits, vegetables, fish, and pork because elders consume these items more often than younger people do) and reductions in the total amount of food consumed per capita (because elders eat smaller amounts of food).[70]

Gender has also been found to influence eating habits. In some cultures women are prohibited from eating specific foods or are expected to serve the largest portions and best pieces of food to the men. In other societies food preference is related to gender. Some people in the United States consider steak to be a masculine food and salad to be a feminine one; or that men drink beer and women drink white wine. Research has shown that gender differences affect how the brain processes satiation responses to chocolate, suggesting that men and women may vary in the physiological regulation of food intake—perhaps accounting for some food preferences.[71]

A person's state of health also has an impact on what is eaten. A chronic condition such as lactose intolerance or a disease such as diabetes requires an individual to restrict or omit certain foods. An individual who is sick may not be hungry or may find it difficult to eat. Even minor illnesses may result in dietary changes, such as drinking ginger ale for an upset stomach or hot tea for a cold. Those who are trying to lose weight may restrict foods to only a few items, such as grapefruit or cabbage soup, or to a certain category of foods, such as those low in fat or carbohydrates. Those who are exceptionally fit, such as students or professional athletes, may practice other food habits, including carbohydrate loading or consumption of high-protein bars. In many cultures, specific foods are often credited with health-promoting qualities, such as ginseng in Asia or chicken soup in Eastern Europe. Corn in American Indian culture may be selected to improve strength or stamina. Well-being is not limited to physiological conditions. Spiritual health is equally dependent on diet in some cultures where the body and mind are considered one entity. A balance of hot and cold or yin and yang foods may be consumed to avoid physical or mental illness. (See Chapter 2, "Traditional Health Beliefs and Practices.")

The final factor in consumer food choice is variety. The omnivore's paradox states that humans are motivated psychologically to try new foods. Further, the desire for new flavors may also have a physiological basis. Sensory-specific satiety (unrelated to actually ingesting and digesting food) results when the pleasure from a certain food flavor decreases after a minute or two of consumption. Introduction of a new food, or even the same food with new added seasoning, arouses the enjoyment in eating again, encouraging the search for new flavor stimuli.[72] In addition, hunger increases the probability that a

Another aspect of food as entertainment is competitive eating as televised sport. Elite eaters can make more than $50,000 a year in winnings, with records such as 46 dozen oysters in ten minutes, 8.4 pounds of baked beans in 2 minutes, 47 seconds, and 11 pounds of cheesecake in 9 minutes.[123]

Old age is a cultural concept; among some American Indians and Southeast Asians, individuals become elders in their 40s.[124]

***Lactose intolerance,* the inability to digest the milk sugar lactose, develops as a person matures. It is believed that only 15 percent of the adult population in the world (those of northern European heritage) can drink milk without some digestive discomfort.**

Research on sensory-specific satiety suggests people eat less food when consuming a monotonous meal, and may overeat and gain weight when abundant variety is available.[8]

The Japanese say that for every new food a person tries, life is extended seventy-five days.

new food will be liked.[73,74] Marketers take advantage of the innate human drive for diet diversity by continual reformulation and repackaging of processed food products to attract consumers.

Interest in the foods of other regions or cultures is associated the desire for new taste experiences, and also with increased income and educational attainment. Wealth permits experimentation and education can increase wealth. Nutritional knowledge, also affected by educational attainment, includes the health-promoting benefits of dietary diversity.

One study reported that college students were more likely to try a new fruit, vegetable, or grain product if information on the nutritional benefits were provided.[75] Some researchers have found that attitudes about the healthfulness of certain foods is important in food selection, and parents may purchase foods they consider healthy for their children even if they would not select those items for themselves.[76] The nutrition knowledge of the person who plans meals in the home impacts food selection for all household members.[77] It has been suggested, however, that choices are more often influenced by beliefs regarding nutritional quality than the actual nutritional value or health consequences of a food.[72,78,79,80] Whether accurate or misinformed, nutritional knowledge does not always translate into knowledge-based food choice: a poll found that six in every ten consumers check nutrition labels frequently for calories and fat content, but nearly half of those who read the information still choose items for taste even when they determine that the item is bad for them.[81,82]

What is evident after exploring the consumer food choice model is that many influences on individual food habits are interrelated. An inborn preference for foods high in sugar, fat, and salt can encourage the consumption of items specifically processed to enhance those taste experiences. Such foods are often convenient, and items such as soft drinks and sandwich meats may be less costly than fruit juice or fresh pork or beef (though certainly some processed items are more expensive than homemade equivalents). Advertisers exploit the need for convenience and the desire to try new foods. A person may be aware of nutrition messages encouraging reduction of sugars and fats in the diet, as seen in the Dietary Guidelines for Americans, but this nutrition knowledge is often overridden by the primary factor in consumer food choice: taste.

Furthermore, influences on choice may change for each person as she or he matures. Food selection in infants (within the dietary domain of available foods provided by parents) is based almost exclusively on taste factors, with a strong resistance to new items. Children become more interested in self-expression as they grow and become sensitive to family and peer pressure. Young adults continue to be concerned with taste and self-expression, to which cost and convenience are typically added, especially in families with children. In middle age, increased income may mitigate cost issues; and in old age, health problems may become a more significant factor in food choice than even taste.

Nutrition and Food Habits

The Need for Cultural Competency

In recent years, the significance of culturally based food habits on health and diet has been recognized, and the need for intercultural competencies in the areas of nutrition research, assessment, counseling, and education has been cited.[83] The Campinha-Bacote Model of Competence outlines a process for cultural competency in health care, involving steps from cultural awareness, cultural knowledge, cultural skill, cultural encounter, and cultural desire.[84,85]Accurate data collection for required for assessment and educationis dependent on respect for different values and a trusting relationship between respondent and researcher; effective intercultural communication is a function of understanding and accepting a client's perspective and life experience.[47] New standards of nutrition care issued by professional accreditation organizations reflect similar guidelines.[86,87] Looking toward the future, it has been proposed that health care professionals should move beyond the theoretical concepts foundational to cultural sensitivity and relevance to the practicalities of cultural competency. Language skills, managerial expertise, and leadership are needed to guide diverse communities in healthy lifestyle changes, to serve hard-to-reach populations, and to effect change in the health care system.

Diversity in the U.S. Population

The growing need for cultural competency is evident in current demographic trends. Since the 1970s, the United States has moved increasingly toward a cultural plurality, where no single ethnic group is a majority. In 1980, only Hawaii and the District of Columbia had plurality populations. Since that time, California, New Mexico, and Texas have joined the list. Pluralities also exist in several metropolitan area populations, including Chicago, Houston, Los Angeles, Miami, New York City, and Philadelphia. Other urban areas such as Dallas and Las Vegas are approaching plurality population status, with less than the national average of whites. Nationwide, demographers estimate that non-Hispanic whites will become less than 50 percent of the total population by the year 2050.

This change can be seen in the dramatic difference in projected ethnic group growth from 2000 to 2050 (see Figure 1.5). Although the total population is expected to increase by about 49 percent during that period, more than 90 percent of the growth will be in minority populations. Gains for the Asian population are expected at more than four times the national average, and more than three times the national average for the Hispanic population. The black population is expected to grow by 71 percent, and the group with the largest projected increase (217 percent) is "all other ethnic groups," which includes the smaller populations of American Indians, Native Alaskans, Native Hawaiians, and other Pacific Islanders, as well as the quickly growing numbers of people of two or more ethnicities.

In actual numbers, Hispanics surpassed African Americans as the largest U.S. minority population in 2008, representing more than than 15 percent of the total population, whereas blacks or African Americans make up approximately 13 percent. Asians are the third largest minority at approximately 4 percent of the total U.S. population. Much smaller numbers of Pacific Islanders and Native Americans were reported, 0.1 percent and 0.9 percent, respectively. Notably, many U.S. ethnic populations have an average age significantly lower than that of the total population. Predicted demographic changes are often seen first among children and young adults.[88]

This profile of the general U.S. population is notably different than that for health care professionals, who are mostly white. Among registered dietitians, 85 percent were white in 2004. Although

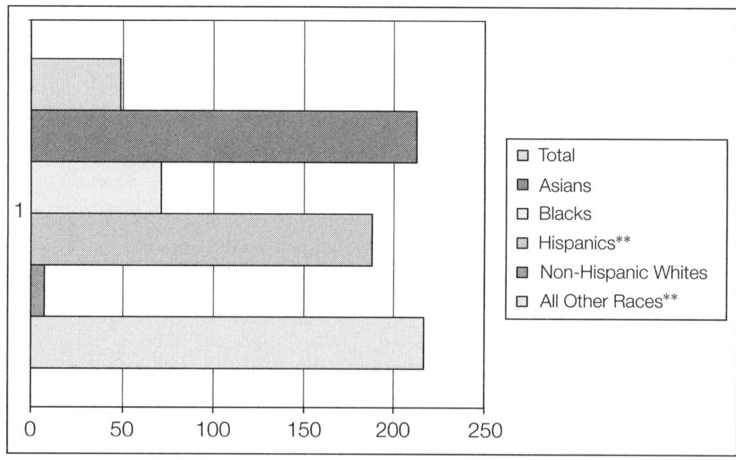

Figure 1.5
Projected U.S. population growth, by percentage, 2000–2050.
Source: U.S. Census Bureau. 2004. U.S. Interim Projections by Age, Sex, Race, and Hispanic Origin http://www.census.gov/ipc/www/usinterimproj/

Asians/Pacific Islanders were well represented at 5 percent, African Americans and Hispanics were significantly under represented at 5 percent and 3 percent, respectively.[89] Researchers note that clients from minority populations prefer to receive health care in settings with minority health care providers; that minority health care providers are more likely to work in underserved areas; and that people from minority groups are more likely to participate in research studies when the investigator is from the same ethnic background.[90]

"Respect for diverse viewpoints and individual differences" is an American Dietetic Association value.

As of 2004, only seven states had populations that were less than 10 percent minority: Iowa, Maine, New Hampshire, North Dakota, Vermont, and West Virginia.

Diversity in the Canadian Population

The Canadian census is conducted differently from the U.S. count. Canadians in 2006 were asked to list their ethnicity in an open-ended question, and multiple responses listing one or more ethnicities were accepted. Over 200 different ethnicities were identified. This has provided a broader picture of ancestry, particularly because single responses and multiple responses were reported separately. For example, of the over 1.3 million Aboriginals (including Native American Indians, Métis—people of mixed Aboriginal and non-Aboriginal heritage—and Inuit), 565,040 listed this ethnicity as a single response and 800,020 listed it as part of a multiple response. A separate question inquired if the census respondent was a member of a visible minority, defined by the Employment Equity Act as "persons, other than Aboriginal peoples, who are non-Caucasian in race or non-white in color." The act specifically lists Chinese, South Asians (i.e., Asian Indians, Pakistanis, Sri Lankans), blacks, Arabs/West Asians, Filipinos, Latin Americans, Japanese, Koreans, and Pacific Islanders. Immigration growth in Canada has dramatically exceeded overall population growth in recent years: immigrants in Canada represented more than 2/3 of the population growth from 2001 to 2006—more than 19 percent of the total population—and the nation is second only to Australia (22 percent) in proportion of foreign-born citizens. Of greater importance, Canadian immigration patterns have shifted during the past three decades. Recent immigrants include almost 60 percent from Asia and 20 percent from the Caribbean. Chinese, South Asian, and black groups are the three largest minority populations, though the fastest growing populations are Arabs/West Asians and Koreans. Nearly all (94 percent) recent immigrants to Canada have settled in urban areas, particularly Toronto, Vancouver and Montreal. Other urban areas with disproportionately large recent immigrant populations include Calgary, Edmonton, and Ottawa-Hull.[91]

Ethnicity and Health

Health is not enjoyed equally by all in the United States. Disparities in mortality rates, chronic disease incidence, and access to care are prevalent among many U.S. ethnic groups (Table 1.2). Poor health status in the United States is also associated with poverty (see Cultural Controversy—Does Hunger Cause Obesity? later in this chapter), low educational attainment, and immigrant status: Immigrant health has been found initially better than similar U.S.-born populations in some research, and is shown to decline with length of stay.[92,93]

Acculturation to the majority culture is believed to be a significant factor in health independent of socioeconomic status. First noted in heart disease rates, modernization has also been linked to increased blood cholesterol levels, increased blood pressure levels, obesity, type 2 diabetes, and some cancers.[18,92,94,95,96] The stress of adaptation to the pressures of a fast-paced society is believed to be significant.[96] Hereditary predisposition to developing certain health conditions most probably plays a significant role. It is important to note, however, that acculturation is difficult to define accurately and is not an inherent risk factor in itself.[97] Some changes in diet—such as a reduction in pickled food intake associated with stomach cancer or increased availability of fruits and vegetables—can be beneficial. Better educational opportunities and health care services can also promote health.

The specific impact of ethnicity on health status is not well delineated through the limited data available. Researchers caution that using race or ethnicity in research can be misleading.[98] For example, the Human Genome project states that there is no genetic basis for use of the term *race*. Hamilton explains that 99.9 percent of all humans have the same genes. Race is simply a category used to describe groups of individuals.[99] Many studies do not explain how participants are categorized. Individuals may self-select differently than investigators, and self-identity may change over time. Even official classifications may vary and change. In the United States, the Office of Management and Budget is responsible for defining the categories used in all government work, including the Census. In 1997, the standards were revised to include five classifications for "race": American Indian or Native American, Asian, Black or African American, Native Hawaiian or other Pacific Islander, and White. Prior to the revision, there were only four groupings because Asians were combined with Pacific Islanders. Additionally, the two categories for ethnicity were expanded in 1997 (ethnic members may be of any race): Hispanic or Latino, and Not Hispanic or Latino. These changes from earlier definitions can lead to difficulties in interpreting data trends. Further, the factor of ethnicity is not sufficiently separated from socioeconomic status in many studies, calling into question whether a stated finding is due to ethnicity or whether it is due to income, occupation, or educational status. For example, evaluation of the incidence of type 2 diabetes in the Black Women's study indicates a strong relationship between individual and neighborhood socioeconomic status and type 2 diabetes even when controlling for factors such as education and income.[100]

Nevertheless, ethnicity data suggesting risks and disparities can be useful to health care providers as long as the caveats above are considered and care is taken to avoid stereotyping a patient by group membership. For example, the CDC reports that "After adjusting for population age differences, 2004–2006 national survey data for people aged 20 years or older indicate that 6.6 percent of non-Hispanic whites, 7.5 percent of Asian Americans, 10.4 percent of Hispanics, and 11.8 percent of non-Hispanic blacks had diagnosed diabetes. Among Hispanics, rates were 8.2 percent for Cubans, 11.9 percent for Mexican Americans, and 12.6 percent for Puerto Ricans."[101]

Projected rates for 2020 show an even higher disparity: the diagnosis for whites is expected to increase 27 percent, while the increase for African Americans may be as high as 50 percent, and for

Starch Foods

Starchy foods form the foundation of nearly all diets.

© Jupiter Images/comstock.com

A. Rice and rice-like grains are eaten by millions worldwide and come in many varieties, including short-grain (far right), long-grain (in scoop), and wild rice (long, black grains in lower left mix). Rice products, such as noodles and papers, are also common.

© Becky Luigart-Stayner/CORBIS

B. Wheat is popular in drier regions, typically eaten as bread, as pasta, and in cereal form, such as couscous (lower bag) and bulgar (upper bag).

© Danny Lehman/CORBIS

C. Corn is an important New World starch, traditionally prepared as flat breads (including these tortillas), as dumplings, in steamed packets, in stews, and as gruel.

© Wolfgang Kaehler/CORBIS

© Andrew Unangst/Spirit/Corbis

D. In tropical areas, fruit and root vegetables are significant sources of starch, including breadfruit (upper basket), cassava (lower basket), yams (lower left corner), taro root (left upper corner), lotus root (cut root with hollow spaces in center), sweet potatoes (ruby orange roots on right), and burdock root (long, pencil-thin roots). Potatoes are more important in cooler climates.

© Siede Preis/Photodisc/Getty Images

E. Acorns are a starch consumed in some Native American, European, Middle Eastern, and Asian cultures.

Protein Foods

Protein foods include a wide variety of meat products, dairy foods, fish, and shellfish, as well as numerous legumes.

© Chris Shorten/Cole Group/Photodisc/Getty Images

A. Sausages are eaten in nearly every culture. They come in hundreds of types and make use of miscellaneous cuts and leftovers, such as blood (dark red links).

© 2002 Wisconsin Department of Tourism

B. Dairy products, including yogurt and cheese, are available in even more varieties. Yet many cultural groups consume only limited amounts of milk or other dairy foods.

Courtesy of United Soybean Board

C. Soy products, such as soy milk and bean curd (known as tofu or *tobu*) are especially significant in the diet of many Asians.

© Pat O'Hara/CORBIS

D. It is estimated that nearly 30 percent of the population worldwide is dependent on fish, such as this dried salmon.

© Comstock/Jupiter Images

E. Legumes, such as beans, peas, and lentils, are eaten daily in many cultures.

© S. Meltzer/PhotoLink/Getty Images

F. Nuts are an extra source of protein in some regions, including many European, Middle Eastern, Indian, and traditional Native American cuisines.

Vegetables

Vegetables are featured in the cuisines of almost all cultures.

© Royalty-Free PhotoDisc/Getty Images

A. Greens, such as this mizuna and bok choy, are especially common.

© Jack Star/PhotoLink/Getty Images

B. Root, tubers, and bulb vegetables, such as beets, carrtos (above), celeriac, *gobo,* Jerusalem artichokes, onion, radishes, turnips, and water chestnuts, are prevalent in many regions.

© Dan Lamont/CORBIS

C. Mushrooms, fungi that are eaten as vegetables, are usually edible, but can be highly toxic. Types include shiitake (large, dark brown cap), oyster mushrooms (yellow, funnel-shaped), lobster mushrooms (orange, knobby), and black cloud ears (dark, curly fungus).

top, © Royalty-Free PhotoDisc/Getty Images; middle, © R.F. Images; bottom, © Cole Group/Getty Images

D. Examples of vegetables grown on bushes or vines include eggplants, which come in many shapes (from round to oblong, to long-thin) and colors (white, striped, green, orange, purple); chile peppers, which vary in heat from mild to mouth-searing; and numerous types of tomatoes.

Fruits

Fruit is a favorite worldwide.

© Dean Uhlinger/Royalty-Free CORBIS

A. Some regions have only a few fruit types available, such as certain desert areas where prickly pear cactus fruit is a specialty.

© Andrew Unangst/Royalty-Free CORBIS

B. Temperate regions have a broader selection of fruits, including apples, citrus fruits, berries, and more unusual fruit such as pomegranates.

© L. Hobbs/PhotoLink/Getty Images

C. The seeds of some fruits are more important than the flesh, such as these cacao pods, which are the source of chocolate. Coffee, nutmeg, tamarind, and vanilla are the other examples.

© Wolfgang Kaehler/CORBIS

D. The variety of fruit in the tropics is extensive, including breadfruit (knobby, green), *casimiroa,* (smooth, round, green), guavas, mangoes, papaya, and plantains.

© Kevin R. Morris/CORBIS

E. Durian, another tropical specialty, has a strong odor similar to rotting onions esteemed by its fans.

© E. Carey/Cole Group/Getty Images

F. The fruits of palm trees, which include coconuts (above), dates, palmyra fruit, and *pejibaye,* add variety in warmer regions worldwide.

Table 1.2 Health Disparities: Prevalence of Selected Risk Factors and Chronic Diseases among Minority Populations in the United States

Men

	American Indian (n = 751)		Black (n = 3,218)		Hispanic (n = 1,535)		Asian (n = 1,655)	
Risk factors/ chronic diseases	percent	(95 percent CI*)	percent	(95 percent CI)	percent	(95 percent CI)	percent	(95 percent CI)
Obesity	40.1	(36.2–44.0)	26.5	(24.4–28.6)	26.6	(24.1–29.2)	2.7	(1.7–4.1)
Cardiovascular diseases	16.4	(13.6–19.7)	9.9	(8.7–11.3)	7.4	(6.0–9.1)	7.5	(5.6–10.1)
Hypertension	38.5	(34.6–42.5)	34.5	(32.3–36.7)	20.5	(18.2–23.0)	16.1	(13.7–18.9)
High cholesterol	37.1	(32.5–41.9)	31.4	(29.0–33.9)	35.7	(31.9–39.7)	31.4	(27.6–35.6)
Diabetes	16.8	(14.1–19.9)	11.6	(10.2–13.1)	7.1	(6.0–8.5)	4.8	(3.6–6.4)
No. risk factors/ chronic diseases†								
0	11.7	(8.8–15.5)	24.8	(22.5–27.3)	25.4	(21.8–29.4)	36.3	(32.3–40.5)
1	26.1	(22.2–30.4)	30.5	(27.9–33.2)	34.6	(30.7–38.8)	37.1	(33.1–41.4)
2	26.4	(22.4–30.9)	22.9	(20.7–25.3)	20.0	(17.0–23.4)	19.3	(15.6–23.8)
≥ 3	35.7	(31.2–40.5)	21.7	(19.7–24.0)	19.9	(16.9–23.3)	7.2	(5.6–9.2)

Women

	American Indian (n =1,040)		Black (n =7,735)		Hispanic (n= 2,722)		Asian (n= 2,549)	
Risk factors/ chronic diseases	percent	(95 percent CI*)	percent	(95 percent CI)	percent	(95 percent CI)	percent	(95 percent CI)
Obesity	37.7	(34.4–41.1)	37.6	(36.1–39.2)	28.4	(26.4–30.6)	3.1	(2.3–4.1)
Cardiovascular diseases	13.0	(11.0–15.4)	9.4	(8.5–10.3)	5.6	(4.8–6.6)	5.5	(4.4–6.9)
Hypertension	36.8	(33.7–40.1)	40.9	(39.4–42.5)	22.4	(20.7–24.3)	17.6	(15.6–19.7)
High cholesterol	33.5	(30.0–37.2)	34.2	(32.5–35.8)	28.9	(26.5–31.5)	23.3	(20.5–26.3)
Diabetes	19.7	(17.2–22.4)	14.5	(13.4–15.7)	8.4	(7.4–9.5)	4.7	(3.8–5.8)
No. risk factors/ chronic diseases†								
0	17.2	(14.3–20.4)	22.7	(21.1–24.4)	35.9	(32.9–38.9)	57.8	(54.5–60.9)
1	27.6	(24.3–31.2)	28.4	(26.7–30.1)	30.2	(27.6–33.1)	25.8	(22.7–29.2)
2	21.9	(18.9–25.2)	22.2	(20.8–23.7)	18.4	(16.2–20.7)	11.6	(9.6–14.0)
≥ 3	33.3	(29.8–37.1)	26.7	(25.1–28.3)	15.5	(13.7–17.5)	4.8	(3.6–6.3)

* Confidence interval.

† Includes obesity, current smoking, cardiovascular diseases, hypertension, high cholesterol, and diabetes.

Source: Centers for Disease Control and Prevention (2003). Health Status of American Indians Comparison with Other Racial/Ethnic Populations—Selected States, 2001–2002. Morbidity and Mortality Weekly, 52 (47): 1148–1152.

Latinos, 107 percent.[102] Closer inspection reveals considerable variation within these broader ethnic designations as well. In Hawaii, native Hawaiians have more than twice the diabetes prevalence rate of the Korean and Filipino populations and more than three times that of the Chinese population; Japanese, Asian Indians, and Samoans have all demonstrated rapidly increasing rates after moving to the United States compared to other Asian/Pacific Islander groups. Nowhere is the difference as great as in Native Americans/Alaska Natives, where the prevalence rate for diabetes ranges from 1.9 percent in Alaska Inuit (older than age 20) to approximately 70 percent for Pima Indians in the Southwest (older than age 45), who have the highest rates of diabetes in the world. Disparities in morbidity and mortality from type 2 diabetes also exist, though it is noteworthy that much of this difference is eliminated when patients achieve the same degrees of glycemic control.

Research comparing both ethnic and racial factors in the development of end stage renal disease (ESRD) among Hispanics have demonstrated a much higher incidence of ESRD than in non-Hispanic whites. The high incidence of diabetes in this population is speculated to be a contributing factor.[103,104] Recent evaluation of health disparities for Hispanics indicate that access to medical care and health insurance is significantly lower for these individuals, which may certainly contribute to their higher prevalence of ESRD.[105]

The variable role of ethnicity is also seen in U.S. infant mortality trends. Although dramatic declines have occurred since 1950, these gains are not evenly distributed throughout the population. The gap between white and black infant deaths has increased; the mortality rate for African Americans (14.1 per 1,000 live births) is double than for the national average of 6.9 deaths per 1,000 live births. Much of the discrepancy is attributed to disorders resulting from prematurity, low birth weight, sudden infant distress syndrome (SIDS), and complications of pregnancy.[106]

Furthermore, a growing number of studies have documented inequalities in health care treatment for certain ethnic groups. Preventive care, such as immunizations and cholesterol screenings, lags behind the U.S. average, and clinical care disparities abound. For example,African Americans are much less likely than whites to

CULTURAL CONTROVERSY—Does Hunger Cause Obesity?

One of the most perplexing problems in nutrition education and policy is why socioeconomic status is associated with overweight and obesity in the United States. Rates of overweight, defined as a body mass index (BMI)* over 25 but below 30, and obesity, defined as a BMI over 30, have doubled in Americans since the late 1970s. Risk for overweight and obesity is highest in the persons with the lowest incomes and education levels regardless of ethnic heritage, and the risk declines parallel to socioeconomic improvement in most studies. Additionally, overweight and obesity rates are higher in all other ethnic groups (except for Asians) than in whites. Since poverty rates are also higher for all other ethnic groups (in some cases more than three times the rate for whites), it may be that socioeconomic status contributes to some of the disparity in the risk of overweight and obesity between ethnic groups.[125,126]

Researchers suggest that food insecurity in households that do not have enough to eat sometimes or often, or do not have enough of preferred foods to eat, may lead to overweight and obesity through overconsumption of inexpensive, less nutritious foods high in fats or sugar. First postulated by a physician in 1995, it was observed that in the cycle of food assistance, where monthly allocations run out and food shortages occur episodically, a person may compensate by eating larger portions of higher-calorie foods when available.[127] Further research has strengthened the hypothesis, finding that high-energy density diets (those that include more fast foods, snacks, and desserts than fruits, vegetables, and lean protein) are cheaper, more palatable, and more filling than healthier choices.[126,128] As with obesity in adults, obesity in children has been found to be associated with lower household incomes, lower education levels of parents, and consumption of high-energy density foods; and family meals, which improve quality of dietary intake in adolescents (including reductions in snacking), are significantly more frequent in higher-income families.[129,130] Biological factors, such as the taste preference for sweets and fats; psychological factors, including the comfort provided in such items; and an obesigenic environment that promotes consumption of energy-dense items in super-size quantities may be other variables. Further, it is uncertain as yet whether hunger and food insecurity drive overweight and obesity, or whether overweight and obesity cause hunger and food insecurity.[131]

*Weight in kilograms divided by the square of height in meters.

have renal transplants for kidney disease or coronary artery bypass surgery for heart disease, but they are significantly more likely than whites to have lower limbs amputated due to diabetic neuropathy and gastrostomy tubes used on elder patients. Further, though it is believed that health care access and low health insurance rates may be a factor in these differences, studies show even insured new immigrants and ethnic patients with comprehensive government benefits often receive unequal treatment.[107]

As these examples suggest, ethnicity can be a significant factor in the development of certain disease conditions, the way they are experienced, and how they are ultimately resolved. (See Chapter 2 for further information.) The explosive growth of ethnic groups in the U.S. population since the mid-1980s, the rapid movement toward cultural pluralism, and the undeniable connection between heritage and health evidence the urgent need for cultural competency among American health care providers.

Intercultural Nutrition

The study of food habits has specific applications in determining nutritional status and implementing dietary change. Even the act of obtaining a diet record has cultural implications. (See Chapter 3, "Intercultural Communication.") Questions such as what was eaten at breakfast, lunch, and dinner not only ignore other daily meal patterns but also make assumptions about what constitutes a meal. Snacks and the consumption of food not considered a meal may be overlooked. Common difficulties in data collection, such as under- or over-reporting food intake, may also be culturally related to the perceived status of an item, for example, or portion size estimates may be an unknown concept, complicated by the practice of sharing food from other family members' plates. Terminology can be particularly troublesome. Words in one culture may have different meanings in another culture or even among ethnic groups within a culture.

Stereotyping is another pitfall in culturally sensitive nutrition applications, resulting from the overestimation of the association between group membership and individual behavior. Stereotyping occurs when a person ascribes the collective traits associated with a specific group to every member of that group, discounting individual characteristics. A health professional knowledgeable about cultural food habits may inadvertently make stereotypical assumptions about dietary behavior if the individual preferences of the client are neglected. Cultural competency in nutrition implies not only familiarity with the food habits of a particular culture, but recognition of intraethnic variation within a culture as well.

Researchers suggest that health care providers working in intercultural nutrition become skilled in careful observation of client groups, visiting homes, neighborhoods, and markets to learn where food is purchased, what food is available, and how it is stored, prepared, served, and consumed. Participation in community activities, such as reading local newspapers and attending neighborhood meetings or events, is another way to gather relevant information. Informant interviewing reveals the most data about a group; individual members of the group, group leaders, and other health care professionals serving the group are potential sources.[108,109] Combining qualitative approaches such as in-depth, open-ended interviews with clients and quantitative measures through questionnaires is one of the most culturally sensitive methods of obtaining data about a group. Qualitative information obtained through the interviews should alert the researcher to nutrition issues within the group and guide development of the assessment tool; the quantitative results should confirm the data provided through the interview in a larger sample. (See Chapter 3 for more information.)

Cultural perspective is particularly important when evaluating the nutritional impact of a person's food habits. Ethnocentric assumptions about dietary practices should be avoided. A food behavior that on first observation is judged detrimental may actually have limited impact on a person's physical health. Sometimes other moderating food habits are unrecognized. For instance, a dietitian may be concerned that an Asian patient is getting insufficient calcium because she eats few dairy products. Undetected sources of calcium in this case might be the daily use of fermented fish sauces or broths rich in minerals made from vinegar-soaked bones.

Likewise, a food habit that the investigator finds repugnant may have some redeeming nutritional benefits. Examples include the consumption of raw meat and organs by the Inuits, which provides a source of vitamin C that would have otherwise been lost during cooking, and the use of mineral-rich ashes or clay in certain breads and stews in Africa and Latin America.

Acculturation is so complex that it has been difficult to develop accurate assessments for use in health care and research. Neither U.S. nativity nor number of years in residence has proved completely indicative, and it has been suggested that acculturation is sometimes based more on ethnic stereotyping than on cultural differences.[132,133]

Sometimes culturally based food habits have vital nutritional benefits. One example is the use of corn tortillas with beans in Mexico. Neither corn nor beans alone supplies the essential amino acids (chemical building blocks of protein) needed to maintain optimum health. Combined, they provide complete protein.

In addition, physiological differences among populations can affect nutritional needs. The majority of the research on dietary requirements has been conducted on young, white, middle-class American men. Extrapolation of these findings to other populations should be done with caution.

Thus, diet should be carefully evaluated within the context of culture. One effective method is to classify food habits according to nutritional impact: (1) food use with positive health consequences that should be encouraged, (2) neutral food behaviors with neither adverse nor beneficial effects on nutritional status, (3) food habits unclassified due to insufficient culturally specific information, and (4) food behaviors with demonstrable harmful affects on health that should be repatterned.[110] When diet modification is necessary, it should be attempted in partnership with the client and respectful of culturally based food habits. Compliance is associated with an approach that is congruent with the client's traditional health beliefs and practices. (See Chapter 2 for more information.) A recent study evaluating women's beliefs about weight gain in pregnancy found that black women indicated that a lower amount should be gained than the recommendations and that prepregnancy weight had no effect on how much should be gained.[111] Having this information could certainly impact the content and approach for nutrition counseling given during pregnancy. In another example, educators developed a food guide for Caribbean Islanders living in the United States that grouped cultural foods into three categories: growth, protection, and energy, reflecting client-group perceptions of how food affects health.[112]

The American Paradox

Food habits are so intrinsic to culture that food terminology is often employed to describe pluralism. Melting pot suggests a blending of different ethnic, religious, and regional groups to produce a smooth, uniform identity; stew implies a cooking of various populations to achieve a bland sameness with only just a touch of cultural integrity; and tossed salad allows for maintenance of cultural identity, randomly mixed and coated with a glistening unity. A more accurate metaphor for the American population is the omnivore's paradox. The nation was founded by immigrants, and most citizens today are proud of a heritage that, to paraphrase the inscription on the Statue of Liberty, accepts the tired, the poor, and the huddled masses yearning to be free. Yet many Americans are also suspicious of cultural difference and comfortable with what is familiar. (See Figure 1.6.) The same can be said for food habits in the United States.

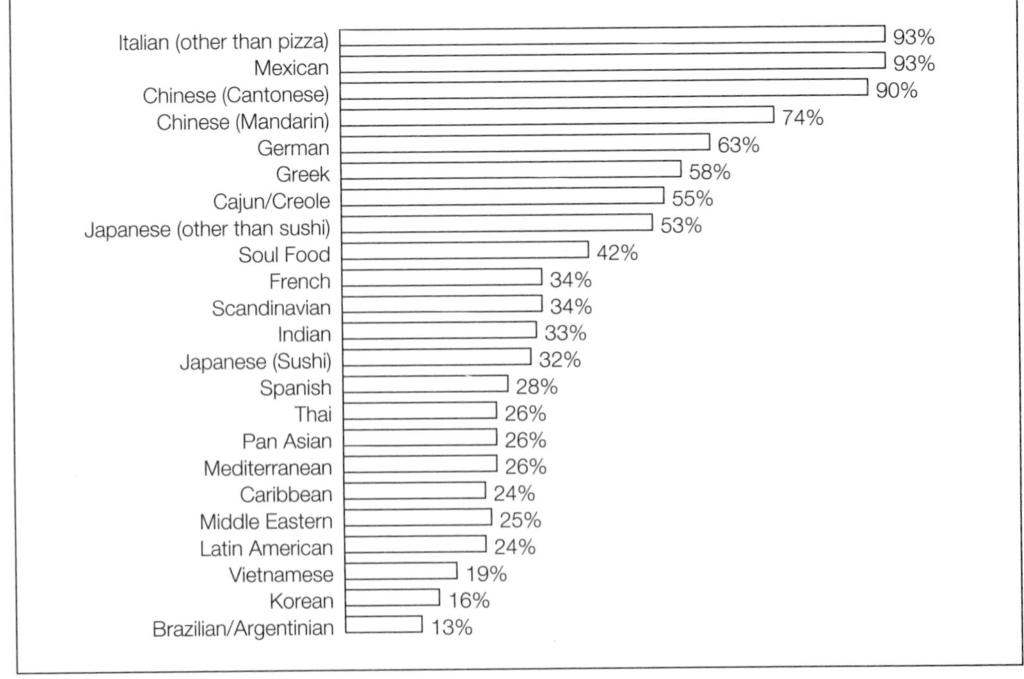

Figure 1.6
Percentage of consumers in the United States who have tried a cuisine at least once, 1999.
Source: Based on the National Restaurant Assoc. Ethnic Cuisines II survey.

The American paradox, in culinary terms, is that although foods from throughout the world are available, and often affordable, consistency and conservatism are also needed. At one end of the spectrum, people who are exposed to new foods through travel and those who crave new taste experiences have driven the rapidly expanding market for imported fruits, vegetables, and meat products, cheeses, and condiments. The growth in ethnic restaurants has far surpassed that of the restaurant industry as a whole in recent decades. A survey found that Italian, Mexican, Japanese (sushi), Thai, Middle Eastern, and Caribbean fare have grown most in popularity in recent years, while interest in French, German, Scandinavian, and soul food has declined.[113] One of the most recent developments is the success of fast-casual ethnic restaurant chains, such as Chipotle, Curry in a Hurry, L&L Hawaiian Barbecue, Mama Fu's, and Pho Hoa. At the other end of the American continuum of cuisine, some people find considerable satisfaction in the uniformity of a meat-and-potatoes diet. A national trends survey found "plain" American food most well liked by respondents (66 percent).[114]

In response to the ambivalence produced by the American paradox, the rising interest in new foods and the continued desire for familiar flavors, ethnic fare is often adapted to American tastes and standardized for national consumption. Spicing is reduced, protein elements (particularly meats and cheeses) are increased beyond traditional ratios of protein to starch or vegetable, more desserts and sweets are offered, and items considered distasteful to the American majority are eliminated. In considering the three most popular ethnic cuisines in the United States, it is unlikely a consumer will find roasted kid at an Italian restaurant, 1,000-year-old eggs at a Chinese take-out counter, or tripe soup at a Mexican drive-up window. Many Americans are convinced that spaghetti with meatballs, fortune cookies, and nachos are authentic dishes, yet they are all items created in the United States for American preferences.[114] Even cultural foods prepared at home from cookbooks are often modified for preparation in American kitchens with American ingredients, losing much of their original content and context. Only in ethnic, religious, and regional enclaves largely isolated from outside influences are traditional food habits maintained. Otherwise, over time, even significant symbolic practices can lose their meaning under the pressure of acculturation. For example, a study of Chinese Americans living in California found that while many attempted to balance their diet between hot and cold foods, few understood the yin-yang principles behind the practice.[115]

© Becky Luigart-Stayner/Terra/Corbis

▲ ***Asian tofu is the main ingredient in this vegetarian adaptation of shepherd's pie, a traditional British entrée popular in the United States.***

In many ways, U.S. cooking is founded on adaptive processes. Hamburgers, hot dogs, and fried chicken are clearly derived from other cultural fare, yet they are changed through resolution of the American paradox. Cheese melted over burgers on a sesame seed bun, chili con carne poured over frankfurters, and cornmeal-crusted chicken served with cream gravy and buttermilk biscuits are nearly unrecognizable compared to their European and African origins. And while the tamale pie in Texas, the saimin noodles in Hawaii, the tofu lasagna in a vegetarian home, and the tuna croissant sandwich in the

One example of a multicultural culinary creation is the California roll, the addition of avocado to traditional Japanese crab sushi. It is called "American sushi" in Japan.

CULTURAL FOODS IN THE UNITED STATES: A TIMELINE*

Pre-17th century

- Regional American Indian cuisines develop.

1500s

- Columbian Exchange: New World foods from the Caribbean and Central/South America (corn, potatoes, tomatoes, chili peppers, peanuts, vanilla, chocolate, etc.) are brought to Europe, Africa, Asia; Old World foods (wheat, rice, sugar, beef, pork, apples, etc.) introduced to the Caribbean and Central/South America.
- Ponce de Leon discovers Florida and most likely brings tomatoes to North America.

1620s

- British traditional midday meal introduced, with meat, fowl, or fish as its centerpiece served with cornbread or biscuit. Steamed or boiled pudding is the first course; dessert of fruit pie or cake follows. It is eaten with a knife, spoon, and fingers.
- First Thanksgiving occurs in 1621 at Plymouth colony, a ten-day celebration combining European and Native-American hunting and harvest feast traditions featuring fowl and venison.
- Dutch colonists at New Amsterdam (present-day Manhattan) introduce coleslaw, doughnuts, cookies, and waffles.

1660s

- Yams, watermelon, okra, black-eyed peas, eggplant, and sesame seeds brought with African slaves who also introduce the New World foods peanuts and chili peppers to North America.

1680s

- German Mennonites settle in Pennsylvania, creating Pennsylvania Dutch fare and popularizing dishes such as scrapple, apple butter, and funnel cakes.
- William Penn founds first brewhouse in Philadelphia, featuring English-style ales.

1760s

- England takes control of Canada from France: French Canadians migrate to New England (Franco-Americans) and Louisiana (Cajuns), bringing fish stews, pork pates, boudin sausages, French toast, and other specialties.
- An English plantation owner in New Smyrna, Florida imports 1,500 indentured servants from Italy, Greece, and Minorca to work his indigo fields, who in turn bring eggplant, lemons, and olives to the region.

1770s

- Boston Tea Party occurs; coffee takes hold as a protest beverage.
- Thomas Jefferson experiments with crops found in Europe, such as rice, broccoli, cauliflower, eggplant, savoy cabbage, and olives.

1790s

- Pineapples introduced to what is now Hawaii by the Spanish.
- *American Cookery* by Amelia Simmons in 1796 is first American cookbook; includes recipes for stuffed turkey, a "tasty Indian pudding," "pomkin" pudding (pie), "American citron" (watermelon) preserves, and cornmeal johnnycakes or hoecakes.

1800s

- First shipment of bananas arrives in the United States.
- First recipe for tomato-based ketchup published in 1812, called "love-apple or tomato catchup."

1820s

- First regional American cookbook published in 1824, *The Virginia Housewife, or Methodical Cook,* by Mary Randall, with recipes for Southern specialties; also foreign dishes, such as ropa vieja and "gaspacho" (Spain), polenta and vermicelli (Italy), curry "after the East Indian manner," and "gumbo—a West Indian dish" (Caribbean).

*REFERENCES FOR TIMELINE

Davidson, A. 1999. The Oxford companion to food. New York: Oxford University Press.
Hess, K. 1992. The Carolina rice kitchen. Columbia, SC: University of South Carolina Press.
Katz, S. H. (Ed.). Encyclopedia of food and culture. New York: Charles Scribner's Sons 2003.
Trager, J. *The Food Chronology.* New York: Henry Holt and Company, 1995.

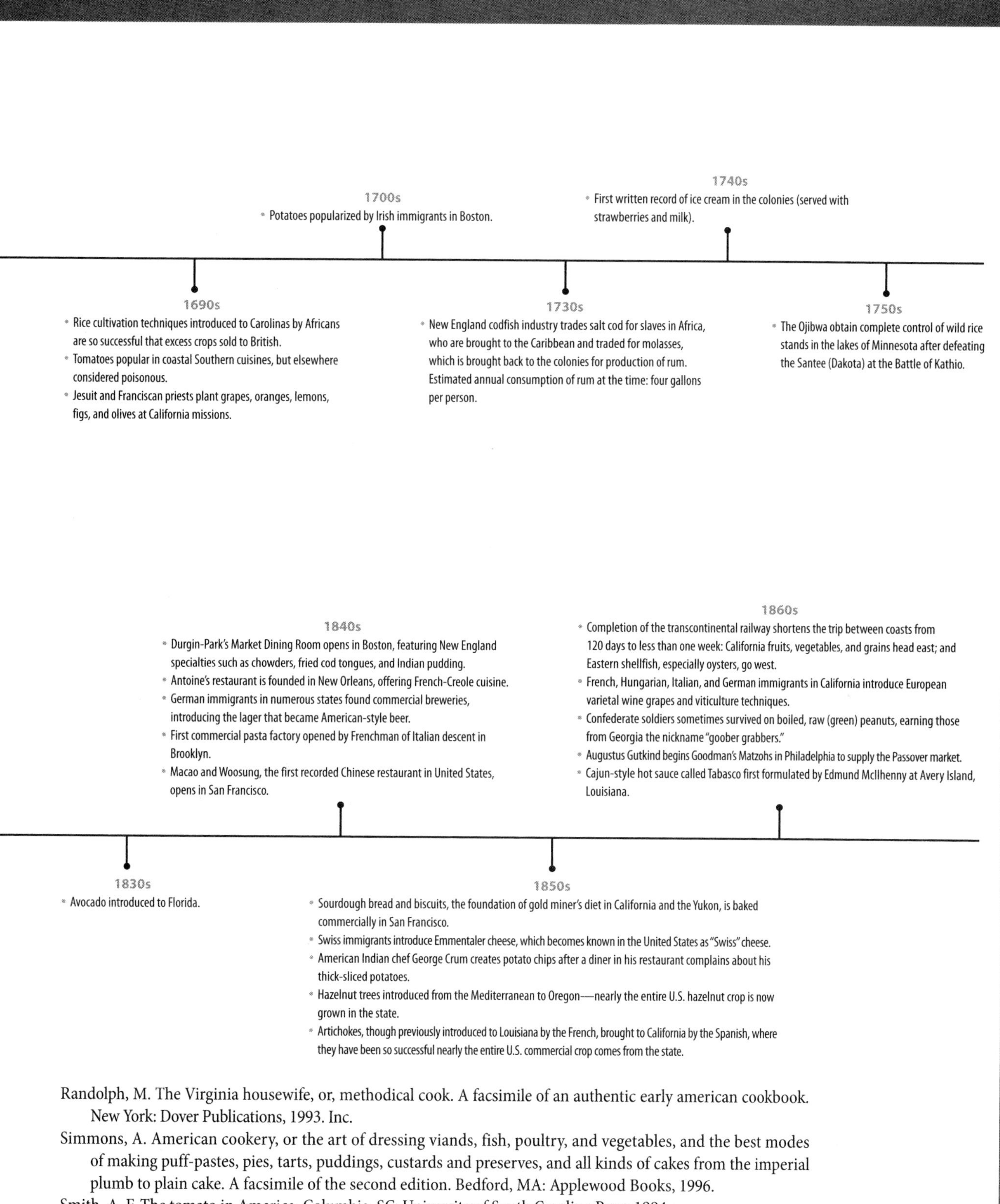

Randolph, M. The Virginia housewife, or, methodical cook. A facsimile of an authentic early american cookbook. New York: Dover Publications, 1993. Inc.

Simmons, A. American cookery, or the art of dressing viands, fish, poultry, and vegetables, and the best modes of making puff-pastes, pies, tarts, puddings, custards and preserves, and all kinds of cakes from the imperial plumb to plain cake. A facsimile of the second edition. Bedford, MA: Applewood Books, 1996.

Smith, A. F. The tomato in America. Columbia, SC: University of South Carolina Press 1994.

CULTURAL FOODS IN THE UNITED STATES: A TIMELINE* *(Continued)*

1870s

- *Jewish Cookery,* first cookbook on the subject in the United States, by Esther Levy published in Philadelphia.
- Chinese Pekin ("Peking") ducks imported by New York farms.
- Acceptance of four-tined fork makes using a knife to eat outmoded; American-style use (transferring the fork from right hand to left when cutting foods) is established.
- Buffalo (a Plains Indian staple), which numbered 30 to 70 million, are reduced to approximately 1,500 animals, in part due to prestige of smoked buffalo tongue in urban areas.
- Navel oranges introduced to California from Brazil, by way of the U.S. Department of Agriculture (USDA) in Washington, D.C.

1880s

- Luchow's restaurant opens in New York City, popularizing dishes found in German-American homes and local beer gardens, such as smoked eel, bratwurst, weinerschnitzel, spatzle, and German-style beers.
- Oscar F. Mayer, a German-American butcher, opens sausage shop in Chicago, later selling wieners to grocery stores throughout the region.
- B. Manischewitz Co. begins production of kosher products in Cincinnati.
- Italian immigrants from Naples introduce spaghetti made with olive oil and tomato paste.

1890s

- Asian immigrants move into San Joaquin valley of California, planting large tracts of land. By the 1940s, Japanese farmers are growing two-thirds of all vegetables in the state, creating shortages when they are interned during World War II.
- Chili powder, combining ground dried chili pepper bits with other seasonings, such as cumin and oregano, is invented in Texas—though attribution is uncertain. Chili stews of beef or goat popular in the region.

1900s

- Chop suey, a Chinese-American vegetable and meat dish that may have come originally from southern China (tsap seui) is popular in "chow-chows" (Chinese restaurants) in California and New York.
- Foods prepared in the "French-fashion" are popular, particularly among the upper classes who can afford to employ cooks knowledgeable in their preparation and dine at expensive restaurants.
- Pistachio tree from Middle East introduced in California and Texas.
- Coca-cola, combining extracts from African kola nuts, South American coca leaves, and fruit syrups, goes on sale in Atlanta as a fountain drink.
- The Kellogg brothers at a Seventh-Day Adventist spa create cereal flakes as a substitute for meat—a year later they add malt sweetener to increase appeal.
- Loma Linda Foods begins production of health breads and cookies for patients at the Seventh-Day Adventist Loma Linda Hospital.
- Broccoli introduced to California by Italian immigrants.
- World's Fair in St. Louis popularizes German hamburger sandwiches and frankfurters (later dubbed *hot dogs*).
- First American pizzeria opened by Italian immigrant in New York City.

1950s

- Trader Vic's restaurants in California popularize Polynesian food, such as luau dishes and pupu platters, as well as the mai tai cocktail, claimed invented by the owner.
- USDA publishes Basic 4 Food Guide.

1960s

- Beef consumption reaches 99 pounds per person in United States, surpassing pork consumption for the first time.
- Frieda's Finest founded to market specialty produce using samples and recipes in supermarkets; popularizes items such as Chinese gooseberries (renamed kiwifruit), Jerusalem artichokes (as sunchokes), radicchio, spaghetti squash, blood oranges, cactus pears, and other items.
- Julia Child debuts her cooking show, *The French Chef,* on public television, demystifying gourmet cooking and promoting French cuisine.
- The first Taco Bell fast-food restaurant opens in Downey, California.
- Benihana of Tokyo opens teppanyaki-style restaurant in New York.
- Term *soul food* coined for traditional African American cuisine.
- In Atlanta, Lester G. Maddox is ordered by the federal government to serve African Americans at his Pickrick Restaurant—he closes the business instead.
- Catfish farming introduced in Arkansas, dramatically increasing production and popularizing the fish nationwide.
- *Foods of the World* cookbooks from Time-Life Books begin with publication of *The Cooking of Provincial France* by M.F.K. Fisher—the series introduces international cuisine through 20 specialized volumes (African cooking to wines and spirits) followed in the 1970s with seven volumes on regional American fare.

1970s

- Falafel stands and restaurants proliferate with increased immigration of Middle Easterners.
- Chez Panisse restaurant opens in Berkeley, California, emphasizing fresh, locally grown ingredients, leading to development of a new California cuisine and promoting regional fare nationally.
- Nissin Foods USA founded in California to market instant noodle products popular in Japan, such as Top Ramen.
- Vietnamese refugees open small restaurants in California, Texas, and other locations, featuring traditional pho, sandwiches, and other items.
- Small numbers of immigrants arrive from Thailand, and many open restaurants, introducing fish sauces such as nuoc mam and noodle dishes, including phad thai.
- Paul Prudhomme opens K-Paul Louisiana Kitchen in New Orleans, popularizing Cajun cooking nationwide, one of the first regional food trends.

1910s

- George Washington Carver extols the virtues of peanuts, soybeans, and sweet potatoes; he popularizes peanut butter, formerly considered a food for the sick and aged.
- U.S. pasta production increases when imported supplies from Italy are cut off during World War I.
- American cheese first processed in Chicago by J.L. Kraft & Bros. (Canadian Mennonite immigrants) by melting bits of cheddar with an emulsifier to produce a smooth, mild cheese-like food.
- The fortune cookie created in California.

1920s

- LaChoy Food Products founded to sell canned and jarred bean sprouts.
- Polish baker Harry Lender opens first bagel plant outside New York, and popularity begins to spread beyond Eastern European enclaves.
- Aplets candy, based on the recipe for Turkish delight, invented by two Armenian immigrants in Washington state.
- Marriott Corp. gets its start as a root beer, tamale, and chili con carne stand in Washington, D.C.
- The Russian Tea Room opens in New York, popularizing blinis, caviar, tea in samovars, and other Russian specialties.
- Colombo Yogurt is founded by Armenian immigrants in Massachusetts.

1930s

- Fritos corn chips first marketed in Texas based on a *tortillas fritas* (fried tortilla strips) recipe purchased from a Mexican restaurant owner.
- Spam is created, becoming a status food in Hawaii and the best-selling canned meat worldwide.
- Goya Foods is founded in New York by Spanish immigrants to import olives and olive oil, later tapping into the growing Latino food market.

1940s

- Influx of Greek immigrants seeking asylum in areas such as New York, Detroit, and Chicago popularize items such as souvlaki and gyros in family-run restaurants and street stands.
- Ed Obrycki's Olde Crab House in Baltimore converts from tavern to restaurant serving Maryland specialties such as soft-shell crab and crab cakes.
- Domestic servants and some housewives take jobs to support the U.S. war effort during World War II, leading to an increased consumption of convenience foods.
- *The Gentleman's Companion, Being an Exotic Cookery Book or, Around the World with Knife, Fork and Spoon* by Charles Baker, a two-volume set, published in 1946, describing dishes and drinks from throughout Europe and Asia—a second two-volume set on the foods and beverages of South America published in 1951.
- Balducci's specialty food shop (founded as a vegetable stand in 1916 by an Italian immigrant) opens in New York, offering an international assortment of foods from Europe, Asia, and Latin America as well as regional specialties, such as rattlesnake and Cajun andouille.
- The McDonald brothers offer franchises of their hamburger stand, founded in 1940 in Pasadena, California.

1980s

- Ethiopian restaurants become popular in some cities where immigrants have settled, introducing items such as injera and berbere.
- Yaohan supermarkets of Japan open in California catering to Asian population and offering ingredients such as bean sprouts, daikon, seaweed, pickled plums, fresh fish, and prepared items, including sushi.
- Korean immigrants, especially in Los Angeles, introduce Korean barbecue, kimchi, and other specialties through restaurants and markets.
- Fresh fugu fish (which can be highly toxic) is imported for first time for use in American-Japanese restaurants under F.D.A. supervision.

1990s

- Salsa becomes the favorite U.S. condiment when sales exceed those of ketchup.
- The term *fusion food* is used for combining the ingredients and preparation techniques of two or more cultural cuisines, such as Thai chicken pizza.
- Chicken consumption per capita first tops beef consumption.
- USDA/DHHS release first version of the Food Pyramid.
- The Food Network begins television broadcasting.
- Spanish tapas restaurants become trendy.

2000s

- Americans consume an average of one tortilla per person each day—representing 30 percent of all bread sales, nearly equal to white bread.
- There are more Chinese restaurants in the United States than McDonald's, Wendy's, and Burger King restaurants combined.
- $1 out of every $7 in grocery purchases is spent on ethnic items in 2005.
- Wine is neck-and-neck with beer as favorite U.S. alcoholic beverage.

DISCUSSION STARTERS: WHO ARE YOU? AND WHAT DO YOU EAT?

- Write a short description of your cultural identity. What is your race? What is your ethnicity? What about your parents and your grandparents? Where is your family originally from? Think about your high school friends and classmates. What was their race? Their ethnicity?
- Now, write a description of what you eat. What are your favorite foods? When living at home, what foods did your family typically eat? If your parents cooked meals, what would they typically cook?

Now, form groups of 3–4 and share your descriptions with each other. Imagine that your instructor asks you to a "potluck," a social gathering where everyone invited is supposed to bring something for everyone else to eat. What you want the other members of your group to bring? What foods might they bring that you would like to try?

university cafeteria are not authentic ethnic fare, they are authentic American foods. It is the unexpected and exciting ways in which the familiar and the new are combined that make the study of food habits in the United States such a pleasurable challenge.

REVIEW QUESTIONS

1. Define food and food habits. How does the omnivore's paradox influence a person's food choices and food habits?
2. List four factors that may influence an individual's choice of foods. Pick one and explain how this factor influences food choices.
3. Define the terms culture and acculturation. Describe an example of a change in food habits that may reflect acculturation.
4. Describe the flavor principles, core foods, and meal patterns of your family's diet.
5. Which of the factors described by the consumer food choice model currently influence your food choices? Which factors do you think will stay the same and which do you think will change as you age?

REFERENCES

1. *U.S. Census Bureau, 2007 American Community Survey Table 52.* Population by Selected Ancestry Group and Region: 2007 accessed 12/3/10.
2. Camarota SA. *Immigrants in the United States—2007: A snapshot of America's foreign-born population.* Washington, DC: Center for Immigration Studies. http://www.cis.org/immigrants_profile_2007. Accessed 12/3/10.
3. Fischler C. 1988. Food, self, and identity. *Social Science Information; 1988: 27*: 275–292.
4. Rozin P. Selection of food by rats, humans, and other animals. In J.S. Rosenblatt, R.A. Hinde, E. Shaw, & C. Beer (Eds.), *Advances in the Study of Behavior.* New York: Academic Press. 1976.
5. Sadella, E., & Burroughs, J. 1981. *Profiles in eating: Sexy vegetarians and other diet-based stereotypes.* Psychology Today (October), 51–57.
6. Stein, R.I., & Nemeroff, C.J. 1995. Moral overtones of food: Judgments of others based on what they eat. *Personality and Social Psychology Bulletin, 21*, 480–490.
7. Larson N, Story M. A review of environmental influences on food choices. *Ann Behav Med. 2009; 38 Suppl 1*: S56–73.
8. Rozin, P. 1996. The socio-cultural context of eating and food choice. In H.L. Meiselman, & H.J.H. MacFie (Eds.), *Food Choice, Acceptance and Consumption.* London: Blackie Academic & Professional.
9. Shepard, R., & Raats, M.M. 1996. Attitudes and beliefs in food habits. In H.L. Meiselman, & H.J.H. MacFie (Eds.), *Food Choice, Acceptance and Consumption.* London: Blackie Academic & Professional.
10. McComber, D.R., & Postel, R.T. 1992. The role of ethnic foods in the food and nutrition curriculum. *Journal of Home Economics, 84*, 52–54, 59.
11. Wansink, B., Cheney, M.M., & Chan, N. 2003. Exploring comfort food preferences across age and gender. *Physiology & Behavior, 79*, 739–742.
12. Heldke, L. 2005. But is it authentic? Culinary travel and the search for the "genuine article." In *The Taste Culture Reader*, C. Korsmeyer (Ed.). New York: Berg.
13. Anderson, E.N. 2005. *Everyone eats: Understanding food and culture.* New York: New York University Press.
14. Helman, C. (2007). *Culture, health and illness.* Oxford, England: Oxford University Press.
15. Andrews M, Backstrand J, Boyle J, Campinha-Bacote J, et al. Theoretical basis for Transcultural Care. *J Transcult Nurs. 2010; 21(supplement)*: 53S–136S.
16. Harnack L, Story M., & Holy Rock, B. 1999. Diet and physical activity patterns of Lakota Indian adults. *Journal of the American Dietetic Association, 99*, 829–835.
17. Satia-Abouta, J.A., Patterson, R.E., Neuhouser, M.L., & Elder, J. 2002. Dietary acculturation: Applications to nutrition research and dietetics. *Journal of the American Dietetic Association, 102*, 1105–1118.
18. Franzen L, Smith C. Acculturation and environmental change impacts dietary habits among adult Hmong. *Appetite. 2009; 52*: 173–83.
19. Mezzich JE, Caracci G, Fabrega H, Kirmayer LJ. Cultural Formulation Guidelines. *Transcultural Psychiatry. 2009; 46*: 383–405.
20. Sodowsky, G.R., & Carey, J.C. 1988. Relationships between acculturation-related demographics and cultural attitudes of an Asian-Indian immigrant group. *Journal of Multicultural Counseling and Development, 16*, 117–136.
21. Bookins, G.K. 1993. Culture, ethnicity, and bicultural competence: Implications for children with chronic illness and disability. *Pediatrics, 91*, 1056–1061.
22. Meleis, A.I., Lipson, J.G., & Paul, S.M. 1992. Ethnicity and health among five Middle Eastern ethnic groups. *Nursing Research, 42*, 98–103.

23. Jelliffe, D.B. 1967. Parallel food classifications in developing and industrialized countries. *American Journal of Clinical Nutrition, 20,* 279–281.
24. Passim, H., & Bennett, J.W. 1943. Social process and dietary change. In *The Problem of Changing Food Habits.* Washington, DC: National Research Council Bulletin.
25. Mintz, S., & Schlettwein-Gsell, D. 2001. Food patterns in agrarian societies: The "core-fringe-legume hypothesis." A dialogue. *Gastronomica, 1,* 41–59.
26. Billings, J., & Sherman, P.W. 1998. Antimicrobial functions of spices: Why some like it hot. *Quarterly Review of Biology, 73,* 3–49.
27. Rozin, P., & Schiller, P. 1980. The nature and acquisition of a preference for chile peppers by humans. *Motivation and Emotion 4,* 77–101.
28. Rozin, E., & Rozin, P. 2005. Culinary themes and variations. In *The Taste Culture Reader,* C. Korsmeyer (Ed.). New York: Berg.
29. Rozin E. Flavor principles: Some applications. In *The Taste Culture Reader,* C. Korsmeyer (Ed.). New York: Berg.
30. Fischler C. Food, self, and identity. *Soc Sci Info. 1988; 27:* 275–292.
31. Douglas, M. 1972. Deciphering a meal. *Daedalus, 101,* 61–81.
32. Sobal, J. 1999. Social change and foodways. In *Proceedings of the Cultural and Historical Aspects of Food Symposium.* Corvallis: Oregon State University.
33. Gaytan, M.S. 2004. Globalizing resistance: Slow food and local imaginaries. *Food, Culture & Society, 7,* 97–116.
34. Waters A. In *the green kitchen: techniques to learn by heart.* 2010. New York: Clarkson-Potter.
35. Lowenberg, M.E. 1970. Socio-cultural basis of food habits. *Food Technology, 24,* 27–32.
36. Schwabe, C.W. 1979. *Unmentionable cuisine.* Charlottesville: University of Virginia Press.
37. Menzel, P., & D'Alusio, F. 1998. *Man eating bugs: The art and science of eating insects.* Berkeley, CA: Ten Speed Press.
38. Drewnowski, A. 2002. Taste, genetics and food choices. In H. Anderson, J. Blundell, & M. Chiva (Eds.), *Food Selection from Genes to Culture.* Levallois-Perret, France: Danone Institute.
39. Munoz, K.A., Krebs-Smith, S.M., Ballard-Barbash, R., & Cleveland, L.E. 1997. Food intakes of US children and adolescents compared with recommendations. *Pediatrics, 100,* 323–329.
40. McCrory, M.A., Saltzman, E., Rolls, B.J., & Roberts, S.B. 2006. A twin study of the effects of energy density and palatability on energy intake of individual foods. *Physiology & Behavior, 87,* 451–459.
41. Anderson, C.H. 1995. Sugars, sweetness and food intake. *American Journal of Clinical Nutrition, 62 (Supp.),* 195S–201S.
42. Anderson, E.N. 2005. *Everyone eats: Understanding food and culture.* New York: New York University Press.
43. Drewnowski, A., & Gomez-Carneros, C. 2000. Bitter taste, phytonutrients and the consumer: A review. *American Journal of Clinical Nutrition, 72,* 1424–1435.
44. Crystal, S.R., & Bernstein, I.L. 1998. Infant salt preference and mother's morning sickness. *Appetite, 30,* 297–307.
45. Delisle H. Findings on dietary patterns in different groups of African origin undergoing nutrition transition. *Appl Physiol Nutr Metab. 2010; 35:* 224–228.
46. Liu A, Berhane Z, Tseng M. Improved dietary variety and adequacy but lower dietary moderation with acculturation in Chinese women in the United States. *J Am Diet Assoc. 2010; 110:* 457–62.
47. Moreland, K., Wing, S., & Roux, A.D. 2002. The contextual effect of the local food environment on resident's diets: The artherosclerosis risk in communities study. *American Journal of Public Health, 92,* 1761–1768.
48. U.S. Dept. of Agriculture, Economic Research Service, *Household Food Security in the United States, 2007, Economic Research Report Number 66;* November 2008; http://www.ers.usda.gov/publications/err66/.
49. Cassidy, C.M. 1994. Walk a mile in my shoes: Culturally sensitive food-habit research. *American Journal of Clinical Nutrition, 59 (Suppl.),* 190S–197S.
50. Jetter, K.M., & Cassady, D.L. 2005. *The availability and cost of healthier food items.* (University of California Agricultural Issues Center AIC Issues Brief 29, March).
51. Wilson TA, Adolph AL, Butte NF. Nutrient adequacy and diet quality in non-overwieght and overweight Hispanic children of low socioeconomic status: the Viva la Familia Study. *J Am Diet Assoc. 2009; 109:* 1012–1022.
52. Zenk SN, Schulz AJ, Odoms-Young AM. How neighborhood environments contribute to obesity. *Am J Nurs. 2009; 109:* 61–64.
53. Block, J.P., Scribner, R.A., & DeSalvo, K.B. 2004. Fast food, race/ethnicity and income: A geographic analysis. *American Journal of Preventative Medicine, 27,* 211–217.
54. Lewis, L.B., Sloan, D.C., Nascimento, L.M., Diamant, A.L., Guinyard, J.J., Yancey, A.K., & Flynn, G.G. 2005. African Americans' access to healthy food options in south Los Angeles restaurants. *American Journal of Public Health, 95,* 668–673.
55. Moore LV, Diez Roux AV, Nettleton JA, Jacobs DR, Franco M. *Am J Epidemiol. 2009; 170:* 29–36.
56. Larson NI, Story MT, Nelson MC. Neighborhood environments: disparities in access to healthy foods in the US. *Am J Prev Med. 2009; 36:* 74–81.
57. Energy Information Administration. 2002. *Cooking trends in the United States: Are we really becoming a fast food country?* U.S. Department of Energy (Nov. 25). http://eia.doe.gov/emeu/recs/cookingtrends/cooking.html
58. Rollins BY, Belue RZ, Francis LA. The beneficial effect of family meals on obesity differs by race, sex, and household education: the national survey of children's health, 2003–2004. *J Am Diet Assoc. 2010; 110:* 1335–9.
59. Beydoun MA, Powell LM, Wang Y. Reduced away-from-home food expenditure and better nutrition knowledge and belief can improve quality of dietary intake among US adults. *Public Health Nutr. 2009; 12:* 369–81.

60. United Press International. *Number of US restaurants decline. August 10, 2010.* Available from: http://www.upi.com/Business_News/2010/08/21/Number-of-US-restaurants-declines/UPI-80231282400173/. Accessed 12/6/10.
61. Jekanowski, M.D., & Binkley, J.K. 2000. Food spending varies across the United States. *Food Review, 23,* 38–51.
62. New Strategist Publications. 2005. *Who's buying groceries* (3rd ed.). Ithaca, NY: Author.
63. Lannon, J. 1986. How people choose food: The role of advertising and packagings. In C. Ritson, L. Gofton, & J. McKenzie (Eds.), *The Food Consumer.* New York: Wiley.
64. Folta, S.C., Goldberg, J.P., Economos, C., Bell, R., & Meltzer, R. 2006. Food advertising targeted at school-age children: A content analysis. *Journal of Nutrition Education and Behavior, 38,* 244–248.
65. Sykes, D. 2003. Food as pleasure: Other-directedness in food ads. *Journal for the Study of Food and Society, 6,* 49–56.
66. Henry AE, Story M. Food and beverage brands that market to children and adolescents on the internet: a content analysis of branded web sites. *J Nutr Ed and Behav. 2009; 41:* 353–9.
67. Duquesne D. *Food porn: love it or hate it?* Huffington Post. December 1, 2010. Available from: http://www.huffingtonpost.com/daphne-duquesne/food-porn_b_790549.html#s196042. Accessed: 12/6/2010.
68. Gardyn, R. 2003. Teen food fetishes: Average male teenager eats five times a day, female eats four times a day. *American Demographics, March,* p. 2.
69. Contento, I.R., Williams, S.S., Michaela, J.L., & Franklin, A.B. 2006. Understanding the food choice process of adolescents in the context of family and friends. *Journal of Adolescent Health, 38,* 575–582.
70. Blisard, N., Lin, B.H., Cromartie, J., & Ballenger, N. 2002. America's changing appetite: Food consumption and spending to 2020. *Food Review, 25,* 1–9.
71. Smeets, P.A., de Graaf, C., Stafleu, A., van Osch, M.J., Nievelstein, R.A., & van der Grond, J. 2006. Effect of satiety on brain activation during chocolate tasting in men and women. *American Journal of Clinical Nutrition, 83,* 1297–1305.
72. Rozin, P. 1996. The socio-cultural context of eating and food choice. In H.L. Meiselman, & H.J.H. MacFie (Eds.), *Food Choice, Acceptance and Consumption.* London: Blackie Academic & Professional.
73. Appleton, K.M., Gentry, R.C., & Shepherd, R. 2006. Evidence of a role for conditioning in the development of liking for flavours in humans in everyday life. *Physiology & Behavior, 87,* 478–486.
74. Birch, L.L. 1999. Development of food preferences. *Annual Review of Nutrition, 19,* 41–62.
75. Martins, Y., Pelchat, M.L., & Pliner, P. 1997. "Try it; it's good for you": Effects of taste and nutrition information on willingness to try novel foods. *Appetite, 28,* 89–102.
76. Sealy YM. Parents' perceptions of food availability: implications for childhood obesity. *Soc Work Health Care. 2010; 49:* 565–580.
77. Variyan, J.N., & Golan, E. 2002. New information is reshaping food choices. *Food Review, 25,* 13–18.
78. Aikman, S.N., Min, K.E., & Graham, D. 2006. Food attitudes, eating behavior, and the information underlying food attitudes. *Appetite, 47,* 111–114.
79. House, J., Su, J., & Levy-Milne, R. 2006. Definitions of healthy eating among university students. *Canadian Journal of Dietetics Practice and Research, 67,* 14–18.
80. Shepard, R., & Raats, M.M. 1996. Attitudes and beliefs in food habits. In H.L. Meiselman, & H.J.H. MacFie (Eds.), *Food Choice, Acceptance and Consumption.* London: Blackie Academic & Professional.
81. Woolcott, D.M. 2002. Impact of information and psychosocial factors on nutrition behavior change. In H. Anderson, J. Blundell, & M. Chiva (Eds.), *Food Selection from Genes to çulture.* Levallois-Perret, France: Darone Institute.
82. Goody CM, Drago L. Introduction: Cultural competence and nutrition counseling. In: *Cultural Food Practices.* Chicago IL: American Dietetic Association, 2010.
83. Campinha-Bacote J. A model and instrument for addressing cultural competence in health care. *J Nurs Educ. 1999; 38:* 203–207.
84. Andrews, M. M., & Boyle, J. S. (Eds.). (2008). *Transcultural concepts in nursing care* (5th ed.). Philadelphia, PA: Lippincott, Williams & Wilkins.
85. Skipper A, Young LO, Mitchell B. 2008 accreditation standards for dietetic education. *J Am Diet Assoc. 2008; 108*: 1732–1735.
86. American Diabetes Association of Diabetes Educators. Cultural sensitivity and diabetes education: recommendations for diabetes educators. *Diabetes Educ. 2007; 33:* 41–44.
87. US Census Bureau. *USA Statistics in Brief.* Available from: http://www.census.gov/compendia/statab/2010/files/racehisp.html. Accessed 12/6/2010.
88. Foundation/Commission on Dietetic Registration 2004 Dietetics Professionals Needs Assessment. *Journal of the American Dietetic Association, 105,* 1348–1355.
89. Pamies, R.J., Hill, G.C., Watkins, L., McNamee, M.J., & Colburn, L. 2006. Diversity and the healthcare workforce. In *Multicultural Medicine and Health Disparities,* D. Satcher, & R.J. Pamies (Eds). New York: McGraw-Hill.
90. Statistics Canada. *2006 Census Analysis.* Available from: http://www12.statcan.gc.ca/census-recensement/2006/as-sa/index-eng.cfm. Accessed: 12/6/2010.
91. Goel, M.S., McCarthy, E.P., Phillips, R.S., & Wee, C.C. 2004. Obesity among US immigrant subgroups by duration of residence. *Journal of the American Medical Association, 292,* 2860–2867.
92. Singh, G.K., & Siahpush, M. 2002. Ethnic-immigrant differentials in health behaviors, morbidity, and cause-specific mortality in the United States: An analysis of two national data bases. *Human Biology, 74,* 83–109.
93. Delisle H. Findings on dietary patterns in different groups of African origin undergoing nutrition transition. *Appl Physiol Nutr Metab. 2010; 35:* 224–228.

94. Liu A, Berhane Z, Tseng M. Improved dietary variety and adequacy but lower dietary moderation with acculturation in Chinese women in the United States. *J Am Diet Assoc. 2010; 110:* 457–462.
95. Perez-Escamilla R. Dietary quality among Latinos: is acculturation making us sick? *J Am Diet Assoc. 2009; 109:* 988–91.
96. Steffen, P.R., Smith, T.B., Larson, M., & Butler, L. 2006. Acculturation to Western society as a risk factor for high blood pressure: A meta-analytic review. *Psychosomatic Medicine, 68,* 386–397.
97. Palinkas, L.A., & Pickwell, S.M. 1995. Acculturation as a risk factor for chronic disease among Cambodian refugees in the United States. *Social Science and Medicine, 40,* 1643–1653.
98. Kaplan, J.B., & Bennett, T. 2003. Use of race and ethnicity in biomedical publication. *Journal of the American Medical Association, 289,* 2709–2716.
99. Hamilton, J. A. (2008). Revitalizing difference in the Hap Map: Race and contemporary human genetic variation research. *Journal of Law and Medical Ethics, Fall,* 471–477.
100. Krishnan S, Cozier YC, Rosenberg L, Palmer JR. Socioeconomic status and incidence of type 2 diabetes: results from the Black Women's Health Study. *Am J Epidemiol. 2010; 171:* 564–70.
101. Centers for Disease Control. 2007. *National diabetes fact sheet: General information and national estimates on diabetes in the United States, 2007.* Atlanta, GA: U.S. Department of Health and Human Services, Centers for Disease Control and Prevention. Accessed 12/7/10.
102. Dagago-Jack, S., & Gavin, J.R. 2006. Diabetes. In *Multicultural Medicine and Health Disparities,* D. Satcher, & R.J. Pamies (Eds.) New York: McGraw-Hill.
103. Collins AJ, Foley RN, Herzog C, et al. United States Renal Data System 2008 Annual Data Report Abstract. *Am J Kidney Dis. 2009; 53 (1 Suppl):* vi–374.
104. Peralta CA, Risch N, Lin F, Shlipak MG, Reiner A, Ziv E, Tang H, Siscovick D, Bibbins-Domingo K. The Association of African Ancestry and elevated creatinine in the Coronary Artery Risk Development in Young Adults (CARDIA) Study. *Am J Nephrol. 2010; 31(3):* 202–8.
105. Lora CM, Daviglus ML, Kusek JW, Porter A, Ricardo AC, Go AS, Lash JP. Chronic kidney disease in United States Hispanics: a growing public health problem. *Ethn Dis. 2009; 19:* 466–72.
106. Centers for Disease Control. *Eliminate disparities in infant mortality.* 2010. Available from: http://www.cdc.gov/omhd/amh/factsheets/infant.htm. Accessed 12/7/2010.
107. Douglas MK, Pacquiao DF. (Eds). 2010. Core curriculum in trnascultural nursing and health care [Supplement]. *J Transcult Nursing, 21 (Suppl. 1).*
108. Stang J, Bayerl CT, American Dietetic Association. Position of the American Dietetic Association: child and adolescent nutrition assistance programs. *J Am Diet Assoc. 2010; 110:* 791–99.
109. Nelms MN, Habash D. Nutrition assessment: foundation of the nutrition care process. IN: Nelms M, Sucher K, Lacey K, Long S. *Nutrition Therapy and Pathophysiology.* 2e. Belmont CA: Cengage Learning, 2011.
110. Jelliffe, D.B., & Bennett, F.J. 1961. Cultural and anthropological factors in infant and maternal nutrition. *Proceedings of the Fifth International Congress of Nutrition, 20,* 185–188.
111. Groth SW, Kearney MH. Diverse women's beliefs about weight gain in pregnancy. *J Midwifery Womens Health. 2009; 54:* 452–7.
112. Stowers, S.L. 1992. Development of a culturally appropriate food guide for pregnant Caribbean immigrants in the United States. *Journal of the American Dietetic Association, 92,* 331–336.
113. National Restaurant Association. 2000. *Non-traditional ethnic cuisines gain in popularity.* August 24. Available online: http://www.restaurant.org/pressroom/pressrelease/?id=126.
114. Sloan, A.E. (Ed.) 2005. Top 10 global food trends. *Food Technology, 59,* 20–32.
115. Chau, P., Lee, H.S., Tseng, R., & Downes, N.J. 1990. Dietary habits, health beliefs, and food practices of elderly Chinese women. *Journal of the American Dietetic Association, 90,* 579–580.
116. Walker, A. 2005. In the absence of food: A case of rhythmic loss and spoiled identity for patients with percutaneous endoscopic gastrostomy feeding tubes. *Food, Culture & Society, 8,* 161–180.
117. Rozin, P., Fallon, A., & Augustoni-Ziskind, M. 1985. The child's conception of food: The development of contamination sensitivity to "disgusting" substances. *Developmental Psychology, 21,* 1075–1079.
118. Sutherland, L.L. 2002. Ethnocentrism in a pluralistic society. *Journal of Transcultural Nursing, 13,* 274–281.
119. Pause, B.M., Rogalski, K.P., Sojka, B., & Ferstl, R. 1999. Sensitivity to androstenone in female subjects is associated with an altered brain response to male body odor. *Physiology & Behavior, 68,* 129–137.
120. Rozin, P., Haidt, J., & McCauley, C.R. 1993. Disgust. In M. Lewis, & J.M. Haviland (Eds.), *Handbook of Human Emotions.* New York: Guilford.
121. U.S. Department of Labor/U.S. Department of Labor Statistics. 2006. *100 years of U.S. consumer spending.* Available online: http://www.bls.gov/opub/uscs/home.htm.
122. Ayala, G.X., Mueller, K., Lopez-Madurga, E., Campbell, N.R., & Elder, J.P. 2005. Restaurant and food shopping selections among Latino women in Southern California. *Journal of the American Dietetic Association, 105,* 38–45.
123. Wilkins, J. 2005. *This black widow has quite a bite.* San Diego Union-Tribune, August 1 p. D3.
124. Yang, F. M., & Levkoff, S. E. (2005). Ageism and minority populations: Strengths in the face of challenge. *Generations, 29*(3), 42–48.
125. DeNavas-Walt, C., Proctor, B.D., & Lee, C.H. U.S. *Census Bureau. 2005. Income, Poverty, and Health Insurance in the United States: 2004.* Washington, DC: U.S. Government Printing Office.
126. Drewnowski, A., & Spector, S.E. 2004. Poverty and obesity: The role of energy density and energy costs. *American Journal of Clinical Nutrition, 79,* 6–16.

127. Dietz, W.H. 1995. Does hunger cause obesity? *Pediatrics, 95,* 766–767.
128. Jetter, K.M., & Cassady, D.L. 2005. *The availability and cost of healthier food items.* (University of California Agricultural Issues Center AIC Issues Brief 29, March).
129. Kumanyika, S., & Grier, S. 2006. Targeting interventions for ethnic minority and low-income populations. *The Future of Children, 16,* 187–207.
130. Drewnowski A. Obesity, diets, and social inequalities. *Nutr Rev. 2009 May; 67 Suppl 1:* S36–9.
131. Scheier, L.M. 2005. What is the hunger-obesity paradox? *Journal of the American Dietetic Association, 105,* 883–886.
132. Schwartz, S.J., Pantin, H., Sullivan, S., Prado, G., & Szapocznik, J. 2006. Nativity and years in receiving culture as markers of acculturation in ethnic enclaves. *Journal of Cross Cultural Psychology, 37,* 345–353.
133. Hunt, L.M., Schneider, S., & Comer, B. 2004. Should "acculturation" be a variable in health research? A critical review of research on US Hispanics. *Social Science & Medicine, 59,* 973–986.

CINCINNATI STATE LIBRARY

Traditional Health Beliefs and Practices

CHAPTER 2

Health and illness in America are usually considered the specialty of mainstream biomedicine. Furthermore, health promotion is based on scientific findings of researchers regarding diet, exercise, and lifestyle issues such as smoking cessation and stress management; disease is treated according to the latest technologies. In reality, health care is pluralistic in the United States, as well as in most other cultures. Many people in the United States never consult a physician or allied health care provider when physical or emotional symptoms occur, relying on home remedies and popular therapies found readily on the Internet rather than seeking professional help. Complementary and alternative medicine (CAM) is popular with many Americans. The National Center for Complementary and Alternative Medicine recently published data gathered from the National Health Interview Survey and estimates that approximately 38 percent of all adults and approximately 12 percent of all children used some form of CAM during the year of 2007. This is a significant increase since the last report in 2002. The top five CAM therapies included natural products, deep breathing, meditation, chiropractic and osteopathic interventions, and massage.[1] Consumer spending on such practices and products has more than tripled in the past decade, from $11 billion annually to nearly $40 billion.[2] When biomedical care is sought, it is often in conjunction with these other systems. The term *integrative medicine* is used when there is a combination of conventional and CAM treatments that have demonstrated scientific evidence of safety and effectiveness.[1]

Culture determines how a person defines health, recognizes illness, and seeks treatment. Traditional health beliefs and practices can be categorized in various ways: through the etiology of illness (due to personal, natural, social, or supernatural causes) or by the cures that are employed (the use of therapeutic substances, physical forces, or magico-religious interventions). There is no consensus, however, on these classifications. In this chapter, home remedies, popular approaches such as folk and alternative traditions, and professional systems (including U.S. biomedicine, traditional Chinese medicine, and ayurvedic medicine) are reviewed within the cultural context of health and illness. Specific beliefs and practices are detailed in the following chapters on each American ethnic group.

Biomedicine is the term used to describe the conventional system of health care in the United States and other Western nations based on the principles of the natural sciences, including biology, physiology, and biochemistry.

Ayurvedic medicine is the ancient Asian Indian system of healing.

Worldview

Cultural Outlook

Each cultural group has a unique outlook on life, based on a common understanding and ranking of values. These standards typically represent what is considered worthy in a life well lived. They are a collective expression of preferences and priorities—not absolutes—and individuals within a society may hold a spectrum of beliefs. However, expectations about personal and public

Table 2.1 Comparison of Common Values

Majority American Culture	Other Cultural Groups
Mastery over nature	*Harmony with nature*
Personal control over the environment	*Fate*
Doing—activity	*Being/becoming*
Time dominates	*Personal interaction dominates*
Human equality	*Hierarchy/rank/status*
Individualism/privacy	*Group welfare*
Youth	*Elders*
Self-help	*Birthright inheritance*
Competition	*Cooperation*
Future orientation	*Past or present orientation*
Informality	*Formality*
Directness/openness/honesty	*Indirectness/ritual/"face"*
Practicality/efficiency	*Idealism*
Materialism	*Spiritualism/detachment*

Source: Adapted from E. Randall-David, *Strategies for Working with Culturally Diverse Communities.* Association for the Care of Children's Health, 19 Mantua Rd., Mt. Royal, NY 08061. Copyright 1989.

conduct, assumptions regarding social interaction, and assessments of individual behavior are determined by this cultural outlook, or worldview. This perspective influences perceptions about health and illness as well as the role of each within the structure of a society.[3,4,5] Majority American values, which are shared by most whites and to some degree by many other ethnic groups in the United States, emphasize individuality and control over fate (Table 2.1). Personal accountability and self-help are considered cultural cornerstones. One study found that 82 percent of American consumers believe they are directly responsible for their own health.[6] Most other cultures worldwide believe that fate—including the will of God, the actions of supernatural agents, or birthright (i.e., astrological alignment or cosmic karma)—is a primary influence in health and illness. Although most cultures have complex practices regarding the maintenance of health, the concept of preventative health care, such as annual checkups, is unknown in some cultures where fate dominates.

The concept of preventive health care, such as annual checkups provided by a biomedical professional, is unknown in some cultures where fate is believed to determine health.

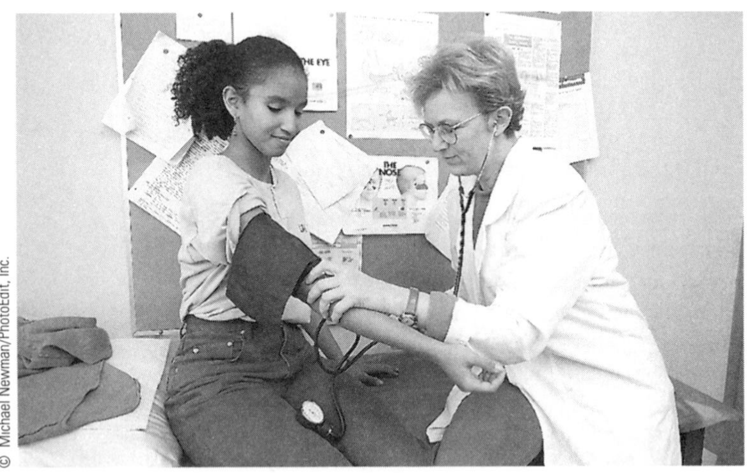

© Michael Newman/PhotoEdit, Inc.

The significance of fate often coincides with differences in perceptions of time. Many Americans place great value on promptness and schedules; they are also future oriented, meaning they are willing to work toward long-term goals or make sacrifices so that they or their children will reap rewards in the future. The majority members in the United States are also monochronistic, with a preference for concentrating on one issue or task at a time in a sequential manner. Many other cultural groups live in the present and are often polychronistic, or comfortable doing many things at once. A Mexican American who is talking with his grandmother while fixing an appliance and watching a baseball game on television is unlikely to cut the visit short just because he has a medical appointment. Immediate interests and responsibilities, including interpersonal relationships, are more important than being on time. A few cultures, such as certain Native American nations, are past oriented, living according to historical direction.

Most majority Americans are very task oriented and desire direct participation in their health care; they feel best when they can do something. Other cultures place a greater value on being and feel comfortable with inactivity. Self-worth is based more on personal relationships than on accomplishments. The expectation is that the health care practitioner will take responsibility for treatment. The whole idea of the provider–client partnership may be alien to Asians, who often expect to be fully directed in their care. While many Americans value patient autonomy and confidentiality, other cultural groups, such as Middle Easterners, believe that the family should be involved in all health care decisions—the welfare of the group outweighs that of the individual.

Americans consider honest, open dialogue essential to effective communication, and informality is usually a sign of friendliness. Many other cultures prefer indirect communication techniques and expect a formal relationship with everyone but intimate family members. In cultures where identity with a group is more significant than individuality, social status and hierarchy are respected, which can have an impact on the practitioner–client relationship (see Chapter 3, "Intercultural Communication," for more information).

Worldview is especially evident in serious, life-and-death health care decisions. Southeast Asians may appear indifferent to a terminally ill family member and have little interest in prolonging life because of a faith in reincarnation. Some African Americans distrust white-American health care recommendations regarding do-not-resuscitate orders in part because they contradict the critical role of faith in African American healing. An Orthodox Jewish patient may believe that physicians are mandated to preserve life and that any person who assists death through denial of sustaining care is a murderer; a non-Orthodox Jew may believe that no one should endure unrelievable pain and thus dying should not be prolonged. Middle Easterners traditionally demand that everything be done to keep a person alive because death is in God's hands, and one must never give up hope. Mexican-American family members might view death as part of God's plan for a relative; they might be against anything that would quicken death, or they may expect the practitioner to make the decision.[7,8,9]

Most health care situations are not cases of life or death, and worldview affects many other, less catastrophic aspects of health and illness as well. It is useful to examine the biomedical worldview and understand the perspective of most U.S. health care providers before learning about other traditional health beliefs and practices. Comparisons between biomedicine and other medical systems can reveal areas of potential disagreement or conflict regarding how and why illness occurs and the expectations for treatment before working with a client. Compliance increases with clinical approaches that concur with the client's worldview.[10,11]

Some majority Americans find eating a meal a disruption of daily tasks; others adhere to strict meal schedules. In polychronistic societies, meals are usually leisurely events, a chance to enjoy the blessings of food in the company of family and friends.

Biomedical Worldview

Biomedicine is a cultural subdivision of the American majority worldview. It shares many beliefs with the dominant outlook but differs in a few notable areas.[12,13] There are certainly exceptions to the biomedical worldview within certain specialties, and by some practitioners, yet many of the underlying assumptions are culture-specific. The tendency is for health care providers to enforce their beliefs, practices, and values upon clients, sometimes unknowingly because they are unaware of cultural differences, but more often because they believe their ideas are superior. This process is called *cultural imposition*, and it impacts nearly all client care.[14]

Some researchers have noted that although the biomedical community often calls clients whose cultural background differs from the majority "hard to reach," this term is equally applicable to health professionals who refuse to provide culturally appropriate care.[67]

Relationship to Nature

Biomedicine adheres to the concept of mastery over nature. Practitioners are soldiers in the war on cancer (or other conditions). They fight infection, conquer disease, and kill pain. Technology is considered omnipotent; its tools are the arsenal used to battle pain and illness.

One factor in this approach is the attitude that health can be measured numerically and that there are standardized definitions of disease. Blood and urine analyses, X-rays, scans, and other diagnostic tests are used to define whether a patient is within normal physical or biochemical ranges. Figures falling within designated parameters mean the patient is functioning normally; if the data are too high or too low, the patient is in an abnormal state that may indicate disease. Diagnosis occurs independent of the idiosyncratic characteristics of the individual, usually without consideration of cultural factors such as ethnic background or religious faith. Symptoms occurring outside numerical confirmation are frequently determined to be of psychosomatic origins.[12]

Personal Control or Fate?

The conventional U.S. medical system leaves little room for chance or divine intervention. Scientific rationality dictates that there is a biomedical cause for every condition, even if it is as yet undiscovered. Each individual inherits a certain physiological constitution and has a personal responsibility to make the choices that prevent illness. Receiving immunizations and getting regular checkups are biomedical ways in which an individual can preserve health. Being obese, smoking cigarettes, consuming immoderate amounts of alcohol, and failing to manage stress are biomedical examples of how an individual may endanger health. A person who behaves in a manner believed to cause disease is often stigmatized.

When a person is ill, the biomedical assumption is that he or she will reliably comply with therapy, and that treatment, if undertaken correctly by the patient, will alleviate the condition. The onus of cure is dependent on personal behavior. From the patient's perspective, there is the presumption that health care professionals will provide mistake-free care. Malpractice suits filed when care was less than perfect have led to extensive charting and record keeping in the U.S. biomedical system.

State of Being

The number of adults over sixty-five years in the United States is expected to double by the year 2050; figures among some ethnic groups, such as African Americans, Asian Americans, Native Americans and Latinos, show even greater growth.

Congruent with the value placed on personal control, biomedical patients are expected to be active partners in their cure. Complacency and noncompliance are disliked by biomedical practitioners. Changes in lifestyle can help preserve health; taking medications and completing therapeutic regimens can relieve symptomatic pain or cure disease. The biomedical emphasis is on doing, not being. Other worldviews may expect client passivity and acceptance of adverse conditions. Clients are the recipients of healing, not participants in healing.

Role of the Individual

Similar to the American majority worldview, individuality is honored in U.S. biomedicine, and client confidentiality is nearly inviolate. Individuals are seen as a single, biological unit, not as members of a family or a particular cultural group. It is assumed that a person desires privacy, and clients are sometimes encouraged by providers to keep medical matters quiet, even if it means withholding information from relatives. Treatment typically is focused on each client, in keeping with the beliefs of personal responsibility and the provider–patient partnership.

Human Equality

A fundamental premise in American biomedicine is that all patients deserve equal access to care, although, practically speaking, cost, location, and convenience prevent many patients from receiving adequate health services. This is a relatively unique perspective; most other societies deliberately ration health care through assessing physical status (e.g., a young person may receive services denied to a terminally ill older person) or through socioeconomic status (e.g., the wealthy can purchase care; the poor are left to whatever society offers).

The biomedical worldview on human equality differs substantially from the mainstream American outlook in one way, however. A hierarchy of biomedical professionals is strictly observed in the United States, with physicians having the highest status and allied health professionals substantially less. Health care workers outside the professional system, such as clerical and custodial workers, and those beyond the reach of biomedicine, such as folk healers, are accorded even lower standing. Deference to those of superior rank is expected. The client is typically inferior to biomedical professionals within this hierarchy.

Aging

Biomedicine supports the majority American worldview in its value on youthfulness. Many aspects of health care practice are dedicated to postponing the aging process, from plastic surgery to the technological prolongment of life. The fear of aging is so pervasive in the U.S. culture that it influences health care outside the conventional biomedical system as well. Numerous alternative traditions promise everlasting youth through the use of certain products. The emphasis on youthfulness is in direct conflict with other cultural worldviews that honor the wisdom that comes with aging and that hold high esteem for elders.

Perceptions of Time

Biomedicine is future-oriented, with a focus on what can be done today so that the client will be better tomorrow. Often treatments are unpleasant, invasive, and even painful at the moment of their application, yet the hope is that they will benefit the client in the future. Long-term management of disease and illness–prevention strategies such as diet is even more oriented toward future benefits.

Although being on time for appointments and taking medications when scheduled are valued in clients, biomedical practitioners are notorious for their disrespect of the client's time. Clients are frequently asked to arrange non-emergency consultations weeks or even months in advance and may be kept waiting on the days of their appointments.

Degree of Formality/Degree of Directness

The established biomedical hierarchy, as well as the emphasis on timeliness, is often reflected in the degree of informality observed in dialogue between provider and patient. The provider often addresses the client by his or her first name, yet expects the patient to use formal titles in return. The provider usually spends limited time on small talk and attempts to get quickly to the problem; the expectation is the patient will also use direct approaches. Extensive jargon without explanation is often employed.

Biomedical practitioners value honest, open communication with patients because it enhances their ability to diagnose and treat disease, and it assists in issues such as informed consent. Other cultural worldviews, however, value indirect or intuitive communication with health care practitioners (see Chapter 3, for more information). Some cultures also believe that the family, not the patient, should be told about serious conditions.[15]

Materialism or Spirituality?

Each disease, from the biomedical viewpoint, has its own physiological characteristics: a certain cause, specific symptoms, expected test results, and a predictable response to treatment. To many biomedical health care providers, an illness isn't real unless it is clinically significant; emotional or social issues are the domain of other specialists. Biomedicine differs from most traditional health care approaches in the recognition of the mind–body duality. Nearly all other cultures consider the mind and body as a unified whole. Somatization refers to the expression of emotions through bodily complaints. In biomedical culture, somatic symptoms are often interpreted as a maladaptive emotional response, yet they are the most common presentation of psychological distress in patients worldwide.[16] In folk medicine and some alternative traditions, the emotional needs of the patient are addressed through physical therapies. Spiritual intervention is frequently sought concurrently.

The separation between physical and emotional or psychological health is so embedded in American culture that no English word exists to even express the concept of mind–body unity.

What Is Health?

Cultural Definitions of Health

Meaning of Health

The World Health Organization (WHO) describes health as "a state of complete physical, mental, and social well-being, not merely an absence of disease or infirmity." Although comprehensive from a biomedical perspective, this definition does not fit the worldview of many cultural groups, because it ignores the natural, spiritual, and supernatural dimensions of health.

Most Native Americans believe that health is achieved through harmony with nature, which includes the family, the community, and the environment. Africans also emphasize a balance with nature and believe that malevolent environmental forces such as those of nature, God, the living, or the dead may disrupt a person's energy and bring illness. Many African Americans, Latinos, Middle Easterners, and some southern Europeans attribute health to living according to God's will. Gypsies maintain health through avoiding contact with non-Gypsies, who are considered inherently polluted. Most Asians believe that health is dependent on their relationship to the universe and that a balance between polar elements, such as yin and yang, must be maintained. Some Southeast Asians are concerned with pleasing their ancestor spirits, who may cause accidents or sickness when angry. Pacific Islanders believe that fulfilling social obligations is essential to health and that disharmony with family or village members can result in illness. Asian Indians consider mind, body, and soul to be interconnected and believe that spirituality is as important to health as a good diet or getting

The word health comes from the Anglo-Saxon term *hal,* meaning "wholeness."

proper rest (see individual chapters on each ethnic group for more details).

Health in other cultures is less dependent on symptoms than on the ability to accomplish daily responsibilities. Among Koreans, there is a strong desire to avoid burdening their children with their health problems. Mexican men may ignore physical complaints because it is considered weak and unmanly to acknowledge pain. Even within a single culture, socioeconomic differences may contribute to the definition of health; daily aches are tolerated when a weekly paycheck is essential.[12]

Health Attributes

As health is defined culturally, so are the characteristics identified with health. Physical attributes are most commonly associated with well-being, including skin color, weight maintenance, and hair sheen. Normal functioning of the body, such as regular bowel movements, routine menstruation, and a steady pulse, is expected, as is the use of arms, legs, hands, and the senses. Undisturbed sleep and appropriate energy levels also suggest good health.

In ayurvedic medicine, a distinction is made between general health and optimal health.

Behavioral norms within the context of marriage, family, and community are sometimes considered a sign of well-being. It is the cultural specifics of health characteristics that tend to vary. Healthy hair in the United States is advertised as clean, shiny, and flake-free. In many cultures, oily hair is the norm, and dandruff is not a significant concern. Americans count on a single, strong pulse of about 72 beats per minute when resting, while in other medical systems there is more than one pulse of importance to health, and these pulses are a primary diagnostic tool in illness. Pregnancy is a medical condition in the United States warranting regular exams by biomedical professionals, whereas in many societies pregnancy is a normal aspect of a healthy woman's cycle, and prenatal care is uncommon. Generally speaking, Americans expect to be content in their lives; many other cultures have no such assumptions and do not link happiness with well-being.

In traditional Chinese medicine, 15 separate pulses are identified, each associated with an internal organ and each with its own characteristics.

Body Image

One area of significant cultural variation regarding health is body image. Perceptions of weight, health, and beauty differ worldwide. In the United States, there is significant societal pressure to be thin. Although there is no scientific concurrence on the definition of ideal or even healthy weight for individuals, being overweight is usually believed to be a character flaw in the majority American culture. Even health care professionals reportedly make moral judgments about obesity, depicting overweight persons as weak-willed, ugly, self-indulgent, and fair game for ridicule.[17,18,19] The health risks associated with being overweight cause some providers to presume ill health in their obese clients. Thinness corresponds to the biomedical worldview regarding mastery of nature, the idea that the intellect can control the appetite.[20]

Historically, thinness has been associated with a poor diet and disease. In many cultures today, including those of some Africans, Caribbean Islanders, Filipinos, Mexicans, Middle Easterners, American Indians, and Pacific Islanders, being overweight is a protective factor that is indicative of health as well as an attribute of beauty. Many overweight African American women, for example, are less concerned about weight issues and more satisfied with their bodies than overweight white women.[21,22] A larger ideal body image is the norm for most black men and women regardless of age, education, or socioeconomic status.[23,24] Some black Caribbean Islanders and Puerto Rican women also report a larger body size as attractive to family and peers when compared to Anglo, Eastern European-American, and Italian-American women[25] Some Hispanic women value a heavier profile for themselves, and even if they opt for a slimmer body personally, they may prefer plump children.[26,27] Researchers have found that some young African American and Latina women purposefully contest the majority culture emphasis on thinness, substituting self-acceptance and nurturance.[28]

Researchers have found that attitudes about weight sometimes change when an immigrant enters a culture with different perceptions regarding health and beauty. More acculturated Hispanic women and children were more likely to choose a thinner figure as ideal than those who were less acculturated, ideal body image for Samoan women in Hawaii varied with whether they identified with Western or non-Western culture, and Puerto Ricans living on the mainland United States expressed a desire for thinness that is between that of their country of origin and that of the majority culture in their new homeland.[29,30] Among some Native Americans, ideal body size has changed over time. Elders are more likely than younger adults to prefer a heavier profile, and children demonstrate a desire for even thinner bodies.[31,32,33] Some studies also suggest the pressure to be thin may be impacting young persons more than adults: the percentage of normal-weight teens engaging in unhealthy weight control behaviors did not vary by ethnicity in a national examination of high school students, and another study found Asian, Hispanic, and American Indian adolescent girls reported similar numbers of weight-related concerns as white girls; however, African American girls had fewer weight issues.[34,35,36] Some Native American schoolchildren express a high level of body dissatisfaction, and concerns about overweight were high in a cohort of third-grade children, with Latinas and African American girls reporting the same or greater level of body dissatisfaction than white and Asian girls.[37,38]

Health Maintenance

Health Habits

Just as with health attributes, there are some broad areas of intercultural agreement on health habits. Nearly all people identify a good diet, sufficient rest, and cleanliness as necessary to preserving health. It is in the definitions of these terms that cultural variations occur. For example, majority Americans typically identify three meals each day as a good diet. Asians may indicate a balance of yin and yang foods is a requirement. Middle Easterners may be concerned with sufficient quantity, and Asian Indians may be concerned with religious purity of the food. To most Americans, keeping clean means showering daily, while some Filipinos bathe several times each day to maintain a proper hot-cold balance.

In the United States the National Center for Health Statistics describes the trends for dietary intake. These include that during 2007–2008, the average energy intake for men was 2,504 kcal and for women was 1,771 kcal. The average carbohydrate intake of 47–50 percent total kcal is slightly reduced from previous years whereas protein intake was slightly higher.[39] Some cultural groups would find these data irrelevant to health status. Macronutrient intake may not be associated with disease prevention, and dietary supplement use may not be familiar. Physical labor is often a factor in preserving health, but recreational exercise is rare throughout much of the world. Alcohol consumption is prohibited by several religions. Preventative care is unusual in many cultures.

Culturally specific health practices differ particularly in those beliefs passed on within families. A small survey of U.S. college students from many backgrounds revealed notable variation in health habits beyond general concepts regarding diet, sleep, physical activity, and cleanliness.[39,40] Dressing warmly (Eastern European, French, French Canadian, Iranian, Irish, Italian, Swedish) and avoiding going outdoors with wet hair (Eastern European, Italian) were listed by some. Daily doses of cod liver oil (British, French, French Canadian, German, Norwegian, Polish, Swedish) or molasses (African American, French, French Canadian, German, Irish, Swedish) as a laxative were frequently reported maintenance measures. Natural amulets were traditionally worn in some families to prevent illness, such as camphor bags (Austrian, Canadian, Irish) or garlic cloves (Italian). Faith was important to many of the students, expressed as blessing of the throat (Irish, Swedish) and wearing holy medals (Irish), as well as daily prayer (Canadian, Ethiopian).

Health-Promoting Food Habits

Food habits are often identified as the most important way in which a person can maintain health. Nearly all cultures classify certain foods as necessary for strength, vigor, and mental acuity. Some also include items that create equilibrium within the body and soul.

General dietary guidelines for health usually include the concepts of balance and moderation. In the United States, current recommendations include a foundation of complex carbohydrates in the form of whole grains, vegetables, and fruits; supplemented by smaller amounts of protein foods such as meats, legumes, and dairy products; and limited intakes of fats, sugar, salt, and alcohol. The Chinese system of yin-yang encourages a balance of those foods classified as yin (items that are typically raw, soothing, cooked at low temperatures, white or light green in color) with those classified as yang (mostly high-kcalorie foods, cooked in high heat, spicy, red-orange-yellow in color), avoiding extremes in both. Some staple foods, such as boiled rice, are believed to be perfectly balanced and are therefore neutral.

In ancient China, nutritionists were ranked highest among health professionals.

The neutral category of foods is usually expanded by those Asians who prefer biomedicine and believe that traditional health systems are unsophisticated.[68]

© Mitch Hrdlicka/Photodisc/Getty Images

▲ *The health value of specific foods varies culturally. In the United States, it is said that milk builds strong bones. But in certain Native American cultures, milk is considered a weak food. Some Latinos believe milk is only good for children.*

Although which foods are considered yin or yang vary regionally in China, the concept of keeping the body in harmony through diet remains the same, usually adjusted seasonally to compensate for external changes in temperature and for physiological conditions such as age and gender (see Chapter 11, "East Asians").

Aspects of the yin-yang diet theory are found in many other Asian nations, and a similar system of balance focused on the hot-cold classification of foods is practiced in the Middle East, parts of Latin America, the Philippines, and India. Hot-cold concepts developed out of ancient Greek humoral medicine that identified four characteristics in the natural world (air-cold, earth-dry, fire-hot, water-moist) associated with four body humors: hot and moist (blood), cold and moist (phlegm), hot and dry (yellow/green bile), and cold and dry (black bile). Applied to daily food habits, this system usually focuses on only the hot and cold aspects of food (defined by characteristics such as taste, preparation method, or proximity to the sun) balanced to account for personal constitution and the weather. In Lebanon, it is believed that the body must have time to adjust to a hot food before a cold item can be eaten. In Mexico, the categorization of hot and cold foods is related to a congruous relationship with the natural world. Asian Indians associate a hot-cold balanced diet with spiritual harmony.

Ritualistic cannibalism, especially when the heart or liver of a brave and worthy enemy was consumed, is an extreme example of the sympathetic qualities of food.

Quantity of food is often associated with health as well. Some African Americans, for example, traditionally eat heavy meals, reserving light foods for ill and recuperating family members. In the Middle East ample food is necessary for good health, and a poor appetite is sometimes regarded as an illness in itself. As discussed previously, being overweight is frequently associated with well-being in some cultures.

In addition to balance and moderation, specific foods are sometimes identified with improved strength or vitality. In the United States, milk is considered to build strong bones, carrots are considered to improve eyesight, and candy is considered to provide quick energy. Chicken soup, a traditional tonic among Eastern European Jews, has become a well-accepted cure-all. Navajos consider milk to be a weak food, but meat and blue cornmeal are strong foods. Asians call strengthening items *pu* or *bo* foods, including protein-rich soups with pork liver or oxtail in China, and bone marrow or dog-meat soup in Korea. Puerto Ricans drink eggnog or malt-type beverages to improve vitality.

The sympathetic quality of a food, meaning a characteristic that looks like a human body part or organ, accounts for many health food beliefs. The properties of a food entering the mouth are incorporated into physical traits. Some Italians drink red wine to improve their blood, and American women sometimes eat gelatin (which is made from animal hooves) to grow longer, stronger fingernails. Throughout Asia and parts of the United States, ginseng, which is a root that resembles a human figure, is believed to increase strength and stamina. Other foods are believed to prevent specific illnesses. Americans, for instance, are urged to eat cabbage-family (cruciferous) vegetables to reduce their risk of certain cancers. Oatmeal (high in soluble fiber) and fish (high in omega-3 fatty acids) have both been promoted as preventing heart disease.

Some cultures believe that fresh foods prepared at home are healthiest and, in the United States, the popularity of locally grown items and organic foods (those produced without the use of chemical additives or pesticides) has increased in recent years. Vegetarianism, macrobiotics, customized diets that account for an individual's food sensitivities or allergies, and very low-fat or low-carbohydrate diets are a few of the other ways in which health is promoted by some people through food habits.

Disease, Illness, and Sickness

Cultural Definitions of Disease, Illness, and Sickness

When health is diminished, a person experiences difficulties in daily living. Weakness, pain and discomfort, emotional distress, or physical debilitation may prevent an individual from fulfilling responsibilities or obligations to the family or society. Researchers call this experience illness, referring to a person's perceptions of and reactions to a physical or psychological condition, understood within the context of worldview. In biomedical culture, illness is caused by disease, defined as abnormalities or malfunctioning of body organs and systems. The term *sickness* is used for the entire disease–illness process. When an individual becomes sick, questions such as how did the illness occur, how are the symptoms experienced, and how is it cured arise—answered primarily through cultural consensus on the meaning of sickness.

Becoming Sick

During the onset of a sickness, physical or behavioral complaints make a person aware that a problem exists. The development may be slow, and the symptoms may take time to manifest into a disease condition. Or symptoms may occur suddenly, and it is quickly obvious that illness is present.

Except in emergencies, an individual usually seeks confirmation of illness first from family or friends. Symptoms are described and a diagnosis is sought. A knowledgeable relative is often the most trusted person in determining whether a condition is cause for concern and whether further care should be pursued; in many cultures, a mother or grandmother is the medical expert within a family. This is a major step in social legitimization of the sickness. If others agree that the person is ill, then the individual can adopt a new role within the family or community—sick person. In this capacity, the sick person is excused from many daily obligations regarding work and family, as well as social and religious duties. A reprieve from personal responsibility for well-being is also given, with care provided by relatives, healers, or health professionals. The role of sick person provides a socially accepted, temporary respite from the physical and psychological burdens of everyday life, with the understanding that sickness is not a permanent condition and that recovery should occur.[12,40]

Explanatory Models

When unexpected events happen, there is a human need to explain the origins and causes of seemingly random occurrences. Explanatory models congruent with a culture's worldview are used to account for why good or evil happens to a person or a community and to calm individual fears of being victimized. In sickness, the explanatory model details the cause of disease, the ways in which symptoms are perceived and expressed, the ways in which the illness can be healed and prevented from reoccurring, and why one person develops a sickness whereas another remains healthy.[41,42,43]

The etiology of sickness is of central concern because the reason an illness occurs often determines the patient's outlook regarding the progression and cure of the sickness. In biomedical culture, three causes of disease are identified: (1) immediate causes, such as bacterial or viral infection, toxins, tumors, or physical injury; (2) underlying causes, including smoking, high cholesterol levels, glucose intolerance, or nutritional deficiencies; and (3) ultimate causes, such as hereditary predisposition, environmental stresses, obesity, or other factors.[44] The causes of illness are generally more complex. Four theories on the etiology of sickness prevalent in most societies have been described (Figure 2.1): those originating in the patient, those from the natural world, those from the social world, and those due to supernatural causes.[12] It is important to note that in no society do all persons subscribe to any single cause,

Macrobiotics is a Japanese diet based on brown rice, miso soup, and vegetables that was popularized in Europe as promoting health in the 1920s. Serious nutritional deficiencies have been identified in infants and toddlers on this restricted diet.[69]

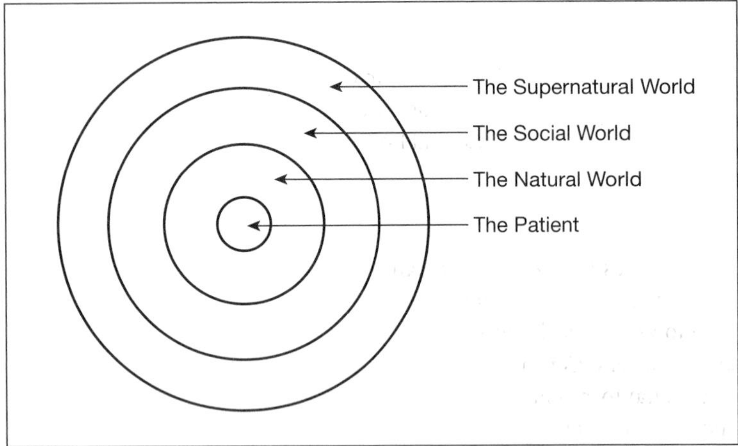

Figure 2.1
Client theories on the etiology of sickness.
Source: From C.G. Helman, *Culture, Health, and Illness* (2nd ed.). © 2000 Hodder Arnold.

and there is considerable variation in degree of belief, intraculturally as well as interculturally. Additionally, believing in a cause does not necessarily result in a practice associated with a cause.

Sickness Due to the Patient. First are those sicknesses that develop within the individual patient, usually attributable to a person's constitution or lifestyle; that is, an individual has a genetic or psychological vulnerability to illness or disease, either an inherited weakness or emotional susceptibility, such as depression. Sickness may also be due to lifestyle choices. A person in the United States may be blamed for a heart attack if he has become overweight, eats fatty foods, smokes tobacco, and never exercises. A person who fails to wear a seat belt and then is injured in an auto accident may also be found at fault. Responsibility for sickness falls primarily on the patient, although in many other cultures when a person's actions are unfavorable to health, it is outside forces that are thought to actually cause an illness or accident in retaliation for the offense.

Sickness Due to the Natural World. Etiology in the natural world includes environmental elements such as the weather, allergens, smoke, pollution, and toxins. Wind or bad air is of particular concern in many cultural groups, including some Arabs, Chinese, Italians, Filipinos, and Mexicans, because it can enter the body through pores, orifices, or wounds in the body, causing illness. Viruses, bacteria, and parasites are natural biological agents of sickness. Humoral systems, which associate various body humors with natural elements (as described previously), connect illness and disease with disharmony in the environment. Astrology, which determines an individual's fate (including health status) through planetary alignment at the time of birth, is another natural world phenomenon. Injuries due to natural forces, such as lightning or falling rocks, are sometimes categorized with this group; however, many cultures believe such accidents are actually the result of supernatural injunction.

Regarding the 1918 influenza epidemic, the *New York Post* reported that epidemics are the punishment that nature inflicts for the violation of her laws and ordinances.[70]

Sickness Due to the Social World. Sickness attributed to social causes occurs around interpersonal conflict within a community. It is common to blame an enemy for pain and suffering. Inadvertent or purposeful malevolence is the source of illness and disease in many cultures. Among the most common causes is the evil eye.

It is widely believed in parts of Africa, Asia, Europe, Greece and the Middle East, India, Latin America, and some areas of the United States that an individual who stares (especially with envy) can project harm on another person, even if the gaze is unintentional. Children are believed to be particularly vulnerable to the evil eye, resulting in colic, crying, hiccups, cramps, convulsions, and seizures. Among adults, the evil eye can cause headache, malaise, complications in pregnancy and birth, impotence and sterility in men, and insanity.[45] Protections against the evil eye include such practices as placing a red bag filled with herbs on an infant's crib in Guatemala; knotting black or red string around children's wrists in India; leaving children unwashed and unadmirable in Iran; wearing a charm in the form of a black hand (*mano negro*) in Puerto Rico; and painting a house white and blue to blend with the sky, thus avoiding notice, in Greece. Eastern European Jews wear a red ribbon; Sephardic Jews wear a blue ribbon. In Scotland, a fragment of the Bible is kept on the body, and in Muslim areas of Southeast Asia, a piece of the Koran is worn.[40]

Conjury is another frequent social cause of sickness. A person with imputed powers to manipulate the natural or supernatural world directs illness or injury toward an individual, or sells the magic charms or substances necessary for a normal person to inflict harm. Conjury is practiced by witches (called *brujos* or *brujas* in Spanish), sorcerers, root doctors, herb doctors, voodoo or hoodoo doctors (see Chapter 8, "Africans"), underworld men and conjure men, most of whom obtain their powers from the devil or other evil spirit. For example, a conjurer might sprinkle graveyard dust under a person's feet, causing him to waste away, known as fading in rural African American tradition. A bundle of sticks placed in the kitchen will cause illness in people who consume food prepared there. A brujo can cause illness in Latinos through contagious magic, using bits of a person's hair or fingernails when casting a spell. Native American conjury often uses animals or natural phenomena (such as lightning) to attack a victim, or causes natural objects to be inserted into the body, resulting in pain. The bewitched Native American may behave in inexplicable, disruptive ways and may be abandoned by the community if considered incurable and unable to change undesirable conduct.[40,46] There is often an overlap between sickness attributed to the social world and that caused by supernatural forces.

Sickness Due to the Supernatural World. In the supernatural realm, sickness is caused by the actions of gods, spirits, or the ghosts of ancestors. The will of God is a prominent factor in illness and disease suffered by many Jews, Christians, and Muslims. Sickness is sometimes considered a punishment for the violation of religious covenants, and other times it is viewed as simply a part of God's unknowable plan for humanity. Even those persons who do not follow a specific faith may ascribe illness to fate, luck, or an act of God. Some Africans, Asians, Latinos, Middle Easterners, American Indians, and Pacific Islanders believe that malevolent spirits can attack a person, causing illness. For example, among Cambodians, death can occur when the nightmare spirit immobilizes a victim through sitting on his or her chest and causing extreme fright.[47] In other situations spirit possession takes place. An evil spirit inhabits the body of a victim who then exhibits aberrant behavior, such as incoherent speech or extreme withdrawal. Many Southeast Asians associate caretaker spirits with body organs and life forces that may desert a person when angered or frightened, leaving that individual vulnerable to sickness. In addition, the ghosts of ancestors usually protect their living relatives from harm, but may inflict pain and illness when ignored or insulted.

One of the most common causes of sickness in many cultures is soul loss, when the soul detaches from a person's body, usually due to emotional distress or spirit possession. The symptoms typically include general malaise, listlessness, depression, a feeling of suffocation, or weight loss. If left untreated, soul loss can lead to more serious illness.

Folk Illnesses

Inasmuch as sickness is culturally sanctioned and explained through culture-specific models, it follows that each culture recognizes different disorders. Certain symptoms, complaints, and behavioral changes are associated with specific conditions and are termed folk illnesses or culture-bound syndromes. Examples of such sicknesses are numerous, such as cases of soul loss experienced by some Asians, Latinos (who call it *susto* or *espanto*), Native Americans, Pacific Islanders, and Southeast Asians. Muso, experienced by young Samoan women as mental illness, and sudden unexpected nocturnal death syndrome (SUNDS) suffered by Cambodians (see the previous section, "Sickness Due to the Supernatural World") are cases of folk illness due to evil spirits. Strong emotions, particularly fright or anger, cause many folk conditions, such as stroke precipitated in *bilis* or *colera* in some Guatemalans, the cooling of the blood and organs in *ceeb* among the Hmong, or the stomach and chest pain of *hwabyung* in some Koreans. Psychological distress is often expressed through somatic complaints in some cultures; for instance, an Asian Indian may present symptoms of extreme stress as burning on the soles of the feet, or a depressed Asian Indian man may experience *dhat*, the loss of semen.

Diet-related folk illnesses are common. High blood and low blood among some African Americans are examples. Depending on the cultural group, imbalance in the digestive system results in numbness of the extremities (*si zhi ma mu*) in some Chinese; nausea and the feeling of a wad of food stuck in the stomach (*empacho*) among Mexicans; and paralysis in some Puerto Ricans (*pasmo*). The eating disorders anorexia nervosa (a fear of fat and failure to maintain body weight resulting in a weight 15 percent or more below that recommended) and bulimia nervosa (binge eating followed by the use of self-induced vomiting, laxatives, enemas, or medications to reduce calorie intake, or the use of excessive exercise or fasting) are

Other names for the evil eye in the United States include the bad eye, the look, the narrow eye, and the wounding eye. Those who are victims have been blinked, eye bitten, forelooked, or overlooked.

Fear of the evil eye is mentioned in Talmudic writings, the Bible, and the Koran.

▲ *Body image is one area where the viewpoint of the client may vary from the biomedical assumptions of the health care provider. Obesity in some cultures is still considered desirable since it indicates wealth and having adeqate amounts of foods.*

sometimes described as a culture-bound syndrome in the United States and other westernized nations, associated with issues such as the drive to thinness, body image, maturity, and control.[48,49,71] In the case of anorexia, it is usually the biomedical culture that identifies the symptoms as a disease state. Many anorectics do not consider themselves ill or in need of medical intervention. Such differences in the definition of sickness account for why some conditions, such as anorexia or other folk illnesses, are difficult to cure with biomedical approaches. Effective treatment of many sicknesses depends on agreement between the patient and the practitioner regarding how the illness has occurred, the meaning of the symptoms, and how the sickness is healed.[42]

Healing Practices

Biomedicine has increasingly focused on disease to the exclusion of illness. Biomedical health professionals attempt to diagnose and cure the structural and functional abnormalities found in patient organs or systems. In contrast, healing addresses the experience of illness, alleviating the infirmities of the sick patient even when disease is not evident. Healing responds to the personal, familial, and social issues surrounding sickness.

Seeking Care

When sickness occurs, a person must make choices regarding healing. Professional biomedical care, if available, is usually initiated when the onset of symptoms is acute or an injury is serious. Nearly all cultural groups recognize the value of biomedicine in emergencies.

Choice of care often depends on the patient's view of the illness in cases when the sickness is not life-threatening. In these situations, home remedies are generally the first treatment applied.[33,45] Therapies may be determined by the patient alone or in consultation with family members, friends, or acquaintances. If the remedies are ineffective, if other people encourage further care, or if the individual experiences continued disruption of work, social obligations, or personal relationships, professional advice may be sought. The type of healer chosen depends on factors such as availability, cost, previous care experiences, referrals by relatives or friends, and how the patient perceives the problem. If the patient suffers from a folk illness, a folk healer may be sought immediately because biomedical professionals are considered ignorant about such conditions. Otherwise, biomedical care may be undertaken, independently or simultaneously with other approaches. A study of Taiwanese patients revealed that in acute illness, biomedical care was initially sought; but, if treatment was ineffective, traditional Chinese medical practitioners were employed; if there was no progress in healing, another traditional specialist would be tried; and if the patient was still afflicted, sacred healers would be sought. In chronic or recurrent sickness, biomedical, traditional Chinese medicine, and spiritual approaches would be attempted concurrently.[10] The use of multiple approaches is particularly common when there are concerns that a condition is culture specific.[50,51,52,53]

Research suggests that large numbers of Americans obtain health care outside the biomedical system for minor and major illnesses.[1,54] As many as one-third to one-half of patients with intractable conditions (i.e., back pain, chronic renal failure, arthritis, insomnia, headache, depression, gastrointestinal problems), terminal illnesses such as cancer or acquired immune deficiency syndrome (AIDS), and eating disorders seek unconventional treatment. Nearly all do so without the recommendation of their biomedical doctor, integrating multiple therapies on their own.

Biomedicine is rejected by some people because their experience with care has been impersonal, costly, inconvenient, or inaccessible. Further, conventional treatments may have been painful or harmful. Some clients believe that biomedical professionals are hostile or uninterested in ethnic health issues.[43,55] Often, the biomedical approach is incongruent with the patient's perspective on sickness. The health care professional may express disdain for explanations of etiology that disagree with the biomedical viewpoint or dismiss complaints with no discernible clinical diagnosis as insignificant. Folk healers and other alternative practitioners can provide an understanding of illness within the context of the patient's worldview and can offer care beyond the cure of disease, including sincere sympathy and renewed hope.

CULTURAL CONTROVERSY—Botanical Remedies

More than 80 percent of the world's population uses herbal remedies to treat illness and optimize health. Technically, an herbal medicine contains only leafy plants that do not have a woody stem.[72] A more comprehensive term is botanical, including all therapeutic parts of all plants, from the root (e.g., ginseng), the bark (e.g., willow), the sap (e.g., from aloe), the gum (e.g., frankincense) or oil (e.g., from nutmeg), the flowers (e.g., echinacea), the seeds (e.g., gingko biloba), to the fruit (e.g., bilberries). Botanical remedies often use the whole plant, which practitioners claim is superior to using a single active extract because other components in the plant may work together synergistically in the preparation to enhance the therapeutic value and to buffer any side effects. For this same reason, plants are often combined in formulary mixtures, particularly in traditional Chinese medicine.

Most consumers select botanical remedies instead of biomedicine because they believe they are safer and more effective than prescription drugs or they are treating chronic conditions for which biomedicine has little to offer in the way of relief. Some proponents note that botanicals have been used for centuries and that reported deaths each year number in only the hundreds, compared to over 100,000 deaths due to prescription drug–induced conditions.[57,73] The key word, however, is *reported*. The Dietary Supplement Health and Education Act (DSHEA) passed by Congress in 1994 defines dietary supplements as separate from food and drugs and thus outside the scope of federal monitoring. Manufacturers are exempt from regulations requiring that complaints, injuries, or deaths due to consumption of their product be reported to the Food and Drug Administration (FDA). Though the FDA retains the right to protect the public from harmful products, the burden of proof is on the government to prove that a particular botanical remedy is unsafe. Many manufacturers have voluntarily adopted good manufacturing processes, and the American Herbal Products Association has created a botanical safety rating system that classifies herbs as (1) safe when consumed appropriately; (2) restricted for certain uses; (3) use only under the supervision of an expert; and (4) insufficient data to make a safety classification.

Unfortunately, the explosive, unregulated growth of the industry has resulted in numerous problems. Of particular concern is the interaction of botanicals when used with biomedical therapies. For example, gingko biloba reduces the effectiveness of some prescription drugs, such as certain antacids and antianxiety medications, while potentiating others, including anticoagulants, antidepressants, and antipsychotics.[74] Some botanicals can react adversely with anesthesia, and others can interact with radiation therapy.[75,76] Further, natural products can be adulterated with pesticides, heavy metals (such as mercury), or prescription drugs (such as warfarin or alprazolam).[77] (Tables of selected botanical remedies for each ethnic group are available through the Food and Culture book companion website.)

Healing Therapies

No consensus prevails among researchers on the classification of what is called unconventional, alternative, or folk medical care. Home remedies (i.e., herbal teas, megavitamins, relaxation techniques), popular therapies (i.e., chiropractic, homeopathy, hypnosis, massage), and professional practices (i.e., those that require extensive academic training in conventionally recognized medical systems, such as biomedicine, traditional Chinese medicine, and ayurvedic medicine) include a variety of treatments that fall into three broad categories: (1) administration of therapeutic substances, (2) application of physical forces or devices, and (3) magico-religious interventions.[13,56] Most patients use unconventional therapies without the supervision of a biomedical doctor or any other kind of health care provider. Popular and professional practitioners, when consulted, may use one or several of these treatments in healing a patient.

Administration of Therapeutic Substances. Biomedical pharmaceutical and diet prescriptions are two of the most common types of therapeutics in this category, which also includes over-the-counter medications, health food preparations, and prepackaged diet meals, as well as vitamins and mineral supplements. In a 2007 survey, it was estimated that more than 38 percent of all adult Americans used complementary and alternative medicine, including high doses of vitamins, during the previous month.[1] Home remedies and health practitioners other than biomedical professionals often emphasize the use of botanical medicine, which includes whole plants or pieces (particularly herbs), and occasionally animal parts, such as antlers or organs, or certain powdered mineral elements. In many cultures,

The person most likely to use complementary and alternative medicine in the United States is a middle- to upper-class white or Asian woman who is well educated in consumer health issues and lives in an urban area in one of the Pacific states.[5]

Medications using digitalis, opiates, and salicylates, common today as biomedical therapeutics, were first used by folk healers.

© Mitch Hrdlicka/Photodisc/Getty Images

▲ *Ginseng, a root found in both North America and Asia, is one of the top ten common herbal remedies used in the nation. It reputedly promotes health through increased strength and vitality, and may be taken specifically to treat digestive upset, anxiety, or sexual impotence.*

healers specialize in the use of herbal preparations; often they are elder men or women with intimate knowledge of the natural environment. Root doctors in the American South, and the proprietors of *botánicas* (herbal pharmacies) found in some Latino neighborhoods are a few examples. In addition to folk healing, both traditional Chinese medicine and ayurvedic medicine make extensive use of botanical medicine (see the chapters on each American ethnic group for more details).

Naturopathic doctors trained in the United States attend a four-year program including many biomedical disciplines. Doctors of chiropractic (DC) are the third largest category of health care practitioners in the United States, following physicians and dentists. Osteopaths are licensed to prescribe medications and perform surgery as doctors of osteopathy (DO) in all 50 states.

Homeopathy also prescribes therapeutic substances, such as botanical medicine, diluted venom, or bacterial solutions, and biomedical drugs. Originating in Germany, homeopathy is based on the concept that symptoms in illness are evidence that the body is curing itself, and acceleration or exaggeration of the symptoms speeds healing. One primary tenant is that "like cures like." Naturopathic medicine also focuses on helping the body to heal itself, usually through noninvasive natural treatments (including some physical manipulations, as the following section describes), although biomedical drugs and surgery are used in certain cases. Nutritional therapy, based on whole foods and dietary supplements, is the foundation of naturopathic health maintenance and healing.

Kur (spa) therapy is popular throughout Europe, particularly in Germany.

Application of Physical Forces or Devices. Manipulations of the body operate on the shared premise that internal functioning improves with minor adjustments of physical structure. Chiropractic theory states that misalignments of the spine interfere with the nervous system, interrupting the innate intelligence that regulates the body, resulting in disease and disorder. Osteopathic medicine proposes that blood and lymph flow, as well as nerve function, improves through manipulation of the musculoskeletal system, particularly the correction of posture problems, mobilization of bone joints, and spine alignment. Health problems are treated through restoration of mobility and suppleness.

Several Asian healing therapies can be classified as the application of physical forces or devices and are practiced by family members in the home, by folk healers, by specialists in the therapies, or by traditional Chinese medical physicians. Massage therapy, acupressure, and pinching or scratching techniques are used to release the vital energy flow through the twelve meridians of the body identified in traditional Chinese medicine, primarily by relieving muscle tension so that oxygen and nutrients can be delivered to organs and wastes removed. Coining is a related practice in which a coin or spoon is rubbed across the skin instead of pressing or pinching specific points. Acupuncture is similar to acupressure in that it attempts to restore the balance of vital energy in the body along the meridians, but it differs in that it stimulates specific junctures through the insertion of nine types of very fine needles. The needles do not cause bleeding or pain. Acupuncture is considered useful in correcting conditions where too much heat (yang) is present in the body. In conditions of too much cold (yin), another technique is preferred, called moxibustion, in which a small burning bundle of herbs (e.g., wormwood) or a smoldering cigarette is touched to specific locations on the meridians to restore the balance of energy. A similar method is cupping, the placement of a heated cup or a cup with a scrap of burning paper in it, over the meridian points.[57]

Application of electricity is used in various electrotherapies, primarily to stimulate muscle or bone healing, especially in sports medicine. Biofeedback also uses small electric pulses to teach a person how to consciously monitor and control normally involuntary body functions, such as skin temperature and blood pressure, to alleviate health problems, which include insomnia, gastrointestinal conditions, and chronic pain.

Hydrotherapy involves the application of baths, showers, whirlpools, saunas, steam rooms, and poultices to relieve the discomforts of back pain, muscle tension, arthritis, hypertension, cirrhosis of the liver, asthma, bronchitis, and head colds. In addition to the hydrotherapeutic qualities, the mineral content of the water is considered stimulating.

Magico-religious Interventions. Spiritual healing practices are associated with nearly all religions and cosmologies. They typically fall into two divisions: those actions taken by the individual, and those taken on behalf of the individual by a sacred healer.

In Western religious traditions, God has power over life and death. Sickness represents a breach between humans and God. Healing is integrally related to salvation because both mend broken ties.[85] Living according to God's will is necessary to prevent illness, and prayer is the most common method of seeking God's help in healing. Roman Catholics, for example, make appeals to the saints identified with certain afflictions—St. Teresa of Avila for headaches, St. Peregrine for cancer, St. John of God for heart disease, St. Joseph for the terminally ill, and St. Bruno for those who are possessed are just a few examples. Pilgrimages to the shrines of these saints are made for special petitions. In Eastern religions, health is determined mostly by correct conduct in this and past lives, as well as in the virtuous behavior of ancestors. Religious offerings are made regularly; for instance, Hindus choose a personal deity to worship daily at a home shrine. Improper actions leading to disharmony within a person, family, community, or the supernatural realm can cause sickness. Healing occurs through restoration of balance, often including offerings to the deities or spirits of the living and dead who have been offended.

Individual healing practices developed out of religious ritual include meditation, a contemplative process of focused relaxation; yoga, the control of breathing and use of systematic body poses to restrain the functions of the mind and promote mind–body unity; and visualization or guided imagery, induced relaxation and targeted willing away of health problems. Each concentrates the power of mind on reducing health risks, such as stress, high blood pressure, and decreased immune response, or on alleviating specific medical conditions. Hypnotherapy works in a similar manner; although it is generally done with the aid of a hypnotherapist, self-hypnosis can be learned for personal use.

In many cases, the spiritual skills of the individual are inadequate for the problem, and the help of a sacred healer is sought. These health practitioners generally work through interventions with the supernatural world, which may include prayers, blessings, chanting or singing, charms, and conjury, as well as the use of therapeutic substances (i.e., herbal remedies) and application of physical cures (i.e., the laying on of hands). Faith healers, most of whom get their healing gifts from God, are common among many Christian groups. Some are affiliated with certain denominations and rites, such as the Cajun *traiteurs* of Louisiana, who specialize in treating one or two ailments through prayers and charms associated with Catholicism.[58,59] Others, such as the sympathy healers of the Pennsylvania Dutch who practice powwowing (also known by the German name *Brauche* or *Braucherei*), are considered the direct instruments of God.[60]

Persons with a spiritual calling are often employed to treat illness. *Neng* among the Hmong, Mexican *curanderos* (or *curanderas*), practitioners of voodoo in the American South, and *espiritos* or *santeros* (or *santeras*) in the Caribbean may communicate with the spirits or saints to heal their patients. Ceremonial invocation is the primary therapy, although charms and spells to counteract witchcraft and botanical preparations to ease physical complaints are used as well.

Eighty percent of respondents in a study on faith and healing in the southeastern United States said they believe God acts through physicians to cure illness.[78]

▼ Acupuncture attempts to restore the balance of vital energy in the body through inserting and manipulating needles.

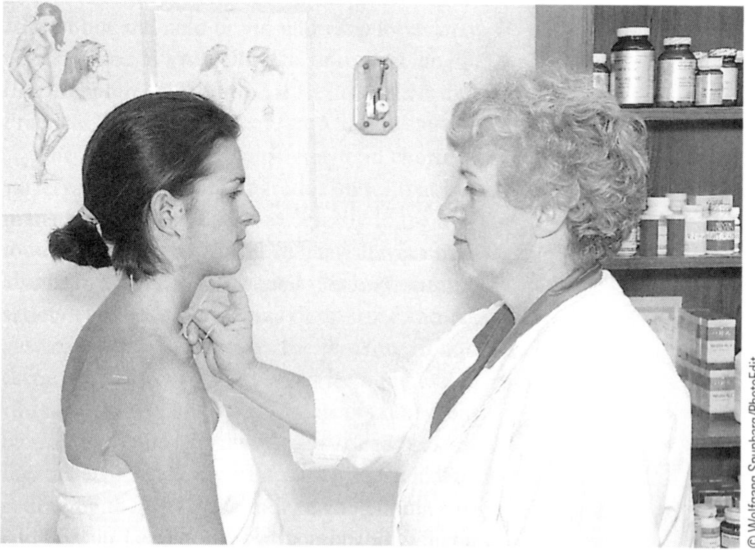

© Wolfgang Spunbarg/PhotoEdit

A study of physician beliefs about health and religion found that although 91 percent of respondents said knowledge of a client's faith practices is important in care, only 32 percent ask about religious affiliation.[79]

When Navajos dream frequently of death, it is usually considered a sign of serious illness.

Consultation with a Native American medicine man may take hours to complete; some healing ceremonies take a week to perform and may cost thousands of dollars.[80]

Shamans, called medicine men among many Native American groups, are sacred healers with exceptional powers. They originated in Russia an estimated 20,000 years ago and spread throughout the world to the indigenous cultures of the Americas, Southeast Asia, Indonesia, Polynesia, and Australia. Remote tribal groups found in Africa, India, and Korea have similar healers. A shaman is a composite priest, magician, and doctor; the position is passed on from generation to generation, or through a calling that could include fainting spells or convulsive fits due to attacks by spirits. Shamans typically complete lengthy apprenticeships and are initiated through a series of trials simulating death and rebirth. In shamanic systems, sickness is due to spiritual crisis, and healing emphasizes strengthening of the soul through redirection of the life forces or, in cases of serious illness, retrieval of the soul, which may have been stolen by evil spirits. Shamanic practices include visualization techniques to create harmony between the patient and the universe, singing, chanting, prognostication, dream analysis, and séances. Shamans are often expert herbalists.[61,62]

Pluralistic Health Care Systems

The enduring popularity of traditional health beliefs and practices is based in cultural congruency. Healing sickness, with or without the services of an expert provider, takes place according to a patient's worldview. Humans value what validates their beliefs and discount anything that differs, regardless of statistical data or scientific claims; they give disproportionate credence to persons they like and respect.

Medical Pluralism

Medical pluralism is the term for the consecutive or concurrent use of multiple health care systems.[44] Although it is often assumed that ethnic minorities, the poor, the less educated, or recent immigrants are most likely to rely on traditional folk medicine, studies report that the use of healers in some groups increases with education and income level. Further, acculturation is not associated with a rise in the use of biomedical services.[43] Medical pluralism is widespread in the United States.

Biomedical Healing

Clients using traditional health practices are generally seeking to alleviate the difficulties experienced in illness through understandable, flexible, and convenient treatment from a warm and caring provider. The personal relationship with the healer is as important as the actual therapy (see Chapter 3).

Biomedical professionals often operate in partnership, knowingly or not, with unconventional health care practitioners. A small study of prostate cancer patients reported that 37 percent used alternative therapies in conjunction with radiation treatment. A separate survey found that their physicians believed only 4 percent of these patients used any other health care practices.[54] Researchers suggest care is optimized when providers work together rather than at cross purposes. Patients are sometimes put in the confusing position of choosing between biomedical and traditional systems that contradict or reject each other.[52]

Studies suggest that some unconventional therapies are effective, benefiting the patient physiologically or psychologically, and should be accepted as complementary to biomedical approaches.[63,64] Cooperative monitoring by a biomedical professional can also detect those few instances when a home remedy or popular practice is harmful to the patient. Furthermore, biomedical health care providers can adopt certain healing strategies. Understanding the patient's perspective on illness and attending to differences in the patient–provider explanatory models is one approach.[10] Recommending alternate, experimental biomedical programs in cases of advanced chronic disease is another.[56] A more comprehensive methodology is offered through transcultural nursing theory, developed to provide cultural congruent care that is beneficial, satisfying, and meaningful to clients.[14] Transcultural nursing theory identifies three modes of effective care: (1) cultural care preservation and/or maintenance, (2) cultural care accommodation and/or negotiation, and (3) cultural care repatterning or restructuring.

Cultural care preservation and/or maintenance is used when a traditional health belief or practice is known to be beneficial in its effect and is encouraged by the provider. Cultural care accommodation and/or negotiation is accomplished between the provider and the patient (or the patient's family) when there is an expectation for care that is outside biomedical convention. Cultural care repatterning or restructuring occurs when both provider and patient agree that a habit is harmful to health, and a cooperative plan is developed to introduce a new and different lifestyle. Applied to food habits, the culture care theory acknowledges that some traditional beliefs and practices regarding diet have beneficial or neutral consequences, some have unknown consequences, and some may be deleterious to the health of a client (see Chapter 1, "Food and Culture").

In addition to specific provider approaches, the health care setting can also help promote biomedical healing through services desired by clients in a comfortable, welcoming atmosphere. Some health care organizations are forming therapeutic alliances with community folk healers or combining non-Western practices with biomedicine in an integrative approach. At a minimum, research on traditional practices used by clients and information about alternative community health resources should be available to providers; staff should be encouraged to keep up on current trends through continuing education, diversity training, and refresher workshops. Other useful steps include increasing care accessibility by taking education and services into client communities when possible and providing flexible, non-traditional hours for appointments. Clients also feel more at ease with a staff representative of the community, so to some degree, this preference should be accommodated.[65,66] Successful biomedical healing is dependent on the intercultural knowledge and sensitivity of the health provider and setting. Care must be undertaken in cooperation with the patient, as well as the patient's

DISCUSSION STARTERS: WHAT IS YOUR "WORLDVIEW" OF HEALTH CARE?

Culture determines how each of us defines health, recognizes illness, and considers medical treatment of illness. Reflecting on how we view health and healthcare can tell us a lot about our cultures. Answer each of the following questions. Don't worry about justifying your beliefs. All you want to do here is reflect on and write down your beliefs. There are no right or wrong answers, as long as the answers honestly express your beliefs.

- **Which most determines a person's health:** the decisions and actions of the person her- or himself? The decisions and actions of one's family and others with whom one has been brought up? The decisions and actions of the government (whether national or local or both)? The will of God? The actions of some other supernatural agency (please identify)? Astrological alignment of the stars and planets? Karma? Fate? Something else?
- **Who is most responsible for a person's healthcare:** the healthcare practitioner? The person her- or himself (or, if a child, the parents or adult caregivers)? Someone else?
- **If you became seriously ill or injured and needed dramatic treatment, such as surgery or radiation treatment, who would most influence your decision(s) regarding that treatment:** you alone? Your family? Your church, synagogue, tabernacle, temple, mosque, or other religious congregation? Others?
- **If a loved one is terminally ill—that is, dying from a disease or injury—which of the following would be most important to you:** trying everything possible to cure your loved one or at least, prolonging that loved one's life? Doing everything possible to relieve any pain that loved one might be enduring? Praying with that loved one and sustaining her or his faith? Helping your loved one die in a way that she or he wants, which could involve *not* employing life-sustaining measures?
- **Which of the following is the most important way you tell whether you are healthy or not:** by whether or not you have any symptoms of disease? By whether or not you can accomplish daily tasks and meet daily responsibilities? By how you look in the mirror and/or how others perceive how you look?

In small groups (3-4), compare answers for about 10 minutes. Restrain any impulse to criticize. Your task is to gather as much information as possible about each person's worldview relating to healthcare in the short time that you have. Remember, there are no right or wrong answers. Different cultures exhibit different worldviews.

When members of your group have completed sharing answers with each other, individually write a reflection on what you've learned and how you feel about it (maximum one page).

family and any concurrent traditional providers in use by the patient. Healing should not be the sole domain of home remedies, popular health care approaches, or alternative medical systems. Medical pluralism offers the opportunity for biomedicine to heal sickness through coordinated client care, with understanding and appreciation for the therapeutic value of traditional health beliefs and practices.

REVIEW QUESTIONS

1. If you become ill, how might your worldview influence your expectations about your illness and its treatment?
2. How does biomedicine in the United States reflect the majority culture?
3. How might another culture's definition of health differ from that of the World Health Organization: "a state of complete physical, mental, and social well-being, not merely an absence of disease or infirmity"?
4. Describe three ways that diet may be used to promote or maintain health, using specific examples of foods and practices.
5. What is meant by "folk illnesses" or "culture-bound syndromes"? Using one example, explain how effective treatment for the condition would differ from the conventional biomedical approach.

REFERENCES

1. Barnes P. M., Bloom B., Nahin R. CDC National Health Statistics Report #12. *Complementary and Alternative Medicine Use Among Adults and Children: United States, 2007.* December 2008.
2. Nahin R. L., Barnes P. M., Stussman B.A., et al. *Costs of complementary and alternative medicine (CAM) and frequency of visits to CAM practitioners: United States, 2007.* (PDF) CDC National Health Statistics Report #18. 2009.
3. Kavanagh, K., Absalom, K., Beil, W., & Schliessmann, L. 1999. Connecting and becoming culturally competent: A Lakota example. *Advances in Nursing Science, 21,* 9–31.
4. Randall-David, E. 1989. *Strategies for working with culturally diverse communities and clients.* Bethesda, MD: Association for the Care of Children's Health.
5. Schilling, B., & Brannon, E. 1986. *Cross-cultural counseling: A guide for nutrition and health counselors. FNS #250.* Alexandria, VA: U.S. Department of Agriculture/ U.S. Department of Health and Human Services.
6. Yankelovich, Inc. 2006. Food for Life Study. In *The Yankelovich MONITOR.* Chapel Hill, NC.
7. Blackhall, L.J., Murphy, S.T., Frank, G., Michel, V., & Azen, S. 1995. Ethnicity and attitudes toward patient autonomy. *Journal of the American Medical Association, 274,* 820–825.
8. Klessig, J. 1992. The effects of values and culture on life-support decisions. *Western Journal of Medicine, 157,* 316–322.
9. Uche-riffin, Ndidi F. 1994. Perceptions of African American women regarding health care. *Journal of Cultural Diversity, 1,* 32–35.
10. Kleinman, A. 1980. *Patients and healers in the context of culture.* Berkeley: University of California Press.
11. Kumanyika, S. K., & Morssink, C.B. 1997. Cultural appropriateness of weight management programs. In *Overweight and weight management: The health professional's guide to understanding and practice.* Gaithersburg, MD: Aspen.
12. Helman, C. G. 2000. *Culture, health and illness: An introduction for health professionals* (4th ed). London: Butterworth-Heinemann.
13. Spector, R. E., 2004. *Cultural diversity in health and illness* (6th ed.). Upper Saddle River, NJ: Pearson Education, Inc.
14. Leininger, M. M. 1991. Becoming aware of types of health practitioners and cultural imposition. *Journal of Transcultural Nursing, 2,* 36.
15. Braun, U. K., Beyth, R. J., Ford, M. E., & McCullough, L. B. (2008). Voices of African American, Caucasian and Hispanic surrogates on the burdens of end-of-life decision making. *Journal of General Internal Medicine, 23,* 267–274.
16. Ots, T. 1990. The angry liver, the anxious heart and the melancholy spleen: The phenomenology of perceptions in Chinese culture. *Culture, Medicine, and Psychiatry, 14,* 21–58.
17. Cassell, J. A. 1995. Social anthropology and nutrition. A different look at obesity in America. *Journal of the American Dietetic Association, 95,* 424–427.
18. Ceballos N..; Czyzewska M Body image in Hispanic/ Latino vs. European American adolescents: implications for treatment and prevention of obesity in underserved populations. *J Health Care Poor Underserved. 2010 ; 21:* 823–38.
19. Gracia-Arnaiz M. Fat bodies and thin bodies. Cultural, biomedical and market discourses on obesity. *Appetite. 2010; 55:* 219–25.
20. DeGarine, I., & Pollack, N. J. 1995. *Social aspects of obesity.* Australia: Gordon & Breach.
21. Neumark-Sztainer, D., Croll, J., Story, M., Hannan, P. J., French, S.A., & Perry, C. 2002. Ethnic/racial differences in weight-related concerns and behaviors among adolescent girls and boys: Findings from Project EAT. *Journal of Psychosomatic Research, 53,* 963–974.
22. York-Crowe, E. E., & Williamson, D. A. 2005. Health and appearance concerns in young Caucasian and African-American women. *Eating and Weight Disorders, 10,* e38–e44.
23. Greenberg, D. R., & LaPorte, D. J. 1996. Racial differences in body type preferences of men for women. *International Journal of Eating Disorders, 19,* 275–278.
24. Davis D. S.; Sbrocco T; Odoms-Young A; Smith DM. Attractiveness in African American and Caucasian women: is beauty in the eyes of the observer? *Eat Behav. 2010; 11:* 25–32.
25. Mossavar-Rahmani, Y., Pelto, G. H., Ferris, A. M., & Allen, L. H. 1996. Determinants of body size perceptions and dieting behavior in a multiethnic group of hospital staff women. *Journal of the American Dietetic Association, 96,* 252–256.

26. Contento, I. R., Basch, C., & Zybert, P. 2003. Body image, weight and food choices of Latina women and their young children. *Journal of Nutrition Education and Behavior, 35,* 236–248.
27. Crawford, P. B., Gosliner, W., Anderson, C., Strode, P., Becerra-Jones, Y., Samuels, S., Carroll, A. M., & Ritchie, L. D. 2004. Counseling Latina mothers of preschool children about weight issues: Suggestions for a new framework. *Journal of the American Dietetic Association, 104,* 387.
28. Rubin, L. R., Fitts, M.L., & Becker, A. E. 2003. "Whatever feels good in my soul": Body ethics and aesthetics among African American and Latina women. *Culture, Medicine and Psychiatry, 27,* 49–75.
29. Olvera, N., Suminski, R., & Power, T.G. 2005. Intergenerational perceptions of body image in Hispanics: Role of BMI, gender, and acculturation. *Obesity Research, 13,* 1970–1979.
30. Wang, C. Y., Abbot, L., Goodbody, A. K., & Hui, W. T. 2002. Ideal body image and health status in low-income Pacific Islanders. *Journal of Cultural Diversity, 9,* 12–22.
31. Gittelsohn, J., Harris, S. B., Thorne-Lyman, A. L., Hanley, A. J., Barnie, A., & Zinman, B. 1996. Body image concepts differ by age and sex in an Ojibway-Cree community in Canada. *Journal of Nutrition, 126,* 2990–3000.
32. Lynch, W. C., Heil, D. P., Wagner, E., Havens, M. D. Ethnic differences in BMI, weight concerns, and eating behaviors: comparison of Native A merican, White, and Hispanic adolescents. *Body Image. 2007; 4:* 179–90.
33. Rinderknect, K., & Smith, C. 2002. Body-image perceptions among urban Native American youth. *Obesity Research, 10,* 315–327.
34. Robinson, T. N., Chang, J. Y., Haydel, K. F., & Killen, J. D. 2001. Overweight concerns and body dissatisfaction among third-grade children: The impacts of ethnicity and socioeconomic status. *Journal of Pediatrics, 138,* 181–187.
35. Neumark-Sztainer, D., Croll, J., Story, M., Hannan, P. J., French, S. A., & Perry, C. 2002. Ethnic/racial differences in weight-related concerns and behaviors among adolescent girls and boys: Findings from Project EAT. *Journal of Psychosomatic Research, 53,* 963–974.
36. Talamayan, K. S., Springer, A. E., Kelder, S. H., Gorospe, E. C., & Joye, K. A. 2006. Prevalence of overweight misperception and weight control behaviors among normal weight-adolescents in the United States. *Scientific World Journal, 6,* 365–376.
37. Davis, S. M., & Lambert, L. C. 2000. Body image and weight concerns among Southwestern American Indian preadolescent schoolchildren. *Ethnicity and Disease, 10,* 184–194.
38. Wright J. D., Wang C-Y. *Trends in intake of energy and macronutrients in adults from 1999–2000 through 2007–2008. NCHS data brief, no 49.* Hyattsville, MD: National Center for Health Statistics. 2010.
39. Seo D. C., Torabi, M. R., Jiang, N., Fernandez-Rojas, X., Park, B. H. Cross-cultural comparison of lack of regular physical activity among college students: Universal versus transversal. *Int J Behav Med. 2009; 16:* 355–9.
40. Spector, R. E., 2004. *Cultural diversity in health and illness* (6th ed.). Upper Saddle River, NJ: Pearson Education.
41. Kleinman, A., Eisenberg, L., & Good, B. (2006). Culture, illness and care: Clinical lessons from anthropologic and cross-cultural research. *FOCUS, 4(1),* 140–149.
42. Kleinman, A., Eisenberg, L., & Good, B. 1978. Culture, illness, and care: Clinical lessons from anthropologic and cross-culture research. *Annals of Internal Medicine, 88,* 251–258.
43. Douglas, M. K., Pacquiao, D. F. (eds). 2010; Core curriculum in transcultural nursing and health care. *Journal of Transcultural Nursing. 21(Suppl 1).*
44. Clark, M. M. 1983. Cultural context of medical practice. *Western Journal of Medicine, 139,* 806–810.
45. Hand, W. D. 1980. *Magical medicine.* Berkeley: University of California Press.
46. Graham, J. S. 1976. The role of the curandero in the Mexican American folk medicine system in West Texas. In W. D. Hand (Ed.), *American Folk Medicine.* Berkeley: University of California Press.
47. Adler, S. R. 1995. Refuge stress and folk belief: Hmong sudden deaths. *Social Science and Medicine, 40,* 1623–1629.
48. Keel, P. K., & Klump, K. L. 2003. Are eating disorders culture-bound syndromes? Implications for conceptualizing care. *Psychological Bulletin, 129,* 747–769.
49. American Psychiatric Association. Practice guideline for the treatment of patients with eating disorders. *Am J Psychiatry. 2000; 157 (suppl):* 1–39.
50. Kim, C., & Kwok, Y. S. 1998. Navajo use of native healers. *Archives of Internal Medicine, 158,* 2245–2249.
51. Ma, G. 1999. Between two worlds: The use of traditional and Western health services by Chinese immigrants. *Journal of Community Health, 24,* 421–437.
52. Marbella, A. M., Harris, M. C., Diehr S., Ignace, G., & Ignace, G. 1998. Use of Native American healers among Native American patients in an urban Native American health center. *Archives of Family Medicine, 7,* 182–185.
53. Poss, J. E., Jezewski, M. A., & Stuart, A. G. 2003. Home remedies for type 2 diabetes used by Mexican Americans in El Paso, Texas. *Clinical Nursing Research, 12,* 304–323.
54. Kao, G.D., & Devine, P. 2000. Use of complementary health practices by prostate carcinoma patients undergoing radiation therapy. *Cancer, 88,* 615–619.
55. Hughes, C. K., & Higuchi, P. 2004. Ka Lokahi Wahine: A culturally based training for health professionals. *Pacific Health Dialog, 11,* 166–169.
56. Murray, R. H., & Rubel, A. J. 1992. Physicians and healers—Unwitting partners in health care. *New England Journal of Medicine, 326,* 61–64.
57. Kittler, P. G. 2006. Complementary and alternative medicine. In *Nutrition Therapy and Pathophysiology,* M. Nelms, Sucher K.P., & S. Long (Eds.). Belmont, CA: Cengage.
58. Brandon, E. 1976. Folk medicine in French Louisiana. In W. D. Hand (Ed.), *American Folk Medicine.* Berkeley: University of California Press.
59. Leistner, C. G., Hirschfield, L. Cajun and creole food practices. In: Goody CM, Drago L.(Eds). *Introduction: Cultural competence and nutrition counseling.* In: *Cultural Food Practices.* Chicago IL: American Dietetic Association, 2010.
60. Yoder, D. 1976. Hohman and Romanus: Origins and diffusion of Pennsylvania German powwow manual.

In W. D. Hand (Ed.), *American Folk Medicine.* Berkeley: University of California Press.

61. Balzer, M.M. 1987. Behind shamanism: Changing voices of Siberian Khanty cosmology and politics. *Social Science and Medicine, 24,* 1085–1093.
62. Sheikh, A. A., Kunsendorf, R. G., & Sheikh, K. S. 1989. Healing images: From ancient wisdom to modern science. In A. A. Sheikh & K. S. Sheikh (Eds.), *Eastern and Western Approaches to Healing: Ancient Wisdom and Modern Knowledge.* New York: Wiley.
63. Oh, B., Butow, P., Mullan, B., Beale, P., Pavlakis, N., Rosenthal, D., Clarke S. The use and perceived benefits resulting from the use of complementary and alternative medicine by cancer patients in Australia. *Asia Pac J Clin Oncol. 2010; 6:* 342–9.
64. Lunny, C. A., Fraser, S. N. The use of complementary and alternative medicines among a sample of Canadian menopausal-aged women. *J Midwifery Womens Health. 2010 Jul; 55 (4):* 335–43.
65. Kaptchuk, T. J., & Millar, F. G. 2005. Viewpoint: What is the best and most ethical model for the relationship between mainstream and alternative medicine: Opposition, integration, or pluralism? *Academic Medicine, 80,* 286–290.
66. Yehieli, M., & Grey, M. A. 2005. *Health matters: A pocket guide for working with diverse cultures and underserved populations.* Yarmouth, ME: Intercultural Press.
67. Kumanyika, S. K., & Morssink, C. B. 1997. Cultural appropriateness of weight management programs. In *Overweight and weight management: The health professional's guide to understanding and practice.* Gaithersburg, MD: Aspen.
68. Anderson, E. N. 1987. Why is humoral medicine so popular? *Social Science and Medicine, 25,* 331–337.
69. Dagnelie, P. C., & van Stavern, W. A. 1994. Macrobiotic nutrition and child health: Results of a population-based, mixed longitudinal cohort study in the Netherlands. *American Journal of Clinical Nutrition, 59 (Suppl.),* 1187S–1196S.
70. Garrett, L. 1994. *The Coming Plague: Newly Emerging Diseases in a World Out of Balance.* New York: Farrar, Strauss & Giroux.
71. Tsai, G. 2000. Eating disorders in the Far East. *Eating and Weight Disorders, 5,* 183–187.
72. DeBusk, R. M. 1999. Herbal medicines: A primer. On the Cutting Edge: *Diabetes Care and Education, 20,* 4–5.
73. Lazarou, J., Pomeranz, B. H., & Corey, P. N. 1998. Incidence of adverse drug reactions in hospitalized patients: A meta-analysis of prospective studies. *Journal of the American Medical Association, 279,* 1200–1205.
74. Bressler, R. 2005. Interactions between Gingko biloba and prescription medications. *Geriatrics, 60,* 30–33.
75. King, A. R., Russett, F.S., Generali, J.A., Grauer, D. W. Evaluation and implications of natural product use in preoperative patients: a retrospective review. *BMC Complement Altern Med. 2009; 13;* 9: 38.
76. Sagar, S. M. Can the therapeutic gain of radiotherapy be increased by concurrent administration of Asian botanicals? *Integr Cancer Ther. 2010; 9:* 5–13.
77. Colson, C. R., & De Broe, M. E. 2005. Kidney injury from alternative medicines. *Advances in Chronic Kidney Diseases, 12:* 261–275.
78. Mansfield, C. J. 2002. The doctor as God's mechanic? Beliefs in the Southeastern United States. *Social Science & Medicine, 54,* 399–409.
79. Chibnall, J. T., & Brooks, C. A. 2001. Religion in the clinic: The role of physician beliefs. *Southern Medical Journal, 94,* 374–379.
80. Gurley, D., Novins, D. K., Jones, M. C., Beals, J., Shore, J. H., & Manson, S. M. 2001. Comparative use of biomedical services and traditional healing options by American Indian veterans. *Psychiatric Services, 52,* 68–74.

Intercultural Communication

No matter whether a person is interacting with people from diverse cultural groups at work and in social settings or traveling to another country for business or pleasure, he or she needs intercultural communication skills to successfully negotiate daily life. Intercultural communication is a specialty in itself. The field encompasses language and the context in which words are interpreted, including gestures, posture, spatial relationships, concepts of time, the status and hierarchy of persons, the role of the individual within a group, and the setting. This chapter presents a broad and limited overview of intercultural communication concepts, as well as information useful in nutrition counseling with individuals or with groups in educational programs. Later chapters on each American ethnic group provide more specific details as reported by researchers and practitioners familiar with cultural communication characteristics.

The Intercultural Challenge

Researchers use an iceberg analogy (Figure 3.1) to describe how a person's cultural heritage can impact communication.[1] Ethnicity, age, and gender are the most visible personal characteristics affecting dialogue—the so-called tip of the iceberg. Beneath the surface, but equally influential, may be degree of acculturation or assimilation, socioeconomic status, health condition, religion, educational background, group membership, sexual orientation, or political affiliation.

Most people are comfortable conversing with those who are culturally similar to themselves. Communication is sometimes described as an action chain,[2] meaning that one phrase or action leads to the next: in the United States, a person who extends her hand in greeting expects the other person to take her hand and shake it, or when a person says "thank you," a "you're welcome" should follow. Communication comprises a whole series of unwritten expectations regarding how a person should reciprocate, and such expectations are largely cultural in origin. If a person understands the communication action chain and responds as expected, a successful relationship can develop. When a person does not respond as expected,

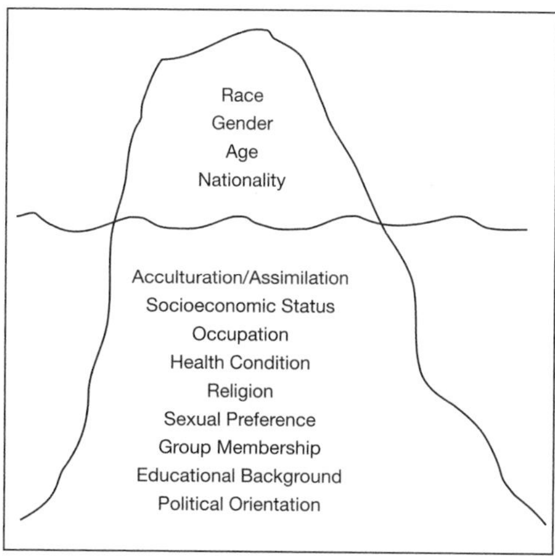

Figure 3.1
Iceberg model of multicultural influences on communication.
Source: Adapted with permission of Sage, from *Effective Communication in Multicultural Health Care Settings*, G. L. Kreps & E. N. Kunimoto, 6. Copyright © 1994. Permission conveyed through Copyright Clearance Center, Inc.

"Observe the nature of each country; diet; customs; the age of the patient; speech; manners; fashion; even his silence. . . . One has to study all these signs and analyze what they portend." Hippocratic writings, fifth century B.C.E.

communication can break down, and the relationship can deteriorate.[1]

When meeting a person for the first time, the only data that speakers usually have to work with come from their own cultural norms. They use these norms to predict how that person will respond to their words and what conversational approaches are appropriate. They may also use social roles to determine their communication behavior. Furthermore, speakers modify their words and actions as they get to know a person individually, observing personal cues about communication idiosyncrasies that vary from cultural or social customs.[3] An employee, for example, may make certain assumptions about a supervisor based on ethnicity, gender, age, and especially occupational status and then make adaptations. For example, an employee may start out calling his boss "Mrs. Smith" as a sign of respect for her position but use the more informal "Sue" when she requests that he call her by her first name.

Interpersonal relationships between two individuals are based mostly on personal communication preferences; group interactions commonly depend on cultural or social norms. Misinterpretations at the cultural or social levels of communication are more likely to occur because they are more generalized. As mentioned in Chapter 1, stereotyping occurs if a person overestimates another individual's degree of association with any particular cultural or social group. Only about one-third of the people from any given group are believed to actually behave in ways typified by the group as a whole.[3] Assumptions about how a person of a different cultural heritage should communicate can elicit certain types of reactions based on norms in that person's culture: stereotyping can become a self-fulfilling prophecy. For instance, Asians have a reputation for being restrained in conversation compared to the typical American approach. A nutrition education provider teaching a prenatal care program may respond to more aggressive speakers during a group meeting, failing to actively involve a Vietnamese-American participant. The Vietnamese-American woman may feel the provider is disinterested in her questions or comments, becoming even less likely to offer input at the next meeting, reinforcing the provider's perception that all Asians are quiet.

The challenge is to increase familiarity with cultural communication behaviors, while remaining aware of personal cues and moving toward an interpersonal relationship as quickly as possible.

Intercultural Communication Concepts

Communication uses symbols to represent objects, ideas, or behaviors. Thoughts, emotions, and attitudes are translated into language and nonverbal actions (i.e., gestures, posture, eye contact) to send messages from one person to another.

The two components of the message are the content and the relationship between the speaker and the receiver. Depending on the situation, the content or the relationship may assume greater prominence in interpretation of meaning. Messages that violate cultural expectations may be accurate in content but have a negative impact on the relationship. If the message consistently offends the receiver, the relationship will deteriorate and the message will be disregarded. For example, if a health professional advises a Chinese client to increase calcium intake through increased milk consumption, it is a high-content message but does little to acknowledge the role of milk in the Chinese diet. Is the client lactose-intolerant? Does the client like milk? Does the client classify milk as health promoting or as a cause of illness? How does milk fit into the balance of the diet according to the client? Unless the provider gains an understanding of how the client conceptualizes the situation, the content of the message may be ignored because the client assumes disinterest or even disrespect for personal beliefs and expectations. Thus, the provider–client relationship is weakened. Messages that demonstrate respect for the individuality of the receiver are called personal messages and these improve relationships; those that are disrespectful are termed object messages and these often degrade relationships. Communication occurs in a continuum between personal and object messages.[1]

Verbal messages are most useful for communicating content, whereas nonverbal messages usually convey information about relationships.

If the nonverbal message is consistent with the verbal message, it can build the relationship and help the receiver correctly interpret the meaning intended by the speaker. When the nonverbal message is inconsistent with the verbal message, both the relationship and the content are undermined. Successful communication is dependent on both verbal and nonverbal skills, each significantly impacted by cultural considerations.

© The New Yorker Collection 1980 Peter Steiner from cartoonbank.com. All Rights Reserved.

Figure 3.2
Cultural context is so embedded in the individual that many people believe it is innate and assume all other people share the same background.
Source: ID: 38626, Published in March 10, 1980.

Verbal Communication

The abstract nature of language means it can only be correctly interpreted within context. The cultural aspects of context are so embedded that a speaker often believes they are innate—that is, that all other people must communicate according to the same expectations. Context includes issues common to cultural worldview, such as the role of the individual in a group and perceptions of power, authority, status, and time. In addition, context in communication also encompasses the significance of affective and physical expression (termed low- or high-context) and level of tolerance for uncertainty and ambiguity (called uncertainty avoidance). Verbal communication occurs within these cultural premises, often operating at an unconscious level in the speaker (Figure 3.2).

Low- and High-Context Cultures

The context in verbal communication varies culturally. Conversational context can be defined as the affective and physical cues a speaker uses to indicate meaning, such as tone of voice, facial expression, posture, and gestures.

In most Western cultures, messages usually concern ideas presented in a logical, linear sequence. The speaker tries to say what is meant through precise wording, and the content of the language is more objective than personal along the continuum of personal and object messages. This communication style is termed low-context because the actual words are more important than who is receiving the message, how the words are said, or the nonverbal actions that accompany them. Communication in a low-context culture is so dependent on words that the underlying meaning is undecipherable if wording is chosen poorly or deliberately to mislead the recipient. Nearly every American has also experienced the obtuse professional languages of attorneys or scientists who fail to convey their message in common, everyday English. The Swiss, Germans, and Scandinavians are examples of low-context cultures.[2]

In cultures with a high-context communication style, most of the meaning of a message is found in the context, not in the words. In fact, the wording used may be vague, circuitous, or incomplete. The content of the language is more personal than objective, dependent on the relationship between speaker and listener. Attitudes and feelings are more prominent in the conversation than thoughts. Communication in high-context cultures is analogous to the expression "reading between the lines." Misunderstandings easily occur if either participant is unfamiliar with the meaning of the nonverbal signifiers being used, such as small eye movements or sounds that are made when in agreement, or disagreement, or

▼ ***Only the person sending the message knows the meaning of the message: the person receiving the message must use what she knows about cultural and social norms, as well as what she knows about the speaker personally, to interpret the message.***

© Tom McCarthy/PhotoEdit, Inc.

One area of conflict often found between blacks and whites is due to differences in high- and low-context communication. African Americans tend to be more high-context than white Americans, using cognitive, affective, and physical responses that may appear disruptive or overly emotional to whites.[90]

when upset. For example, Caucasian Americans tend to squirm a bit when uncomfortable with a conversation, while the Japanese will quickly suck in air. High-context cultures are most prevalent among homogeneous populations with a common understanding of the affective and physical expression used in sending the message (see the following "Nonverbal Communication" section). Asian, Middle Eastern, and Native American cultures are very high-context. Latino societies are moderately high-context. American culture is thought to be toward the low end, but more middle-context than many European societies.

In low-context cultures, communication is usually explicit, straightforward, and unambiguous. The focus is on the speaker, who uses words to send messages that are often intended to persuade or convince the receiver. In high-context cultures, indirect communication is preferred. Implicit language is used, and many qualifiers are added; nonverbal cues are significant to interpreting the message. The locus of conversation is the receiver; the speaker makes adjustments in consideration of the listener's feelings.[3] Low-context listeners are often impatient with high-context speakers, wondering when the speaker will get to the point of the conversation. Low-context listeners also frequently miss the affective and physical expression in the message.

Health care situations are often extremely low-context. The conversation is focused on the provider who delivers a verbal message to the client with little consideration for the nonverbal message. The communication is high on content and low on relationship. Clients from high-context cultures are likely to be dissatisfied, even offended, by such impersonal, objective interactions. Communication problems may not be evident to a low-context clinician until the client leaves and never returns.

One way to patronize a person is to speak in a low-context mode, elaborating beyond what is needed for understanding.

Interactions may range from low- to high-context depending on the situation, regardless of the overall cultural preference. In uncomfortable or embarrassing situations, a low-context communicator may be very sensitive and indirect. In high-context cultures, direct language is frequently used in intimate relationships.

Even within low-context cultures, intimate conversations are usually highly contextual; words and phrases may be significantly shortened or abbreviated—just a look may be enough for understanding.

As an example of low- versus high-context communication in different situations, consider a researcher presenting current nutritional data on spinach to a group of other professionals. She will probably speak in a relatively monotone voice and use scientific jargon. She will present her points in a sequential manner, support her thesis with examples, and then restate her ideas in the conclusion. She will probably stand erect and limit the expressive use of her hands and face. The message is almost entirely in the content of the words she says. In contrast, this same woman might behave very differently when feeding her reluctant toddler spinach for dinner. She might smile and make yummy sounds as she offers him a spoonful or pretends the spinach is a plane coming in for a landing in his mouth. She might give him a spoonful of meat or potato, then try the spinach again. She might even dance around his high chair a little or hum a few bars of the old cartoon theme song about a sailor who liked spinach. She doesn't try to get him to eat spinach by explaining its nutrient content, as she did at her meeting. The message is nonlinear and not dependent on the content of the words she uses. This is not to say that a health care provider should burst out in song when working with a client from a high-context culture. But it does suggest that indirect, expressive approaches may be more effective in some intercultural clinical, educational, or counseling settings. Identification of a culture as either low- or high-context provides a general framework for communication but may be affected by other situational factors.

Individuals and Groups

The relationship of the individual to the group is determined in part by whether a culture is low- or high-context. In low-context cultures, the individual is typically separate from the group, and self-realization is an important goal. Self-esteem is dynamic, based on successful mastery or control of a situation. In high-context cultures, the individual is usually defined by group association, and a person desires oneness with the group, not individuation. A mutual dependency exists and self-esteem is based on how well a person can adjust to a situation.[3] Individualism is a prominent characteristic in Australia, Canada, Great Britain, New Zealand, the Netherlands, and the United States. Collectivism is especially valued in the nations of Denmark, Ghana, Guatemala, Indonesia, Nigeria, Panama, Peru, El Salvador, Sierra Leone, Taiwan, Thailand, and Venezuela.

In societies emphasizing individuality, a person must communicate to gain acceptance by the group, whether it is the family, the workplace, or the community. Communication is used to establish the self within an individual or group relationship. When meeting someone new, the action chain in the conversation is flexible, with few expectations. The two people may focus on one or the other speaker and often delve into personal preferences, such as favorite restaurants or sports teams. When group identity is the focus of a society, there is no need for a person to seek acceptance from the group or to communicate individuality. Silence is highly valued. Interactions between strangers tend to be ritualized, and if the action chain is broken, communication cannot continue. The expectation is that each speaker will indicate group affiliation and that such identity conveys all the information needed to know that person.

The role of the individual within the group can have an impact on health care delivery. Within more group-oriented cultures, greater participation is required of their members in matters of health and illness, and it may be expected that relatives will participate in giving patient histories, overseeing physical exams, or making decisions regarding treatment.[4] Middle Easterners expect to go to the hospital with an ill family member to provide care. Next of kin is determined along bloodlines among Latinos, and care decisions are often the responsibility of a grandmother or mother instead of the spouse. Koreans prefer the whole family make decisions regarding treatment for a terminally ill patient. Some American Indians are so strongly associated with the group it is difficult for them to communicate individual needs.

Uncertainty Avoidance

Related to the role of the individual in a group is tolerance for uncertainty and ambiguity. Some groups exhibit great discomfort with what is unknown and different; these are defined as high uncertainty avoidance cultures. Members of these cultures may become anxious about behavior that deviates from the norm; high uncertainty avoidance cultures desire consensus. Argentina, Belgium, Chile, Columbia, Costa Rica, Croatia, Egypt, France, Greece, Guatemala, Israel, Japan, Korea, Mexico, Panama, Peru, Portugal, Turkey, Serbia, and Spain are stronger in uncertainty avoidance than the United States, as are most African and other Asian nations. They typically have a history of central rule and complex laws that regulate individual action on behalf of the group.[3]

Cultures with low or weaker uncertainty avoidance include Canada, Denmark, Great Britain, Hong Kong, India, Indonesia, Jamaica, the Netherlands, the Philippines, Sweden, and the United States. People from these nations are usually curious about the unknown and different. They are more informal, willing to accept dissent within a group, and open to change.

It is important to distinguish the differences between risk avoidance and uncertainty avoidance. A person from a high uncertainty avoidance culture may be quite willing to take familiar risks or even new risks in order to minimize the ambiguity of a situation. But in general, risks that involve change and difference are difficult for people with strong uncertainty avoidance; this is especially a concern when changes threaten acceptance by the group. For example, researchers suggest African-American women may resist certain preparations or seasonings if family members object or if the foods might undermine ethnic identity. Furthermore, weight loss may be avoided if being thin means the potential loss of a peer group that values a larger figure.[5]

Working with family or peers in a group setting to effect dietary change may be more successful for persons with a low tolerance for uncertainty, especially when the positive value of change is accepted and group consumption patterns are modified.

Power, Authority, and Status

The perception of power, or power distance, can strongly influence communication patterns. In low-context cultures, where individuality is respected, power or status is usually attributed to the role or job that a person fulfills. Power distance is small. People are seen as equals, differentiated by their accomplishments. It is common for an individual to question directions or instructions; the belief is that a person must understand why before a task can be completed. A client may desire a full explanation of a condition and expected outcomes before undertaking a specific therapy. In many high-context cultures, where group identification is esteemed, superiors are seen as fundamentally different from subordinates. Authority is rarely questioned. For example, a health care provider counseling

The Inuit conception of time is governed by the tides—one set of tasks is done when the tide is out, another when the tide comes in.

The Arabs say, "Bukra insha Allah," which means, "Tomorrow, if God wills."

To many Chinese, the gift of a clock means the same as saying, "I wish you were dead." Each tick is perceived as a reminder of mortality.

an Ethiopian patient with type 2 diabetes may believe that a culturally sensitive approach is to ask him about his perceptions of the disease. What does he call it? How does he think it can be cured? Unknown to the provider, the Ethiopian man has a large power distance, and he assumes that the provider is the expert. Why would she ask such questions of him? Doesn't she know what she is doing? He expects her to provide all the answers, with little participation from him. He may even become uncooperative or fail to return for a follow-up visit because he questions her expertise.

Although there is usually some combination of both small and large power distance tendencies in a culture, one tendency will predominate. Some countries with small power distance include Austria, Canada, Denmark, Germany, Great Britain, Ireland, Israel, the Netherlands, New Zealand, Sweden, and the United States. Those with larger power distance include most African, Asian, Latino, and Middle Eastern cultures, including (but not exclusively) Egypt, Ethiopia, Ghana, Guatemala, India, Malaysia, Nigeria, Panama, Saudi Arabia, and Venezuela. Client empowerment, particularly in setting goals and objectives, may be resisted by people from groups who come from cultures with a larger power distance; maximum personal responsibility may be preferred by people from groups with a smaller power distance.

Several cultures are gender oriented as well. In masculine cultures, power is highly valued. Some, such as Germany, Hong Kong, and the United States, are considered masculine due to their aggressive, task-oriented, materialistic culture. Others, such as Italy, Japan, Mexico, and the Philippines, are characterized as masculine because sex roles are strongly differentiated. Men are accorded more authority in masculine societies. In more feminine cultures, quality of life is important; men and women share more equally in the power structure. More feminine countries can be task oriented and materialistic, such as Denmark, the Netherlands, Norway, and Sweden; but hard work and good citizenship are seen as benefiting the whole society, not a hierarchy of superiors within an organization or nation. Workplace relationships and an obligation to others are characteristically emphasized. Nearly all other nations combine masculine and feminine power qualities. Strong gender orientation can cause conflict. A health care team, for instance, may include several experts, often both women and men. An Italian-American patient might show little respect for the female members and may ignore their directions unless restated by one of the male practitioners, regardless of his area of expertise. Even women from masculine societies may find it difficult to accept the authority of a female provider. Generally, masculine-oriented U.S. health care practitioners can communicate more successfully with most ethnic clients by using a less assertive, less autocratic approach that includes compromise and consensus.

Time Perception

Being on time, sticking to a schedule, and not taking too much of a person's time are valued concepts in the United States, but these values are unimportant in societies where the idea of time is less structured. Low-context cultures tend to be monochronistic, meaning that they are interested in completing one thing before progressing to the next. Monochronistic societies are well suited to industrialized accomplishments. Polychronistic societies are often found in high-context cultures. Many tasks may be pursued simultaneously, but not to the exclusion of personal relationships. Courtesy and kindness are more important than deadlines in polychronistic groups.[2] Exceptions occur, however. The French have a relatively low-context culture but are polychronistic; the Japanese can become monochronistic when conducting business transactions with Americans.

People who are single-minded often find working with those performing multiple tasks frustrating. Monochronistic persons may see polychronistic behaviors, such as interrupting a face-to-face conversation for a phone call or being late for an appointment, as rude or contemptuous. Yet no disrespect is intended, nor is it believed that polychronistic persons are less productive than monochronistic people. In fact, multitasking, the ability to do many things simultaneously, is valued in many organizations today.

Nonverbal Communication

High-context cultures place great emphasis on nonverbal communication in the belief that body language reveals more about what a person is thinking and feeling than words do. Yet customs about touching, gestures, eye contact, and spatial relationships vary tremendously among cultures, independent of low- or high-context communication style. As discussed previously, such nonverbal behavior can reinforce the content of the verbal message being sent, or it can contradict the words and confuse the receiver. Successful intercultural communication depends on consistent verbal and nonverbal messages. During personal and group interactions, persons move together in a synchronized manner. Barely detectable motions, such as the tilt of the head or the blink of an eye, are imitated when people are in sync and communicating effectively. The ways in which a person moves, however, are usually cultural and often unconscious. Although body language is closely linked to ethnicity, most people believe that the way they move through the world is universal.

Misinterpretations of nonverbal communication subtleties are common and often inadvertent. More than 7,000 different gestures have been identified, and meaning is easily misunderstood when awareness of differences is limited.[6]

Touching

Touching includes handshakes, hugging, kissing, placing a hand on the arm or shoulder, and even unintentional bumping. In China, for example, touching between strangers, even handshaking if one person is male and the other female, is uncommon in public. Orthodox Jewish men and women are prohibited from touching unless they are relatives or are married. To Latinos, touching is an expected and necessary element of every relationship. The abrazo, a hug with mutual back patting, is a common greeting. Touching norms frequently vary according to attributes such as gender, age, or even physical condition. In the United States, it is acceptable for an adult to pat the head of a child, but questionable with another adult. Women may kiss each other, men and women may kiss, but men may not publicly kiss other men. It is admirable to take the arm of an elderly person crossing the street but rude to do so for a healthy young adult.[7]

Cultures in which touching is mostly avoided include those of the United States, Canada, Great Britain, Scandinavia, Germany, the Balkans, Japan, and Korea. Those in which touching is expected include the Middle East and Greece, Latin America, Italy, Spain, Portugal, and Russia. Cultures that fall in-between are those of China, France, Ireland, and India, as well as those in Africa, Southeast Asia, and the Pacific Islands. Health care professionals should take careful note of cultural touching behaviors. Vigorous handshaking is often considered aggressive behavior, and a reassuring hand on the shoulder may be insulting.[6] Of special mention are attitudes about the head. Many cultures consider the head sacred, and an absent-minded pat or playful cuff to the chin may be exceptionally offensive. Conversely, persons from cultures in which frequent touching is the norm may be insulted by reticence to a hug or a kiss and may be unaware of legal issues regarding inappropriate touching in the United States.

Gesture, Facial Expression, and Posture

Gestures include overt movements such as waving hello or goodbye or standing to indicate respect when a person enters the room, as well as more indirect motions such as handing an item to a person or nodding the head in acknowledgment. Facial expression includes deliberate looks of attention or questioning and unintentional wincing or grimacing. Even smiling has specific cultural connotations.

Confusion occurs when movements have significantly different meanings to different people. Crossed arms are often interpreted as a sign of hostility in the United States, yet do not have similar negative associations in the Middle East, where it is a common stance while talking. The thumbs-up gesture is obscene in Afghanistan, Australia, Nigeria, and many Middle Eastern nations. The crooked-finger motion used in the United States to beckon someone is considered

Anthropologists speculate that the handshake, the hug, and the bow with hands pressed together all originated to demonstrate that a person was not carrying weapons.

In Japan, the small bow used in greetings and departures is a sign of respect and humility. The inferior person in the relationship always bows lower and longer than the person in the superior position.

Many Asians completely avoid touching strangers, even in transitory interactions, such as returning change after a purchase. This can be offensive to persons who consider physical contact a sign of acceptance, such as African Americans and Latinos.

▼ *Touching norms frequently vary according to attributes such as ethnicity, gender, age, or physical condition.*

© Michael Newman/PhotoEdit

The one-finger salute is considered an insult in many cultures. This gesture dates back to Roman times, when it was called *digitus impudicus* (the "impudent finger").

Koreans say, "A man who smiles a lot is not a real man."

The Japanese demonstrate attentiveness by closing their eyes and nodding.

lewd in Japan; is used to call animals in Croatia, Malaysia, Serbia, and Vietnam; and is used to summon prostitutes in Australia and Indonesia. To many Southeast Asians, it is an insolent or threatening gesture.[7,8]

Asians find it difficult to directly disagree with a speaker and may tilt their chins quickly upward to indicate "no" in what appears to Americans to be an affirmative nod. Some Asian Indians, Greeks, Turks, and Iranians shake their heads back and forth to show agreement and nod up and down to express disagreement. Puerto Ricans may smile in conjunction with other facial expressions to mean "please," "thank you," "excuse me," or other phrases. The Vietnamese may smile when displeased. In India, smiles are used mostly between intimates; an Asian Indian client may not know what to make of a health care provider who smiles in a friendly fashion.

Good posture is an important sign of respect in nearly all cultures. Slouching or putting one's feet up on the desk are generally recognized as impolite. In many societies, the feet are considered the lowest and dirtiest part of the body, so it is rude to point the toe at a person when one's legs are crossed or to show the soles of one's shoes.

Eye Contact

The subtlest nonverbal movements involve the eyes. Rules regarding eye contact are usually complex, varying according to issues such as social status, gender, and distance apart. Most Americans consider eye contact indicative of honesty and openness, yet staring is thought to be rude. To Germans, direct eye contact is an indication of attentiveness. African Americans may be uncomfortable with prolonged eye contact, but may also find rapid aversion insulting. In general, blacks tend to look at a person's eyes when speaking and look away when listening. To Filipinos, direct eye contact is an expression of sexual interest or aggression. Among Native Americans, direct eye contact is considered rude, and averted eyes do not necessarily reflect disinterest. When Asians and Latinos avoid eye contact, it is a sign of respect. Middle Easterners believe that the minute motions of the eyes and pupils are the most reliable indication of how a person is reacting in any situation.[7]

Spatial Relationships

Each person defines his or her own space—the surrounding area reserved for the individual. Acute discomfort can occur when another person stands or sits within the space identified as inviolate. Middle Easterners prefer to be no more than two feet from whomever they are communicating with so that they can observe their eyes. Latinos enjoy personal closeness with friends and acquaintances. African Americans are likely to be offended if a person moves back or tries to increase the distance between them. Intercultural communication is most successful when spatial preferences are flexible.

In addition to distance, the way a person is positioned affects communication in some cultures. It is considered rude in Samoan and Tongan societies, for instance, to speak to a person unless the parties are positioned at equal levels, for example, both sitting or both standing.

Role of Communication in Health Care

Health care providers in the United States take pride in their technical expertise and mastery of knowledge. They spend years understanding biochemical and physiological processes, laboratory assessments, diagnostic data, and therapeutic strategies; yet little of that time is devoted to how valuable information is effectively communicated to the client or members of the health care team. Skills are needed for successful communication with these and other participants, such as extended family members or traditional health practitioners, despite possible differences in language, ethnicity, religious affiliation, gender, age, educational background, occupation, health beliefs, or other cultural factors.

Words are the primary tool of the clinician following diagnosis. Whereas the surgeon depends on the scalpel, most other providers rely on language to inform and guide patients in the treatment and lifestyle changes necessary to maintain or improve health.[9] The surgeon has significant control within the surgical setting; in most cases the patient is not even conscious. In contrast, the clinician interacts directly with a patient who has independent, sometimes contradictory, ideas about health, illness, and treatment. The provider can control only her or his side of the conversation; if the words are ineffective, the client may reject recommended medications or therapies. Although the actions of the surgeon are generally limited to the patient, the advice of the health care provider often impacts not only the patient, but also the patient's family. Dietary modifications in particular may have long-term implications; if cultural food habits are changed, the new ways of eating may be passed on for generations.

© Steve Kaufman/Documentary/Corbis

▲ *Different cultural expectations regarding nonverbal behaviors, such as eye contact and spatial preferences, can lead to misunderstandings in intercultural communication.*

Interaction Between Provider and Client

In the time-pressured and cost-constrained setting of health care delivery, object messages are more common than personal messages, and content is considered more relevant than the relationship. Typically, the health care professional relies on the client to provide accurate, detailed information about his or her medical history and current symptoms so that the appropriate diagnosis and treatment can be determined. The client depends on the practitioner to explain any medical condition in terms that are understandable and to describe treatment strategies and expectations clearly. This basic conversation is repeated between providers and their clients daily; it is the essence of clinical health care. In practice, however, this common interaction between provider and client greatly underestimates the complexity of intercultural communication. Confidence and caring that is established between health care providers and patients can also contribute to the results of the overall health outcome.[10]

Numerous barriers to the sharing and understanding of knowledge can prevent successful communication in the health care setting. For example, a client may be fearful or in pain when seeking help, more focused on immediate discomfort than on conversing clearly with the provider. During times of stress, a client is also more likely to use her or his mother tongue than English if it is a second language.[11] The provider often assumes the role of the expert, leaving little room for participation of the client as the authority on what he or she is experiencing physically or emotionally.[12,13] The provider may rely on medical jargon because it is difficult to interpret many terms without extensive explanations or oversimplification.[14,15] A provider may be most concerned with the technical aspects of a health problem and inadvertently ignore the interpersonal dimensions of the relationship with the patient or may be too rushed to express care and compassion.[1,16] Furthermore, cultural communication customs may interfere directly with the trust and respect necessary for effective health care.[17]

Researchers in effective communication have identified five ways in which misunderstandings occur that are applicable to the health care setting:[3,18]

1. A provider can never fully know a client's thoughts, attitudes, and emotions, especially when the client is from a different cultural background.
2. A provider must depend on verbal and nonverbal signals from the client to learn what the client believes about health and illness, and these signals may be ambiguous.
3. A provider uses his or her own cultural understanding of communication to interpret verbal and nonverbal signals from the client, which may be inadequate for accurate deciphering of meaning in another cultural context.
4. A provider's state of mind at any given time may bias interpretation of a client's behavior.
5. There is no correlation between what a provider believes are correct interpretations of a client's signals or behaviors and the accuracy of the provider's belief. Misunderstandings of meaning are common.

Health care providers and consumers depend on their abilities to communicate sensitively and effectively with one another to relieve discomfort, save lives, and promote health. Ineffective communication in health care can, and often does, result in unnecessary pain, suffering, and death (p. 8).[1]

© Jeff Dunn/Index Stock/Photolibrary

▲ *The client depends on the practitioner to explain any medical condition in terms that are understandable and to describe treatment strategies and expectations clearly.*

The results of ineffective communication in health care can be serious. Noncompliance issues are among the most important for the clinician.

Patients may reject recommendations or fail to return for follow-up appointments because they are dissatisfied with their relationship with their health care provider.[12,19,20] Patients with diabetes who perceived discriminatory behavior from their health provider due to race, age, socioeconomic status, or gender suffered more symptoms and had higher levels of hemoglobin A1C (a blood test for three-month average sugar levels) than other patients.[21] Conversely, patients who received treatment by health professionals in accordance with their desired care reported significantly better dietary management of their diabetes in another study.[22] Patients often report better outcomes with traditional healers than with biomedical practitioners because there is more time spent on explanation and understanding of the condition.[23] Development of the interpersonal relationship with the practitioner is crucial to a patient's understanding and accepting treatment strategies, particularly if recommendations conflict with cultural perceptions regarding health and illness.[24]

Participatory research on a Lakota Indian reservation reported that intercultural connections were directly related to the investment of time and commitment to establishing and pursuing meaningful dialogue.[39]

Being involved with decision-making was significantly associated with adherence to medical advice for whites, whereas being treated with dignity improved adherence for racial/ethnic minorities in one survey.[91]

Responsibilities of the Health Care Provider

Although communication requires the active participation of at least two persons, the health care provider has certain responsibilities in interactions with clients. The provider often assumes the superordinate position in the relationship because she is accorded that status by the client or because the client is distracted by pain or discomfort. In that role, it is the practitioner's obligation to understand what is said by the client and to provide the client with the information needed to participate in treatment. This may require that the clinician be familiar with cultural norms, listen carefully and seriously to the client (observing personal cues), and take action based on what is said by the client. Caring and considered communication can empower the client within the relationship and improve treatment efficacy.[1,7,25]

It is believed intercultural communication awareness occurs in four stages. First is unconscious incompetence, when a speaker misunderstands communication behaviors but doesn't even know misinterpretation has occurred. The second stage is conscious incompetence, when a speaker is aware of misunderstandings but makes no effort to correct them. Third is conscious competence, when a speaker considers his or her own cultural communication characteristics and makes modifications as needed to prevent misinterpretations. The final stage is unconscious competence, when a speaker is skilled in intercultural communication practices and no longer needs to think about them during conversation.[26]

Successful Intercultural Communication

Effective intercultural communication begins when the speaker is mindful of his or her own communication behaviors and is sensitive to misinterpretations that may result from them. Practitioner knowledge about a culture does not necessarily facilitate effective care without awareness of cultural differences and personal biases.[27,28,36] A willingness to listen carefully to a client without assumptions or bias and to recognize that the client is the expert when it comes to information about his or her experience is requisite to successful health care interactions.

The mnemonic CRASH has been suggested as a useful way to remember the components to cultural competency that underlie effective care: C—consider Culture in all patient–practitioner interactions; R—show Respect and avoid gratuitous familiarity and affection; A—Assess/Affirm intracultural differences due to language skills, acculturation, and other factors, recognizing each individual as an expert on his or her health beliefs and practices; S—be Sensitive to issues that may be offensive or interfere with trust in the relationship, and show Self-awareness regarding personal biases that may cause miscommunication; and H—demonstrate Humility, apologizing quickly and accepting responsibility for communication missteps.[29] With CRASH in mind, the health professional can begin mastery of the communication skills needed to promote understanding and acceptance from clients of many cultural backgrounds.

Intercultural Communication Skills

Reading about culturally based communication differences is an intellectual undertaking. Actually applying intercultural communication concepts is much more challenging. Successful face-to-face interactions require understanding cultural communication expectations and being familiar with the idiosyncratic style of the other person. Often there is contradictory information to assimilate. The Japanese, for example, come from a culture that is male dominated, assertive, and achievement oriented, yet also high in uncertainty avoidance, emphasizing consensus and ritualistic communication practices. Should an explicit or implicit approach be used when speaking with someone of Japanese heritage? Clearly, much depends on the circumstances surrounding the conversation and the particular people involved. A few guidelines on applied intercultural communication skills are a beginning. Numerous books, articles, and courses on health care communication are available to supplement this brief overview.

Name Traditions

Determine how clients prefer to be addressed. Americans are among the most informal worldwide, frequently calling strangers and acquaintances by their given names. Nearly all other cultures expect a more respectful approach. This can include the use of title or prefixes (Mr., Mrs., Miss, Ms., Dr., Sir, Madam, etc.), use of surname, and proper pronunciation. Never use "Dear," "Honey," "Sweetie," "Fella," "Son," or other endearments or pejorative terms in place of proper names.

Name order is often different from the United States pattern of title, given name, middle name, and surname. In many Latin American countries a married woman uses her given name and her maiden surname, followed by *de* ("of") and her husband's family surname. For instance, a married woman named Ana would use her maiden name of Lopez followed by de plus her husband's family surname of Perez. She would then use the name Ana Lopez de Perez. Children typically use their given name, their father's surname, and then their mother's family surname. This causes confusion because while most Latino men prefer to be addressed by their father's family surname, it is often the name that sequentially is placed on the "middle name" line of forms. Persons reading the form often assume the mother's family surname is a Latino's last name.[30]

Middle Easterners use their title, given name, and surname. They may also use *bin* (for men) or *bint* (for women), meaning "of" a place or "son/daughter of." This is not to be confused with the given name "Ben" (although it is pronounced similarly). For example, Abdel Al-Fakeeh *bin* Saud literally means Abdel son of Saud. If the grandparent of a Middle Easterner is well known, his or her last name may be added to the name order, following the surname, with another *bin*. The Vietnamese, Hmong, and Cambodians place the surname first, followed by the given name (although many make the switch to the American name order with acculturation). The Chinese and Koreans use a similar system, including a generation name following the family name and before the given name (the generation name is sometimes hyphenated with the given name, or the two names are run together). For example, a man whose name is Yung and has a generation name of Tsing and a surname of Wai would be referred to as Yung Tsingwai. Married women in China and Korea do not take their husband's surname. In Japan, the surname is followed by *san*, meaning "Mr." or "Ms." Given names are only used among close friends and intimates. Hindus in India do not traditionally use surnames. A man goes by his given name, preceded by the initial of his father's given name. A woman follows the same pattern, using her father's initial until marriage, at which time she uses her given name, followed by her husband's given name. Muslims

One researcher reports that the use of cultural food potlucks among hospital staff members has facilitated cultural understanding among employees of diverse backgrounds.[92]

Studies suggest that the provider–client relationship improves when the client is also trained in communication skills. A meeting with coaches to guide clients in formulating appropriate questions and to practice negotiation techniques before an appointment has been successful in some health care settings.[41]

in India use the Middle Eastern order, and Christian Indians use the U.S. pattern.[31] There are numerous other name traditions and preferences of address. When in doubt, it is best to ask.

Appropriate Language

Use unambiguous language when working with clients who are limited in English. Choose common terms (not necessarily simple words), avoiding those with multiple meanings, such as "to address," which may mean to talk to someone, to give a speech, to send an item, or to consider an issue. Vague verbs, such as "get," "make," and "do," may cause confusion. Use specific verbs, such as "purchase," "complete," or "prepare," when directing clients.

Slang and idioms may have no meaning in another culture. Many new English speakers interpret words literally. *How's it going?* makes no sense if one does not understand the meaning of *it* in this context.[30] Sports analogies, including *score* and *strike out,* are indecipherable if the game is unfamiliar. Phrases that suggest a mental picture, such as *run that by me* or *easy as pie* or *dodge a bullet,* are barriers to comprehension.[32] Medical jargon is also likely to interfere with communication.[15] Some persons with limited English skills are embarrassed to admit that they do not understand what is being said or to ask that something be repeated. When comprehension is critical, respectfully request clients to repeat instructions in their own words, or ask that they demonstrate a skill so that misunderstandings can be corrected.

Avoid asking questions that can be answered with a simple yes or no. For example, "Do you understand?" will often prompt a positive response in practitioner–client conversations, regardless of comprehension level. It is better to ask leading questions—for example, "What confuses you?" or "Tell me what you don't understand."[30] In some Asian cultures it is impossible to say "no" to a request. The Japanese, for instance, have developed many ways to avoid a negative response, such as answering "maybe," countering or criticizing the question, issuing an apology, remaining silent, or leaving the room.[33]

A professional interpreter reports that unintended results due to limited language skills can be confusing, insulting, or even comic. A friendly physician meant to ask one of his clients, "Cuantos años tiene usted?" ("How old are you?") He mispronounced the word as anos, however, saying, "How many anuses do you have?"[11]

The direct, explicit communication style of majority Americans is predicated on the assumption that each person is saying what he or she means. This can cause difficulties when conversing with persons for whom negotiation is standard practice. For example, a practitioner offers coffee or tea to an Iranian client. She refuses, so he sits down and begins the discussion. The client is upset because she was being polite and had expected the health care professional to ask again, and then insist that she have something to drink.

Typically, Americans not only believe what is said initially; they also consider answers absolute. In some cultures it is acceptable to make a commitment, then decide later to change the terms of the agreement or decline altogether. People from these groups assume that one cannot predict intervening events or future needs. Further, there can be differences in what is accepted as "truth." In the United States, truth is considered objective, supported by immutable facts. In many other cultures, truth is subjective, often based on emotions.[30,34] A Filipino-American client may report to the practitioner that she has not lost any weight on her low-calorie diet this week. But when she gets on the scale, she weighs three pounds less. Asked about it, the client explains she is frustrated because she gained one-half a pound yesterday and does not feel she is progressing. Understanding that the definitions of truth vary culturally can help explain some miscommunication.

Use of an Interpreter

Language can be the most difficult of all intercultural communication barriers to overcome. It is estimated that nearly 14 million people living in the United States have poor English skills.[35] Many are recent immigrants, and others view their stay in the United States as temporary and therefore see no need to learn English.[10]

According to Title VI of the Civil Rights Act of 1964, all persons in the United States are guaranteed equal access to health care services regardless of national origin, which has been interpreted to mean that there can be no discrimination based on language. Programs not offering access to persons with limited English skills may lose federal funding, including Medicare and Medicaid reimbursements. However, the regulation is vague and difficult to enforce. Compliance is mostly complaint driven. At present, few medical institutions have adequate professional interpreters available to meet the needs of their non-English-speaking clientele.

Unfortunately, many health care providers resort to nonprofessional interpreters, such as the client's family or friends, to facilitate communication. The inadequacies of such interpretations are numerous. Patients may be reluctant or embarrassed to discuss certain conditions in front of their relatives, or family members may decide that the information provided by the practitioner isn't really needed by the patient, so they do not interpret accurately. Untrained interpreters are often unfamiliar with medical terminology. One study indicated that 23 to 52 percent of phrases were misinterpreted by nonprofessionals; for example, "laxative" was the term used for diarrhea, and "swelling" was confused with getting fat. The interpreter tended to ignore questions about bodily functions altogether.[34] Even totally bilingual individuals may not be familiar with all dialects; the terms used in one part of a country may be very different from those used in another region. Ethical issues arise when children are used as interpreters. Children may be frightened of medical procedures, and dependence on a child for communication can invert family dynamics, causing unnecessary intergenerational stress.

Issues regarding informed consent, patient safety, and noncompliance occur when interpretations are inadequate. Some health care providers attempt to use their personal skills in a foreign language, believing it is better to try to communicate directly than to lose some control through interpreters. Although conversing in the language of the client is often greatly appreciated, it is important for the provider not to overestimate fluency. Obtaining the services of a professional interpreter is warranted in all but emergency situations when a delay could be life threatening.[10]

When using an interpreter, the practitioner should speak directly to the client, and then watch the client rather than the interpreter during interpretation. If the nonverbal response doesn't fit the comment, confirmation with the interpreter can ensure that the meaning is clear. Sometimes an interpreter may appear to answer for the patient; the interpreter may be very familiar with the patient's history based on previous interpretations for other health providers. Conversely, it may take an interpreter considerably longer to interpret a comment than it takes to say it in English, in part because certain cultural interpretations and explanations may be necessary.[37] The technique of back interpretation, meaning that instructions are repeated back to the clinician, can prevent miscommunication and open the conversation to any further questions by the client. Providers can increase effective communication through an interpreter by using a positive tone of voice and avoiding a judgmental or condescending attitude. Short, direct phrases—avoiding metaphors or colloquialisms—and repeating important information more than once can improve client understanding.

In areas where interpreters are unavailable, telephone interpretation services may be an alternative. The AT&T Language Line is available throughout the United States and offers interpreters trained in medical terminology in most major languages.[35]

Successful communication in the interpreter–client relationship is also dependent on intercultural skills. It has been suggested that translators who build rapport and trust are more effective than those who are emotionally detached.[75]

Intercultural Counseling

Practitioner attitude toward outcomes is perhaps the most important element of successful intercultural counseling. A health care provider cannot be open minded if objectives are completely preplanned. Participation in a relaxed, give-and-take exchange can reveal issues of primary concern to the client. An invitation to share stories, for example, may address concerns that cannot be expressed directly. Mutual commitment to shared goals can be developed by attentive listening to client needs and learning about client expectations.[25,38,39] This collaboration in defining and achieving outcomes is the difference between advocacy and manipulation. Effective intercultural counseling is an ongoing process of practice and refinement, requiring an open attitude, cultural knowledge, and intercultural communication skills.[39,40]

A study on patient participation in decision-making found that 96 percent of clients want care choices offered to them and want their opinions sought. However, 52 percent preferred to leave the final decision to their physician. Well-educated white women were most likely to want shared decision-making; African Americans and Hispanics were least likely.[93]

Pre-Counseling Preparation

Researchers have made many recommendations regarding effective intercultural communication. The basic competencies needed by practitioners include (1) information transfer—the verbal and nonverbal ability to convey object messages; (2) relationship development and

Table 3.1 Checklist for Intercultural Nutrition Counseling

Attitudes	Knowledge	Skills
• I am open-minded and willing to be a learner instead of the expert when it comes to the client's life experiences and ways of knowing. • I am sincerely interested in different cultural perspectives on reality, and I can respect cultural worldviews other than my own. • I can tolerate the ambiguities of intercultural communication. • I can accept that some interactions will be uncomfortable or unfamiliar to me. • I am patient; I attempt to understand the ideas and feelings of the client.	• I understand that although some cultural influences on communication are readily apparent, others are hidden and require development of a personal relationship so that salient factors in communication and compliance can be identified. • I know that body language can provide significant information about the client's concerns and feelings; the relationship can improve or deteriorate through nonverbal communication. • I understand that modification of culturally held beliefs and behaviors can have significant, long-term effects on the client and the client's family. Attempts to force change may result in ineffective communication and noncompliance by the client. • I am familiar with cultural food habits of my clients. • I have learned about traditional health beliefs and practices.	• I explain diet rationale in common terms within the context of a client's worldview, including concepts regarding the cause, prevention, and treatment of illness; I set realistic goals with the client. • I emphasize the continuation of positive cultural food habits and recommend modification of only those food habits that may be detrimental to the client's health. I avoid personal bias. • I attempt to send nonverbal messages consistent with my verbal messages. • I engage in effective intercultural communication with all participants in the health care system to help clients through illness and improve health through supportive personal relationships, through cooperation with families, and through the gathering and sharing of relevant data with other health care professionals.

Source: Adapted from Cassidy, C.M. 1994. Walk a mile in my shoes: Culturally sensitive food-habit research. *Journal of the American Dietetic Association, 97*, 1288–1292; Gudykunst, W.B., & Nishida, T. 1994. *Bridging Japanese/North American differences*. Thousand Oaks, CA: Sage; Kavanagh, K.H., & Kennedy, P.H. 1992. *Promoting cultural diversity: Strategies for health professionals*. Newbury Park, CA: Sage; Kreps, G.L., & Kunimoto, E.N. 1994. *Effective communication in multicultural health care settings*. Thousand Oaks, CA: Sage; Sanjur, D. 1995. *Hispanic foodways, nutrition, and health*. Boston: Allyn & Bacon.

Health care providers note that age affects intercultural communication because older minority members are often socially isolated and may be unwilling to communicate with health care providers from a different culture.[94]

Demographic data on practitioners show a disproportionate number of whites (among dietitians, 90.5 percent are non-Hispanic whites), suggesting that intercultural counseling will become increasingly prevalent until greater diversity in the health care professions is achieved.[95]

maintenance—the ability to create rapport, establish trust, and demonstrate empathy and respect; and (3) compliance gaining—the ability to obtain client cooperation (Table 3.1).[41]

Practically speaking, a health care provider cannot be expected to become an expert in intercultural communication or to fully understand the communication modes best suited to each of the many clients from different cultural heritages. Most patients living in the United States do not expect to be treated as they would in their homeland. But familiarity with intercultural communication attitudes, knowledge, and skills can greatly enhance health care efficacy.

The In-Depth Interview

The in-depth interview is essential in intercultural counseling to determine many of the iceberg issues that may affect communication and cooperation in health care, including ethnicity, age, degree of acculturation or bicultural adaptation, socioeconomic status, health condition, religious affiliation, educational background, group membership, sexual orientation, or political affiliation. However, a client may believe that personal questions about his background are invasive or unnecessary, especially if he comes from a high-context culture. Direct inquiry may even suggest to the client that the practitioner is incompetent because she cannot determine the problem through indirect methods.

One culturally sensitive approach is the respondent-driven interview, in which simple, open-ended questions by the provider initiate conversation.[12] The client can express her understanding and experience in her own words. The practitioner exerts little control over the flow of the response, yet elicits data through careful prompting.[23,42] Useful questions to ask during the conversation include these:

- What do you call your problem? What name do you give it?
- What do you think caused it?
- Why did it start when it did?

- What does your sickness do to your body? How does it work?
- Will you get better soon, or will it take a long time?
- What do you fear about your sickness?
- What problems has your sickness caused for you personally? for your family? at work?
- What kind of treatment will work for your sickness? What results do you expect from treatment?
- What home remedies are common for this sickness? Have you used these home remedies?

Furthermore, information should be requested about traditional healers:

- How would a healer treat your sickness? Are you using that treatment?

For nutritional assessment within the context of the client's condition, questions about food habits are appropriate:

- Can what you eat help cure your sickness or make it worse?
- Do you eat certain foods to keep healthy? To make you strong?
- Do you avoid certain foods to prevent sickness?
- Do you balance eating some foods with other foods?
- Are there foods you will not eat? Why?

Learning about how a client understands his illness, including expectations about how the illness will progress, what the provider should do, and what he has established as therapeutic goals, allows the provider to compare her own view of the illness and to resolve any discrepancies that might interfere with care. Demonstrating sincere interest in cultural health beliefs through an open-ended conversation can elicit the information needed to begin assessment and to determine the most effective approaches for each individual.

Intercultural Nutrition Assessment

Several difficulties in the collection and analysis of cultural health data have emerged in recent years. Researchers have discovered that standardized tools can introduce systematic bias into results or provide misleading information when used with different cultural populations.

Generalized approaches to the use of the 24-hour food recall, food frequency forms, and nutrient databases can produce large errors in assessment.[12,43–45] Cultural unfamiliarity with concepts, such as fiber; terminology differences, such as using one word for several foods or not having a name for a certain category of food (e.g., there is no American Indian word for vegetables), or grouping foods by different categories (e.g., by medicinal properties or status); lack of differentiation between meals and snacks; checking "phantom foods" (those not actually consumed) when not enough traditional items are available for selection; and translation mistakes (e.g., literal translations of food names, use of a brand name for a generic item, or use of the name for a traditional food for a similar American item) are a few ways collected data can be invalidated. Frequent consumption of mixed dishes can result in omission of some nutrient sources (as when rice is prepared with dried peas or beans, yet reported as rice by some Caribbean Islanders) and over- or underestimation of intake due of complications in portion-size estimates. Tremendous variability in the amounts of food eaten has been reported between individuals and cultural groups.[34] Nutritional variety may be artificially reduced if the composition data for a cultural food prepared in significantly different ways among subgroups are used without allowances for recipe modifications. Using food composition data for similar foods when specific listings on a cultural item are unavailable can lead to intake miscalculations if numerous substitutions are necessary.

Food lists derived from data on the U.S. population as a whole may miss significant dietary nutrient sources in subgroups. A study comparing a generalized food frequency questionnaire with ethnic-specific tools developed for African Americans, whites, and Mexican Americans found improvement in assessing total fat, vitamin A, and vitamin C intake with the modified forms. Calculations increased by 6 to 7 percent for fat sources, 1 to 3 percent for vitamin A, and 1 to 2 percent for vitamin C when twenty-eight cultural food items were added to the seventy-four food items typical of the total population.[46] Similar research also reported a modest increase in nutrient estimates for Hispanic, Chinese, and

A study of Chinese-American schoolchildren found that USDA food composition databases supplied information on only 65 percent of the 120 different food items identified in the three-day diet records.[96]

Research conducted with African American women found that an interviewer-administered food frequency questionnaire was more feasible, and less burdensome for staff and respondents, than multiple twenty-four-hour dietary recalls.[97]

Japanese respondents.[47] A review of food frequency questionnaires in minority populations, however, found that the number of published examples was so few, and the methodologies were so varied, that no conclusions regarding how best to develop valid, reliable instruments could be reported. Questions regarding how extensive questionnaires should be (especially number of included foods), whether food groupings need to be modified, and how portion size should be standardized remain unanswered.[48]

Other assessment tools may be questionable in intercultural settings as well. A model used to determine health attitudes among whites was found unreliable when used with Mexican Americans, even after operational adjustments for cultural differences.[49] Contradictions were found, for example, between self-administered and interviewer-directed questionnaire responses; participants were more likely to express disagreement about an item when completed individually than when asked about it by the interviewer. Cultural attitudes regarding pleasing authorities are believed to have influenced respondent answers, calling into question the use of the interview as a valid tool for gathering data in this ethnic group. In another study on low-literacy populations the opposite was found; self-reported data on food frequency questionnaires were found unreliable when compared to comments made by respondents during follow-up interviews.[50,98]

Standardized height and weight growth curves have not been validated for all ethnic groups and should be applied cautiously with cultural variation in mind.

© David Young-Wolff/PhotoEdit

A review of acculturation scales and indexes found that many were unsuited for use in dietary interventions or nutrition education programs.[49] Single-item measures of acculturation, including broad questions such as "How long have you lived in the United States?" and "What language do you speak at home?" provide only introductory information about a client and resulted in outcome discrepancies in data collected on dietary fat intake and acculturation in one study.[51] Acculturation scales are more comprehensive but also do not address questions regarding food habit changes. In addition, acculturation scales are typically validated on homogeneous population samples, such as college students or hospital patients, and may not be fully applicable to a more diverse clientele. Food-based assessments are more promising, but they do not provide data on the psychological or social aspects of acculturation and are usually limited to use with a single cultural group.

Furthermore, anthropometric measurement tools are sometimes inappropriate for certain populations. Height and weight growth curves are particularly vulnerable to misinterpretation due to cultural variation, especially among Asian groups. Stature prediction equations for whites were inaccurate for Latinos[52–55] and African Americans.[56] The predictive value of waist-to-hip ratios and body mass index (BMI) may vary in some populations.[57–60] Questions about the standard BMI cutoff for overweight and obesity in Asians have been raised due to a high risk for health problems at lower numbers.[61] When percent body fat data was used instead of BMI for calculating obesity in African Americans and whites, the difference in rates between women was cut in half (with blacks still more at risk than whites), and the gap for men was widened, with far more white men identified as obese.[62] Even physiologic calculations, such as basal metabolic rate equations, may differ culturally.[63,64]

The development of culturally specific techniques and tools is a critical need in nutritional assessment. For an individual client, switching from a quantitative to a more qualitative approach can establish trust and cooperation in initial interviews. The twenty-four-hour recall, for instance, can be conducted in an open-ended manner, requesting simply that all foods consumed the previous day be remembered.[12] This eliminates difficulties with obtaining portion sizes or differentiating meals. The dietitian does not need to predetermine food items or categories. In subsequent meetings, more information, such as frequency and amount of given items, can be requested after explanation of why these types of data are needed.

When working with many clients of a single cultural heritage, it may be useful to prepare quantitative tools based on qualitative research. This approach is most successful when done by an investigator already familiar with the specific group's food culture. A well-intentioned but culturally biased open-ended question (for example, asking a participant to list "any other foods eaten weekly") may not prompt the recall of foods eaten seasonally, thus underestimating a particular nutrient. The burden of negotiating two different cultural food systems should be on the researcher, not the study participant.[47,65] Detailed interviews with individuals can provide information on appropriate language, categories, concepts, and formatting of the instruments helpful in culturally specific nutritional assessment.

Monthly twenty-four-hour recalls of small, representative samples are useful in determining overall consumption patterns, especially where seasonal variation occurs.[34] Focus groups have been found effective in selecting food items to include and quantification measures in the preparation of multicultural food frequency questionnaires.[43,49] Guidance from the targeted population is essential.[49,65] Two questions regarding the accuracy of data collection are suggested: First, how do cultural perceptions about food affect the way in which clients report their intake? Second, how is the report of intake affected by the client's relationship with the interviewer, the setting, and the assessment tool? Answers to these questions can suggest culturally sensitive approaches and improve the validity of collected data.[13,29]

Access to cultural foods composition data and culturally specific anthropometric and physiological measurements is more problematic. Requests for recipes can be used to expand current databases, although this technique may be too time consuming to complete with every client. Being mindful that data analysis is often approximate and that standardized measurements may be questionable, dietary modifications should be made carefully and cautiously with all clients from cultural backgrounds other than the American majority.

Intercultural Nutrition Education

The biomedical paradigm emphasizes behavioral change accomplished through one-on-one work with an individual. However, many cultural groups prefer learning about nutrition in settings with family members or peers.[66] For example, researchers have found that while white adolescent girls demonstrate poor outcomes when counseled with their mothers, black adolescent girls show significantly improved outcomes when their mothers participate in weight loss sessions.[67,68]

Successful nutrition education strategies for groups are as dependent on intercultural communication skills as is nutrition counseling with individuals. For example, researchers have described how culture can affect program outcomes in a group weight loss setting.[69] Negative results are possible at any phase of the process, from motivation and attendance to skill acquisition and behavior change (see Figure 3.3). At each point cultural influences may reinforce or contradict the content and context of the education messages conveyed by the health care practitioner. When communication conflict develops, an inexact period of time exists when the client is willing to negotiate toward resolution of the message. If dissatisfaction continues, a poor weight loss outcome results because the person (1) is never motivated to sign up, (2) drops out of the program before completion, (3) attends but never learns skills, or (4) learns skills but does

Figure 3.3
Schematic illustration of how cultural factors might influence participation and outcomes at various phases of the program.

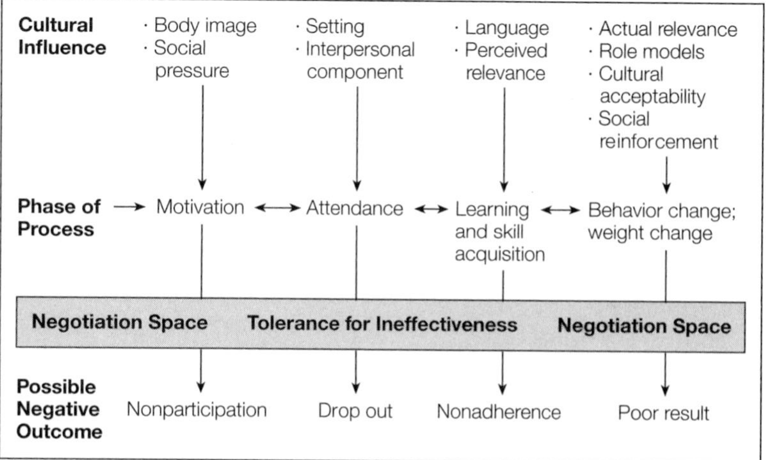

A comparison of food composition tables in nine European nations exemplifies the difficulty in obtaining accurate nutrient data. Differences in definitions, analysis methods, and expression made it impossible to compare local tables with international data.[66]

not apply them in practice. Program designers must do more than superficially modify the program materials and the setting to communicate effectively with a different cultural group. Understanding the cultural health beliefs, attitudes, and values of a target audience; developing education programs within the context of those perceptions; and using culturally appropriate, consistent verbal and nonverbal messages in an accepted medium increase communication efficacy.

Culturally Relevant Program Preparation

Although health education program models typically advise a step-by-step process of planning and execution, the reality is that some aspects of preparation and implementation occur concurrently. A health educator may begin planning with a general idea of goals for a population group, but will probably modify and refine objectives as more information about the target audience is gathered. Ongoing evaluation may suggest better message formats or more suitable influence channels as the effort proceeds. Effective programs are often nonlinear, with each element in planning and implementation connected through feedback and assessment into a continuous improvement loop.

Targeting the Audience

Identification of the target audience in nutrition education efforts is among the most important steps in program planning. Learning about the cultural orientation of the group is imperative; campaigns to change behaviors aimed independently at the individual are usually misdirected.[70] What appear to be significant needs to the health educator may be considered unimportant or too difficult to remedy by mem bers of the target population.

Definitions of health differ widely among cultures (see Chapter 2, "Traditional Health Beliefs and Practices"). A common belief is that illness is a matter of heredity, fate, or punishment by God, or is due to supernatural causes. In a study of health perceptions held by hard-to-reach populations in the United States, it was found that though individuals believed that lifestyle might impact acute infection, there was almost no association between diet or exercise and chronic disease; respondents had limited motivation to improve health behaviors because they felt they had little personal control over their health.[71] The role of the individual within the group can also affect responsibility for health maintenance; in some cultures, the extended family is held accountable for the health of each member.

Content of the message can be critical. For example, one study on the effects of public education efforts to reduce bulimic eating behaviors revealed that some women learned about vomiting as a weight-control method from the campaign.[99]

Demographic information about the target audience can guide program development. Primary language should be identified, as well as gender, average age, socioeconomic status, educational attainment, religious affiliation, and other iceberg factors in communication. Assessment of acculturation or bicultural adaptation is equally important. The more culturally homogeneous the target population, the more appropriate are the messages that can be created.[72,78] In many cases, the larger, heterogeneous audience may be stratified into smaller, segmented target groups that share similar cultural beliefs and attitudes.

Involving members of the targeted audience in program planning is one of the best ways to determine cultural orientation.[73] Of special note is the role of community leaders or spokespersons in the process. Seeking the respect, trust, and endorsement of influential persons within the target audience for a particular nutrition education program can open intercultural communication channels otherwise limited to the formal interactions reserved for strangers.[39,74,75] The educator establishes a relationship with the group through asking for permission to present the health message to its members.

Goals and Objectives

The next step in intercultural program planning is to define clear and realistic goals and objectives within the cultural context of the target audience. Even culturally sensitive education messages do not necessarily translate into sustained modification of food habits without follow-up support, and overly ambitious expectations are a common reason for failure.[76,77] Nevertheless, strategies emphasizing continuation of positive cultural dietary patterns or portion control rather than elimination of certain foods are reportedly of interest to African Americans, Asian Americans, Latinos, and Native Americans, as well as whites: one study found a barrier to eating healthfully was that participants believed they would have to give up their cultural heritage and conform to the dominant culture.[69,71,79] Programs coordinating objectives with cultural beliefs about the role of food in health, such as balancing yin and yang foods in Chinese meals, can reinforce dietary change.[49] Consulting health care practitioners in the targeted community can provide information on local needs and concerns useful in defining achievable goals and objectives.

Triangulation

An especially useful step in the program design process is triangulation, a method of confirming congruence between data collected on the target audience and proposed program goals and objectives (Figure 3.4)[80] Triangulation means that information gathered through one source or method is used to confirm and extend information gathered through other sources and methods. In addition to corroboration, triangulation can provide improved understanding of local issues and perspectives.

In the triangulation pilot program, community nutritionists were interviewed to help define the target population of young African American women and for direction in program development. Next, target group women participated in focus groups and were asked to discuss benefits and obstacles to healthy eating. A final step surveyed community resources on the availability of quality food products. When researchers compared data from the three qualitative studies, they found that their target audience was confirmed as appropriate, that the nutritionists had correctly identified the need for culturally relevant skill-building messages, and that there was a barrier to achieve program goals and objectives due to the lack or excessive cost of fresh and frozen foods. The triangulation process provided concrete data on target population needs and credible communication channels, directing program planning toward culturally relevant interventions and resource development.

Developing the Message

It is believed that the more fundamental the health message is in relation to a group's survival, safety, or social needs, the more effective it will be interculturally.[1] The message must satisfy the individual's need to gain knowledge or offer a solution to a perceived problem before it is worth the person's time to process the information.[70] Messages should be as direct and explicit as allowed within cultural norms.[81] Language relevant to the group should be used in development of the message, and translation of existing materials should be avoided to prevent inappropriate phrasing and terminology. Common words used by the target audience are effective, although it is important that they not be used in an insincere or condescending way. Written materials should be brief and prepared at the reading level of the target population.

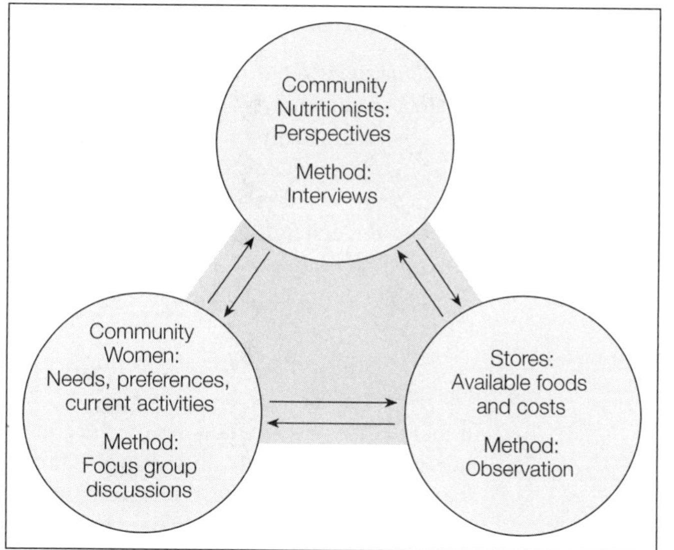

Figure 3.4
Triangulation.
Reprinted from Journal of the American Dietetic Association, Volume 99 (6), J. Goldberg et al, "Using 3 Data Sources and Methods to Shape a Nutrition Campaign," pp. 717–722, Copyright © 1999 The American Dietetic Association, with permission from Elsevier.

Marketing experts recognize many cultural groups are high-context communicators and have greater abilities than the white American majority culture to send and receive messages through nonverbal modes.[75] Body language must be culturally congruent with the verbal message for successful communication to occur. The use of pictures, cartoons, and photographic images can symbolically enhance content meaning of a message in a high-context culture, as well as aid target populations with mixed English language skills or reading abilities.

Educational messages are most effective when they are more personal than objective; the emotional dimension is as important as the content. Many researchers recommend the universally accepted format of storytelling to deliver the message.[82,83] Actors and other celebrities are especially suited to recounting personal experiences about health issues. Stories can transcend many cultural boundaries; if a message is targeted toward one cultural group yet applicable to many audiences, a spokesperson identified with the intended target group also may have broader appeal when using a narrative approach.

The four barriers to healthy eating identified by African American women in the triangulation focus groups were taste, cost, time, and lack of information such as recipes, shopping tips, and a chart comparing healthful and unhealthful choices.[80]

A pilot test of the message with targeted audience members can improve success. Focus groups can be especially useful in assessing cultural appropriateness of education materials and in identifying any resistance triggers inadvertently included in the message.[70,71,84]

Implementation Strategies

Dissemination of a nutrition education message should include analysis of cultural influence channels and media preferences, development of an effective marketing mix, and evaluation of the program. Whether a person actually hears, sees, and understands a message is dependent on frequency, timing, and accessibility. Exciting, informative, culturally appropriate messages fail if they never reach the target audience.[70,76]

Influence Channels

Influence channels are the ways in which message materials are transmitted to the target audience. They include television, video, computers, radio, magazines, newspapers, newsletters, direct mail, and telephones. Each cultural group demonstrates distinct media-use patterns and is best approached through those influence channels. Oral traditions are strong among some populations, while written messages are favored by others. It is estimated that many African Americans (97 percent) listen to the radio for more than thirty hours each week; Latinos enjoy television programs focusing on family and relationship issues, watching on average fifteen hours of Spanish programming and ten hours of English programming each week.[72,85] Asians watch about six hours of native-language television weekly. In a study of Native Americans in California, a majority of respondents reported that they would prefer receiving nutrition information through newsletters (69 percent) or videotapes (67 percent); approximately one-quarter indicated they would like to receive a visit from a health professional, and only 6 percent selected a workshop with family and friends.[79] Computer-based, interactive nutrition education programs are an emergent educational tool, particularly suitable for audiences with low literacy or limited English language skills, as was found in a study of rural Appalachian women.[86] The Internet has become a useful technology, offering twenty-four-hour access to health education materials and easy access to group support through bulletin boards and chat rooms, and individualized therapy through e-mail.[87–89]

Cultural icons incorporated into educational messages should be selected with care. For example, the owl represents wisdom in some American Indian cultures; in others, the owl is a symbol for death.[82]

Mass media campaigns are believed to influence a change in health behavior by about 10 percent of the targeted audience, which can be a significant number in a large campaign.

According to marketing experts, the most effective presentation of a message requires a combination of pictures, sounds, and words in the broadcast and print media.[72] The use of multiple influence channels and frequent repetitions of the message at times when the target audience is listening or watching is also important. Beyond the mass media, health fairs, neighborhood clinics, farmers' markets and grocery stores, traditional healers, churches, schools, food banks or social service centers, carnivals, and sporting events are a few of the locations where culturally relevant nutrition education materials can be successfully distributed on a smaller scale. Low-income, hard-to-reach whites, blacks, and Latinos express interest in nonjudgmental small-group support meetings similar to Alcoholics Anonymous and Tupperware-style home meetings with food samples and cooking demonstrations as settings for nutrition education programs.[71]

Marketing Mix

The four Ps of the marketing mix are product, price, placement, and promotion.[1] They refer to a well-developed message (product) that advances program goals and objectives at minimal economic or psychological cost to target audience members (price) and presents this message in a method congruent with target audience media preferences (placement) in such a way that the target audience members are encouraged to become more involved in the program, either through phone numbers for further information or through attendance at group meetings (promotion). Attention to all four areas of the marketing mix assures that the health care message is fully accessible to the target audience.

Evaluation

Process evaluation keeps track of progress throughout the program, especially the identification of larger community conditions that may be presenting barriers to dissemination of the message. Summative evaluation is used to assess program results after completion of the effort. Evaluation data are useful in refining intercultural nutrition education strategies both during implementation and in future programs. Publication of culturally sensitive nutrition education program results greatly benefits other health professionals and their clients through shared knowledge about intercultural communication techniques and tools.

DISCUSSION STARTERS: HOW DOES DR. PETROCELLI IMPROVE HIS PATIENT RELATIONS?

Consider this situation: Dr. Petrocelli is a neighborhood doctor in a large U.S. city. His office has been in this urban neighborhood for forty years. Originally, the neighborhood was ethnically fairly homogenous: primarily European American and African American, but increasingly, over the last decade, more people of various different ethnicities (Asian, Hispanic, and Middle Eastern), mostly immigrants to the United States, have moved into the neighborhood and are seeking health care. Dr. Petrocelli recognizes that sometimes communications with his newer patients are strained, and he worries that he is not fully understanding them and they are not fully understanding him. He's considering several different options:

- Hire an interpreter. But in order to pay this person's salary, he will need to increase the fees for his patients.
- Continue to rely on patients' relatives or friends to translate for him. And continue to worry if the translations are accurate.
- Start holding an in-depth interview with each of his new patients to get to know them better. But these interviews will take a lot of time and could mean that either he will have to extend his hours—and pay his staff overtime—or set a moratorium on taking any new patients until he has completed the interviews. He worries that such a moratorium could mean that some people living in the neighborhood might not receive adequate healthcare.

First individually and then in small groups, brainstorm which of these options might be the best—or perhaps better yet, brainstorm a solution to Dr. Petrocelli's problem that combines options or offers a completely different option.

REVIEW QUESTIONS

1. Why is communication with another person or group described as an action chain? Give an example of an action chain that might occur when you meet (1) a friend, (2) your new boss, and (3) a young child.
2. Why is it important to become familiar with other cultures' communication behaviors? Give three examples of nonverbal communication behaviors.
3. What is meant by low- or high-context and uncertainty avoidance in describing verbal communications? Name one culture that is low-context and one culture that is high-context. How may an individual's relationship to the group differ between high- and low-context cultures? How does uncertainty avoidance differ from risk avoidance? Give an example.
4. What would be the culturally appropriate verbal address for when you meet the following individuals: an African American, a Latin American, a Vietnamese, and an Asian?

REFERENCES

1. Kreps, G.L., & Kunimoto, E.N. 1994. *Effective communication in multicultural health care settings.* Thousand Oaks, CA: SAGE.
2. Hall, E.T. 1977. *Beyond culture.* Garden City, NY: Anchor/Doubleday.
3. Gudykunst, W.B. 2003. *Bridging differences: Effective intergroup communication* (4th ed.). Thousand Oaks, CA: SAGE.
4. Gostin, L.O. 1995. Informed consent, cultural sensitivity, and respect for persons. *Journal of the American Medical Association, 274,* 844–845.
5. Kumanyika, S.K., & Morssink, C.B. 1997. Cultural appropriateness of weight management programs. In *Overweight and Weight Management: The Health Professional's Guide to Understanding and Practice.* Gaithersburg, MD: Aspen.
6. Axtell, R.E. 1991. *Gestures: The do's and taboos of body language around the world.* New York: Wiley.
7. Eubanks RL, McFarland M,Mixer SJ, Munoz C, Pacquiao DF, Wenger AFZ. Cross Cultural Communication. In: Douglas MK, Pacquiao DF (eds). *Core Curriculum in transcultural nursing and health care [Supplement]. Journal of Transcultural Nursing, 21 (Suppl.1),* 2010.
8. Dresser, N. 2005. *Multicultural manners: Essential rules of etiquette for the 21st century.* New York: John Wiley & Sons.
9. Tumulty, P. 1970. What is a clinician and what does he do? *New England Journal of Medicine, 283,* 20–24.
10. Woloshin, S., Bickell, N.A., Schwartz, L.M., Gany, F., & Welch, H.G. 1995. Language barriers in medicine in the United States. *Journal of the American Medical Association, 273,* 724–728.
11. Haffner, L. 1992. Translation is not enough: Interpreting in a medical setting. *Western Journal of Medicine, 157,* 255–259.

12. Brookins, G.K. 1993. Culture, ethnicity, and bicultural competence: Implications for children with chronic illness and disability. *Pediatrics, 91,* 1056–1062.
13. Cassidy, C.M. 1994. Walk a mile in my shoes: Culturally sensitive food-habit research. *American Journal of Clinical Nutrition, 59 (Suppl.),* 190S–197S.
14. Kreps, G.L. 1990. Communication and health education. In E.B. Ray & L. Donohew (Eds.), *Communication and Health.* Hillsdale, NJ: Erlbaum.
15. Sevinc, A., Buyukberber, S., & Camci, C. 2005. Medical jargon: Obstacle to effective communication between physicians and patients. *Medical Principles and Practice, 14,* 292.
16. Cricco-Lizza, R. 2006. Black non-Hispanic mother's perceptions about the promotion of infant-feeding methods by nurses and physicians. *Journal of Obstetrics and Gynecology, 35,* 173–180.
17. Fong, A.K.H. 1991. Educating clients: Successful communication creates change. *Journal of the American Dietetic Association, 91,* 289–290.
18. Beck, A. 1988. *Love is never enough.* New York: Harper & Row.
19. Lynch, J. 1990. Organ donation & approaching the African American family. *Contemporary Dialysis & Nephrology, 11,* 21–22.
20. Uche-Griffin, Ndidi, F. 1994. Perceptions of African American women regarding health care. *Journal of Cultural Diversity, 1,* 32–35.
21. Piette, J.D., Bibbins-Domingo, K., & Schillinger, D. 2006. Health care discrimination, processes of care, and diabetes patients' health status. *Patient Education and Counseling, 60,* 41–48.
22. Anderson, C.H. 1995. Sugars, sweetness and food intake. American *Journal of Clinical Nutrition, 62 (Suppl.),* 195S–201S.
23. Kleinman, A., Eisenberg, L., & Good, B. 1978. Culture, illness, and care: Clinical lessons from anthropologic and cross-culture research. *Annals of Internal Medicine, 88,* 251–258.
24. MacGregor, K., Handley, M., Wong, S., Sharifi, C., Gjeltema, K., Schillinger, D., & Bodenheimer, T. 2006. Behavior-change action plans in primary care: A feasibility study of clinicians. *Journal of the American Board of Family Medicine, 19,* 215–223.
25. Street, R.L. Jr., Gordon, H.S., Ward, M.M., Krupat, E., & Kravitz, R.L. 2005. Patient participation in medical consultations: Why some patients are more involved than others. *Medical Care, 43,* 960–969.
26. Howell, W. 1982. *The emphatic communicator.* Belmont, CA: Wadsworth.
27. Purnell, L.D., & Paulanka, B.J. 2003. *Transcultural health care: A culturally competent approach* (2nd ed.). Philadelphia: F.A. Davis Company
28. Reimann, J.O., Talavera, G.A., Nunez, S.M., & Velasquez, R.J. 2004. Cultural competence among physicians treating Mexican Americans who have diabetes: A structural model. *Social Science & Medicine, 59,* 2195–2205.
29. Rust, G., Kondwani, K., Martinez, R., Dansie, R., Wong, W., Fry-Johnson, Y., Woody, R.D.M., Daniels, E.J., Herbert-Carter, J., Aponte, L., & Strothers, H. 2006. A CRASH-course in cultural competence. *Ethnicity & Disease, 16,* S3–29, S3–36.
30. Dresser, N. 2005. *Multicultural manners: Essential rules of etiquette for the 21st century.* New York: John Wiley & Sons.
31. Morrison, T., Conaway, W.A., & Borden, G.A. 1994. *Kiss, bow, or shake hands: How to do business in sixty countries.* Holbrook, MA: Adams.
32. Harris, P.R., & Moran, R.T. 1987. *Managing cultural differences* (2nd ed.). Houston, TX: Gulf.
33. Usunier, J.C. 1993. *International marketing.* New York: Prentice Hall.
34. Hankin, J.H., & Wilkins, L.R. 1994. Development and validation of dietary assessment methods for culturally diverse populations. *American Journal of Clinical Nutrition, 59 (Suppl.),* 198S–200S.
35. Woloshin, S., Bickell, N.A., Schwartz, L.M., Gany, F., & Welch, H.G. 1995. Language barriers in medicine in the United States. *Journal of the American Medical Association, 273,* 724–728.
36. Ebden, P., Bhatt, A., Carey, O.J., & Harrison, B. 1988. *The bilingual consultation. Lancet,* 13, 347.
37. Muecke, M.A. 1983. In search of healers—Southeast Asian refugees in the American health care system. *Western Journal of Medicine, 139,* 835–840.
38. Street, R.L. Jr., Gordon, H.S., Ward, M.M., Krupat, E., & Kravitz, R.L. 2005. Patient participation in medical consultations: Why some patients are more involved than others. *Medical Care, 43,* 960–969.
39. Kavanagh, K., Absalom, K., Beil, W., & Schliessmann, L. 1999. Connecting and becoming culturally competent: A Lakota example. *Advances in Nursing Science, 21,* 9–31.
40. Nunez, A., & Robertson, C. 2006. Cultural competency. In *Multicultural Medicine and Health Disparities,* D. Satcher & R.J. Pamies (Eds.), New York: McGraw-Hill.
41. Ruben, B.D. 1990. The health caregiver–patient relationship: Pathology, etiology, treatment. In E.B. Ray, & L. Donohew (Eds.), *Communication and Health.* Hillsdale, NJ: Erlbaum.
42. Anderson, C.H. 1995. Sugars, sweetness and food intake. *American Journal of Clinical Nutrition, 62 (Suppl.),* 195S–201S.
43. Forsythe, H.E., & Gage, B. 1994. Use of a multicultural food-frequency questionnaire with pregnant and lactating women. *American Journal of Clinical Nutrition, 59 (Suppl.),* 203S–206S.
44. Loria, C., Arroyo, D., & Briefel, R. 1994. Cultural biases influencing dietary interviews with Mexican Americans: The Hanes experience. *American Journal of Clinical Nutrition, 59 (Suppl.),* 290S.
45. Teufel, N.I. 1997. Development of culturally competent food-frequency questionnaires. *American Journal of Clinical Nutrition, 65 (Suppl.),* 1173S–1178S.
46. Borrud, L.G., McPherson, R.S., Nichaman, M.Z., Pillow, P.C., & Newell, G.R. 1989. Development of a food frequency instrument: Ethnic differences in food sources. *Nutrition and Cancer, 12,* 201–211.
47. Block, G., Mandel, R., & Gold, E. 2004. On food frequency questionnaires: The contribution of

open-ended questions and questions about ethnic foods. *Epidemiology, 15,* 216–221.

48. Coates, R.J., Clark, W.S., Eley, J.W., Greenberg, R.S., Huguley, C.M., & Brown, R.L. 1990. Race, nutritional status, and survival from breast cancer. *Journal of the National Cancer Institute, 82,* 1684–1692.
49. Satia-Abouta, J.A., Patterson, R.E., Neuhouser, M.L., & Elder, J. 2002. Dietary acculturation: Applications to nutrition research and dietetics. *Journal of the American Dietetic Association, 102,* 1105–1118.
50. Bettin, K.J. 1994. A food frequency questionnaire design for a low literacy population. *American Journal of Clinical Nutrition, 59 (Suppl.),* 289S.
51. Norman, S., Castro, C., Albright, C., & King, A. 2004. Comparing acculturation models in evaluating dietary habits among low-income Hispanic women. *Ethnicity & Disease, 14,* 399–404.
52. Netland, P.A., & Brownstein, H. 1985. Anthropometric measurements for Asian and Caucasian elderly. *Journal of the American Dietetic Association, 85,* 221–223.
53. Freimer, N., Echenberg, D., & Kretchmer, N. 1983. Cultural variation—Nutritional and clinical implications. *Western Journal of Medicine, 139,* 928–933.
54. Gardner, W.E. 1994. Mortality. In N.W.S. Zane, D.T. Takeuchi, & K.N.J. Young (Eds.), *Confronting Critical Health Issues of Asian and Pacific Islander Americans.* Thousand Oaks, CA: SAGE.
55. Chumlea, W.C., Shumei, S.G., Wholihan, K., Cockram, D., Kuczmarski, R.J., & Johnson, C.L. 1998. Stature prediction equations for elderly non-Hispanic white, non-Hispanic black, and Mexican-American persons developed from NHANES III data. *Journal of the American Dietetic Association, 98,* 137–142.
56. Hoerr, S.L., Nelson, R.A., & Lohman, T.R. 1992. Discrepancies among predictors of desirable weight for black and white obese adolescent girls. *Journal of the American Dietetic Association, 92,* 450–453.
57. Slattery ML, Ferucci ED, Murtaugh MA, Edwards S, Ma KN, Etzel RA, Tom-Orme L, Lanier AP. Associations among body mass index, waist circumference, and health indicators in American Indian and Alaska Native adults. *Am J Health Promot. 2010; 24:* 246–54.
58. Yatsuya H, Folsom AR, Yamagishi K, North KE, Brancati FL, Stevens J; Atherosclerosis Risk in Communities Study Investigators. Race- and sex-specific associations of obesity measures with ischemic stroke incidence in the Atherosclerosis Risk in Communities (ARIC) study. *Stroke. 2010; 41:* 417–25.
59. Huxley R, Mendis S, Zheleznyakov E, Reddy S, Chan J. Body mass index, waist circumference and waist:hip ratio as predictors of cardiovascular risk–a review of the literature. *Eur J Clin Nutr. 2010; 64:* 16–22.
60. Lusky, A., Lubin, F., Barell, V., Kaplan, G., Layani, V., & Wiener, M. 2000. Body mass index in 17-year-old Israeli males of different ethnic backgrounds: National or ethnic-specific references? *International Journal of Obesity and Related Metabolic Disorders, 24,* 88–92.
61. McNeely, M.J., & Boyko, E.J. 2004. Type 2 diabetes prevalence in Asian Americans: Results of a national health study. *Diabetes Care, 27,* 66–69.
62. Crawley, J., & Burkhauser, R.V. 2006. Beyond BMI: The value of more accurate measures of fatness and obesity in social science research. Working Paper No. 12291. *NBER Bulletin on Aging and Health, June.* Cambridge, MA: National Bureau of Economic Research.
63. Case, K.O., Brahler, C.J., & Heiss, C. 1997. Resting energy expenditures in Asian women measured by indirect calorimetry are lower than expenditures calculated from prediction equations. *Journal of the American Dietetic Association, 97,* 1288–1292.
64. Liu, H.Y., Lu, Y.F., & Chen, W.J. 1995. Predictive equations for basal metabolic rate in Chinese adults: A cross-validation study. *Journal of the American Dietetic Association, 95,* 1403–1408.
65. Teufel, N.I. 1997. Development of culturally competent food-frequency questionnaires. *American Journal of Clinical Nutrition, 65 (Suppl.),* 1173S–1178S.
66. Deharveng, G., Charrondiere, U.R., Slimani, N., Southgate, D.A.T., & Riboli, E. 1999. Comparison of nutrients in the food composition tables available in the nine European countries participating in EPIC. *European Journal of Clinical Nutrition, 53,* 60–79.
67. Brownell, K.D., Kelman, J.H., & Stunkard, A.J. 1983. Treatment of obese children with and without their mothers: Changes in weight and blood pressure. *Pediatrics, 71,* 515–523.
68. Kumanyika, S.K., & Morssink, C.B. 1997. Cultural appropriateness of weight management programs. In *Overweight and Weight Management: The Health Professional's Guide to Understanding and Practice.* Gaithersburg, MD: Aspen.
69. Kreuter, M.W., Sugg-Skinner, C., Holt, C.L., Clark, E.M., Haire-Joshu, D., Fu, Q., Booker, A.C., Steger-May, K., & Bucholtz, D. 2005. Cultural tailoring for mammography and fruit and vegetable intake among low-income African-American women in urban public health centers. *Preventative Medicine, 41,* 53–62.
70. Brown, J.D., & Einsiedel, E.F. 1990. Public health campaigns: Mass media strategies. In E.B. Ray & L. Donohew (Eds.), *Communication and Health.* Hillsdale, NJ: Erlbaum.
71. White, S.L., & Maloney, S.K. 1990. Promoting healthy diets and active lives to hard-to-reach groups: Market research study. *Public Health Reports, 105,* 224–231.
72. Rabin, S. 1994. How to sell across cultures. *American Demographics, 16,* 56–57.
73. Williams, J.H., Auslander, W.F., de Groot, M., Robinson, A.D., Housto, C., & Haire-Joshu, D. 2006. Cultural relevancy of a diabetes prevention nutrition program for African American women. *Health Promotion and Practice, 7,* 56–67.
74. Massaro, E., & Claiborne, N. 2001. Effective strategies for reaching high-risk minorities with diabetes. *The Diabetes Educator, 27,* 820–826, 828.
75. Preloran, H.M., Browner, C.H., & Leiber, E. 2005. Impact of interpreter's approach on Latina's use of amniocentesis. *Health Education & Behavior, 32,* 599–612.
76. Donohew, L. 1990. Public health campaigns: Individual message strategies and a model. In E.B. Ray, & L. Donohew (Eds.), *Communication and Health.* Hillsdale, NJ: Erlbaum.

77. Elshaw, E.B., Young, E.A., Saunders, M.J., McGurn, W.C., & Lopez, L.C. 1994. Utilizing a 24-hour dietary recall and culturally specific diabetes education in Mexican Americans with diabetes. *Diabetes Educator, 20,* 228–235.
78. Chew, T. 1983. Sodium values of Chinese condiments and their use in sodium-restricted diets. *Journal of the American Dietetic Association, 82,* 397–401.
79. James, D.C. 2004. Factors influencing food choices, dietary intake, and nutrition-related attitudes among African-Americans: Application of a culturally sensitive model. *Ethnicity & Health, 9,* 349–367.
80. Goldberg, J., Rudd, R.E., & Deitz, W. 1999. Using 3 data sources and methods to shape a nutrition campaign. *Journal of the American Dietetic Association, 99,* 717–722.
81. Randall-David, E. 1989. *Strategies for working with culturally diverse communities and clients.* Bethesda, MD: Association for the Care of Children's Health.
82. Hodge, F.S., Paqua, A., Marquez, C.A., & Geishirt-Cantrell, B. 2002. Utilizing traditional storytelling to promote wellness, in American Indian communities. *Journal of Transcultural Nursing, 13,* 6–11.
83. Esquivel, G.B., & Keitel, M.A. 1990. Counseling immigrant children in the schools. *Elementary School Guidance & Counseling, 24,* 213–221.
84. Kiefer, E.C., Willis, S.K., Odoms-Young, A.M., Guzman, J.R., Allen, A.J., Two Feathers, J., & Loveluck, J. 2004. Reducing disparities in diabetes among African-American and Latino residents of Detroit: The essential role of community planning focus groups. *Ethnicity & Disease, 14 (Suppl. 1),* S27–37.
85. Mogelonsky, M. 1998. Watching in tongues. *American Demographics, April, 48.*
86. Tessaro, I., Rye, S., Parker, L., Trangsrud, K., Mangone, C., McCrone, S., & Leslie, N. 2006. Cookin' up health: Developing a nutrition intervention for a rural Appalachian population. *Health Promotion and Practice, 7,* 252–257.
87. Archer WR, Batan MC, Buchanan LR, Soler RE, Ramsey DC, Kirchhofer A, Reyes M. Promising practices for the prevention and control of obesity in the worksite. *Am J Health Promot. 2011; 25:* e12–26.
88. Chou HK, Lin IC, Woung LC, Tsai MT. Engagement in E-Learning Opportunities: An Empirical Study on Patient Education using Expectation Confirmation Theory. *J Med Syst. 2010 Nov 23.* [Epub ahead of print]
89. Kaufman N. Internet and information technology use in treatment of diabetes. *Int J Clin Pract Suppl. 2010; 166: 41–6.* Review.
90. Gilbert, S.E., & Gay, G. 1989. Improving the success in school of poor black children. In B.J. Robinson Shade (Ed.), *Culture, Style, and the Educative Process.* Springfield, IL: Thomas.
91. Beach, M.V., Sugarman, J., Johnson, R.L., Arbelaez, J.J., Duggan, P.S., & Cooper, L.A. 2005. Do patients treated with dignity report higher satisfaction, adherence, and receipt of preventative care? *Annals of Family Medicine, 3,* 331–338.
92. Burner, O.Y., Cunningham, P., & Hattar, H.S. 1990. Managing a multicultural nurse staff in a multicultural environment. *Journal of Nursing Administration, 20,* 30–34.
93. Levinson, W., Kao, A., Kuby, A., & Thisted, R.A. 2005. Not all patients want to participate in decision making. A national study of public preferences. *Journal of General Internal Medicine, 20,* 531–535.
94. Wood, J.B. 1989. Communicating with older adults in health care settings: Cultural and ethnic considerations. *Educational Gerontology, 15,* 351–362.
95. *Commission on Dietetic Registration.* Available from: http://www.cdrnet.org/certifications/rddtr/rddemo.htm. Accessed January 11, 2011.
96. Wang, M.C., & Sabry, Z.I. 1994. A culturally-sensitive method for the assessment of diet in a group of American Chinese children. *American Journal of Clinical Nutrition, 59 (Suppl.),* 290S.
97. Yanek, L.R., Moy, T.F., & Becker, D.M. 2001. Comparison of food frequency and dietary recall methods in African-American women. *Journal of the American Dietetic Association, 101,* 1361–1364.
98. Jerome, N.W. 1997. Culture-specific strategies for capturing local dietary intake patterns. *American Journal of Clinical Nutrition, 65,* 1166S–1167S.
99. Swartz, L. 1987. Illness negotiation: The case of eating disorders. *Social Science and Medicine, 24,* 613–618.

CHAPTER 4

Food and Religion

The function of religion is to explain the inexplicable, thus providing humans with a sense of comfort in a chaotic world. Food, because it sustains life, is an important part of religious symbols, rites, and customs, those acts of daily life intended to bring about an orderly relationship with the spiritual or supernatural realm.

In the Western world, Judaism, Christianity, and Islam are the most prevalent religions, whereas Hinduism and Buddhism are common in the East. The Western religions, originating in the Middle East, are equated with the worship of a single God and the belief that that God is omnipotent and omniscient. It is for God to command and for humankind to obey. This life is a time of testing and a preparation for life everlasting, when humans will be held accountable to God for their actions on Earth.

The Eastern religions of Hinduism and Buddhism developed in India. Unlike the Western religions, they do not teach that God is the lord and maker of the universe who demands that humankind be righteous. Rather, the principal goal of the Indian religions is deliverance, or liberation, of the immortal human soul from the bondage of the body. Moreover, nearly all Indian religions teach that liberation, given the right disposition and training, can be experienced in the present life.

This chapter discusses the beliefs and food practices of the world's major religions. Other religions of importance to specific cultures are introduced in the following chapters on each ethnic group. As with any description of food habits, it is important to remember that religious dietary practices vary enormously, even among members of the same faith. Many religious food practices were codified hundreds or thousands of years ago for a specific locale and, consequently, have been reinterpreted over time and to meet the needs of expanding populations.

As a result, most religions have areas of questionable guidelines. For example, fish without scales are banned under kosher food laws. Are sturgeon, which are born with scales but lose them as they mature, considered fit to eat for Jews? Orthodox Jews say no, whereas many Conservative and Reform Jews say yes: smoked sturgeon can be found at almost any Jewish deli. Hindus, who avoid fish with "ugly forms," identify those fish that are undesirable according to local tradition. In addition, religious food practices are often adapted to personal needs. Catholics, encouraged to make a sacrifice during Lent, traditionally gave up meat but today may choose pastries or candy instead. Buddhists may adopt a vegetarian diet only during the period when as an elder they become a monk or nun. Because religious food prescriptions are usually written in some form, it is tempting to see them as being black-and-white. Yet they are among the most variable of culturally based food habits. See Table 4.1.

U.S. law now prohibits the census from including mandatory questions regarding religion. Independent national survey data often differ from religious group records regarding membership.

A study on religious affiliation in 2008 found the percentage of self-identified Christians in the United States had declined in the past decade, whereas the percentage that adhere to non-Christian religions remained relatively constant. The biggest gain was seen in those who practice no religion, nearly doubling from 8 percent to 15 percent. Younger persons, men, Asians, and those living in the West were most likely to say they have a secular outlook (see Figure 4.1).[38]

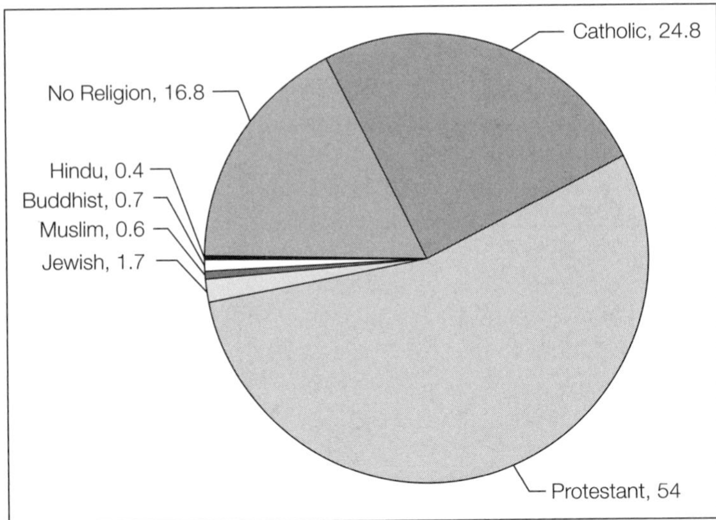

Figure 4.1
Self-described religious affiliation in the United States by percentage—2008.
Source: U.S. Religious Landscape Survey Religious Affiliation: Diverse and Dynamic February 2008, Pew Forum on Religion & Public Life http://religions.pewforum.org/pdf/report-religious-landscape-study-full.pdf 1/10/11

In the 2001 Canadian census, 43 percent of the population was identified as Roman Catholic, and another 30 percent was listed as Protestant. It is estimated that there are 579,640 Muslims (2 percent of the population), 479,620 Eastern Orthodox Christians, 329,995 Jews, 300,345 Buddhists, 297,200 Hindus, and 278,415 Sikhs. Sixteen percent of the population adheres to no religion.

Western Religions

Judaism

The Jewish religion, estimated to be 4,000 years old, started when Abraham received God's earliest covenant with the Jews. Judaism was originally a nation as well as a religion. However, after the destruction of its capital, Jerusalem, and its main sanctuary, the Temple of Solomon, by the Romans in 70 C.E., it had no homeland until the birth of Israel in 1948.

During the Diaspora (the dispersion of Jews outside the homeland of Israel), Jews scattered and settled all over the ancient world. Two sects of Judaism eventually developed: the Ashkenazi, who prospered in Germany, northern France, and the eastern European countries; and the Sephardim, originally from Spain, who now inhabit most southern European and Middle Eastern countries. Hasidic Jews are observant Ashkenazi Jews who believe salvation is to be found in joyous communion with God as well as in the Bible. Hasidic men are evident in larger U.S. cities by their dress, which includes long black coats and black or fur-trimmed hats (worn on Saturdays and holidays only), and by their long beards with side curls.

The cornerstone of the Jewish religion is the Hebrew Bible, particularly the first five books of the Bible, the Pentateuch, also known as the books of Moses, or the Torah. It consists of Genesis, Exodus, Leviticus, Numbers, and Deuteronomy. The Torah chronicles the beginnings of Judaism and contains the basic laws that express the will of God to the Jews. The Torah not only sets down the Ten Commandments, but also describes the right way to prepare food, give to charity, and conduct one's life in all ways. The interpretation of the Torah and commentary on it are found in the Talmud. The basic tenet of Judaism is that there is only one God, and His

Table 4.1 Common Religious Food Practices

	ADV	BUD	EOX	HIN	JEW	MOR	MUS	RCA
Beef		A		X				
Pork	X	A		A	X		X	
All Meat	A	A	R	A	R		R	R
Eggs/dairy	0	0	R	0	R			
Fish	A	A	R	R	R		R	
Shellfish	X	A	0	R	X			
Alcohol	X			A		X	X	
Coffee/tea	X					X	A	
Meet and dairy at the same meal					X			
Leavened foods					R			
Ritual slaughter of meats					+		+	
Moderation	+	+				+	+	

Note: ADV, Seventh-Day Adventist; BUD, Buddhist; EOX, Eastern Orthodox; HIN, Hindu; JEW, Jewish; MOR, Mormon; MUS, Muslim; RCA, Roman Catholic. X, prohibited or strongly discouraged; A, avoided by most devout; R, some restrictions regarding types of foods or when foods are eaten observed by some adherents; 0, permitted, but may be avoided at some observances; +, practiced.

will must be obeyed. Jews do not believe in original sin (that humans are born sinful) but rather that all people can choose to act in a right or wrong way. Sin is attributed to human weakness. Humans can achieve, unaided, their own redemption by asking for God's absolution (if they have sinned against God) or by asking forgiveness of the person they sinned against. The existence of the hereafter is recognized, but the main concern in Judaism is with this life and adherence to the laws of the Torah. Many Jews belong to or attend a synagogue (temple), which is led by a rabbi, who is a scholar, teacher, and spiritual leader. In the United States, congregations are usually classified as Orthodox, Conservative, or Reform, although American Jews represent a spectrum of beliefs and practices. The main division among the three groups is their position on the Jewish laws. Orthodox Jews believe that all Jewish laws, as the direct commandments of God, must be observed in all details. Reform Jews do not believe that the ritual laws are permanently binding but that the moral law is valid. They believe that the laws are still being interpreted and that some laws may be irrelevant or out of date, and they observe only certain religious practices. Conservative Jews hold the middle ground between Orthodox and Reform beliefs.[1]

Immigration to the United States

In the early nineteenth century, Jews, primarily from Germany, sought economic opportunities in the New World. By 1860, there were approximately 280,000 Jews living in the United States. Peak Jewish immigration occurred around the turn of the century (1880–1920); vast numbers of Jews moved from eastern Europe because of poverty and pogroms (organized massacres practiced by the Russians against the Jews before World War II).

During the Great Depression, Jews continued to immigrate into the United States, primarily to escape from Nazi Germany. Their numbers were few, however, because of restrictions in the immigration quota system. Today Jews continue to come to the United States, especially from Russia. Some come from Israel as well. The Jewish population in the United States was over 6.5 million adults and children in 2010, according to data compiled from local federations; close to half of U.S Jews live in the northeastern region of the nation.[2] Large populations are found in New York, California, and Florida. Most Jews in the United States are Ashkenazi: 10 percent identify themselves as Orthodox, 34 percent as Conservative, 29 percent as Reform, and the rest are not affiliated with a specific congregation.

Kashrut, the Jewish Dietary Laws

Some people in the United States believe that Jewish food consists of dill pickles, bagels and lox (smoked salmon), and chicken soup. In actuality, the foods Jews eat reflect the regions where their families originated. Because most Jews in the United States are Ashkenazi, their diet includes the foods of Germany and eastern Europe. Sephardic Jews tend to eat foods similar to those of southern Europe and the Middle Eastern countries, whereas Jews from India prefer curries and other South Asian foods. All Orthodox and some Conservative Jews follow the dietary laws, kashrut, that were set down in the Torah and explained in the Talmud.

Since 2003, sales of kosher foods have increased by 68 percent. More than 50 percent of kosher food is purchased by non-Jews, including Muslims, Seventh-Day Adventists, vegetarians, and those with food allergies, in part due to their reputation for safety and purity.[25,40]

Kosher or kasher means "fit" and is a popular term for Jewish dietary laws and permitted food items. *Glatt* kosher means that the strictest kosher standards are used in obtaining and preparing the food. Kashrut is one of the pillars of Jewish religious life and is concerned with the fitness of food. Many health-related explanations have been postulated about the origins of the Jewish dietary restrictions; however, it is spiritual health, not physical health or any other factor, that is the sole reason for their observance. Jews who keep kosher are expressing their sense of obligation to God, to their fellow Jews, and to themselves. The dietary laws governing the use of animal foods can be classified into the following categories:

1. ***Which animals are permitted for food and which are not:*** Any mammal that has a completely cloven foot and also chews the cud may be eaten, and its milk may be drunk. Examples of permitted, or clean, animals are cattle deer, goats, oxen, and sheep. Unclean animals include swine, carnivorous animals, and rabbits. Clean birds must have a crop, gizzard, and an extra talon, as do chickens, ducks, geese, and turkeys. Their eggs are also considered clean. All birds of prey and their eggs are unclean and cannot be eaten. Among fish, everything that has fins and scales is permitted; everything else is

Most gelatin is obtained from processed pig tissues. Kosher, gelatin-like products are available.

In 2002, the first kosher food in outer space was served to astronaut Ilan Ramon on the space shuttle Columbia.

unclean. Examples of unclean fish are catfish, eels, rays, sharks, and all shellfish. As discussed previously, Orthodox rabbis consider sturgeon a prohibited food; Conservative authorities list it as kosher. Caviar, which comes from sturgeon, is similarly disputed.[2] All reptiles, amphibians, and invertebrates are also unclean.

Tevilah is the ritual purification of metal or glass pots, dishes, and utensils through immersion in the running water of a river or ocean. China Porcelain and ceramic items are exempt.

2. ***Method of slaughtering animals:*** The meat of permitted animals can be eaten only if the life of the animal is taken by a special process known as *shehitah*. If an animal dies a natural death or is killed by any other method, it may not be eaten. The *shohet* (person who kills the animal) must be a Jew trained and licensed to perform the killing, which is done by slitting the neck with a sharp knife, cutting the jugular vein and trachea at the same time. This method, which is quick and painless, also causes most of the blood to be drained from the carcass.[3]
3. ***Examination of the slaughtered animal:*** After the animal is slaughtered, it is examined by the *shohet* for any blemishes in the meat or the organs that would render the animal *trefah*, meaning unfit for consumption. Disease in any part of the animal makes the whole animal unfit to eat.
4. ***Forbidden parts of a permitted animal:*** Two parts of the animal body are prohibited. Blood from any animal is strictly forbidden; even an egg with a small bloodspot in the yolk must be discarded. *Heleb* (fat that is not intermingled with the flesh, forms a separate solid layer that may be encrusted with skin or membrane, and can easily be peeled off) is also proscribed. The prohibition against *heleb* only applies to four-footed animals.
5. ***The preparation of the meat:*** For meat to be kosher, the *heleb*, blood, blood vessels, and sciatic nerve must be removed. Much of this work is now done by the Jewish butcher, although some Jewish homemakers still choose to remove the blood. This is known as koshering, or kashering, the meat. It is accomplished in five steps: First, the meat is soaked (within seventy-two hours after slaughter) in tepid water for thirty minutes; second, it is drained on a slanted, perforated surface so that the blood can drain easily; third, the meat is covered with kosher salt for at least one hour; fourth, the salt is rinsed from the meat; and finally, the meat is rinsed repeatedly to make sure all blood and salt are removed.[4] The liver cannot be made kosher in the ordinary way because it contains too much blood. Instead, its surface must be cut across or pierced several times, then it must be rinsed in water, and finally it must be broiled or grilled on an open flame until it turns a grayish-white color.
6. ***The law of meat and milk:*** Meat (*fleischig*) and dairy (*milchig*) products may not be eaten together. It is generally accepted that after eating meat a person must wait six hours before eating any dairy products, although the period between is a matter of custom, not law.[3] Only one hour is necessary if dairy products are consumed first. Many Jews are lactose intolerant and do not drink milk. However, other dairy items such as cheese, sour cream, and yogurt are often included in the diet (Kosher cheese must be made with rennet obtained from a calf killed according to the Jewish laws of slaughtering). Separate sets of dishes, pots, and utensils for preparing and eating meat and dairy products are usually maintained. Separate linens and washing implements are often employed. Eggs, fruits, vegetables, and grains are pareve, neither meat nor dairy, and can be eaten with both. Olives are considered dairy foods, prohibited with meat, if they are prepared using lactic acid.
7. ***Products of forbidden animals:*** The only exception to the rule that products of unclean animals are also unclean is honey. Although bees are not fit for consumption, honey is kosher because it is believed that it does not contain any parts from the insect.
8. ***Examination for insects and worms:*** Because small insects and worms can hide on fruits, vegetables, and grains, these foods must be carefully washed twice and examined before being eaten. Kosher-certified prepackaged produce is available from some suppliers. A processed food product (including therapeutic dietary formulas) is considered kosher only if a reliable rabbinical authority's name or insignia appears on the package. The most common insignia is a *K*, permitted by the U.S. Food and Drug Administration (FDA), indicating rabbinical supervision by the processing company. Other registered symbols include those found in Figure 4.2.

Because it is not known how much salt remains on the meat after rinsing, Orthodox Jews with hypertension are often advised to restrict their meat consumption.

The prohibition of the sciatic nerve is based on the biblical story of Jacob's nighttime fight with a mysterious being who touched him on the thigh, causing him to limp. Because the nerve is difficult to remove, the entire hindquarter of the animal is usually avoided.

 The Union of Orthodox Jewish Congregations
New York, New York

 O.K. (Organized Kashrut) Laboratories
Brooklyn, New York

 Kosher Supervision Service
Teaneck, New Jersey

 Asian-American Kashrus Services
San Rafael, California

 The Heart "K" Kehila Kosher
Los Angeles, California

 Chicago Rabbinical Council
Chicago, Illinois

 Orthodox Vaad of Philadelphia
Philadelphia, Pennsylvania

 Vaad Hakahrus of Dallas, Inc.
Dallas, Texas

 Vaad Harabonim of Greater Detroit and Merkaz
Southfield, Michigan

 Orthodox Rabbinical Council of S. Florida
(Vaad Harabonim De Darom Florida)
Miami Beach, Florida

 Vaad Horabonim of Massachusetts
Boston, Massachusetts

 Vaad Hakashrus of the Orthodox Jewish
Council of Baltimore
Baltimore, Maryland

 Atlanta Kashruth Commission
Atlanta, Georgia

 Vaad Hakashrus of Denver
Denver, Colorado

 Vaad Harabonim of Greater Seattle
Seattle, Washington

 Kashruth Council Orthodox Division
Toronto Jewish Congress
Willowdale, Ontario, Canada

 Montreal Vaad Hair
Montreal, Canada

 Vancouver Kashruth
British Columbia, Canada

Figure 4.2
Examples of kosher food symbols.

Religious Holidays

The Sabbath. The Jewish Sabbath, the day of rest, is observed from shortly before sundown on Friday until after nightfall on Saturday. Traditionally, the Sabbath is a day devoted to prayer and rest, and no work is allowed. All cooked meals must be prepared before sundown on Friday because no fires can be kindled on the Sabbath. Challah, a braided bread, is commonly served with the Friday night meal. In most Ashkenazi homes the meal would traditionally contain fish or chicken or cholent, a bean and potato dish that can be prepared Friday afternoon and left simmering until the evening meal on Saturday. Kugel, a pudding, often made with noodles, is a typical side dish.

Rosh Hashanah. The Jewish religious year begins with the New Year, or Rosh Hashanah, which means "head of the year." Rosh Hashanah is also the beginning of a ten-day period of penitence that ends with the Day of Atonement, Yom Kippur. Rosh Hashanah occurs in September or October; as with all Jewish holidays, the actual date varies from year to year because the Jewish calendar is based on lunar months counted according to biblical custom and does not coincide with the secular calendar. For this holiday, the challah is baked in a round shape that symbolizes life without end and a year of uninterrupted health and happiness. In some communities, the challah is formed like a bird representing God's protection. Apples are dipped in honey, and a special prayer is said for a sweet and pleasant year. Some families traditionally consume the head of a fish or of a sheep, with the wish that God's will for them is to be at the head, not the tail, of any undertakings in the upcoming year. Foods with Hebrew names similar to other words may also be eaten, such as beets (similar to *remove* and used to pray that enemies be removed). On the second night a new fruit, one that hasn't been consumed for a long period of time, is enjoyed with a prayer for a year of plenty. Often, the fruit is a pomegranate, which reputedly contains 613 seeds, the same as the number of commandments listed in the Torah. No sour or bitter foods are served on this holiday, and special sweets and delicacies, such as honey cakes, are usually prepared.

Two breads, or one bread with a smaller one braided on top, are usually served on Fridays, the beginning of the Sabbath, symbolic of the double portion of manna (nourishment) provided by God to help sustain the Israelites when they wandered in the desert for forty years after their exodus from Egypt.

Yom Kippur, the Day of Atonement. Yom Kippur falls ten days after Rosh Hashanah and is the holiest day of the year. On this day, every Jew atones for sins committed against God

In poor Ashkenazi homes, gefilte (filled) fish became popular on the Sabbath. Similar to the concept of meatloaf, it is made by extending the fish through pulverizing it with eggs, bread, onion, sugar, salt, and pepper, then stewing the balls or patties with more onions.

and resolves to improve and once again follow all the Jewish laws. Yom Kippur is a complete fast day (no food or water; medications are allowed) from sunset to sunset. Everyone fasts, except boys under thirteen years old, girls under twelve years old, persons who are very ill, and women in childbirth. The meal before Yom Kippur is usually bland to prevent thirst during the fast. The meal that breaks the fast is typically light, including dairy foods or fish, fruits, and vegetables.

Sukkot, Feast of Tabernacles. Sukkot is a festival of thanksgiving. It occurs in September or October and lasts one week. On the last day, Simchat Torah, the reading of the Torah (a portion is read every day of the year) is completed for the year and started again. This festival is very joyous, with much singing and dancing. Orthodox families build a sukkah (hut) in their yards and hang fruit and flowers from the rafters, which are built far enough apart so that the sky and stars are visible. Meals are eaten in the sukkah during Sukkot.

Hanukkah, the Festival of Lights. Hanukkah is celebrated for eight days, usually during the month of December, to commemorate the recapture of the Temple in Jerusalem in 169 C.E. Families celebrate Hanukkah by lighting one extra candle on the menorah (candelabra) each night so that on the last night all eight candles are lit. Traditionally, potato pancakes, called latkes, are eaten during Hanukkah. Other foods cooked in oil, such as doughnuts, are sometimes eaten as well.

Purim. Purim, a joyous celebration that takes place in February or March, commemorates the rescue of the Persian Jews from the villainous Haman by Queen Esther. It is a mitzvah (good deed) to eat an abundant meal in honor of the deliverance. The feast should include ample amounts of meat and alcoholic beverages. Customarily, people dress in disguise for the day to hide from Haman, to add surprise to gift giving, or to hide from God in order to overindulge in anonymity. A food closely associated with Purim is hamantaschen (literally, "Haman's pockets," but usually interpreted to mean Haman's ears clipped in the humiliation of defeat). A hamantasch is a triangular-shaped pastry filled with sweetened poppy seeds or fruit jams made from prunes or apricots. Another pastry associated with Purim is kreplach (a triangular or heart-shaped savory pastry stuffed with seasoned meat or cheese and then boiled like ravioli). Purim challah (a sweet bread with raisins) and fish cooked for the holiday in vinegar, raisins, and spices are often served. Seeds, beans, and cereals are offered in remembrance of the restricted diet eaten by the pious Queen Esther.

Passover. Passover, called *Pesach* in Hebrew, is the eight-day festival of spring and of freedom. It occurs in March or April and celebrates the anniversary of the Jewish exodus from Egypt. The Passover seder, a ceremony carried out at home, includes readings from the seder book, the Haggadah, recounting the story of the exodus, of the Jews' redemption from slavery, and of the God-given right of all humankind to life and liberty. A festive meal is a part of the seder; the menu usually includes chicken soup, matzo balls, and meat or chicken. When Moses led the Jews out of Egypt, they left in such haste that there was no time for their bread to rise. Today, *matzah*, a white-flour cracker, is the descendant of the unleavened bread or bread of affliction. During the eight days of Passover, no food that is subject to a leavening process or that has come in contact with leavened foods can be eaten. The forbidden foods are wheat, barley, rye, and oats. Wheat flour can be eaten only in the form of *matzah* or *matzah* meal, which is used to make matzo balls. In addition, beans, peas, lentils, maize, millet, and mustard are also avoided. No leavening agents, malt liquors, or beers can be used.

Because milk and meat cannot be mixed at any time, observant Jewish families have two sets of special dishes, utensils, and pots used only for Passover. The entire house, especially the kitchen, must be cleaned and any foods subject to leavening removed before Passover. It is customary for Orthodox Jews to sell their leavened products and flours to a non-Jew before Passover. It is very important that all processed foods, including wine, be prepared for Passover use and be marked "Kosher for Passover."

The seder table is set with the best silverware and china and must include candles, kosher wine, the Haggadah, three pieces of *matzot* (the plural of *matzah*) covered separately in the folds of a napkin or special Passover cover, and a seder plate. The following items go on the seder plate:

Ashkenazi Jews traditionally avoided pepper during Passover because it was sometimes mixed with bread crumbs or flour by spice traders.

In some Sephardic homes, matzah is layered with vegetables and cheese or meat for the Passover meal.

1. *Z'roah.* *Z'roah* is a roasted shank bone, symbolic of the ancient paschal lamb in Egypt, which was eaten roasted.
2. *Beitzah.* *Beitzah* is a roasted egg, representing the required offering brought to the Temple at festivals. Although the egg itself was not sacrificed, it is used in the seder as a symbol of mourning. In this case, it is for the loss of the Temple in Jerusalem.
3. *Marror.* *Marror* are bitter herbs, usually horseradish (although not an herb), symbolic of the Jews' bitter suffering under slavery. The *marror* is usually eaten between two small pieces of matzot.
4. *Haroset.* *Haroset* is a mixture of chopped apple, nuts, cinnamon, and wine. Its appearance is a reminder of the mortar used by the Jews to build the palaces and pyramids of Egypt during centuries of slavery. The *haroset* is also eaten on a small piece of matzah.
5. *Karpas.* A green vegetable, such as lettuce or parsley, is placed to the left of the *haroset*, symbolic of the meager diet of the Jews in bondage. It is dipped into salt water in remembrance of the tears shed during this time. It also symbolizes springtime, the season of Passover.
6. A special cup, usually beautifully decorated, is set on the seder table for Elijah, the prophet who strove to restore purity of divine worship and labored for social justice. (Elijah is also believed to be a messenger of God, whose task it will be to announce the coming of the Messiah and the consequent peace and divine kingdom of righteousness on Earth.)

© Michael Newman/PhotoEdit, Inc.

▲ *Typical seder meal.*

Shavout, Season of the Giving of the Torah. The two-day festival of Shavout occurs seven weeks after the second day of Passover and commemorates the revelation of the Torah to Moses on Mount Sinai. Traditional Ashkenazi foods associated with the holiday include blintzes (extremely thin pancakes rolled with a meat or cheese filling, then topped with sour cream), kreplach, and knishes (dough filled with a potato, meat, cheese, or fruit mixture, then baked).

Fast Days

There are several Jewish fast days in addition to Yom Kippur (see Table 4.2). On Yom Kippur and on Tisha b'Av, the fast lasts from sunset to sunset and no food or water can be consumed. All other fast days are observed from sunrise to sunset. Most Jews usually fast on Yom Kippur, but other fast days are observed only by Orthodox Jews. Extremely pious Jews may add personal fast days

The Torah prohibits the drinking of wine made by non-Jews because it might have been produced for the worship of idols. Some Orthodox Jews extend the prohibition to any grape product, such as grape juice or grape jelly.

Table 4.2 Jewish Fast Days

Tzom	Day after Rosh Hashanah	In memory of Gedaliah, who ruled after the First Temple was destroyed
Yom Kippur	10 days after Rosh Hashanah	Day of Atonement
Tenth of Tevet Seventeenth of Tamuz	December July }	Commemorate an assortment of national calamities listed in the Talmud
Ta'anit Ester	Eve of Purim	In grateful memory of Queen Esther, who fasted when seeking divine guidance
Ta'anit Bechorim	Eve of Passover	Gratitude to God for having spared only the the first-born of Israel; usually only first-born son fasts
Tisha b'Av	August	Commemorates the destruction of the First and Second Temples in Jerusalem

Cottage cheese is associated with Shavout because the Israelites were late in returning home after receiving the Ten Commandments and the milk had curdled. Many dishes served during the holiday contain cheese fillings.

on Mondays and Thursdays.[5] All fasts can be broken if it is dangerous to a person's health; those who are pregnant or nursing are exempt from fasting.

Additional information about Jewish dietary laws and customs associated with Jewish holidays can usually be obtained from the rabbi at a local synagogue. The Union of Orthodox Jewish Congregations of America also publishes a directory of kosher products.

Nutrition Status

Although Judism is a religion but in many ways Jewish people are also considered an ethnic group. Few studies have been conducted to determine the nutritional status of Jews but certain physiological conditions and medical disorders have been reported to have a higher incidence in Jews. It is estimated that 60 percent to 80 percent of Ashkenazi Jewish people are lactose intolerant.[6] Research has identified a genetic predisposition to inflammatory bowel disease in Ashkenazi Jews (two to eight times more common).[7]

Christianity

Throughout the world, more people follow Christianity than any other single religion. The three dominant Christian branches are Roman Catholicism, Eastern Orthodox Christianity, and Protestantism. Christianity is founded on recorded events surrounding the life of Jesus—believed to be the Son of God and the Messiah—chronicled in the New Testament of the Bible. The central convictions of the Christian faith are found in the Apostles' Creed and the Nicene Creed. These creeds explain that people are saved through God's grace, through the life and death of Jesus, and through his resurrection as Christ.

The commemoration of the Last Supper is called Corpus Christi, when Jesus instructed his disciples that bread was his body and wine his blood. In Spain and many Latin American countries, Corpus Christi is celebrated by parading the bread (called the "Host") through streets covered with flowers.

St. Valentine's Day traditions may date back to Lupercalia, a Roman festival held in mid-February, at which a young man would draw the name of a young woman out of a box to be his sweetheart for a day.

For most Christians, the sacraments mark the key stages of worship and sustain the individual worshiper. A sacrament is an outward act derived from something Jesus did or said, through which an individual receives God's grace. The sacraments observed, and the way they are observed, vary among Christian groups. The seven sacraments of Roman Catholicism, for example, are baptism (entering Christ's church), confirmation (the soul receiving the Holy Ghost), Eucharist (partaking of the sacred presence by sharing bread and wine), marriage (union of a man and woman through the bond of love), unction (healing of the mind, spirit, and body), reconciliation (penance and confession), and ordination of the clergy.

Roman Catholicism

The largest number of persons adhering to one Christian faith in the United States are Roman Catholics (approximately 67 million in 2009, according to parish records).[8] The head of the worldwide church is the pope, considered infallible when defining faith and morals. The seven sacraments are conferred on the faithful.[9]

Although some Roman Catholics immigrated to the United States during the colonial period, substantial numbers came from Germany, Poland, Italy, and Ireland in the 1800s and from Mexico and the Caribbean in the twentieth and twenty-first century. There are small groups of French Catholics in New England (primarily in Maine) and in Louisiana. In addition, most Filipinos and some Vietnamese people living in the United States are Catholics.

Feast Days. Most Americans are familiar with Christmas (the birth of Christ) and Easter (the resurrection of Christ after the crucifixion). Other Christian feast days celebrated in the United States are New Year's Day, the Annunciation (March 25), Palm Sunday (the Sunday before Easter), the Ascension (forty days after Easter), Pentecost Sunday (fifty days after Easter), the Assumption (August 15), All Saint's Day (November 1), and the Immaculate Conception (December 8).

Holiday fare depends on the family's country of origin. For example, the French traditionally serve *bûche de Noël* (a rich cake in the shape of a Yule log) on Christmas for dessert, while the Italians may serve panettone, a fruited sweet bread (see individual chapters on each ethnic group for specific foods associated with holidays).

Fast Days. Fasting permits only one full meal per day at midday. It does not prohibit the taking of some food in the morning or evening; however, local custom as to the quantity and quality of this supplementary nourishment varies. Abstinence forbids the use of meat, but not of eggs, dairy products, or condiments made of animal fat and is practiced on certain days and in conjunction with fasting. Only Catholics older than the age of fourteen and younger than the age of sixty are required to observe the dietary laws.[10]

The fast days in the United States are all the days of Lent, the Fridays of Advent, and the Ember Days (the days that begin each season), but only the most devout fast and abstain on all of these dates. More common is fasting and abstaining only on Ash Wednesday and Good Friday. Before 1966, when the U.S. Catholic Conference abolished most dietary restrictions, abstinence from meat was observed on every Friday that did not fall on a feast day. Abstinence is now encouraged on the Fridays of Lent in remembrance of Christ's sacrificial death.

Some older Catholics and those from other nations may observe the pre-1966 dietary laws. In addition, Catholics are required to avoid all food and liquids, except water, for one hour before receiving communion.

Eastern Orthodox Christianity

The Eastern Orthodox Church is as old as the Roman Catholic branch of Christianity, although not as prevalent in the United States. In the year 300 C.E., there were two centers of Christianity, one in Rome and the other in Constantinople (now Istanbul, Turkey). Differences arose over theological interpretations of the Bible and the governing of the church, and in 1054 the fellowship between the Latin and Byzantine churches was finally broken. Some of the differences between the two churches concerned the interpretation of the Trinity (the Father, the Son, and the Holy Ghost), the use of unleavened bread for the communion, the celibacy of the clergy, and the position of the pope. In the Eastern Orthodox Church, leavened bread, called phosphoron, is used for communion, the clergy are allowed to marry before entering the priesthood, and the authority of the pope of Rome is not recognized.

The Orthodox Church consists of fourteen self-governing churches, five of which—Constantinople, Alexandria (the Egyptian Coptic Church), Antioch, Jerusalem, and Cyprus—date back to the time of the Byzantine Empire. Six other churches represent the nations where the majority of people are Orthodox (Russia, Rumania, Serbia, Bulgaria, Greece, and the former Soviet state of Georgia). Three other churches exist independently in countries where only a minority profess the religion (Poland, Albania, and the Sinai Monastery). Additionally, there are four churches considered autonomous, but not yet self-governing: Czech Republic/Slovakia, Finland, China, and Japan.[11] The Orthodox Church in America was constituted in 1970.[12] The beliefs of the Orthodox churches are similar; only the language of the service differs.

The first Eastern Orthodox Church in America was started by Russians on the West Coast in the late 1700s. It is estimated that nearly 3 million persons in the United States are members of the Eastern Orthodox religion, with the largest following (1,500,000) being Greek.[8] Most Eastern Orthodox churches in the United States recognize the patriarch of Constantinople as their spiritual leader.

Feast Days. All the feast days are listed in Table 4.3. Easter is the most important holiday in the Eastern Orthodox religion and is celebrated on the first Sunday after the full moon after March 21, but not before the Jewish Passover. Lent is preceded by a pre-Lenten period lasting ten weeks before Easter or three weeks before Lent. On the third Sunday before Lent (Meat Fare Sunday), all the meat in the house is eaten. On the Sunday before Lent (Cheese Fare Sunday), all the cheese, eggs, and butter in the house are eaten. On the next day, Clean Monday, the Lenten fast begins. Fish is allowed on Palm Sunday and on the Annunciation Day of the Virgin Mary. The Lenten fast is traditionally broken after the

Courtesy of Grossich and Bond, Inc.

▲ *Italian-American Catholics often serve panettone, a sweet bread with dried fruits, on feast days, especially Christmas.*

Lent is the forty days before Easter; the word originally meant spring. The last day before Lent is a traditional festival of exuberant feasting and drinking in many regions where Lenten fasting is observed. In France and in Louisiana, it is known as Mardi Gras; in Britain, Shrove Tuesday; in Germany, Fastnacht; and throughout the Caribbean and in Brazil, Carnival.

The Ethiopian church is an Orthodox denomination similar to the Egyptian Coptic Church. Timkat (Feast of the Epiphany) is the most significant Christian holiday of the year, celebrating the baptism of Jesus. Beer brewing, bread baking, and eating roast lamb are traditional.

Table 4.3 Eastern Orthodox Feast Days

Feast Day	Date*
Christmas	Dec. 25 or Jan. 7
Theophany	Jan. 6 or Jan. 19
Presentation of our Lord into the Temple	Feb. 2 or Feb. 15
Annunciation	Mar. 25 or Apr. 7
Easter	First Sunday after the full moon after Mar. 21
Ascension	40 days after Easter
Pentecost (Trinity) Sunday	50 days after Easter
Transfiguration	Aug. 6 or Aug. 19
Dormition of the Holy Theotokos	Aug. 15 or Aug. 28
Nativity of the Holy Theotokos	Sept. 8 or Sept. 21
Presentation of the Holy Theotokos	Nov. 21 or Dec. 4

*Date depends on whether the Julian or Gregorian calendar is followed.

Koljivo, boiled whole-wheat kernels mixed with nuts, dried fruit, and sugar, must be offered before the church altar three, nine, and forty days after the death of a family member. After the koljivo is blessed by the priest, it is distributed to the friends of the deceased. The boiled wheat represents everlasting life, and the fruit represents sweetness and plenty.

The red Easter egg symbolizes the tomb of Christ (the egg) and is a sign of mourning (the red color). The breaking of the eggs on Easter represents the opening of the tomb and belief in the resurrection.

midnight services on Easter Sunday. Easter eggs in the Eastern Orthodox religion range from the highly ornate (eastern Europe and Russia) to the solid red ones used by the Greeks.

Fast Days. In the Eastern Orthodox religion there are numerous fast days (see Table 4.4). Further, those receiving Holy Communion on Sunday abstain from food and drink before the service. Fasting is considered an opportunity to prove that the soul can rule the body. On fast days, no meat or animal products (milk, eggs, butter, and cheese) are consumed. Fish is also avoided, but shellfish is generally allowed. Older or more devout Greek Orthodox followers do not use olive oil on fast days, but will eat olives.

Protestantism

The sixteenth century religious movement known as the Reformation established the Protestant churches by questioning the practices of the Roman Catholic Church and eventually breaking away from its teachings. The man

Table 4.4 Eastern Orthodox Fast Days and Periods

Fast Days

Every Wednesday and Friday except during fast-free weeks:

- Week following Christmas until Eve of Theophany (12 days after Christmas)
- Bright Week, week following Easter
- Trinity Week, week following Trinity Sunday

Eve of Theophany (Jan. 6 or 18)

Beheading of John the Baptist (Aug. 29 or Sept. 27)

The Elevation of the Holy Cross (Sept. 14 or 27)

Fast Periods

Nativity Fast (Advent): Nov. 15 or 28 to Dec. 24 or Jan. 6

Great Lent and Holy Week: 7 weeks before Easter

Fast of the Apostles: May 23 or June 5 to June 16 or 29

Fast of the Dormition of the Holy Theotokos: Aug. 1 or 14 to Aug. 15 or 28

Dates depend on whether the Julian or Gregorian calendar is followed.

primarily responsible for the Reformation was Martin Luther, a German Augustinian monk who taught theology.[13] He started the movement when, in 1517, he nailed a document containing 95 protests against certain Catholic practices on the door of the castle church in Wittenberg, Germany. He later broadened his position. A decade later, several countries and German principalities organized the Protestant Lutheran Church based on Martin Luther's teachings.

Luther placed great emphasis on the individual's direct responsibility to God. He believed that every person can reach God through direct prayer without the intercession of a priest or saint; thus, every believer is, in effect, a minister. Although everyone is prone to sin and inherently wicked, a person can be saved by faith in Christ, who by his death on the Cross atoned for the sins of all people. Consequently, to Luther, faith was all-important and good works alone could not negate evil deeds. Luther's theology removed the priest's mystical function, encouraging everyone to read the Bible and interpret the Scriptures. The beliefs taught by Martin Luther established the foundation of most Protestant faiths.

Other reformers who followed Luther are associated with specific denominations. In the mid-sixth century, John Calvin developed the ideas that led to the formation of the Presbyterian, Congregationalist, and Baptist churches; John Wesley founded the Methodist movement in the eighteenth century. Other denominations in the United States include Episcopalians (related to the English Anglican Church started under King Henry VIII); Seventh-Day Adventists; Jehovah's Witnesses; Disciples of Christ; Church of Jesus Christ of Latter Day Saints (Mormons); Church of Christ, Scientist (Christian Scientists); and Friends (Quakers).

The most significant food ordinance in Protestant churches is the Eucharist, also called Communion, or the Lord's Supper. However, other than a liquid and a consecrated bread-like morsel being offered, there is little consistency in celebration of this ordinance. It can signify an encounter with the living presence of God, a remembrance of the Passover Seder attended by Jesus, a continuity of tradition through community, or an individual spiritual experience. Though wine is traditional, many churches switched to grape juice during Prohibition and continue this temperance practice. Some churches offer the wine/juice in a single cup which is shared, while others provide small, individually filled cups. Many liturgical churches, such as the Lutheran church, offer wafers similar to Catholic practice. Others, such as Methodists, often use a bread pellet. Some organize their members to bake bread (of any type), and many denominations simply cut up white bread of some sort.[14] The primary holidays of the Protestant calendar are Christmas and Easter. The role of food is important in these celebrations; however, the choice of items served is even more varied than Communion practices, determined by family ethnicity and preference rather than religious rite. Fasting is also uncommon in most Protestant denominations. Some churches or individuals may use occasional fasting, however, to facilitate prayer and worship. Only a few of the Protestant denominations, such as the Mormons and the Seventh-Day Adventists, have dietary practices integral to their faith.

Mormons. The Church of Jesus Christ of Latter Day Saints is a religion that emerged in the U.S. during the early 1800s. Its founder, Joseph Smith Jr., had a vision of the Angel Moroni, who told him of golden plates hidden in a hill and the means by which to decipher them. The resulting Book of Mormon was published in 1829, and in 1830 a new religious faith was born.

The Book of Mormon details the story of two bands of Israelites who settled in America and from whom certain Native Americans and Pacific Islanders are descended.[15] Christ visited them after his resurrection, and they thus preserved Christianity in its pure form. The tribes did not survive, but the last member, Moroni, hid the nation's sacred writings, compiled by his father, Mormon.

The Mormons believe that God reveals himself and his will through his apostles and prophets. The Mormon Church is organized along biblical lines. Members of the priesthood are graded upward in six degrees (deacons, teachers, priests, elders, seventies, and high priests). From the priesthood are chosen, by the church at large, a council of twelve apostles, which constitutes a group of ruling elders; from these, by seniority, a church president rules with life tenure. There is no paid clergy. Sunday services are held by groups of Mormons, and selected church members give the sermon.[16]

To escape local persecution, Brigham Young led the people of the Mormon Church to Utah in 1847. Today, Utah is more than 80 percent Mormon, and many Western states have significant numbers of church members, with an estimated total number of adherents over 5.8 million in 2010.[8] The main branch of the church is headquartered in Salt Lake City, but a smaller branch, the Reorganized Church of Jesus Christ of Latter Day Saints, is centered in Independence, Missouri. All Mormons believe that Independence will be the capital of the world when Christ returns.

Joseph Smith, through a revelation, prescribed the Mormon laws of health, dealing particularly with dietary matters.[15] These laws prohibit the use of tobacco, strong drink, and hot drinks. Strong drink is defined as alcoholic beverages; hot drinks mean tea and coffee. Many Mormons do not use any product that contains caffeine. Followers are advised to eat meat sparingly, and to base their diets on grains, especially wheat. In addition, all Mormons are required to store a year's supply of food and clothing for each person in the family. Many also fast one day per month (donating to the poor the money that would have been spent on food).

Loma Linda Foods began as a bakery in 1906 providing whole-wheat bread and cookies to the Adventist patients and staff of Loma Linda University Medical Center in southern California.

The American breakfast cereal industry is the result of the dietary and health practices of the Seventh-Day Adventists. In 1886, Dr. John Kellogg became director of the Adventists' sanitarium in Battle Creek, Michigan, and in his efforts to find a tasty substitute for meat, he invented corn flakes.

Seventh-Day Adventists. In the early 1800s, many people believed that the Second Coming of Christ was imminent. In the United States, William Miller predicted that Christ would return in 1843 or 1844. When both years passed and the prediction did not materialize, many of his followers became disillusioned. However, one group continued to believe that the prediction was not wrong but that the date was actually the beginning of the world's end preceding the coming of Christ. They became known as the Seventh-Day Adventists and were officially organized in 1863.[17]

The spiritual guide for the new church was Ellen G. Harmon, who later became Mrs. James White. Her inspirations were the result of more than 2,000 prophetic visions and dreams she reportedly had during her life. Mrs. White claimed to be not a prophet but a conduit that relayed God's desires and admonitions to humankind.

There were nearly a million Seventh-Day Adventists in the United States in 2010 and more than 16 million worldwide.[8] Besides the main belief in Christ's advent, or second coming, the Seventh-Day Adventists practice the principles of Protestantism. They believe that the advent will be preceded by a monstrous war, pestilence, and plague, resulting in the destruction of Satan and all wicked people; the Earth will be purified by holocaust. Although the hour of Christ's return is not known, they believe that dedication to his work will hasten it.

The church adheres strictly to the teachings of the Bible. The Sabbath is observed from sundown on Friday to sundown on Saturday and is wholly dedicated to the Lord. Food must be prepared on Friday and dishes washed on Sunday. Church members dress simply, avoid ostentation, and wear only functional jewelry. The church's headquarters are in Tacoma Park, Maryland, near Washington, D.C., where they were moved after a series of fires ravaged the previous center in Battle Creek, Michigan.

Each congregation is led by a pastor (more a teacher than a minister), and all the churches are under the leadership of the president of the general conference of Seventh-Day Adventists. Adventists follow the Apostle Paul's teaching that the human body is the temple of the Holy Spirit. Many of Mrs. White's writings concern health and diet and have been compiled into such books as The Ministry of Healing, Counsels on Health, and Counsels on Diet and Foods.[18–20]

Adventists believe that sickness is a result of the violation of the laws of health. One can preserve health by eating the right kinds of foods in moderation and by getting enough rest and exercise. Overeating is discouraged. Vegetarianism is widely practiced because the Bible states that the diet in Eden did not include flesh foods. Most Adventists are lacto-ovo-vegetarians (eating milk products and eggs, but not meat). Some do consume meat, although they avoid pork and shellfish. Mrs. White advocated the use of nuts and beans instead of meat, substituting vegetable oil for animal fat, and using whole grains in breads. Like the Mormons, the Adventists do not consume tea, coffee, or alcohol and do not use tobacco products. Water is considered the best liquid and should be consumed only before and after the meal, not during the meal. Meals are not highly seasoned, and hot spices such as mustard, chili powder, and black pepper are avoided. Eating between meals is discouraged so that food can be properly digested. Mrs. White recommended that five or six hours elapse between meals.

Islam

Islam is the second largest religious group in the world. Although not widely practiced in the United States, Islam is the dominant religion in the Middle East, northern Africa, Pakistan, Indonesia, and Malaysia. Large numbers of people also follow the religion in parts of sub-Saharan Africa, India, Russia, and Southeast Asia.

Islam, which means "submission" (to the will of God), is not only a religion but also a way of life.[21] One who adheres to Islam is called a Muslim, "he who submits." Islam's founder, Mohammed, was neither a savior nor a messiah but rather a prophet through whom God delivered his messages. He was born in 570 C.E. in Mecca, Saudi Arabia, a city located along the spice trade route. Early in his life, Mohammed acquired a respect for Jewish and Christian monotheism. Later, the archangel Gabriel appeared to him in many visions. These revelations continued for a decade or more, and the archangel told Mohammed that he was a prophet of Allah, the one true God. Mohammed's teachings met with hostility in Mecca, and in 622 he fled to Yathrib. The year of the flight (hegira) is the first year in the Muslim calendar. At Yathrib, later named Medina, Mohammed became a religious and political leader. Eight years after fleeing Mecca, he returned triumphant and declared Mecca a holy place to Allah.

The most sacred writings of Islam are found in the Qur'an (sometimes spelled *Koran* or *Quran*), believed to contain the words spoken by Allah through Mohammed. It includes many legends and traditions that parallel those of the Old and New Testaments, as well as Arabian folk tales. The Qur'an also contains the basic laws of Islam, and its analysis and interpretation by religious scholars have provided the guidelines by which Muslims lead their daily lives.

Muslims believe that the one true God, Allah, is basically the God of Judaism and Christianity but that his word was incompletely expressed in the Old and New Testaments and was only fulfilled in the Qur'an. Similarly, they believe that Mohammed was the last prophet, superseding Christ, who is considered by Muslims a prophet and not the Son of God. The primary doctrines of Islam are monotheism and the concept of the last judgment—the day of final resurrection when all will be deemed worthy of either the delights of heaven or the terrors of hell.

Mohammed did not institute an organized priesthood or sacraments but instead advocated the following ritualistic observances, known as the Five Pillars of Islam:

1. Faith, shown by the proclamation of the unity of God, and belief in that unity, as expressed in the creed, "There is no God but Allah; Mohammed is the Messenger of Allah."
2. Prayer, *salat,* performed five times daily (at dawn, noon, mid-afternoon, sunset, and nightfall), facing Mecca, wherever one may be; and on Fridays, the day of public prayer, in the mosque (a building used for public worship). On Fridays, sermons are delivered in the mosque after the noon prayer.
3. Almsgiving, *zakat*, as an offering to the poor and an act of piety. In some Islamic countries, Muslims are expected to give 2.5 percent of their net savings or assets in money or goods. The money is used to help the poor or to support the religious organization in countries where Islam is not the dominant religion. In addition, *zakat* is given to the needy on certain feast and fast days (see the next section on dietary practices for more details).
4. Fasting, to fulfill a religious obligation, to earn the pleasure of Allah, to wipe out previous sins, and to appreciate the hunger of the poor and the needy.
5. Pilgrimage to Mecca, *hadj*, once in a lifetime if means are available. No non-Muslim can enter Mecca. Pilgrims must wear seamless white garments; go without head covering or shoes; practice sexual continence; abstain from shaving or having their hair cut; and avoid harming any living thing, animal, or vegetable.

There are no priests in Islam; every Muslim can communicate directly with God, so a mediator is not needed. The successors of the prophet Mohammed and the leaders of the Islamic community were the caliphes (or *kalifah*). No caliphes exist today. A mufti, like a lawyer, gives legal advice based on the sacred laws of the Qur'an. An imam is the person appointed to lead prayer in the mosque and deliver the Friday sermon.

The following prominent sects in Islam have their origin in conflicting theories on the office of caliph (caliphate): (1) Sunni, who form the largest number of Muslims and hold that the

If one is unable to attend a mosque, the prayers are said on a prayer rug facing Mecca.

The Kaaba, in Mecca, is the holiest shrine of Islam and contains the Black Stone given to Abraham and Ishmael by the Archangel Gabriel. During the *hadj*, each pilgrim touches the stone and circles the shrine.

No one claiming title to the office of caliphate has been recognized by all Muslim sects since its abolition by the Turkish government in 1924 following the fall of the Ottoman Empire. The role of the caliphate in modern Islam is uncertain.

The status of fish varies by sect. Most Muslims consider anything from the sea halal; however, some, such as Shiites, eat only fish with scales.[25]

caliphate is an elected office that must be occupied by a member of the tribe of Koreish, the tribe of Mohammed. (2) Shi'ia, the second largest group, who believe that the caliphate was a God-given office held rightfully by Ali, Mohammed's son-in-law, and his descendants. The Shiites (followers of Shi'ia Islam) are found primarily in Iran, Iraq, Yemen, and India. (3) The Khawarij, who believe that the office of caliph is open to any believer whom the faithful consider fit for it. Followers of this sect are found primarily in eastern Arabia and North Africa. (4) The Sufis, ascetic mystics who seek a close union with God now, rather than in the hereafter. Only 3 percent of present-day Muslims are Sufis, and many remain outside mainstream Islam.[22,23]

It is estimated that nearly 2.5 million Muslims live in the United States; many came from the Middle East. Most are Sunnis, with only a small percentage of Shiites, though there is some crossover in worship and religious celebrations.[24] In addition, some African Americans believe Allah is the one true God; they follow the Qur'an and traditional Muslim rituals in their temple services. The movement was originally known as the Nation of Islam and its adherents identified as Black Muslims. A split in the Nation of Islam resulted in one faction of Black Muslims becoming an orthodox Islamic religion called the World Community of Al-Islam in the West. It is accepted as a branch of Islam. The other Black Muslim faction has continued as the Nation of Islam under the leadership of Louis Farrakhan.

Islamic laws consider eating to be a matter of worship, and Muslims are encouraged to share meals.

© Nigel Blythe/Cephas Picture Library/Alamy

Halal, Islamic Dietary Laws

In Islam, eating is considered to be a matter of worship. Muslims are expected to eat for survival and good health; self-indulgence is not permitted. Muslims are advised against eating more than two-thirds of their capacity, and sharing food is recommended. Food is never to be thrown away, wasted, or treated with contempt. The hands and mouth are washed before and after meals. If eating utensils are not used, only the right hand is used for eating, as the left hand is considered unclean.

Permitted or lawful foods are called halal. Allah alone has the right to determine what may be eaten, and what is permitted is sufficient—what is not permitted is unnecessary.[25] Unless specifically prohibited, all food is edible. Unlawful or prohibited (haram) foods listed in the Qur'an include:

1. All swine, four-footed animals that catch their prey with their mouths, birds of prey that seize their prey with their talons, and any by-products of these animals, such as pork gelatin or enzymes used in cheese making. If the source of any by-product is in question, it is avoided.
2. Improperly slaughtered animals (including carrion). An animal must be killed in a manner similar to that described in the Jewish laws, by slitting the front of the throat; cutting the jugular vein, carotid artery, and windpipe; and allowing the blood to drain completely. In addition, the person who kills the animal must repeat at the instant of slaughter, "In the name of God, God is great." Fish and seafood are exempt from this requirement.

Some Muslims believe that a Jew or a Christian can slaughter an animal to be consumed by Muslims as long as it is done properly. Others will eat only kosher meat, while some abstain from meat altogether unless they know it is properly slaughtered by a Muslim or can arrange to kill the animal themselves. Meat of animals slaughtered by people other than Muslims, Jews, or Christians is prohibited. Meat from an animal that was slaughtered when any name besides God's was mentioned is also prohibited.[26]

3. Blood and blood products.
4. Alcoholic beverages and intoxicating drugs, unless medically necessary. Even foods that have fermented accidentally are avoided. The drinking of stimulants, such as coffee and tea, is discouraged, as is smoking; however, these prohibitions are practiced only by the most devout Muslims.

A Muslim can eat or drink prohibited food under certain conditions, such as when the food is taken by mistake, when it is forced by others, or there is fear of dying by hunger or disease. The term for a food that is questionably halal or haram is *mashbooh*, and when in doubt, a Muslim is encouraged to avoid the item. Foods that combine halal items with haram items, such as baked goods made with lard or pizza with bacon, ham, or pork sausage topping, are also prohibited.[27] Muslims vary in their observance of the halal diet, however, with the strictest adherence found among the most orthodox believers.[28] Foods in compliance with Islamic dietary laws are sometimes marked with symbols registered with the Islamic Food and Nutrition Council of America (IFNCA) (Figure 4.3), signifying the food is fit for consumption by Muslims anywhere in the world.[29]

Figure 4.3 Examples of halal food symbols.

Feast Days

The following are the feast days in the Islamic religion:

1. Eid al-Fitr, the Feast of Fast Breaking—the end of Ramadan is celebrated by a feast and the giving of alms.
2. Eid al-Azha, the Festival of Sacrifice—the commemoration of Abraham's willingness to sacrifice his son, Ishmael, for God. It is customary to sacrifice a sheep and distribute its meat to friends, relatives, and the needy.
3. Shab-i-Barat, the night in the middle of Shaban—originally a fast day, this is now a feast day celebrated mostly in non-Arab nations, often marked with fireworks. It is believed that God determines the actions of every person for the next year on this night.
4. Nau-Roz, New Year's Day—primarily celebrated by the Iranians, it is the first day after the sun crosses the vernal equinox.
5. Maulud n'Nabi—the birthday of Mohammed.

NEW AMERICAN PERSPECTIVES—Islam

My name is Hafsabibi Mojy and I am a student at San Jose State University. I also work as Nutrition Technician in a local medical center. Although I immigrated from India to the United States in 1994, I am an observant Muslim, especially when it comes to Islamic food laws. I will shop at special stores in order to buy halal meats, which for me include beef, goat, chicken, veal, and turkey and are more expensive then meat from the supermarket. Most Americans I have met do not know what halal means, and they often think I am a vegetarian, probably because I come from India. It is common for me to call myself a "meat-eating vegetarian" as I only eat meat at home or places where halal is available, but otherwise I call myself vegetarian when I am at places where halal food is not available.

I do fast from sunrise to sunset during Ramadan. It is not hard, but it takes a few days in the beginning to get back to the rhythm of fasting. To me, personally, I get thirsty more than hungry. My favorite Islamic holiday is Eid Al-Adha or Feast of Sacrifice. It is the most important feast of the Muslim calendar and lasts for three days. It concludes the Pilgrimmage to Mecca. The feast reenacts prophet Ibrahim's obedience to God by sacrificing a cow or ram. The family eats about a third of the meal and donates the rest to the poor. My favorite foods include all Indian meat curry dishes, Biryani (spicy rice pilaf with meat or chicken), and kababs.

Health care that includes dietary modifications may interfere with Islamic food laws. Say, for example, a patient is on a clear liquid diet at a hospital . . . there will be very few choices left to give to a Muslim patient as Jell-O will be excluded (Jell-O, considered a clear liquid because it is liquid at body temperature, contains gelatin which is usually derived from animals and hence will be a non-halal item). Some "vegetarian" dishes contain a chicken-broth base and hence will be considered non-halal. Even a vegetarian burger can have wine in it and will be considered non-halal as all alcohol products are also prohibited under Islamic food laws.

Some devout Muslims also avoid land animals without external ears, such as snakes and lizards.

The month of Ramadan can fall during any part of the year. The Muslim calendar is lunar but does not have a leap month; thus, the months occur at different times each solar year.

Women who wish to undertake voluntary fasts must seek permission from their husbands.

In 2006, a French humanitarian group was accused of racism when it deliberately served "identity soup" made with pork parts to needy Muslims in the country.

Feasting also occurs at birth, after the consummation of marriage, at Bismillah (when a child first starts reading the Qur'anic alphabet), after circumcision of boys, at the harvest, and at death.

Fast Days

On fast days, Muslims abstain from food, drink, smoking, and coitus from dawn to sunset. Food can be eaten before the sun comes up and again after it sets. Fasting is required of Muslims during Ramadan, the ninth month of the Islamic calendar. It is believed that during Ramadan, "the gates of Heaven are open, the gates of Hell closed, and the devil put in chains." At sunset, the fast is usually broken by taking a liquid, typically water, along with an odd number of dates.

All Muslims past the age of puberty (fifteen years old) fast during Ramadan. A number of groups are exempt from fasting, but most must make up the days before the next Ramadan. They include sick individuals with a recoverable illness; people who are traveling; women during pregnancy, lactation, or menstruation; elders who are physically unable to fast; insane people; and those engaged in hard labor. During Ramadan, it is customary to invite guests to break the fast and dine in the evening; special foods are eaten, especially sweets. Food is often given to neighbors, relatives, and needy individuals or families.

Muslims are also encouraged to fast six days during Shawwal, the month following Ramadan; the tenth day of the month of Muhurram; and the ninth day of Zul Hijjah, but not during the pilgrimage to Mecca. A Muslim may fast voluntarily, preferably on Mondays and Thursdays. Muslims are not allowed to fast on two festival days: Eid al-Fitr and Eid al-Azha; or on the days of sacrificial slaughter: Tashriq—the twelfth, thirteenth, and fourteenth days of Dhu-al-Hijjah. It is also undesirable for Muslims to fast excessively (because Allah provides food and drink to consume) or to fast on Fridays.

Eastern Religions

Hinduism

Hinduism is considered the world's oldest religion, and, like Judaism, it is the basis of other religions such as Buddhism. Although Hinduism was once popular throughout much of Asia, most Hindus now live in India, its birthplace. The common Hindu scriptures are the Vedas, the Epics, and the Bhagavata Purana. The Vedas form the supreme authority for Hinduism. There are four Vedas: the Rigveda, the Samaveda, the Yajurveda, and the Aatharvaveda.

The goal of Hinduism is not to make humans perfect beings or life a heaven on Earth but rather to make humans one with the Universal Spirit or Supreme Being. When this state is achieved, there is no cause and effect, no time and space, no good and evil; all dualities are merged into oneness. This goal cannot be reached by being a good person, but it can be obtained by transforming human consciousness or liberation, *moksha*, into a new realm of divine consciousness that sees individual parts of the universe as deriving their true significance from the central unity of spirit. The transformation of human consciousness into divine consciousness is not achieved in one lifetime, and Hindus believe that the present life is only one in a series of lives, or reincarnations. Hindus believe in the law of rebirth, which postulates that every person passes through a series of lives before obtaining liberation; the law of karma, that one's present life is the result of what one thought or did in one's past life. In each new incarnation, an individual's soul moves up or down the spiritual ladder; the goal for all souls is liberation.

There is one Supreme Being, Brahman, and all the various gods worshiped by men are partial manifestations of him. Hindus choose the form of the Supreme Being that satisfies their spirit and make it an object of love and adoration. This aspect of worship makes Hinduism very tolerant of other gods and their followers; many different religions have been absorbed into Hinduism.

The three most important functions of the Supreme Being are the creation, protection, and destruction of the world, and these functions have become personified as three great gods: Brahma, Vishnu, and Siva (the Hindu triad or trinity). The Supreme Being as Vishnu is the protector of the world. Vishnu is also an avatara, meaning he can take on human forms whenever the world is threatened by evil. Rama and Krishna are regarded as two such embodiments and are also objects of worship.

Hindus believe that the world passes through repeating cycles; the most common version of

CULTURAL CONTROVERSY—Meat Prohibitions

Scientists have calculated that animal protein comprised over 50 percent of the total daily calories consumed by prehistoric peoples, a far higher amount than what most Americans eat today.[41] Further, it is believed that only 1 percent of the world population refuses to eat all types of meat, poultry, and fish and that total vegans, who avoid all animal products, equal only one-tenth of 1 percent. Humans favor protein foods. Nevertheless, many cultures impose some restrictions on what meats may be consumed, mostly in accordance with prevailing religious dietary laws. The devout of each faith see little reason to ask why a particular food is prohibited. It is considered presumptuous or sacrilegious for humans to question the directives of God or church.

This has not deterred researchers from speculating on the rationale of meat taboos. Some have investigated the whole field of taxonomy and how animals are classified as different or unnatural, thus abominable due to their physical characteristics. The Jewish prohibition against pork, for example, seems to be because pigs do not chew their cud, marking them as dissimilar from other animals with cloven hooves.[42] This theory is supported by the omnivore's paradox and the psychological need for food familiarity. Others have focused on the use of the term *unclean* in relation to biblical and Qur'anic pork prohibitions, claiming that pork consumption is unhealthful. Many researchers discard this theory because it is thought that ancient populations could not have made the association between eating pork and the slow development of diseases, such as trichinosis, not to mention that other animals that carry fatal illnesses (e.g., spongiform encephalitis or mad cow disease) are not avoided.

The socioecological theory for why certain meats are avoided suggests that if an animal is more valuable alive than dead or, conversely, if it does not fit well into the local ecology or economy, consumption will be prohibited.[43] Religious dietary codes often reinforce preexisting food practices and prejudices. When reviewing the history of pork in the Middle East, for example, archeological records show it was part of the ancient diet. But by 1900 B.C.E., pork had become unpopular in Babylonia, Egypt, and Phoenicia, coinciding with an expanding population and deforestation of the region. Pigs compete with humans for food sources. Additionally, they do not thrive in hot, dry climates. Cows, goats, and sheep, on the other hand, can graze over large areas and survive on the cellulose in plants unavailable to human metabolism. And they need no protection from the sun. The nomadic Hebrews were unlikely to have herded pigs in their early history, and by the time they settled there was a broad aversion to pigs by many Middle Easterners. The first followers of Mohammed were also pastoral people, which may explain why the only explicitly prohibited animal flesh in Islam is pork.

The socioeconomic theory is useful in examining other meat prohibitions. In India, where beef is banned for Hindus, cattle are the primary power source in rural farming communities due to the expense of tractors. Further, cattle provide dung that is dried to produce a clean, slow-burning cooking fuel, and cows provide milk for the dairy products important in some vegetarian fare. Even dead cows serve a purpose, providing the very poor with scavenged meat to eat and skins to craft leather products. The value of cattle in India is reinforced by religious custom. Horsemeat in Europe is a different example. Though horse consumption was frequent in early Europe, other cultures who used the animals for travel and cavalry often banned it. Asian nomads who roamed on horseback consumed horse milk and blood but only ate the flesh in emergencies. It was avoided by the Romans and most Middle Easterners (prohibited for Jews and by custom among Muslims). During the eighth century, when European Christian strongholds came under attack from Muslim cavalry in the south and mounted nomads from the west, Pope Gregory III recognized the need for horses in the defense of the church. He prohibited horsemeat as "unclean and detestable." However, horse consumption was never entirely eliminated, especially during times of hardship, and gradually religious restrictions were eased. By the nineteenth century, horsemeat had regained favor, especially in France and Belgium, where it is a specialty item today. Despite the initial need for horsepower, the religious prohibition was unsustainable over time because it contradicted prevailing food traditions.

the creation is connected to the life of Vishnu. From Vishnu's navel grows a lotus, and from its unfolding petals is born the god Brahma, who creates the world. Vishnu governs the world until he sleeps; then Siva destroys it, and the world is absorbed into Vishnu's body to be created once again.

The principles of Hinduism are purity, self-control, detachment, truth, and nonviolence. Purity is both a ceremonial goal and a moral ideal. All rituals for purification and the elaborate rules regarding food and drink are meant to lead to purity of mind and spirit. Self-control governs both the flesh and the mind.

Hinduism does not teach its followers to suppress the flesh completely but rather to regulate its appetites and cravings. The highest aspect of self-control is detachment. Complete liberation

The Seven Social Sins according to Gandhi are politics without principle; wealth without work; pleasure without conscience; knowledge without character; commerce without morality; science without humanity; and worship without sacrifice.[44]

Ganesh got his elephant head when he angered his father, Siva, who cut off his human head. When his mother pleaded with Siva to replace his head, Siva used the head of a nearby elephant. Hindus honor Ganesh through offerings of the foods he favored.

Yoga means yoke, as in yoking together or union.

Some Hindu worshipers break coconuts on the temple grounds to symbolize the spiritual experience. The hard shell is a metaphor for the human ego, and once it is cracked open, the soft, sweet meat representing the inner self is open to becoming one with the Supreme Being.[45]

from this world and union with the divine are not possible if one clings to the good or evil of this existence. Pursuit of truth is indispensable to the progress of humans, and truth is always associated with nonviolence, ahimsa. These principles are considered the highest virtues. India's greatest exponent of this ideal was Mahatma Gandhi, who taught that nonviolence must be practiced not only by individuals, but also by communities and nations.

One common belief of Hinduism is that the world evolved in successive stages, beginning with matter and going on through life, consciousness, and intelligence to spiritual bliss or perfection. Spirit first appears as life in plants, then as consciousness in animals, intelligence in humans, and finally bliss in the supreme spirit. A good person is closer to the supreme spirit than a bad person is, and a person is closer than an animal. Truth, beauty, love, and righteousness are of higher importance than intellectual values (e.g., clarity, cogency, subtlety, skill) or biological values (e.g., health, strength, vitality). Material values (e.g., riches, possessions, pleasure) are valued least.

The organization of society grows from the principle of spiritual progression. The Hindu lawgivers tried to construct an ideal society in which people are ranked by their spiritual progress and culture, not according to their wealth or power. The social system reflects this ideal, which is represented by four estates, or castes, associated originally with certain occupations. The four castes are the Brahmins (teachers and priests), the Kshatriyas (soldiers), the Vaisyas (merchants and farmers), and the Sundras (laborers). Existing outside social recognition are the *dalits*, or untouchables (e.g., butchers, leather workers), a group of persons who do not fall into the other four categories; although this designation was outlawed by the Indian government in 1950, it is estimated that untouchables still constitute almost 14.4 percent of the population.[30]

The four castes are represented as forming parts of the Creator's body: respectively, his mouth, arms, thighs, and feet. The untouchables were supposedly created from darkness that Brahma discarded in the process of creation. The castes also conform to the law of spiritual progression, in that the most spiritual caste occupies the top and the least spiritual the bottom. The Hindus believe that nature has three fundamental qualities: purity, energy, and inertia. Those in whom purity predominates form the first caste, energy the second caste, and inertia the third and fourth castes. Each caste should perform its own duties, follow its hereditary occupation, and cooperate with the others for the common welfare. People's good actions in this life earn them promotion to a higher caste in the next life.

There are thousands of subdivisions of the four main castes. The subcastes often reflect a trade or profession, but some scholars contend that the latter was imposed on the former. In reality, the subcaste is very important to daily life, whereas what major caste one belongs to makes little difference to non-Brahmins (see Chapter 14, "South Asians"). The ideal life of a Hindu is divided into four successive stages, called asramas. The first stage is that of the student and is devoted entirely to study and discipline. The guru becomes an individual's spiritual parent. After this period of preparation, the student should settle down and serve his or her marriage, community, and country. When this active period of citizenship is over, he or she should retire to a quiet place in the country and meditate on the higher aspects of the spirit (become a recluse). The recluse then becomes a sannyasi, one who has renounced all Earthly possessions and ties. This stage is the crown of human life. The goals of life are dharma (righteousness), *artha* (worldly prosperity), *kama* (enjoyment), and *moksha* (liberation). The ultimate aim of life is liberation, but on their way to this final goal people must satisfy the animal wants of their bodies, as well as the economic and other demands of their families and communities. However, all should be done within the moral law of dharma. Adherence to dharma reflects a unique aspect of Hinduism, namely, that practice is more important than belief. There are no creeds in Hinduism; it is the performance of duties associated with one's caste or social position that make a person a Hindu.[31]

Common practices in Hinduism include rituals and forms of mental discipline. All Hindus are advised to choose a deity on whose form, features, and qualities they can concentrate their mind and whose image they can worship every day with flowers and incense. The deity is only a means of realizing the Supreme Being by means of ritualistic worship. Externally, the deity is worshiped as a king or honored guest. Internal worship consists of prayer and meditation. Mental discipline is indicated by the word *yoga*. Along with mental discipline, yoga has come to mean a method of restraining the functions of the mind and their physiological consequences.

Hindus can be divided into three broad sects according to their view of the Supreme Being. They are the Vaishnava, the Saiva, and the Sakta, who maintain the supremacy of Vishnu, Siva, and the Sakti (the female and active aspects of Siva), respectively. Different sects are popular in different regions of India. Many Hindus do not worship one God exclusively. Vishnu may be worshiped in one of his full embodiments (Krishna or Rama) or partial embodiments. In addition, there are hundreds of lesser deities, much like saints. One is Siva's son, the elephant-headed Ganesh, who is believed to bring good luck and remove obstacles.

It is estimated that in 2008 there are nearly 2.3 million Hindus in the United States. This number was based on the U.S. Census of South Asian Indians, but it was adjusted by the percentage of the Hindu population in India.[32] It was assumed that the percentage of Hindus in the United States would be similar to that found in India. A small percentage of non–Indian Americans have become followers of the Hindu religion. The International Society for Krishna Consciousness, founded in 1966 by devotees of a sixteenth-century Bengali ascetic, has the largest number of converts.

Hindu Dietary Practices

In general, Hindus avoid foods believed to hamper the development of the body or mental abilities. Bad food habits will prevent one from reaching mental purity and communion with God. Dietary restrictions and attitudes vary among the castes.

The Laws of Manu (dating from the fourth century C.E.) state that "no sin is attached to eating flesh or drinking wine, or gratifying the sexual urge, for these are the natural propensities of men; but abstinence from these bears greater fruits." Many Hindus are vegetarians.[33,34] They adhere to the concept of ahimsa, avoiding inflicting pain on an animal by not eating meat. Although the consumption of meat is allowed, the cow is considered sacred and is not to be killed or eaten. If meat is eaten, pork as well as beef is usually avoided. Crabs, snails, crocodiles, numerous birds (e.g., crows, doves, domesticated fowl, ducks, flamingos, parrots, vultures, and woodpeckers), antelopes, camels, boars, bats, porpoises, and fish with ugly forms (undefined) should also be rejected. In addition, the laws make many other recommendations regarding foods that should be avoided, including foods prepared by certain groups of people (e.g., actors, artists, carpenters, cobblers, doctors, eunuchs, innkeepers, musicians, prostitutes, liars, spies, and thieves); foods that have been contaminated by a person sneezing or through contact with a human foot, clothing, animals, or birds; milk from an animal that has recently given birth; and water from the bottom of a boat. No fish or meat should be eaten until it has been sanctified by the repetition of mantras offering it to the gods. Pious Hindus may also abstain from alcoholic beverages. Garlic, turnips, onions, leeks, mushrooms, and red-hued foods, such as tomatoes and red lentils, may be avoided. Despite such lengthy prohibitions, Hindus exert considerable personal discretion regarding taboo foods.[35]

Intertwined in Hindu food customs is the concept of purity and pollution. Complex rules regarding food and drink are meant to lead to purity of mind and spirit. Pollution is the opposite of purity and should be avoided or ameliorated. To remain pure is to remain free from pollution; to become pure is to remove pollution. Certain substances are considered both pure in themselves and purifying in their application. These include the products of the living cow—milk products, dung, and urine—and water from sources of special sanctity, such as the Ganges River. Pure and purifying substances also include

According to legend, Vishnu rested on a 1,000-headed cobra between the creation and destruction of the world. During the festivities of Naga Panchami, snakes are venerated at Hindu temples, and milk is offered to cobras to prevent snakebite.

Hindus are encouraged to practice moderation—they are advised not to eat too early, not to eat too late, and not to eat too much.

Students studying the Vedas and other celibates are usually vegetarians and may restrict irritating or exciting foods, such as honey.

▼ *The numerous religious holidays and secular events celebrated in India include feasting, which serves to distribute food throughout the community.*

© United Nations/J. Isaac

materials commonly employed in rituals, such as turmeric and sandalwood paste. All body products (e.g., feces, urine, saliva, menstrual flow, and afterbirth) are polluting. Use of water is the most common method of purification because water easily absorbs pollution and carries it away.

Feast Days

Water is the beverage of choice at meals. Standing water is easily defiled if it is touched by a member of a lower caste; flowing water is so pure that even an untouchable standing in it does not pollute it.

The Hindu calendar marks eighteen major festivals every year. Additional important feast days are those of marriages, births, and deaths. Each region of India observes its own special festivals; it has been said that there is a celebration going on somewhere in India every day of the year. All members of the community eat generously on festive occasions, and these may be the only days that very poor people eat adequately. Feasting is a way of sharing food among the population because the wealthy are responsible for helping the poor celebrate the holidays.

One of the gayest and most colorful of the Hindu festivals is Holi, the spring equinox and the celebration of one of Krishna's triumphs. According to legend, Krishna had an evil uncle who sent an ogress named Holika to burn down Krishna's house. Instead, Krishna escaped and Holika burned in the blaze. It is traditional for Indians to throw colored water or powder at passersby during this holiday.

The ten-day celebration of Dusshera in late September or early October commemorates the victory of Prince Rama (one of Vishnu's embodiments) over the army of the demon Ravana. It is also a grateful tribute to the goddess Durga, who aided Rama. The first nine days are spent in worshiping the deity, and the tenth day is spent celebrating Rama's victory.

In southern India the rice harvest is celebrated in the festival called Pongal—new rice is cooked in milk, and when it begins to bubble, the family shouts, "Pongal!" ("It boils!").

Divali, celebrated throughout India in November, marks the darkest night of the year, when souls return to Earth and must be shown the way by the lights in the houses. For many, Divali is also the beginning of the new year, when everyone should buy new clothes, settle old debts and quarrels, and wish everyone else good fortune.

The Hindu calendar is lunar; thus, its religious holidays do not always fall on the same day on the Western calendar. Every three to five years the Hindu calendar adds a thirteenth leap month (a very auspicious period) to reconcile the months with the seasons.

Fast Days

In India, fasting practices vary according to one's caste, family, age, sex, and degree of orthodoxy. A devoutly religious person may fast more often and more strictly than one who is less religious. Fasting may mean eating no food at all or abstaining from only specific foods or meals. The fast days in the Hindu calendar include the first day of the new and full moon of each lunar month; the tenth and eleventh days of each month; the feast of Sivaratri; the ninth day of the lunar month Cheitra; the eighth day of Sravana; days of eclipses, equinoxes, solstices, and conjunctions of planets; the anniversary of the death of one's father or mother; and Sundays.

Buddhism

Siddhartha Gautama, who later became known as Buddha (the Enlightened One), founded the Eastern religion of Buddhism in India in the sixth century B.C.E. Buddhism flourished in India until 500 C.E., when it declined and gradually became absorbed into Hinduism. Meanwhile, it had spread throughout southeastern and central Asia. Buddhism remains a vital religion in many Asian countries, where it has been adapted to local needs and traditions.

Buddhism was a protestant revolt against orthodox Hinduism, but it accepted certain Hindu concepts, such as the idea that all living beings go through countless cycles of death and rebirth, the doctrine of karma, spiritual liberation from the flesh, and that the path to wisdom includes taming the appetites and passions of the body. Buddha disagreed with the Hindus about the methods by which these objectives were to be achieved. He advocated the Middle Way between asceticism and self-indulgence, stating that both extremes in life should be avoided. He also disagreed with the Hindus on caste distinctions, believing that all persons were equal in spiritual potential.

The basic teachings of Buddha are found in the Four Noble Truths and the Noble Eightfold Path.[36] The Four Noble Truths are as follows:

1. *Dukkha*—The Noble Truth of Suffering: Suffering is part of living. Persons suffer when they experience birth, old age, sickness, and death. They also suffer when they fail to obtain what they want. At a deeper level, *dukkha* embodies other concepts, such as imperfection, emptiness, and impermanence.
2. *Samudaya*—The Noble Truth of the Cause of Suffering: This is the arising of *dukkha*. Suffering is caused by a person's cravings for life, which cause rebirth. It is manifested by

an attachment to pleasure, wealth, power, and even ideals and beliefs.

3. *Nirodha*—The Noble Truth of the Cessation of Suffering: This is the cessation of *dukkha.* A person no longer suffers if all cravings are relinquished.
4. *Magga*—The Noble Truth to the Path Leading to the Cessation of Suffering: This is the Eightfold Path. It is a middle way between the search for happiness through pursuit of pleasure and the search for happiness through self-mortification and asceticism. By following this path (right view, right thought, right speech, right action, right livelihood, right effort, right mindfulness, and right concentration), craving is extinguished and deliverance from suffering ensues.

The third and fourth phases of the Eightfold Path (Figure 4.4), right speech and right action, have been extended into a practical code of conduct known as the Five Precepts. These are (1) abstain from the taking of life, (2) abstain from the taking of what is not given, (3) abstain from all illegal sexual pleasures, (4) abstain from lying, and (5) abstain from consumption of intoxicants because they tend to cloud the mind.

The person who perfects Buddha's teachings achieves nirvana, a state of calm insight, passionlessness, and wisdom. In addition, the person is no longer subject to rebirth into the sorrows of existence. Because the ideal practice of Buddhism is impractical in the turmoil of daily life, Buddhism has encouraged a monastic lifestyle. The ideal Buddhists are monks, following a life of simplicity and spending considerable time in meditation. They own no personal property and obtain food by begging. They are usually vegetarians and are permitted to eat only before noon. The monk confers a favor or merit (good karma) on those who give him food.

There are numerous sects in Buddhism and two great schools of doctrine: Theravada (also known as Hinayana) Buddhism, which is followed in India and Southeast Asia; and Mahayana Buddhism, which is followed in China, Japan, Korea, Tibet, and Mongolia.[36,37] Theravada Buddhism is primarily a spiritual philosophy and system of ethics. It places little or no emphasis on deities, teaching that the goal of the faithful is to achieve nirvana. In Mahayana, a later form of Buddhism, Buddha is eternal and

FIGURE 4.4
The Buddhist Wheel of Law—the spokes represent the Eightfold Path.

cosmic, appearing variously in many worlds to make known his truth, called dharma. This has resulted in a pantheon of Buddhas who are sometimes deified and, for some sects, a hierarchy of demons. Some sects promise the worshiper a real paradise rather than the perfected spiritual state of nirvana.

The number of Buddhists in the United States was over 2.1 million in 2008[38] majority are immigrants from Japan, China, and Southeast Asia and their descendants. Based on number of meditation centers, a majority are believed to be Mahayana.[39] A small number of non-Asians also have more recently converted to Buddhism. Vajrayana Buddhism, a Tibetan Mahayana sect also known as Tantric Buddhism, and Zen Buddhism, a Chinese sect that spread to Japan around the year 1200, have gained followers, especially in the West.

In both Theravada and Mahayana temples, worshipers may offer food at the altar, such as apples, bananas, grapes, oranges, pineapples, candy, rice, dried mushrooms, and oil.

A Zen Buddhist monastery, Tassajara, located in central California, is famous for its vegetarian restaurant and popular cookbook. Macrobiotics is not associated with Zen Buddhism (see Chapter 2 "Traditional Health Beliefs and Practices").

Dietary Practices

Buddhist dietary restrictions vary considerably depending on the sect and country. Buddhist doctrine forbids the taking of life; therefore, many followers are lacto-ovo-vegetarians (eating dairy products and eggs, but no meat). Some eat fish, and others abstain only from beef. Others believe that if they were not personally responsible for killing the animal, it is permissible to eat its flesh.

Buddhist monks in Tibet carve sculptures in butter (as high as fifteen feet) and parade them during an evening in March, lit by lanterns, for Chogna Choeba, the Butter Lamp Festival. Afterward they are dismantled and thrown in the river, symbolic of the impermanence of life.

Feasts and Fasts

Buddhist festivals vary according to region. From July to October, Buddhist monks are directed to remain in retreat and meditate, coinciding with

DISCUSSION STARTERS: DOES YOUR RELIGION AFFECT YOUR EATING HABITS?

Explore your own religious and cultural dietary restrictions. Even if you are not religious, you can probably identify the major religions of your culture, a religion or multiple religions of your parents, grandparents, or great grandparents. It's probable that, unless you have made a conscious decision to change your dietary habits from your childhood, you still adhere to at least some of the dietary practices of your family's religion. Answer the following:

- What foods and/or drink are prohibited by your religion or you were prohibited from consuming when growing up?
- Are there certain times during the year when your religion directs you—or your family directed you as a child—to avoid certain foods and drink or maybe even to fast (not eat at all)?
- Are there certain times during the year when your religion directs you—or when, as a child, you were directed—to consume particular kinds of food or drink?
- Do you observe certain feast days during the year, days when you are supposed to eat a lot?

After answering the questions above, seek out others with whom to compare your answers. Don't call just on friends. Contact someone at a local mosque, synagogue, temple, church, or other religious meeting place, and ask to interview someone about her or his religion. Be sure to explain that you are student studying food and religion. Another way of finding someone of a different culture and religion to interview is to look for local restaurants serving foods of particular ethnic groups: Thai, Chinese, Vietnamese, Korean, Japanese, Mexican, Middle Eastern, Indian, Ethiopian, Cuban, Caribbean, Greek, Italian, French, Cajun, African American, or another culture. Contact the owner or manager to request an interview, again making sure to explain who you are and your reason for wanting to interview her or him. Because this person may be busy with work, she or he may not be able to talk with you right away. When meeting with the person for the interview, show her or him the questions above and your answers. Compare your answers with those that your interviewee gives to those same questions.

the rainy season and the sprouting of rice in the fields. The first day of retreat is a time for worshipers to bring gifts of food and articles of clothing to the monks; the retreat ends with *pravarana*, the end of the rainy season, when worshipers once again offer gifts to the monks, invite them to a meal, and organize processions. On three separate days (which vary according to the regional calendar), Mahayana Buddhists commemorate the birth of Buddha, his enlightenment, and his death; Theravada Buddhists celebrate Magha Puja, the Four Miracles Assembly, in February or March when Buddha appointed the first Buddhist brotherhood of monks at a coincidental meeting of 1,250 disciples at a shrine. In April or May they honor the Buddha on a single holiday called Vesak. Buddhist monks may fast twice a month, on the days of the new and full moon. They also do not eat any solid food after noon.

REVIEW QUESTIONS

1. What are the basic tenets of Western and Eastern religions?
2. Pick two of the following religions and describe the dietary laws for food preparation and consumption, and any additional laws for holy days: Judaism, Hinduism, and Islam.
3. List the Five Pillars of Faith in Islam and the Four Noble Truths and Noble Eightfold Path in Buddhism.
4. Describe and compare the roles of fasting in Islam and Hinduism, using examples of fasting practices in each faith.

REFERENCES

1. Raphael, M.L. 2003. *Judaism in America.* New York: Columbia University Press.
2. Greenberg, B. 1989. *How to run a traditional Jewish household.* Northvale, NJ: Aronson.
3. Himelstein, S. 1990. *The Jewish primer.* New York: Facts on File.
4. Shwide-Slavin C. Jewish Food Practices. In: Goody CM, Drago L. (eds.). *Cultural Food Practices. Diabetes Care and Education Dietetic Practice Group.* Chicago IL: American Dietetic Association, 2010.
5. Oxford University Press. 2002. *The Oxford Dictionary of the Jewish Religion.* R.J.Z. Werblowsky & G. Weigoder (Eds.). New York: Author.
6. Lloyd ML, Olsen WA. Disaccharide malabsorption. In: Haubrich WS, Schaffner F, Berk JE, eds. *Bockus Gastroenterology.* 5th ed. Philadelphia, PA: Saunders; 1995: 1087–1100.
7. Shugart YY, Silverberg MS, Duerr RH, Taylor KD, Wang MH, Zarfas K, Schumm LP, Bromfield G, Steinhart AH, Griffiths AM, Kane SV, Barmada MM, Rotter JI, Mei L, Bernstein CN, Bayless TM, Langelier D, Cohen A, Bitton A, Rioux JD, Cho JH, Brant SR. An SNP linkage scan identifies significant

Crohn's disease loci on chromosomes 13q13.3 and, in Jewish families, on 1p35.2 and 3q29.*Genes Immun. 2008 Mar; 9*(2): 161–7.

8. National Council of Churches of Christ in the USA. 2010 . *Yearbook of American and Canadian Churches.* E.W. Lindner (Ed.). Nashville: Abingdon Press.
9. Eagan, J.F. 1995. *Restoration & renewal: The church in the third millennium.* Kansas City, MO: Sheed & Ward.
10. Clancy, P.M.J. 1967. *Fasting and abstinence. In The New Catholic Encyclopedia.* New York: McGraw-Hill.
11. Fairnbairn, D. 2002. *Eastern orthodoxy through Western eyes.* Louisville, KY: Westminster John Knox Press.
12. Smart, N. 1998. *The world's religions* (2nd ed.). New York: Cambridge University Press.
13. Kolb, R. 2004. Martin Luther. In *The Encyclopedia of Protestantism,* H.J. Hillerbrand (Ed.). New York: Rutledge.
14. Sack, D. 2000. *Whitebread Protestants: Food and religion in American culture.* New York: St. Martin's Press.
15. Newell, C. 2000. *Latter days: A guided tour through six billion years of Mormonism.* New York: St. Martin's Press.
16. Douglas, D. 2004. Mormonism. In *The Encyclopedia of Protestantism,* H.J. Hillerbrand (Ed.). New York: Rutledge.
17. Greenleaf, F. 2004. Seventh-Day Adventists. In *The Encyclopedia of Protestantism,* H.J. Hillerbrand (Ed.). New York: Rutledge.
18. White, E.G.H. 1905. *The ministry of healing.* Hagerstown, MD: Review and Herald Publishing.
19. White, E.G.H. 1923. *Counsels on health.* Hagerstown, MD: Review and Herald Publishing.
20. White, E.G.H. 1938. *Counsels on diet and foods.* Hagerstown, MD: Review and Herald Publishing.
21. Frager, R. 2002. *The wisdom of Islam: An introduction to the living experience of Islamic belief and practice.* Haupage, NY: Godsfield Press.
22. Denny, F.M., & Mamiya, L.H. 1996. Islam in the Americas. In *The Muslim Almanac,* A.A. Nanji (Ed.). New York: Gale Research, Inc.
23. Waines, D. 1995. *An introduction to Islam.* New York: Cambridge University Press.
24. Pew Report, 2007, *Muslim Americans: Middle Class and mostly Mainstream.* Available from: http://pewresearch.org/pubs/483/muslim-americans. Accessed: January 17, 2011.
25. Regenstein, J.M., Chaudry, M.M., & Regenstein, C.E. 2003.The kosher and halal food laws. *Comprehensive Reviews in Food Science and Food Safety, 2,* 111–127.
26. Chaudry, M.M. 1992. Islamic food laws: Philosophical basis and practical implications. *Food Technology, 46,* 92–93, 104.
27. Kulkarni KD. Asian Indian and Pakistani Food Practices. In: Goody CM, Drago L. (eds.). *Cultural Food Practices. Diabetes Care and Education Dietetic Practice Group.* Chicago IL: American Dietetic Association, 2010.
28. Eliasi, J.R., & Dwyer, J.T. 2002. Kosher and halal: Religious observances affecting dietary intakes. *Journal of the American Dietetic Association, 102,* 911–913.
29. Hussaini, M.M. 1993. *Islamic dietary concepts and practices.* Bedford Park, IL: Islamic Food and Nutrition Council of America.
30. Censusindia.gov.in. *"Census of India—India at a Glance: Scheduled Castes & Scheduled Tribes Population."* http://www.censusindia.gov.in/Census_Data_2001/India_at_Glance/scst.aspx. Retrieved 1/10/2010.
31. Flood, G. 1996. *An introduction to Hinduism.* New York: Cambridge University Press.
32. *Hindu American Foundation.* Available from: http://www.hafsite.org/resources/hinduism_101/hinduism_demographics. Accessed: January 10, 2011.
33. Achaya, K.T. 1994. *Indian food: A historical companion.* Delhi: Oxford University Press.
34. Pandit, B. 2005. *The Hindu mind: Fundamentals of Hindu religion and philosophy for all ages.* Glen Ellyn, IL: Dharma Publishing.
35. Kilara, A., & Iya, K.K. 1992. Food and dietary practices of the Hindu. *Food Technology, 46,* 94–102, 104.
36. Crosby, K. 2004. Theraveda. In *Encyclopedia of Buddhism.* R.E. Buswell (Ed.). New York: Macmillan Reference.
37. Schopen, G. 2004. Mahayana. In *Encyclopedia of Buddhism,* R.E. Buswell (Ed.). New York: Macmillan Reference.
38. Pew Forum on Religion and Public Life. *US Religious Landscape Survey Religious Affiliation 2008.* Available from: http://religions.pewforum.org/pdf/report-religious-landscape-study-full.pdf. Accessed: January 10, 2011.
39. Morreale, D. 1998. Everything has changed in Buddhist America. In *The Complete Guide to Buddhist America,* D. Morreale (Ed.). Boston: Shambhala.
40. Barrow K. *More people choosing Kosher for health.* New York Times. April 13, 2010.
41. Cordain, L., Miller, J.B., Eaton, S.B., Mann, N., Holt, S.H.A., & Speth, J.D. 2000. Plant-animal subsistence ratios and macronutrient energy estimations in worldwide hunter-gatherer diets. *Journal of Clinical Nutrition, 71,* 682–692.
42. Douglas, M. 1966. *Purity and danger: An analysis of concepts of pollution and taboo.* New York: Praeger.
43. Harris, M. 1998. *Good to eat: Riddles of food and culture.* Long Grove, IL: Waveland Press.
44. Pandit, B. 2005. *The Hindu mind: Fundamentals of Hindu religion and philosophy for all ages.* Glen Ellyn, IL: Dharma Publishing.
45. Dresser, N. 2005. Multicultural manners: *Essential rules of etiquette for the 21st century.* New York: John Wiley & Sons.

CHAPTER

Native Americans

It has been suggested that native Hawaiians be included as Native Americans; however, the history and culture of these peoples are substantially different from those of Native Americans of the U.S. mainland, Alaska, and Canada, so they are discussed in Chapter 12, "Southeast Asians and Polynesians." Native American cultures of Mexico, Central America, and South America are considered in Chapters 9, "Mexicans and Central Americans" and 10, "Caribbean Islanders and South Americans."

The U.S. Census uses the phrase "American Indians and Alaska Natives" (abbreviated AI/AN), and notes that approximately 15 percent of this population is also of Hispanic ethnicity.[90]

The designation *Native American*, which includes the greatest number of ethnic groups of any minority population in the United States, is a term for the indigenous people of the Americas. It is used for both American Indians and Alaska Natives, who are comprised of American Indians, Inuits, and Aleuts. Each of the approximately 400 American Indian and Alaska Native nations has its own distinct cultural heritage.

Over 5 million persons of Native American heritage lived in the United States in 2009, according to U.S. Census figures, representing 1.6 percent of the total population.[1] The vast majority of American Indians and Alaska Natives today live west of the Mississippi River. Roughly half live in rural areas, either on government reservations or on nearby farms. Native American ethnic identity varies tremendously, from tenacious maintenance of heritage to total adoption of the majority culture.

Traditional Native American foods have made significant contributions to today's diet in the United States. Corn, squash, beans, cranberries, and maple syrup are just a few of the items Native Americans introduced to European settlers. Historians question whether the original British colonists would have survived their first years in America without the supplies they obtained and the cooking methods they learned from the Native Americans. The diet of Native Americans has changed dramatically from its origins, yet recent renewed interest in American Indian and Alaska Native culture has prevented the complete disappearance of many traditional foods and food habits (Figure 5.1). This chapter reviews both the past and present diet of Native American ethnic groups.

Cultural Perspective

History of Native Americans

Settlement Patterns

It is hypothesized that the Native Americans came to North America approximately 20,000 to 50,000 years ago across the Bering Strait, which links Asia to Alaska, although some evidence suggests earlier migrations may have occurred. Archaeological research provides little insight into the settlement patterns and diversification of Native American culture in the years before European contact in the 1600s. Furthermore, the Native American languages were entirely verbal, so written historical records are nonexistent. There are, consequently, enormous gaps in what is known of early Native American societies.

Observations of Native Americans by white settlers have been well documented. These commentaries identified three major centers of American Indian culture during the seventeenth century. In the Southeast the sophisticated social organization of the Cherokees, Chickasaws, Choctaws, Creeks, and Seminoles led the Europeans to call them the "Five Civilized Tribes." The Iroquois, in what is now New York State, ruled a democratic confederacy of five nations. Religion and the arts flourished in Pueblo communities adjacent to the Rio Grande and Little Colorado rivers in the Southwest.

The introduction of horses, firearms, and metal knives changed the lifestyles of many nations, especially those that used the new tools to exploit the resources of the Great Plains. This initial interaction between white settlers and Plains Indians resulted in the development of the stereotype of the buffalo-hunting horseman with feathered headdress who came to represent all American Indian ethnic groups in the white imagination.

European diseases and the massacre of whole nations reduced the numbers of both Native American individuals and ethnic groups. In addition, many American Indians were forced to migrate west to accommodate white expansion. The hardships of involuntary relocation and the deaths caused by illness and assault may have caused the extinction of nearly one-quarter of all Native American ethnic groups.

Native American lands dwindled as white settlers moved westward. By the late nineteenth century, the majority of Native Americans lived on lands held in trust for them by the U.S. government, called federal reservations. Still others resided in state reservation communities. Although they were not required by law to live on reservations, there were few other viable Native American communities.

The Bureau of Indian Affairs (BIA) took over the administration of the reservations near the turn of the twentieth century. It established a program of cultural assimilation designed to bring the Native American residents into mainstream U.S. society. Before the 1930s, Native American children were sent to off-reservation boarding schools where white values were encouraged. Later, public reservation schools attempted similar indoctrination. The BIA program usually failed to force Native Americans to accept white values, however. The Native Americans changed their dress, occupation, and social structure, but they did not fully assimilate. In many cases their religious beliefs were strengthened, and their involvement in crafts, music, and dance was deepened to support their ethnic identity.

Current Demographics

Many Native Americans left the reservations for the employment opportunities available during World War II. Some joined the armed services, where they became fluent in English and the ways of the majority society. Others took war-related industry jobs. In the 1950s and 1960s, the BIA Employment Assistance Program was a major factor in the continuing out-migration of Native Americans from the reservations to the cities. By 2000, two-thirds of all American Indians and Alaska Natives resided in the farms, towns, and urban areas outside of reservations

Photo by Laurie Macfee

▲ *Traditional Native American foods: Some typical foods include beans, berries, corn, fish, jerky, maple syrup, squash, and tomatoes.*

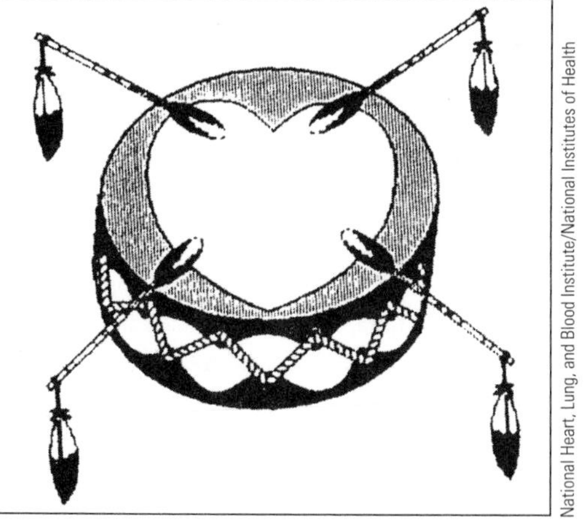

National Heart, Lung, and Blood Institute/National Institutes of Health

Figure 5.1

The logo for the Building Healthy Hearts for American Indians and Alaska Natives Initiative depicts the drum as a focal point in tribal traditional life. The drum is a living thing to be treated with deep respect, and the act of drumming is connected to the heartbeat. The shape of the drum represents the circle of life, reminding us that we are all linked with one another. The drumsticks symbolize the four directions, which in turn are associated with the four stages of life: east, the newborn; south, the adolescent; west, the adult; and north, the elder.

During the early nineteenth century, the Cherokee had a written language, a bilingual newspaper, a school system, a court system, and a Cherokee Nation constitution.[115] They were a prosperous tribe; many owned black slaves.

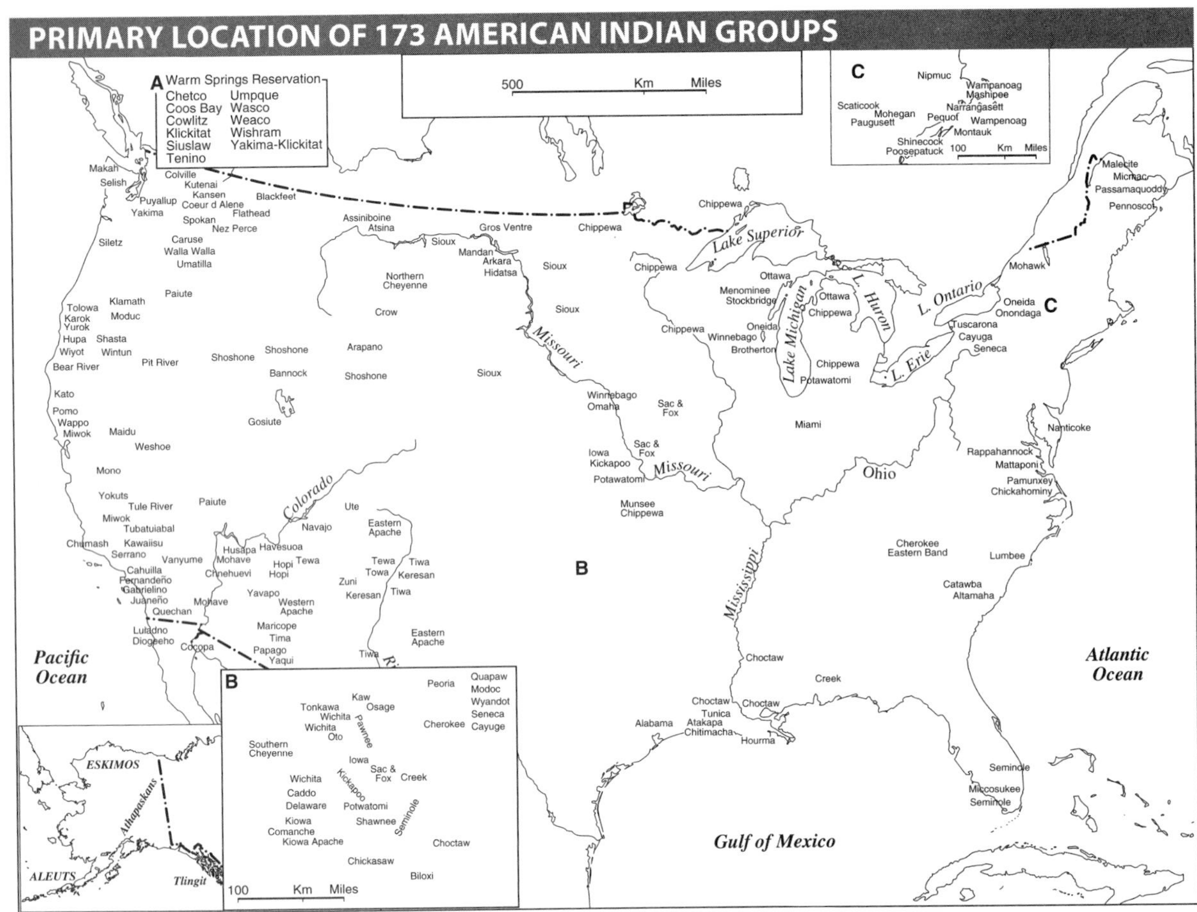

Figure 5.2
Native American nations in the United States.
Source: Adapted and reprinted by permission of the publisher from *Harvard Encyclopedia of American Ethnic Groups* edited by Stephan Thernstrom, Ann Orlov, and Oscar Handlin, p. 61, Cambridge, Mass.; The Belknap Press of Harvard University Press, Copyright © 1980 by the President and Fellows of Harvard College.

(Figure 5.2). More recent statistics indicate that 60% live in metropolitan areas.[2,3] Among the states, California has the highest number of resident American Indians (fewer than 700, 000), followed by Oklahoma and Arizona. In 2009, the largest ethnic groups within the states of Alaska, Montana, North Dakota, Oklahoma, and South Dakota were Alaska natives and Native Americans/American Indians.[2]

Large, urban Native American populations are found in Los Angeles, New York City, Phoenix, Tulsa, Anchorage, Oklahoma City, Albuquerque, and Tucson. The remainder of the American Indian and Alaska Native population resides in rural areas, including reservations where approximately 1.9 million individuals reside. Many first-generation urban Native Americans maintain close ties with the reservation of their ethnic group and travel often between the city and tribal land. Members of the second generation living in urban regions are more likely to think of the city as their permanent home.

Socioeconomic Status

The socioeconomic status of Native Americans declined drastically with the forced migrations of the nineteenth century. Even those Native American nations that were agriculturally self-sufficient suffered when relocated to regions with poor growing conditions. Further, there were few native occupations that were valued in the job market outside the reservations. BIA education efforts were generally unsuccessful, and most Native Americans did not begin to find employment until World War II and the development of the BIA Employment Assistance Program. The Indian Self-Determination and Education Act of 1975 was enacted to promote Native American participation in government and education but economic improvement continues to be slow. Approximately 26 percent of American Indians and Alaska Natives work in management and professional occupations.[3] The Apaches and Dakota have been active in ranching and are known for their expertise in the rodeo circuit.[4,6] In the

Southwest, small-scale agriculture and livestock grazing are still important among some Hopi, Pueblo, and Navajo peoples; and traditional crafts such as weaving, pottery, and silversmithing are significant.[7,8,9] Some Alaska Natives combine part-time paid employment with subsistence living and often find jobs in the fishing and forestry industries.[10–12]

The overall poverty rate for Native Americans (nearly 32 percent) in 2008 was more than double that of the general population; however, significant tribal differences are seen. The median family income for American Indian and Alaska Natives in 2008 was $33,627. Most Alaska Natives have rates somewhat higher than the U.S. average (15 to 23 percent), while poverty levels among the Apache, Navajo, and Sioux approached or exceeded three times the national rates (34 to 39 percent). Rates of formal education have been improving, yet 24 percent of all adult Native Americans have not finished high school or obtained an equivalency diploma. Fourteen percent of American Indians and Alaska Natives at least a bachelor's degree and approximately 50,500 American Indians and Alaska Natives have at least an advanced graduate degree.[3]

Native American Organizations

Few Native American neighborhoods develop in urban areas. Native Americans who settle in the cities usually arrive as individuals or small family groups and typically do not live near others of their nation. Sometimes longtime city residents exhibit a sense of superiority over recent arrivals from the reservations. In general, the difficulties of adjusting to urban white society stimulate many young Native Americans to identify not only with Native Americans of other nations but also with people from other ethnic groups.

Activists for Native American rights often come from the cities and are not always supported by Native Americans who live on tribal lands. Native American organizations have done much to maintain Indian identity. Most areas with large Native American populations have their own clubs and service associations. Organizations to promote ethnic identity have been founded by the Navajo, Pueblos, Tlingit, Haida, and Pomo. Other groups such as athletic clubs and dance groups serve the social needs of the pan–Native American community.

Worldview

Harmony best describes the Native American approach to life. Each individual strives to maintain a balance among spiritual, social, and physical needs in a holistic approach. Only what is necessary for life is taken from the natural environment; the belief is that the Earth should be cared for and treated with respect. Generosity is esteemed and competitiveness is discouraged, yet individual rights are also highly regarded. Personal autonomy is protected through the principle of noninterference. Among the Navajo, for example, an individual would never presume to speak for another, even a close family member. For most Native Americans, time is conceptualized as being without beginning or end, and the culture is present-oriented, meaning that the needs of the moment are emphasized over the possible rewards of the future.

Religion

Traditional Native American religions vary from an uncomplicated belief in the power of a self-declared evangelist to elaborate theological systems with organized hierarchies of priests. Yet they all share one characteristic: The religion permeates all aspects of life. Rather than a separate set of beliefs practiced at certain times in specific settings, religion is an integral part of the Native American holistic worldview. Religious concepts influence both the physical and emotional well-being of the individual.

Efforts to increase prosperity on reservations include utilization of natural resources and establishment of gambling operations, which are legal on tribal lands.

Many Native American nations have rejected all attempts at Christian conversion, especially in the Southwest. The Navajo, Arizona Hopi, Rio Grande Pueblos, Potawatomi, Lakota, and Dakota have retained most of their native religious values and rituals, such as sweat lodge purification rites. Other religions unique to Native Americans emerged after European contact, such as the Drum Dance cult and the Medicine Bundle religions, which combine spiritual elements from several different ethnic groups. A Paiute visionary, Wovoka, founded the Ghost Dance religion in the late 1880s, which prophesied an end to white domination through prayer, abstinence from alcohol, and ritual dancing.[13] In addition, religions mixing Christianity with traditional beliefs have been popular since the late nineteenth century; the Native American Church has been especially successful. Other groups claim Christian

© Lionel Delevingne/Stock, Boston, Inc.

▲ *The primary social unit of Native Americans is the extended family, which includes all close and distant kin.*

fellowship but continue to practice native religions as well. Finally, many Native Americans now adhere to Roman Catholicism or some form of Protestantism, especially in urban areas where churches have been established to serve all Native American congregations. In Alaska, some Alaska Natives have become adherents of Russian Eastern Orthodoxy.

Family

The primary social unit of Native Americans is the extended family. Children are valued highly, and there is great respect for elders. All blood kin of all generations are considered equal; there is no differentiation between close and distant relatives. Aunts and uncles are often considered like grandparents, and cousins are viewed as brothers or sisters. Even other tribal members are sometimes accepted as close kin. In many Native American societies an individual without relatives is considered poor.

Many Native American nations are matrilineal, meaning that lineage is inherited from the mother. Traditionally, property was often passed down through women in these tribes, and decision-making often rested with an elder woman in the family.[14,15] Today, even in matrilineal systems, the men are the family providers and heads of the household; women are typically in charge of domestic matters. Due to the respect for the individual within most Native American groups, men and women hold equal standing. Native American children are expected to assist their parents in running the home.

Traditional Health Beliefs and Practices

In Native American culture, health reflects a person's relationship to nature, broadly defined as the family, the community, and the environment. Every illness is due to an imbalance with supernatural, spiritual, or social implications. Treatment focuses on the cause of the imbalance, not the symptoms, and is holistic in approach. The sick individual is at odds with the universe, and community and family support is focused on restoring harmony, not curing the disease.[16]As explained by the Cherokee medicine man, Sequoyah, "Indian medicine is a guide to health, rather than a treatment. The choice of being well instead of being ill is not taken away from an Indian."[17] Traditional Native American medicine is concerned with physical, mental, and spiritual renewal through health maintenance, prevention of illness, and restoration of health.

Many causes account for illness. Some Navajos believe that witchcraft, through agents such as animals, lightning, and whirlwinds; transgressions committed at ceremonial occasions; or evil spirits (especially ghosts) may cause fainting, hysteria, or other conditions. Witchcraft may also take the form of intrusive objects, causing pain where the object is inserted, as well as emaciation. Possession by a spirit may dislodge the soul, resulting in a feeling of suffocation (or possession may be a sign of a gift for healing). Soul loss may also cause mental disorders. Violation of a taboo, whether an actual breach by an individual or contact with evil objects that have committed mythical breaches, results in general seizures.[18,19] Traditionally, these beliefs are shared by many other Native Americans as well. For example, some Iroquois believe in a similar list of reasons for illness, adding that unfulfilled dreams or desires may also be a contributing factor,[20] and some Inuit believe sleep paralysis occurs when the soul is attacked by malevolent spirits or through witchcraft.[21]

Some Native Americans reject the concept that poor nutrition, bodily malfunctions, or an infection by a virus or bacteria can cause sickness. An evil external source is often identified instead. Some Dakotas, for instance, blame type 2 diabetes on disease-transmitting foods provided by whites with the intention of eliminating all Native Americans.[22] An outbreak of serious respiratory infections due to the Hanta virus was explained by Navajo healers as being due to rejection of traditional ways and adoption of convenience foods.[23] Some Native Americans attribute alcoholism to soul loss and the cultural changes due to domination by white society.[15]

Small bags of herbs (called "medicine bundles" by certain Plains Indians), fetishes, feathers, or symbols may be worn to protect against malevolent forces. Fetishes are used when an animal that has been harmed or killed causes an ailment; a fetish in the form of the animal is rubbed on the afflicted body part with appropriate chants.

Traditional healers often specialize in their practice. Navajo medicine men and women usually exert a positive influence in preventing disharmony through rituals such as the sweat bath to promote peace. They also have negative powers, which can be used to counteract witchcraft or evil acts by a person's enemies. Diagnosticians may be called on to identify the cause of an illness through stargazing or listening (if crying is heard, the patient will die). Hand motions or trembling also may be involved, sometimes including painting with white-, blue-, yellow-, and black-colored sand to produce a picture of magical healing power. Other traditional Navajo practitioners are singers, who cure with sacred chanting ceremonies, and healers, who have specific responsibility for care of the soul. Among the Oneida, dreamers have the ability to see the future and diagnose illness.[24] In many Native American groups, herbalists, often women, assist in the treatment of illness through the ceremonial collection and application of wild plant remedies.[25] (See "Therapeutic Uses of Foods" later in this chapter for examples.)

Among California Native Americans, illness was treated first with home remedies. If that proved ineffective, non-sacred healers such as herbalists or masseuses were contacted. If the patient still did not improve, a diviner would be consulted for a diagnosis. If spiritual or supernatural intervention was needed, a shaman (medicine man) was employed. Consultation with native healers is often concurrent with seeking Westernized health care.[15,26,27]

Traditional Food Habits

The traditional food habits of Native Americans were influenced primarily by geography and climate. Each Native American nation adopted a way of life that allowed it to maximize indigenous resources. Many were agriculturally based societies, others were predominantly hunters and gatherers, and some survived mainly on fish. Most of each day was spent procuring food.

Ingredients and Common Foods

Indigenous Foods

Archaeological records and descriptions of America by European settlers indicate that Native Americans on the East Coast enjoyed an abundance of food. Fruits, including blueberries, cranberries, currants, grapes, persimmons, plums, and strawberries, as well as vegetables, such as beans, corn, and pumpkins, are mentioned by the New England colonists. They describe rivers so full of life that fish could be caught with frying pans, sturgeon so large they were called "Albany beef," and lobster so plentiful that they would pile up along the shoreline after a storm. Game included deer, moose, partridge, pigeon, rabbit, raccoon, squirrel, and turkey. Maple syrup was used to sweeten foods. Farther south, Native Americans cultivated groundnuts (*Apios americana*, or Indian potatoes) and tomatoes, and collected wild Jerusalem artichokes (a starchy tuber related to the sunflower). Native Americans of the Pacific Northwest collected enough food, such as salmon and fruit, during the summer to support them for the rest of the year. Peoples of the plains hunted buffalo, and those of the northeastern woodlands gathered wild rice; nations of the Southwest cultivated chili peppers and squash amid their corn (see Table 5.1).

Native Americans not only introduced whites to indigenous foods but shared their methods of cultivation and food preparation as well. Legend has it that the Pilgrims nearly starved despite the plentiful food supply because they were unfamiliar with the local foods. One version of the tale is that Squanto, the sole surviving member of the Pautuxets (the other members had succumbed to smallpox following contact with earlier European explorers), saved the Pilgrims, who were mostly merchants, by teaching them to grow corn. He showed them the Native American method of planting corn kernels in mounds with a fish head for fertilizer and using the corn stalks as supports for beans.

Foods Introduced from Europe

Foods introduced by the Europeans, especially the French Jesuits in the North and the Spanish in the South, were well accepted by Native Americans. Apples, apricots, carrots, lentils, peaches, purslane, and turnips were some of the more successful new foods. Settler William Penn noted that he found peaches in every large Native American farm he encountered barely one hundred years after they had been introduced to the Iroquois. The Europeans also brought rye and wheat. However, few Native American nations replaced corn with these new grains.

Table 5.1 Indigenous Foods of the Americas*

Fruits	Berries (blackberries, blueberries, cranberries, gooseberries, huckleberries, loganberries, raspberries, strawberries), cactus fruit (*tuna*), cherimoya, cherries (acerola cherries, chokecherries, ground-cherries), grapes (e.g., Concord), guava, mamey, papaya, passion fruit (*granadilla*), pawpaw, persimmon (American), pineapple, plums (American, beach), soursop (*guanabana*), zapote (*sapodilla*)
Vegetables	Avocado, bell peppers (sweet peppers, pimento), cactus (*nopales, nopalitos*), chayote (*christophine, chocho, huisquil, mirliton,* vegetable pear), pumpkins, squash, tomatillo, tomatoes
Tubers/roots	Arrowroot, cassava (*yuca, manioc, tapioca*), groundnut, Indian breadroot, Jerusalem artichoke, jicama, malanga (*yautia*), potatoes, sweet potatoes
Grains/cereals	Amaranth, corn (maize), quinoa, wild rice
Nuts/seeds	Brazil nuts, cashews, hickory nuts, pecans, pumpkin seeds (*pepitas*), sunflower seeds, walnuts (black)
Legumes	Beans (green beans, most dried beans), peanuts
Poultry	Turkey
Seasonings/flavorings	Allspice, chile peppers (e.g., hot and sweet), chocolate (cocoa), maple syrup, sassafras (*filé* powder), spicebush, vanilla

* Foods native to North, Central, or South America. Some items not indigenous to the United States (e.g., pineapple, potatoes) were popularized only after acceptance in Europe and introduction by European settlers. Other foods (e.g., avocado, jicama, tomatillo) have become more common in the United States with the growing Latino population.

Livestock made a much greater impact on Native American life than did the new fruits and vegetables. Cattle, hogs, and sheep reduced the Native Americans' dependence on game meats. The Creeks and Cherokees of the Southeast fed their cattle on corn and fattened their suckling pigs and young lambs on apples and nuts. The Powhatans of Virginia fed their hogs peanuts, then cured the meat over hickory smoke. Lamb and mutton became staples in the Navajo diet after the introduction of sheep by the Spanish. In addition, the Europeans brought horses and firearms, which made hunting easier, and metal knives and iron pots, which simplified food preparation. They also introduced the Native Americans to distilled spirits.

Staples

The great diversity of Native American cultures has resulted in a broad variety of cuisines. The cooking of one region was as different from that of another as French food is from German food today. Native American cooking featured local ingredients and often reflected the need to preserve foods for future shortages. The only staple foods common to many, though not all, Native American nations were beans, corn, and squash. The cultural food groups are listed in Table 5.2.

Regional Variations

Native American fare has been divided by regions into five major types: northeastern, southern, plains, southwestern, and Pacific Northwest/Alaska Native. Although each area encompasses many different Native American nations, they share similarities in foods and food habits.

Northeastern. The northeastern region of the United States was heavily wooded, with numerous freshwater lakes and a long Atlantic coastline. It provided the local Native Americans, including the Iroquois and Powhatan, with abundant indigenous fruits, vegetables, fish, and game. Most nations also cultivated crops such as beans, corn, and squash. Many of the foods associated with the cooking of New England have their origins in northeastern Native American recipes. The clambake was created when the Narragansett and the Penobscot steamed their clams in beach pits lined with hot rocks and seaweed. Dried beans were simmered for days with maple syrup (the precursor of Boston baked beans). The dish that today is called succotash comes from a stew common in the diet of most Native Americans; it combined corn, beans, and fish or game. In the Northeast it was usually flavored with maple syrup. Clam chowder, codfish

Table 5.2 Cultural Food Groups

Group	Comments	Common Foods	Adaptations in the United States
Protein Foods			
Milk/milk products	High incidence of lactose intolerance among Native Americans with a high percentage of Native American heritage.	No common milk products in traditional diets.	Powdered milk and evaporated milk are typical commodity products, usually added to coffee, cereal, and traditional baked goods; ice cream is popular with some groups. Some reports have been made of frequent milk consumption.
Meat/poultry/fish/eggs/legumes	Meat is highly valued, considered healthful. Meats are mostly grilled or stewed, preserved through drying and smoking. Beans are an important protein source.	*Meat:* bear, buffalo (including jerky, *pemmican*), deer, elk, moose, opossum, otter, porcupine, rabbit, raccoon, squirrel. *Poultry and small birds:* duck, goose, lark, pheasant, quail, seagull, wild turkey. *Fish, seafood, and marine mammals:* abalone, bass, catfish, clams, cod, crab, eel, flounder, frogs, halibut, herring, lobster, mussels, olechan, oysters, perch, red snapper, salmon, seal, shad, shrimp, smelts, sole, sturgeon, trout, turtle, walrus, whale. *Eggs:* bird, fish. *Legumes:* many varieties of the common bean (kidney, navy, pinto, etc.), *tepary* beans.	Beef is well accepted; lamb and pork are also popular. Canned and cured meats (bacon, luncheon meat) may be common if income is limited. Game is rarely eaten. Meats remain a favorite food. Chicken eggs are commonly eaten.
Cereals/Grains	Corn is primary grain; wild rice is available in some areas.	Cornmeal breads (baked, steamed), hominy, gruels, corn tortillas, *piki*, toasted corn; wild rice.	Wheat has widely replaced corn; store-bought or commodity breads and sugared cereals are common. Cakes, cookies, pastries are popular.
Fruits/Vegetables	Indigenous plants are major source of calories in diet of some Native American nations. Fruits and vegetables are either gathered or cultivated; fruit is a popular snack food.	*Fruit:* blackberries, blueberries, buffalo berries, cactus fruit (*tuna*), chokeberries, cherries, crab apples, cranberries, currants, elderberries, grapes, groundcherries, huckleberries, persimmons, plums, raspberries, salal, salmonberries, strawberries (beach and wild), thimbleberries, wild rhubarb. *Vegetables:* camass root, cacti (*nopales*), chile peppers, fiddleheads, groundnuts, Indian breadroot, Jerusalem artichokes, lichen, moss, mushrooms, nettles, onions, potatoes, pumpkin, squash, squash blossoms, sweet potatoes, tomatoes, wild greens (cattail, clover, cow parsnip, creases, dandelion, ferns, milkweed, pigweed, pokeweed, saxifrage, sunflower leaves, watercress, winter cress), wild turnips, *yuca* (cassava).	Apples became common after European introduction. Apples, bananas, oranges, peaches, pineapple have been well accepted; canned fruits are popular. Wild berries are still gathered in rural areas. Some traditional vegetables are eaten when available. Green peas, string beans, instant potatoes are common commodity items. Intake of vegetables is low; variety is limited. Potato chips and corn chips often are popular as snacks.
Additional Foods			
Seasonings		Chiles, garlic, hickory nut cream, onions, peppermint, sage, salt, sassafras, seaweed, spearmint, and other indigenous herbs and spices.	
Nuts/seeds	Nuts and seeds are often an important food source; acorns are sometimes a staple.	Acorn meal, black walnuts, buckeyes, chestnuts, hazelnuts, hickory nuts, mesquite tree beans, pecans, peanuts, *piñon* nuts (pine nuts), pumpkin seeds, squash seeds, sunflower seeds, seeds of wild grasses.	
Beverages	Herbal teas often consumed for enjoyment, therapeutic, or spiritual value.	Teas of buffalo berries, mint, peyote, rose hip, sassafras, spicebush, sumac berries, *yerba buena;* honey and water.	Coffee, tea, soft drinks are common beverages. Alcoholism is prevalent.
Fats/oils	Traditional diets vary in fat content, from extremely low in the mostly vegetarian cooking of California and Nevada Indians to very high in the primarily animal-based fare of Native Alaskans.	Fats rendered from buffalo, caribou, moose, and other land mammals; seal and whale fat.	Butter, lard, margarine, vegetable oils have replaced rendered fats in most regions; seal and whale fat are still consumed by Inuits and Aleuts.
Sweeteners	Consumption of sweets is low in traditional diets.	Maple syrup, other tree saps, honey.	Sugar is primary sweetener; candy, cookies, jams, and jellies are popular.

A Native American hotel chef, George Crum, is attributed with the invention of potato chips in 1853. Today Americans consume an average of seventeen pounds of potato chips per person each year.

balls, brown bread, corn pudding, pumpkin pie, and the dessert known as Indian pudding are all variations of northeastern Native American recipes. In addition to clams, the Native Americans of the region ate lobster, oysters, mussels, eels, and many kinds of salt- and freshwater fish.

Game, such as deer and rabbit, was eaten when available. Wild ducks, geese, and turkeys were roasted with stuffings featuring crab apples, grapes, cranberries, or local mushrooms. Corn, as the staple food, was prepared in many ways, such as roasting the young ears; cooking the kernels or meal in soups, gruels, and breads; steaming it in puddings; or preparing it as popcorn. Pumpkins and squash were baked almost daily, and beans were added to soups and stews. Local green leafy vegetables were served fresh. Sweets included cherries stewed with maple syrup, cranberry pudding, crab apple sauce, and hazelnut cakes.

Southern. The great variety of foods found in the northeastern region of the United States was matched by the plentiful fauna and lush flora of the South. Oysters, shrimp, and blue crabs washed up on the warm Atlantic beaches during tropical storms. The woodlands and swamplands teemed with fish, fowl, and game, including bear, deer, raccoon, and turtle, as well as ample fresh fruit, vegetables, and nuts. The Native Americans of this region, such as the Cherokee, Creek, and Seminole, were accomplished farmers, growing crops of beans, corn, and squash.

When Africans were first brought as slaves to America, they were often housed at the periphery of farms. Initially a great deal of interaction took place between blacks and local Native Americans, who taught them how to hunt the native game without guns and to use the indigenous plants. Some of the Native American cooking techniques were later introduced into white southern cuisine by African-American cooks, and many of the flavors typical of modern southern cooking come from traditional Native American foods. Hominy (dried corn kernels with the hulls removed) and grits (made of coarsely ground hominy) were introduced to the settlers by the Native Americans. The chicken dish known as Brunswick stew is an adaptation of a southern Native American recipe for squirrel. The Native Americans also made sophisticated use of native plants for seasoning, and they thickened their soups and stews with sassafras.

The staple foods of corn, beans, and squash were supplemented with the indigenous woodland fruits and vegetables. Blackberries, gooseberries, huckleberries, raspberries, strawberries, crab apples, grapes, groundcherries, Jerusalem artichokes, leafy green vegetables, persimmons (pounded into a paste for puddings and cakes), and plums were some of the numerous edible native plants. Tomatoes and watermelons were added after introduction by the Spanish into Florida. The Native Americans of the South also used beechnuts, hazelnuts, hickory nuts, pecans, and black walnuts in their cooking. The thick, creamlike oil extracted from hickory nuts was used to flavor corn puddings and gruels, and a traditional Cherokee specialty was kanuche, a soup made from the nuts (often with the addition of corn, hominy, or rice), is still popular today. Honey was the sweetener used most frequently, and it was mixed with water for a cooling drink. Teas were made from mint, sassafras, or spicebush (Lindera benzoin), and during the summer "lemonade" was made from citrus-flavored sumac berries.

[SAMPLE MENU]

A Traditional Northeastern Indian Meal

Iroquois Soup[a,b] or Duck with Wild Rice[a]

Bannocks[a,c]

Indian Pudding[a,b] or Maple Popcorn Balls[a,c]

Strawberry Juice

[a]Cox, B., & Jacobs, M. 1991. *Spirit of the harvest: North American Indian cooking.* New York: Stewart, Tabori & Chang.

[b]*RecipeSource* at http://www.recipesource.com/ethnic/americas/native/

[c]*Recipe Goldmine* at http://www.recipegoldmine.com/worldnativeam/nativeam.html

Plains. The Native Americans who lived in the area that is now the American Midwest were mostly nomadic hunters, following the great herds of bison across the flat plains for sustenance. The land was rugged and generally unsuitable for agriculture. Those nations that settled along the fertile Mississippi and Missouri River valleys, however, developed farm-based societies supported by crops of beans, corn, and squash.

Bison meat was the staple food for most plains nations such as the Arapaho, Cheyenne, Crow, Dakota, and Pawnee. The more tender cuts were roasted or broiled, while the tougher ribs, joints, and other bones with marrow were prepared in stews and soups. Pieces of meat, water, and sometimes vegetables would be placed in a hole in the ground lined with cleaned buffalo skin. The stew would then be stone boiled: rocks that had been heated in the fire would be added to the broth until the mixture was thoroughly cooked. All parts of the bison were eaten, including the liver and kidneys (which were consumed raw immediately after the animal was slaughtered), udder, tongue, and hump. Extra meat was preserved by cutting it into very thin strips and then dehydrating it in the sun or over the fire. This tough, dried meat would keep for several years and was known as jerked buffalo or jerky. The jerky would be pulverized and mixed with water or corn gruel, or, in emergencies, eaten dry. Most often it was shredded and mixed with bison fat and berries, then formed into cakes called pemmican.

When bison were unavailable, the plains nations would hunt deer, rabbit, and game birds. Fresh leafy green vegetables were consumed in season, and root vegetables, such as wild onions, prairie turnips (*Psoralea esculenta,* also called breadroot, tipsin, and timpsila), and Jerusalem artichokes, were eaten throughout the year. Wild rice, a native aquatic grass with an earthy, nutty flavor, was collected in the northern parts of the Midwest. It was served with bison, venison, or duck and used as a stuffing for grouse, partridge, and duck. Wild rice is believed to have traditionally provided as much as 25 percent of the total Ojibwa diet, and it was customarily prepared with maple syrup.[28] Blackberries, shadberries (also called Juneberries or saskatoon berries), cherries, crab apples, grapes, persimmons, and plums were available in some areas, but the most popular fruit was the scarlet buffalo berry (*Shepherdia canadensis*), so called because it was often served in sauces for bison meat or dried for pemmican. In addition, berries were traditionally boiled with bison suet and/or blood to make a thick pudding called wojapi.

Southwestern. Some of the oldest Native American settlements in North America were located along the river valleys of the arid Southwest. Despite the semi-desert conditions, many Native Americans such as the Hopi, Pima, Pueblo, and Zuni lived in pueblo (Spanish for "town" or "village") communities and were mostly farmers, cultivating beans, chili peppers, corn, and squash. Others, including the Apache and Navajo, were originally roving hunters and gatherers. After the Spanish introduced livestock, some of these nomadic groups began to raise sheep. Mutton has since become associated as a traditional staple food of the region.

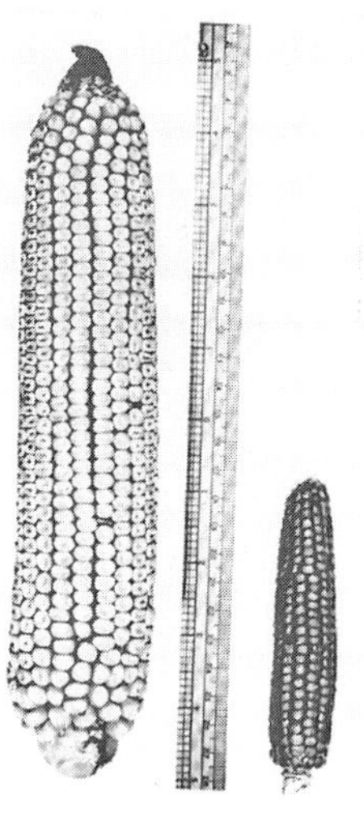

▲ ***Modern Indian corn compared to Indian corn circa 500 C.E.***

From A Pictorial History of the American Indian Copyright © 1956 by Oliver La Farge Reprinted by permission of Frances Collin, Literary Agent

[SAMPLE MENU]

A Traditional Southeastern Indian Meal

Cherokee Pepper Pot Soup[a, b] or Brunswick Stew[a]

Corn Pone[a, b, c]

Huckleberry Honey Cake[a, c, d] or Grape Dumplings[a, c]

Honey Water

[a]Cox, B., & Jacobs, M. 1991. *Spirit of the harvest: North American Indian cooking.* New York: Stewart, Tabori & Chang.

[b]*RecipeSource* at http://www.recipesource.com/ethnic/americas/native/

[c]*Cherokee North Carolina, Recipes* at http://www.cherokee-nc.com/recipes_main.php

[d]*Recipe Goldmine* at http://www.recipegoldmine.com/worldnativeam/nativeam.html

Until the arrival of livestock, the diet of the region was predominantly plant based, providing a nourishing diet when supplemented with small game such as rabbit and turkey. Corn was the primary food, and at least five different colors of corn were cultivated. Each color symbolized one of the cardinal points for the Zuni, and each had its own use in cooking. White corn (East) was ground into a fine meal and used in gruels and breads. Yellow corn (North) was roasted and eaten in kernel form or off the ear. The rarer red (South), blue (West), and black (the nadir, the lowest point beneath the observer) corn was used mostly for special dishes, such as the lacy flat Hopi bread made from blue cornmeal, known as piki. Multicolored corn represented the zenith. The Hopi also attached importance to the color of corn and cultivated twenty different varieties. In many areas, corn was prepared in ways similar to those of the northern Mexican Native Americans—tortillas (flat, griddle-fried cornmeal bread), pozole (hominy), and the tamale-like chukuviki (stuffed cornmeal dough packets). Juniper ash (considered a good source of calcium and iron) was often added to cornmeal dishes for flavoring.[29]

Beans were the second most important crop of the southwestern region. Many varieties were grown, including the domesticated indigenous tepary beans and pinto beans from Mexico. Both squash and pumpkins were commonly consumed, and squash blossoms were fried or added to soups and salads. Squash and pumpkin seeds were also used to flavor dishes, and chile peppers were used as vegetables and to season stews. Cantaloupes (also known as muskmelons) were also grown after they were introduced by the Spanish.

When crops were insufficient, the southwestern Native Americans relied on wild plants, and to add variety, tender amaranth greens were eaten in summer. Piñon seeds (also called pine nuts) flavored stews and soups. Both the fruit (tunas) and the pads (nopales) of the prickly pear cactus were eaten, as were the pulp and fruit of other succulents, such as yucca (the starchy fruit known today as "Navajo bananas"). A unique food popular with some Apaches was the root of the mescal plant, another desert succulent. It would be baked for hours in a covered, stone-heated pit until it developed a soft, sticky texture and a flavor similar to molasses.[5] The beans of the mesquite tree were a staple in some desert regions; they were ground into a flour and used in gruels, breads, and sun-baked cakes.

[SAMPLE MENU]

A Traditional Great Plains Indian Meal

Pemmican[a, b, c]

Stuffed Pumpkin[a] or Buffalo/Bison Stew[b]

Broiled Jerusalem Artichokes[a]

Chokecherry Pudding[a,b] or *Wojapi* (pudding)[b]

Peppermint Tea[b]

[a]Cox, B., & Jacobs, M. 1991. *Spirit of the harvest: North American Indian cooking.* New York: Stewart, Tabori & Chang.

[b]*Native Tech: Indigenous Foods and Traditional Recipes* at http://www.nativetech.org/recipes/index.php

[c]*Pemmican: Recipes, Stories and Stores* at http://w4.lns.cornell.edu/~seb/pemmican.html

Northwest Coast/Alaska Natives. This culinary region incorporates a diverse geographic area. The climate of the Pacific Northwest coast is temperate. The luxuriantly forested hills and mountain slopes abound with edible plants and game, and the sea supplies fish, shellfish, and marine mammals. Farther north, in Alaska and Canada, the growing season shortens to only a few summer months, and temperatures in the winter regularly plunge to minus fifty degrees Fahrenheit. Two-thirds of Alaska is affected by permafrost, and the vast stretches of tundra are inhospitable to humans.

The Native American ethnic groups inhabiting this region include Indians as well as Inuits (Eskimos) and Aleuts, known as Alaska Natives. Native American nations such as the Tlingit and Kwakiutl inhabit the northwest coastal area and some interior Alaskan regions. The Aleuts live on the thousand-mile-long chain of volcanic islands that arch into the Pacific from Alaska called the

Aleutians. The Inuits, including the Yupik and Inupiat, live in the northern and western areas of Alaska, as well as in Canada, Greenland, and Siberia.

The Native Americans of the Northwest Coast had no need for agriculture. Food was plentiful, and salmon was their staple. The fish were caught annually in the summer as they swam upstream to spawn. They were roasted over the fire when fresh, and the eggs, known today as red caviar, were a favorite treat when dried in the sun into chewy strips. Extra fish were smoked to preserve them for the winter. In addition, cod, clams, crabs, halibut, herring, shrimp, sole, smelt, sturgeon, and trout were consumed. Ocean mammals such as otter, seal, and whale were also hunted. Bear, deer, elk, and mountain goats were eaten, as were numerous wild fowl and game birds.

Despite the abundant fish and game, wild plants made up more than half the diet of the Northwest Coast Native Americans. More than 100 varieties of indigenous fruits, vegetables, and even lichen were consumed, including acorns, blackberries, blueberries, chokecherries (*Prunus virginiana*), desert parsley, hazelnuts, huckleberries, mint, raspberries, salal (*Gaultheria shallon*), and strawberries. Camass roots (*Camassia quamash*), a bulb related to the hyacinth, were roasted or dried by many Indians of the region. Fresh greens were also popular.

In contrast to the plenty of the Northwest Coast, the diet of many Alaska Natives was often marginal. The Inuits and Aleuts were usually seminomadic, traveling as necessary to fish and hunt. Fish and sea mammals, such as seal, walrus, and whale, were the staple foods. Arctic hare, caribou, ducks, geese, mountain goats, moose, musk oxen (hunted to extinction in Alaska by the 1870s), polar bear, and mountain sheep were consumed when available. Some items were boiled, but many were eaten raw due to the lack of wood or other fuel. The fat of animals was especially valued as food. Muktuk (also called muntak or onattak), still a commonly consumed item, consists of chunks of meat with the layer of fat and skin attached. Muktuk is typically frozen before use. Walrus or whale muktuk can also be preserved by rolling it in herbs (with no salt) and fermenting it in a pit for several months to make a treat known as kopalchen. Akutok, a favored dish, was a mixture of seal oil, berries, and caribou fat. Even the stomachs of certain game were examined for edible undigested foods, such as lichen in

[SAMPLE MENU]

A Traditional Southwestern Meal

Green Chili Stew [a, b, c, d]

Blue Corn Bread[b, c, d] **or Frybread**[a, b, c, d]

Pueblo Piñon/Feast Day Cookies[a, d] **or Navajo Peach Crisp**[a, b]

[a]Cox, B., & Jacobs, M. 1991. *Spirit of the harvest: North American Indian cooking.* New York: Stewart, Tabori & Chang.

[b]Keegan, M. 1996. *Southwest Indian cookbook.* Santa Fe, NM: Clear Light Publishers.

[c]*Traditional Native American Recipes from The Cooking Post* at http://www.cookingpost.com/recipe.cfm

[d]*Recipe Goldmine* at http://www.recipegoldmine.com/worldnativeam/nativeam.html

© Bettmann/CORBIS

▲ ***Inuit women preparing a dead seal for butchering.***

[SAMPLE MENU]

A Northwest Coast Indian Meal

Barbecued Clams[a] or Fresh Salmon[c]

Elk Stew with Acorn Dumplings[a, b]

Steamed Fiddlehead Ferns[a, b]

Whipped Raspberry (Soup)[a, b]

[a]Cox, B., & Jacobs, M. 1991. *Spirit of the harvest: North American Indian cooking.* New York: Stewart, Tabori & Chang.

[b]*Astray Recipes* at http://www.astray.com/recipes/

[c]*Kwakuitl Recipes* at http://www.hallman.org/indian/recipe.html

elk and clams in walruses. The limited selection of wild plants included willow shrubs, seaweed, mosses, lichen, a few blueberries, salmonberries, and cranberries. Leaves from an aromatic bush known as bog shrub (*Ledum palustre*) were brewed to make tundra tea (also called Hudson Bay tea), a beverage still popular today.

Other Native American Cuisines. Many traditional Native American diets do not fit conveniently within the five major regional cuisines. Among them was the fare of the population found in what is now Nevada and parts of California, called Digger Indians by the first whites to encounter them because they subsisted mostly on dug-up roots, such as Indian breadroot, supplemented by small game and insects. In central California, numerous nations, such as the Miwoks and Pomos, had an acorn-based diet. Acorns contain tannic acid, a bitter-tasting substance that is toxic in large quantities. To make the acorns edible, Native American women would first crack and remove the hard hull, grind the meat into a meal, add water to make a dough, and then leach the tannic acid from the dough by repeatedly pouring hot water through it. Acorns were sometimes leached in sandy-bottomed streams as well.

In the rugged northern mountains and plains lived nations such as the Blackfeet, Crows, Shoshones, and Dakotas, who were nomadic hunters of game. Although many may have hunted bison at one time, they were limited to the local bear, deer, moose, rabbits, wildfowl, and freshwater fish when the expansion of other Native Americans and whites into the Midwest pushed them northward and westward. Wild plants added variety to their diet. For example, the Nez Percés baked camass roots in a covered pit with heated rocks, which caramelized the starch providing a sweet, onion-like flavor. The cooked roots were made into gruel or dough for bread.[30]

Meal Composition and Cycle

Daily Patterns

Traditional meal patterns varied according to ethnic group and locality. In the Northeast, one large, hearty meal was consumed before noon, and snacks, such as soup, were available throughout the day. In some tribes, no specific meal time was standard. The men were served first, and stood or sat while they consumed the meal in silence. Women and children ate next.

Serving two meals per day was more common in the Southwest. The women would rise before dawn to prepare breakfast, eaten at sunrise. The afternoon was spent cooking the evening meal, which was eaten before sunset. Two meals per day was also the pattern among the Native Americans of the Pacific Northwest.

In regions with limited resources, meals were often monotonous. The two daily meals of the southwestern Native Americans, for example, regularly consisted of cornmeal gruel or bread and boiled dehydrated vegetables. No distinction was made between morning or evening menus. Other dishes such as game, fresh vegetables, or fruit were included when seasonally available. The single meal of the northeastern Native Americans often included roasted game; the Northwest Coast Native Americans frequently included some form of salmon twice a day, in addition to the many local edible greens and roots.

Food was simply prepared. It was roasted over the fire or in the ashes or cooked in soups or stews.

The northeastern and Northwest Coast Native Americans steamed seafood in pits; southwestern Native Americans baked cornmeal bread in adobe ovens called hornos. (After the introduction of hogs, flat breads were commonly fried in lard.) Seasonal items were preserved by drying them in the sun or smoking them over a fire; for meat, fish, and oysters, special wood was often used to impart a distinctive flavor. Other foods were ground into a meal or pounded into a paste. In Alaska, meats, greens, and berries were preserved in fermented (aged) blubber. All nations liked sweets, but they were limited to fruits and dishes flavored with maple syrup, honey, or other indigenous sweeteners.

Special Occasions

Many Native American religious ceremonies were accompanied by feasts. Among the northeastern Iroquois, seasonal celebrations were held for the maple, planting, strawberry, green corn, harvest, and New Year's festivals. The southern nations held an elaborate Green Corn Festival in thanks for a plentiful summer harvest. No one was allowed to eat any of the new corn until the ceremony was complete. Each home was thoroughly cleaned, the fires were extinguished, and all old pieces of pottery and clothing were replaced with newly made items. The adult men bathed and purged themselves with an emetic. When everything and everyone were thoroughly clean in body and spirit, a central fire was lit by rubbing two sticks together, and each hearth fire was relit with its flames. The feasting on new corn then began. Amnesty was granted for all offenses except murder, and the festival signified the beginning of a new year for marriages, divorces, and periods of mourning.

Role of Food in Native American Culture and Etiquette

Historically, many Native American nations, especially in the inland regions, experienced frequent food shortages. As a result, food is valued as sacred, and, in the holistic worldview of most Native American groups, food is also considered a gift of the natural realm. In some nations elaborate ceremonies accompanied cultivation of crops, and prayers were offered for a successful hunt.

The men in many nations were traditionally responsible for hunting or the care of livestock. The job of food gathering, preparation, and storage usually belonged to the women, who also made the cooking utensils, such as watertight baskets or clay pots.[8,24] In predominantly horticultural societies, both men and women were frequently involved in cultivation of the crops. Among the nations of the Northeast, the men ate first, followed by women and children. In the Southwest, men prepared the game they caught and served it to the women.

© Bettmann/CORBIS

▲ *Baking bread in a southwestern outdoor oven.*

Sharing food is an important aspect of most Native American societies today. Food is usually offered to guests, and in some tribes it is considered rude for a guest to refuse food. It is also impolite to eat in front of others without sharing.[31] Any extra food is often given to members of the extended family. In some nations of the Southwest, meals are prepared and eaten communally. Each woman makes a large amount of one dish and shares it with the other families, who in turn share what they have prepared. Many Native Americans find the idea of selling food inconceivable; it is suggested that this is one reason there are few restaurants featuring Native American specialties.

Therapeutic Uses of Food

The role of food in spiritual and physical health is still important for many Native Americans, and many food plants provide medicine in some form. Corn is significant in some healing ceremonies. Cornmeal may be sprinkled around the bed of a patient to protect him or her against further illness. Corn pollen may be used to ease heart palpitations, and fine cornmeal is rubbed on children's rashes. Navajo women drink blue cornmeal gruel to promote the production of milk after childbirth, and Pueblo women use a mixture of water and corn ear smut (*Usti Lago maydis*, a kind of fungus) to relieve diarrhea and to cure irregular menstruation. A similar drink was given to Zuni women to speed childbirth and to prevent postpartum hemorrhaging. Corn silk tea was used as a diuretic and was prescribed for bladder infections.[32]

Numerous other indigenous plants are used by Native Americans for medicinal purposes. For example, agave leaves (from a succulent common in the Southwest) are chewed as a general tonic, and the juice is applied to fresh wounds. Another succulent, yucca, was considered a good laxative by the Hopis. Pumpkin pastes soothe burns. Chili peppers are used in compresses for arthritis and applied directly to warts. Infusions are used for many remedies, such as wild strawberries or elderberry flowers for diarrhea and mint tea to ease colic, indigestion, and nausea. The Ojibwas boiled blackberry roots to prevent miscarriages and sumac fruit and roots to stop bleeding. Traditionally, maple sugar lozenges were used for sore throats. Bitter purges and emetics are administered because they are distasteful and repugnant to any evil spirits that might cause illness.[33]

Food restrictions are still common during illness. Depending on the nation, many Native Americans believe that cabbage, eggs, fish, meat, milk, onions, or organ meats should be eliminated from a patient's diet. Conversely, some foods may be considered important to maintain strength during sickness, such as meat among the Seminole in Florida,[34] and both meat and blue cornmeal among the Navajo.[35] The Navajo may avoid sweets during pregnancy to prevent having a weak infant.[16] Some foods are prohibited after childbirth, such as cod, halibut, huckleberries, and spring salmon for Nootka women of the Northwest Coast.

Native Americans found many plants had psychotherapeutic properties. They were used to relax and sedate patients, to stupefy enemies, and to induce hypnotic trances during religious ceremonies. The opiates in the roots of California poppies dulled the pain of toothache, for example. Lobelia was smoked as an antispasmodic for asthma and bronchitis. In the Southwest, knobs from the peyote cactus were used to produce hallucinations, sometimes in combination with other intoxicants. Jimsonweed (*Datura stramonium*) was traditionally used to keep boys in a semiconscious state for twenty days so that they could forget their childhood during Algonquin puberty rites.[36] It is still used today by some Native Americans for medicinal and ritual purposes.

Contemporary Food Habits

Native American ethnic identity is changing. Traditional beliefs and values are often in direct conflict with those of the majority society, and Native Americans' self-concept has undergone tremendous changes in the process of acculturation. Three transitional adaptations of members of the Ojibwa (also called Chippewa) tribe, identified by three different lifestyles, may serve as a model for the adaptations made by members of other Native American groups.[37] The first stage of adaptation is traditional; during this stage parents and grandparents speak the Ojibwa language at home, practice the Midewiwin religion, and participate in Native American cultural activities such as feasts and powwows. The second stage is more acculturated. English is the primary language, although some Ojibwa also is spoken. Catholicism is the preferred religion, and the family is involved in activities of the majority society. In the third Ojibwa lifestyle—the pan-traditional stage—the family speaks either English or Ojibwa exclusively; practices a religion that is a combination of Native American and Christian beliefs, such as the Native American Church; and is actively involved in activities of both traditional Native American and white societies.

Adaptation of Food Habits

Food habits reflect changes in Native American ethnic identity. Many Native Americans eat a diet that includes few traditional foods. Others are consciously attempting to revive the foods and dishes of their ancestors.

Ingredients and Common Foods

When Native Americans were uprooted from their lands and their known food supplies, many immediately became dependent on the foods provided to the reservations. One study evaluating the diets of Havasupai Indians living on a reservation in Arizona found that 58 percent of the subjects ate only foods purchased or acquired on the reservation during the twenty-four-hour recall period.[38] Commodity foods currently include items such as canned and chopped meats, poultry, fruit juices, peanut butter, eggs, evaporated and powdered milk, dried beans, instant potatoes, peas, and string beans. Researchers report that many of these foods, such as kidney beans, noodles, and peanut butter, are discarded by the Navajo; powdered milk may also be rejected because it is disliked or is considered a weak food suitable only for infants or the elderly.[19] On some reservations, large supermarkets provide a selection of foods similar to that found throughout the United States; however, on more remote reservations and in many rural areas access to markets is very limited.[18,39] Other sources of food include gardening (reportedly practiced by between 43 and 91 percent of rural Native Americans), fishing, hunting, gathering indigenous plants, and raising livestock. One study of California Miwok, for instance, reports that 67 percent of respondents recall that their grandparents harvested wild greens, nuts, berries, and mushrooms, and that 47 percent of respondents continue to supplement their diet this way.[40] A national Canadian survey reported that 66 percent of aboriginal peoples obtained some of their meat, poultry, and fish through hunting and fishing; 10 percent obtained most in this way; and 5 percent obtained all this way.[41] These supplementary food sources are limited by seasonality and, in some cases, by state and federal hunting and fishing laws.[39] Native Americans living in urban areas have the same access to food as other city dwellers.

Over the years, traditional foods were lost and substitutions were made. For example, beef is a commonly accepted substitute for game among many Native American ethnic groups. Fry bread is another example. It is a flat bread made from wheat flour typically fried in lard, and has been prepared in the Southwest for about one hundred years and in other regions for even less time. Though made from ingredients introduced by the Europeans, it is one of the items most often identified as "traditional" among Indians throughout the nation,[42] and it is often served at Native American festivals. It has been suggested that the substitution of Western foods, especially commodity items, in the preparation of traditional Indian dishes has adversely affected the nutritional value of these foods.[43]

Traditional foods make up less than 25 percent of the daily diet among the Hopi. Older Hopi women lament the fact that younger Hopi are no longer learning how to cook these dishes.[44] In a study of Cherokee women, the degree of Native American heritage of the woman in charge of the food supply in the home directly affected the consumption of traditional foods in that home.[45] Corn and corn products, such as hominy, were among the most popular traditional foods; game meat, hickory nuts, raspberries, and winter squash were the least commonly served. Among Cherokee teenagers, traditional items such as fry bread, bean bread (corn bread with pinto beans), and chestnut bread (made from chestnuts and cornmeal) were well accepted. Although more than 80 percent of the adolescents were familiar with typical Cherokee dishes, including native greens and game meat (bear, deer, groundhog, rabbit, raccoon, squirrel, and wild boar), these foods were rarely eaten.[46] Pima consume traditional items, such as tepary beans and cactus stew, mostly at community get-togethers.[47] A small sample of children from four different Native American communities recorded that only 7 out of 1,308 items listed in food recalls for the study were traditional.[48] California Miwok list mostly southwestern items such as beans, rice, and tortillas as those they most associate with Native American foods and recall numerous items eaten by their grandparents but not consumed now, such as squirrel, rabbit, deer, acorn mush, and certain insects. Access to wild game is limited due to hunting restrictions. Navajo women eat traditional foods infrequently, with the exception of fry bread, mutton, and tortillas. Blue cornmeal mush (with ash), hominy, and sumac berry pudding are a few of the native dishes consumed occasionally.[49] Dakota women of all ages take pride in traditional foods but prepare them only when it is convenient or for special occasions. A study of Sencotan Indians in British Columbia found that marine foods, especially salmon, retain social and economic importance within the community.[50] Among some Baffin Inuit, traditional foods (e.g., sea and land mammals, fish) make up about one-third of energy intake. Men were found to eat more traditional items than women.[51] A study of Yupik Inuit found that younger respondents (ages 14–19) ate significantly fewer traditional foods than did the oldest respondents (ages 40–81).[52]

Broader efforts to preserve traditional and adapted Native American food traditions are

Table 5.3 America's Top Ten Endangered Foods

Chapalote Corn	Considered the original cultivated corn with small ears, coffee-colored kernels, and a flinty flavor
Chiltepin Pepper	Pea-sized, very hot wild chile pepper native to the Southwest considered the ancestor of most varieties used today—drought, diminishing habitat, and unscrupulous harvesting threatens this chile with extinction
Eulachon Smelt	Pacific Northwest source of oil that has suffered serious declines in population—further, traditional methods of processing are gradually being lost
Gulf Coast Sheep	Introduced to the Southeast by the Spanish in the 1500s, this breed adapted well to the humid conditions of the region, providing excellent meat and wool—newer breeds are lessening their popularity
Java Chicken	One of the first chicken breeds introduced to the United States, these birds now number only about 100
Marshall Strawberry	An heirloom fruit discovered in Massachusetts in 1890 but grown commercially in the Pacific Northwest—very intense flavor
Native American Sunflowers	Indigenous plants cultivated by the Native Americans for seeds and oil—brought to Europe as an oil source—popularity of the oil led to growing a single "improved" variety, and this one type is now susceptible to numerous rust diseases—few sources of the original seeds remain
Pineywoods Cattle	A foraging breed introduced by the Spanish to Florida, particularly suited to conditions in the South, and popular with early Native American ranchers—approximately 200 animals are left
Seminole Pumpkin	A pear-shaped pumpkin grown on vines that use trees for support—found in the Everglades but rarely cultivated today
White Abalone	The deepest inhabitant of West Coast Abalone, it has neared extinction due to the popularity of its sweet meat—now being bred in a recovery program designed to save it

Source: Nabhan, G.P., & Rood, A. 2004. Renewing America's Food Traditions (RAFT): Bringing cultural and culinary mainstays of the past into the new millennium. *Flagstaff, AZ: Center for Sustainable Agriculture at Northern Arizona University.*

also underway. Notably, the group Renewing America's Food Traditions (RAFT), a coalition of organizations dedicated to bringing the foods of the past into the present, have listed over 700 endangered food items (see Table 5.3). Support of communities attempting to recover and conserve food traditions is their primary goal.[53]

Meal Composition and Cycle

Little has been reported regarding current Native American meal patterns. It is assumed that three meals per day has become the norm, especially in families without income constraints. Meals consumed by Native Americans vary considerably among regions. In the text, *Cultural Food Practices*, the diet for Northern Plains Indians' is described as one centered around meats and starches with a limited variety of fruits and vegetables. The access to commodity food supplements also influences food choices.[18] Navajos still use traditional cooking methods but have also adapted to using more fat and salt in food preparation.[18] Navajo women were found to eat fry bread or tortillas, potatoes, eggs, sugar, and coffee most frequently. Fried foods were preferred for breakfast, and lunch and dinner consisted of one boiled meal and one fried or roasted meal. Pimas in Arizona prefer eggs, bacon or sausage, and fried potatoes for breakfast, while Southwest specialties, such as tacos, tamales, and chili con carne, are common at other meals. A study comparing Indians in New England living on reservations to those living in urban areas found baking and boiling remain favored preparation methods by respondents living on reservations, whereas urban residents were more likely to fry items. Grilled meats and smoked fish were also common ways of cooking.[42]

Special Occasions

Numerous traditional celebrations are maintained by Native American tribal groups.[18] Among the largest is the five-day Navajo Nation Fair held each

Labor Day weekend; the Pawnee Veteran's Day Dance and Gathering where ground meat with pecans and corn with yellow squash are served; the Miccosukee Arts Festival and the Seminole Fair in Miami, where alligator meat is featured; the Iroquois Midwinter Festival held in January to mark the new year; the Upper Mattaponi Spring Festival in Virginia over Memorial Day weekend; the three-day Creek Nation Festival and Rodeo; the Yukon International Storytelling Festival at which wild game such as caribou and musk ox are available; and the Apache Sunrise Ceremony which features gathered foods such as amaranth leaves and the pulp and fruit from the saguaro and prickly pear cacti. More local festivities are also common. Pueblo Feast Days are observed in honor of the Catholic patron saint of each village with a soup of posole (hominy) and beef or pork ribs. Northwest Coast potlatches are common in the spring, featuring herring roe, fish or venison stews, euchalon, salmon, and other traditional foods. In addition, there are all-Indian festivals that draw attendees from throughout the country, such as O'Odham Tash Indian Days in Casa Grande, Arizona; the Red Earth Festival in Oklahoma City; and the Gallup Intertribal Indian Ceremonial in New Mexico. Native Americans may also eat traditional foods on special occasions such as birthdays, but for holidays of the majority culture, other foods are considered appropriate. For example, turkey with all the trimmings is served by Dakotas for Thanksgiving and Christmas.[54]

Nutritional Status

Nutritional Intake

Research on the nutritional status of Native Americans is limited. Severe malnutrition was documented in the 1950s and 1960s, including numerous cases of kwashiorkor and marasmus. Today, lower socioeconomic status and higher unemployment contribute to an inadequate diet for some Native Americans. In general, however, recent changes in morbidity and mortality figures suggest that Native Americans have transitioned from the conditions associated with under-consumption, such as infectious diseases, to conditions associated with overconsumption, including obesity, type 2 diabetes, and cardiovascular disease.[55]

Studies of current Alaska Native eating habits suggest that diets high in refined carbohydrates (starchy and sugary foods) and fat, and low in fruits and vegetables, are common. The protein and nutrient of Alaska Natives has declined during the past several decades, as many foods obtained through hunting and gathering were replaced by processed, canned, and packaged items.[56] The estimated carbohydrate content of the Alaska Native diet before contact with Westerners was exceptionally low (3 to 5 percent of daily calories) due to a dependence on sea mammals and fish. Within only a few generations, that figure had increased to 50 percent of total calories, much of it from low-nutrient–density foods.[57,58] Research on Alaska Natives has also shown low intakes of calcium, iron, phosphorus, magnesium, zinc, vitamins A, C, D, and E, riboflavin, and folic acid, as well as fiber, omega-6, and omega-3 fatty acids. Traditional Alaska Native diets have been found lower in fat and carbohydrates, and higher in protein, phosphorus, potassium, iron, zinc, copper, magnesium, manganese, selenium, and several vitamins, including A, D, E, riboflavin, and B_6.[59–62]

A similar transition occurred in the diets of American Indians in other parts of the nation; and today, refined carbohydrates are prominent in the diet. Studies have identified white breads, tortillas, potato chips, French fries, and candy as the top contributors of energy for many Native Americans.[48,63–66] Of particular note is the consumption of sweetened beverages, including soda and fruit-flavored drinks, estimated to be 15 to 27 percent of all carbohydrates consumed, and as much as 17 percent of daily calories.[63,66–70] High-fat foods, including fried foods and processed meats and beef dishes, are another significant source of energy.

A low intake of fruits and vegetables is prevalent in the diets of both Native American adults and adolescents.[63,64] Of Indians in California, 60 percent said they had not eaten any fruit the previous day, and 28 percent reported they had not consumed any vegetables.[71] Among the Lakotas of South Dakota, nearly 60 percent stated they ate fruit only two to eight times a month; nearly half reported eating vegetables (including potatoes) over five times weekly.[72] A study of Catawba showed 47 percent ate less than one fruit daily, and 87 percent consumed only one vegetable or less.[73] The vegetable most often consumed in a study of Native American women in Oklahoma was French fries; and only tomatoes, tossed salad, green beans, potato salad, and mashed potatoes were also mentioned in the list of the top fifty-three items most often eaten.[68] Barriers to increased consumption of fruit and vegetables included cost, availability, and quality.

Native American nutrient intake has been compared to that of the general population in a few studies. It is notable that among both children and adults, few differences are found. One large study in three states reported that median intakes of vitamins A and C and folate were low for both American Indians and the total population.[74] A comparison of dietary calcium intake among African American, Native American, and white women in North Carolina found that although whites ate significantly more high-calcium foods than did Native Americans, none of the respondents were consuming recommended levels of the mineral through their food.[5,75] Other researchers who investigated rural Native American and white children in Oklahoma suggest that their diets are more influenced by factors such as poverty and living in a rural area than by cultural or structural issues related to race or ethnicity.[66] Nutrient deficiencies in Native American adults and children may occur, but with the exception of Alaska Natives, most studies suggest dietary adequacy similar to that of the total U.S. population.

Life expectancy has improved over recent years, yet disparities are still found. The average life expectancy is approximately 2.4 years less when compared to the U.S. population. "American Indians and Alaska Natives die at higher rates than other Americans from tuberculosis (600% higher), alcoholism (510% higher), motor vehicle crashes (229% higher), diabetes (189% higher), unintentional injuries (152% higher), homicide (61% higher) and suicide (62% higher)."[76]

Native American mothers are more likely to be younger than the general population and less likely to be married. Thirty percent of American Indian and Alaska Native women do not receive prenatal care during the first trimester of pregnancy. American Indian and Alaska Native infants die at a rate of 8.5 per every 1,000 live births, as compared to 6.8 per 1,000 for the U.S. all races population (2000–2008 rates).[4,75] Postnatal mortality rates are nearly 60 percent higher for American Indians and Alaska Natives compared to the total population, and sudden infant death syndrome is particularly problematic.[21] Breast-feeding has traditionally been considered the proper way to feed infants among most Native Americans. Among the Navajo in one study, elders reported that breast-fed infants were better able to hear traditional teachings and were better disciplined. It also demonstrated that the children are loved. Eighty-one percent of subjects initiated breast-feeding. Most added infant formula within the first week and used this combined feeding practice for more than five months.[77] Other estimates of breast-feeding show rates of 24 to 62 percent.[78,79] Baby-bottle tooth decay, due most often to extended use of a bottle with formula, milk, juice, or soda, affects more than one-half of all Native American and Alaska Native children.

Overweight and obesity are prevalent among Native Americans. National data suggest obesity rates of 37 to 40 percent among American Indian adults in 2008, approximately 20 percent above the average for the total population.[4] In another survey, the highest obesity rates were found in Native Americans living in Alaska, and the lowest in the Pacific Northwest.[80] Though definitions of overweight and obesity vary in other research, the trends are consistent. Group-specific studies report 83 percent of Havasupai subjects and 60 percent of Seminoles were identified as obese; 63 percent of Navajo women and 33 to 50 percent of Navajo men were overweight; and 61 percent of Indians residing in Oklahoma were overweight or obese.[81,82] In some groups, such as the Inuits, overweight was traditionally desirable as a visible demonstration of wealth during times of privation.[10] Among the Navajos, elder men and women prefer a heavier body shape.[83] Studies of Native American school children and adolescents show consistently higher weight-for-height ratios than for white, black, or Hispanic populations, while body mass index (BMI) rates over the 85 percentile varied, up to 50 percent for girls, and nearly 50 percent for boys.[78,84–87]

Figures on obesity contradict some nutritional intake data, which show the caloric intake of many Native Americans to be normal or less than the recommended dietary allowances. Metabolic differences in obese Native Americans may be a factor,[55,74,115] or lower rates of energy expenditure through exercise may contribute: American Indians and Alaska Natives report lack of leisure-time physical activity at higher rates than any other U.S. ethnic group.[80,88] However, a study of Hualapai women of Arizona indicated that the daily caloric intake of obese women was significantly higher than for the non-obese women. Sweetened beverages and alcoholic beverages accounted for the differences. Researchers investigating Zuni adolescents suggest that underreporting of foods and/or alcohol may account for low reported energy intakes.[63,68,89] Dieting behaviors among adult Native American women mostly involve healthy approaches such as eating more fruits and vegetables and exercising more, according to one study, although skipping meals, fasting, and disordered eating such as self-induced vomiting were also mentioned; 10 percent engaged in binge eating. A national survey of Native American youth revealed that over 40 percent reported binge-eating behavior with vomiting rates of 4 to 6 percent. Frequent dieting was also common. All disordered eating occurred more frequently among overweight respondents.[90,91]

The incidence of type 2 diabetes mellitus, especially among some Native Americans of the Plains and Southwest, is estimated to be between two and four times that of the general population. More than 16 percent of American Indians and Alaska Natives are estimated to have diabetes which is the highest rate of any ethnic group in the United States.[4] The Pima Indians are believed to have the highest rate of type 2 diabetes in the world, affecting 70 percent of all adults over the age of forty-five. Rates of the disease among children are also increasing substantially, and it has been noted that acanthosis nigricans (a patchy darkening of the skin) is an independent marker for insulin resistance in Native American youngsters.[92–94] The death rate from type 2 diabetes is more than three times as high for Native Americans as for the total population.[95] Notably, diabetes was rare among Native Americans fifty years ago.[96]

One theory for the high rates of type 2 diabetes among Native Americans is genetic predisposition.[97] However, a comparison of type 2 diabetes among Pima Indians living in Arizona and those living in Mexico found rates in Arizona to be more than five times those found in Mexico, suggesting that genetic predisposition alone does not account for high prevalence in the United States and that a Westernized environment may be a factor.[98] A recent study describes a significantly lower plasma insulin level in Pima Indians living in Mexico when compared to the U.S. Pima population even before diabetes is diagnosed. As the authors state: "This finding underscores the importance of lifestyle factors as protecting factors against insulin resistance in individuals with a high propensity to develop diabetes".[99] Higher rates of diabetes are found among Alaska Natives who have significant increased intake of non-indigenous protein (i.e., beef, chicken), carbohydrates (i.e., white bread, potatoes or rice, soft drinks), and fat (i.e., butter, shortening), combined with a lower intake of native foods such as salmon, caribou, berries, and seal oil; and higher rates are found among Pima Indians who consume an Anglo diet when compared to Pima who consume a traditional diet.[100,101] Some researchers suggest the difference is due to the dietary change from indigenous starches to the refined flours and sugars of the adapted diet. Traditional starches are harder to digest and absorb, leading to lower blood sugar levels and insulin responses that may be protective in the development of diabetes.[102] Other researchers suggest the increased intake of fat in the modern Native American diet may be responsible for the increase in type 2 diabetes.[103] (See Cultural Controversy feature.)

Associated with obesity and type 2 diabetes is a dramatic increase in the prevalence of heart disease among Native Americans during the last twenty-five years; heart disease is now the leading cause of death. Rates of cardiovascular disease for Native Americans have surpassed those of the total population in many locations, and are often more fatal. Additional risk factors include high rates of cigarette smoking, alcohol consumption, elevated blood lipid levels, obesity, and hypertension.

Chronic kidney disease is also a concern, particularly associated with type 2 diabetes, with incidence up to twenty times higher than in the general population. American Indians and Alaska Natives die at higher rates than other Americans from tuberculosis, with estimates being up to 600% higher.[2–4]

The incidence of alcoholism among Native Americans has decreased in recent years, but it still remains a significant medical and social problem. High unemployment rates and loss of tribal integrity, ethnic identity, and self-esteem are frequently cited as reasons for substance abuse among both reservation and urban Native Americans. The rate for alcohol-related deaths is more than eight times that of the general U.S. population.[3–4]

Counseling

A survey of Native American nurses identified attitudes, skills, and knowledge needed to serve Native Americans successfully. They listed being open-minded, avoiding ethnocentrism, and using intercultural communication skills, especially the ability to listen carefully and to provide respectful silence. Learning about the Native American worldview, traditional health beliefs, differences between nations, and the history of each group was considered essential to effective interaction.[18,104] Of particular importance is recognition of diversity within Native American groups and understanding of local culture.

Access to biomedical health care may be limited for some Native Americans because of low income or inadequate transportation. Many are also limited to care through the Indian Health Service (IHS), which has a federal trust to provide health services to all Native Americans who are members of tribes recognized by the U.S. government. It began operation under the BIA in 1924 and was placed under what is now called the Department of Health and Human Services in 1955. It operates forty-three hospitals and works in collaboration with state, tribal, and private health care facilities to provide comprehensive services. Under the Indian Self-Determination and Education Assistance Act of 1975, tribes are given the option of staffing and managing IHS programs in their communities. However, the number of Native American health professionals is limited, and intercultural care is the norm rather than the exception in many facilities.

Some Native Americans hold beliefs that cause them to avoid biomedical treatment in

CULTURAL CONTROVERSY—Type 2 Diabetes, Thrifty Genes, and Changing Theories

Scientific theory sometimes takes on a life of its own, existing in the public memory long past the time it has been modified or disproved. Such is the case with the thrifty gene hypothesis. In the 1960s when researchers first proposed that certain Native Americans might be predisposed to developing diabetes mellitus (DM) due to a feast-or-famine metabolism compromised by a modern diet, type 1 DM had not yet been fully differentiated from type 2 DM. Four decades later, our understanding of the disease has improved, revealing a complexity that defies simple explanations.[86]

What is now known as type 1 DM results from a complete lack of insulin in the body and is not correlated to lifestyle. Type 2 DM is more common; it is associated with normal or reduced levels of insulin and the inability to use insulin efficiently, and it is closely correlated with obesity and limited physical activity. Both types of diabetes result in high levels of blood glucose that over time can cause lifelong disability and death. Diagnosis of type 2 DM has increased by more than 65 percent in the general adult population over the past ten years. Long considered a disease of middle age, type 2 DM was rarely diagnosed in urban pediatric clinics as recently as the early 1990s, but type 2 DM is now increasing rapidly as a proportion of all newly diagnosed cases of diabetes in children.[63] Among Native Americans, prevalence rates vary from between 4 and 70 percent of adults over the age of forty-five, depending on group, with the lowest rates seen in Alaska and the highest found in the Southwest. In a few Native American groups a unique genetic mutation has been identified as the cause, as in the Oji-Cree of Canada, who have type 2 DM rates of 40 percent among adults.[49]

Yet attempts to find a universal thrifty gene that predisposes other Native Americans to the disease have been unsuccessful. It was initially argued that a metabolism adapted to a diet plentiful at times (feast periods) interspersed with food shortages (famine periods) could not cope with the constant abundance of the modern diet, resulting in high insulin responses and type 2 DM. The theory collapsed when researchers were unable to find a specific mechanism that causes an ethnic predisposition to developing diabetes. Further, it has been difficult to totally account for high prevalence rates through differences in diet and activity levels between diabetic and non-diabetic members in certain Native American groups.[28, 32, 113, 123, 127, 128]

These data beg the question: Why do some Native Americans develop type 2 DM at rates many times above that of the white U.S. population? Researchers propose that the development of type 2 DM is much more complex than previously thought, including factors in three domains: political-economic (such as ongoing stress, unavailability of healthy foods, and barriers to health care access), etiological (both genetic and non-genetic), and cultural (including traditional health beliefs, values regarding body image, norms about exercise, etc.).[106] Of particular interest are theories regarding the roles of historical trauma (similar to posttraumatic stress syndrome)[15] and prenatal adaptation to malnutrition and gestational diabetes in Native American mothers resulting in increased rates of type 2 DM.[6] Genes alone, thrifty or not, cannot completely explain what has become a problem of epidemic proportions in much of the Native American community.

general. For example, pregnancy is often considered a healthy state, and Native American women may not seek prenatal care for this reason. Some older Native Americans report fear of non–Native American providers, and others find biomedical physicians too negative because of impersonal care. Disclosure of risks may also be regarded as negative, in violation of a positive approach to life.[105] Some Native Americans may be angry about their condition and may blame illnesses such as diabetes on a Western conspiracy. Some are suspicious of federal and state government and may be hesitant to sign any forms.[15]

One social worker reports that her clients at a dialysis center in Arizona are so accustomed to friends and family members with kidney failure that they fatalistically expect a similar outcome for themselves and are sometimes relieved when it finally happens.[106] Research shows that many Native Americans believe that renal failure, amputations, and blindness are inevitable consequences of diabetes.[96] Traditional attitudes about time may cause delays in seeking care—the importance of finishing a current project or commitment may outweigh that of keeping an appointment with a health provider.

In general, both verbal and nonverbal communication with Native Americans must take into account cultural traditions. Among the Navajo, the patient should be asked directly for his medical history, because even family members may believe that they have no right to speak for another. However, family members are often consulted in making medical decisions.

It is estimated that more than one-half of Native Americans speak their native language in addition to English. Native American languages are primarily verbal, and some Native Americans may experience difficulties with written information or instructions. Further, some English words, such as *germ,* may not exist in Native American languages.[23] It has been reported, for example, that increasing vegetables in the diet of some Native Americans has met with resistance because the closest equivalent Native American word for vegetable is *weeds.*[107] Many Native Americans are comfortable with periods of silence in a conversation, using the time to compose their thoughts or to translate responses. A yes or no response may be considered a complete answer to a question, and a Native American may answer, "I don't know," if he or she thinks that a question is inappropriate or does not wish to discuss the topic. Stories or metaphors may be used to make a point.[15] Information may be withheld until she feels she can trust the provider. Saving face and avoiding conflict are crucial, and a Native American may not ask questions during an interview because that would suggest that the health care provider was not communicating clearly. Among the Navajo, direct questioning by the provider suggests that the practitioner is unknowledgeable or incompetent. Open-ended questions are preferred. The very concept of a dietary interview may be interpreted by some Native Americans as interference with their personal autonomy. Emphasis on how personal health promotes the welfare of the client's family and community can be effective.[108]

Nonverbal communication is very sophisticated among some Native Americans. It has been reported that some children may not be taught to speak until other senses are developed.[109] A Native American client may expect the practitioner to intuit the problem through nonverbal techniques rather than through an interview.[110] Although a smile and a handshake are customary, a vigorous handshake may be considered a sign of aggressiveness. Native Americans often sit at a distance, and direct eye contact may be considered rude—the health professional should not interpret averted eyes as evidence of disinterest. Quiet, unhurried conversations are most conducive to successful interaction.

Researchers note that one-on-one diet education in a clinical setting is often ineffective with Native Americans.[39] Counseling can be improved through recognition of the strong oral tradition found in most Native American groups. The preferred learning style for many Native Americans is in an interactive, informal, and cooperative setting. Talking with clients, instead of to clients, can improve efficacy. For many Native Americans, it is the relationship established with the caregiver, not the content of the conversation, that is important. The sharing of personal stories, in particular, can elicit information and address issues that are uncomfortable for clients to address directly. In group settings, talking circles can be effective because they facilitate communication and demonstrate equality among all participants.[79,108,111] In the text *Cultural Food Practices,* it is noted that visual learning is key so the use of food models, pictures or videos may be successful as teaching tools.[18] Developing a nutrition model

PRACTITIONER PERSPECTIVES—Native American

LORRAINE WHITEHAIR, RD, MPH, RN, CDE

Worked at Indian Health Services for over ten years,

Tribal RD for a number of years, and currently works for CDC

I was born in Oljato, Utah, a rural Navajo reservation located 150 miles northeast of Farmington, New Mexico. After graduating from high school, I worked as a nurses' aid in a small, fifteen-bed hospital near my home in Utah. Later, I completed a bachelor's degree in nutrition science at the University of Utah and a dietetic internship and a master's degree in public health at the University of California, Berkeley. As a registered dietitian I have worked with the Navajo nation for approximately twelve years.

Briefly, how would you describe the Navajo foods?

Traditionally, during the fifties, sixties, and early seventies, gardening and sheepherding was our livelihood. We hauled all our water and used wood-burning stoves for cooking. My parents told us of Navajo native foods when they were young, some of which I have never seen. My mother told me she spent many days grinding corn for the winter. One stew made by parents was *ad Alth ta' nash besh,* meaning boiled with several mixtures. It had melon seeds, local wild green plants, and a variety of plant seeds, and wild onions, and was thickened with corn meal. When I was young we grew watermelon, cantaloupe, fresh corn, and summer and winter squash. Other favorites included boiled mutton backbone in green chili stew, grilled mutton ribs, liver and greater omentum (the fatty tissue that covers the stomach) in an open-face sandwich with onions or green chili in a tortilla. We may not have had as much mutton as we wanted, but we had plenty of fruits and vegetables during the summer months.

Traditional food eaten today includes corn prepared many different ways from fresh to dried or ground into meal in varying degrees of texture. One recipe for fresh corn is called "Kneel down Bread" because it is prepared in a sitting position. A one-fourth cup of the ground corn is placed in corn leaves, wrapped and tied together. It is then placed in a large hole that is dug in the ground and is cooked by building a fire on top. All parts of the sheep and goat are used for food. This includes the head, intestine, organs, skin, hoofs, and blood. Blood sausage and liver sausage are very popular. The stuffed intestine is made by wrapping the greater omentum with the intestine. It is then grilled on hot coals until crispy.

Most elderly like simple, unmixed foods. Older people dislike marinated meats. They prefer the meats cooked plain. I would say Navajo native food is fairly bland, although men like to eat with very hot chili and raw, fresh onions.

How much has Western food culture influenced the current diet?

In the past three to four decades our diet has changed significantly. Our level of physical activity also decreased at the same time. Popular foods now include spam, canned meat, cold cuts, ramen noodles, chips, candy, sodas, and flavored fruit drinks.

Are there any foods that would be difficult or easy to modify in the diet?

Yes. Fried bread would be hard to modify. People enjoy the texture of bread cooked in hot oil, which makes the recipe difficult to alter. Intestine wrapped in greater omentum is the one that could be easy to modify. Replace the intestinal fat with thick sliced vegetables like onions, carrots, cabbage, green peppers, and celery. Ramen noodles, a non-traditional food, are easier to modify as well—I suggest to my clients that they add a variety of vegetables (frozen and fresh) and muscles from beef, chicken, or pork and decrease the serving size, since the noodles contain fat. People seem to like it with fresh ginger as well.

What advice would you give to new RDs working with Native Americans?

Learn about the history of the people. Learn about the way of life then and now. The Navajo people value family and community. There is a saying which states, "Know who you are and where you came from." That is to know your clan system. Introductions are a very important part of interaction with people on the Navajo reservation. As time consuming as it may be, establishing good introductions helps establish trust, kinship, and credibility. Making a good introduction about your family clan (family background) is a cultural practice, and it is believed to help make business meetings successful. For elderly clients you may address them as your mother (*Shi ma'*) or your father (*Shi yes ah'*) even though they are not related to you.

For the middle aged and elderly population, living in harmony with the surroundings is important. Parents and grandparents spend a great deal of time talking about being respectful of kinship, elderly people, and nature, including land and living creatures. We are to look toward living in harmony with our surroundings. Illness is often thought of as a result of an interruption of harmony.

We enjoy laughing. There is great pleasure in tastefully teasing when appropriate. We have many jokes about life on the reservation, our first boarding school experiences, interaction with people in towns, schools, and medical clinics, and meeting with public health officials. We have many animal stories that are funny, and they also teach. For example, I use this story to help teach my clients about diabetes:

PRACTITIONER PERSPECTIVES—Native American *(continued)*

Two Birds

Two birds, Jay and Woody, lived near the Interstate 40 highway fifty miles west of Albuquerque, New Mexico. Jay loves to pick up foods that are left by humans who were traveling back on Interstate 40 highway from Albuquerque, New Mexico. The most common foods that were left along the highway were french fries, hamburgers, shakes, pies, cookies, ice cream, fried chicken, biscuits, potato chips, fried bread, and candy bars. Jay always invited Woody to help him eat the leftover foods that were thrown near the highway. He would tell Woody about the "delicious" food he had for the day. He would say, "You do not have to hunt for your food. It is there along the road." Woody told Jay the food he was eating was not good for him. Woody faithfully hunted for his food supply. He worked to store his food daily so that he would have enough for an emergency during the winter months. Daily, Woody flew long distances to get his food. Meanwhile, Jay had been steadily gaining weight. He also had difficulty flying moderate distances. In the fall both birds wanted to visit distant relatives near Albuquerque. Jay was not able to complete the trip to Albuquerque. He experienced shortness of breath and exhaustion half way to Albuquerque. Jay and Woody decided that Woody should go alone to see their distant relatives. In early spring Jay became very ill after a flu he had developed following a large snow storm. He had frequency of urination, tiredness, sores on his toes not healing, and blurred vision. His doctor told him he had diabetes. Jay took the news very hard. He did not want to change his diet. Woody begged Jay to eat more healthfully. In the end Jay's diet changed back to eating whole grains, nuts, and seeds to help control his blood sugars. He also had to retrain his wings to fly longer distances daily, which helped bring his good cholesterol (HDL) in excellent ranges. His doctor was very happy to see Jay living more healthfully.

that tells the story of changes to traditional foodways and resulting consequences, using culturally appropriate narrative and imagery, has also been recommended (Figure 5.1).[112]

A client may have misconceptions about biomedicine. Because the Navajo health system classifies illness by cause rather than by symptom, clients may have difficulty understanding the necessity for a physical exam or medical history. There may also be the expectation that medication can cure illness and that an injection is needed for every disorder.[19] Very low compliance rates among some Native Americans on special diets have been reported.[107] A Native American may be confused about recommendations for weight loss because slimness is associated with disease or witchcraft.

Native American clients may follow traditional health practices instead of, or in conjunction with, biomedical therapies.[18] On one Navajo reservation, it was found that 90 percent of clients visiting a biomedical clinic had first sought traditional care.[113] A study of Native American veterans found a greater use of traditional therapies in areas where biomedical facilities were less accessible.[20] Other research suggests that though combining traditional and biomedical approaches is common, up to 40 percent of some American Indians use only traditional healing.[114] Some Indians believe a biomedical practitioner is needed to cure any "white man's disease," and traditional healers, such as herbalists and medicine men and women, or shamans, will frequently decline to treat conditions unfamiliar to them, referring patients to Westernized health care instead.[43,107,115] Compliance is most effective when traditional practices are accepted and encouraged as an integral part of complete care.

An in-depth interview is necessary to determine not only ethnic identity but also degree of

DISCUSSION STARTERS: WHO ARE YOU? AND WHAT DO YOU EAT?

After reading about the different regional, traditional Native American diets, decide which one you think would be the healthiest. Why? Then, decide which one you think would taste the best to you. Why? Once you've considered the traditional diets, reflect on the differences between the traditional diet you see as the healthiest and the contemporary diet of many Native Americans. Consider the following questions:

- What needs to be done to improve the diets of many Native Americans today?
- What are the major cultural obstacles in improving contemporary Native American diets?

Imagine that you are a member of a research institution that has been asked to recommend dietary policy for a contemporary Native American community. In a small group, compose a list of your major recommendations for such a policy that the majority of your group can agree to. Also provide advice to health care professionals serving that community about how to implement those dietary changes.

acculturation. Traditional medical beliefs and customs, if practiced, should be acknowledged. Personal dietary preferences are of special importance due to the variety of Native American foods and food habits. Note taking may be considered exceptionally rude, so memory skills or a tape recorder may be preferable during the interview.

REVIEW QUESTIONS

1. In Native American culture, what is considered the cause of illness? How may this influence the treatment of a medical disorder, like type 2 diabetes?
2. Pick two regional classifications of traditional Native American cuisine. Describe the similarities and differences between these two classifications in food and their preparation.
3. Describe three therapeutic uses of corn and one therapeutic use of a non-corn item by Native Americans.
4. What is Indian fry bread? Is it a traditional food? Why or why not?
5. What factors may have increased the incidence of type 2 diabetes among Native Americans?

REFERENCES

1. U.S. Census Bureau, Administration and Customer Services Division. *USA Statistics in Brief—Race and Hispanic Origin.* Available from: http://www.census.gov/compendia/statab/2010/files/racehisp.html
2. Public Information Office. 2010. *Facts for Features: American Indian and Alaska Native Heritage Month, November 2010.* Washington, DC: U.S. Census Bureau.
3. US Health and Human Services. Office of Minority Health. *American Indian/Alaska Native Profile.* Available from: http://minorityhealth.hhs.gov. Accessed: January 22, 2011.
4. Barnes PM, Adams PF, Powell-Griner E. Health *Characteristics of the American Indian or Alaska Native Adult Population: United States, 2004–2008.* Available from: http: //www.cdc.gov/nchs/data/nhsr/nhsr020.pdf. Accessed: January 20, 2011.
5. Birchfield, D.L. 2000. Apaches. In *Gale Encyclopedia of Multicultural America,* R.V. Dassanowsky & J. Lehman (Eds.). Farmington Hills, MI: Gale Group.
6. Birchfield, D.L. 2000. Sioux. In *Gale Encyclopedia of Multicultural America,* R.V. Dassanowsky & J. Lehman (Eds.). Farmington Hills, MI: Gale Group.
7. Birchfield, D.L. 2000. Navajos. In *Gale Encyclopedia of Multicultural America,* R.V. Dassanowsky & J. Lehman (Eds.). Farmington Hills, MI: Gale Group.
8. Birchfield, D.L. 2000. Pueblos. In *Gale Encyclopedia of Multicultural America,* R.V. Dassanowsky & J. Lehman (Eds.). Farmington Hills, MI: Gale Group.
9. French, E., & Hanes, R.C. 2000. Hopis. In *Gale Encyclopedia of Multicultural America,* R.V. Dassanowsky & J. Lehman (Eds.). Farmington Hills, MI: Gale Group.
10. Jones, J.S. 2000. Inuit. In *Gale Encyclopedia of Multicultural America,* R.V. Dassanowsky & J. Lehman (Eds.). Farmington Hills, MI: Gale Group.
11. Benson, D.E. 2000. Tlingit. In *Gale Encyclopedia of Multicultural America,* R.V. Dassanowsky & J. Lehman (Eds.). Farmington Hills, MI: Gale Group.
12. Kawagley, O. 2000. Yupiat. In *Gale Encyclopedia of Multicultural America,* R.V. Dassanowsky & J. Lehman (Eds.). Farmington Hills, MI: Gale Group.
13. Hanes, R.C., & Hillstrom, L.C. 2000. Paiutes. In *Gale Encyclopedia of Multicultural America,* R.V. Dassanowsky & J. Lehman (Eds.). Farmington Hills, MI: Gale Group.
14. Birchfield, D.L. 2000. Choctaw. In *Gale Encyclopedia of Multicultural America,* R.V. Dassanowsky & J. Lehman (Eds.). Farmington Hills, MI: Gale Group.
15. Yehieli, M., & Grey, M.A. 2005. *Health matters: A pocket guide for working with diverse cultures and underserved populations.* Boston: Intercultural Press.
16. Purnell, L.D., & Paulanka, B.J. 2003. *Transcultural health care: A culturally competent approach,* 2nd ed. Philadelphia: FA Davis Company.
17. Garrett, J.T. 1990. Indian health: Values, beliefs, and practices. In M.S. Harper (Ed.), *Minority aging: Essential curricula content for selected health and allied health professions.* Health Resources and Services Administration, Department of Health and Human Services, DHHS Publication No. HRS (P-DV-90-4). Washington, DC: U.S. Government Printing Office.
18. Brown TL, Zephier E, Johnson ML.American Indian Food Practices. In *Cultural Food Practices,* Goody CM, Drago L (eds). Chicago: American Dietetic Association/American Diabetes Association. 2010.
19. Kunitz, S.J., & Levy, J.E. 1981. Navajos. In A. Harwood (Ed.), *Ethnicity and Medical Care.* Cambridge, MA: Harvard University Press.
20. Wilson, C.S. 1985. Nutritionally beneficial cultural practices. *World Review of Nutrition and Diet, 45,* 68–96.
21. Castor, M.L., Smyser, M.S., Taualii, M.M., Park, A.N., Lawson, S.A., & Forquera, R.A. 2006. A nationwide population-based study identifying health disparities between American Indians/ Alaska Natives and the general populations living in select urban counties. *American Journal of Public Health, 96,* 1478–1484.
22. Lang, G.C. 1985. Diabetes and health care in a Sioux community. *Human Organization, 44,* 251–260.
23. Plawecki, H.M., Sanchez, T.R., & Plawecki, J.A. 1994. Cultural aspects of caring for Navajo Indian clients. *Journal of Holistic Nursing, 12,* 291–306.
24. Heisey, A.W., & Hanes, R.C. 2000. Oneidas. In *Gale Encyclopedia of Multicultural America,* R.V. Dassanowsky & J. Lehman (Eds.). Farmington Hills, MI: Gale Group.
25. Hillstrom, L.C., & Hanes, R.C. 2000. Nez Perce. In *Gale Encyclopedia of Multicultural America,* R.V. Dassanowsky & J. Lehman (Eds.). Farmington Hills, MI: Gale Group.
26. Kim, C., & Kwok, Y.S. 1998. Navajo use of native healers. *Archives of Internal Medicine, 158,* 2245–2249.
27. Sanchez, T.R., Plaweck, J.A., & Plawecki, H.M. 1996. The delivery of culturally sensitive health care to Native Americans. *Journal of Holistic Nursing, 14,* 295–307.

28. Roy, L. 2000. Ojibwa. In *Gale Encyclopedia of Multicultural America.* R.V. Dassanowsky & J. Lehman (Eds.). Farmington Hills, MI: Gale Group.
29. Christensen, N.K., Sorenson, A.W., Hendricks, D.G., & Munger, R. 1998. Juniper ash as a source of calcium in the Navajo diet. *Journal of the American Dietetic Association, 98,* 333–334.
30. Gunderson, M. 2003. *The Food Journal of Lewis & Clark: Recipes for an Expedition.* Yankton, SD: History Cooks Publishing.
31. Woolf, N., Conti, K.M., Johnson, C., Martinez, V., McCloud, J., & Zephier, E.M. 1999. *Northern Plains Indian food practices, customs, and holidays.* Chicago: American Dietetic Association/American Diabetes Association.
32. Kavasch, E.B., & Baar, K. 1999. *American Indian healing arts.* New York: Bantam.
33. Vogel, V.J. 1981. American Indian medicine. In G. Henderson & M. Primeaux (Eds.), *Transcultural Health Care.* Menlo Park, CA: Addison-Wesley.
34. Joos, S.K. 1980. Diet, obesity, and diabetes mellitus among the Florida Seminole Indians. *Florida Science, 43,* 148–152.
35. Schilling, B., & Brannon, E. 1986. *Cross-cultural counseling: A guide for nutrition and health counselors.* FNS #250. Alexandria, VA: U.S. Department of Agriculture/ U.S. Department of Health and Human Services.
36. Emboden, W.A. 1976. Plant hypnotics among North American Indians. In W.D. Hand (Ed.), *American Folk Medicine.* Los Angeles: University of California Press.
37. Primeaux, M., & Henderson, G. 1981. American Indian patient care. In G. Henderson & M. Primeaux (Eds.), *Transcultural health care.* Menlo Park, CA: Addison-Wesley.
38. Vaughan, L.A., Benyshek, D.C., & Martin, J.F. 1997. Food acquisition habits, nutrient intakes, and anthropometric data of Havasupai adults. *Journal of the American Dietetic Association, 97,* 1275–1282.
39. Wiedman, D. 2005. American Indian diets and nutritional research: Implications of the Strong Heart Dietary Study, Phase II, for cardiovascular disease and diabetes. *Journal of the American Dietetic Association, 105,* 1874–1880.
40. Ikeda, J., Dugan, S., Feldman, N., & Mitchell, R. 1993. Native Americans in California surveyed on diets, nutrition needs. *California Agriculture, 47,* 8–10.
41. Wein, E.E. 1995. Evaluating food use by Canadian aboriginal peoples. *Canadian Journal of Physiological Pharmacology, 73,* 759–764.
42. Di Noia, J., Schinke, S.P., & Contento, I.R. 2005. Dietary patterns of reservation and non-reservation Native *American youths. Ethnicity & Disease, 15,* 705–712.
43. Williams, D.E., Knowler, W.C., Smith, C.J., Hanson, R.L., Roumain, J., Saremi, A., Kirska, A.M., Bennett, P.H., & Nelson, R.G. 2001. The effect of Indian or Anglo dietary preference on the incidence of diabetes in Pima Indians. *Diabetes Care, 24,* 811–816.
44. Kuhnlein, H.V., Calloway, D.H., & Harland, B.F. 1979. Composition of traditional Hopi foods. *Journal of the American Dietetic Association, 75,* 37–41.
45. Story, M., Bass, M.A., & Wakefield, L.M. 1986. Food preferences of Cherokee teenagers in Cherokee, North Carolina. *Ecology of Food and Nutrition, 19,* 51–59.
46. Story, M., Snyder, P., Anliker, J., Cunningham-Sabo, L., Weber, J.L., Kim, R., Platero, H., & Stone, E.J. 2002. Nutrient content of school meals in elementary schools on American Indian reservations. *Journal of the American Dietetic Association, 102,* 253–256.
47. Smith, C.J., Nelson, R.G., Hardy, S.A., Manahan, E.M., Bennett, P.H., & Knowler, W.C. 1996. Survey of the diet of Pima Indians using quantitative food frequency assessment and 24-hour recall. *Journal of the American Dietetic Association, 96,* 778–784.
48. Lyttle, L.A., Dixon, L.B., Cunningham-Sabo, L., Evans, M., Gittelsohn, J., Hurley, J., Snyder, P., Stevens, J., Weber, J., Anliker, J., Heller, K., & Story, M. 2002. Dietary intakes of Native American children: Findings from the Pathways Feasibility Study. *Journal of the American Dietetic Association, 102,* 555–558.
49. Heck, K.E., Schoendorf, K.C., & Parker, J. 1999. Are very low birthweight births among American Indians and Alaska Natives underregistered? *International Journal of Epidemiology, 28,* 1096–1101.
50. Mos, L., Jack, J., Cullon, D., Montour, L., Alleyne, C., & Ross, P.S. 2004. The importance of marine foods to a near-urban first nation community in coastal British Columbia, Canada: Toward a risk-benefit assessment. *Journal of Toxicology and Environmental Health, 67,* 791–808.
51. Kuhnlein, H.V. 1995. Benefits and risks of traditional food for indigenous peoples: Focus on dietary intakes of Arctic men. *Canadian Journal of Physiological Pharmacology, 73,* 765–771.
52. Bersamin, A., Luick, B.R., Ruppert, E., Stern, J.S., & Zidenberg-Cherr, S. 2006. Diet quality among Yup'ik Eskimos living in rural communities is low: The Center for Alaska Native Health Research Pilot Study. *Journal of the American Dietetic Association, 106,* 1055–1063.
53. Nabhan, G.P., & Rood, A. 2004. *Renewing America's Food Traditions (RAFT): Bringing Cultural and Culinary Mainstays of the Past into the New Millenium.* Flagstaff, AZ: Center for Sustainable Agriculture at Northern Arizona University.
54. Bass, M.A., & Wakefield, L.M. 1974. Nutrient intake and food patterns of Indians on Standing Rock Reservation. *Journal of the American Dietetic Association, 64,* 36–41.
55. Compher, C. 2006. The nutrition transition in American Indians. *Journal of Transcultural Nursing, 17,* 217–223.
56. Jackson, M.Y. 1986. Nutrition in American Indian health: Past, present, and future. *Journal of the American Dietetic Association, 86,* 1561–1565.
57. Murphy, N.J., Schraer, C.D., Thiele, M.C., Bulkow, L.R., Doty, B.J., & Lanier, A.P. 1995. Dietary change and obesity associated with glucose intolerance in Alaska Natives. *Journal of the American Dietetic Association, 95,* 676–682.
58. Risica, P.M., Nobmann, E.D., Caulfield, L.E., Schraer, C., & Ebbesson, S.O. 2005. Springtime macronutrient intake of Alaska Natives of the Bering Straits

region: The Alaska Siberia Project. *International Journal of Circumpolar Health, 64,* 222–233.

59. Kuhnlein, H.V., Receveur, O., Soueida, R., & Egeland, G.M. 2004. Arctic indigenous peoples experience the nutrition transition with changing dietary patterns and obesity. *Journal of Nutrition, 134,* 1447–1453.
60. Kuhnlein, H.V., Soueida, R., & Receveur, O. 1996. Dietary nutrient profiles of Canadian Baffin Island Inuit differ by source, season, and age. *Journal of the American Dietetic Association, 96,* 155–162.
61. Nakano, T., Fediuk, K., Kassi, N., & Kuhnlein, H.V. 2005. Dietary nutrients and anthropometry of Dene/Metis and Yukon children. *International Journal of Circumpolar Health, 64,* 147–156.
62. Nobmann, E.D., & Lanier, A.P. 2001. Dietary intake among Alaska Native women resident of Anchorage, Alaska. *International Journal of Circumpolar Health, 60,* 123–137.
63. Cole, S.M., Teufel-Shone, N.I., Ritenbaugh, C.K., Yzenbaard, R.A., & Cockerham, D.L. 2001. Dietary intake and food patterns of Zuni adolescents. *Journal of the American Dietetic Association, 101,* 802–806.
64. deVera, N. 2003. Perspectives on healing foot ulcers by Yaquis with diabetes. *Journal of Transcultural Nursing, 14,* 39–47.
65. Story, M., Snyder, P., Anliker, J., Cunningham-Sabo, L., Weber, J.L., Kim, R., Platero, H., & Stone, E.J. 2002. Nutrient content of school meals in elementary schools on American Indian reservations. *Journal of the American Dietetic Association, 102,* 253–256.
66. Stroehla, B.C., Malcoe, L.H., & Velie, E.M. 2005. Dietary sources of nutrients among rural Native American and white children. *Journal of the American Dietetic Association, 105,* 1908–1916.
67. Bersamin, A., Luick, B.R., Ruppert, E., Stern, J.S., & Zidenberg-Cherr, S. 2006. Diet quality among Yup'ik Eskimos living in rural communities is low: The Center for Alaska Native Health Research Pilot Study. *Journal of the American Dietetic Association, 106,* 1055–1063.
68. Teufel, N.I., & Dufour, D.L. 1990. Patterns of food use and nutrient intake of obese and non-obese Hualapai Indian women in Arizona. *Journal of the American Dietetic Association, 90,* 1229–1235.
69. Taylor, C.A., Keim, K.S., & Gilmore, A.C. 2005. Impact of core and secondary foods on nutritional composition of diets in Native-American women. *Journal of the American Dietetics Association, 105,* 413–419.
70. Wharton, C.M., & Hampl, J.S. 2004. Beverage consumption and risk of obesity among Native Americans in Arizona. *Nutrition Reviews, 62,* 153–159.
71. Ikeda, J.P., Murphy, S., Mitchell, A., Flynn, N., Mason, I.J., Lizer, A., & Lamp, C. 1998. Dietary quality of Native American women in rural California. *Journal of the American Dietetic Association, 98,* 812–813.
72. Harnack, L., Story, M., & Holy Rock, B. 1999. Diet and physical activity patterns of Lakota Indian adults. *Journal of the American Dietetic Association, 99,* 829–835.
73. Costacou, T., Levin, S., & Mayer-David, E. 2000. Dietary patterns among members of the Catawba Indian nation. *Journal of the American Dietetic Association, 100,* 83–835.
74. Stang, J., Zephier, E.M., Story, M., Himes, J.H., Yeh, J.L., Welty, T., & Howard, B.V. 2005. Dietary intakes of nutrients thought to modify cardiovascular risk from three groups of American Indians: The Strong Heart Dietary Study, Phase II. *Journal of the American Dietetics Association, 105,* 1895–1903.
75. Bell, R.A., Quandt, S.A., Spangler, J.G., & Case, D. 2002. Dietary calcium intake and supplement use among older African American, white, and Native American women in a rural southeastern community. *Journal of the American Dietetic Association, 102,* 844–847.
76. Indian Health Service. *Facts on Indian Health Disparities.* 2006. Available from: http://info.ihs.gov/Files/DisparitiesFacts-Jan2006.pdf.Accessed: January 22, 2011.
77. Wright, A.L., Clark, C., & Bauer, M. 1993. Maternal employment and infant feeding practices among the Navajo. *Medical Anthropology Quarterly, 7,* 260–280.
78. Story, M., Strauss, K., Zephier, E., & Broussard, B. 1998. Nutritional concerns in American Indian and Alaska Native children: Transitions and future directions. *Journal of the American Dietetic Association, 98,* 170–176.
79. Houghton, M.D., & Graybeal, T.E. 2001. Breastfeeding practices of Native American mothers participating in WIC. *Journal of the American Dietetic Association, 101,* 245–247.
80. Denny, C.H., Holtzman, D., & Cobb, N. 2003. Surveillance for health behaviors of American Indians and Alaska Natives. Findings from the Behavioral Risk Factor Surveillance System, 1997–2000. Morbidity and Mortality Weekly Report. *CDC Surveillance Summaries, 52,* 1–13.
81. Bursac, Z., & Campbell, J.E. 2003. From risky behaviors to chronic outcomes: Current status and Healthy People 2010 goals for American Indians in Oklahoma. *Journal of the Oklahoma Medical Association, 96,* 569–573.
82. Vaughan, L.A., Benyshek, D.C., & Martin, J.F. 1997. Food acquisition habits, nutrient intakes, and anthropometric data of Havasupai adults. *Journal of the American Dietetic Association, 97,* 1275–1282.
83. White, S.L., & Maloney, S.K. 1990. Promoting healthy diets and active lives to hard-to-reach groups: Market research study. *Public Health Reports, 105,* 224–231.
84. Bernard, L., Lavallee, C., Gray-Donald, K., & Delisle, H. 1995. Overweight in Cree schoolchildren and adolescents associated with diet, low physical activity, and high television viewing. *Journal of the American Dietetic Association, 95,* 800–802.
85. Caballero, B., Himes, J.H., Lohman, T., Davis, S.M., Stevens, J., Evans, M., Going, S., Pablo, J. Pathways Study Research Group. 2003. Body composition and overweight prevalence in 1704 school children from 7 American Indian communities. *American Journal of Clinical Nutrition, 78,* 308–312.
86. Jackson, M.Y. 1993. Height, weight and body mass index of American Indian schoolchildren 1990–1991. *Journal of the American Dietetic Association, 93,* 1136–1140.
87. Zephier, E., Himes, J.H., & Story, M. 1999. Prevalence of overweight and obesity in American Indian school children and adolescents in the Aberdeen area: A population study. *International Journal of Obesity and Related Metabolic Disorders, 23 (Suppl. 2),* S28–S30.

88. Liao, Y., Tucker, P., Okoro, C.A., Giles, W.H., Mokdad, A.H., Harris, V.B., Division of Adult and Community Health, National Center for Chronic Disease Prevention and Health Promotion, Centers for Disease Control and Prevention (CDC). 2004. REACH 2010 surveillance for health status in minority communities—United States, 2001–2002. Morbidity and Mortality Weekly Report. *CDC Surveillance Summaries, 27,* 1–36.
89. Wiedman, D. 2005. American Indian diets and nutritional research: Implications of the Strong Heart Dietary Study, Phase II, for cardiovascular disease and diabetes. *Journal of the American Dietetic Association, 105,* 1874–1880.
90. Neumark-Sztainer, D., Story, M., Resnick, M.D., & Blum, R.W. 1997. Psychosocial concerns and weight control behaviors among overweight and nonoverweight Native American adolescents. *Journal of the American Dietetic Association, 97,* 598–604.
91. Sherwood, N.E., Harnack, L., & Story, M. 2000. Weight-loss practices, nutrition beliefs, and weight-loss program preferences of urban American Indian women. *Journal of the American Dietetic Association, 100,* 442–446.
92. Copeland, K., Pankratz, K., Cathey, V., Immohotichey, P., Maddox, J., Felton, B., McIntosh, R., Parker, D., Burgin, C., & Blackett, P. 2006. Acanthosis Nigricans, insulin resistance (HOMA) and dyslipidemia among Native American children. *Journal of the Oklahoma Medical Association, 99,* 19–24.
93. Fagot-Campagna, A., Pettitt, D.J., Englegau, M.M., Burrow, N.R., Geiss, L.S., Valdez, R., Beckles, G.L., Saaddine, J., Gregg, E.W., Williamson, D.F., & Narayan, K.M. 2000. Type 2 diabetes among North American children and adolescents: An epidemiological review and a public health perspective. *Journal of Pediatrics, 136,* 664–672.
94. Acton KJ, Burrows, Wang J, Geiss LS. Diagnosed diabetes among American Indians and Alaska Natives ages <35 years: United States, 1994–2004. *MMWR Morb Mortal Weekly Rep. 2006; 55:* 1201–1203.
95. Warne, D. 2006. Research and educational approaches to reducing health disparities among American Indians and Alaska Natives. *Journal of Transcultural Nursing, 17,* 266–271.
96. Parker, J.G. 1994. The lived experience of Native Americans with diabetes within a transcultural nursing experience. *Journal of Transcultural Nursing, 6,* 5–11.
97. Baier, L.J., & Hanson, R.L. 2004. Genetic studies of the etiology of type 2 diabetes in Pima Indians: Hunting for pieces to a complicated puzzle. *Diabetes, 53,* 1181–1186.
98. Schulz, L.O., Bennett, P.H., Ravussin, E., Kidd, J.R., Kidd, K.K., Esparza, J., & Valencia, M.E. 2006. Effects of traditional and western environments on prevalence of type 2 diabetes in Pima Indians in Mexico and the U.S. *Diabetes Care, 29,* 1866–1872.
99. Esparza-Romero J, Valencia ME, Martinez ME, Ravussin E, Schulz LO, Bennett PH. Differences in insulin resistance in Mexican and U.S. Pima Indians with normal glucose tolerance. *Clin Endocrinol Metab. 2010 Nov; 95*(11):E358–62. Epub 2010 Jul 28.
100. Murphy, N.J., Schraer, C.D., Thiele, M.C., Bulkow, L.R., Doty, B.J., & Lanier, A.P. 1995. Dietary change and obesity associated with glucose intolerance in Alaska Natives. *Journal of the American Dietetic Association, 95,* 676–682.
101. Williams, D.E., Knowler, W.C., Smith, C.J., Hanson, R.L., Roumain, J., Saremi, A., Kirska, A.M., Bennett, P.H., & Nelson, R.G. 2001. The effect of Indian or Anglo dietary preference on the incidence of diabetes in Pima Indians. *Diabetes Care, 24,* 811–816.
102. Brand, J.C., Snow, B.J., Nabham, G.P., & Traswell, A.S. 1990. Plasma glucose and insulin responses to traditional Pima Indian meals. *American Journal of Clinical Nutrition, 51,* 416–420.
103. Gittelsohn, J., Wolever, T.M., Harris, S.B., Harris-Giraldo, R., Hanley, A.J., & Zinman, B. 1998. Specific patterns of food consumption and preparation are associated with diabetes and obesity in a Native Canadian community. *Journal of Nutrition, 128,* 541–547.
104. Weaver, H.N. 1999. Transcultural nursing with Native Americans: Critical knowledge, skills, and attitudes. *Transcultural Nursing, 10,* 197–202.
105. Carrese, J.A., & Rhodes, L.A. 2000. Bridging cultural differences in medical practice: The case of discussing negative information with Navajo patients. *Journal of General Internal Medicine, 15,* 92–96.
106. Juarez, C.A.B. 1990. Professional practice at a Native American community. *Contemporary Dialysis and Nephrology, 11,* 23–28.
107. Jackson, M.Y., & Broussard, B.A. 1987. Cultural challenges in nutrition education among American Indians. *Diabetes Educator, 13,* 47–50.
108. Hodge, F.S., Paqua, A., Marquez, C.A., & Geishirt-Cantrell, B. 2002. Utilizing traditional storytelling to promote wellness, in American Indian communities. *Journal of Transcultural Nursing, 13,* 6–11.
109. Wilson, C.S. 1985. Nutritionally beneficial cultural practices. *World Review of Nutrition and Diet, 45,* 68–96.
110. Galanti, G.A. 2004. *Caring for patients from different cultures,* 3rd Ed. Philadelphia: University of Pennsylvania Press.
111. Kavanagh, K., Absalom, K., Beil, W., & Schliessmann, L. 1999. Connecting and becoming culturally competent: A Lakota example. *Advances in Nursing Science, 21,* 9–31
112. Conti, K.M. 2006. Diabetes prevention in Indian country: Developing nutrition models to tell the story of food-system change. *Journal of Transcultural Nursing, 17,* 234–245.
113. Gurley, D., Novins, D.K., Jones, M.C., Beals, J., Shore, J.H., & Manson, S.M. 2001. Comparative use of biomedical services and traditional healing options by American Indian veterans. *Psychiatric Services,52,* 68–74.
114. Novins, D.K., Beals, J., Moore, L.A., Spicer, P., Manson, S.M., & AISUPERPFP Team. 2004. Use of biomedical services and traditional healing options among American Indians: Sociodemographic correlates, spirituality, and ethnic identity. *Medical Care, 42,* 670–679.
115. Conley, R.J. 2000. Cherokees. In *Gale Encyclopedia of Multicultural America,* R.V. Dassanowsky & J. Lehman (Eds.). Farmington Hills, MI: Gale Group.

CHAPTER

Northern and Southern Europeans

Some of the largest American ethnic groups come from northern and southern Europe (see the map in Figure 6.1). Immigrants from these regions began arriving in what is now the United States in the sixteenth century and are still coming, significantly influencing American majority culture. Many foods and food habits we consider to be American were introduced by these settlers. The northern European idea of a meal, consisting of a large serving of meat, poultry, or fish with smaller side dishes of starch and vegetable, was quickly adopted and expanded in the United States to include even bigger portions of protein foods. Adaptations of some southern European specialties have become commonplace American fare. Each ethnic group from northern and southern Europe has brought a unique cuisine that was combined with indigenous ingredients—blended with the cooking of Native Americans, other Europeans, and Africans; and flavored with the foods of Latinos, Asians, and Middle Easterners—to form the foundation of the typical American diet. This chapter discusses the traditional foods and food habits of Great Britain, Ireland, France, Italy, Spain, and Portugal and examines their contributions to the cooking of the United States.

Northern Europeans

Great Britain includes the countries of England, Scotland, Wales, and Northern Ireland. Ireland is now a sovereign country. Although quite northern, the climate is temperate due to the warming influence of the Gulf Stream, and the lowlands are suitable for growing crops.

Just across the English Channel is France, regarded for centuries as the center of Western culture politically, as well as in the arts and sciences. Its capital, Paris, is one of the world's most beautiful and famed cities. France contains some of the best farmland in Europe, and three-fifths of its land is under cultivation. It is especially well known for premium wine production.

In 1607, people from Great Britain began immigrating to what has become the United States. They brought with them British trade practices and the English language, literature, law, and religion. By the time the United States gained independence from Britain, the British and their descendants constituted one-half of the American population. They produced a culture that remains unmistakably British-flavored, even today. The French came to the United States later and in smaller numbers, yet they have made significant regional contributions. Belgium, situated northeast of France, shares many French food habits but has had very little influence on the American diet due to the relatively low number of immigrants to the United States. The traditional foods of northern Europe and their influence on American cuisine are examined in the next section.

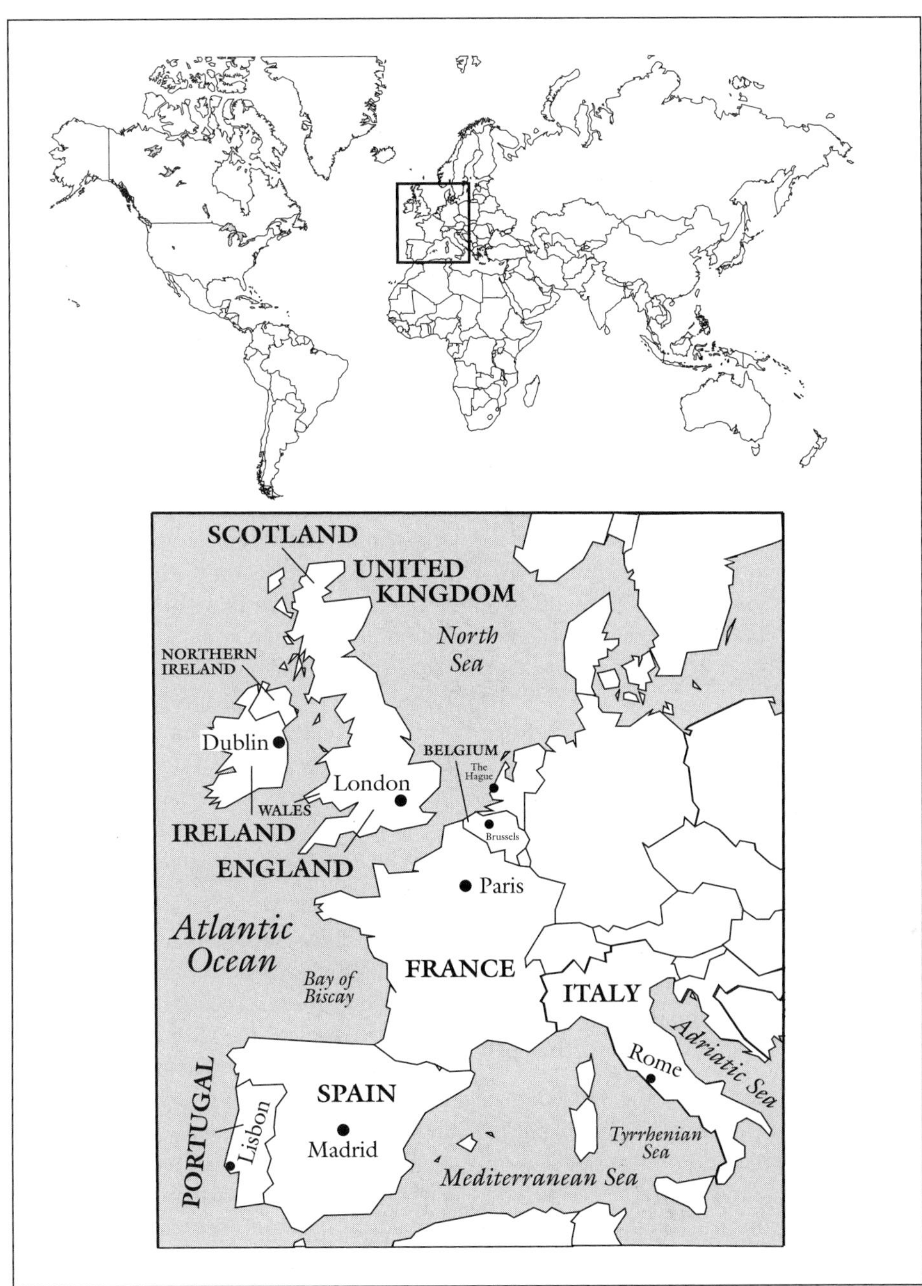

Figure 6.1
Northern and southern Europe.

Cultural Perspective

History of Northern Europeans in the United States

Immigration Patterns

Great Britain. The British who immigrated in the seventeenth century settled primarily in New England, Virginia, and Maryland. Although many originally came to avoid religious persecution, such as the Puritans in New England and the Catholics in Maryland, most later immigrants earned their passage to America by signing on as indentured servants.

By the eighteenth century, British immigration had slowed. After independence, British immigration to the United States further declined due to American hostility and disapproval by the British government. However, reported arrivals of British in the nineteenth century increased substantially. Early in the century, most immigrants were families from rural areas of southern and western England and Wales. In the latter half of the century, the majority of immigrants were from large English towns; many were seasonal unskilled workers who repeatedly returned to Britain.

It is said there have been Scots in America as long as there have been Europeans on the continent. More than one hundred towns and cities in the United States bear Scottish names, and it has been estimated that 1.5 million Scots immigrated to America. Although the majority of Scots came during the eighteenth and nineteenth centuries, 400,000 immigrated between 1921 and 1931, when Scotland suffered a severe economic depression. The Scottish settled over most of the United States and were often professionals or skilled laborers.

Although British immigration did not decline in the early twentieth century, the United States was no longer the country of first choice for those leaving Great Britain. During the Great Depression in the 1930s, more British people returned to Britain than came to America. After World War II, an increase in immigration was attributable to British war brides returning to the United States with their American husbands. Since the 1970s, British immigration has been constant at about 10,000 to 20,000 persons per year.

The Scots have been stereotyped as thrifty. "Scotch" tape was so named with the hope it would suggest an economical product.[1]

Canadian census data from 2001 reported 5.9 million of English ancestry (almost 20 percent of the total population), 4.6 million of French heritage (over 15 percent of the total population), 4.1 million of Scottish heritage (almost 14 percent of the total population), and 3.8 million of Irish background (about 12 percent of the total population).

Ireland. The first Irish people to immigrate in substantial numbers to the United States were the descendants of Scottish Presbyterians who had settled in Northern Ireland in the seventeenth century. Large-scale immigration began in the eighteenth century, and by 1775 there were an estimated 250,000 Scotch Irish living in the American colonies. Most of the immigration was the result of an economic depression brought on by a textile slump in Ireland.

Initially, the Scotch Irish settled in Pennsylvania. Before long the direction of Scotch-Irish immigration was westward to the frontier, first up the Delaware River and then beyond the Susquehanna into the rich farmlands of the Cumberland Valley. The Scotch Irish played an important role in the settlement of the trans-Allegheny region and eventually clustered around the site of Pittsburgh and in other areas of southwestern Pennsylvania. They also settled in the frontier regions of western Maryland, the Shenandoah Valley of Virginia, and the backcountry of Georgia.

Irish Catholics started to arrive in the United States by 1820, and their immigration reached an apex between 1840 and 1860, when approximately 2 million people arrived. The impetus to leave Ireland was not only religious persecution but also repeated crop failures. The potato blight that destroyed their principal crop in 1845 resulted in death by starvation of 1 million Irish people.

The Irish Catholics were the first great ethnic minority in American cities, and their early history set the pattern for later minority immigrant groups. They settled in the northeastern cities and were at the bottom of the socioeconomic ladder. The Scotch Irish, who were often of relatively high economic standing and Protestant, found it fairly easy to move into mainstream American society; however, the Irish Catholics were often stereotyped as drunkards, brawlers, and incompetents. The Irish achieved success with painful slowness. For many, their first entry into the American mainstream came by way of city politics.

France. Immigration directly from France has been the smallest, yet most constant, of that from any European country, but the return rate has been high. Most of the estimated 1 million persons who have immigrated to the United States from France have been middle-class and skilled and have come for economic opportunity.

A smaller number came because of religious persecution. More than 12,000 Huguenots (French Protestants) settled in the American colonies in the eighteenth century. They were considered to be excellent skilled workers. Generally, French people who settled in the United States were eager to assimilate and able to do so because they were economically successful.

Few pockets of French culture remain in the United States, with the exception of southern Louisiana, originally a French holding, and northern New England. However, the Frenchness of these areas is probably due more to the influence of French Canadian immigration than to direct French immigration. French Canadians are the descendants of explorers and settlers who came from France, primarily Normandy and Brittany, during the seventeenth century. They established New France in what is today known as Canada. When the English gained control of Canada, many French Canadians moved to the United States; in some instances they were deported from Canada. Most settled in northern New England, especially Maine, and their descendants are known as Franco-Americans. Others from Acadia (Nova Scotia, New Brunswick, Prince Edward Island, and part of Maine) relocated, often not by choice, to central and southern Louisiana; their descendants are known as Cajuns.

Current Demographics and Socioeconomic Status

British and Irish. The British assimilated into American mainstream society easily. Distinct groups from specific regions of Great Britain can still be found, however. For example, Cornish immigrants of the nineteenth century were often miners, and their descendants are still living in certain old mining regions, such as Grass Valley, California; Butte City, Montana; and the areas around Lake Superior. The Welsh who immigrated in the nineteenth century were miners and mill workers. They settled in the mid-Atlantic and midwestern states, especially Ohio and Pennsylvania; many were Baptist, Calvinist, or Methodist. Remnants of Welsh communities in the United States still celebrate St. David's Day (the feast day of the patron saint of Wales) and the annual festival of the National Gymanfa Ganu Association (an assembly that sings Welsh folk songs). It is estimated that there were nearly 29 million Americans of English or British ancestry in 2008.

Close to 5 million Irish Catholics have immigrated to the United States, and in 2008 there were approximately 36 million Americans who claimed Irish descent. Although they started out on a lower economic rung than other older immigrant groups, they are now scattered throughout the occupational structure.

In the 1950s, the Irish were overrepresented as clergymen, firefighters, and police officers. Today, there are disproportionately more in law, medicine, and the sciences; and Irish Catholics are more likely than other whites to attend college and to pursue advanced graduate degrees.[2] Although Irish Catholics have to some extent assimilated into mainstream American society, they still remain an identifiable ethnic group.

Photo by Laurie Macfee

▲ *Traditional foods of northern Europe. Some typical foods include apples, bacon, beef, cheese, cream, French bread, oatmeal, salt cod, and tripe.*

Persons identifying themselves as of Scottish and Scotch Irish heritage totaled over 9.3 million in the 2008 census estimate. They are well assimilated, though pockets of Scotch-Irish populations can still be found in certain Appalachian communities.

French. Over 9.5 million Americans listed French as their ancestry in the 2008 U.S. Census survey. An estimated fewer than 40 percent of French Americans are descended from immigrants who came to the United States directly from France.[3]

More than 2.2 million people of French-Canadian descent live in the United States as of 2005, many of whom make their home in the Northeast.[4] The French Canadians who settled in the New England states worked in factories that processed textiles, lumber, and bricks. Since 1950 there has been an increase in the percentage of Franco-Americans holding white-collar jobs, but they still lag behind other ethnic groups economically. Compared to the French who immigrated directly from France and assimilated rapidly into American culture, the descendants of French Canadians have clung to their heritage, including language, customs, and religious affiliation.

In Louisiana, more than 800,000 people are of Acadian ancestry, with smaller populations

The Welsh honor the patron saint of Wales, St. David, by wearing a leek stalk on the lapel. St. David identified his followers by placing a leek in the brim of their caps.

located along sections of the Gulf Coast, and one small colony remaining in northeastern Maine.[5] The Cajuns settled in rural and inaccessible areas of southern Louisiana, the bayous, and along the Mississippi River. Primarily farmers, fishermen, and herders, they were self-sufficient and kept to themselves. Today they are still rural, but their occupations reflect local economic conditions.

Worldview

Religion

British. Nearly all early British immigrants to America were Protestant. Although many came to escape persecution by the Church of England, others maintained this faith and established congregations throughout the American colonies. The Church of England in the United States became the Episcopal Church during the late eighteenth century.

British ethnicity was often expressed through religious affiliation, particularly with the Episcopal, Methodist, Baptist, and Quaker faiths. Many immigrants established distinctively English congregations, but within a generation most became indistinguishable from other American churches. Today, Americans of British descent participate in most U.S. faiths.

Irish. Religion is a cornerstone of Irish Catholic society, and in the United States it is centered on the parish. Over time, the Catholic Church in America came to be dominated by the Irish, often to the resentment of other Catholic immigrants. The church spared no effort to aid its members; it established schools, hospitals, and orphanages across the United States. The church helped to bridge the cultural gap for many Irish immigrants through advice, job placement, savings clubs, and temperance societies. Today, religion plays a less important role in Irish Catholic life, although the role of the Irish in the church is still significant.

French. Among French Americans, Franco-Americans, and Acadians, the Catholic Church provided the nucleus of the community, gave it stability, and helped preserve the language and traditions of the people. The church today still plays a central role.

New England Puritans and English Quakers were among the first in the United States to promote free public education.

Family

British. The immigrant English family formed the model for the typical American family. It included a father, a mother, and their children. This family group sometimes resided near other relatives, but more often established solitary households. The father was in charge of the public and business aspects of the family, while the mother controlled the domestic and social responsibilities. Traditionally, the oldest children in the home were well educated and were sent to private schools, if affordable. Such an education was considered an investment in the future, and children were expected to continue the family business and to maintain the family's social position. Due to the similarities between the British family and the emerging American family, new immigrants from England assimilated quickly. It was very common for British immigrants to marry non-English spouses.

Irish. Many of the characteristics of the Irish family in the nineteenth century persisted into the twentieth century. Irish Catholics tended to marry at a late age, have large families, and divorce rarely. Today, however, most first- and second-generation Irish Catholics are likely to marry outside their group and, with increasing frequency, outside Catholicism. Traditionally, the father was the breadwinner in the Irish Catholic family, but the mother's position was a strong one. Daughters were often as well educated as sons. The Irish people's relatively egalitarian attitudes toward sex roles may be responsible for the high concentration of Irish American women in professional jobs and white-collar work.

French. The Franco-Americans in New England maintained many French traditions through their continued contact with French relatives in Quebec. They had little desire to acculturate. During the 1930s, due to the Great Depression and new laws restricting reentry into the United States, the bond to Canada weakened, which diminished new French Canadian immigration. Today the descendants of the French Canadians speak French infrequently and often marry outsiders. Family ties are still strong, but, as with the Cajuns, family size has decreased. Franco-American women have traditionally had higher status and more authority than their counterparts in France.

Until the twentieth century, Cajuns lived in rural areas in extended family households with as many as ten or twelve children per couple. The whole family worked as a unit, and decisions that affected the group were made jointly by all the adults. Until 1945, many Cajuns were illiterate and spoke only Cajun French. The use of Cajun French was prohibited by the public schools in 1921; as a result, many younger Cajuns today do not speak or understand French. The average family size today is smaller, and there is more marriage outside the community, but Cajuns still retain strong ties to their families.

Traditional Health Beliefs and Practices. Many of what are characterized as American majority cultural beliefs regarding health originated in northern Europe. For example, when students were surveyed on family home remedies,[6] those of British, Irish, and French descent shared similar health maintenance practices such as a good diet, plentiful sleep, and daily exercise. Fresh air, cleanliness, and keeping warm and dry were also commonly mentioned. The Irish traditionally wear protective religious medallions.

Among the British and the Irish there is the more generalized belief that good health is dependent on proper attitude (which includes religious faith) and a rigorous, regular lifestyle. Many northern Europeans associate a moderate diet with maintaining bowel regularity, and laxative use is common. Stomach ailments may be explained as due to food that is too spicy, spoiled, or incompatible (causing an allergic reaction).

The traditional French lifestyle, which features leisurely meals and little structured exercise, presents a paradox to researchers. Although the French consume more total saturated fat and cholesterol than Americans, their death rate from heart disease is less than one-half of that in the United States. Scientists speculate that some other protective factor in the French diet or lifestyle may account for this discrepancy, such as the higher intake of wine or more frequent walking. Genetic factors do not appear to be a cause. Studies comparing the French with Americans of French ancestry have not been reported.

The French Canadians who settled in Louisiana brought numerous traditional remedies. Today, Americans of French descent in Louisiana, including the Cajuns and urban dwellers of all socioeconomic groups, often use home remedies and consult folk healers. Salves of whiskey and camphor or sheep's tallow and turpentine are considered beneficial for colds. Tobacco smoke is blown into the ears to cure earaches. A string of garlic is tied around the neck of a baby with worms. Red flannel pouches filled with camphor or asafetida are worn to prevent illness. It should be noted that some Americans of French descent living in the region may also consult practitioners of voodoo for health problems (see Chapter 8, "Africans" for more details).

Traditional Food Habits

The influence of France on the food habits of Great Britain and Ireland and vice versa has led to many similarities in the cuisines of these countries, although the ingredients of southern French cooking differ in that they are more like those of Mediterranean countries. The influence of these northern European cuisines on American foods and food habits has been extensive.

Ingredients and Common Foods

Staples and Regional Variations

Great Britain and Ireland. Animal products are of key importance in Great Britain and Ireland. Some form of meat, poultry, or fish is present at most meals, in addition to eggs and cheese. In Britain and Ireland, lamb is a commonly eaten meat, as is roast beef, which is often made for Sunday dinner with Yorkshire pudding (a popover cooked in meat drippings). Pork is often served as sausages (bangers) and bacon. Various game birds are also eaten. The cultural food groups list (Table 6.1) includes a more complete detailing of ingredients.

The British and Irish diets also contain a variety of seafood. A well-known fast-food item is fish and chips. The fish is battered and deep-fried, served with fried potatoes, and seasoned with salt and malt vinegar. Salt-dried fish, including ling, cod, and pollack, was traditionally served with a white sauce and potatoes for meatless days by Irish Catholics.[51] Preserved fish is also found as an appetizer or at breakfast. Examples are smoked Scottish salmon and kippers, which are a salted and smoked fish.

Dairy products and eggs also play an important role in the diet of the British and Irish. Eggs are traditionally served for breakfast, and cheese is the key ingredient in the traditional ploughman's lunch served in pubs. It consists of a piece of cheddar cheese, bread, pickled onions, and a pint of beer. Other cheeses produced in England are the slightly nutty Cheshire and

A "pub," or public house, is a bar that serves beer, wine, hard liquor, and light meals. The British pub is often the place where friends and family meet to socialize.

Oatcakes, called bannocks, were traditionally eaten to celebrate the pagan Celtic holiday of Beltane on May 1. One section was burnt or covered with ash; the unlucky person who received the marked portion was sacrificed (in more recent times the victim would leap through a small bonfire three times instead).

Table 6.1 Cultural Food Groups: Northern European

Group	Comments	Common Foods	Adaptations in the United States
Protein Foods			
Milk/milk products	The English and Irish drink milk as a beverage. Cheese is eaten daily.	Cheese (cow, sheep, and goat milk), cream, milk, sour cream, yogurt	
Meat/poultry/fish/ eggs/legumes	Meat, poultry, or fish is usually the centerpiece of the meal. Meats are generally roasted or broiled in Great Britain; also prepared as stews or in pies. Smoked, salted, or dried fish is popular in England.	*Meat:* beef (roasts; variety cuts such as brains, kidneys, liver, sweetbreads, tongue, and tripe), horsemeat, lamb, oxtail, pork, rabbit, snails, veal, venison *Poultry and small birds:* chicken, duck, goose, partridge, pheasant, pigeon, quail, thrush, turkey *Fish and shellfish:* anchovies, bass, clams, cod, crab, crawfish, haddock, herring, lobster, mackerel, mullet, mussels, oysters, perch, pike, pompano, salmon, sardines, scallops, shad, shrimp, skate, sole, sturgeon, trout, whiting *Eggs:* poultry and fish *Legumes:* kidney beans, lentils, lima beans, split peas	The Irish consume more animal protein.
Cereals/Grains	Wheat bread usually accompanies the meal. In Britain and Ireland oatmeal or porridge is common for breakfast.	Barley, hops, oats, rice, rye, wheat	Corn and corn products are consumed more.
Fruits/Vegetables	Potatoes are frequently eaten in Ireland. Arrowroot starch is used as thickener, and tapioca (from cassava tubers) is eaten.	*Fruits:* apples, apricots, cherries, currants, gooseberries, grapes (many varieties), lemons, melons, oranges, peaches, pears, plums, prunes, raisins, raspberries, rhubarb, strawberries *Vegetables:* artichokes, asparagus, beets, brussels sprouts, cabbage, carrots, cauliflower, celery, celery root, cucumbers, eggplant, fennel, green beans, green peppers, kale, lettuce (many varieties), leeks, mushrooms (including chanterelles, cèpes), olives, onions, parsnips, peas, potatoes, radishes, salsify, scallions, sorrel, spinach, tomatoes, turnips, truffles, watercress	Native and transplanted fruits and vegetables, such as bananas, blueberries, okra, and squash, were added to the diet.
Additional Foods			
Seasonings	British and Irish dishes emphasize naturalness of foods with mild seasoning, served with flavorful condiments or sauces used to taste. French dishes are often prepared with complementary sauces or gravies that enhance food flavor.	Angelica (licorice-flavored plant), bay leaf, capers, chile peppers, chives, chocolate, chutney, cinnamon, cloves, coffee, cognac, fennel seeds, garlic, ginger, horseradish, juniper berries, mace, malt vinegar, marjoram, mint, mustard, nutmeg, oregano, paprika, parsley, pepper (black, white, green, and pink), rosemary, saffron, sage, shallots, sweet basil, Tabasco sauce (and other hot sauces), tarragon, thyme, vanilla, Worcestershire sauce	Cajun and Creole cooking are highly spiced. Stews are thickened with filé powder (sassafras).
Nuts/seeds	Nuts especially popular; used primarily in desserts.	*Nuts:* almonds (sweet and bitter), chestnuts, filberts (hazelnuts), pecans, walnuts (including black) *Seeds:* sesame	
Beverages	Alcoholic beverages consumed as part of the meal.	Beer (ale, stout, bitters), black and herbal tea (mint, anise, chamomile, etc.), cider, coffee, gin, hot chocolate, liqueurs, port, sherry, whiskey, wine (red, white, champagne, and fruit/vegetable)	
Fats/oils	Butter used extensively in cooking of northern and central France; olive oil more common in southern regions of the country.	Butter, goose fat, lard, margarine, olive oil, vegetable oil, salt pork	
Sweeteners		Honey, sugar	Molasses and maple syrup are used as sweeteners. Irish Americans use more sugar than members of other groups.

Stilton, a blue cheese. In Ireland, a market for hand-crafted farmhouse cheeses has developed over the past decade. They include both fresh, soft cheeses and aged types (often flavored with herbs or other seasonings), from cow's or goat's milk. One cheese that has gained international acclaim is Cashel blue, which is produced in Tipperary, Ireland.

Devonshire, England, is known for its rich cream products, such as double cream (which has twice as much butter fat as ordinary cream) and clotted cream, a slightly fermented, thickened cream. It is often spread on scones, biscuits made with baking powder.

Though not the main focus of the meal, breads are not overlooked. In Ireland, soda bread, a bread made with baking soda instead of yeast, was traditionally prepared every day to accompany the meal and remains popular today. Another version was made of cornmeal. Wheat flour is commonly used for baking, and oatmeal is eaten as a porridge for breakfast in Scotland or used in making bread and biscuits throughout Britain and Ireland. Biscuits, or biskcake, in England can refer to bread, cake, cookies, crackers, or what are known in the United States as biscuits. Scottish shortbread is an example of a sweet, buttery biscuit.

Fruits and vegetables are limited to those that grow best in cool climates. Potatoes, brought to Ireland from the New World in the seventeenth century, are the mainstay of the Irish diet and are found in British fare as well. Potatoes are found in stews or pies, such as stobhach Gaelach, an Irish stew of lamb's neck, and shepherd's or cottage pie, a meat pie made of leftover ground meat and onions and topped with mashed potatoes. Mashed potatoes are often just referred to as *mash*, as in bangers and mash (sausages and mashed potatoes). Some side dishes made of potatoes are boxty, a type of potato pancake or dumpling; bubble and squeak, a dish made of leftover cabbage and potatoes chopped and fried together; and colcannon, mashed and seasoned boiled white vegetables with onions or leeks. Berries are popular in puddings, pies, and jams. Kitchen gardens are still found in many areas, providing tomatoes, cucumbers, watercress, and other items. Farmers' markets, featuring fresh local produce, are increasingly popular.

One unusual vegetable of the region is seaweed. Laver is a purple seaweed (called nori in Japan) that is a specialty in Wales and parts of northern coastal England. It is customarily boiled into a gelatinous paste, then mixed with oatmeal and formed into patties that are fried. Known as laverbread, these cakes are traditionally served at breakfast with bacon. Dulse, a red algae eaten in Ireland, can be consumed fresh, but is usually

© Photodisc/PhotoLink/Getty Images

▲ *Fish and chips are classic pub food in England and Ireland, typically served with malt vinegar.*

[SAMPLE MENU]

An Irish Pub Supper

Steak/Beef and Guinness Pie[a,b]

Brown Bread[a,b]

Apple Crumble[a,b]

Stout

[a]Johnson, M.M. 2006. *The Irish pub cookbook*. San Francisco: Chronicle Books.

[b]*Recipes from the Irish Kitchen* at http://www.littleshamrocks.com/Irish-Food-Recipes.html

dried, then chewed like beef jerky for a snack, or flaked and added to soups or warm milk.

The most common beverages consumed by adults in Ireland and England are tea, beer, and whiskey. Tea, which has become synonymous with a meal or break in the afternoon, was introduced to England in 1662 by the wife of Charles II. Drunk with most meals and as a refreshment, strong black tea is preferred, served with milk and sugar. Frequently consumed alcoholic beverages include beer and whiskey. The British and the Irish do not drink the bottom-fermented style of beer common in the United States. Instead, in Britain the pubs usually serve bitters, an amber-colored, top-fermented beer, strongly flavored with hops, while in Ireland, a favorite is stout, a dark, rich beer that can provide substantial calories to the diet. Both beers are served at cellar temperature and are naturally carbonated.

Whiskey is made in both Ireland and Scotland, but the Irish are usually credited with its invention and name. In Ireland, it is distilled from mashed, fermented barley. Scotch, or Scotch whisky (spelled without an e), is distilled from a blend of malted whiskey (in which the barley germinated before fermentation) and unmalted whiskey. Scotch is traditionally a much stronger, smokier-tasting beverage than Irish whiskey. Other alcoholic drinks popular in Britain are gin, Port (a brandy-fortified wine made in Portugal), and Sherry (a fortified wine from Spain). A less common beverage but still popular in some regions is mead, a type of honey wine made from the fermentation of honey and water. The Welsh prefer a stronger, highly spiced variety called metheglyn.

France. The cooking of France has traditionally been divided into classic French cuisine (*haute or grande* cuisine) and provincial or regional cooking.[7] Classic French cooking is elegant and formal, mostly prepared in restaurants using the best ingredients from throughout the country. Provincial cuisine is simpler fare made at home or local cafes featuring fresh local ingredients. Broadly speaking, butter and cream enriches many dishes in the northeastern and central regions of the country, while lard, duck fat, and goose fat flavor foods in the northwest and south central areas.[8] In the southeast, olive oil is prominent. Seafood and lamb are specialties of the north, while pork is common in the regions bordering Belgium and Germany. Beef and veal are favorites in the central areas, and in the southernmost regions near Spain, fish is a specialty. Cold weather fruits and vegetables are featured in northern dishes, while temperate, Mediterranean produce is the mainstay in southern areas. In the north, foods are subtly seasoned. In the south, garlic flavors many dishes.

The ancestors of most French Americans are from two of France's northern provinces, Brittany and Normandy. Brittany, known as Bretagne, is located in the northwest; its shores are washed by the English Channel and the Bay of Biscay. Seafood, simply prepared, is common, and delicate Belon oysters are shipped throughout France. Mutton and vegetables from the region are said to have a naturally salty taste because of the salt spray. Apples are the prevalent fruit, and cider is widely exported.

Located along the English Channel and east of Brittany is Normandy, also known for its seafood and apples. Calvados, an apple brandy, is thought to be the mother of applejack, an alcoholic apple drink used to clear the palate during meals in Louisiana. Another alcoholic drink produced in

▼ *Tea time in Great Britain has become an afternoon meal, with small sandwiches, scones (on the second rack of the silver tray), and an assortment of cookies and pastries.*

Courtesy of Grossich and Partners

the region is Bénédictine, named after the Roman Catholic monks who still make it at the monastery in Fecamp. Normandy is also renowned for its rich dairy products; its butter is considered one of the best in France. Camembert, a semisoft cheese with a mild flavor, and Pont-l'Évêque, a hearty aromatic cheese, are produced in the area. Dishes from Normandy are often prepared with rich cream sauces. Crêpes, very thin, unleavened pancakes, originated in this region; they are typically served topped with sweet or savory sauces or rolled with meat, poultry, fish or seafood, cheese, or fruit fillings.

Champagne, bordered by the English Channel and Belgium, has a cuisine influenced by the Germanic cultures. Beer is popular, as are sausages, such as andouille and andouillette, large and small intestinal casings stuffed with pork or lamb stomach. Charcuterie, cold meat dishes such as sausages, pâtés, and terrines, which often are sold in specialty stores, are especially good from this region. Pâté is a spread of finely ground, cooked, seasoned meats. A terrine is commonly made with leftover meats cut into small pieces, mixed with spices and a jelling substance, then baked in a loaf pan. Throughout the world, Champagne is probably best known for its naturally carbonated wines. Only sparkling wines produced in this region can be legally called champagne in France.

The province that borders Germany, Alsace-Lorraine, has been alternately ruled by France and Germany. One of its principal cities is Strasbourg. Many German foods are favored in the region, such as goose, sausages, and sauerkraut. Goose fat is often used for cooking, and one of the specialties of the area is pâté de fois gras, pâté made from the enlarged livers of force-fed geese. Another famous dish is quiche Lorraine, pie pastry baked with a filling of cream, beaten eggs, and bacon. Alsace-Lorraine is a wine-producing area; its wines are similar to the German Rhine wines but are usually not as sweet. Distilled liquors produced in the region are kirsch, a cherry brandy, and the brandy eau de vie de framboise, made from raspberries.

Located south of Normandy and Brittany in the west-central part of France is Touraine, the province that includes the fertile Loire valley. Along the river one can see the beautiful chateaux or palaces built by the French nobility. Known as the "garden of France," Touraine produces some of the finest fruits and vegetables in the country. A dry white wine produced in the area is Vouvray. In the north-central region is the area surrounding the city of Paris called the Ile-de-France, the home of classic French cuisine. Some of the finest beef and veal, as well as a variety of fruits and vegetables, are produced in this fertile region. Brie, semisoft and mild flavored, is the best-known cheese of the area. Dishes of the Ile-de-France include lobster à l'américaine, lobster prepared with tomatoes, shallots, herbs, white wine, and brandy; potage St. Germain, pea soup; filet de bœuf béarnaise, filet of beef with a béarnaise sauce; and tarte tatin, an upside-down apple and caramel tart.

Located southeast of Paris is Burgundy, one of the foremost wine-producing regions of France. Burgundy's robust dishes start to take on the flavor of southern France; they contain garlic and are often prepared with olive oil. Dijon, a principal city, is also the name of the mustards of the region, prepared with white wine and herbs. Dishes of the area are escargot, or snails (raised on grape vines) cooked in a garlic butter and served in the shell; coq au vin, rooster or chicken cooked in wine; and bœuf bourguignon, a hearty red wine beef stew. In Burgundy, the red wines are primarily made from the pinot noir grape and the white wines from the chardonnay grape. The great wines of the area are usually named after the villages in which they are produced; for example, Gevrey-Chambertin, Vosne-Romanée, and Volnay. Cassis, a black currant liquor, is also produced in the region, and brandy from Cognac is a specialty. To the east, along the border with Switzerland, is the mountainous Franche-Comte region, known for its exceptionally tender and flavorful Bresse chicken.

The other major wine-producing region of France is Bordeaux, which is also the name of its principal city. Famous for its hearty dishes, the term *à la bordelaise* can mean (1) prepared in a seasoned sauce containing red or white wine, marrow, tomatoes, butter, and shallots; (2) use of mirepoix, a finely minced mixture of carrots, onions, and celery seasoned with bay leaves and thyme; (3) accompanied by cèpes, large fleshy mushrooms; or (4) accompanied by an artichoke and potato garnish. A red Bordeaux wine is full-bodied and made primarily from the cabernet sauvignon grape. (In Great Britain, Bordeaux is called claret.) Among the wines produced are St. Julien, Margaux, Graves, St. Emilion, Pomerol, and Sauternes, a sweet white dessert wine.

Colcannon was customarily served for the harvest dinner and on Halloween in Ireland. For Halloween, coins were wrapped and buried in the dish so the children could find them as they ate.

The term *honeymoon* originated with the European custom of newlyweds drinking mead for the first lunar month following their wedding.

Tomatoes were introduced to Europe in 1523 from the New World. Reaction was mixed; some people thought they were poisonous, while others believed they brought luck. Tomato-shaped pin cushions developed from the latter superstition.

Belgians are renowned for their beers. One specialty ale is lambic, a fruity brew distinctive for its use of unmalted (raw) wheat and open-tank fermentation with wild yeast.

Fresh cream in France, called fleurette, is often added to sauces or whipped for dessert. Also popular is crème fraîche, cream that is fermented until it is thickened and slightly tangy.

In the south of France is Languedoc, famous for cassoulet, a complex dish containing duck or goose, pork or mutton, sausage, and white beans, among other ingredients. Provence, located on the Mediterranean Sea, is a favorite vacation spot because of its warm Riviera beaches. Provence is also known for the large old port city of Marseilles, its perfumes from the city of Grasse, and the international film festival in Cannes.

The cooking of Provence is similar to that of Italy and Spain. Staple ingredients are tomatoes, garlic, and olive oil; à la Provençal means that a dish contains these three items. Other common food items are seafood from the Mediterranean, artichokes, eggplant, and zucchini. Popular dishes from the region are bouillabaisse, the famed fish stew made with tomatoes, garlic, olive oil, and several types of seafood, seasoned with saffron, and usually served with rouille, a hot red pepper sauce; ratatouille, tomatoes, eggplant, and zucchini cooked in olive oil; salade Niçoise, a salad originating in Nice, containing tuna, tomatoes, olives, lettuce, other raw vegetables, and sometimes hard-boiled eggs; and pan bagna, a French bread sandwich slathered with olive oil and containing a variety of ingredients, such as anchovies, tomatoes, green peppers, onions, olives, hardboiled eggs, and capers. One unique specialty item in the region associated with haute cuisine is black truffles. This costly, pungent underground fungus flavors or garnishes many classic French dishes.

Cooking Styles. Although the ingredients used in the countries on the opposite sides of the English Channel are not substantially different, their cooking styles vary greatly. British and Irish food is described as simple and hearty fare that developed out of rural, seasonal traditions.[9] French cuisine is admired for its fresh ingredients, attention to detail, and technical proficiency—and it is imitated around the world.

Great Britain and Ireland. Both the British and the Irish take pride in the naturalness of their dishes and their ability to cook foods so the flavors are enhanced rather than obscured. In recent years, the eating local movement (see Chapter 15, "Regional Americans," for more details) and government programs promoting regional specialties have lead to a renewed interest in traditional fare.[10]

Meat is usually roasted or broiled, depending on the cut, and lightly seasoned with herbs and spices. Strong-flavored condiments such as Worcestershire sauce (flavored with anchovies, vinegar, soy, garlic, and assorted spices) on roast beef or mint jelly on lamb are often served. Chutneys, highly spiced fruit or vegetable pickles originally from India, are also popular. Leftover meat is finely chopped, then served in a stew, pie, or pudding. Offal, parts of the animal often discarded, such as lamb's brains, pig's tail, and calf's heart, have become trendy items in England, appreciated for both their traditional heritage and the ecological/ethical value of using the whole animal.

While most Americans think of pies and puddings as being sweet desserts, in Britain and Ireland this is not necessarily the case. A pie is a baked pastry consisting of a mixture of meats, game, fish, and vegetables, or fruit, covered with or enclosed in a crust. A Cornish pasty is an individual pillow-shaped pie filled with meat, onions, potatoes, and sometimes fruit. Another well-known British dish is steak and kidney pie.

[SAMPLE MENU]

A French Lunch

***Pâté*[a,b*] and Baguette**

***Quiche* (Seafood, Onion, or Lorraine)[a,b]**

Green Salad

Selection of Cheeses (e.g., *Brie, Pont-l'Evêque*)

Fresh Fruit or *Tarte aux Pommes* (Apple Tart)[a,b]

Wine

[a]Child, J., Bertholle, L., & Beck, S. 2001. *Mastering the art of French cooking* (Vol. I). New York: Knopf.

[b]*French Food and Cook* at http://www.ffcook.com

*Can be purchased at a specialty cheese shop or delicatessen.

Pudding is a steamed, boiled, or baked dish that may be based on anything from custards and fruits to meat and vegetables. An example of a sweet pudding is plum pudding, which is served traditionally at Christmas. It is a steamed dish of suet, dried and candied fruit, and other ingredients. Trifle is a layered dessert made from custard, pound cake, raspberry jam, whipped cream, sherry, and almonds.

France. Classic French cuisine implies a carefully planned meal that balances the texture, color, and flavor of the dishes, similar to the harmony found in musical compositions or paintings.[45] The soul of French cooking is its sauces, often painstakingly prepared from stocks simmered for hours to bring out the flavor. A white stock is made from fish, chicken, or veal; and a brown stock is made from beef or veal.[11]

Sauces are subtly flavored with natural ingredients, such as vegetables, wine, and herbs. They must never overwhelm the food, but rather complement it. The five basic sauces are espagnole (or brown sauce), made with brown stock, mirepoix, and roux (thickening agent made from flour cooked in butter or fat drippings); velouté, made with white stock, roux, onions, and spices; béchamel (or cream sauce), made with white stock, milk, and roux; tomato, made with white stock, tomatoes, onions, carrots, garlic, and roux; and, hollandaise, which combines egg yolks and drawn butter (béarnaise is similar, flavored with tarragon). Examples of classic cold sauces are mayonnaise and vinaigrette, a mixture of vinegar and oil.

Some common rules in preparing French dishes are (1) never mix sweet and sour flavors in the same dish; (2) never serve sweet sauces over fish; (3) do not under- or overcook food; (4) with the exception of salad and fruit, do not serve uncooked food; (5) always use the freshest, best-tasting ingredients; and (6) wine is an integral part of the meal and must complement the food.

French breads and pastries are particularly noteworthy. Breads are typically made with white flour, shaped into long loaves (e.g., thin baguettes), rounds, braids, or rings, then baked in a wood-fired oven. Sweeter breads, such as eggy brioche or buttery, flaky croissants, are between breads and pastries. Specialty doughs, such as cream puff pastry, multilayered puff pastry, and the classic sponge cake génoise, are used to create the numerous desserts of France, such as cakes, petits fours (small, bite-size pastries), and tarts. Chocolate, fresh fruits, and pastry cream thickened with egg yolks enrich these pastries.

© K Sanchez/Cole Group/Getty Images

▲ *French breads are consumed at every meal, and include baguettes, braids, rings, and sweeter versions such as brioche (with little topknots).*

In recent years, classic French cuisine has merged with a rediscovery of regional fare to create what is known as nouvelle cuisine (new cuisine).[12] The practice of nouvelle cuisine has popularized French cooking worldwide and influenced the development of local specialties with an emphasis on fresh ingredients. An appreciation for the cooking of other nations, especially those of Asia, has occurred in France, and many dishes now use foreign seasonings or use exotic techniques and presentations.

Meal Composition and Cycle

Daily Pattern

Great Britain and Ireland. In Britain, four meals are traditionally served each day—breakfast, lunch, tea, and an evening meal (dinner). In the nineteenth and early twentieth centuries, breakfast was a very substantial meal, consisting of oatmeal; bacon, ham, or sausage; eggs (prepared several ways); bread fried in bacon grease; toast with jam or marmalade; grilled tomatoes or mushrooms; and possibly smoked fish or deviled kidneys. All this was washed down with tea. Today, in Scotland, oatmeal is usually eaten for breakfast, while in England, packaged breakfast cereals are often eaten during the week, and the more extensive breakfast is reserved for weekends and special occasions.

The first eating chocolate was introduced by the British in 1847. Europeans now consume twenty to twenty-five pounds of chocolate per person each year—twice the amount eaten by U.S. citizens.

Potted is an English term for fish, meat, poultry, or game pounded with lard or butter into a coarse or smooth pâté, then preserved in jars or pots. *Deviled* describes a dish prepared with a spicy hot sauce or seasoning.

Lunch was originally a hearty meal and still is on Sundays, but during the week it is squeezed in between work hours. It may include a meat pie, fish and chips, or a light meal at the pub with a pint of bitters or stout. Both Sunday lunch and the weekday dinner are much like a U.S. dinner. The meals consist of meat or fish, vegetable, and starch. The starch is often potatoes or rice, and bread also accompanies the meal. Dessert (often called "pudding") follows the main course.

In the late afternoon in Britain and Ireland, most people take a break and have a pot of tea and a light snack. In some areas a high tea is served. This can be a substantial meal that includes potted meat, fish, shrimp, ham salad, salmon cakes, fruits, and a selection of cakes and pastries. High tea is associated with working-class or rural families who have maintained the custom of a large lunch, with high tea serving as dinner. It is thought that the upper British classes add the term *high* to tea as a dinner when it is served occasionally in place of dinner as a novelty or to children as an informal substitute for dinner. Whether snack or meal, the British often just call it tea.

In the small town of Palmiers, the city council has banned any ready-made or mass-produced food (e.g., frozen pizza) from the local school cafeteria to "ensure our kids stay healthy, teach them the taste of proper French food, and help keep our small farmers in business."[14]

In Britain, a piece of mincemeat pie is eaten at midnight on New Year's Eve while making a wish for the upcoming year.

France. The French eat only three meals a day—breakfast, lunch, and dinner. Second helpings are uncommon, and there is very little snacking between meals. Breakfast, in contrast to the British meal, is very light, consisting of a croissant or French bread with butter and jam, and strong coffee with hot milk or hot chocolate. The French breakfast is what is known in the United States as a continental breakfast. Lunch is traditionally the largest meal of the day and, in some regions of France, businesses close at midday for two hours so people can return home to eat. The meal usually starts with an appetizer (hors d'oeuvre) such as pâté. The main course is a meat, fish, or egg dish accompanied by a vegetable and bread. If salad is eaten, it is served after the main course. Dessert at home is usually cheese and fruit. In a restaurant, ice cream (more like a fruit sherbet or sorbet), cakes, custards, and pastries are served in addition to fruit and cheese. Wine is served with the meal and coffee after the meal. Dinner is similar, but traditionally a lighter meal with a starter course of soup or appetizer, then a main dish, followed by a cheese course. However, meal patterns are changing in France. The popularity of fast foods and shorter lunch periods is resulting in a more American pattern of smaller lunch followed by larger dinner.

Etiquette. The fork is not passed from the right hand to the left hand when cutting food in England and Ireland. Instead the fork remains in the left hand, and the knife in the right. The two are often used together to scoop food onto the fork. All dishes are passed to the left. When not eating, the hands should be placed in one's lap. In Ireland, a small plate to the left of the setting is used for placing potato peelings.[13]

France is similar to Great Britain in the use of the fork and knife. For example, lettuce in a salad should not be cut, but folded into a small, easy-to-eat packet. The French also pass dishes to the left. They do not usually use bread plates but place their portions of bread directly on the table. In contrast to England and Ireland, it is considered impolite to put your hands on your lap. The wrists should be rested on the table with the hands in view. Chocolates are appropriate gifts to bring to a dinner. In England, a bottle of champagne is appreciated, while in Ireland a bottle of wine to serve with the meal is also common. In France, a dessert-style wine or after-dinner liqueur is the best beverage to give.

Special Occasions. Christmas and Easter are the most important Christian holidays celebrated in England, Ireland, and France. Ireland and France are predominantly Catholic countries and tend to observe all the holy days of obligation and patron saints' days. France commemorates the beginning of the French Revolution on July 14, Bastille Day.

Great Britain and Ireland. The British celebrate Christmas by serving hot punch or mulled wine; roast beef, goose, turkey, or ham; plum pudding and mincemeat (or just "mince") pies; and, afterward, port with nuts and dried fruit. The plum pudding is traditionally splashed with brandy and then flamed before being served. Mincemeat pies were originally prepared with seasoned, ground meats, suet, and fruit, but today they are usually made with only dried and candied fruit, nuts, and spices. Boxing Day, the day after Christmas, is when friends and relatives visit one another.

Foods served at Easter include hot cross buns and Shrewsbury simnel. In ancient times, the cross on the buns is believed to have symbolized both sun and fire; the four quarters represented the seasons. Today the cross represents Christ and the resurrection. Shrewsbury simnel is a rich spice cake topped with twelve decorative balls of marzipan originally representing the astrological signs. (It is also served on Mother's Day.) Another holiday celebrated throughout Great Britain is New Year's Day on January 1.

The Scottish traditionally eat haggis on New Year's Eve. Haggis is a sheep's stomach stuffed with a pudding made of sheep's innards and oatmeal. After it is served, adult diners drench their portions with Scotch whisky before eating. It is also the traditional entrée (served with "neeps," mashed turnips, and "tatties," mashed potatoes) on Burns's Night, commemorating the national poet Robert Burns (January 25). St. Patrick's Day began as a religious commemoration for the patron saint of Ireland. The Irish-American custom of eating a corned beef and cabbage meal on March 17 is now as popular in Ireland as it is in the United States.[15]

France. In France, the main Christmas meal is served after mass on the night of December 24. Two traditional dishes are a boudin noir and boudin blanc, also known as black pudding and white pudding (dark blood sausage or a light-colored one made from veal, chicken, or pork with milk) and a goose or turkey with chestnuts. In Provence, the Christmas Eve meal is meatless, usually cod, but the highlight is that it is followed by thirteen desserts.

On Shrove Tuesday (Mardi Gras), the French feast on pancakes, fritters, waffles, and various biscuits and cakes. During Lent, no eggs, fat, or meat are eaten. Dishes served during Lent often contain cod or herring. Cod is also the traditional dish served on Good Friday; in some regions, lentils are eaten to wash away one's sins. Easter marks the return of the normal diet, and eggs are often served hard-boiled (also colored), in omelets, or in breads and pastries. French toast (croûtes dorée) is a traditional Easter dish. Also common are pies filled with minced meats.

Therapeutic Uses of Food. Most northern Europeans share a belief that a good diet is essential to maintaining health. Traditional home remedies for minor illnesses include chicken soup, tea with honey or lemon or whiskey, hot milk, or hot whiskey with cloves. Practices less common today are taking sulfur with molasses as a laxative and regular use of cod liver oil. Some Irish Americans may use senna (Cassia actifolia) weekly to cleanse their bowels.[16,17]

© Robert Brenner/PhotoEdit, Inc.

▲ *St. Patrick's Day in Savannah, Georgia.*

Home remedies popular with Americans of French descent include infusions made from magnolia leaves, elderberry flowers, sassafras, or citronella, which are prescribed for colds. Sore throats are treated by gargling herbal teas or hot water with dissolved honey, salt, and baking soda. Sassafras tea is used to cleanse the blood, and garlic is ingested for worms.

Contemporary Food Habits in the United States

Adaptations of Food Habits

Ingredients and Common Foods

British and Irish. Many United States dishes have their origins in Great Britain. The Puritans, adapting Native American fare, made a pudding with cornmeal, milk, molasses, and spices. Today, this is called Indian pudding. Pumpkin pie is just a custard pie to which the Native American squash, pumpkin, is added. Apple pie has been so well accepted that we say, "American as apple pie," despite its English heritage. Syllabub, a milk and wine punch drunk in the American South at Christmastime, is also an English recipe.

French. French cooking has had less influence on everyday American cooking (except for French fries), but there are probably few cities that don't have a French restaurant (which may or may not be owned by a French immigrant).

French Americans adapted their cuisine to the available ingredients and other ethnic cooking styles. The best example of this is found in Louisiana, where Creole and Cajun cooking developed. Creole cooking is to Cajun cooking what

Robert Burns once wrote that haggis was the "great chieftain o' the puddin' race." Nevertheless, Scottish government officials recommended in 2006 that haggis be served to children no more than once a week due to its high fat and sodium content.

The members of the French Foreign Legion are nicknamed "boudin," because the red blanket roll they wear across their chests resembles a black pudding.

Gumbo is the African Bantu word for okra.

The name for the popular Cajun music style, zydeco, is derived from the French term for green bean, haricot (pronounced "ar-ee-ko") because it is snappy, like a bean.

French grande cuisine is to provincial cooking. Some dishes may sound typically French, such as the fish stew known as bouillabaisse, but this is made with fish from the Gulf of Mexico, not from the Mediterranean. Even the coffee is slightly different, flavored with the bitter chicory root.

Ingredients for Cajun cooking reflect the environment of Louisiana: Bayou Cajun foods are from lake and swamp areas, whereas prairie Cajun dishes are found in inland areas. Fish and shellfish abound, notably crawfish, crabs, oysters, pompano, redfish, and shrimp, to name just a few. Shellfish is commonly eaten raw on the half shell (oysters) or boiled in a spicy mixture. Gumbo and jambalaya are often made with seafood. Gumbo is a thick, spicy soup made with a variety of seafood, meat, and vegetables. It is thickened with either okra or filé powder and then ladled over rice. Jambalaya, also a highly seasoned stew made with a combination of seafood, meats, and vegetables, was brought to New Orleans by the Spanish. Originally made only with ham (jambon), it was later modified. The base for these stews and gravies is roux; however, the Cajun roux is unique in that the flour and fat (usually vegetable oil) are cooked very slowly until the mixture turns brown and has a nutlike aroma and taste.

Other key ingredients in Cajun cooking are rice (which has been grown in Louisiana since the early 1700s), red beans, tomatoes, chayote squash, eggplant, spicy hot sauce, and a variety of pork products. One of the better known hot sauces, Tabasco, is produced in the bayous of southern Louisiana from fermented chile peppers, vinegar, and spices. A deep-fried rice fritter, calas, is the Louisiana version of a doughnut. Other rice dishes are red beans and rice and dirty rice. Dirty rice derives its name from the fact that its ingredients, bits of chicken gizzards and liver, give the rice a brown appearance. Cajun boudin sausages are a specialty. Boudin blanc is made with pork and rice; boudin rouge has pork blood added to it. Cochon de lait, a suckling pig roasted over a wood fire, is prepared at Cajun festivals in central Louisiana. Fricot is a popular soup made with potatoes and sausage or shredded meat. Cracklings, known as gratons, are bite-size bits of fried pork skin (often with meat attached) popular in some regions.

Pecan pralines are a famous New Orleans candy. Pecans are native to Louisiana; pralines are large, flat patties made from brown sugar, water or cream, and butter. Another confection eaten often with coffee is beignets, round or square puffed French doughnuts dusted with powdered sugar. French toast, or pain perdu, is another French specialty that was transported to New Orleans and is now familiar to most Americans.

The cuisine of French Americans in New England tends to be traditionally French, but it is influenced by common New England foods and food habits. Franco-Americans use more herbs and spices than other New Englanders and take time to prepare the best-tasting food. Traditional French dishes are pork pâté, called creton by the Franco-Americans, and the traditional Yule log cake (bûche de Noël) served at Christmas. Franco-American cuisine offers numerous soups and stews. One of the most elaborate of the stews, which is also called a pie, is cipate, known as cipaille, si-pallie, six-pates, and sea pie in some areas. A typical recipe calls for chicken, pork, veal, and beef, plus four or five kinds of vegetables layered in a heavy kettle, covered with pie crust. It is slowly cooked after chicken stock has been added through vents in the crust.

Maple syrup is commonly used. One unique breakfast dish is eggs poached in the syrup. Maple syrup is also served over bread dumplings or just plain bread. Franco-Americans appreciate wine and distilled spirits. One unusual combination of both is caribou, a mixture of white whiskey (a distilled, colorless liquor) and red wine, which is drunk on festive occasions. (See also Chapter 15.)

***La boucherie:* French-speaking Cajuns in Louisiana maintain the hog-butchering traditions of their past. Before the days of refrigeration, everyone in the community helped prepare the meat and lard. Participants went home with fresh pork cuts and spicy sausages called boudin. *La boucherie* continues today at many Cajun festivals.**

Courtesy of the Louisiana Office of Tourism

Meal Composition and Cycle

British and Irish. American food habits have been greatly influenced by British and Irish immigrants. Meal patterns and composition are very similar to those in Great Britain. The typical meal of a meat, poultry, or fish main dish served with vegetable and starch side dishes, and often bread, continues to this day. Though English Americans also consumed a hearty breakfast that often included ham or bacon and eggs, in more recent years, time constraints and health concerns have changed this pattern on weekdays; weekend breakfast sometimes reverts to the more British-style meal.

Festive meals also reflect the British and Irish influence. A traditional Christmas dinner includes roast turkey or ham, stuffing, and mashed vegetables. For dessert, a pie is customary, often mincemeat. Two holidays Americans think of as being typically American, Thanksgiving and Halloween, are actually of British and Irish origin. Thanksgiving combined the tradition of an old British harvest festival with the Pilgrims' celebration of surviving in their new environment. In Great Britain and Ireland, Halloween, or All Hallow's Eve, is believed to have originated in ancient times. Ghosts and witches were thought likely to wander abroad on Halloween night.

French. Americans of French descent have adopted the U.S. meal cycle with the main meal in the evening. In Louisiana, the best-known celebration is Mardi Gras, culminating on Shrove Tuesday, just before the beginning of Lent. In New Orleans there are parades, masquerading, and general revelry; the festival reaches its climax at a grand ball before midnight. After this day and night of rich eating and grand merriment, the forty days of fasting and penitence of Lent begin. In the Cajun countryside, Mardi Gras is celebrated with run: Men on horseback ride from farmhouse to farmhouse collecting chickens and sausages to add to a community gumbo. Participants enjoy beer, boudin, and faire le maque ("make like a monkey," or clowning around) at each stop. During the rest of the year, Cajuns sponsor many local festivals, such as the crawfish, rice, and yam festivals.

Franco-Americans, like their French ancestors, serve meat pies on religious holidays. The special pie for Easter has sliced hard-boiled eggs laid down on the bottom crust and then a layer of cooked meat topped with well-seasoned pork and beef meatballs. For Christmas, tourtière, a pie made with simmered seasoned pork, is eaten cold after midnight Mass.

Nutritional Status

The influence of the British and French on American cuisine is undoubtedly one reason the U.S. diet is high in cholesterol and fat, and low in fiber and complex carbohydrates. Current research in Europe suggests continuing similarities. A recent survey of dietary habits found that consumption of potatoes, animal protein, processed foods, margarine and butter, and sweets is relatively high in the United Kingdom.[18] In France, consumption of added animal fats and oils is high.[19] Estimated prevalence of overweight and obesity (body mass index [BMI] >25) is 57 percent for women and 66 percent for men in England; 48 percent for women and more than 66 percent for men in Ireland; and 41 percent for women and nearly 66 percent for men in France.[20] Although few studies have been conducted on the nutritional status of Americans who are of French, Irish, and British descent, it is assumed that they have the same nutritional advantages and disadvantages as the general U.S. population.

Nutritional Intake. Very little has been reported on the diets of northern Europeans living in the United States. A study to determine differences in mortality from coronary heart disease examined Irish brothers—one group in Ireland, one group living in the United States (Boston)—and a third control group of first-generation Irish Americans in Boston.[21] Although there was no significantly different relative risk for death from heart disease among the three groups, it was found that their diets varied significantly. The Boston brothers and the first-generation Irish Americans had a higher intake (as a percentage of caloric intake) of animal protein, total fat (more vegetable and less animal), sugar, fiber, and cholesterol, and a lower intake of starch. The Irish brothers had a higher caloric intake than the Boston brothers and the first-generation Irish Americans, yet their relative weight was significantly lower.

It is commonly assumed that a high rate of alcoholism prevails among Irish Americans; however, little has been reported confirming or

Crawfish are also known as crayfish (especially in New Orleans), crawdads, crawdaddy crab (in the Great Lakes area), clawfish, and mudbugs, among others. They are small crustaceans that look like miniature lobsters, found in the fresh waters of Louisiana, Lake Michigan, California, and the Pacific Northwest.

Cornish pasties are still popular in parts of the country where immigrants from Cornwall came to work in the mines, such as the Upper Peninsula in Michigan, where May 24 was declared Pasty Day in 1968.

A study in France found that consumption of mineral water provided up to 25 percent of total daily intake of calcium and 6 to 7 percent of magnesium in participants.[16]

Descendants of southern Europeans may have a higher incidence of lactose intolerance than other European groups. Alcohol consumption among the Basques in Spain is high, especially for men.[39]

An Italian proverb states that after age forty, a person can "expect a new pain every morning."

Some Italians believe that wine mixed with milk in the stomach causes too much acid, so milk is avoided at meals and consumed mostly with snacks.

Historically, *Mac* before a family name meant "son of," whereas *O* signified "descended from."

Some Cajuns think that being thin means a person is puny or unattractive.[27] Cajuns may believe that milk and fish should not be consumed at the same meal.

The market for processed spaghetti sauce in the United States is more than $500 million annually.

refuting this belief. One comparative study found that Irish-American men had higher rates of excessive drinking, and more physical, psychological, sexual, and/or occupational problems with alcohol misuse than did Puerto Rican men.[22]

Prevalence for hereditary hemochromatosis, which may be treated with a low-iron diet (and avoidance of alcohol and foods or supplements high in vitamin C), are higher northern Europeans (estimated to be 0.54 percent in whites).[23] Though it is hypothesized that the gene for the disease is Celtic in origin,[24] a study of French Canadians also noted a high prevalence rate.[25] (It should be noted that hemochromatosis are often undetected in these populations, and may not show overt clinical symptoms until middle or late adulthood.) A study on the causes of inherited chylomicronemia indicates that the frequency of lipoprotein lipase deficiency is very high among French Canadians.[26] Franco-Americans may also have high rates of this genetic defect, leading to elevated triglycerides and the necessity of a very low-fat diet.

Counseling. Studies of people in France suggest that the French are likely to undertake many activities at once, change plans frequently, ignore schedules, and communicate indirectly with enthusiastic body language.[28] Direct, intense eye contact is important. Among the English and Irish, less eye contact between strangers or acquaintances is common. The English are low context in their communication style, yet prefer understatement and somewhat indirect discussion compared to Americans. The Irish are high context, direct, and often effusive. The French are very low context and can be exceptionally direct.[13] Most northern Europeans use a quick, light handshake when greeting one another. It is important to note, however, that most Americans of British, Irish, or French descent are completely acculturated.

Americans of British and Irish descent are often stoic in the face of illness and reserved in the communication of their symptoms. Some Irish believe that the best way to stay healthy is to avoid doctors unless very ill. The British, Irish, and French all tend to be more formal than Americans and politeness is expected. Socioeconomic status and religious practice are likely to have greater impact on foods and food habits than country of origin. The in-depth personal interview should reveal any notable ethnically based preferences.

NEW AMERICAN PERSPECTIVES—Irish

JOHN CASEY, Retired

I came to the U.S. in 1956 from Ireland when I was twenty-six years old. I first lived in New York City where plenty of other Irish live. When I left Ireland, it wasn't as well-off as it is now, and food was not plentiful, and it was mostly grown locally. You raised pigs and killed two a year, and that provided the bacon for the rest of a year. The foods we ate every day were bread, butter, milk, and eggs. My dad owned a food shop, so we had a bit more of other foods. When I came to America, I was overwhelmed by the amount of food available and all the different types. I had never had juice with breakfast, didn't know what a grapefruit was—thought it was a very big lemon. Other foods that I tried for the first time were watermelon, turkey, hamburgers, corn on the cob, and french fries. I like all of them except the watermelon.

In Ireland the main meal of the day was lunch, and what we usually had was all boiled together, like a New England boiled dinner but without as much meat and usually no meat. On sick days we got toast and tea. But the bread was only toasted on one side. When I first got toast here, it was toasted on both sides, and I wasn't sure if you buttered both sides as well. The three biggest holidays in Ireland are Christmas, Easter, and St. Patrick's Day. Christmas was the biggest feast day—bacon, eggs, and sausage for breakfast and for the main meal we had goose with dressing and mashed potatoes, plus custard for dessert. For Easter we often had mutton, and the children—if it was affordable—got chocolate Easter eggs, just like here. St. Patrick's Day wasn't as much fun because it fell during Lent, and the pubs were closed.

I eat a lot of different foods now, more than I did when I was younger, and I like Chinese and German food, but I miss Irish bacon. My grandchildren are still trying to get me to order different flavors of ice cream, but I will only eat vanilla. When I first came to America, Ireland didn't have enough food, and Americans ate too much. Today, both the Irish and Americans eat too much.

Southern Europeans

Southern European countries lie along the Mediterranean Sea and include Italy, southern France, Spain, and Portugal. Italy, shaped like a boot, sticks out into the Mediterranean and includes the island of Sicily, which lies off the boot toe. Italy is separated from the rest of Europe by the Alps, which form its northern border. Spain, located to the west of France (the Pyrenees Mountains form a natural border between the two countries), occupies the majority of the Iberian Peninsula. Portugal sits on the western end of the peninsula and includes the Azore and Madeira Islands located in the Atlantic Ocean (the Cape Verde Islands were formerly a Portuguese territory, but they gained independence in 1975). Most of southern Europe enjoys a warm Mediterranean climate except in the cooler mountainous regions.

Immigration to the United States from southern Europe has been considerable, primarily from the poorer regions of southern Italy. Many Americans, even those of non-European descent, enjoy Italian cuisine in some form. The foods of Spain and Portugal are similar to those of Italy and France due to the shared climate and history of Greek and Roman influence in the region, but their preparations differ. The following section reviews the traditional diets of Italy, Spain, and Portugal. The influence of these cuisines on U.S. fare is also discussed.

Photo by Laurie Macfee

▲ *Traditional foods of southern Europe. Some typical foods include almonds, artichokes, basil, cheese, eggplant, garlic, chickpeas, olive oil, olives, onions, pasta, prosciutto, salt cod, sweet bread, and tomatoes.*

Cultural Perspective

History of Southern Europeans in the United States

Immigration Patterns. The majority of immigrants from southern Europe were Italians, who swelled the population of U.S. cities on the eastern seaboard during the late nineteenth and early twentieth centuries. Next in number were the Portuguese, primarily from the Azore Islands. Smaller numbers of Spanish immigrants have been reported.

Italians. According to immigration records, more than 5 million Italians have settled in the United States. The majority came from the poorer southern Italian provinces and from Sicily between 1880 and 1920. Although earlier immigrants from northern Italy settled on the West Coast of the United States during the gold rush, most of these later immigrants settled in the large industrial cities on the East Coast. Many Italians who arrived faced discrimination and hostility, and, in response, formed concentrated communities within urban centers, often called Little Italies.[29] Several cities still boast Italian neighborhoods such as the North End in Boston and North Beach in San Francisco.

Many Italians came to the United States for economic reasons; more than one-half of the immigrants, mostly men, returned to their homeland after accumulating sufficient money. Peasants in their native land, Italians in the United States often became laborers in skilled or semiskilled professions, especially the building trades and the clothing industry. Immigration from Italy fell sharply after World War I; however, more than 500,000 Italians have immigrated since World War II.

Spaniards. More than one-quarter of a million people from Spain have immigrated to the United States since 1820. However, the majority of the Spanish-speaking population in the United States comes from the U.S. acquisition of Spanish territories and the immigration of people from Latin American countries (see Chapters 9 and 10 for more detail).

The earliest Spanish settlers arrived during colonial times, establishing populations in what is now Florida, New Mexico, California, Arizona, Texas, and Louisiana. A majority were from the poorest regions of southern Spain and the Canary Islands.[30] Half of all other Spanish immigrants to the United States came later in the nineteenth and early twentieth centuries, due to depressed economic conditions in Spain. In 1939, after the fall of the second Spanish republic, a small number of refugees immigrated for political reasons.

Additional Spanish immigrants were from the Basque region, located in northeastern Spain on the border with France (there are also French Basques). The Basques are thought to be one of the oldest surviving ethnic groups in Europe; they lived in their homeland before the invasion

There are more than 1.2 million Canadians of Italian ancestry, according to the 2001 census figures. Boise, Idaho, is considered the Basque capital of the United States.

Among the Basques, it is said that the Devil once came to the region to learn their language, Euskera, so that he could entrap the inhabitants. He gave up after seven years when he was only able to mastert wo words: *bai* and *ez* (yes and no).

of the Indo-Europeans around 2000 B.C.E. Their language, Euskera, is not known to be related to any other living language. Though the earliest Basque immigrants to the United States were fishermen and whalers who probably arrived before Columbus, most came in the mid-nineteenth century, arriving first in California for the Gold Rush, then spreading north and east throughout the West. Many emigrated from South America, where they had first settled, and were listed as Chileans (the umbrella term used for all South Americans at the time). An accurate estimate of their numbers is impossible.[31]

Portuguese. Beginning in the early nineteenth century, two waves of Portuguese immigrants arrived in the United States. Early immigrants were primarily from the Azore Islands and Cape Verde Islands, and they often located in the whaling ports of New England and Hawaii. They were followed in the 1870s by immigrants hoping to escape poverty. They arrived with little education and few skills, but were willing to do farm labor in California and Hawaii, and work in the service trades of northeastern cities.

After World War II, a small number of Portuguese from Macao, a Portuguese settlement on the coast of China near Hong Kong, settled in California. They were well educated and many held professional jobs. A much more significant number of Portuguese, more than 150,000, entered the United States after 1958, again mostly from the Azore Islands, following a series of volcanic eruptions that devastated the region. It has been estimated that the Portuguese currently have one of the higher rates of new arrivals to the United States among European groups.

Current Demographics and Socioeconomic Status

Italians. In the United States in 2008 there were nearly 18 million Americans of Italian descent, most of whom live in or around major cities. Economic conditions improved during the 1980s in Italy, and immigration from the nation slowed significantly. However, there are still nearly 500,000 immigrants born in Italy who make their homes in the United States today.

Economically, Italian Americans shared in the general prosperity after World War II, and today most are employed in white-collar jobs or as skilled laborers. Four generations of Italians living in the United States have been identified. The elderly living in urban Italian neighborhoods are one group, those who are middle age and living in either urban or suburban settings are the second group, the well-educated younger Italian Americans of subsequent generations living in mostly suburban areas are the third, and the very recent immigrants from Italy are the fourth.[32] These groupings can be expected to change as each grouping ages and with increased assimilation: only 20 percent of Italian Americans born after 1940 married other Italian Americans.

Spaniards. People who report Spanish, Spaniard, or Spanish-American heritage were over 1.1 million in the 2005 U.S. Census estimates and are now grouped with Hispanics in the U.S. Census. Seven percent were born in Spain. Most are well-integrated into their communities, although larger populations are found in New York and Tampa, Florida. A distinctive group of Isleños, descendants of Canary Island immigrants, is found in southern Louisiana. The Basques settled mostly in the rural regions of California, Nevada, Idaho, Montana, Wyoming, Colorado, New Mexico, and Arizona and became ranchers. Some Basque immigrants, however, were drawn to the mining jobs of West Virginia and the rubber and steel plants of Ohio, Illinois, Michigan, and Pennsylvania. Though the 2008 census estimates report 58,000 Basque Americans, it is thought that this number may underrepresent the total population, which may be as high as 100,000.[31] Today most Basque descendants are involved in some aspect of animal husbandry or small business; few have entered other professions. Newer Basque communities now exist in Connecticut and Florida because jai alai (a Basque sport) facilities were established there.

Portuguese. As of 2008 over 1.4 million Americans were of Portuguese descent. In 2000, fifty thousand claimed Cape Verdean ancestry, and four thousand reported Azore Islands heritage. (Immigrants from the Cape Verde and Azore Islands and those from Madeira may not feel Portuguese. Instead, they identify with their island or city of origin.) Initially, the Portuguese Americans on the West Coast were farmers and ranchers, but eventually their descendants moved into professional, technical, and administrative positions.[33] On the East Coast, the descendants of the Portuguese who settled in the whaling ports now make up a significant part of the fishing industry, though only 3 percent of all Portuguese Americans work in this occupation. The percentage of Portuguese families living in poverty is half that of the U.S. average.

Worldview

Religion

Italians. In Italy, the Roman Catholic Church was a part of everyday life. Immigrants to the United States, however, found the church to be more remote and puritanical, as well as staffed by the Irish. The church responded by establishing national parishes (parishes geared toward one ethnic group with a priest from that group), that helped immigrants adjust to the United States. Some religious festivals, part of daily spiritual life in Italy, were transferred to the United States and are still celebrated today, such as the Feast of San Gennaro in New York's Little Italy.

Spaniards. Most Spaniards are Roman Catholic. The Jesuit Order was founded in Basque country and has significantly influenced Basque devotion. Basque Americans are involved in their parishes, and there is the expectation that religion is part of daily life and sacrifice.

Portuguese. The Roman Catholic Church also helped the Portuguese ease into the mainstream of U.S. life. Local churches and special parishes often sponsor traditional religious festas that include Portuguese foods, dances, and colorful costumes.

Family

Italians. The social structure of rural villages in southern Italy was based on the family, whose interests and needs molded each individual's attitudes toward the state, church, and school. The family was self-reliant and distrusted outsiders. Each member was expected to uphold family honor and fulfill familial responsibilities. The father was head of the household; he maintained his authority with strict discipline. The mother, although subordinate, controlled the day-to-day activities in the home and was often responsible for the family budget. Once in the United States, the children broke free of parental control due to economic necessity. Although sons had always been allowed some independence, daughters soon gained freedom, as well, because they were expected to work outside the home like their brothers. Education eventually also changed the family. Early immigrants repeatedly denied their children schooling, sending them to work instead. However, by 1920, education was considered an important stepping stone for Italian Americans.

Spaniards. In the traditional Spanish family, the father spent much of his time working and socializing outside the home, while the mother devoted her life to her children. Typically, one daughter would choose not to marry and would care for her parents as they aged. In the United States, Spanish-American families are usually limited to immediate members, although the obligation to parents remains stronger than for most Americans. An elder may live part of the year with one child, then part of the year with another child. Independent living and retirement homes are also common. The Basque family was customarily an extended one. Basques in Spain are prohibited from marrying non-Basques, but in the United States many Basques marry other nationalities. Basques accept all members who marry into their families.

Spanish women hold unique status among southern Europeans. Class distinctions are more important than gender when it comes to educational and professional attainment. The Basque women are historically recognized for their equality. Since ancient times, their duties have been as valued as those of men, and jobs are often not gender-specific.

Portuguese. Like the Italians, the Portuguese have close family solidarity and have had some success in maintaining the traditional family structure. Grown sons and daughters often live in close proximity to their parents, and family members try to care for the sick at home. Family structure is threatened, however, when women must work or generational values change. Men tend to dominate the family, and, as a result, some Portuguese-American women marry outside the group.

Traditional Health Beliefs and Practices. Traditional Italian health beliefs include concepts common in the American majority culture as well as concerns associated with folk medicine. Fresh air is believed necessary to health, and some older Italian Americans maintain that the heavy air of the United States is considered unhealthy compared to the light air of Italy. Well-being is

A tortilla in Spain is an egg omelet, not the cornmeal flatbread eaten by the Mexicans. It is believed that the Spanish called the Mexican bread by that name because of its similar shape.

defined as the ability to pursue normal, daily activities. There is the expectation that health declines with age.

Some Italian Americans believe that illness is due to contamination (through an unclean or sick person) or heredity (blood). Older immigrants may also think that sickness occurs because of drafts (surgery may be avoided so that organs will not be exposed to air), the suppression of emotions (i.e., anxiety, fear, grief), or supernatural causes. Some Italians believe that a minor illness can be attributed to the evil eye and that serious conditions result from being cursed by a malicious person or God.[34,35,6] Saints may be implored for protection, good luck charms worn, or, more traditionally, the practices of a *maghi* (witch) used to avoid illness. Although many Italians do not profess a belief that God punishes sin with a curse, there is often a fatalistic approach to terminal illness as being the result of God's will. Italian Americans, especially those of older generations, sometimes believe that problems in pregnancy are due to diet,. Unsatisfied cravings are reputed to cause deformities, and if a woman does not eat a food she smells, she may suffer a miscarriage. Little has been reported regarding Spanish and Portuguese health practices.

Recenty, research has related the traditional food habits of the Mediterrean diet pattern to a lower incidence of coronary heart disease, various types of cancer, and other diseases. It appears that the tradition diet of southern Europe is protective against several chronic diseases.[36]

Olive oil is labeled according to method of processing and the percent of acidity, from extra virgin to virgin (or pure) olive oil. In the United States, only the oils derived from the first press of the olives can be called virgin or extra virgin depending on its acidity (extra virgin is lower); a blend of olive oil, produced by by refining that does not alter it's fat structure, and virgin olive oils developed to reduce acidity and must be labeled "pure."

Traditional Food Habits

Although the foods of the southern European countries are similar, as detailed in the Cultural Food Groups list (Table 6.2), there are notable differences in preparation and presentation. Many Americans think of Italian cooking as consisting of pizza and spaghetti. In reality, these dishes are only a small part of the regional cuisine of southern Italy, the original homeland of most Italian Americans. Spanish food is mistakenly equated with the hot and spicy cuisine of Mexico. Although Mexico was a colony of Spain, the foods and food habits of the two countries differ substantially. Portugal and Spain have very similar cuisines, but most of the Portuguese immigrants to the United States are from the Azore Islands and the island of Madeira. Their diet was less varied than that of the mainland Portuguese.

Ingredients and Common Foods

Foreign Influence. The Phoenicians and Greeks, who settled along the Mediterranean coast in ancient times, are believed to have brought the olive tree and chickpeas (garbanzo beans) to the region. In addition, fish stew, known as bouillabaisse in France and zuppa di pesce alla marinara in Italy, may be of Greek origin. The Muslims brought eggplants, lemons, oranges, sugar cane, rice, and a variety of sweetmeats and spices. Marzipan, a sweetened almond paste used extensively in Italian desserts, and rice flavored with saffron, as in the northern Italian dish risotto alla Milanese, are both believed to have Muslim origins. In Spain the Muslim influence is also seen in saffron-seasoned rice and in the use of ground nuts in sauces, candies, and other desserts.

It was the food of the New World colonies, however, that shaped much of Italian, Spanish, and Portuguese cuisine. Chocolate, vanilla, tomatoes, avocados, chili peppers, pineapple, white and sweet potatoes, corn, many varieties of squash, and turkey were brought back from the Americas. The tomato is of particular importance to the character of southern European cooking. Asian ingredients have had a significant impact on the fare of Portugal and, to a lesser degree, the dishes of Spain and Italy. From India and the Far East came coconuts, bananas, mangoes, sweet oranges, and numerous spices, such as pepper, nutmeg, cinnamon, and cloves.

Staples

Italy. Although the cooking styles and ingredients vary from region to region in Italy, some general statements can be made about ingredients. Pasta is the quintessential dish throughout the nation. It is prepared fresh, from dough made with the addition of eggs, or dried, from a dough made without eggs. It is traditionally served three ways: with sauce (*asciutta*), in soup (*en brodo*), or baked (*al forno*). There are literally hundreds of pasta shapes, such as thin, round strips that include spaghetti (from the Italian word for string) and capelli d'angelo (angel hair); flat strips such as linguini and fettucini (ribbon); tubular forms, such as macaroni, penne, and the larger manicotti; and sheets such as lasagna and pappardelle. There are additional forms, such as spirals (e.g., fusilli, rotelle), shells (conchiglie), little ears (orechiette), bowties (farfalle), and small barley- or rice-shaped orzo. The most common pasta in Italy is tagliatelle, a medium-width flat noodle.[8]

In the north, fresh pastas are more common, and stuffed versions made with bits of

Table 6.2 Cultural Food Groups: Southern European

Group	Comments	Common Foods	Adaptations in the United States
Protein Foods			
Milk/milk products	Most adults do not drink milk but do eat cheese. Dairy products are often used in desserts. Many adults suffer from lactose intolerance.	Cheese (cow, sheep, buffalo, goat), milk	It is assumed that second- and third-generation southern Europeans drink more milk into their adulthood than their ancestors did.
Meat/poultry/fish/eggs/legumes	Dried salt cod is eaten frequently. Small fish, such as sardines, are eaten whole, providing substantial dietary calcium.	*Meat:* beef, goat, lamb, pork, veal (and most variety cuts *Poultry:* chicken, duck, goose, pigeon, turkey, woodcock *Fish:* anchovies, bream, cod, haddock, halibut, herring, mullet, salmon, sardines, trout, tuna, turbot, whiting, octopus, squid *Shellfish:* barnacles, clams, conch, crab, lobster, mussels, scallops, shrimp *Eggs:* chicken *Legumes:* chickpeas, fava and kidney beans, lentils, lupine seeds, white beans	More meat and less fish are eaten than in Europe.
Cereals/Grains	Bread, pasta, or grain products usually accompany the meal.	Cornmeal, rice, wheat (bread, farina, a variety of pastas)	
Fruits/Vegetables	Fruit is often eaten as dessert. Fresh fruits and vegetables are preferred.	*Fruit:* apples, apricots, bananas, cherries, citron, dates, figs, grapefruit, grapes, lemons, medlars, peaches, pears, pineapples, plums (prunes), pomegranates, quinces, oranges, raisins, Seville oranges, tangerines *Vegetables: arugala*, artichokes, asparagus, broccoli, cabbage, cardoon, cauliflower, celery, chicory, cucumber, eggplant, endive, escarole, fennel, green beans, lettuce, kale, kohlrabi, mushrooms, mustard greens, olives, parsnips, peas, peppers (green and red), pimentos, potatoes, *radicchio*, swiss chard, tomatoes, turnips, zucchini	First- and second-generation southern Europeans generally eat only fresh fruits and vegetables. Fruit and vegetable consumption tends to reflect general American food habits by the third generation.
Additional Foods			
Seasonings	Dishes using similar ingredients in Italy, Spain, and Portugal often differentiated by distinctive use of herbs and spices. Seasoning in Azore Islands and Cape Verde Islands is usually very mild.	Basil, bay leaf, black pepper, capers, cayenne pepper, chocolate, chervil, cinnamon, cloves, coriander, cumin, dill, fennel, garlic, leeks, lemon juice, marjoram, mint, mustard, nutmeg, onion, oregano, parsley (Italian and curley leaf), rosemary, saffron, sage, tarragon, thyme, vinegar	
Nuts/seeds	Nuts commonly used in desserts and added to some entrees and side dishes.	Almonds, hazelnuts, *pignolis* (pine nuts), walnuts, lupine seeds	
Beverages		Coffee, chocolate, liqueurs, port, Madeira, sherry, flavored sodas (e.g., *orzata*), tea, wine	
Fats/oils	Olive oil flavors numerous dishes; used for deep-frying in Spain.	Butter, lard, olive oil, vegetable oil	Use of olive oil has decreased.
Sweeteners		Honey, sugar	

meat, cheese, and vegetables, such as ravioli, are especially popular. Pasta in the north is also frequently topped with rich cream sauces. In the agriculturally poorer south, the pasta is generally dried, and it is usually served unfilled with a tomato-based sauce.

Other broad differences are that northern fare uses more butter, dairy products, rice, and meat than the south, which is notable for the use of olive oil, more fish, and more beans and vegetables, such as artichokes, eggplants, bell peppers, and tomatoes. Garlic is found throughout the nation, though it is more popular in the north. Other seasonings common to all of Italy are parsley, basil, and oregano. Anise, cinnamon, nutmeg, mace, and cloves are also used in many dishes.

The Italians eat more rice than any other Europeans. Thomas Jefferson supposedly smuggled rice out of Italy to the United States, where his first attempts to cultivate it were unsuccessful.

Espresso, which may mean "made expressly for you," is made from finely ground dark roast coffee through which water is forced by steam pressure. Cappuccino is espresso topped with frothy steamed milk.

Spain. The rugged terrain in Spain is suitable for raising small animals and crops, such as grapes and olives. Spain is the largest producer of olives in the world. Entrées usually feature eggs, lamb, pork, poultry, or dried and salted fish (especially cod, called bacalao). Eggs are consumed day and night. They are enjoyed fried in olive oil, often topped with migas (fried bread crumbs combined with garlic, bacon, and ham). Tortilla española (potato omelet) is perhaps the national dish, eaten as appetizers, entrees, snacks, and as a filling for bocadillos (sandwiches). Sausages, such as the paprika- and garlic-flavored chorizo and the blood sausage called morilla, are common. Serrano (meaning "from the mountains") ham is a salty, dry-cured meat served in paper-thin slices that has gained worldwide acclaim. Seafood is popular in coastal regions. Meats are often combined with vegetables in savory stews. Each region has its own recipe for paella, which typically includes saffron-seasoned rice topped with chicken, mussels, shrimp, sausage, tomatoes, and peas. Cocido, a stew of chickpeas, vegetables (e.g., cabbage, carrots, potatoes), and meats (e.g., beef, chicken, pork, meatballs, sausages), also varies from area to area, but is always served in three courses. The strained broth with added noodles is eaten first, followed by a plate of the boiled vegetables, and concluded with a plate of cooked meats. Crusty bread is served with the meal.

"Cods' tongues" are a specialty enjoyed by all southern Europeans. Though they are an especially succulent strip of meat from inside the fish's mouth, they are not actually tongues.

The name *gazpacho* may have come from the vinegar-and-water drink called posca, reportedly offered to Christ on the cross.

Garlic and tomatoes flavor many Spanish dishes, for example, gazpacho, a refreshing pureed vegetable soup that is usually served cold, and zarzuela (meaning "operetta"), a fresh seafood stew. Olive oil is also a common ingredient used in almost all cooking, even deep-frying pastries, such as the ridged, cylindrical doughnuts known as churros. Sauces accompany many dishes. Alioli is made from garlic pulverized with olive oil, salt, and a little lemon juice. It is served with grilled or boiled meats and fish. Another popular sauce, called romescu, is sometimes mixed with alioli to taste by each diner at the table. It combines pureed almonds, garlic, paprika, and tomatoes with vinegar and olive oil. Fruit, particularly oranges, is popular for dessert, sometimes served in custard. One favorite is membillo, a quince paste served with slices of a salty sheep's-milk cheese known as Manchego. Spain's best-known dessert is flan, a sweet milk-and-egg custard topped with caramel. Wine usually accompanies the meal. Sangria, made with red or white wine and fresh fruit juices, is served chilled in the summer. Spain is probably most famous in the United States for its Sherries, which are wines fortified with added brandy. Sherries can be dry or sweet and are categorized by the length of time they are aged. They are often described as having a nutty flavor.

© M Lamotte/Cole Group/Getty Images

▶ *Pasta comes in dozens of forms in Italy, including thin strings, flat ribbons, tubes, spirals, sheets, and shapes that resemble wheels, bowties, little ears, hats, rice, and other items. It is found fresh or dried.*

Portugal. Portuguese fare shares some similarities in ingredients with Spanish cuisine, but a more generous addition of herbs and spices, including cilantro, mint, and cumin, distinguishes the cooking. Fish dominates the diet of the

Portuguese; they are said to have as many recipes for bacalhau (dried salt cod) as there are days in the year.[37] Sardines are often grilled or cooked in a tomato and vegetable sauce. Lamprey is a popular food in northern Portugal, where it is often prepared with curry-like seasonings. Shellfish, such as clams, are often combined with pork or other meats in stewed dishes. Chouriço, similar to the Spanish pork sausage, chorizo, and linguiça, a pork and garlic sausage, are often eaten at breakfast. Other typical dishes are cacoila, a stew made from pig hearts and liver, then served with beans or potatoes; isca de figado, beef liver seasoned with vinegar, pepper, and garlic, then fried in olive oil or lard; and assada no espeto, meat roasted on a spit. A common soup is caldo verde, or green broth, made from kale or cabbage and potatoes. A unique dry soup of bread moistened with oil or vinegar and topped with anything from meat, chicken, or shellfish and vegetables is called açordu. Fava beans, chickpeas, and lupine seeds (tremocos) are added to some dishes. Rice and fried potatoes are so popular they are often served together. Crusty country breads and, in the north, a cornmeal bread called broa also accompany the meal. Portuguese sweet breads, pan doce, and doughnuts, malassadas, are also specialties. Desserts often feature fruit, such as bananas, grapes, and figs, as well as eggs and almonds. Puddings, custards, and sponge cakes are popular.

Regional Variations

Italy. Some of the regional specialties in the northern area of Lombardy, around Milan, are risotto, a creamy rice dish cooked in butter and chicken stock, flavored with Parmesan cheese and saffron; polenta, cornmeal mush (thought to have been made originally from semolina wheat), often served with cheese or sauce; and panettone, a type of fruitcake. Veal is very popular, served in the stew known as osso buco and in veal piccata (chops that are pounded very thin, then breaded and pan-fried, topped with lemon juice, capers, and minced parsley). The cheeses of the region include Gorgonzola, a tangy, blue-veined cheese made from sheep's milk, and Bel paese, a soft, mild-flavored cheese. The area is also known for its aperitifs, such as bittersweet Vermouth.

Venice, located on the east coast and known for its romantic canals (although the city actually consists of 120 mud islands), has a

[SAMPLE MENU]

An Italian Lunch

Crostini[a,b,c]

Spaghetti con Cozze **(Spaghetti with Mussels)**[a,c]

Chicken *Saltimboca*[b,c]

Sauteed Spinach

Biscotti **and** ***Espresso***[b,c]

[a]Hazin, M. 2004. *Marcella says...* New York: HarperCollins Publishers.

[b]Casella, C. 2005. *True Tuscan*. New York: HarperCollins Publishers.

[c]Marion Batali at http://www.mariobatali.com/

◀ *Olives and olive oil are found in numerous southern European dishes. Spain is the primary producer of olives worldwide.*

© Photodisc/PhotoLink/Getty Images

[SAMPLE MENU]

Spanish *Tapas*

***Croquetas*[a,b]**

Spanish Potato *Tortilla* (omelet)[a,b]

***Empanadas*[a,b]**

***Gambas* (grilled shrimp)[a,b]**

Fried Almonds, Pieces of Cheese, Sausage Bites

Sherry, Beer, or *Sangria*[a,b]

[a]Von Bremzen, A. 2005. *The new Spanish table*. New York: Workman Publishing.

[b]Tapas recipes: http://www.tapas-recipes.com/tapas-recipes.html

Linguiça comes from the Portuguese word meaning "tongue," a reference to the shape of the sausage.

Sweets were a traditional source of income for Portuguese convents, and the names of many pastries reflect this past, including papas-de-anjo (angel puffs) and gargantas de friera (nun's wattles).[37]

Italian folklore has it that basil can only develop its full flavor if the gardener curses daily at it.

cuisine centered on seafood. Its best-known dish is scampi, made from large shrimp seasoned with oil, garlic, parsley, and lemon juice. Inland is Verona, famous for its delicate white wines, such as Soave. Turin, the capital of the western province of Piedmont, is known for its grissini, the slender breadsticks popular throughout Italy, and bagna cauda (meaning "hot bath"), a dip for raw vegetables consisting of anchovies and garlic blended into a paste with olive oil or butter. A summer favorite is vitello tonnato, braised veal served cold with a spicy tuna sauce. Located on the northwest coast of Italy, Genoa is known for its burrida, a fish stew containing octopus and squid, and pesto, an herb, cheese, and nut paste (usually made with basil), which has become popular in the United States.

Moving westward, the city of Bologna is the center of a rich gastronomic region known as Emilia-Romagna. Pasta favorites of the area include lasagne verdi al forno, spinach-flavored lasagna noodles baked in a ragu (a meat sauce typically made with four different meats and red wine), and a white sauce, flavored with cheese; and tortellini, egg pasta stuffed with bits of meat, cheese, and eggs, served in soup or a rich cream sauce. It is traditionally served on Christmas Eve. A similar stuffed pasta is cappelleti, named for its shape, a little hat. Cured meats are a specialty of the region, including salami and sopresseta (similar to salami but rougher textured); mortadella, a pork sausage (similar to American bologna); pancetta (a salt-cured bacon); prosciutto, a raw, smoked ham (served thinly sliced, often as an appetizer with melon or fresh figs); and culatello (a milder and creamier ham than prosciutto). Parmesan cheese, a sharply flavored cow's milk cheese with a finely grained texture, also comes from the area, as does aceto balsamico di Modena (or di Reggio Emilia), a vinegar made from the white Trebbiano wine grapes. When labeled tradiziolone, it means the vinegar has been twice fermented and aged in wood casks for at least twelve years, which intensifies and sweetens the flavor, and thickens it into a syrupy consistency. Those labeled condimento are imitations of vinegar blends and reduced aging.

Florence, the capital of Tuscany, has a long history of culinary expertise. In 1533, Catherine de' Medici (the Medici family ruled Florence) married into the royal family of France. She is often credited with introducing Italian fare—at the time the most sophisticated cuisine in Europe—to France. Florence is renowned for its green noodles (colored by bits of spinach) served with butter and grated Parmesan cheese, called fettucini Alfredo. The term alla Fiorentina refers to a dish garnished with or containing finely chopped spinach. Whole grilled fish and wild game dishes are popular. Rosemary flavors many dishes of the region. Tuscany is also famous for its full-bodied red wine, Chianti, and its use of chestnuts, which are featured in a cake eaten at Lent called castagnaccio alla Fiorentina.

Rome, the capital of Italy, has its own regional cooking and is probably best known for fettucine Alfredo, long, flat egg noodles mixed with butter, cream, and grated cheese. Another dish is saltimbocca (meaning "jumps in the mouth")—thin slices of veal rolled with ham and cooked in butter and Marsala wine. Gnocchi, which are dumplings, are eaten throughout Italy, but in Rome they are made out of semolina and baked in the oven. Fried artichokes are popular at Easter time, as is roast baby lamb or kid. Pecorino romano is the hard sheep's milk cheese of Rome, similar to Parmesan but with a sharper flavor.

The capital of Campania in southern Italy is Naples, considered the culinary capital of the South. Pasta is the staple food, and a favorite way of serving it is simply with olive oil and garlic, or mixed with beans, in the soup pasta e fagiole. Pizza is native to Naples and is said to date back to the sixteenth century, perhaps originating with toppings for the savory flatbread known as foccacia. Another form of pizza is calzone, which is pizza dough folded over a filling of cheese, ham, or salami, then baked or fried. The area's best-known cheeses are Mozzarella, an elastic white cheese originally made from buffalo milk; Provolone, a firm smoked cheese; and Ricotta, a soft, white, unsalted cheese made from sheep's milk and often used in desserts. Sicily and other regions of southern Italy use kid and lamb as their principal meats. It is sometimes prepared alla cacciatore (hunter's style), with tomatoes, olives, garlic, wine, or vinegar (and sometimes anchovies)—a method also used with wild boar, venison, and chicken. Along the coast, fresh fish, such as tuna and sardines, are used extensively; baccala, dried salt cod, is often served on fast days. The North African influence shows up in Sicily in the use of couscous, called cuscus in Italy, which is commonly served with fish stews. Southern Italy's cuisine is probably best known for its desserts. Many examples can be found in Italian-American bakeries and espresso bars: cannoli, crisp, deep-fried tubular pastry shells filled with sweetened ricotta cheese, shaved bittersweet chocolate, and citron; cassata, a cake composed of sponge cake layers with a ricotta filling and a chocolate- or almond-flavored sugar frosting; gelato, fruit or nut (e.g., black currant or pistachio) ice cream; and granita, intensely flavored ices. Spumoni is chocolate and vanilla ice cream with a layer of rum-flavored whipped cream containing nuts and fruits. Another popular sweet is zeppole, a deep-fried doughnut covered with powdered sugar. The sweet white wine fortified with grape spirits, Marsala, is also a specialty in the region. It develops a deep-tawny color when aged.

Spain. The cooking of Spain can be divided broadly by preparation methods. In the north, stewing is most common. In the central regions, roasting is favored. Although deep-fried foods are found in every region of the nation, they are especially popular in the southern regions.[38]

© Cole Group/Getty Images

▲ *Fish and shellfish are a favorite in Italy, Spain, and Portugal.*

Most Spanish dishes prepared in the United States reflect the cooking of Spain's southern region, with its seafood, abundant fruits and vegetables, and Muslim influence. Fried fish, arroz negro (rice blackened with squid ink), and salmorejo (a fresh tomato soup thickened with bread crumbs and garnished with Serrano ham and hard-boiled egg) are popular dishes. Central Spain has a more limited diet; roast suckling pig and baby roast lamb are favorites. Garlic soup starts many meals. In the northwest, fish is common, and often fills empanadas (small pastry turnovers). Octopus flavored with paprika is a specialty. In the Basque provinces lamb is the primary meat, and charcoal-grilled lamb is a specialty. Seafood, such as bacalao al pil-pil (dried salt cod cooked in olive oil and garlic), bacalao a la vizcaina (dried salt cod cooked in a sauce of onions, garlic, pimento, and tomatoes), and

Forks were originally two-pronged—the three-pronged fork was created in Italy for eating pasta.

The art of making ice cream is credited to the Chinese, who brought it to India; from there it spread to the Persians and Arabs. The Muslims brought it to Italy, and it was a Sicilian, Francisco Procopio, who introduced ice cream to Paris in the 1660s. The British discovered it soon after and later brought ice cream to America.

During the nineteenth century, Madeira wine was sent to other European nations in the holds of ships where it became very hot. Instead of ruining the wine, it aged it more quickly—Madeira that had circumnavigated the globe twice became popular in England. Today, it is heated during aging to simulate voyage conditions.

angulas (tiny eel spawn cooked with olive oil, garlic, and red peppers), is a favorite in some Basque areas. Other popular dishes include garlic soup, babarrun gorida (red beans with chorizo), and pipperrada vasca (eggs with peppers). Simple rice puddings or fruit compotes are typical desserts.

Portugal. Though Portuguese cuisine varies from north to south, from hearty soups and stews to a more refined, lighter style of entrée, the largest regional differences occur between the mainland and the islands. The foods of the Madeiras, the Azores, and the Cape Verde Islands include tropical ingredients imported from both Africa and the Americas. In Madeira, which attracts many tourists from throughout Europe, avocados, cherimoya, guava, mango, and papaya are featured in its dishes. Corn is common, as is couscous. Honey cakes and puddings reflect the influence of other European nations. In the Azores and Cape Verde Islands, fare varies significantly from island to island and even city to city. Bananas, corn, cherimoya, passion fruit, pineapples, and yams are prominent. Açorda d'azedo is one specialty—a mixture of cornbread, vinegar, onions, garlic, saffron, and a little lard boiled together and eaten for breakfast. Beef is the preferred meat, and seafood such as cockles, limpets, crab, lobster, and octopus is eaten in many areas. Little fat or oil is added to dishes; and spicing is mild, often limited to onion, garlic, salt, and pepper. Tea is the preferred beverage. Portugal is famous for its rich sweet wines: Port (from the northern region) and Madeira (from the islands), which are fortified with grape spirits at the start of fermentation. They can be consumed young, or aged for forty or more years, becoming drier, more nutty, and smoother in flavor. They are popular with dessert or as after-dinner drinks.

Meal Composition and Cycle

Daily Patterns

Italy. A traditional Italian breakfast tends to be light, including coffee with milk (caffe latte), tea, or a chocolate drink, accompanied by bread and jam. Lunch is the main meal of the day and may be followed by a nap. It usually starts with an appetizer course of antipasti, such as ham, sausages, pickled vegetables, and olives; or crostini, crispy slices of bread with various toppings, such as tomatoes or cheese. Next is minestra (wet course), usually soup, or asciutta (dry course) of pasta, risotto, or gnocchi. The main course is fish, meat, or poultry, roasted, grilled, pan-fried, or stewed. It is served with a starchy or green vegetable, followed by a salad. Bread is served with the meal, often with olive oil and balsamic vinegar for dipping. Dessert often consists of fruit and cheese; pastries or biscotti (crunchy twice-baked cookie slices) and ice creams are served on special occasions. Dinner is served at about 7:30 p.m. and is a lighter version of lunch. Wine usually accompanies lunch and dinner. Coffee or espresso is enjoyed after dinner, either at home or in a coffeehouse. Marsala may be served with cheese before the meal for a light appetizer course, or after dinner. It is also often used in the preparation of desserts. One such sweet, now prepared all over Europe, is zabaglione, a wine custard.

Spain. By U.S. standards, the Spanish appear to eat all the time. The traditional pattern, four meals plus several snacks, is spread across the day. A light breakfast (*desayuno*) of coffee or chocolate, bread, or churros is eaten about 8:00 a.m., followed by a midmorning breakfast around 11:00 a.m. of grilled sausages, fried squid, bread with tomato, or an omelet. A light snack, tapas, is consumed around 1:00 p.m. as a prelude to a three-course lunch (*comida*) at around 2:00 p.m., consisting of soup or salad, fish or meat, and dessert, which is often followed by fruit and cheese. Many businesses close for several hours in the afternoon to accommodate lunch and a nap. Tea and pastries (*merienda*) are eaten between 5:00 and 6:00 p.m. and more tapas are enjoyed at 8:00 or 9:00 p.m. Finally supper, including three light courses such as soup or omelets and fruit, is consumed between 10:00 p.m. and midnight. Tapas are usually served in bars and cafés and are accompanied by Sherry or wine; the variety of tapas is tremendous; it is not unusual for more than twenty kinds to be offered on a menu. They are differentiated from appetizers in Spain in that they are strictly finger foods, such as olives, almonds, croquetas (fried croquettes with fish, ham, cheese, etc.), stuffed mushrooms, shrimp, sausage bits, pieces of cheese, and other small bites. The evening meal may be skipped if a substantial number of tapas are eaten at night. The main meal of the day is lunch, which is generally eaten at home and consists of three hefty courses.

Portugal. Portuguese meal patterns are similar to those of Spain, often starting out with the day around 8:00 a.m. with espresso coffee and a roll with marmalade, or pastel de nata, a cinnamon-flavored custard tart in puff pastry.

A morning coffee break, including coffee served with hot milk, is followed by lunch in the early afternoon. This is traditionally the largest meal of the day, and even in urban areas often includes several courses. Unlike the Spanish pattern, the evening meal in Portugal is usually eaten earlier. As in Spain, red wine usually accompanies the meal.

Etiquette. Italy, Spain, and Portugal share many etiquette rules. The fork remains in the left hand, and the knife remains in the right hand. The knife can be used to help scoop food onto the fork. Bread is not served with butter and should be placed on the edge of the main plate, or next to it on the table. Manners regarding the consumption of pasta include using your fork to twirl the pasta against the edge of the plate or bowl (never use a spoon to help with this), and never slurping. Bread may be used daintily to soak up extra sauce, but should not be used to mop the plate. When not eating, the hands should be kept above the table with the wrists resting on the edge.[13]

When in someone's home, or at a hosted meal, never start eating until the host has said buòn appetito (in Italy), buen apetito (in Spain), or bom appetite (in Portugal). In Italy, when invited to someone's home, it is considered rude to discuss any serious topic before a meal is shared. Chocolates are considered a good hostess gift when invited for dinner in all three nations. In Italy, wine is appreciated if enough is brought for all guests; wine should be avoided as a gift in Spain and Portugal, where hosts have likely chosen favorites to accompany the meal.

Special Occasions

Italy. Italy celebrates few national holidays, probably because of its divided history. Most festas are local and honor a patron saint. Other significant religious holidays are usually observed by families at home, although some cities, such as Venice, have a public pre-Lenten carnival. In some areas of the United States where southern Italians predominate, St. Joseph, the patron saint of Sicily, is honored during Lent. Breads in the shape of a cross blessed by the parish priest, pasta with sardines, and other meatless dishes are featured. Among Italian Americans, it is traditional to serve seven seafood dishes on Christmas Eve. During the Easter holidays, Italian American bakeries sell an Easter bread with hard-boiled eggs still in their shells braided into it. Special desserts may accompany the holiday meal, such as panettone, amaretti (almond macaroons), and torrone (nougats) at Christmastime and cassata at Easter. Colored, sugar-coated Jordan almonds, which the Italians call confetti (meaning "little candies"), are served at weddings.

© PhotoLink/Photodisc/Getty Images

Traditional Italian lunches are large, often ending with fruit and cheese. On special occasions, pastries, such as twice-baked cookies called biscotti, or ice cream are served with coffee or espresso.

Legend has it that zabaglione was created to increase male vigor by a Franciscan monk who tired of hearing the confessional complaints from women about their tired husbands.

The word tapas means "lids," and the first tapas were pieces of bread used to cover wine glasses to keep out flies.

Spain. The most elaborate of Spanish festivals is Holy Week, the week between Palm Sunday and Easter. It is a time of Catholic processions; confections and liqueurs such as coffee, chocolate, and anisette (licorice flavored) are served. Holiday sweets include tortas de aceite, which are cakes made with olive oil, sesame seeds, and anise; cortados rellenos de cidra, or small rectangular tarts filled with pureed sweetened squash; torteras, or large round cakes made with cinnamon and squash and decorated with powdered sugar; and yemas de San Leandro, which is a sweet made by pouring egg yolks through tiny holes into boiling syrup. It is often served with marzipan.

Special dishes are also prepared for Christmas and Easter. The Basques eat roasted chestnuts and pastel de Navidad, or individual walnut and raisin pies, at Christmas; an orange-flavored doughnut, called causerras, is featured on Easter. At New Year's, it is customary for the Spanish (and the Portuguese) to eat twelve grapes or raisins at the twelve strokes of midnight to bring luck for each month of the coming year.

Portugal. Christmas Eve typically features two meals in Portugal: dinner and a post–midnight Mass buffet in the early hours of Christmas morning. Dinner often includes a casserole of bacalhau and potatoes, as well as meringue cookies known as suspiros (sighs). The buffet offers

In the 1600s, the Spanish were the first to add sugar to the bitter chocolate beverage native to Mexico. Its popularity spread quickly through Europe, even though certain clerics tried to ban it due to its association with the "heathen" Aztecs. Although the Spanish are fond of chocolate, it is used mostly as a beverage and is rarely added to pastries or confections.

In the Portuguese town Amarante, the Festa de São Gonçalo is held the first weekend in June. Dating back to pagan times, it is traditional for unmarried men and women to exchange phallus-shaped cakes as tokens of their affection for each other.

In rural areas of Portugal, people traditionally collect medicinal plants for home remedies on Quinta-Feira da Espiga (Ear of Wheat Thursday or Ascension Day), the fortieth day after Easter.

The Spanish American Isleños of Louisiana marinate shrimp in vinegar with olives and onions, make almond-honey nougat, and use ample olive oil (instead of the butter and lard favored in nearby Cajun cooking).

mostly finger foods, such as fried cod puffs and sausages.

In the United States, the Holy Ghost (Spirit) Festival is the most popular and colorful social and religious event in the Portuguese community. It is not widely celebrated in Portugal and probably came to the United States with immigrants from the Azores. Although the origins of the event are obscure, it is believed to date back to Isabel (Elizabeth) of Aragon, wife of Portugal's poet-king, Dom Diniz (1326). One story is that the festival derives its character from the belief that because Isabel was particularly devoted to the Holy Ghost, she wanted to give an example of charity in the annual distribution of food to the poor.

The week-long festival is usually scheduled sometime after Easter and before the end of July. It is held at the local church or Hall of the Holy Ghost (also called an IDES Hall). The main event of the festival takes place on the last day, Sunday, with a procession to the church and the crowning of a queen after the service. The donated food (originally distributed to needy persons on Sunday afternoon but now often served at a free community banquet) is blessed by the priest. The most traditional foods at the feast are a Holy Ghost soup of meat, bread, and potatoes, and a sweet bread called massa sovoda. The bread is sometimes shaped like little doves, called pombas. Also celebrated in the United States is the Feast of the Most Blessed Sacrament, which was started in New Bedford, Massachusetts, by four Madeirans in gratitude for their salvation from a shipwreck en route to the United States in 1915. It attracts over 150,000 visitors on the first weekend in August for music, dancing, and traditional foods such as linguiça, bacalhau, fava beans, assada no espeto, and cacoila. The Festa de Sennor da Pedra is held later in the month. This Azore Islands tradition includes a parade and similar traditional foods. Other festivities not mentioned here are associated with the Madeiran cult of Our Lady of the Mount (a shrine on the island of Madeira).

Therapeutic Uses of Food

Little has been reported on the therapeutic uses of food by Americans of southern European descent. Some Italians, particularly older immigrants, categorize foods as being heavy or light, wet or dry, and acid or nonacid.[35] Heavy foods, such as fried items and red meats, are considered difficult to digest; light foods, including gelatin, custards, and soups, are regarded as easy to digest and appropriate for people who are ill. Wet and dry refers to how foods are prepared (with or without ample broth or fluid), as well as to their inherent qualities. For example, leafy greens such as escarole, spinach, and cabbage are considered wet. A wet meal is served once a week by some Italian Americans to "cleanse out the system." Wet meals, especially soups, are considered necessary when a person is sick because illness is associated with dryness in the body. Citrus fruits, raw tomatoes, and peaches are thought to be acidic foods that may cause skin ailments and are avoided if such conditions exist.

Other Italian beliefs about foods are that liver, red wine, and leafy vegetables are good for the blood and that too many dairy products make the urine hard (kidney stones). A clove of garlic may be eaten each day to prevent respiratory infections, and a raw egg or dandelion greens may be consumed for strength and vitality.[13] Both balsamic vinegar and olive oil, which are served with bread at meals, are believed to be health-promoting foods in Italy.

Contemporary Food Habits in the United States

Adaptations of Food Habits

It is generally assumed that second- and third-generation Americans of southern European descent have adopted the majority American diet and meal patterns, preserving some traditional dishes for special occasions. These assimilated Americans consume more milk and meat but less fish, fresh produce, and legumes than their ancestors. Olive oil is still used often, although not exclusively; pasta remains popular with Italians.

Nutritional Status

Nutritional Intake. Little research has been conducted on the nutritional intake of southern European Americans. It can be assumed that they suffer from dietary deficiencies and excesses similar to those of the majority of Americans. A study of elderly Portuguese immigrants in Cambridge, Massachusetts, found that dinner was the main meal of the day, and the subjects had moderate intake of breads and grains and low intake of fruits, vegetables, and dairy products. Although dairy intake was low, many of the subjects ate sardines, a rich source of calcium. The subjects reported low consumption of sweets and alcohol, although the researcher stated that the Americanized Portuguese diet tends to be high in sugar and fat.[40] One study comparing body weights of American and Italian women with polycystic ovary syndrome found that though the BMIs for the American women were significantly higher, the total calorie intake and dietary constituents were similar, suggesting unknown genetic or lifestyle components may play a role.[42]

According to a survey of European dietary habits, a majority of the population in Italy consumes more plant products than protein, and approximately equal amounts of both are consumed in Spain.[18] In addition, meat consumption is highest in the northern regions of these nations, and lowest in the southern areas.[19] In general, the Mediterranean diet, which is typified in southern Italy and Spain, has been characterized as health promoting due to a high intake of complex carbohydrates, a high intake of protective phytochemicals, and a low intake of fat with a higher proportion of monounsaturated fats from olive oil as compared to saturated animal fats.[43,44] The greater emphasis on grains, legumes, vegetables, and fruits; lower intakes of meat and dairy foods; and promotion of wine in moderation differentiate the Mediterranean diet from that recommended by U.S. health officials.[45] However, a study by the Italian Association for Cancer Research has found that cancer rates increased as food habits changed in Italy; pasta consumption has fallen, and meat intake has quadrupled since 1950; changes toward a more westernized diet are found in Spain and Portugal as well.[41,46] Rates of overweight and obesity in Italian women are 35 percent, but are over 53 percent in men. In Spain, rates exceed 60 percent for men and 46 percent in women; in Portugal, overweight and obesity in women approach 50 percent, and in men, 60 percent.[20]

Counseling. The conversational style of southern Europeans is animated, warm, and expressive. Feelings are more important than objective facts in a discussion. Shaking hands with everyone in the room in greeting and leaving is appropriate; some men include pats on the back, and women may quickly embrace or kiss on the cheeks. Eye contact among elders tends to be frequent and quick, whereas younger people may prefer steady eye contact. Touching is very common, especially between members of the same sex. It has been noted that Italian-American clients are open, willing to detail symptoms with their health-care professional, and expressive with chronic pain—although some women may demonstrate high levels of modesty and may resist discussing personal topics.[34,6] Italian Americans may seek medical advice from family and friends before consulting a health professional. They express preference for providers who are warm and empathetic (*simpatico*) and disdain those who are perceived as arrogant and unapproachable (*superbo*).

Recent Italian immigrants or those who are elders may be very concerned about the qualities of their blood or may have many gastrointestinal complaints. There may be confusion regarding hypertension, which is considered high or too much blood, and anemia or low blood pressure, which is associated with low blood.[35]

Dietary requirements should be carefully detailed for some Italian Americans. Restrictions recommended for clients with diabetes may be ignored if daily social activities (i.e., coffee and pastries with friends) must be modified. Language difficulties may occur among elders or new immigrants.

Information regarding the counseling of Spanish Americans or Portuguese Americans is limited. The people of Spain and Portugal are traditionally high context communicators and very polychronic, though many urban residents have more western monochronic viewpoints. A quick handshake is the customary greeting, and clients from southern Europe will typically sit and stand closer to each other than many Americans prefer. Direct eye contact is important.[13] A high rate of illiteracy has been reported in the Portuguese-American population (40 percent of surveyed elders; 15 percent of recent immigrants). This should be taken into consideration when preparing educational materials. An in-depth interview can be used to assess the client's degree of acculturation and traditional health practices, if any. Personal food preferences should be determined.

DISCUSSION STARTERS: LET'S OPEN A PUB!

Americans often have an inaccurate view of British and Irish pubs. Many of us tend to identify these pubs with our American bars, but in fact, pubs are much more like American "bar and grills," "sports taverns," and restaurants that serve beer and wine. Most British and Irish pubs serve hot lunches and dinners as well as alcoholic beverages. Traditional British pub fare includes fish and chips (what Americans call french fries), shepherd's pie or cottage pie (beef or mutton, mashed potatoes, maybe green peas, and a potato crust on top or cooked in a pie crust), steak or steak and kidney pie, bangers and mash (sausages and mashed potatoes), Yorkshire pudding (a batter such as a pancake batter, covered with beef gravy), Quaker pudding (a grayish spiced pudding), Cornish pastry, and mince pies. The name "pub" is short for "public house," and historically, these pubs functioned as local meeting places and served to strengthen cultural ties within the community.

Imagine that you plan to open an "authentic" British and Irish pub in your college community. In a small group, decide what your food menu should include. Remember that you will need to balance your effort to be authentic with your need to attract college students and serve students who eat only vegetarian meals.

REVIEW QUESTIONS

1. Summarize the immigration patterns of northern and southern Europeans.
2. Describe the American majority cultural beliefs regarding health, and the origins of these beliefs.
3. Describe the traditional food habits of England, Ireland, and Italy. List five of your favorite foods. Do any of these foods have their roots in Europe? Describe your typical meal cycle and meal composition. Are these similar to those of Europe?
4. What is the difference between Cajun and Creole cooking? What are the origins of both styles of cooking?
5. Compare and contrast the immigrant experiences of the Irish and Italians.
6. How did the new world foods (tomatoes, potatoes, corn, etc.) influence European foodways?
7. Why is Mediterranean diet considered healthy?

REFERENCES

1. Hess, M.A. Scottish and Scotch Irish Americans. In *Gale Encyclopedia of Multicultural America,* R.V. Dassanowsky & J. Lehman (Eds.). Farmington Hills, MI: Gale Group.
2. Ryan, E., O'Keane, C., & Crowe, J. 1998. Hemochromatosis in Ireland and HFE. *Blood Cells, Molecules, and Diseases, 24,* 428–432.
3. Hillstrom, L.C. 2000. French Americans. In *Gale Encyclopedia of Multicultural America,* R.V. Dassanowsky & J. Lehman (Eds.). Farmington Hills, MI: Gale Group.
4. Fedunkiw, M. 2000. French-Canadian Americans. In *Gale Encyclopedia of Multicultural America,* R.V. Dassanowsky & J. Lehman (Eds.). Farmington Hills, MI: Gale Group.
5. Heimlich, E. 2000. Acadians. In *Gale Encyclopedia of Multicultural America,* R.V. Dassanowsky & J. Lehman (Eds.). Farmington Hills, MI: Gale Group.
6. Spector, R.E., 2004. *Cultural Diversity in Health and Illness* (6th ed.). Upper Saddle River, NJ: Pearson Education, Inc.
7. Fisher, M.F.K. 1968. *The cooking of provincial France.* New York: Time-Life.
8. Zibart, E. 2001. *The ethnic food lovers companion: Understanding the cuisines of the world.* Birmingham: Menasha Ridge Press
9. Bailey, A. 1969. *The cooking of the British Isles.* New York: Time-Life.
10. Ffrench, H.H., & VisitBritain. 2005. United Kingdom: A flavorful adventure. In *Culinary Cultures of Europe,* D. Goldstein & K. Merkle (Eds.). Strasbourg, France: Council of Europe Publishing.
11. Claiborne, C., & Franey, P. 1970. *Classic French cooking.* New York: Time-Life
12. Poulain, J.P. 2005. French gastronomy, French gastonomies. In *Culinary Cultures of Europe,* D. Goldstein & K. Merkle (Eds.). Strasbourg, France: Council of Europe Publishing.
13. Foster, D. 2000. *The global etiquette guide to Europe.* New York: John Wiley & Sons.
14. Henley, J. 2000, January 16. *French town decrees local fare for students.* San Jose Mercury News, 5 AA.
15. Wilson, S.A. 2003. People of Irish heritage. In *Transcultural Health Care* (2nd ed.), L.D. Purnell & B.J. Paulanka (Eds.). Philadelphia: FA Davis Company.
16. Galan, P., Arnaud, M.J., Czernichow, S., Delabroise, A.M., Preziosi, P., Bertrais, S., Franchisseur, C., Maurel, M., Favier, A., & Hercberg, S. 2002. Contribution of mineral waters to the dietary calcium and magnesium intake in a French adult population. *Journal of the American Dietetic Association, 102,* 1658–1662.
17. Zibart, E. 2001. *The ethnic food lovers companion: Understanding the cuisines of the world.* Birmingham: Menasha Ridge Press.
18. Slimani, N., Fahey, M., Welch, A.A., Wirfalt, E., Stripp, C., Bergstrom, E., Linseisen, J., Schulze, M.B., Bamia, C., Chloptsios, Y., Veglia, F., Panico, S., Bueno-de-Mesquita, H.B., Ocke, M.C., Brustad, M., Lund, E., Gonzalez, C.A., Barcos, A., Berglund, G., Winkvist, A., Mulligan, A., Appleby, P., Overvad, K., Tjonneland, A., Clavel-Chapelon, F., Kesse, E., Ferrari, P., Van Staveren, W.A., & Riboli, E. 2002. Diversity of dietary patterns observed in the European Prospective

Investigation into Cancer and Nutrition (EPIC) project. *Public Health Nutrition, 5,* 1311–1328.

19. Linseisen, J., Bergstrom, E., Gafa, L., Gonzalez, C.A., Thiebaut, A., Trichopoulou, A., Tumino, R., Navarro Sanchez, C., Martinez Garcia, C., Mattisson, I., Nilsson, S., Welvh, A., Spencer, E.A., Overvad, K., Tjonneland, A., Clavel-Chapelon, F., Kesse, E., Miller, A.B., Schulz, M., Botsi, K., Naska, A., Sieri, S., Sacerdote, C., Ocke, M.C., Peeters, P.H., Skeie, G., Engeset, D., Charrondiere, U.R., & Slimani, N. 2002. Consumption of added fats and oils in the European Prospective Investigation into Cancer and Nutrition (EPIC) centres across 10 European countries as assessed by 24-hour recalls. *Public Health Nutrition, 5,* 1227–1242.
20. International Obesity Task Force. 2010. *Overweight and obesity among adults in the European Union.* Available online: http://www.iaso.org/site_media/uploads/AdultEU27March2010notonwebyetupdatev2.pdf
21. Kushi, L.H., Lew, R.A., Stare, E.J., Curtis, R.E., Lozy, M., Bourke, G., Daly, L., Graham, I., Hickey, N., Mulcahy, R., & Kevaney, J. 1985. Diet and 20-year mortality from coronary heart disease: The Ireland-Boston diet-heart study. *New England Journal of Medicine, 312,* 811–818.
22. Johnson, P.B. 1997. Alcohol-use-related problems in Puerto Rican and Irish-American males. *Substance Use and Misuse,* 32, 169–179.
23. Phatak, P.D., Sham, R.L., Raubertas, R.F., Dunnigan, K., O'Leary, M.T., Braggins, C., & Cappuccio, J.D. 1998. Prevalence of hereditary hemochromatosis in 16031 primary care patients. *Annals of Internal Medicine, 129,* 954–961.
24. Ryan, E., O'Keane, C., & Crowe, J. 1998. Hemochromatosis in Ireland and HFE. *Blood Cells, Molecules, and Diseases, 24,* 428–432.
25. Girouard, J., Giguere, Y., Delage, R., & Fousseau, F. 2002. Prevalence of HFE gene C282Y and H63D mutations in a French-Canadian population of neonates and in referred patients. *Human Molecular Genetics, 11,* 185–189.
26. Ma, Y., Murthy, V., Roderer, G., Monsalve, M.V., Clarke, L.A., Normand, T., Julien, P., Gagne, C., Lambert, M., Davignon, J., Lupien, F.J., Brunzell, J., & Hayden, M.R. 1991. A mutation in the human lipoprotein lipase gene as the most common cause of familial chylomicronemia in French Canadians. *New England Journal of Medicine, 324,* 1761– 1766.
27. Leistner, C.G. 1996. Cajun and Creole food practices, customs, and holidays. Chicago: American Dietetic Association/American Diabetes Association Hall, E.T., & Hall, M.R. 1990. *Understanding cultural differences: Germans, French, and Americans.* Yarmouth, ME: Intercultural Press.
28. Hall, E.T., & Hall, M.R. 1990. *Understanding cultural differences: Germans, French, and Americans.* Yarmouth, ME: Intercultural Press.
29. Pozzetta, G. 2000. Italian Americans. In *Gale Encyclopedia of Multicultural America,* R.V. Dassanowsky & J. Lehman (Eds.). Farmington Hills, MI: Gale Group.
30. Colahan, C. 2000. Spanish Americans. In *Gale Encyclopedia of Multicultural America,* R.V. Dassanowsky & J. Lehman (Eds.). Farmington Hills, MI: Gale Group.
31. Shostak, E. 2000. Basque Americans. In *Gale Encyclopedia of Multicultural America,* R.V. Dassanowsky & J. Lehman (Eds.). Farmington Hills, MI: Gale Group.
32. Harwood, A. 1981. *Ethnicity and medical care.* Cambridge, MA: Harvard University Press.
33. Norden, E.E. 2000. Portuguese Americans. In *Gale Encyclopedia of Multicultural America,* R.V. Dassanowsky & J. Lehman (Eds.). Farmington Hills, MI: Gale Group.
34. Hillman, S.M. 2003. People of Italian heritage. In *Transcultural Health Care* (2nd ed.), L.D. Purnell & B.J. Paulanka (Eds.). Philadelphia: FA Davis Company.
35. Ragucci, A.T. 1981. Italian Americans. In A. Harwood (Ed.), *Ethnicity and medical care.* Cambridge, MA: Harvard University Press.
36. de Lorgeril M, Salen P. The Mediterranean diet: rationale and evidence for its benefit. *Curr Atheroscler Rep.2008 Dec; 10*(6): 518–22.
37. Pessoa e Costa, A. 2005. Portugal: A dialogue of cultures. In *Culinary Cultures of Europe,* D. Goldstein & K. Merkle (Eds.). Strasbourg, France: Council of Europe Publishing.
38. Valverde Villena, D. 2005. Spain: Agape and conviviality at the table. In *Culinary Cultures of Europe,* D. Goldstein & K. Merkle (Eds.). Strasbourg Cedex, France: Council of Europe Publishing.
39. Aranceta, J., Perez, C., Gondra, J., Gonzalez de Gaideano, L., & Saenz de Buruaga, J. 1993. Fat and alcohol intake in the Basque Country. *European Journal of Clinical Nutrition, 47,* S66–S70.
40. Poe, D.M. 1986. *Profile of Portuguese elderly nutrition participants: Demographic characteristics, nutrition knowledge and practices.* Master's thesis, MGH Institute of Health Professions, Boston, MA.
41. Fernandez San Juan, P.M. 2006. Dietary habits and nutritional status of school aged children in Spain. *Nutritión Hospilataria, 21,* 374–378.
42. Carmina, E., Legro, R.S., Stamets, K., Lowell, J., & Lobo, R.A. 2003. Differences in body weight between American and Italian women with polycystic ovary syndrome: Influence of the diet. *Human Reproduction, 18,* 2289–2293.
43. Ferro-Luzzi, A., & Branca, F. 1995. Mediterranean diet, Italian-style: Prototype of a healthy diet. *American Journal of Clinical Nutrition, 61* (Suppl.), 1338S–1354S.
44. Visioli, F., Bogani, P., Grande, S., Detopoulou, V., Manios, Y., & Galli, C. 2006. Local food and cardio-protection: The role of phytochemicals. *Forum of Nutrition, 59,* 116–129.
45. Willet, W.C., Sacks, F., Trichopoulou, A., Drescher, G., Ferro-Luzzi, A, Helsing, E., & Trichopoulos, D. 1995. Mediterranean diet pyramid: A cultural model for healthy eating. *American Journal of Clinical Nutrition, 61* (Suppl.), 1402S–1406S.
46. Marquez-Vidal, P., Ravasco, P., Dias, C.M., & Camilo, M.E. 2006. Trends of food intake in Portugal, 1987–1999: Results from the National Health Surveys. *European Journal of Clinical Nutrition, 60,* 1414–1422.

CHAPTER

Central Europeans, People of the Former Soviet Union, and Scandinavians

The European settlers from central Europe, the former Soviet Union (FSU), and Scandinavia were some of the earliest and largest groups to come to the United States. Though many arrived as early as the 1600s and most had come before the beginning of the twentieth century, the upheavals of two world wars and the collapse of the Soviet Union have led to continuous immigration from these regions during the last century (see the map in Figure 7.1).

The influence of immigrants from central Europe, the FSU (especially Russia), and Scandinavia on American majority culture, especially in the area of cuisine, is substantial. Bread baking, dairy farming, meat processing, and beer brewing are just a few of the skills these groups brought with them. Their expertise permitted the expansion of food production and distribution that encouraged nationwide acceptance of their ethnic specialties, leading to the creation of a typical American cuisine. This chapter focuses on the traditional and adapted foods and food habits of Germans, Poles, and other central European groups; Russians and other FSU populations; and Danes, Swedes, and Norwegians.

Eastern Europe is the term sometimes used to define the region that is also called "European Russia" (the western half of the country; east of the Ural Mountains is known as Siberia or "Asian Russia"). Before the breakup of the USSR, "eastern Europe" was sometimes used to describe those countries under Soviet control (e.g., Czechoslovakia, Hungary, East Germany).

Central Europeans and the People of the FSU

Central Europe stretches from the North and Baltic Seas, south to the Alps, and east to the Baltic States. It includes the nations of Germany, Austria, Hungary, Romania, the Czech Republic, Slovakia, and Poland, as well as Switzerland and Liechtenstein. Most of the countries share common borders; Austria, Hungary, the Czech Republic, Romania, and Slovakia are situated south of Germany and Poland. Switzerland, an isolated nation, is surrounded by Germany, Austria, France, Italy, and Liechtenstein. The climate of central Europe is harsher and colder than that of southern Europe, but much of the land is fertile.

The FSU includes the Commonwealth of Independent States or CIS (the Russian Federation, Armenia, Azerbaijan, Belarus, Georgia, Kazakstan, Kyrgyzstan, Republic of Moldavia, Tajikistan, Turkmenistan, Ukraine, and Uzbekistan) and the Baltic States (Estonia, Latvia, and Lithuania), extending east to the border with China and the Pacific Ocean. Its vast geography includes the Arctic and parts of the Middle East. Except in the southern republics, the harsh winters of the region affect agricultural capacity.

The large number of immigrants from central Europe and parts of the FSU made significant contributions to the literature, music, and cuisine of the United States. Many central European foods have become standard American fare. Imagine a baseball game without hot dogs and beer or a picnic without potato salad. This section explores these and other food customs of central Europe and the FSU and their impact on the American diet.

Figure 7.1
Central Europe, Scandinavia, former Soviet Union.

Cultural Perspective

History of Central Europeans and Russians in the United States

Immigration Patterns

Germans. For almost three centuries, Germans have been one of the most significant elements in the U.S. population. According to U.S. 2008 Census figures, one in every six Americans is of German descent, making this the largest ethnic group in the nation. Germans are also one of the least visible of any American group.

The earliest German settlement in the American colonies was Germantown, Pennsylvania, founded in 1681. By 1709 large-scale immigration began, primarily from the Palatinate region of southwestern Germany. Many of the immigrants, who were mostly of Amish,

Mennonite, or other religious minority faiths seeking freedom from discrimination, settled in Pennsylvania. The majority were farmers who steadily pushed westward searching for new lands for their expanding families. Those in Pennsylvania, Ohio, and Indiana became known as the "Pennsylvania Dutch." Immigration dropped off after 1775, but an economic crisis in Europe once again prompted numerous Germans to come to the United States. Approximately 5 million Germans immigrated to the United States between 1820 and 1900. Like the earlier settlers, most were farmers who arrived with their families, although by the end of the century there were increasing numbers of young, single people who were agricultural laborers and servants. Many of these settled in the Mississippi, Ohio, and Missouri River valleys, the Great Lakes area, or the Midwest. Most Germans avoided the southern United States, but there are sizable German settlements in Texas and New Orleans.

Small numbers of Schwenkfelders from southern Germany, members of a pacifist religious sect similar to Quakers, settled in Pennsylvania in the mid-1700s. They introduced crocus flowers, source of the spice saffron.

The word Dutch is a corruption of Deutsch, meaning "German," and has nothing to do with the Netherlands.

A third significant phase of immigration began after the turn of the twentieth century, when approximately 1.5 million Germans arrived. Many were unmarried industrial workers seeking higher pay, and others were the descendants of Germans who had settled in ethnically isolated colonies in Russia as early as the sixteenth century. Discrimination and the revolution of 1917 led to their departure. Most of these third phase immigrants joined growing numbers of second- and third-generation Germans living in urban areas. Cities with considerable German populations included Cleveland, New York, Toledo, Detroit, Chicago, Milwaukee, and St. Louis. German Russians, however, tended to settle in rural areas, especially in Colorado.

During the 1930s, many of the German immigrants were Jewish refugees. After World War II, displaced persons of German descent and East German refugees made up the sizeable German immigrant group who settled in the United States.

▼ *Traditional foods of central Europe and the FSU. Some typical foods include beets, cabbage, ham, herring, kasha, potatoes, rye bread, sausages, and sour cream.*

Photo by Laurie Macfee

Poles. Poles have arrived in the United States continuously since 1608. The largest wave of immigration occurred between 1860 and 1914, mostly for economic reasons. The early phase was dominated by Poles (approximately 500,000) from German-ruled areas of Poland (Pomerania and Poznan) and by Poles who worked in western Germany. German Poles often became part of the German or Czech communities or established farming settlements in the Southwest and Midwest.

The number of Polish immigrants from Germany began to decline after 1890, but the slack was taken up by the arrival of more than 2 million Poles from areas under Russian and Austrian rule. The German Poles left their homeland to become permanent settlers, but the Russian and Austrian Poles came as temporary workers. Although 30 percent returned to Poland, many eventually moved back to the United States permanently. The Austrian and Russian Poles tended to settle in the rapidly developing cities of the Mid-Atlantic and Midwestern states, especially Chicago, Buffalo, and Cleveland.

Polish emigration after World War I was usually not for economic reasons. Most (more than 250,000) left because of political dissatisfaction: government instability and dictatorship in the 1920s and 1930s, the German invasion and occupation from 1939 to 1945, and the pro-Soviet communist government after 1945. Many settled in urban areas in which there were substantial existing Polish populations. More recently, small numbers of younger Poles have taken advantage of the freedom resulting from post-Communist rule to come to the United States.

Other Central Europeans. Nearly 4 million Austrians, Hungarians, Czechs, Slovaks, and Swiss have come to the United States, primarily during the late nineteenth and early twentieth centuries, for economic and political reasons.

Austrian immigration patterns are not entirely known because Austrians and Hungarians were classified as a single group in U.S. statistics until 1910. More than 2 million Austrians are believed to have come to the United States searching for economic opportunities in the decade following 1900. Most were unskilled, and many were fathers who left families in Austria with the hopes of making their fortune. Many Austrians never found the advancement they were seeking, and approximately 35 percent returned home. A second, smaller wave of immigration occurred in the 1930s, when 29,000 well-educated, urban Austrian Jews fled Hitler's arrival.

The first group of Hungarians arriving in the United States was several thousand political refugees following the revolution of 1848. Most were men—well educated, wealthy, and often titled. Later Hungarian immigrants who arrived at the turn of the century were often poor, young, single men who found job opportunities in the expanding industrial workplace. Many worked in the coal mines of eastern Ohio, West Virginia, northern Illinois, and Indiana. Cities that developed large Hungarian populations were primarily located in the Northeast and Midwest. More than 50,000 additional Hungarians entered the United States as refugees after World War II and the 1956 uprising against the communist government. They first settled in the industrial towns populated by earlier Hungarian immigrants, but many, mostly professionals, soon moved to other cities that offered better jobs.

Czech immigrants initially tended to be farmers or skilled agricultural workers who settled in the states of Nebraska, Wisconsin, Texas, Iowa, and Minnesota, often near the Germans. Later Czech immigrants were skilled laborers; they settled in the urban areas of New York, Cleveland, and especially Chicago.

The majority of the Slovak immigrants were young male agricultural workers who arrived before World War II. Those who decided to remain in the United States later sent for their wives and families. The majority settled in the industrial Northeast and Midwest; they labored in coal mines, steel mills, and oil refineries.

Immigrants from Switzerland came to the United States for economic opportunities. The majority arrived prior to World War I, seeking jobs as artisans or professionals in the urban areas of New York, Philadelphia, Chicago, Cincinnati, St. Louis, San Francisco, and Los Angeles.

Another group without national boundaries found throughout central Europe (as well as in northern and southern Europe) is the Gypsies, also known as Roma. Gypsy immigrants to the United States are not counted in U.S. Census figures. There are an estimated 100,000 to 1,000,000 Gypsies in America, from a variety of Gypsy groups and speaking different dialects.[1] Though they originate from numerous European countries, the majority living in the United States are believed to have come from central Europe.

The Gypsies are an insular ethnic group found throughout the world. When they first arrived in Europe in the 1300s from India, they were mistaken for Egyptians. Their name derives from this error. Those from eastern and southeastern Europe are the Rom, and some Gypsies prefer the name Roma.

Russians and People of the FSU. Russian immigrants originally came to Alaska and the West Coast, rather than to the eastern states. Most of their settlements were forts or outposts used to protect their fur trade and to shelter missionaries. When Russia sold Alaska to the United States in 1867, half of the settlers returned home and many of the others moved to California. Subsequent immigration was primarily to the East Coast, although some Russians (Molokans, followers of a religion that had rejected the Russian Eastern Orthodox Church) immigrated to the West Coast in the early twentieth century.

Russians, mainly impoverished peasants seeking a better life, began to arrive in large numbers during the 1880s. Over 1.5 million were Jews seeking freedom from persecution as well as economic opportunity. A second wave of immigrants came after the 1917 revolution, when more than 2 million people fled the country; 30,000 settled in the United States. After World War II, only small numbers of Soviet refugees, primarily Jews, were allowed to emigrate. Following the breakup of the Soviet Union, nearly 200,000 Russians settled in the United States between 1990 and 1993. The settlement patterns of Russians are similar to those of other immigrants from central Europe. For the later wave of immigrants, the port of entry was New York City. Many remained in New York, and others settled in nearby industrial areas that offered employment in the mines and factories.

The largest populations of immigrants from the FSU are from the Ukraine, Lithuania, and Armenia. Lengthy Russian domination of the region hinders estimates of the total numbers in the United States because some settlers were listed as Russians in immigration figures. It is believed that the first significant number of Lithuanians, approximately 300,000, arrived in the United States following the abolition of serfdom in 1861.[2] Nearly 30,000 refugees fleeing Soviet control came following World War II. The largest influx of Ukrainian immigrants occurred in the 1870s, when almost 350,000 men were recruited to work the Pennsylvania mines as strikebreakers. A majority settled in that state, though smaller numbers found factory work in Ohio, New York, and Michigan. Later immigrants, including 80,000 Ukrainians displaced by World War II, favored the urban centers of the Northeast.

Significant Armenian immigration began in 1890, when immigrants came for economic opportunity. A second wave of Armenians from Turkey who were seeking escape from persecution arrived following the two world wars. More than 60,000 Armenian refugees have come since the 1980s, settling primarily in Los Angeles, with smaller numbers joining the older American communities in Boston, New York, Detroit, Chicago, and the agricultural region of Fresno, California.

Current Demographics and Socioeconomic Status

Germans. There are more than 50 million Americans of German heritage in the nation today, according to 2008 census estimates. Wisconsin-Minnesota-North Dakota-South Dakota-Nebraska-Iowa is considered the German belt; however, only the Pennsylvania Dutch, the rural-dwelling Germans from Russia who settled in the Midwest, and a few concentrated communities in Texas retain some aspects of their cultural heritage.[3]

Germans differ little from the national norms demographically, although they are slightly higher in economic achievement and are generally conservative in attitudinal ratings. The high degree of German acculturation is attributed to their large numbers, their occupations, and the time of their arrival in the United States. Furthermore, entry of the United States into World War I created a storm of anti-German feeling in America. German-composed music was banned, German-named foods were renamed, and German books were burned. As a result, German Americans rapidly assimilated, abandoning the customs still common in other ethnic groups, such as ethnic associations and use of their oral and written language.

Poles. Polish Americans form one of the largest ethnic groups in the United States today. In 2008, it was estimated that there were close to 9 million Americans of Polish descent; many still live in the urban areas of the Northeast and upper Midwest where their ancestors originally settled. Economically, the third-generation Polish American has moved modestly upward, but the majority of Polish Americans still live just below or solidly at middle-class level. Forty-five percent of the males have white-collar jobs, and 40 percent of working Polish women are semiskilled or unskilled laborers. Poles have been active in the formation and leadership of U.S. labor unions. More recent immigrants usually possess higher occupational skills and educational backgrounds than earlier immigrants.

Other Central Europeans. There is continued confusion over Austrian ethnicity, dating back to changing national boundaries and names. It is believed that although only 772,000 Americans identify themselves as being of Austrian descent in the 2008 U.S. Census figures, as many as 4 million U.S. citizens may actually be of Austrian ancestry.[4] Though early immigrants settled mostly in the Northeast, the largest populations of Austrian Americans are now found in New York, California, and Florida. At the turn of the century, Austrians were involved in clothing and tailoring, mining, and the food industry, including bakeries, meatpacking operations, and restaurants. Today, Austrians are found in a diverse range of occupations.

In the 2008 U.S. Census figures, more than 1.5 million Hungarian Americans were estimated to be in the United States. Most settled originally in the Northeast, but younger generations have migrated to California and Texas, while many Hungarian retirees have moved to Florida.[5] Economically, the Hungarians differ little from other central European immigrants. Most live in urban areas and work mostly in white-collar occupations. First- and second-generation Hungarian Americans encouraged their children to become engineers, a science that was respected by the Hungarian aristocracy at the turn of the century.

Nearly 1,600,000 Americans of Czech descent were identified in the 2008 U.S. Census estimates. Most Czechs now live in cities or rural non-farm areas[6] and are very acculturated. Cities and states with large Czech populations are California, Chicago, Iowa, Minnesota, Nebraska, New York City, Texas, and Wisconsin. Occupationally, only a small number of Czech Americans are still farmers; a majority of Czechs now hold sales, machinist, or white-collar jobs. Many Czechs have been successful in industry, founding businesses that produce cigars, beer, and watches.

There are over 809,000 Americans of Slovak descent, according to the 2008 U.S. Census figures. Actual numbers may exceed 2 million when those originally misidentified as Czechoslovakian or Hungarian are included.[7] The first two generations of Slovaks grew up in tightly knit communities anchored by work, church, family, and social activities. The third and fourth generations have sought higher education, work in white-collar jobs, and live in the suburbs; median family income is far above the national average. Cultural ties are still strong among the later generations of Slovaks.

More than 1 million citizens declared Swiss ancestry in the 2008 U.S. Census estimates. Most Swiss were multilingual and often multicultural when they arrived, assimilating quickly into U.S. culture. The few Swiss who come to the United States today work mostly in the U.S. branches of Swiss companies.[8]

After arrival in the United States, Gypsies retained their tradition of roving; their exact numbers are unknown, and they are a very mobile population, often living in trailer parks. Many renovate apartments and houses to accommodate large social gatherings, then pass the homes on to other Gypsy Americans when they move. It is estimated that approximately half live in the rural areas of the South and West, and half live in urban regions. The cities with the largest concentrations of Gypsies are Los Angeles, San Francisco, New York, Chicago, Boston, Atlanta, Dallas, Houston, Seattle, and Portland. Traditionally tinkers and traders, Gypsies have been very successful at independent trades, such as house painting and asphalt paving, and service work such as body-fender repair and dry cleaning. Gypsies have also entered the car dealership profession in large numbers. Women are a strong presence in the mystical arts, including fortunetelling. The Gypsies divide the urban regions of the United States to minimize competition between Gypsy-owned businesses.

Russians and People of the FSU. In 2008 approximately 3 million Russian Americans were living in the United States. They have mostly moved out of the inner-city settlements to the suburbs, especially in the Northeast. A 20,000-member community of Russian Molokans is in California.[9] Figures from the 2008 census reported close to 1 million Americans of Ukrainian descent, 712,000 of Lithuanian heritage, and 464,000 of Armenian ancestry. In the past decade, one-third of FSU immigrants are from Russia, one-third are from Ukraine, and the remaining third are from all other FSU nations. These recent immigrants have settled in urban areas, including New York, Chicago, Los Angeles, Boston, Detroit, and San Francisco. Today, 40 percent of Ukrainians are found in Pennsylvania, and 60 percent of Armenians are living in California.

Since World War II, the relations between the Russian-American community and American society have largely been dependent on the political relations between the United States and Russia. During the 1950s, anti-Soviet and anticommunist sentiments in the United States caused many Russian Americans to assume a low profile that hastened their acculturation. Since the bulk of Ukrainian and Lithuanian immigration occurred several generations past, most of these populations are assimilated. Armenians, who are typically well educated and English-speaking on arrival, have also found it easy to adapt to U.S. society.

Immigrants recently arriving from Russia and the FSU have come from relatively advanced educational and professional backgrounds. Estimates are nearly half of Russian immigrants have a university degree. Most have professional experience, many as engineers, economists, scientists, or physicians. In addition, many Armenian Americans own their own businesses and have income levels that are well above average; however, higher than average numbers of families who have recently immigrated from Armenia fall below the poverty line.[10]

Many recent refugees from the FSU have been admitted to the United States due to ethnic or religious persecution. In one study, immigrants reported they feared for their safety and felt they could not live openly as Jews due to anti-Semitism.[11]

Based on 2001 census data, it is estimated that there are almost 3 million Canadians of German descent and 817,000 of Polish ancestry. In addition, 1 million Canadians list their heritage as Ukranian, and over 300,000 report Russian origins.

Seventy-three percent of the population in Switzerland speaks Swiss German, 20 percent speaks Swiss French, and 5 percent speaks Swiss Italian; in addition, most Swiss speak one or two other languages.

Many Czech immigrants immediately Americanized their names upon arrival in the United States.[12]

Worldview

Religion

Germans. The majority of German immigrants were Lutheran; a minority were Jewish or Roman Catholic. Today the Pennsylvania Dutch and the rural Germans from Russia faithfully maintain their religious heritage.

Both groups are primarily Protestant, mostly Lutheran or Mennonite. Mennonites are a religious group derived from the Anabaptist movement, which advocated baptism and church membership for adult believers only. They are noted for their simple lifestyle and rejection of oaths, public office, and military service. The Amish, a strict sect of the Mennonites, follow the Bible literally. They till the soil and shun worldly vanities such as electricity and automobiles. Their life centers on Gelassenheit, meaning submission to a higher authority through reserved and humble behavior, and placing the needs of others before the needs of the individual.

Poles. Most Polish immigrants were devout Catholics; they quickly established parish churches in the United States. The Catholic Church is still a vital part of the Polish-American community, although Polish Americans have been found to marry outside the church more than other Catholics.

Central Europeans. Austrians are mostly Roman Catholic and have been actively involved in promoting Catholicism in America. In 1829 the Leopoldine Stiftung was founded in Austria to collect money throughout Europe to introduce religion to the U.S. frontier, resulting in more than 400 Catholic churches established in the East, the Midwest, and in what was known as "Indian country." There are also small numbers of Austrian Jews. The majority of Hungarians are Catholic, although in the United States, nearly 25 percent are Protestant.

In Europe most Czechs were Roman Catholics, but one-half to two-thirds of nineteenth century Czech immigrants from rural areas left the church and were considered free thinkers who believed in a strong separation of church and state. Subsequent generations now belong to a variety of faiths. Religion is still an important factor in the lives of Slovaks. Most are Roman Catholics who attend services regularly. First- and second-generation Slovaks usually send their children to parochial schools supported by the ethnic parish.

Traditional spirituality for the Gypsies is derived from Asian Indian religions, such as Hinduism and Zoroastrianism (see Chapter 2, "Traditional Health Beliefs and Practices," regarding eastern religions, and Chapter 14, "Asian-Indians and Pakistanis" for more information on Asian Indian faiths). While traditions and customs vary by tribe and to a certain degree by the host culture, Gypsies are thought to be united in their worldview, called romaniya. Many believe in God, the devil, ghosts, and predestination. Most of all they adhere to the concept that persons and things are either pure or polluted. Gypsy culture is structured to preserve purity and to avoid contamination through contact with non-Gypsies. Some Gypsy Americans are Christians, often members of fundamentalist congregations, and several churches have specifically combined Gypsy spiritual concepts with Christian practices.

Russians and People of the FSU. Except for the Soviet Jews, the primary organization of the Russian-American community today is the Russian Orthodox Church. Religion has always played a central role in the Russian community, and the Orthodox Church has tried to preserve the culture. However, the largest branch of the Eastern church, officially known as the Orthodox Church in America (formerly the Russian Orthodox Church outside Russia), now includes people from other central European and FSU countries, and the Russian traditions have been deemphasized.

Among Ukrainian Americans, more belong to the Roman Catholic Church than to the Eastern Orthodox faith. Lithuanians are also predominantly Roman Catholic; however, there are small numbers of Protestants, Jews, and Eastern Orthodox followers. Most Armenians are members of the Armenian Apostolic Church (an Eastern Orthodox faith noted for allowing its members to make decisions on issues such as birth control and homosexuality without religious influence), although some Americans of Armenian descent are Protestants or members of the Armenian Rite of the Roman Catholic Church.

Family

Germans. The traditional German family was based on an agricultural system that valued large families in which every member worked in the fields to support the household. Even when

German immigrants moved to urban areas, family members were expected to help out in the family business. Most German families today are assumed to have adopted the smaller American nuclear configuration. The exception may be among the Pennsylvania Dutch, particularly the Amish, who continue to have large families of seven to ten children. It is not unusual for an Amish person to know as many as seventy-five first cousins or for a grandparent to have thirty-five grandchildren.[13] Many Amish families are finding it difficult to maintain traditional values due to growing contact with the majority culture through suburban sprawl.

Poles. Traditionally the Polish-American family was patriarchal, and the father exerted strong control over the children, especially the daughters. The mother took care of the home, and, if the children worked, it was near the home or the father's workplace. Since the 1920s, the overwhelming majority of Polish-American families have been solely supported by the father's income; wives and children have rarely worked.

Other Central Europeans. Tight nuclear families typify traditional Austrian households. Although the father is in charge of family finances, it is the mother who rules home life. Assimilation in the United States has led to a deterioration of the nuclear family, including an increased divorce rate. Traditional Czech and Hungarian families were male dominated and included many relatives. In the United States, participation in church activities, fraternal societies, and political organizations often served to replace the extended family for both men and women. The role of women has become less circumscribed; children are typically encouraged to pursue higher education and professional careers. Family ties are strong among the Slovaks. Parents are respected; they are frequently visited and cared for in their old age. Weddings are still a major event, although they are not celebrated for several days, as they once were.

Gypsies customarily maintain extended families, although in the United States more nuclear families have been established. When traveling was common, multifamily groups (smaller than tribes) would temporarily band together. Affiliation with this group, called a *kumpania*, often continues today. The father is in charge of all public matters, but women may make most of the family income and manage all money matters. Women also retain some power through their ability to communicate with the supernatural world. Usually Gypsies do not date, and arranged marriages are still common.

Russians. Traditionally, Russians lived in very large family groups with women legally dependent on their husbands. This structure changed, however, with the education and employment opportunities offered to women during the communist rule of the Soviet Union. Most women worked, and families became smaller. Even when employed fulltime, however, women remained responsible for all household chores. In the United States, Russian family structure has shrunk even further. Russian couples have significantly fewer children than the national average for American families. Education is emphasized, especially if it can be obtained at a Russian-language school. Many first-generation immigrants attempted to maintain ethnic identity by restricting their children to spouses from their immediate group, but marriage to non-Russians is now the norm.

Since many of the Ukrainian and Lithuanian immigrants in the nineteenth century were men, most were forced to intermarry with other ethnic groups. The men dominated the household, the women ran the home, and the extended family was the norm. Ukrainian and Lithuanian families have since moved toward a more typically American composition with just two working parents and children. In many ways, Armenian homes are also similar to the average U.S. household. Both parents usually work, and education is a high priority. Nearly 70 percent of second-generation Armenian Americans obtain a college degree. It has been noted, however, that most Armenian children retain respect for elders after acculturation, which is uncommon in other cultures, and tight-knit families have allowed many Armenians to pass on traditional customs.[14]

Traditional Health Beliefs and Practices. German biomedicine makes extensive use of botanical remedies, though continued use is not documented in German Americans. A study of German Americans elders in Texas showed that many believe illness is caused by infection or stress-related conditions.[15] Some Germans

The Amish and Mennonites are referred to as the "Plain People." The more liberal and worldly members of the Lutheran and Reformed Churches are called the "Gay Dutch," "Fancy Dutch," or "Church People."

In 301 C.E., Armenia became the first nation to adopt Christianity as its state religion.

AP/World Wide Photos

▲ *Eastern Orthodox priest blessing the Coca-Cola plant in Moscow, 1996.*

believe sickness is an expected consequence of strenuous labor. Health is maintained by dressing properly, avoiding drafts, breathing fresh air, exercising, doing hard work, and taking cod liver oil. A few respondents mentioned the importance of religious practices and that suffering from illness is a blessing from God. Numerous home remedies are common (see "Therapeutic Uses of Foods" section for more information).

The Pennsylvania Dutch traditionally believed a hearty diet high in meats, dairy products and grain foods was important for maintaining good health. Many use home remedies, homeopathic preparations, and healers to treat illness. Sympathy healing is especially well developed. This traditional folk practice uses charms, spells, and blessings to cure the symptoms of disease. It is called either "powwowing" (though not related to Native American beliefs and practices) or by its German name, *Brauche* or *Braucherei.* There is a strong religious foundation to the practice, and the healer acts as God's instrument, requesting God's direct assistance in treatment. Powwow compendiums still in use today offer everything from household tips to cures for warts, burns, toothache, and the common cold. The Amish in particular subscribe to sympathy healing, the laying on of hands to diagnose illness, and reflexology (foot massage thought to benefit other areas of the body, such as the head, neck, stomach, and back), as well as the use of herbal home cures, especially teas.[16–19]

Symptoms of colic in Amish infants are attributed to a condition known as livergrown, which can only be cured with sympathy healing.

Polish American elders in Texas have reported that a shortage of medical supplies in Poland led to the widespread use of faith healers.[15] Although such healing practices are not documented in the United States, many Polish Americans are deeply religious and believe that faith in God and the wearing of religious medals will help to prevent illness. Other health-maintenance beliefs include avoidance of sick people, a healthy diet, sleep, keeping warm, exercise, a loving home, and avoidance of gossip.

Gypsies have unique health beliefs.[10] Health is maintained through *marimé,* a system of purity and pollution that may be related to Asian-Indian beliefs (see Chapter 14). The separation of clean from unclean dictates much of Gypsy life. The body is an example of this dichotomy. The upper body is pure, as are all its secretions, such as saliva. The lower half is impure and shameful. Care is taken to avoid contamination through contact of the upper body by the lower body (only the left hand is used for personal care). Menstrual blood is especially impure. Purity is also maintained by avoiding public places that non-Gypsies (who are considered unclean) frequent and by not touching contaminated surfaces, as well as by the use of disposable utensils, cups, and towels when in impure locations (e.g., hospitals).

Gypsies divide illnesses into those that are due to contact with non-Gypsies, and those that are Gypsy conditions caused by spirits, ghosts, the devil, or breaking cultural rules. Home remedies and Gypsy healers (usually older women versed in medicinal lore) are considered best for Gypsy illnesses. Non-Gypsy conditions are suitable for treatment by non-Gypsy physicians, though a non-Gypsy folk healer such as a *curandero* may be consulted as well.

Natural cures and alternative medicine are used extensively in Russia and the nations of the FSU, and they are often integrated with a biomedical therapy.[18] For example, cupping is used for respiratory illnesses. Saunas, massage, steam baths, and balneotherapy (bathing in mineral springs) are often prescribed in conjunction with biomedical approaches; mud baths may be used for hypertension, and sulfurated hydrogen baths for cardiac ailments. In addition, homeopathic preparations and herbal remedies are popular.[20,11,21,18] Magic and the occult may be used to cure illnesses due to supernatural causes. Psychics and *znakarki* (elder women who whisper charms and sprinkle water with magic powers) may be employed for chronic conditions that biomedicine cannot ameliorate. In the Siberian region of Russia, sickness was traditionally attributed to spiritual crisis, soul loss, evil spirits, breach of taboos, or curses. Treatments used by shamans (magicoreligious healers) included realigning life forces or retrieving the soul through visualization techniques, singing, chanting, prognostication, dream analysis, and séances.[22,23] Russians who do not believe in any occult practices may blame illness on other factors outside their control, including social conflict, political problems, war, poor medical care, and starvation.[24]

Traditional Food Habits

Ingredients and Common Foods: Staples and Regional Variations

The regional variations in central European and FSU cuisine are minor. The exceptions are the foods of the southern CIS nations, such as Armenia. (See "The Cooking of Armenia" in this chapter.) The temperate climate of the region and proximity to the Arabs, Turks, and Greeks have resulted in a cuisine similar to Middle Eastern fare (see Chapter 13, "People of the Balkans and the Middle East"). Ingredients in traditional central European and FSU dishes were dictated by what could be grown in the cold, often damp climate. Common ingredients are potatoes, beans, cabbage and members of the cabbage family, beets, eggs, dairy products, pork, beef, fish and seafood from the Baltic Sea, freshwater fish from local lakes and rivers, apples, rye, wheat, and barley (see Table 7.1 for the cultural food groups). Foods were often dried, pickled, or fermented for preservation—for example, cucumber pickles, sour cream, and sauerkraut.

Bread is a staple item, and there are more than one hundred varieties of bread. Because the climate in central Europe and FSU makes wheat harder to grow, bread is often made with rye and other grains; thus it is darker in color than bread made from wheat flour. Common types are whole wheat, cracked wheat, white, black, rye, pumpernickel, caraway, egg, and potato. Cornmeal breads are found in more southern nations such as Romania. Soft pretzels are a favorite in Germany and in Switzerland; they are sometimes sliced to make sandwiches. Noodles and dumplings abound and are often served as side dishes. Boiled dumplings (called knedliky in Czech, Knödel in German, and kletski in Russian), can be made with flour or potatoes and with or without yeast. Spätzle are tiny dumplings common in southern Germany. They are made by forcing the dough through a large spoon with small holes into the hot water. Stuffed dumplings, filled with meat, liver, bacon, potatoes, or fruit, are called Maultaschen (German), pierogi (Polish), pelmeni (Russian), or varenyky (Ukrainian). Related to the filled dumpling is stuffed pastry dough, which is baked or fried. It is customarily filled with meat or cabbage. Small individual pastries are called pirozhki in Russian, and a large oval pie is known as a pirog (also called a kulebi). One elaborate version, kulebiaka, usually includes a whole fish, such as salmon, with mushroom and rice filling. In Lithuania, lamb-stuffed pockets served with sour cream are called kulduny. A specialty product of Russia is buckwheat (an Asian grain), which has a very distinctive, nutty flavor, especially when it is toasted. The groats (hulled and crushed grains) are prepared in ways similar to rice, especially as side dishes and stuffings.[25] Buckwheat meal is used to make baked goods.

Next to bread, meat is the most important element of the diet. Pork is the most popular. Schnitzel is a meat cutlet, often lightly breaded and then fried. Ham is served fresh or cured. Poland is famous for its smoked ham, and in Germany, Westphalian ham is lightly smoked, cured, and cut into paper-thin slices. Beef is also common. In Germany, Sauerbraten, a marinated beef roast, is the national dish. It is also rolled around various fillings, such as bacon, onions, and pickles, to make Rouladen. Veal is especially popular in Lithuania. Poultry is well liked. Germans often eat roast goose stuffed with onions, apples, and herbs on holidays. In Russia, chicken is stewed on special occasions, and breaded chicken cutlets called kotlety are common. A famous Russian dish is chicken Kiev—breaded, fried chicken breasts filled with herbed butter. Game meats are a favorite in many areas, especially deer, wild boar, and rabbit. A well-known German dish is Hasenpfeffer—hare cooked in red wine with

The Poles say, "A doctor's mistakes are covered in earth."

Gypsy women are traditionally prohibited from touching food, water, or utensils intended for other family members during their period or following childbirth.

Legend has it that it was the Mongols who showed the central Europeans how to broil meat, make yogurt and other fermented dairy products, and preserve cabbage in brine.

Frankfurter means a sausage from the German city of Frankfurt; Weiner means one from Vienna. In the United States the term *hot dog* became popular in the late 1800s because of the sausage's resemblance to a dachshund; stories about a cartoonist inventing the nickname are myth.

Table 7.1 Cultural Food Groups: Central European and Russian/FSU

Group	Comments	Common Foods	Adaptations in the United States
Protein Foods			
Milk/milk products	Dairy items, fresh or fermented, are frequently consumed. Whipped cream is popular in some areas; sour cream popular in other regions.	Milk (cow, sheep) fresh and fermented (buttermilk, sour cream, yogurt), cheese, cream	Milk products are still frequently consumed or increased.
Meat/poultry/fish/eggs/legumes	Meats are often extended by grinding and stewing. Russians tend to eat their meat very well done.	*Meat:* beef, boar, hare, lamb, pork (bacon, ham, pig's feet, head cheese), sausage, variety meats, veal, venison *Fish:* carp, flounder, frog, haddock, halibut, herring, mackerel, perch, pike, salmon, sardines, shad, shark, smelts, sturgeon, trout *Shellfish:* crab, crawfish, eel, lobster, oysters, scallops, shrimp, turtle *Poultry and small birds:* chicken, Cornish hen, duck, goose, grouse, partridge, pheasant, quail, squab, turkey *Eggs:* hens, fish (caviar) *Legumes:* kidney beans, lentils, navy beans, split peas (green and yellow)	Consumption of meat and poultry has increased; use of variety meats has decreased. Sausages and other processed meats are often eaten.
Cereals/Grains	Bread or rolls are commonly served at all meals. Dumplings and kasha are also common. Numerous cakes, cookies, and pastries are popular. Rye flour is commonly used.	Barley, buckwheat, corn, millet, oats, potato starch, rice, rye, wheat	More white bread, less rye and pumpernickel bread are eaten. Breakfast cereals well accepted.
Fruits/Vegetables	Potatoes are used extensively, as are all the cold-weather vegetables. Cabbage is fermented to make sauerkraut. Fruits and vegetables are often preserved by canning, drying, or pickling. Fruit is often added to meat dishes.	*Fruits:* apples, apricots, blackberries, blueberries, sour cherries, sweet cherries, cranberries, currants, dates, gooseberries, grapefruit, grapes, lemons, lingonberries, melons, oranges, peaches, pears, plums, prunes, quinces, raisins, raspberries, rhubarb, strawberries *Vegetables:* asparagus, beets, broccoli, Brussels sprouts, cabbage (red and green), carrots, cauliflower, celery, celery root, chard, cucumbers, eggplant, endive, green beans, kohlrabi, leeks, lettuce, mushrooms (domestic and wild), olives, onions, parsnips, peas, green peppers, potatoes, radishes, sorrel, spinach, tomatoes, turnips	Tropical fruits may be eaten. Greater variety of vegetables consumed; salads popular.
Additional Foods			
Seasonings	Central Europeans tend to season their dishes with sour-tasting flavors, such as sour cream and vinegar.	Allspice, anise, basil, bay leaves, borage, capers, caraway, cardamom, chervil, chives, cinnamon, cloves, curry powder, dill, garlic, ginger, horseradish, juniper, lemon, lovage, mace, marjoram, mint, mustard, paprika, parsley, pepper (black and white), poppy seeds, rosemary, rose water, saffron, sage, savory (summer and winter), tarragon, thyme, vanilla, vinegar, woodruff	Saffron is a popular spice in Pennsylvania Dutch fare.
Nuts/seeds	Poppy seeds are often used in pastries; caraway seeds flavor cabbage and bread.	*Nuts:* almonds (sweet and bitter), chestnuts, filberts, pecans, walnuts *Seeds:* poppy seeds, sunflower seeds	
Beverages	Central Europeans drink coffee; Russians drink tea. Many varieties of beer are produced. Hungarians and Austrians tend to drink more wine than other central European people.	Beer, hot chocolate, coffee, syrups and juices, fruit brandies, herbal teas, milk, tea, kvass, vodka, wine	Soft drinks common.
Fats/oils		Butter, bacon, chicken fat, flaxseed oil, goose fat, lard, olive oil, salt pork, suet, vegetable oil	Commercial salad dressings, non-dairy creamers added to diet.
Sweeteners		Honey, sugar (white and brown), molasses	

black pepper. In Poland, bigos ("hunter's stew"), made of venison, hare, and vegetables (some form of cabbage is always added), is traditional. Geese and duck are widely eaten, also.

In the past, meat was often scarce and expensive; thus many traditional recipes stretched it as far as possible. Dishes common throughout the region consist of seasoned ground meat mixed with a binder such as bread crumbs, milk, or eggs, then formed into patties and fried. In Germany, ground beef (and sometimes pork or veal) is served raw on toast as steak tartare. Ground meat is also used to stuff vegetables (such as stuffed cabbage) or pastry, or is cooked as meatballs, such as Königsberger Klopse topped with capers and a white sauce. Cut-up meat is often served in soups, stews, or one-pot dishes. In Germany a slowly simmered one-pot dish of meat, vegetables, potatoes, or dumplings is called Eintopf. Hungary is known for its gulyás, a paprika-spiced stew known as goulash in the United States. Sweet Hungarian paprika is ground, dried, red chile peppers to which sugar has been added. As chile peppers are a New World food, it is thought that the Hungarians used black pepper to season their food before the discovery of the Americas.

Ground meats are also made into sausages. In Germany there are four basic categories of sausage (Wurst). Rohwurst, similar to American-style liverwurst, is cured and smoked by the butcher and can be eaten as is. Examples include Teewurst, a raw, spiced pork sausage that is spreadable like pâté, and Mettwurst, a mild, sliceable pork sausage. Bruhwurst (the frankfurter or Wienerwurst is one type) is smoked and scalded by the butcher; it may be eaten as is or heated by simmering. Knockwurst, which is like a cold cut, may be smoked and is fully cooked by the butcher. Leberwurst (liverwurst), Blutwurst (blood sausage), and Süize (head cheese) are examples. Bratwurst, similar to sausage links, is sold raw by the butcher and must be panfried or grilled before eating. The Polish are famous for kielbasa, a garlic-flavored pork sausage. In Austria, some sausages are called Wieners. Two popular sausages with both the Czechs and Slovaks are jaternice, made from pork, and jelita, a blood sausage, which can be boiled or fried.

Fresh- and saltwater fish and seafood are often eaten fresh, smoked, or cured. Trout, carp, and eel are popular throughout much of the region. In Germany herring is commonly pickled and eaten as a snack or at the main meal, sometimes as Rollmops, wrapped around a bit of pickle or onion. In Russia, smoked salmon and sturgeon are considered delicacies, as is caviar, which is roe from sturgeon. Caviar is classified according to its quality and source. Beluga, the choicest caviar, is taken from the largest fish and has the largest eggs; its color varies from black to gray. Sevruga and osetra, taken from smaller sturgeon, have smaller eggs and are sometimes a lighter color. Sterlet, or imperial caviar, is from a rare sturgeon with golden roe. The finest caviar is sieved by hand to remove membranes and is lightly salted. Less choice roes are more heavily salted and pressed into bricks. Though some fish is consumed in Poland, it is not a popular food, and is in some cases associated with shortages endured during Soviet rule.[26]

Dairy products are eaten daily. Cheeses may be served at any meal, from the fresh, sweet varieties, such as Lithuanian farmer's cheese, to the strongly flavored aged types like German Limburger. Fresh milk is drunk; butter is the preferred cooking fat. Buttermilk (a thick type called kefir is popular in southern areas of the FSU), sour cream, and fresh cream are also common ingredients in sauces, soups, stews, and baked products. In Austria and Germany, whipped cream is part of the daily diet, served with coffee or pastries.[27]

Traditionally, cold weather fruits and vegetables added variety to the diet. Red and green cabbage is ubiquitous—found fried, boiled, fermented as sauerkraut, and added to stuffings, soups, and stews. Potatoes are equally popular. They are most often boiled, or roasted and sliced. One German specialty found in the northern area of the country is called Himmel und Erde ("heaven and earth"), a boiled dish of potatoes and sliced apples topped with fried bacon and onions. Other root crops such as beets and kohlrabi accompany many meals. Cucumbers are frequently pickled or served dressed with vinegar for a salad. Onions and mushrooms flavor numerous dishes. Wild mushrooms are so popular in Poland that they are often used as a meat substitute on religious fast days.[26] Temperate vegetables, including tomatoes and eggplant, are found in the more southern nations of the FSU, and are now widely available throughout the region. Cauliflower and tomatoes are the favorite vegetables in Germany today.[28] Common

After diners eat smoked eel on pumpernickel bread in some restaurants of northern Germany, the waiter pours inexpensive Schnapps over their hands to rid them of a fishy odor.

The national dish of Switzerland is cheese fondue (chunks of bread dipped into melted cheese). The Swiss are known for their zesty cheeses with holes, such as Emmenthal (the original Swiss cheese) and Gruyère.

Green vegetables as a group are called wloszczyzna in Polish, meaning "Italian commodities," since so many were originally imported from the southern nation.[30]

fruits include apples, cherries, plums, and berries, though imported bananas are a favorite in Russia.[29]

In much of central Europe, sweets are enjoyed daily. They are eaten at coffeehouses in the morning or afternoon, or bought at the local bakery and served as dessert. There are numerous types, such as cheesecakes, coffee cakes, doughnuts, and nut- or fruit-filled individual pastries. Apple, cherry, raspberry, chocolate, almond, and poppyseed are favorite flavors. Austria is reputed to be the home of apple strudel, made from paper-thin sheets of dough rolled around cinnamon-spiced apple pieces. They are also known for Sachertorte, a chocolate sponge cake with apricot or cream filling. Germany is famous for Schwarzwälder Kirschtorte (Black Forest cake), a rich chocolate cake layered with cherries, whipped cream, and Kirsch (cherry liqueur). "Branch" cake, galęziak, which looks similar to a gnarled log, is a popular pastry from Lithuania that is also found in Poland, where is it is known as sękacz or "pyramid" cake. Dobosch torte, a multilayered sponge cake with chocolate filling and caramel topping, is a favorite in Hungary. Though fresh fruits are eaten infrequently, cooked fruits, such as the berry pudding called kisel in Russia, are common desserts throughout the region.

In central Europe, the most common hot beverage is coffee. In Russia, strong tea diluted with hot water from a samovar is consumed instead. A samovar is a brass urn, which may be very ornate, heated by charcoal inserted in a vertical tube running through the urn's center. Although southwestern Germany, Austria, and Hungary produce excellent white wine, the most popular alcoholic drink in the region is beer. The Czechs are known for pilsner beer, which is bitter tasting but light in color and body. German beers can be sweet, bitter, weak, or strong and are typically bottom-fermented (meaning the yeast sinks during brewing). Lager, a bottom-fermented beer that is aged for about six weeks, is the most common type. Bock beer is the strongest flavored, has a higher than average alcohol content, and is sometimes called "liquid bread." Märzenbier, a beer midway between a pilsner and a bock beer, is served at Oktoberfest (see section on "Special Occasions") in Munich. Weissbier is a light, top-fermented beer brewed from wheat, and often mixed with a lemon or raspberry fruit syrup for a refreshing summer beverage. In Russia and other FSU nations, a sour beer fermented from rye bread or beets, called kvass, is popular. It is slightly sweet and fizzy and sometimes flavored with black currant leaves, caraway, mint, or lemon. Mead, a beer-like product fermented from honey, is a Polish specialty. Schnapps, a fruit brandy made from fermented fruit, such as cherries, is popular in Germany. Vodka, which is commonly drunk in Poland and Russia, is a distilled spirit made from potatoes. It is served ice cold and often flavored with seasonings, such as lemon or black pepper. In Poland, the vodka goldwasser contains flakes of pure gold.

Sachertorte was the subject of a famous Viennese court battle regarding who had the rights to claim the original recipe—it was known as the "Sweet Seven Years War" due to the length of the case.[31]

Sharing food is essential to Gypsy culture. The harshest punishment that can be imposed on an individual is to be banned from communal meals.

Meal Composition and Cycle

Daily Patterns

Central Europe. In the past, people of this region ate five or six large meals a day, if they could afford it. The poor, and usually the people who worked the land, had fewer meals, which were often meatless. Today, modern work schedules have changed the meal pattern, resulting in three meals with snacks each day.

In Germany and the countries of central Europe, the first meal of the day is breakfast, which consists of bread served with butter and jam. Sometimes breakfast is accompanied by soft-boiled eggs, cheese, and ham. In Poland, tea served in glasses is the traditional beverage. The tea is sucked through a sugar cube held between the teeth. At midmorning many people have their second breakfast, which may include coffee, tea, or hot chocolate, and pastries, bread, and fruit, or a small sandwich. Lunch is the main meal of the day. In the past, people ate lunch at home, but today they are more likely to go to a cafeteria or restaurant. A proper lunch begins with soup, followed by a fish course, and then one or two meat dishes served with vegetables, and perhaps stewed fruit. Dessert is the final course, usually served with whipped cream. A quicker and lighter lunch may consist of only a stew or a one-pot meal.

A break is taken at mid-afternoon, if time permits. It typically includes coffee or tea and cake or cookies. The evening meal tends to be light, usually including salads and an assortment of pickled or smoked fish, cheese, ham, and sausages eaten with a selection of breads. In Germany this meal is called Abendbrot, meaning "evening bread." However, westernization and shorter lunch hours mean that nearly a third of

EXPLORING GLOBAL CUISINE—Armenia

Legend is that following the flood, Noah's Ark first found land in Armenia and that the nation was founded by his descendants. One of the first countries to adopt Christianity, Armenia has been ruled by a succession of conquerors, from Rome to Russia, and today is surrounded by Muslim nations. Its fare reflects this unique past. Bread is so important there that the Armenian word for it is also used colloquially for meal or food in general.[32] Lamb is a staple, though chicken and beef are also popular. Pork is rarely consumed, dating back to pre-Christian biblical prohibitions. Freshwater fish such as trout and sturgeon (including its roe, caviar) are well liked. Yogurt, known as mahdzoon, is consumed daily in soups, salad dressings, and beverages, as are cheeses such as feta and mozzarella-like string cheese, which are often flavored with spices or herbs. Numerous fruits and vegetables are cultivated, used fresh, dried, and pickled. Apricots, grapes, lemons, persimmons, pomegranates, quince, bell peppers, cabbage, cucumbers, eggplants, okra, squash, and tomatoes are examples. Olives and olive oil are common; however, lamb fat is often preferred for cooking. Dishes are seasoned with onions, garlic, lemon juice, sesame seeds, allspice, basil, cumin, fenugreek, rosemary, and mint. Honey flavors most desserts.[33]

Armenian cuisine has been significantly influenced by neighboring Greeks, Turks, Persians, Syrians, and other Arabs. Shared dishes include the chickpea puree called hummus, tabouli bulgur salad, grilled kebabs, kufta (meatballs), meat turnovers known as boereg, meat- or grain-stuffed vegetables called dolma, and paklava (baklava). Many dishes have a distinctly Armenian twist. Pilaf, also called plov, is preferred with bulgur and vermicelli instead of rice in many areas. Lahjuman, an Armenian pizza made with lamb, vegetables, and feta cheese, is a favorite, as is keshkeg, a lamb or chicken stew that includes whole hulled wheat kernels called zezads.

Dinner begins with a selection of mezze (appetizers) served with the anise-flavored aperitif raki. Soups follow, made with yogurt, eggs, and lemon, or tomatoes, often with added lentils, meatballs, or even fruit. Salads are also served regularly, often with the main course of kebabs, stew, or casserole. Every meal includes bread, such as pita, lavash, or choereg—Armenian yeast rolls. Dessert is usually fruit, with pastries on special occasions. Traditional beverages include coffee, tea, and tahn—yogurt thinned with water and flavored with mint. Armenians are world-acclaimed vintners. Wine and brandy made from grapes, raisins, apricots, or other fruits are frequently consumed.

Germans now eat smaller lunches with a larger meal than Abendbrot in the evening.[28]

Gypsies customarily eat two meals each day, first thing in the morning and in the late afternoon. Meals are typically social occasions, featuring the dishes common in their adopted homeland; stews, fried foods, and unleavened breads are especially popular.

Russia and the FSU. In czarist times, the aristocracy ate four complete meals per day; dinner was the largest. The majority of the population never ate as lavishly or as often. The meal that peasants ate after a long day's work is still the basis of a typical diet in present-day Russia, as well as in some parts of the FSU, including Ukraine. Three hearty meals a day are common, with the largest meal consumed at lunch. Snacking is rare. The staples are bread; soup made from beets (borscht, which is of Ukrainian origin), cabbage (shchi), or fish (ukha); and kasha (cooked porridge made from barley, buckwheat, or millet).[34] In Lithuania, soup is often replaced by salads. Tea, kvass, vodka, or beer usually accompanies meals.

One part of the traditional czarist evening meal, zakuski (meaning "small bites"), is still part of dinner in Russia today. This traditional array of appetizers starts the meal, and may range from two simple dishes, such as pickled herring and cucumbers in sour cream, to an entire table spread with countless hors d'oeuvres. An assortment of zakuski

Many German foods have been adopted in the United States, including lager-style beer, pork or veal sausages, and pretzels.

© Eising/Photodisc/Getty Images

[SAMPLE MENU]

German *Abendbrot*

A Selection of Sausages, Sliced Ham, and Cheeses

(*Westphalian* Ham, *Teewurst*, etc.)*

Herring (Salad) in Cream Saucea,[a,c]*

Pumpernickel Bread, Small Rolls

Potato Salad,[a,b,c] Beet Salad,[a,b,c] Pickles

***Mohnkuchen* (Poppyseed Cake)[a,b]**

Beer or White Wine

[a]Heberle, M.O. 1996. *German cooking: The complete guide to preparing classic and modern German cuisine, adapted for the American kitchen*. New York: HP Books.

[b]*German Recipes* at http://www.myrecipes.com/german-recipes/

[c]*German Recipes* at http://www.ustash.com/recipes/list/33

*Can be purchased at German-style delicatessen

Fast foods, including U.S. hamburger franchises, are very popular throughout the region. In Germany, street stands offering bratwurst and currywurst (sausage with curry seasoning) and French fries are common. Döner kebabs, Turkish-style lamb in pita bread, are another favorite in Germany and in Russia.

usually includes a variety of small, open-faced sandwiches topped with cold, smoked fish; anchovies or sardines; cold tongue and pickles; and ham, sausages, or salami. Caviar, the most elegant of zakuski, is served with an accompanying plate of chopped, hard-boiled eggs and finely minced onions. Other zakuski include marinated or pickled vegetables, hot meat dishes, and eggs served a variety of ways.

Etiquette. Central Europeans tend to be more formal than most Americans. In Germany guests are generally not invited for dinner but may be asked for dessert and wine later in the evening. If you are invited for a meal at a home or restaurant, the invitation may indicate "c.t." (cum tempore), meaning you can arrive up to fifteen minutes late, or "s.t." (sine tempore) meaning to be exactly on time. Do not begin the meal before the host says, "Guten Appetit." In Poland people do not begin a meal until everyone is served and the host says, "smacznego." In the Czech Republic and Slovakia the host says "dobrochot"; in Hungary it is "jo atvadyat." In Russia wait for the host to say "pree yat na vah appeteetah." The word for *hospitality* in Russia is khlebosol'stvo, meaning "bread" and "salt." Bread is traditionally served with butter and a small bowl of salt for dipping to welcome diners.[25]

Appropriate hostess gifts are good quality dessert wines, candies, or pastries. Do not bring vodka in regions where it is commonly served, because it suggests your host does not have enough on hand. In the Czech Republic, Slovakia, and Hungary, Jack Daniels whiskey is particularly appreciated.[35]

The continental European style of eating with the fork in the left hand and the knife in the right is common throughout the region. Do not switch utensils when eating. In Germany, knives are only used when absolutely necessary. Cutting potatoes, pancakes, or dumplings with a knife is an insult to the cook or host because it suggests that these items are tough. When not eating, keep your hands above the table with the wrists resting on the edge. Pass all dishes to the left. Wine and vodka may be poured in tumblers instead of glasses specific for each beverage, and will usually be refilled as soon as they are emptied. Vodka is traditionally consumed in one shot.

Special Occasions. The majority of central European holidays have a religious significance, although some traditions date back to pre-Christian times. The two major holidays in the region are Christmas and Easter. Many of the familiar symbols and activities associated with these holidays, such as the Christmas tree and the Easter egg hunt, were brought to the United States by central European immigrants.

Germany. Germany is a land of popular festivals. Nearly all are accompanied by food and drink. Probably the best-known celebration is Munich's Oktoberfest, which lasts for sixteen days from late September through early October. Founded in 1810 to commemorate the marriage of Prince Ludwig of Bavaria, it is now an annual festival with polka bands and prodigious sausage-eating and beer-drinking.

Advent and Christmas are the holiest seasons in German-speaking countries. The Christmas tree, a remnant of pagan winter solstice rites, is lit on Christmas Eve when the presents, brought not by Santa Claus but by the Christ Child, are opened. The Christmas tree is not taken down until Epiphany, January 6. A large festive dinner is served on Christmas Day, and it is customary for families to visit one another. Foods served during the Christmas season include carp on Christmas Eve and roast hare or goose accompanied by apples and nuts on Christmas Day. Brightly colored marzipan candies in the shape of fruits and animals are traditional Christmas sweets. Other desserts prepared during the season are spice cakes and cookies (Pfeffernüsse and Lebkuchen), fruit cakes (Stollen), cakes in the shape of a Christmas tree (Baumkuchen), and gingerbread houses.

On Easter Sunday, the Easter bunny hides colored eggs in the house and garden for the children to find. Ham and pureed peas are typically served for Easter dinner. Candy Easter eggs and rabbits are also part of the festivities.

Poland. Christmas and Easter are the two most important holidays in Poland, a predominantly Catholic country. On Easter, the festive table may feature a roast suckling pig, hams, coils of sausages, and roast veal. Always included are painted hard-boiled eggs, grated horseradish, and a Paschal lamb sculptured from butter or white sugar. Before the feasting begins, one of the eggs is shelled, divided, and reverently eaten. The crowning glory of the meal is the babka, a rich yeast cake. All the foods are blessed by the priest before being served. On Christmas Eve, traditionally a fast day, a meal of soup, fish, noodle dishes, and pastries is served when the first star of the evening is seen.[26] One popular soup, barszcz Wigilijny, a cousin to Russian borscht, is made with mushrooms as well as beets. Carp is usually the fish served on Christmas Eve. A rich Christmas cake, makowiec, is shaped like a jelly roll and filled with black poppy seeds, honey, raisins, and almonds. Jelly doughnuts, or paczki, are eaten on New Year's Eve, while on New Year's Day bigos, a hearty stew containing meats, cabbage and other vegetables plus dried fruit, is washed down with plenty of vodka.

Austria, Hungary, the Czech Republic, and Slovakia. At Christmastime, the Czechs eat carp four different ways: breaded and fried, baked with dried prunes, cold in aspic, and in a fish soup. The Christmas Eve meal might also include pearl barley soup with mushrooms, as well as fruits and decorated cookies. Christmas dinner features giblet soup with noodles, roast goose with dumplings and sauerkraut, braided sweet bread (vanocka or houska), fruits and nuts, and coffee. Kolaches, round yeast buns filled with poppy seeds, dried fruit, or cottage cheese, are served at the Christmas meal and on most festive occasions. For Easter, a baked ham or roasted kid is served with mazanec (vanocka dough with raisins and almonds shaped into a round loaf).

The Slovaks break the Advent fast on Christmas Eve by eating oplatky, or small, wafer-like Communion breads spread with honey. The meal may contain wild mushroom soup, cabbage and potato dumplings, stuffed cabbage (holubjy), and mashed potato dumplings covered with butter

© Lottie Davies/Digital Vision/Getty Images

▲ *Russian appetizers, called zakuski, often feature blini topped with sour cream, smoked salmon or caviar, and chives.*

In Hungary a gift-giving tradition called komatál includes an exchange of sweets, fruit, and wine. It is customarily done between girls to establish lifelong friendships and to support women after childbirth.[10]

Marzipan, a paste of ground almonds and sugar, is commonly used in desserts and candies throughout central Europe.

Courtesy of Florida State News Bureau

Vanocka, a Christmas bread popular in the Czech Republic.

chicken, or goose accompanied by roast potatoes and stuffed cabbage, followed by desserts of brandied fruits or fruit compote and poppy seed and nut cakes.

In addition to Christmas and Easter, the Austrians celebrate Fasching (a holiday also known in southern Germany). Originating as a pagan ceremony to drive out the evil spirits of winter in which a procession would parade down the main street of a town ringing cowbells, it developed into a multi-day carnival associated with Lent. In Vienna, over 300 sumptuous balls are held during the event, including those hosted by the coffee brewers and the confectioners of the city. Doughnuts, fritters, and other sweets are typical festival food.

In the Polish Easter meal, the dairy products, meats, and pastries symbolize the fertility and renewal of springtime, while the horseradish is a reminder of the bitterness and disappointments in life.

One customary Christmas Eve dish in Poland is karp po zydowsku, chilled slices of carp in a sweet-and-sour aspic with raisins and almonds. It is of Jewish origin, dating to when Poland was a haven for Jews in the fourteenth century.

The term *pascha* comes from the Hebrew word for Passover, Pesach.

and cheese (halusky). A favorite dessert is babalky, pieces of bread that are sliced, scalded, and drained and then rolled in ground poppy seeds, sugar, or honey. Mulled wine usually accompanies this meal, as do assorted poppy seed and nut pastries and a variety of fruits. For Easter, the Slovaks prepare paska, a dessert in the form of a pyramid containing cheese, cream, butter, eggs, sugar, and candied fruits, decorated with a cross. The meal, blessed by the priest on Holy Saturday, includes ham, sausage (klobása), roast duck or goose, horseradish, an Easter cheese called syrek, and an imitation cheese ball made from eggs (hrudka).

In Hungary, the most important religious holiday is Easter. Starting before Lent, pancakes are traditionally eaten on Shrove Tuesday; sour eggs and herring salad are served on Ash Wednesday. During Easter week, new spring vegetables are enjoyed, as well as painted Easter eggs. The Good Friday meal may include a wine-flavored soup, stuffed eggs, and baked fish. The biggest and most important meal of the year is the feast of Easter Eve, which consists of a rich chicken soup served with dumplings or noodles, followed by roasted meat (ham, pork, or lamb), then several pickled vegetables, stuffed cabbage rolls, and finally a selection of cakes and pastries served with coffee. The Christmas Eve meal, which is meatless, usually features fish and potatoes. The Christmas Day meal often includes roast turkey,

Russia and the FSU. Before the 1917 revolution, Russians celebrated a full calendar of religious holidays, including 250 fast days.[25] Fish was significant on days when it was allowed, but at other fast meals, mushrooms were so common as entrees that they became known as forest meat.[29] Today, the observant still does not eat any animal products during fasts (see Chapter 4, "Food and Religion" for the fast days in the Eastern Orthodox Church).

The most significant holiday is Easter, which replaced a pre-Christian festival that marked the end of the bleak winter season. The Butter Festival (maslenitas) precedes the forty days of Lent. One food eaten during this period is blini, raised buckwheat pancakes. Blini can be served with various toppings such as butter, jam, sour cream, smoked salmon, or caviar. Butter is the traditional topping because it cannot be eaten during Lent. Traditional foods served on Easter after midnight Mass are pascha, similar to the Slovak paska but decorated with the letters XB ("Christ is risen"); kulich, a cake made from a very rich, sweet yeast dough baked in a tall, cylindrical mold; and red or hand-decorated hard-boiled eggs. On Pentecost (Trinity) Sunday (fifty days after Easter), kulich left over from Easter is eaten.

Twelve different dishes, representing the twelve apostles, are traditionally served during the Russian and Ukrainian Christmas Eve meal. One of the dishes is kutia, or sochivo, a porridge of wheat grains combined with honey, poppy seeds, and stewed dried fruit consumed when the first stars of Christmas Eve appear. A festive meal is served on Christmas Day. On New Year's Day children receive gifts, and spicy ginger cakes are eaten. A pretzel-shaped sweet bread, krendel, is eaten on wedding anniversaries and name days (saint's days are celebrated as birthdays in the Eastern Orthodox faith).

Therapeutic Uses of Foods. A study of German American elders found that they sometimes use home remedies to treat minor illnesses. Chicken soup is used for diarrhea, vomiting, or sore throat. Tea is taken for an upset stomach, and milk with honey is commonly used for a cough.[15] Traditionally, the Pennsylvania Dutch believe cold drinks are unhealthy and that eating meat three times a day is the cornerstone of a good diet. Herbal teas are consumed for a variety of complaints.[36,19]

Polish American elders reportedly believe that sauerkraut is good for colic, as are tea and soda water. Chamomile tea is used for cramps, tea with dried raspberries and wine for colds, cooked garlic for high blood pressure, and warm beverages (milk, tea, or lemonade) for coughs. Tea with honey and alcoholic spirits may be used to help sweat out an illness.[37,15]

Gypsies have unique health beliefs, many involving food. For example, Gypsies traditionally think that fresh food is the most nourishing and that leftovers are unwholesome. Canned and frozen items may be mistrusted as not being fresh. Many Gypsies believe that non-Gypsies carry disease, and they may insist on using disposable plates and utensils any time they must eat in a public place. Insufficient intake of lucky foods, such as salt, pepper, vinegar, and garlic, can predispose a person to poor health. Home remedies, such as tea with crushed strawberries, asafetida (called "devil's dung"), and ghost vomit (Fuligo septica), are common.[38,10]

Russians and people of the FSU ascribe health benefits to many different foods. Butter is considered good for eyesight, dill for dyspepsia, and honey for flatulence.[23] Respiratory infections may be treated with gogomul, a mixture of egg yolk, sugar, milk, and baking soda.[11] Teas made from raspberry, chamomile, eucalyptus, and cornsilk are used for numerous complaints. Some alcoholic beverages are of particular therapeutic repute. Kvass, the slightly fermented beverage made from bread, is believed to be good for digestion and to cure hangovers. Balsam, flavored vodka, is traditionally used to cure everything from the common cold to alcoholism. More recently, vodka distilled with medicinal herbs, such as ginseng or schizandra, has become a favorite supplement consumed as a shot or added to tea or coffee. Full, hearty meals are considered important in maintaining health.[39]

Most Christmas food traditions in Russia did not survive the Soviet-ruled period when religious ceremonies were discouraged.[29]

Ukrainians set a place at the table for the spirits of dead ancestors at Christmas.

Contemporary Food Habits in the United States

Adaptations of Food Habits

Ingredients and Common Foods. The central European and Russian diet is not significantly different from U.S. fare. Immigrants made few changes in the types of foods they ate after they came to the United States. What *did* change was the quantity of certain foods. Most central European immigrants were not wealthy in their native lands, and their diets had included meager amounts of meat. After immigrating to the United States, they increased the quantity of meat they ate considerably.

[SAMPLE MENU]

A Russian Dinner

***Pirozhki* (Baked Turnovers)[a,b]**

***Shchi* (Cabbage Soup)[a,b]**

***Kotlety* (Chicken Cutlets)[a,b]**

***Kasha* with Mushrooms (Buckwheat Groats)[a,b]**

Fruit Compote

Vodka and Beer

[a]Visson, L. 2004. *The Russian heritage cookbook*. Woodstock, NY: The Overlook Press.

[b]*Russian Cuisine* at http://www.ruscuisine.com/

Malzbier is a German beer (1 percent alcohol) that is considered appropriate for young children and nursing mothers.

Ghost vomit (Fuligo septica) is a myxomycete or slime mold. The thin, yellow, creeping mass is found on rotting wood and other decaying material, where it eats bacteria.

Some Russian-American mothers fear that cold milk may cause illness in their child and leave milk at room temperature for hours to warm it.[39]

The people of eastern Pennsylvania, where there is a large concentration of German Americans, still eat many traditional German dishes adapted to accommodate available ingredients. Common foods include scrapple (also called ponhaus), a pork and cornmeal sausage flavored with herbs and cooked in a loaf pan, served for breakfast with syrup; sticky buns, little sweet rolls thought to be descended from German cinnamon rolls known as Schnecken; schnitz un knepp (apples and dumplings), a one-pot dish made from boiled ham, dried apple slices, and brown sugar, topped with a dumpling dough; boova shenkel, beef stew with potato dumplings; hinkel welschkarn suup, a rich chicken soup brimming with tender kernels of corn; apple butter, a rich fruit spread, very much like a jam; schmierkaes, a German cottage cheese; funnel cake, a type of doughnut; shoofly pie, a molasses pie thought to be descended from a German crumb cake called Streuselkuchen; fastnachts, doughnuts originally prepared and eaten on Shrove Tuesday to use up the fat that could not be eaten during Lent; and sweets and sours, sweet-and-sour relishes, such as coleslaw, crabapple jelly, pepper relish, apple butter, and bread-and-butter pickles, served with lunch and dinner. Saffron crocuses were cultivated in parts of Pennsylvania, and the seasoning was used to color many dishes dark yellow, from soups to a traditional wedding cake known as Schwenkfelder.

Much of German cooking has been incorporated into U.S. cuisine. Many foods still have German names, although they are so common in the United States that their source is unrecognized (e.g., sauerkraut, pretzels, pickles). Other foods contributed by the Germans are hamburgers, frankfurters, braunschweiger (liver sausage), thuringer (summer sausage), liverwurst, jelly doughnuts, and pumpernickel bread. Beer production, especially in Milwaukee, was dominated by the Germans for more than one hundred years. German immigrants created a lager-style beer that is milder, lighter, and less bitter than typical German beer; it can now be described as American-style beer (see also Chapter 15, "Regional Americans").

"Sauerkraut Yankees" was a derogatory nickname for the Pennsylvania Dutch during the Civil War.

On average, Americans drink 29.5 gallons of beer per person annually.[40]

In the 1800s, fried oysters with a glass of Schnapps was a popular Pennsylvania Dutch breakfast for men.

Meal Composition and Cycle. Third- and fourth-generation central European, Russian, and FSU Americans tend to consume three meals a day (with snacks), and meal composition is similar to that of a typical American meal, although more dairy products and sausages may be eaten. Some traditional foods and ingredients may not be available in areas without large central European or Russian populations.

Little dietary acculturation was found in a study of recent immigrants from Russia and other FSU countries.[39] An increased quantity of familiar foods, particularly soured milk, sour cream, kefir, homemade cheeses, cold cuts, and eggs, and a greater variety of fruits (including citrus, bananas, mangoes, pineapples, kiwis, and fruit juices) and vegetables (especially broccoli, cauliflower, spinach, red cabbage, and mixed salads) are consumed in the United States, but are incorporated into traditional dishes and meal patterns. Breakfast cereals are well accepted, and traditional grains such as barley, buckwheat, and millet are consumed less often. Favorite American items include soft drinks, ice cream, yogurt, commercial salad dressings, nondairy cream, and whipping cream. Coffee is consumed more often and soups less often, especially at dinner. Snacking on fruit, milk, bread, pastries, sandwiches, and candy is becoming more common.

Many central European and Russian-American families serve traditional foods at special occasions. On October 11, Polish Americans celebrate Pulaski Day, a national day of remembrance that features a large parade in traditional apparel down the streets of New York City. The Austrians and Czechs typically observe St. Nicholas Day (December 6), when apples and nuts are put in the stockings of well-behaved children, and coal is given to the naughty ones. Hungarian Americans observe three unique holidays in the United States with a combination of patriotic and religious activities. The first is March 15, commemorating the revolution of 1848; the second is August 20, St. Stephen's Day; and the third is the attempted revolt known as the Revolution of 1956 on October 23.

Among the Amish, many national holidays, such as the Fourth of July and Halloween, are not observed. A second day of celebration is added to Christmas, Easter, and Pentecost: the first day is reserved for sacred ceremonies, the second day for social and recreational activities. Most Amish celebrate all holidays quietly with family.

Nutritional Status

Nutritional Intake. Very little has been reported on the nutritional intake of acculturated central European or Russian Americans. Recent European-wide studies show the diets of central Europeans are among the highest in animal products, potatoes, sweets, and refined/processed items in Europe.[41] Consumption of fats and oils of animal origin was highest in Germany: the mean daily intake for added fats and oils was sixty-six grams for men living in Potsdam.[42] Rates of overweight and obesity are 50 percent for German women and nearly 66 percent for German men. Slightly lower rates are found in the Czech Republic, Slovakia, and Hungary. Polish men scored higher than Germans, but Polish women scored lower.[43] Recent immigrants from Russia and other FSU nations were found to consume a diet high in saturated fats, sodium, and sugar.[39] It is reasonable to assume that central European, Russian, and other FSU Americans may be at risk of developing cardiovascular disease and other conditions associated with high-fat items (particularly red meats, processed meats, and dairy products) popular both in their traditional cultures and in U.S. fare.

Recent immigrants from Russian and FSU nations may suffer some nutritional deficiencies due to inadequate consumption of fruits and vegetables. Low intakes of riboflavin and vitamin C are reported.[41] In the Chuvash Republic of Russia, significant dietary selenium deficiency and moderate iron and manganese deficiencies have been reported.[44] High rates of diabetes, hypertension, hyperlipidemia, and cardiovascular disease are found in Russian-speaking immigrants. Tuberculosis and HIV incidence has been increasing dramatically in Russia in recent years and is thus a concern for new immigrants. Nearly 80 percent of immigrants from Russia are thought to be from the regions most affected by the Chernobyl nuclear power accident of 1979. Increases in leukemia and thyroid cancer in this population have been noted.[45]

A study of recent immigrant Russian mothers found strong support for breast feeding. All but one of ninety participants breast fed their infants exclusively or partially for an average of twenty-eight to thirty weeks, regardless of whether the babies had been born in Russia or the United States.[46] This contrasts with health reports from Russia, where breast feeding is discouraged, and infant nutrition has been compromised by inadequate supplies of formula. High rates of iron-deficiency anemia (affecting 25 to 50 percent of children) and endemic goiter have been reported.[45] Since the Chernobyl nuclear power accident, some Russians may resist having x-rays taken because they fear radiation exposure.[11]

Heavy alcohol use has been reported in Russia, certain FSU nations, and Poland.[47–50] Some researchers are concerned that immigrants from these countries may continue similar drinking patterns; however, data are limited.[37,18] Low rates of alcoholism have been reported among Russian-speaking immigrants.[49]

A small study of Gypsies in Boston found high rates of hypertension, type 2 diabetes, and vascular disease, affecting between 80 and 100 percent of the population over age fifty. Approximately 85 percent were obese. Chronic renal insufficiency was also a problem.[51,52] In Europe, researchers have found high rates of dyslipidemia, obesity, and insulin resistance, suggesting a predisposition for metabolic syndrome (for more information on this topic, see Chapter 14).[53,54] Genetic problems due to high rates of consanguinity show Gypsies worldwide at risk for several metabolic conditions, including phenylketonuria, galactokinase deficiency, citrullinaemia, Wilson's disease, and metachromatic leukodystrophy. Rates of both infant mortality and unfavorable birth outcomes are high; overall life expectancy is low.[55]

Counseling. Communication difficulties may occur with recent or older central European or

Courtesy Pennsylvania Dutch Visitors Bureau

▲ *Pennsylvania Dutch funnel cake is an unusual "doughnut" made by pouring the batter through a funnel or from a pitcher into the hot oil in a swirled pattern.*

High rates of gastric cancer in Lithuania are believed to be due in part to a high consumption of salted and cured meats and fish.[56]

Some Russian Americans favor homemade cheeses made with unpasteurized milk, risking Listeria monocytogenes infection.[57]

Eggs are sometimes used raw in uncooked dishes by Russian and FSU Americans, putting them at risk for Salmonella poisoning.

Russian immigrants. Language barriers may necessitate the use of a competent translator.

Good manners and formality are expected in German conversations. Education is respected, and the use of titles is important.[35] Honesty and directness are appreciated. Germans are monochronistic and prefer to deal with one topic at a time. Direct eye contact is a sign of attentiveness and trust. A handshake is used in greeting, but there is little other touching between acquaintances. Poles tend to speak more quietly than Americans and feel uncomfortable with loud behavior. Discussions about politics are avoided. Direct eye contact and a handshake during a greeting and when leaving are appropriate.

German Americans, as well as other central European Americans, emphasize self-reliance and may avoid health care. Polish Americans may be reluctant to express pain and often deny symptoms of illness.[37] Compliance problems develop if medication is seen as a last resort, taken only until symptoms disappear.

Many Amish still speak a German dialect, so an interpreter may be necessary to provide care. The Amish are often reserved and respectful and expect to be treated in a similar manner. Modesty is important to the Amish, and wearing conservative clothing when counseling clients is appropriate. Modern health technology is not in conflict with traditional Amish religious precepts; however, it may be avoided unless absolutely necessary, such as during a medical emergency.[18] Photos are not allowed, though x-rays and other scans are accepted if their necessity is explained. Educational materials should not include pictures of the Amish, nor illustrations representing human faces. A survey of physicians with experience treating the Amish noted more problems with digestive disorders, obesity, and chronic bed-wetting than with non-Amish clients.[36] The doctors' opinion was that the Amish eat a diet higher in fat and salt than the non-Amish, which may contribute to these problems; however, the Amish showed fewer symptoms of heart disease and alcoholism than non-Amish patients. Conditions regarding maternal and child health are common, and hereditary diseases, including phenylketonuria and other metabolic disorders, are more prevalent among the Amish than the U.S. population as a whole.[18]

In working with Gypsy-American clients, health care providers should be aware that English is usually a second language and that illiteracy is common. However, Gypsies are often very adaptive and may use many forms of communication, depending on the situation. Gypsies believe that the measure of a man's worth is in his girth—weight gain is associated with health, weight loss with illness.[1] Gypsies will often not seek treatment until an emergency develops. Numerous family members will come with a client to provide support, often bringing food, because most Gypsies consider food prepared by non-Gypsies impure.[10]

Russians expect formality between acquaintances, and first names are saved for close friends and relatives. Some Russians may initially respond to any question that requires an affirmative or negative answer with a no; in most cases, however, communication is direct, and may be loud or expressive.[18] Direct eye contact is the norm. A quick three kisses on the cheeks or a handshake is used in greeting, and touching becomes more prominent with familiarity. Russians may consider it impolite to sit with the legs splayed or with an ankle resting on the knee. Preventative screenings or prophylactic therapies are uncommon in the FSU, and clients may not understand the need for the procedures.[11] They may not be compliant with medication or lifestyle modifications. Counseling and mental health issues were avoided by elderly Russian immigrants in one study, perhaps due to the social and economic stigma associated with psychological disorders in Russia.[24] These Russian clients expected extensive help with personal problems and information on social services from their primary provider. An increased perception of pain and somatic symptoms are common; many Russian Americans do not believe U.S. physicians understand their ailments or treatments.[21] Culture-specific illnesses, such as "avitaminosis" or "dysbacteriosis," are unfamiliar diagnoses in the United States.[23] Medical care in Russia mainly provides drug therapy regimes mixed with alternative therapies, thus Russians may expect medications in order to believe they are being helped, and they may believe care is incomplete if it does not include alternative treatments. Injections are preferred over oral medications. One study found that patients frequently self-medicate with drugs obtained from the FSU and often cut the dosage of prescriptions obtained in the U.S. because

Gypsies may reject injections for fear that something impure from the outside will contaminate the pure inner body.

Health care providers working with Russians report that some immigrants believe potatoes cause type 2 diabetes.

they are considered "too strong."[11] Russian clients are often very assertive in their requests because aggressive behavior was necessary to receive attention in the Russian health care system. However, some Russian-speaking immigrants may be reluctant to discuss any infectious diseases for fear of being ostracized within their community.

Many Americans of central European and Russian decent are highly acculturated. An in-depth interview can determine any communication preferences or traditional health practices.

NEW AMERICAN PERSPECTIVES—Russian

CLARA SCHMIDT

I emigrated from the Moldova Republic (formerly Moldavia in the former Soviet Union), first to Israel in 1988, and then to the United States to be near my son. I was an English professor in the Soviet Union and Israel. My family was originally from Romania.

In the Soviet Union, we had enough food but not in abundance. Food was very important to us because of previous shortages, and as a result we were forced to finish everything we were served. Food was never thrown away. In the morning we might have tea with a bread sandwich of butter and jam and/or possibly an egg. Snacks were not typical, but it might be a sandwich or fruit. It was something you brought with you if you went out, since you couldn't purchase them. Dinner was at lunch time (1:00–3:00 p.m.) and was the biggest meal of the day. We ate at home, and the meal was served when my father came home. The meal included soup (vegetable, chicken, or beet/cabbage), salad of several types of vegetables, meat, or fish with potatoes/rice, and fruit for dessert. We drank soft drinks with the meal. Supper was from 6:00 to 8:00 p.m. and could include sausage, vegetables, bread, and maybe leftovers, dairy (yogurt, cottage cheese), or cottage cheese pancakes with sour cream.

My diet changed the most after I immigrated to Israel. We ate less soup, less fat, and more fresh fruit and vegetables year round. We ate much less meat and of course no pork. In Israel I learned to love avocados and hummus (chickpea dip), and I still eat them here. In the United States, we started to eat the American way. Dinner, now the main meal, was closer to 5:00 or 6:00 p.m. It might include chicken soup with vegetables, lamb, cooked fish (not fried), and vegetables. One thing I love here is barbequed lamb ribs. Breakfast has changed to cereal/oatmeal, yogurt, and fruit. Lunch is now a sandwich or salad. I have switched from tea to coffee, and it is something I can't do without. Other things I couldn't change in my diet would be limiting fruit, vegetables, and dairy products. For special occasions I still prepare family recipes, like Georgian chicken with walnuts, chicken necks stuffed with chopped liver, and a three-layer sour cream cake with honey and nuts.

My impression of American eating is that some Americans count every calorie they eat, but the food they eat is not tasty, and that most Americans eat a lot of very tasty but unhealthy food. Fast foods taste good, but the amount of fat is awful—seems like it is all meat and fried food. I prefer plain food, but made at home.

Scandinavians

The Scandinavian countries include Sweden, Norway, Denmark, Finland, and Iceland. With the exception of Denmark and the island of Iceland, they are located north of the Baltic and North seas and share common borders with each other and Russia. Most of the population in Scandinavia is concentrated in the warmer southern regions; the harsher northern areas extend above the Arctic Circle. Norway's weather is more moderate than that of Finland and Sweden because its long western coastline is washed in the temperate North Atlantic Drift. Denmark juts into the North Sea to the north of Germany, and its capital, Copenhagen, is directly opposite from Sweden. Numerous Scandinavians have made their homes in the United States. This section reviews the traditional foods of Scandinavia and the Swedish, Norwegian, and Danish contributions to the U.S. diet.

Cultural Perspective

History of Scandinavians in the United States

Legend is that the Norsemen (ancient Scandinavians), renowned seafarers and explorers, first discovered North America and colonized as far west as Minnesota in the thirteenth and fourteenth centuries. The documented presence of Scandinavians in the United States dates back to the seventeenth century. Jonas Bronck, a Dane, arrived in 1629 and bought a large tract of land from the Native Americans that later became known as the Bronx in New York City.

Immigration Patterns. The majority of Scandinavians arrived in the United States in the 1800s, led by the Norwegians and the Swedes. During the nineteenth century, no other country except Ireland contributed as large a proportion of its population to the settlement of North America as Norway. During the 1900s, an additional 363,000 Norwegians, more than 1,250,000 Swedes, 363,000 Danes, and 300,000 Finns entered the United States.

Few Icelanders have made their home in the United States; the 2008 U.S. Census indicates approximately 55, 000 Americans claim Icelandic descent.

The peak years of Scandinavian immigration to the United States were between 1820 and 1930. The population of all the Scandinavian countries had grown substantially, resulting in economies that could not absorb the unemployed and landless agrarian workers. In Sweden, the problem was magnified by a severe famine in the late 1860s. For the Norwegians there was the additional lure of freedom that the United States offered, and the chance for emancipation for the peasant class. Scandinavian immigrants typically settled in homogeneous communities.

Norwegians and Swedes often moved to the homestead states of the Midwest, especially Illinois, Minnesota, Michigan, Iowa, and Wisconsin. One-fifth of all Swedish immigrants settled in Minnesota. Pockets of Finns and Danes also settled in this region, but they were fewer in number. Many Norwegians and Swedes later moved to the Northwest, working in the lumber and fishing industries. The shipping industry attracted some Norwegians to New York City, where they still live in an ethnic enclave in Brooklyn. Although Swedes and Norwegians are often associated with the rural communities of the Midwest, by 1890 one-third of all Swedes lived in cities, and many Norwegians were seeking opportunity in the urban areas. Chicago and Minneapolis still have large Scandinavian populations.

The Danes, in an effort to preserve their ethnicity, developed twenty-four rural communities between 1886 and 1935, in which, for a set number of years, land could be sold only to Danes. The best known of these communities are Tyler, Minnesota; Danevang, Texas; Askov, Minnesota; Dagmar, Missouri; and Solvang, California. Today, most Danes live in cities, primarily on the East or West coasts. The largest concentrations of Danes are in the Los Angeles area and in Chicago.

Following World War I, U.S. immigration from Finland dropped significantly as Finns chose other countries for emigration. As the Finnish population stagnated, ethnic identity became very difficult to maintain. Second- and third-generation Finns are highly acculturated.

The Scandinavians introduced the cast-iron stove to the United States in the early 1800s. Cooking was previously done in fireplaces and brick ovens.

Current Demographics and Socioeconomic Status. According to the 2008 U.S. Census, there are approximately 1.5 million Danes, 4.4 million Swedes, 4.6 million Norwegians, and 688,000 Finns and their descendants now living in the United States. Most Scandinavians assimilated rapidly into U.S. society, rising from blue-collar to white-collar jobs within a few generations.

A majority of Scandinavian immigrants were literate in their own language and often produced local newspapers, periodicals, and books. Education was valued as a way to improve economic standing. The Danes opened folk schools designed to foster a love of learning in their communities, as well as two liberal arts schools. Norwegians and Swedes established many colleges. The Finns founded summer schools where traditional Finnish culture and religion were taught. Women often had educational opportunities in these schools that were unavailable to them in Scandinavia.

Today, many Norwegians and Swedes have continued farming in the Midwest or have taken jobs in the construction industry. However, nearly one-third of Norwegians are employed in management or specialty professions, while many Swedish Americans have moved into engineering, architecture, and education.[20,58] Danes entered a variety of occupations, but they were most prominent in raising livestock and dairying. Urban Danes today are not associated with specific occupations.[59] Finnish American men have been active in fields such as natural resources management, mining engineering, and geology, while women have been attracted to nursing and home economics.[60]

Worldview

Religion. The majority of Scandinavians who immigrated to the United States were Lutheran, though each nationality had its own branch of the church, and within each branch numerous sects were represented. Factionalism was common until many of the Scandinavian and German Lutheran churches joined together to create the Evangelical Lutheran Church in America (ELCA) in 1987. Those Scandinavians who are not Lutheran often belong to other Protestant churches, including Methodist, Mormon, Seventh-Day Adventist, Baptist, Quaker, and Unitarian.

Family. The nuclear Scandinavian family was at the center of rural life. Families were typically large, and the father was head of the household. Kinship ties were strong: families were expected to pay the way for relatives remaining in Scandinavia to come to the United States, where they would be given a room, board, and help in finding employment. The power of the father diminished and family size decreased as the Scandinavian Americans became more integrated in mainstream society. Among the Finns, it has been noted that when both parents work, they may choose to have only one child.[60]

Traditional Health Beliefs and Practices. Information on traditional Scandinavian health beliefs and practices is very limited. Fish was considered necessary for good health. In Norway, cough and cold confectionaries (such as lozenges and pastilles) are very popular over-the-counter remedies. A current Norwegian study also found that 56 percent of cancer patients use herbs and dietary supplements primarily to boost immune function.[61] The Finns believe in natural health care, practicing massage, cupping, and bloodletting. Further, the sauna (a traditional steambath) is reputed to have therapeutic qualities. It is used by Finns when ill and even by midwives during childbirth. It is considered a remedy for colds, respiratory or circulatory problems, and muscular aches and pains.[60]

One health practice in Sweden that has been widely adopted in the United States is Swedish massage, also known as therapeutic massage. It is a deep muscle technique that uses five main strokes to provide relaxation, increase circulation, and promote healing.

Traditional Food Habits

Scandinavian fare is simple and hearty, featuring the abundant foods of the sea and making the best use of the limited foods produced on land. Scandinavian cooking often reflects the preservation methods of previous centuries. Fish was traditionally dried, smoked, or pickled, and milk was often fermented or allowed to sour before being consumed. Scandinavians still prepare a large variety of preserved foods and prefer their food salty. The basics of the Scandinavian diet are given in the cultural food groups list (Table 7.2).

Ingredients and Common Foods: Staples and Regional Variations

The traditional cooking of Scandinavia is hearty. Spices were expensive in the past, and most dishes feature the natural flavors of ingredients with subtle seasoning. Black pepper, onions, and dill are used in many recipes, and juniper berries add interest in others. Caraway, cloves, nutmeg, and cardamom flavor many baked goods.

Scandinavians are probably best known for their use of fish and shellfish, such as cod, herring, mackerel, pike, salmon, sardines, shrimp, and trout. It is so common that in restaurants the lunch special is often listed simply as dagensratt ("fish of the day").[25] In Norway the fish-processing industry is believed to date back to the ninth century, and today, Scandinavian dried salt cod is exported all over the world. Popular fish dishes include salmon marinated in dill, called gravlax; smoked salmon, known as lox; and the many varieties of pickled herring. Fish sticks and fish baked with cheese and breadcrumbs are common homestyle dishes.

Cream and butter are used in many dishes, especially in Denmark, which has a slightly warmer climate than the other Scandinavian nations. White sauce made with milk and minced parsley tops many Danish dishes, such as roast bacon, eel, and herring, as well as boiled and cured meats.[62] One Swedish specialty is known as Jansson's frestelse ("temptation") and includes anchovies and grated potatoes baked in a cream and onion sauce. Considerable quantities of fermented dairy products are used throughout the region, such as sour cream, cheese, buttermilk, and yogurt-like products, including filmjok (Sweden) and skyr (Iceland). Sour cream in particular is added to soups, sauces, and dressings for salads and potatoes. In Norway, it is traditional to prepare an oatmeal porridge known as rømmegrøt with sour cream, and the dish is still found at many festive occasions. Fish, such as fried trout or herring, is also customarily served with a sour cream sauce and boiled potatoes. Denmark is renowned for its cheeses, such as semi-firm, mellow, nutty-tasting Tybo (usually encased in red wax); firm and bland Danbo; semisoft, slightly acidic Havarti; rich, soft Crèma Dania; and Danish blue cheese. Jarlsberg, a Swiss-style cheese, is a well-known Norwegian product, but brunost, a sweet, brown cheese made from cow's milk mixed with goat's milk and served in thin slices on brown bread, is most popular in the nation.[63] Cheese is eaten daily for breakfast, on sandwiches, and as snacks.

Though fish is popular in many parts of Scandinavia, the more inland areas feature many

In Scandinavia they say, "Danes live to eat, Norwegians eat to live, and Swedes live to drink."

Table 7.2 Cultural Food Groups: Scandinavian

Group	Comments	Common Foods	Adaptations in the United States
Protein Foods			
Milk/milk products	Dairy products, often fermented, are used extensively.	Buttermilk, milk, cream (cow, goat, reindeer); cheese, sour cream, yogurt	
Meat/poultry/fish/ eggs/legumes	Fish is a major source of protein, often preserved by drying, pickling, fermenting, or smoking.	*Meat:* beef, goat, lamb, hare, pork (bacon, ham, sausage), reindeer, veal, venison *Fish and shellfish:* anchovies, bass, carp, cod, crab, crawfish, eel, flounder, grayling, haddock, halibut, herring, lobster, mackerel, mussels, oysters, perch, pike, plaice, roche, salmon (fresh, smoked, pickled), sardines, shrimp, sprat, trout, turbot, whitefish *Poultry and small birds:* chicken, duck, goose, grouse, partridge, pheasant, quail, turkey *Eggs:* chicken, goose, fish *Legumes:* lima beans, split peas (green and yellow)	More meat and less fish are consumed.
Cereals/Grains	Wheat is used less than other grains. Rye is used frequently in breads.	Barley, oats, rice, rye, wheat	More wheat used, fewer other grains.
Fruits/Vegetables	Fruits with cheese are frequently served for dessert. Preserved fruits and pickled vegetables are common. Tapioca (from cassava) is eaten.	*Fruits:* apples, apricots, blueberries, cherries, cloudberries, currants, lingonberries, oranges, pears, plums, prunes, raisins, raspberries, rhubarb, strawberries *Vegetables:* asparagus, beets, cabbage (red and green), carrots, cauliflower, celery, celery root, cucumber, green beans, green peppers, nettles, kohlrabi, leeks, mushrooms (many varieties), onions, parsnips, peas, potatoes, radishes, spinach, tomatoes, yellow and white turnips	A greater variety of fruits and vegetables are obtainable in the United States than in Scandinavia but may not be eaten.
Additional Foods			
Seasonings	Savory herbs and spices preferred. Cardamom is especially associated with Scandinavian sweets.	Allspice, bay leaf, capers, cardamom, chervil, cinnamon, cloves, curry powder, dill, garlic, ginger, horseradish, lemon juice, lemon and orange peel, mace, marjoram, mustard, mustard seed, nutmeg, paprika, parsley, pepper (black, cayenne, white), rose hips, saffron, salt, tarragon, thyme, vanilla, vinegar	
Nuts/seeds	Marzipan (sweetened almond paste) is used in many sweets.	Almonds, chestnuts, walnuts	
Beverages		Coffee, hot chocolate, milk, tea, ale, aquavit, beer, vodka, wine, liqueurs	
Fats/oils	Butter is often used.	Butter, lard, margarine, salt pork	
Sweeteners		Sugar (white and brown), honey, molasses	

meat dishes. In the southern regions pork is particularly common. Roast loin stuffed with prunes and apples is a favorite in Denmark, which is also noteworthy for its smoked ham and bacon. Though beef is available, veal is more commonly consumed. Mutton and lamb are used in roasts and stews, such as the Norwegian fårikål (lamb with cabbage), and are also dried or salt-cured, then thinly sliced and used like ham. In the more northern areas reindeer is raised and is a popular meat. Other game meats such as elk, venison, and hare are hunted in the fall and considered a delicacy. Poultry is not especially well liked; however, pickled goose and stuffed fresh goose are eaten, as are certain game birds, such as grouse.

Historically, meat was in limited supply, so it was stretched by chopping it and combining it with other ingredients, resulting in many traditional dishes. Scandinavians eat many vegetables, such as onions and cabbage, stuffed with ground pork, veal, or beef. The Swedes are known for their meatballs served in cream gravy or brown sauce, and the Danes for frikadeller, which are patties of ground pork and veal, breadcrumbs, and onion fried in butter. The Norwegians prepare kjøttkaker, minced beef cakes seasoned with a bit of ginger that are pan-fried, then boiled, and served with brown gravy. Beef hash made with beets and onions is another example. Sausages are also very common.

Cold weather vegetables, including potatoes, cabbage, kale, Brussels sprouts, carrots, celery root, cucumber, beets, turnips, onions, and leeks are widely available. Rutabagas, sometimes called "Nordic oranges" or "swedes," are a customary side dish.[63] Yellow and green split pea soups with pieces of ham or pork are a winter specialty throughout Scandinavia, sometimes served with pancakes. Wild mushrooms are a specialty in some areas of Finland and Norway. Though vegetables were traditionally served cooked, fresh salads have become popular in recent years. Apples, cherries, prunes, and several varieties of berries (particularly lingonberries) are typical fruits, often stewed or made into preserves that are sometimes served with meat.

Bread is a staple food item and is often prepared from rye flour, although wheat, barley, and oats are also used. Scandinavian breads may or may not be leavened, vary in size and shape, and may be white, brown, or almost black in color. Some are crisp, such as Norwegian flatbrød and Swedish knäckebröd, and are similar to hardtack or crackers. A thin, round bread called lefse (also known as lompe in Norway) is made with a potato and wheat flour dough, then cooked on an ungreased griddle. It may be used to wrap a sausage, or be eaten with butter and sugar or jam and folded like a handkerchief. In Sweden a recent innovation is Tunnbrød, a very thin, wheat tortilla-like bread that is sold as fast food, rolled around fillings such as mashed potatoes, sausages, or shrimp salad and condiments such as pickles, mustard, and ketchup. Dumplings are a favorite in Norway, often made from potatoes.

Desserts, whether they are served after a meal or at a coffee break, are rich but not overly sweet. Most are made with butter and also contain cream or sweetened cheese and the spice cardamom. Aebleskivers are spherical Danish pancake puffs, sometimes stuffed with fruit preserves. During Advent, they are filled with whole almonds. Another popular dessert is pancakes or crêpes served with preserved berries or jam. The Scandinavians use almonds, almond paste, or marzipan in desserts as often as Americans use chocolate. One regional dessert specialty is kransekake, a stack of progressively smaller frosted almond pastry rings. The Danes are best known for their pastries or, as they call them, Wienerbrød ("Vienna bread"). The pastries were brought to Denmark by Viennese bakers a century ago when the Danish bakers went on strike. When the strike was over, the Danes improved the buttery yeast dough by adding jam and other fillings. A Finnish specialty is pulla, a braided sweet bread popular at breakfast.

Milk is well liked as a beverage in Scandinavia. Other common drinks are coffee, tea, beer, wine, and aquavit. Aquavit, which means "water of life," is liquor made from the distillation of potatoes or grain. It may be flavored with an herb such as caraway and is served ice cold in a Y-shaped glass. It is downed like a shot and often followed by a beer chaser.

Meal Composition and Cycle

Daily Patterns. The Scandinavians eat three meals a day, plus a coffee break midmorning, late in the afternoon, or after the evening meal. Breakfast is usually a light meal consisting of bread or oatmeal porridge, cold cereal, eggs, small pastries, cheese, fruit, potatoes, or herring. Sour cream or yogurt-like fermented milk may be served to eat with cereal. Fruit soups may be served in the winter. Milk and coffee or tea accompany the meal.

Traditionally, lunch in Denmark is smørrebrød, which means "buttered bread," an open-faced sandwich eaten with a knife and fork. Buttered bread is topped with anything from smoked salmon to sliced boiled potatoes with bacon, small sausages, and tomato slices. Smørrebrød may also be served as a late afternoon or bedtime snack. Today, Danes are as likely to pick up a quick sandwich made with a bagel or Italian bread as they are to eat smørrebrød.

The cooking of Finland mixes Swedish and Russian elements, such as smörgåsbords, pirozhkis (meat turnovers), and blini (thin buckwheat pancakes). Vodka is preferred over aquavit.

Whey, a byproduct of cheese making, was traditionally mixed with water to make a refreshing beverage in Norway, northern Sweden, and Iceland.

When Scandinavians toast, they say "skoal," which probably derives from the word for skull. Ancient Norsemen used the empty craniums of their enemies as drinking vessels.

At one time Øllebrød, a Danish rye bread porridge made with nonalcoholic beer, was a favorite breakfast for children.

Veal Oscar, veal topped with a béarnaise sauce, white asparagus, and lobster or crab, is named after Swedish King Oscar II (1872–1907), a renowned gourmet.

Courtesy of Denmark Cheese Association

▲ *Danish smørrebrød: Danish Fontina and Havarti cheeses, ham, salami, and smoked salmon are a few of the toppings typical of the open-faced sandwiches known in Denmark as smørrebrød.*

Eating lutefisk has become a symbol of ethnic identity in some Scandinavian-American communities.

Midsummer's Day is still observed by many Scandinavian Americans. In some areas of the United States, it has become Svenskarnas Day (Swede's Day), celebrating Swedish culture and solidarity.

St. Urho's Day (March 16) was invented by Finnish Americans as a spoof on St. Patrick's Day. It commemorates the saints driving out the grasshoppers from Finland.

A buffet meal in Sweden is the smörgåsbord (bread and butter table), a large variety of hot and cold dishes arrayed on a table and traditionally served with aquavit. Ritual dictates the order in which foods are eaten at a smörgåsbord.[64] The Swedes start with herring, followed by other fish dishes, such as smoked salmon and fried fins. Next are the meats and salads (pâtés and cold cuts), and the final course before dessert is comprised of hot dishes, such as Swedish meatballs and mushroom omelets. Today, the smörgåsbord is rarely served except at special occasions. More typically, Swedes consume a hot lunch that might include pea soup, brisket or hash, and mashed rutabaga at the school or work cafeteria.[65]

If lunch is light sandwiches, dinner is complete, often including an appetizer, soup, entrée, vegetables, and dessert. Potatoes are usually served with the evening meal. If a hot meal is eaten at lunch, a more informal supper with convenience foods is preferred. Italian items are especially popular. For example, the three most common dishes served in Swedish homes today are boiled Falun sausage (a Swedish specialty made of beef, veal, and pork), spaghetti with meat sauce, and pizza.[65] Milk, beer, or wine accompanies the meal. Coffee or wine may be served with dessert. If a dessert wine is served, coffee will follow the end of the meal.

Etiquette. As with many other Europeans, the fork remains in the left hand, and the knife remains in the right one. Despite the prevalence of sandwiches in Scandinavian menus, only bread is eaten with the hands at the table; sandwiches are consumed using the fork and knife. Pass dishes to the left. When not eating, keep your hands above the table with the wrists resting on the edge. In Finland, it is important to wait for the host to initiate eating. In Norway, a male guest of honor is expected to thank the hosts on behalf of all guests. Wine is expensive throughout Scandinavia, so it is always appreciated as a hostess gift.[35]

Special Occasions. Social occasions are usually marked with food in Scandinavia. In Sweden, conferences and meetings always offer milk and coffee with sweets, such as cinnamon buns, open-faced sandwiches, or fruit. Sandwich cakes, which are layers of bread with fillings such as pâté, ham, and sliced sausages, garnished with mayonnaise, shrimp, and herbs, are served mid-afternoon whenever a crowd gathers for an event.

December is the darkest month of the year in Scandinavia, and Christmas celebrations are a welcome diversion. The Christmas season lasts from Advent (four weeks before Christmas) until January 13, Saint Canute's Day. In Sweden, on the morning of December 13, St. Lucia's Day, the eldest daughter in the home, wearing a long white dress and a crown of lingonberry greens studded with lit candles, serves her parents saffron yeast buns and coffee in bed. However, the climax of the season is on Christmas Eve, when the biggest, richest, and most lavish meal of the year is eaten.

In Norway, traditional foods eaten on Christmas Eve begin with rice porridge sprinkled with sugar and cinnamon. Buried in the dish is one blanched almond; the person who receives it will have good fortune in the coming year. Lutefisk, a dried salt cod soaked in lye, then boiled, is customary in the north. In the east, pork ribs and sausages with cabbage are served; in the west, a dried lamb rib specialty with mashed rutabaga is favored, while in the south, cod or halibut served with a white sauce and green peas is preferred. Boiled potatoes are a common side dish at all the meals. A selection of cookies and cakes complete the meal. Women traditionally demonstrated their proficiency in the kitchen by offering at least seven types of sweets, a custom still followed in many homes today.[63]

In Sweden, a Christmas smorgasbord with twenty or thirty dishes featuring ham, herring, and other traditional fare is served. In Denmark, roast duck, goose, or pork is served with brown gravy. Typical side dishes are red cabbage and caramelized potatoes. Rice pudding with whipped cream and hot cherry sauce is also traditional. In Finland, the meal starts with pickled herring and salmon, then continues with ham, and vegetable casseroles of potatoes, carrots, or turnips. Prunes are usually featured in one dessert, followed by cookies and pies. In recent years, other European Christmas specialties such as Stollen, panettone, and bûche de Noël have also become popular.[66]

Dozens of cookies and cakes are prepared for the Christmas season. The cookies are often flavored with ginger and cloves; the Christmas tree may be hung with gingerbread figures. Deep-fried, brandy-flavored dough, known as klejner, klener, or klenätter, is also popular. The traditional holiday beverage is glögg, a hot alcoholic punch.

Midsummer's Day (June 24) is a popular secular Scandinavian holiday. It features maypoles, bonfires, and feasting. In Sweden, fish accompanied by boiled new potatoes and wild strawberries are eaten; in Norway, rømmegrøt is served; and in Finland, new potatoes with dill and smoked salmon are typical festive fare.

© Susanna Blavarg/Getty Images

▲ *Crawfish are a specialty in Scandinavia, where they are boiled and seasoned with dill, then served chilled at summer festivals with aquavit, vodka, or beer.*

[SAMPLE MENU]

A Swedish Lunch

Yellow Split Pea Soup[a,b]

Swedish Meatballs[a,b]

New Potatoes with Dill[a,b]

***Pepparkakor* (Gingersnaps)[a,b]**

Milk or Beer

[a]Henderson, H. 2005. *The Swedish table*. Minneapolis: University of Minnesota Press.

[b]*Scandinavian Cooking* at http://scandinaviancooking.com

Contemporary Food Habits in the United States

Adaptations of Food Habits

Scandinavians assimilated quickly into American society, yet their diet did not change significantly because many of their food habits are similar to the diet of the American majority, including three meals per day containing ample dairy products and animal protein. Many Scandinavian foods have been adopted by all Americans (see Chapter 15 for more information).

Nutritional Status

Nutritional Intake. Very little has been published on the nutritional status of Americans of Scandinavian descent. In Scandinavia, estimated

DISCUSSION STARTERS: WHICH CUISINES DOMINATE IN U.S. RESTAURANTS?

Whenever Americans go out to eat at an ethnic-based restaurant, they typically end up at an Italian, Asian, or Mexican institution. French restaurants also tend to carry a certain cache with Americans. German and other central European and Scandinavian restaurants are less common, however, and are typically concentrated in areas of the country with a tradition of central European or Scandinavian immigration. Fast-Food Italian, Asian, and Mexican restaurants are also widespread in the U. S., while there are few fast-food Scandinavian, German, or Russian restaurants.

Individually and then in small groups, brainstorm why ethnic cuisines are distributed in this way. What possible reasons could explain why central European and Scandinavian restaurants are not more widespread, even though immigration from these countries has been as large historically as that from Italy, China, and Mexico?

prevalence of overweight and obesity in Finland is nearly 40 percent for women and almost 60 percent for men. Figures are somewhat lower in Denmark, Norway, and Sweden.[43, 67] Finns purportedly have high rates of heart disease, stroke, alcoholism, depression, and lactose intolerance.[60] Because both their traditional diet and the well-accepted typical U.S. diet are high in cholesterol and saturated fat, Scandinavian Americans may be at increased risk of developing cardiovascular disease and other conditions associated with the westernized diet.

Counseling. Scandinavians are often low-context communicators and highly analytical. Emotions are controlled; superficiality and personal inquiries are avoided. Swedes and Finns are comfortable with silence during a conversation. As a rule, Danes are a little more informal than other Scandinavians or northern Europeans and may use first names. Danes and Swedes make and maintain direct eye contact, whereas Norwegians and Finns make direct eye contact intermittently. A brief, firm handshake is used in greetings. Other touching is infrequent, reserved for friends and relatives.

Scandinavians are likely to avoid discussion of illness until necessary. Some may consider sickness indicative of either physical or moral weakness. An in-depth interview should be used to establish any traditional health beliefs of clients, as well as dietary patterns.

REVIEW QUESTIONS

1. Briefly describe the traditional health practices and beliefs of the Russians, Germans, and Scandinavians.
2. What were the common staples of central Europeans, Scandinavians, and people of the former Soviet Union (FSU)? What were their methods of preservation?
3. List two well-known prepared foods associated with Germany, Poland, FSU, one Scandinavian country, and one other central European country. Describe three sausages that can be found in Germany or Poland. List four U.S. foods that are thought to be descended from eastern European countries.
4. What is zakuski in a Russian meal? What foods may be included? What is a smorgasbord in Scandinavian countries? What foods might be included?
5. Describe a traditional Christmas or Easter dessert for three countries in Central Europe, FSU, and one Scandinavian country.
6. Describe the religion, worldview, and food and health beliefs attributed to Gypsies.

REFERENCES

1. Heimlich, E. 2000. Gypsy Americans. In *Gale Encyclopedia of Multicultural America*, R.V. Dassanowsky & J. Lehman (Eds.). Farmington Hills, MI: Gale Group.
2. Granquist, M.A. 2000. Lithuanian Americans. In *Gale Encyclopedia of Multicultural America*, R.V. Dassanowsky & J. Lehman (Eds.). Farmington Hills, MI: Gale Group.
3. Rippley, L.V.J. 2000. German Americans. In *Gale Encyclopedia of Multicultural America*, R.V. Dassanowsky & J. Lehman (Eds.). Farmington Hills, MI: Gale Group.
4. Jones, S. 2000. Austrian Americans. In *Gale Encyclopedia of Multicultural America*, R.V. Dassanowsky & J. Lehman (Eds.). Farmington Hills, MI: Gale Group.
5. Vardy, S.B., & Szendry, T. 2000. Hungarian Americans. In *Gale Encyclopedia of Multicultural America*, R.V. Dassanowsky & J. Lehman (Eds.). Farmington Hills, MI: Gale Group.
6. Jones, S. 2000. Polish Americans. In *Gale Encyclopedia of Multicultural America*, R.V. Dassanowsky & J. Lehman (Eds.). Farmington Hills, MI: Gale Group.
7. Alexander, J.G. 2000. Slovak Americans. In *Gale Encyclopedia of Multicultural America*, R.V. Dassanowsky & J. Lehman (Eds.). Farmington Hills, MI: Gale Group.

8. Schelbert, L. 2000. Swiss Americans. In *Gale Encyclopedia of Multicultural America*, R.V. Dassanowsky & J. Lehman (Eds.). Farmington Hills, MI: Gale Group.
9. Magocsi, P.R. 2000. Russian Americans. In *Gale Encyclopedia of Multicultural America*, R.V. Dassanowsky & J. Lehman (Eds.). Farmington Hills, MI: Gale Group.
10. Sutherland, A. 1992. Gypsies and health care. *Western Journal of Medicine, 157,* 276–280.
11. Lipson, J.G., Weinstein, H.M., Gladstone, E.A., & Sarnoff, R.H. 2003. Bosnian and Soviet refugees' experiences with health care. *Western Journal of Nursing Research, 25,* 854–871.
12. Molinari, C. 2000. Czech Americans. In *Gale Encyclopedia of Multicultural America*, R.V. Dassanowsky & J. Lehman (Eds.). Farmington Hills, MI: Gale Group.
13. Kraybill, D.B. 2000. Amish. In *Gale Encyclopedia of Multicultural America*, R.V. Dassanowsky & J. Lehman (Eds.). Farmington Hills, MI: Gale Group.
14. Takooshian, H. 2000. Armenian Americans. In *Gale Encyclopedia of Multicultural America*, R.V. Dassanowsky & J. Lehman (Eds.). Farmington Hills, MI: Gale Group.
15. Spector, R.E., 2004. *Cultural Diversity in Health and Illness* (6th ed.). Upper Saddle River, NJ: Pearson Education, Inc.
16. Hirschfelder, G., & Schonberger, G.U. 2005. Germany: Sauerkraut, beer and so much more. In *Culinary Cultures of Europe*, D. Goldstein & K. Merkle (Eds.). Strasbourg, France: Council of Europe Publishing.
17. Offner, J. 1998. Pow-wowing: The Pennsylvania Dutch way to heal. *Journal of Holistic Medicine*, 16, 479–486.
18. Yehieli, M., & Grey, M.A. 2005. *Health matters: A pocket guide for working with diverse cultures and understanding populations.* Yarmouth, ME: Intercultural Press.
19. Yoder, D. 1976. Hohman and Romanus: Origins and diffusion of Pennsylvania German powwow manual. In W.D. Hand (Ed.), *American Folk Medicine.* Berkeley: University of California Press.
20. Granquist, M.A. 2000. Norwegian Americans. In *Gale Encyclopedia of Multicultural America*, R.V. Dassanowsky & J. Lehman (Eds.). Farmington Hills, MI: Gale Group.
21. Smith, L. 1996. New Russian immigrants: Health problems, practices, and values. *Journal of the American Dietetic Association, 3,* 68–73.
22. Balzer, M.M. 1987. Behind shamanism: Changing voices of Siberian Khanty cosmology and politics. *Social Science and Medicine, 24,* 1085–1093.
23. Grabbe, L. 2000. Understanding patients from the former Soviet Union. *Family Medicine, 32,* 201–206.
24. Brod, M., & Heurtin-Roberts, S. 1992. Older Russian émigrés and medical care. *Western Journal of Medicine, 157,* 333–336.
25. Zibart, E. 2001. *The ethnic food lover's companion: Understanding the cuisines of the world.*
26. Krzysztofel, K. 2005. Poland: Cuisine, culture and a variety on the Wisla river. In *Culinary Cultures of Europe*, D. Goldstein & K. Merkle (Eds.). Strasbourg, France: Council of Europe Publishing.
27. Hazelton, N.S. 1969. *The cooking of Germany.* New York: Time-Life Books.
28. Hirschfelder, G., & Schonberger, G.U. 2005. Germany: Sauerkraut, beer and so much more. In *Culinary Cultures of Europe*, D. Goldstein & K. Merkle (Eds.). Strasbourg, France: Council of Europe Publishing.
29. Grigorieva, G. 2005. Russian Federation: Rediscovering classics, enjoying diversity. In *Culinary Cultures of Europe*, D. Goldstein & K. Merkle (Eds.). Strasbourg, France: Council of Europe Publishing.
30. Field, M., & Field, F. 1970. *A quintet of cuisines.* New York: Time-Life.
31. Wechsberg, J. 1974. *The cooking of Vienna's empire.* New York: Time-Life.
32. Haik Poghosyan, S. 2005. Armenia: Insights into traditional food culture. In *Culinary Cultures of Europe*, D. Goldstein & K. Merkle (Eds.). Strasbourg, France: Council of Europe Publishing.
33. Uvezian, S. 1974. *The Cuisine of Armenia.* New York: Hippocrene Books.
34. Papashvily, H., & Papashvily, G. 1969. *Russian cooking.* New York: Time-Life.
35. Foster, D. 2000. *The global etiquette guide to Europe.* New York: John Wiley & Sons.
36. Hostetler, J.A. 1976. Folk medicine and sympathy healing among the Amish. In W.D. Hand (Ed.), *American Folk Medicine.* Berkeley: University of California Press.
37. From, M.A. 2003. People of Polish heritage. In *Transcultural Health Care* (2nd ed.), L.D. Purnell & B.J. Paulanka (Eds.). Philadelphia: FA Davis Company.
38. Hancock, I. 1991. Romani foodways: Gypsy culinary culture. *The World and I,* 666–677.
39. Romero-Gwynn, E., Nicholson, Y., Gwynn, D., Kors, N., Agron, P., Flemming, J., Raynard, H., & Screenivasan, L. 1997. Dietary practices of refugees from the former Soviet Union. *Nutrition Today, 32,* 153–156.
40. Beer Institute http://www.beerinstitute.org/statistics.asp?sid=2 Accessed 1/20/2011
41. Slimani, N., Fahey, M., Welch, A.A., Wirfalt, E., Stripp, C., Bergstrom, E., Linseisen, J., Schulze, M.B., Bamia, C., Chloptsios, Y., Veglia, F., Panico, S., Bueno-de-Mesquita, H.B., Ocke, M.C., Brustad, M., Lund, E., Gonzalez, C.A., Barcos, A., Berglund, G., Winkvist, A., Mulligan, A., Appleby, P., Overvad, K., Tjonnelanad, A., Clavel-Chapelon, F., Kesse, E., Ferrari, P., Van Staveren, W.A., & Riboli, E. 2002. Diversity of dietary patterns observed in the European Prospective Investigation into Cancer and Nutrition (EPIC) project. *Public Health Nutrition, 5,* 1311–1328.
42. Linseisen, J., Bergstrom, E., Gafa, L., Gonzalez, C.A., Thiebaut, A., Trichopoulou, A., Tumino, R., Navarro Sanchez, C., Martinez Garcia, C., Mattisson, I., Nilsson, S., Welvh, A., Spencer, E.A., Overvad, K., Tjonneland, A., Clavel-Chapelon, F., Kesse, E., Miller, A.B., Schulz, M., Botsi, K., Naska, A., Sieri, S., Sacerdote, C., Ocke, M.C., Peeters, P.H., Skeie, G., Engeset, D., Charrondiere, U.R., & Slimani, N. 2002. Consumption of added fats and oils in the European

Prospective Investigation into Cancer and Nutrition (EPIC) centres across 10 European countries as assessed by 24-hour recalls. *Public Health Nutrition, 5,* 1227–1242.
43. International Obesity Task Force. 2010. Overweight and obesity among adults in the European Union. Available online (Accessed 01/20/2011): http://www.iaso.org/site_media/uploads/AdultEU27March2010notonwebyetupdatev2.pdf
44. Rabotaev EF, Khokhlova EA. Topical problems of micronutrients deficiency in the Chuvash Republic. *Gig Sanit 2009; Jan–Feb (1)*: 36–8.
45. Ackerman, L.K. 1997. Health problems of refugees. *Journal of the American Board of Family Practice, 10,* 337–348.
46. Knapp, R.B., & Houghton, M.D. 1999. Breast-feeding practices of WIC participants from the former USSR. *Journal of the American Dietetic Association, 99,* 1269–1271.
47. Engs, R.C., Slawinska, J.B., & Hanson, D.J. 1991. The drinking patterns of American and Polish university students: A cross-sectional study. *Drug and Alcohol Dependence, 27,* 167–175.
48. Grogan, L. 2006. Alcoholism, tobacco, and drug use in the countries of Central and Eastern Europe and the former Soviet Union. *Substance Use and Misuse, 41,* 567–571.
49. Shpilko, I. 2006. Russian-American health care: Bridging the communication gap between physicians and patients. *Patient Education and Counseling, 64,* 331–341.
50. Webb, C.P., Bromet, E.J., Bluzman, S., Tintle, N.L., Schwartz, J.E., Kostyuchenko, S., & Havenaar, J.M. 2005. Epidemiology of heavy alcohol use in Ukraine: Findings from the world mental health survey. *Alcohol and Alcoholism, 40,* 327–335.
51. Thomas, J.D. 1985. Gypsies and American medical care. *Annals of Internal Medicine, 102,* 842–845.
52. Thomas, J.D. 1987. Disease, lifestyle, and consanguinity in 58 American Gypsies. *The Lancet, 2,* 377–379.
53. Krajcovicova-Kudlackova, M., Blazicek, P., Spustova, V., & Ginter, E. 2002. Insulin levels in Gipsy minority. *Bratislavské lekárske listy. 105,* 256–259.
54. Krajcovicova-Kudlackova, M., Blazicek, P., Spustova, V., Valachovicova, M., & Ginter, E. 2004. Cardiovascular risk factors in young Gypsy populations. *Bratilavské lekárske listy. 105,* 256–259.
55. Sepkowitz, K.A. 2006. Health of the world's Roma population. *The Lancet, 367,* 1707–1708.
56. Strumylaite, L., Zickeute, J., Dudzevicius, J., & Dregval, L. 2006. Salt-preserved foods and risk of gastric cancer. *Medicina (Kaunas, Lithuania),* 42, 164–170.
57. Gorodetskaya, I., Matl, J., & Chertow, G.M., 2000. Issues in renal nutrition for persons from the former Soviet Union. *Journal of Renal Nutrition, 10,* 98–102.
58. Lovoll, O.S. 2000. Norwegian Americans. In *Gale Encyclopedia of Multicultural America,* R.V. Dassanowsky & J. Lehman (Eds.). Farmington Hills, MI: Gale Group.
59. Nielsen, J.M., & Petersen, P.L. 2000. Danish Americans. In *Gale Encyclopedia of Multicultural America,* R.V. Dassanowsky & J. Lehman (Eds.). Farmington Hills, MI: Gale Group.
60. Wargelin, M. 2000. Finnish Americans. In *Gale Encyclopedia of Multicultural America,* R.V. Dassanowsky & J. Lehman (Eds.). Farmington Hills, MI: Gale Group.
61. Johansen, R., & Toverud, E.L. 2006. Norwegian cancer patients and the health food market—what is used and why? *Tidsskrift for den Norske Laegeforening, 126,* 773–775.
62. Boyhus, E.M. 2005. Denmark: Nation-building and cuisine. In *Culinary Cultures of Europe,* D. Goldstein & K. Merkle (Eds.). Strasbourg, France: Council of Europe Publishing.
63. Notaker, H. 2005. Norway: Between innovation and tradition. In *Culinary Cultures of Europe,* D. Goldstein & K. Merkle (Eds.). Strasbourg Cedex, France: Council of Europe Publishing.
64. Brown, D. 1968. *The cooking of Scandinavia.* New York: Time-Life.
65. Tellstrom, R. 2005. Sweden: From crispbread to ciabatta. In *Culinary Cultures of Europe,* D. Goldstein & K. Merkle (Eds.). Strasbourg, France: Council of Europe Publishing.
66. Makela, J. 2005. Finland: Continuity and change. In *Culinary Cultures of Europe,* D. Goldstein & K. Merkle (Eds.). Strasbourg, France: Council of Europe Publishing.
67. Meyer, H.E., & Tverdal, A. 2005. Development of body weight in the Norwegian Population. *Prostaglandins, Leukotrienes, and Essential Fatty Acids, 73,* 3–7.

Africans

CHAPTER 8

African Americans are one of the largest cultural groups in the United States, including nearly 39 million people in 2008, more than 12 percent of the total American population.[1] The majority are blacks who came originally from West Africa, although some arrived from the Caribbean, Central America, and, more recently, from the famine- and strife-stricken East African nations. A small number of Americans of African heritage are white, primarily immigrants from the nation of South Africa.

Most African Americans are the only U.S. citizens whose ancestors came by force, not choice. Their long history in America has been characterized by persecution and segregation. At the same time, blacks have contributed greatly to the development of American culture. The languages, music, arts, and cuisine of Africa have mingled with European and Native American influences since the beginnings of the nation to create a unique American cultural mix.

African Americans live with this difficult dichotomy. They are in many ways a part of the majority culture because of their early arrival, their large population, and their role in the development of the country. Much of their native African heritage has been assimilated, and their cultural identity results more from their residence in the United States than from their countries of origin. Yet they are often more alienated than other ethnic groups from white American society. This chapter discusses sub-Saharan African cuisines (North African fare is more similar to that of the Middle East; see Chapter 13, "People of the Balkans and the Middle East") and their contributions to U.S. foods and food habits. The historical influence of West African, black slave, and southern cuisines on current African American cuisine is examined.

Cultural Perspective

Africa is the second largest continent in the world and has a population estimated in 2010 at more than 1 billion people. It straddles the equator, and much of its climate is tropical, yet rainfall varies tremendously (see Figure 8.1). Rainforests, grassland savannas, high mountain forests, and temperate zones are found in the far south and along the Mediterranean.

In the north, the Sahara, the largest desert in the world, stretches from the Atlantic to the Red Sea, separating the Arabic northern African nations (Morocco, Algeria, Tunisia, Libya, and Egypt) from the sub-Saharan western, eastern, and southern regions. Numerous ethnic groups have evolved in Africa, and it is estimated that between 800 and 1,700 distinct languages are spoken. Cultural identity is strong. The long history of conflict and conquest on the continent has never completely eliminated tribal affinity; most destabilization in individual nations today arises over ethnic issues.

The terms *African American*, *black*, and *black American* are used interchangeably in research literature. *African American* is usually the preferred term because it emphasizes cultural heritage. However, *black* or *black American* is used by many African Americans who feel these terms more accurately reflect their current identity. Some recent immigrants from Africa resent the use of *African American* by persons who have lived in the United States for generations.

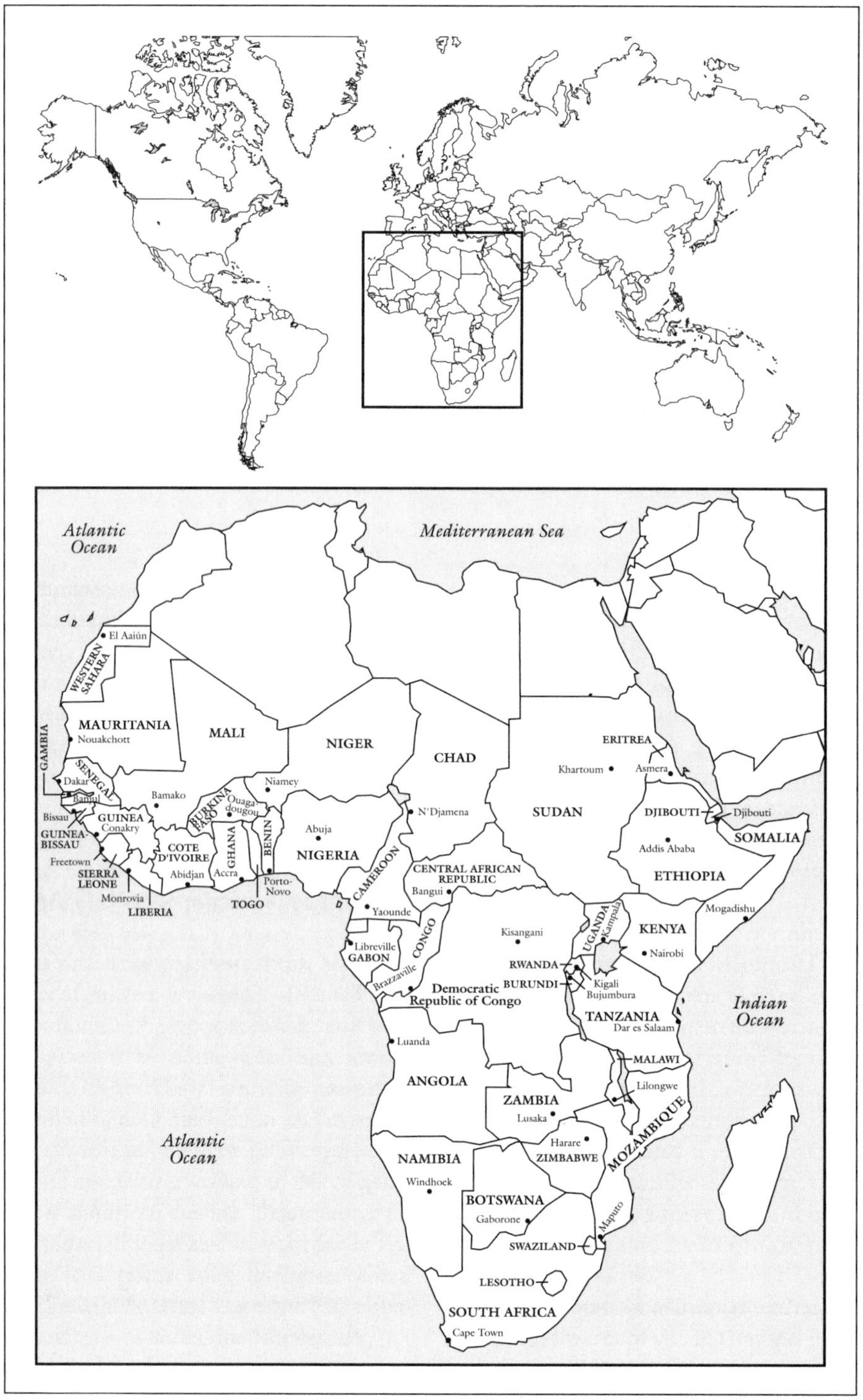

Figure 8.1
Sub-Saharan Africa.

History of Africans in the United States

The arrival of black indentured servants taken forcefully from West Africa preceded the arrival of the *Mayflower* in America. In 1619, Dutch traders sold twenty West Africans to colonists in Jamestown, an early English settlement. More than 425,000 slaves were subsequently imported legally, ancestors of the majority of black Americans residing in the United States today.

Enslavement

The institution of slavery was well established before the first blacks were brought to North America. European slave traders negotiated with African slave suppliers for their human cargo; which West Africans became slaves and which sold slaves depended on intertribal conditions. The European traders kept some records of tribal affiliation, but none of place of origin. It is believed that more than one-half of the slaves in the United States came from the coastal areas of what are now Angola and Nigeria. Others came from the regions that are today Senegal, Gambia, Sierra Leone, Liberia, Togo, Ghana, Benin, Gabon, and the Democratic Republic of Congo (formerly Zaire).

These political identities are relatively recent, however, and the slaves identified with their tribal groups, such as Ashanti, Bambara, Fulani, Ibo, Malinke, or Yoruba, rather than with a specific country or Africa as a whole. The tribal villages of West Africa were predominantly horticultural. Individuals viewed their existence in relation to the physical and social needs of the group. The extended family and religion were the foundations of tribal culture. It was especially difficult for individuals to be separated from their tribe because identity was so closely associated with the group. It is perhaps for this reason that enslaved individuals held on tenaciously to their African traditions. African language, ornamentation (i.e., scarification and teeth filing), and other customs were very threatening to slave owners. New slaves, usually in small groups from eight to thirty, were often housed at the perimeter of plantations until they became acclimated. They learned English in two or three years through contact with Native Americans or white indentured servants. When they became sufficiently acculturated, they would be allowed to work in positions closer to the main plantation house.

This initial period of separation allowed slaves to maintain many cultural values despite exposure to slaves from other tribal groups, indentured servants of different ethnic groups, and the majority culture of the white owners. At the same time, frontier farming was sufficiently difficult that slave owners were quite willing to learn from the slaves' agricultural expertise; therefore, intercultural communication was inevitable. Instead of becoming totally acculturated to the ways of the owners, slaves developed a black Creole, native-born culture during the early slave period, combining both white and West African influences.

After the end of slave importation in 1807, the black Creole population swelled. By law, slavery was a lifelong condition, and the children of female slaves were slaves as well. Although most slaves worked on farms and on cotton, tobacco, sugar, rice, and hemp plantations, many others worked in mines and on the railroads. A large number of slaves were found in the cities doing manual labor and service jobs. It was during this period of rapid growth in the African American population that separate racial group identities began to form in the United States.

Slaves working in kitchens built outside plantation living quarters (due to fire danger) were asked to whistle as they brought food to the main house to prevent them from sampling along the way.

The word *Creole* has numerous definitions, mostly describing Africans and Europeans who moved to the U.S. South and Latin America during the colonial period. *Creole Negro* (typically translated "black Creole") was the term used for Africans who developed their own American culture influenced by the British, Spanish, and French settlers in the region.

Emancipation

The movement to free the slaves began with the American Revolution. Many of the northern states banned slavery from the beginning of independence. By the 1830s, there were 300,000 free persons of color living in urban areas outside of the South. Tension between states that supported slavery (the Confederacy) and those that opposed it (the Union) was one factor that led to the Civil War (called the War between the States in the South) in 1861.

In 1862, President Abraham Lincoln signed the Emancipation Proclamation. Union victory over the Confederacy in 1865 and the subsequent ratification of the Thirteenth Amendment gave all blacks living in the United States their freedom. Some left the South immediately, searching for relatives and a better life. However, most remained in the South because they lacked the skills needed to begin an independent life; for instance, fewer than 2 percent of former slaves were literate. After emancipation, former slave owners continued to exploit African American labor through tenant farming and sharecropping. Under this system, black farmers were perpetually in debt to white landowners. Slowly, as competition for skilled farm labor increased, working conditions for black Americans in the South improved. Literacy rates and political representation increased. At the same time, racial persecution by white supremacists, such as members of the Ku Klux Klan, became more frequent.

At the turn of the twentieth century, depressed conditions in the South and industrial job opportunities in the northern states prompted more than 750,000 African Americans to settle in the Northeast and Midwest. Most were young men, and the majority moved to large metropolitan areas, such as New York, Boston, Detroit, Chicago, and Philadelphia. The influx of southern blacks was resented by both whites and the small numbers of middle-class blacks who had been well accepted in the northern cities. Laws that established racial segregation were enacted for the first time in the early 1900s, resulting in inner-city African American ghettos.

Because of poor economic conditions throughout the country, there was a pause in African American migration north during the Great Depression. The flow increased in the 1940s, and in the following thirty years more than four million African Americans left the South to settle in other regions of the country. This migration resulted in more than a change in regional demographics. It meant a change from a slow-paced rural lifestyle to a fast-paced, high-pressured, urban industrial existence.

In the 1960s, the movement against the injustices of "separate but equal" laws (which permitted segregation as long as comparable facilities, such as schools, were provided for African Americans) gained momentum under the leadership of blacks such as Martin Luther King Jr. Violent riots in city ghettos underscored the need for social reform. Civil rights activism resulted in the repeal of many overtly racist practices and passage of compensatory laws and regulations meant to reverse past discrimination, as typified by federal affirmative action requirements.

Current Demographics

Today, slightly more than one-half of African Americans live in the U.S. South (56 percent), most in suburban areas. Since 1988, more blacks have been moving to the South than to the northern states, reversing the demographic trend northward established in 1900. The remaining African American population is found predominantly in northeastern and midwestern urban areas.[2] The most recent census data indicates that most African Americans live in geographic areas that are predominantly African American or Hispanic.[4] As of 2007, approximately 3.7% percent of immigrants to the United States were of African descent. About one-third of African immigrants in the United States are from West Africa, including the countries of Nigeria or Ghana. Approximately 386,225 immigrants are from East Africa, including the countries of Kenya, Tanzania, and Uganda.[3] In addition, small populations from Eritrea (17,000) and Sudan (19,000) have arrived in recent years, many of whom are refugees seeking escape from ethnic and civil conflicts in the region (some more recent Ethiopian and Somali immigrants are also fleeing war). Less than 1 percent of the black population is of Caribbean or Central American descent.

Socioeconomic Status

African Americans continue to suffer from the discriminatory practices that began with their enslavement, yet it is estimated that 70 percent are making steady economic progress. The black middle class is growing, and the economic gap between blacks and whites is narrowing as blacks increasingly enter fields such as business, health care, and law. Nevertheless, according to 2010 Census data, the poverty rate for African Americans was twice the rate for non-Hispanic white families.[5,6] According to 2007 census data, high school graduation rates are approximately 9 percent lower for blacks than for whites. The African American unemployment rate is more than double that of whites, and median income is substantially lower. In 2009, nearly one in every four black families (24.5 percent) lived below the poverty line, and three times as many black children lived in poverty (30 percent), as did white children (10 percent).[6]

Many African Americans believe they are not completely accepted in U.S. society. Blacks isolated in urban ghettos frequently experience alienation. Frustration, hopelessness, and hostility often result. At the same time, discrimination has promoted ethnic identity among African Americans, due in part to a shared history of persecution. Although Americans of African descent are geographically, politically, and socioeconomically diverse, there is a strong feeling of ethnic unity.

Socioeconomic status for recent immigrants from Africa varies greatly, according to U.S. Census data.[7] Many Nigerians come to the United States for educational opportunity. Nearly 30 percent have advanced college degrees. Nigerians who choose to stay often find employment in academia, whereas others choose to open their own businesses.[8] Most immigrants from Ghana are also well educated, with many obtaining jobs in management and the professions. Ghanaian women often work as teachers, nurses, and secretaries, although some are pursuing business careers.[9] Ethiopians in the United States graduate from high school at rates similar to those of whites, and nearly 60 percent have college degrees. However, many are underemployed, working in low-level service jobs. Approximately 14 percent of Ethiopian families in the United States live below the poverty line. Education levels in immigrants from Eritrea are far lower. Approximately 85 to 90 percent of women in Eritrea were functionally illiterate in 1999.[10] High school graduation rates are far below the U.S. average, and small numbers obtain a college degree. Nevertheless, 23 percent of Eritrean immigrants held jobs in management and the professions (particularly engineering) in 2000. The number of Eritrean families living in poverty was 19 percent. Recent Somali Bantu refugees are largely illiterate with little or no education. In Kenya, those who were employed worked in farming, construction, cleaning, cooking, and other manual labor jobs. Most have never lived with modern conveniences such as electricity or plumbing.[11]

According to 2009 U.S. Census data, many immigrants from Sudan have college or advanced degrees. In addition to jobs in management and the professions, similar numbers are employed in production and transportation, as well as in sales and office work, and average salaries are low. In 1999, 27 percent of Sudanese families lived in poverty. Kenyans who immigrate to the United States are typically well educated, and over half of adults over the age of twenty-five were college graduates. More than 48 percent worked in management or professional jobs, particularly in the technology sector. South Africans are also often well educated. More than 50 percent hold college degrees, and more than 60 percent work in management or professional careers. Other South Africans have come to the United States for entrepreneurial opportunity.[12]

Worldview

Religion

Spirituality was integral to African tribal society, and indigenous religious affiliations were maintained by most slaves, despite attempts to convert blacks to Christianity. Although the first black church, a Baptist congregation, was founded in the 1770s in South Carolina, it was only after U.S. religious groups became involved in the antislavery movement that the black Creole community responded with large numbers of conversions.

Religion is as essential to African American culture today as it was to African society. For many black Americans, the church represents a sanctuary from the trials of daily life. It is a place to meet with other African Americans, to share fellowship and hope. More than 75 percent of African Americans belong to a church. The largest denomination is the National Baptist Convention of the U.S.A. Others with large African American followings include the Methodist Episcopal churches and Pentecostal denominations, such as the Church of God in Christ. A small percentage of African Americans who are Muslim are members of either the World Community of All Islam or the Nation of Islam (see Chapter 4, "Food and Religion," for more information about this religion).

Recent immigrants from Africa adhere to a variety of faiths. Approximately one-quarter of Africans today are Muslim (the fastest growing religion on the continent) and one-quarter are Christian, mostly Protestant. Many Ethiopians and Eritreans follow an Eastern Orthodox faith that is similar to, but separate from, the Egyptian Coptic Church. In addition, it is believed that nearly one-half of the African population participates in traditional tribal religions or combine elements of several faiths.

Family

The importance of the extended family to African Americans has been maintained since tribal times. It was kinship that defined the form of African societies. During the early slave period, the proportion of men to women was two to one, and the family structure included many unrelated members. As the black Creole population increased, nuclear families were established but were often disrupted by the sale or loan of a parent. The extended family provided for dislocated parents and children.

In 2002, 43 percent of African American families were headed by women (compared to 13 percent of white families).[2] The family network often includes grandparents, aunts, uncles, sisters, brothers, deacons or preachers, and friends. Such extended kinship still supports and protects individuals, especially children, from the problems of a discriminatory society. The extended family has been found to be equally valued by both wealthy and poor African Americans.

Most families in Africa today are still extended, and multigenerational homes are typical for recent immigrants to the United States. Children are highly valued, and elders are often responsible for helping to make family decisions. Many immigrants keep in close touch with relatives in Africa and often send money for support.

African societies are highly patriarchal, and in most cases women are expected to be subservient to men. Marriages are often arranged by parents, typically with an exchange of property. In the United States, women work as often as men, but are expected to maintain the house and raise the children in addition to their jobs.[13] In approximately 42 percent of households, grandparents are responsible for child care.[14] Kenyan women who come to the United States are often well educated, and for them the stresses of gender-role changes are not as difficult. However, many Kenyans worry about the loss of traditional values as their children adapt to life in the United States, and conflict between parents and children is common.[15] Intergenerational difficulties are also seen in many Nigerian families where, traditionally, strict obedience was expected of children, who were not allowed to question or contradict their parents.[16] Ghanaians promote cultural traditions from the numerous Ghanaian associations formed to support the immigrant community. Traditionally, these associations in Africa were strictly limited to specific ethnic, cultural, political, or economic groups. Those in the United States, however, generally have more open memberships. Ghanaians located in rural areas may belong to associations found in nearby cities.[8]

Somali Bantus marry young, between the ages of fourteen and sixteen, and women typically have their first child around the age of fifteen. The family group often includes cousins or other kin who have become part of the home. Divorce and remarriage are common.[11]

Traditional Health Beliefs and Practices

Africans view life as energy rather than matter. A person lives transitionally on earth, interacting with all environmental forces, from those of the gods and nature to those exerted by the living and dead. Life events can be influenced by these forces, and a person can, in turn, influence these forces toward good or evil.

Health is maintained through harmony. Disharmony and illness occur when someone (living or dead), the gods, or nature is intentionally malevolent. As described by one African expert:

> Even if it is explained to a patient that he has malaria because a mosquito carrying malaria parasites has stung him, he will still want to know why that mosquito stung him and not another person. The only answer which people find satisfactory to that question is that someone has "caused" (or "sent") the mosquito to sting a particular individual, by means of magical manipulations (p. 169).[17]

One example of how a person may become ill is the evil eye, whereby one person causes illness and misfortune by sending negative energy through an evil gaze.[13] A traditional African healer must first diagnose the illness, determine the supernatural cause of the illness, then dislodge the evil and take measures to prevent reoccurrence. The healer often uses herbs and other natural prescriptions to treat the symptoms and may depend on the spirits of the ancestors to transmit medical knowledge. Bleeding, massage, dietary restrictions, chants, and charms may complete the cure.

The health beliefs and practices of some African Americans reflect traditional African concepts as well as those encountered through early contact with both Native Americans and whites. It is often difficult to determine the origins of a specific practice, and it is also likely that both blacks and whites adhere to similar beliefs. Some Americans of African heritage maintain health by eating three meals each day, including a hot breakfast. Laxatives may be used regularly, and cod liver oil may be taken to prevent colds.

Vicks® VapoRub® may also be ingested for colds. A copper or silver bracelet is sometimes worn for protection; if it is removed, harm will occur. If the skin darkens around the bracelet, illness is impending and precautions, such as more rest and a better diet, should be undertaken.[18,19,20]

One study of black men found that they defined health as more than simply lack of illness. The ability to support their family, fulfill social obligations, and maintain emotional and spiritual well-being was also important. Self-empowerment was one method used to combat the difficulties of racism and poverty thought to undermine good health.[21] Prayer for health is common, practiced by a significant amount of this population.[22] Some African Americans believe that illness is a punishment from God, and many feel that God acts through physicians to heal patients.[17] Stress is frequently cited as the cause for poor health. It is considered by some blacks to be the source of hypertension; likewise, "worriation" results in diabetes.[21,23] Others, especially in the rural South, believe that illness is due to evil spirits or witchcraft. A person may be "hexed," "fixed," "mojoed," or "rooted" by someone with supernatural skills. Healers and conjurers are needed to "fix" or "trick" the evil. The resulting illness can be cured by herbal treatments, incantations, or magical transference. For example, a toad is placed on the head of someone with a headache, and when the toad later dies, the headache will disappear.[18,23]

Best known of traditional healers are the practitioners of voodoo, also called hoodoo. This combination of African and Catholic beliefs is thought to have originated in the Caribbean (see Chapter 10, "Caribbean Islanders and South Americans"); where it is still practiced in the South, it was also likely influenced by European witchcraft.[24] The men and women practitioners can use voodoo magic for good or evil. They cure unnatural illnesses (those of supernatural cause) through casting spells, the use of magic powders and gris-gris, bags worn around the neck with powders, animal bones or teeth, stones, and/or herbs. Other healers include traditional herbalists or root doctors, and spiritual, sympathy, or faith healers who derive their powers from God. A patient may choose to use one or all such healers to treat an illness, and the specialty of one may overlap into another.[25] A root doctor, for example, may apply home remedies or may use charms like a conjurer to remove (or even send) evil. In most cases, healers of all kinds use a holistic approach and spend a great deal of time on a patient, providing a feeling of spiritual as well as physical well-being.

Photo by Laurie Macfee

▲ *Traditional African-American foods. Some typical foods of the Southern black diet include bacon, black-eyed peas, chayote squash, corn, greens, ham hocks, hot sauce, okra, peanuts, watermelon, and sweet potatoes.*

Few African Americans today believe in African witchcraft or employ root doctors. However, the influence of traditional healing practices is still found in the idea that ill health is due to bad luck or fate, in the frequent use of home remedies, and in a preference for natural therapies by some blacks,[13,26,27] as in the popularity of garlic pills found in one study.[21]

Traditional Food Habits

What are traditional African American foods: foods of Africans in the seventeenth and eighteenth centuries, foods of the slaves, or foods of the black South since emancipation? African American cuisine today often includes elements from each of these diets.

Ingredients and Common Foods

Historical Influences

African American foods offer a unique glimpse into the development of a cuisine. Even before West Africans were brought to the United States, their food habits had changed significantly due to the introduction of New World foods such as cassava (*Manihot esculenta*, a tuber that is also called manioc), corn, chiles, peanuts, pumpkins, and tomatoes during the fifteenth and sixteenth centuries. The slaves brought a diet based on these new foods and native West African foods, such as watermelon, black-eyed peas, okra, sesame, and taro. Adaptations and substitutions were made based on available foods. Black cooks added their West African preparation methods to British, French, Spanish, and Native American techniques to produce American southern cuisine, emphasizing fried, boiled, and roasted dishes using pork, pork fat, corn, sweet potatoes, and local green leafy vegetables. The cuisines of other African regions have had little impact at this point on the typical U.S. diet, although recent immigrants may continue to prepare and consume traditional fare.

African Fare

West African. Knowledge of West African food habits before the nineteenth century is incomplete. It is mostly based on the records of North African, European, and U.S. traders, many of whom considered the local cuisine unhealthy. Most West Africans during the slave era lived in preliterate, horticulturally based tribal groups. There was a heavy dependence on locally grown foods, although some items, such as salt and fish (usually salt cured), could be traded at the daily markets held throughout each region.

Historically, staple foods varied in each locality. Corn, millet, and rice were used in the coastal areas and Sierra Leone. Yams were popular in Nigeria. Cassava (often roasted and ground into a flour known as gari) and plantains formed the dietary foundation of the more southern regions, including the Congo and Angola. The arid savanna region of West Africa bordering the Sahara Desert was too dry for cultivation, so most tribes were pastoral, herding camels, sheep, goats, and cattle. In the north, these animals were eaten; in other regions, local fish and game were consumed. Insects such as termites and locusts were consumed in many regions of Africa and are still considered a treat in some areas. Chickens were also raised, though in many tribes the eggs were frequently traded, not eaten, and the chicken itself was served mostly as a special dish for guests. Chicken remains a prestigious meat in many regions today.

There were many similarities in cuisine throughout West Africa. Most foods were boiled or fried, and then small chunks were dipped in a sauce and eaten by hand. Starchy vegetables including yams, plantains, cassava, sweet potatoes, and potatoes were often boiled, then pounded into a paste (called fufu). Each diner formed the dough into bite-sized scoops that were used like spoons to eat stew. Palm oil was the predominant fat used in cooking, giving many dishes a red hue. Peanut oil, shea oil (from the nuts of the African shea tree), and occasionally coconut oil were used in some regions. The addition of tomatoes, hot chili peppers, and onions as seasoning was so common that these items were simply referred to as "the ingredients." Most dishes were preferred spicy, thick, and sticky (mucilaginous).

Legumes were popular throughout West Africa. Peanuts were especially valued and were eaten raw, boiled, roasted, or ground into meal, flour, or paste. Cow peas (*Vigna ungulculata*, neither a standard pea nor a bean—black-eyed peas are one type of cow pea) were eaten as a substitute for meat, often combined with a staple starch such as corn, yams, or rice. Bambara groundnuts, similar to peanuts, were also common. Nuts and seeds were frequently used to flavor and thicken sauces. Mango seeds (called agobono, og bono, or apon), cashews, egusi (watermelon seeds, usually dried and ground), kola nuts, and sesame seeds were popular.

Many varieties of tropical and subtropical fruits and vegetables were available to West Africans, but only a few were widely eaten. Ackee apples, baobab (both the pulp and seeds from the fruit of the baobab tree), guava, lemon, papaya (also called pawpaw), pineapple, and watermelon were the most common fruits. Many dishes included coconut milk. In addition to starchy staple roots and the flavoring ingredients of onions, chili peppers, and tomatoes, the most popular vegetables were eggplant, okra, pumpkin, and the leaves from plants such as cassava, sweet potato, and taro (also called callaloo or cocoyam).

West African cuisine today remains very similar to that of the past. Fish is favored, and little meat is consumed. A mostly vegetarian fare has developed based on regional staples such as beans, yams, and cassava. Gari foto is a popular Nigerian specialty often eaten for breakfast; it combines gari (cassava meal) with scrambled eggs, onions, chiles, and tomatoes and is sometimes served with beans. Stews featuring root vegetables, okra, or peanuts and flavored with small amounts of fish, chicken, or beef are common. Curries are popular in Nigeria, often served with dozens of condiments and garnishes, such as coconut, raisins, chopped dates, peanuts, dried shrimp, and diced fruits. Pili-pili, a sauce of chili peppers, tomatoes, onion, garlic, and horseradish, is usually offered at the table so that each diner may spice dishes to taste.

Deep-fried fish, fried plantain chips, and balls made from steamed rice, black-eyed peas, yams, or peanuts are snack foods available at street stalls in urban areas. The favorite West African sweet is kanya, a peanut candy. Chin-chins, sweet fried pastries topped with sugar and flavoring such as cinnamon or orange zest, are also popular. Bananas are commonly baked and flavored with sugar, honey, or coconut for dessert. Sweetened dough balls prepared from millet or wheat flour are a specialty in many areas; in Ghana they are called togbei ("sheep balls") and are brightly colored with food dye before being deep fried. They are served for special occasions such as birthdays and weddings.

Photo courtesy of the World Health Organization/P. Almasy

▲ *Market scene in Ghana, West Africa.*

[SAMPLE MENU]

A West African Meal

Spicy Fried Plantains[a,b]

Groundnut Chop/Stew over Rice[a,b]

Ginger Beer[a,b] or Green Tea with Mint

Tropical Fruit Salad[a,b]

[a]Jackson, E.A. 1999. *South of the Sahara: Traditional cooking from the lands of West Africa*. Hollis, NH: Fantail.

[b]*The Congo Cookbook* at http://www.congocookbook.com

Ethiopian, Eritrean, Somali, and Sudanese. Mountainous plains and lowland valleys cover much of Ethiopia, Eritrea, and Somalia, and the climate is mostly arid. Though Ethiopia is landlocked, Eritrea and Somalia have lengthy coastal access along the horn of Africa. Millet (including a variety unique to the region called teff), sorghum, and plantains are the staple foods produced, and coffee is the leading export crop. Other foods common to the region include barley, wheat, corn, cabbage, collards, onions, kale, and potatoes, as well as peanuts and other legumes. Enset, a plantain-like plant, is a staple in the high mountainous regions of Ethiopia. Some chicken, fish, mutton, goat, and beef are available.

Historically, Ethiopian cuisine had minimal outside influences, though a large number of Muslims now living in the nation have introduced certain halal dietary practices in some regions (see Chapter 4 for more information on religious dietary practices). More significant has been the Ethiopian Eastern Orthodox religion, which has facilitated the development of vegetarian fare due to the restricted intake of animal proteins. Examples include yataklete kilkil, a garlic- and ginger-flavored casserole, and yemiser selatta, a lentil salad. A mixture of ground legumes called mitin shiro is added to most vegetarian stews. Wat, meaning stew, is the national dish of Ethiopia. Typically thick and spicy, wat may include legumes, meats, poultry, or fish. Milder versions are known as alechas, and most stews can be prepared either way.[28] Doro wat is one popular example, featuring chicken and whole hard-boiled eggs. Yemiser wat is made with red lentils, sega wat with beef, and yebeg wat with lamb. Wat is served with rice or the traditional Ethiopian flat bread called injera. Injera is prepared with a spongy, fermented dough made from teff cooked in a very large circular loaf on a griddle. (Sometimes wat is prepared with added pieces of injera in the stew, known as fitfit.) Another variation of this flat bread, known as kocho, is made with enset.

Ethiopian foods are frequently flavored with a hot spice mixture known as berbere, which includes allspice, cardamom, cayenne, cinnamon, cloves, coriander, cumin, fenugreek, ginger, nutmeg, and black pepper. Niter kebbeh, clarified butter with onions, garlic, ginger, and other spices, is added to many dishes, including kitfo, a raw ground beef specialty. Salted and sweet cheeses are common. Honey (sometimes consumed with the bee grubs) is especially popular as a sweetener and is used in savory dishes such as alechas, and desserts such as baklava, a drier version of the Greek pastry.[29] It is also fermented to make tej, a mead-like beverage. Tella, home-brewed millet or corn beer, is commonly consumed, as is coffee, especially espresso (introduced by the Italians).

Eritrean and Somali food is very similar to Ethiopian, with the exception of more frequent use of seafood. For example, a typical meal includes a spicy stew, often with beef, lamb, kid, or fish, eaten with injera-like breads, known in Somalia as anjeero. The Eritreans often consume their bread with shuro, a thick paste made from chickpeas, onions, tomatoes, and a touch of berbere, similar to Ethiopian mitin shiro. In addition, though many Eritreans belong to the Coptic Eastern Orthodox faith and adhere to the proscriptions on meat, nearly all Somalis are Muslim and follow halal dietary practices. Camel milk is consumed in some areas, and in Somalia, sweetened tea is consumed frequently. In Eritrea, coffee is preferred, and a bitter, fermented barley beverage called sowa is served at most meals. On special occasions a wine similar to tej, called mez, is popular.

Former Italian occupation of the region introduced numerous dishes. Favorites include spaghetti, lasagna, pasta, seafood, and frittata, a scrambled egg dish made with green peppers and onions.[10,28] In some parts of Ethiopia and Eritrea and throughout Somalia, Asian Indian-influenced items, such as curried dishes, unleavened breads such as roti and chapati, and vegetable- or meat-stuffed fritters known as sambosa are also common.[30]

The Sudan, which bridges the desert regions of North Africa and the tropical forests of West and East Africa, has a cuisine reflecting both Middle Eastern and African influences. For example, fava beans or a salad of cucumber and yogurt might be served at the same meal as an okra stew and kisra, the Sudanese staple bread similar to injera.

[SAMPLE MENU]

An Ethiopian Dinner

Kitfo[a,b,c]

Doro Wat[a,b]

Bamia Alich'a[b] ***or* Vegetable *Alecha***[c]

Injera[a,c]

Fruit juice, *Tej*[c], or Ethiopian Coffee

[a]Hafner, D. 2002. *A taste of Africa: Traditional and modern African cooking.* Berkeley, CA: Ten Speed Press.

[b]Harris, J.B. 1998. *The Africa cookbook: Tastes of a continent.* New York: Simon & Schuster.

[c]*The African Cookbook* at http://www.africa.upenn.edu/Cookbook/about_cb_wh.html

East African. The climate and topography of Kenya, Tanzania, and Uganda are well suited to farming and ranching. Cassava, corn, millet, sorghum, peanuts, and plantains combine to form the foundation of the diet. Crops grown for export include coffee, tea, cashews, and cloves. Cattle are raised in the northern plateaus of Kenya; they are considered a gift of the gods (especially among the Maasai tribe), and they indicate wealth. The abundant game animals are also often sacred, although specific taboos vary from region to region. Eating fish and seafood is common along the coast.

The cuisines of East Africa are predominantly vegetarian, influenced in part by Arab, Asian Indian, and British fare. Breads are common at every meal, including chapatis, kitumbua, a rice fritter, and mandazi, slightly sweetened doughnut-like bread. In Kenya, the national dish is ugali, a very thick, doughy cornmeal porridge. Ugali is also found in Tanzania. Mashed beans, lentils, corn, plantains, and potatoes are also popular. Coconut milk, chili peppers, and curry spice blends flavor many dishes. In Uganda, which is inland and less influenced by foreign cuisines, peanuts are a staple food used in everything from stews—such as beef, tomato, and onion stew with peanut butter sauce—to desserts. Plantains are the core food of Tanzania. They are used in soups (with or without beef), stews, fritters, custards, and even wine. Coconut milk is a frequent flavoring, as is curry powder. Throughout the region, dishes made with taro greens or other leafy vegetables and side dishes of local grains and produce, such as eggplant and papaya, round out the cuisine.

South African. South Africa has a very temperate climate favorable to many fruits and vegetables uncommon in the rest of the continent, such as cucumbers, carrots, apricots, tangerines, grapefruit, quinces, and grapes. The cuisine has been strongly influenced by the European settlers of the region, including the Dutch, British, and French. Muslim slaves imported from Malaysia and India have also had a significant impact on South African fare. Mutton, beef, pork, fish, and seafood are popular.

South African meat specialties include sosaties, or skewered, curried mutton; bredie, a mutton stew that may include onions, chiles, tomatoes, potatoes, or pumpkin; frikkadels, braised meat patties; bobotie, a meatloaf flavored with curry and topped with a custard mixture when baked; and biltong, meat strips dried and preserved over smoke. Grape-stuffed chicken or suckling pig is sometimes served for special occasions. Spicy fruit or vegetable relishes called chutney (for more information see Chapter 14, "South Asians"); atjar, or unripe fruit or vegetables preserved in fish or vegetable oil with spices like tumeric and dried chiles; and fresh grated fruit or vegetable salads flavored with lemon juice or vinegar and chiles accompany the dishes.

Sweets are very common. Dried fruits, fruit leathers called planked fruit, and fruit preserves or jams are popular. Many pastries are available, too, such as tarts made with raisins, sweet potatoes, coconut, or custard. Cookies are a favorite. Koeksister are braided crullers that are deep fried and dipped in a cinnamon syrup, and soetkoekies are spice cookies flavored with the sweet wine Madeira.

The Slave Diet

When West Africans were forcefully taken from their tribes, they were not immediately separated from their accustomed foods. Conditions on the slave ships were appalling, but most slave traders did provide a traditional diet for the tribal members on board. The basic staples of each region, plus dried salt cod (which was familiar to most West Africans), were fed to the slaves in minimal quantities. Chili peppers and the native West African malagueta peppercorns were used for seasoning because they were believed to prevent dysentery. It wasn't until the Africans were sold in the United States that significant changes in their cuisine occurred.

The diet of the field workers was largely dependent on whatever foods the slave owners provided. Salt pork and corn were the most common items. Sometimes rice (instead of corn), salted fish, and molasses were included. Greens, legumes, milk, and sweet potatoes were occasionally added. The foods provided, as well as their amount, were usually contingent on local availability and agricultural surplus.

Hunger was common among the slaves. Some slave owners allowed or required their slaves to maintain garden plots or to plant needed vegetables around the periphery of the cotton or tobacco fields. Okra and cow peas from Africa were favored, as well as American cabbage, collard and mustard greens, sweet potatoes, and turnips. Herbs were collected from the surrounding woodlands, and small animals such as opossums, rabbits, raccoons, squirrels, and an occasional wild pig were trapped for supplementary meat. Children would often catch catfish and other freshwater fish.

During the hog-slaughtering season in the fall, variety pork cuts, such as chitterlings

© Bonnie Kamin/PhotoEdit, Inc.

▲ *Many traditional Southern foods, such as fried chicken, corn bread, spicy stews, bean dishes, and simmered greens, reflect West African influences.*

(intestines, pronounced *chitlins*), maw (stomach lining), tail, and hocks, would sometimes be given to slaves. Some slaves were encouraged to raise hogs and chickens. The eggs and the primary pork cuts were usually sold to raise cash for the purchase of luxury foods. Chickens, a prestigious food in West Africa, continued to be reserved for special occasions.

West African cooking methods were adapted to slave conditions. Boiling and frying remained the most popular ways to prepare not only meats but also vegetables and legumes. Bean stews maintained popularity as main dishes. Corn was substituted for most West African regional staple starches and was prepared in many forms, primarily as cornmeal pudding, cornmeal breads known as pone or spoon bread, grits (coarsely ground cornmeal), and hominy (hulled, dried corn kernels with the bran and germ removed). Pork fat (lard) replaced palm oil in cooking and was used to fry or flavor everything from breads to greens. Hot pepper sauces were used instead of fresh chiles for seasoning. No substitutions were available for many of the nuts and seeds used in West African recipes, although peanuts and sesame seeds remained popular.

Food for the slave field workers had to be portable. One-dish vegetable stews were common, as were fried cakes, such as hush puppies (perhaps named because they were used to quiet whining dogs), and the cornmeal cakes baked in the fire on the back of a hoe, called hoecakes. Meals prepared at home after a full day of labor were usually simple.

The slaves who cooked in the homes of slave owners had a much more ample and varied diet. They popularized fried chicken and fried fish. They introduced sticky vegetable-based stews (thickened with okra or the herb sassafras, which when ground is called filé powder), such as the southern specialty gumbo z'herbes, nearly identical to a recipe from the Congo. Green leafy vegetables (simply called greens) became a separate dish instead of being added to stews, but they were still cooked for hours and flavored with meat. Ingredients familiar to West Africans, such as nuts, beans, and squash, were used for pie fillings.

Foods after the Abolition of Slavery

The food traditions of African Americans did not change significantly after emancipation, and they differed little from those of white farmers of similar socioeconomic status. One exception was that pork variety cuts and salt pork remained the primary meats for blacks, while whites switched to beef during the post-Civil War period.

African American Southern Staples

The traditional Southern African American cuisine that evolved from West African, slave, and post-abolition fare emphasizes texture before flavor; the West African preference for sticky foods continues. Pork, pork products, corn, and greens still form the foundation of the diet. The cultural food groups list (Table 8.1) includes other common Southern black foods. (For information about the food habits of blacks from the Caribbean, see Chapter 10; for more information on foods of the South, see Chapter 15, "Regional Americans.")

Pork variety cuts of all types are used. Pig's feet (or knuckles) are eaten roasted or pickled; pig's ears are slowly cooked in water seasoned with herbs and vinegar and then served with gravy. Bits of pork skin (with meat or fat attached) are fried to make cracklings. Chitterlings also are usually fried, sometimes boiled. Sausages and head cheese (a seasoned loaf of meat from the pig's head) make use of smaller pork pieces. Barbecued pork is also common. A whole pig (or just the ribs) is slowly roasted over the fire. Each family has its own recipe for spicy sauce, and each has its opinion about whether the pork should be basted in the sauce or the sauce should be ladled over the cooked meat. Other meats, such as poultry, are also popular.

Occasionally, the small game that was prevalent during the slave period, such as opossum and raccoon, is eaten. More often the meal

Table 8.1 Cultural Food Groups: African American (Southern United States)

Group	Comments	Common Foods	Adaptations in the United States
Protein Foods			
Milk/milk products	Dairy products are uncommon in diet (incidence of lactose intolerance estimated at 60–95 percent of the population). Milk is widely disliked in some studies; well accepted in others. Few cheese or fermented dairy products are eaten.	Milk (consumed mostly in desserts, such as puddings and ice cream), some buttermilk; cheese	Blacks in urban areas may drink milk more often than rural blacks.
Meat/poultry/fish/eggs/legumes	Pork is most popular, especially variety cuts; fish, small game, poultry also common; veal and lamb are infrequently eaten. Bean dishes are popular. Frying, boiling are most common preparation methods; stewed dishes preferred thick and sticky. Protein intake is high.	*Meat:* beef, pork (including chitterlings, ham hocks, sausages, variety cuts) *Poultry:* chicken, turkey *Fish and shellfish:* catfish, crab, crawfish, perch, red snapper, salmon, sardines, shrimp, tuna *Small game:* frogs, opossum, raccoon, squirrel, turtle *Eggs:* chicken *Legumes:* black-eyed peas, kidney beans, peanuts (and peanut butter), pinto beans, red beans	Pork remains primary protein source; prepackaged sausages and lunch meats are popular. Small game is rarely consumed. Variety cuts are considered to be "soul food" and eaten regardless of socioeconomic status or region. Frying is still popular, but more often at evening meal; boiling and baking are second most common preparation methods.
Cereals/Grains	Corn is primary grain product; wheat flour is used in many baked goods. Rice is used in stew-type dishes.	Biscuits; corn (corn breads, grits, hominy); pasta; rice	Store-bought breads often replace biscuits (toasted at breakfast, used for sandwiches at lunch).
Fruits/Vegetables	Green leafy vegetables are most popular, cooked with ham, salt pork, or bacon, lemon and hot sauce; broth is also eaten.	*Fruits:* apples, bananas, berries, peaches, watermelon	Fruits are eaten according to availability and preference; intake remains low.
	Intake of fresh fruits and vegetables is low.	*Vegetables:* beets, broccoli, cabbage, corn, greens (chard, collard, kale, mustard, pokeweed, turnip, etc.), green peas, okra, potatoes, spinach, squash, sweet potatoes, tomatoes, yams	Green leafy vegetables ("greens") are popular in all regions; other vegetables are eaten according to availability and preference; intake remains low.
Additional Foods			
Seasonings	Dishes are frequently seasoned with hot-pepper sauces. Onions and green pepper are common flavoring ingredients.	*Filé* (sassafras powder), garlic, green peppers, hot-pepper sauce, ham hocks, salt pork or bacon (added to vegetables and stews), lemon juice, onions, salt, pepper	
Nuts/seeds	Nuts often used in ways similar to traditional West African dishes, such as nut- or seed-based desserts.	Peanuts, pecans, sesame seeds, walnuts	
Beverages		Coffee, fruit drinks, fruit juice, fruit wine, soft drinks, tea	
Fats/oils		Butter, lard, meat drippings, vegetable shortening	
Sweeteners		Honey, molasses, sugar	Cookies (and candy) are preferred snacks.

[SAMPLE MENU]

A Traditional Black Southern Supper

Fried Chicken[a,b] with Biscuits[a,b]

Macaroni and Cheese[a,b]

Collard Greens[a,b]

Sweet Potato Pie[a,b] or Pound Cake[a,b]

Fruit Juice or Iced Tea

[a]Tillery, C.Q. 1996. *The African American heritage cookbook*. New York: Citadel Press.
[b]*The Chitterling Site* at http://www.chitterlings.com

includes local fish and shellfish, such as catfish, crab, or crawfish. Frog legs and turtle are popular in some areas. Meats, poultry, and fish are often combined in thick stews and soups, such as gumbos (still made sticky with okra or filé powder) that are eaten with rice. They may also be coated with cornmeal and deep-fried in lard, as in Southern-fried chicken and catfish.

The vegetables most characteristic of Southern African American cuisine are the many varieties of greens. Food was scarce during the Civil War, and most southerners were forced to experiment with indigenous vegetation, in addition to cultivated greens such as chard, collard greens, kale, mustard greens, spinach, and turnip greens. Dockweed, dandelion greens, lamb's quarter, marsh marigold leaves, milkweed, pigweed, pokeweed, and purslane were added as acceptable vegetables. Traditionally, the greens are cooked in water flavored with salt pork, fatback, bacon, or ham, plus hot chili peppers (or hot-pepper sauce) and lemon. As the water evaporates, the flavors intensify, resulting in a broth called "pot likker." Both the greens and the liquid are served; hot sauce is offered for those who prefer a spicier dish. Other common vegetables include black-eyed peas, okra, peas, and tomatoes. Onions and green peppers are frequently used for flavoring.

Corn and corn products are as popular in Southern black cuisine today as they were during the slave period. Corn bread and fried hominy are served sliced with butter. Wheat flour biscuits are also served with butter or, in some regions, gravy. Dumplings are sometimes added to stews and greens.

Squash is eaten as a vegetable (sometimes stuffed) and as a dessert pie sweetened with molasses. Sweet potatoes are also used both ways. Other common desserts include bread pie (bread pudding), crumb cake, chocolate or caramel cake, fruit cobblers, puddings, and shortcake, as well as sesame seed cookies and candies.

Meal Composition and Cycle

Daily Patterns

Historically, two meals a day were typical in West Africa, one late in the morning and one in the evening. Snacking was common; in poorer tribes, snacks would replace the morning meal and only dinner would be served. Food was eaten family style or, more formally, the men were served first, then the boys, then the girls, and last the women. Sometimes men gathered together for a meal without women. Mealtimes often were solemn; people concentrated on the attributes of the food, and conversation was minimal.

The West African tradition of frequent snacking continued through the slave period and after emancipation. Meals were often irregular, perhaps due to the variable hours of agricultural labor. The traditional Southern-style meal pattern was adopted as economic conditions for both blacks and whites improved. Breakfast was typically large and leisurely, always including boiled grits and homemade biscuits. In addition, eggs, ham or bacon, and even fried sweet potatoes would be served. Coffee and tea were more common beverages than milk or juice.

Lunch, called dinner, was the main meal of the day. It was eaten at mid-afternoon and featured a boiled entrée, such as legumes or greens with ham, or another stew-type dish. Additional vegetables or a salad may have been served, as well as potatoes and bread or biscuits. Dessert was mandatory and was usually a baked item, not simply fruit. In some homes a full supper of meat, vegetables, and potatoes was served in the evening. Poorer agricultural families often ate only two of these hearty meals a day. Today, few southern African Americans, or whites, continue this traditional meal pattern in full. The Southern-style breakfast might be served just on weekends or holidays, for example. As in the rest of the country, a light lunch has replaced the large dinner on most days, and supper has become the main meal.

Meal traditions vary for recent African immigrants to the United States today. Throughout West Africa, three meals a day are typical, though in some areas, only two meals are consumed during periods of privation, such as just before the harvest.[31] Ethiopians, Eritreans, and Somalis usually eat one or two meals a day, snacking in between. In Eritrea and Somalia, fool, which is a puree made from chickpeas in Eritrea, and pinto beans in Somalia, is a popular breakfast item. Food is typically offered on a communal plate, and individuals use bread to scoop up what is desired. Meals are often joyous and noisy. Three meals each day are common in East Africa, and in Kenya many people also stop for British-style tea in the afternoon. However, in many areas meals are limited to two daily when food is in short supply. Traditionally, men were served first, followed by women, and then children dined. In South Africa, a Westernized pattern of three meals, with dinner the largest, is usually followed.

Special Occasions

Sunday dinner had become a large family meal during the slave period, and it continued to be the main meal of the week after emancipation. It was a time to eat and share favorite foods with friends and relatives, a time to extend hospitality to neighbors.

Many Southern African Americans still enjoy a large Sunday dinner, usually prepared by the mother of the house, who begins cooking in the early morning. The menu would probably include fried chicken, spareribs, chitterlings, pig's feet (or ears or tail), black-eyed peas or okra, corn, corn bread, greens, potato salad, rice, and sweet potato pie. Homemade fruit wines, such as strawberry wine, might also be offered.

© Merritt Vincent/PhotoEdit, Inc.

▲ ***Kwanzaa, the African-American holiday celebrated from December 26 through January 1 each year, culminates with a feast featuring dishes from throughout Africa, the Caribbean, the U.S. South, and other regions where Africans now live.***

Other holiday meals, especially Christmas, feature menus similar to the Sunday meal, but with added dishes and even greater amounts of food. Turkey with cornbread stuffing and baked ham are often the entrees; other vegetable dishes, such as corn pudding, sweet peas, and salads, are typical accompaniments. A profusion of baked goods, including yeast rolls, fruit cakes and cobblers, custard or cream pies, and chocolate, caramel, and coconut cake, round out the meal. Some blacks eat symbolic foods on New Year's Eve, such as fish for motivation, greens for money, black-eyed peas for good luck, and rice for prosperity.[6]

Southern black cuisine is particularly well suited to buffet meals and parties. A pan of gumbo, a pot of beans, or a side of barbecued ribs can be stretched to feed many people on festive occasions. Informal parties to celebrate a birthday, or just the fact that it's Saturday night, are still common. Traditional Southern food is also served at Juneteenth celebrations held in many African American communities to commemorate the emancipation of the slaves.

The African American holiday of Kwanzaa has gained popularity in recent years. Created in southern California in 1966, Kwanzaa recognizes the African diaspora and celebrates the unity of all people of African heritage. It begins on December 26 and runs through New Year's Day. Each day, a new candle is lit to symbolize one of seven principles: unity, self-determination, collective work and responsibility, cooperative economics, purpose, creativity, and faith. The holiday culminates with a feast featuring dishes from throughout Africa, the Caribbean, the U.S. South, and any other region where Africans were transported.

Recent African immigrants may celebrate many religious holidays, especially those associated with the Eastern Orthodox and Islamic faiths. In Nigeria, child-naming ceremonies are particularly important celebrations.[17] A grandmother performs the ritual, offering symbolic foods to the infant, including water (purity), oil (power and health), alcohol (wealth and prosperity), honey (happiness), kola nuts (good fortune), and salt (intelligence and wisdom). Following the tasting, the name is whispered to the child, then announced aloud to the attendees. The family and guests then enjoy a meal together. Many Nigerian Americans continue the custom in the United States.

Role of Food in African American Society and Etiquette

In the American South, food has traditionally been a catalyst for social interaction, and Southern hospitality is renowned. For some blacks, eating is an intimate or a spiritual experience that is shared with others.[6,33,34] Food is lovingly prepared for family and friends, and food is considered an important factor in the cohesiveness of African American society.

In Africa today, sharing food is still an important social activity, often accompanied by loud conversation and gaiety. Food is offered to anyone who is in the home, and in some nations, such as Nigeria, it is common for extended family members to drop by unannounced for a meal.[8] In many urban areas, Western styles of dining are practiced. However, in rural regions meals are often served communally and consumed with the hands. Only the right hand may be used for eating, and the left hand should not touch anything on the table. Although it is usually a sign of respect or affection to feed a bite of food directly into another person's mouth, it is considered exceptionally rude to pass food from one person's hand to another person's hand.[32] In Uganda diners stay seated until all people have finished eating, and sticking one's legs out or leaning on an elbow is not acceptable.[33] In Eritrea, an invitation to coffee means a visit of over an hour, with a minimum of three cups consumed in a ritual that includes the burning of incense.[11]

Therapeutic Uses of Food

Many African Americans maintain health by eating three hearty meals each day, including a hot breakfast. Numerous other beliefs about food and health are noted among small numbers of African Americans living in the rural South. Some of these dietary concepts were brought to other regions of the country during the great migrations and may be found among black elders. The conditions known as "high blood" and "low blood" are one example. High blood (often confused with high blood pressure and high blood sugar levels) is most prevalent and thought to be caused by excess blood migrating to one part of the body, typically the head. High blood is caused by eating excessive amounts of rich foods, sweet foods, or red-colored foods (beets, carrots, grape juice, red wine, and red meat, especially pork). Low blood, associated with anemia, is believed to be caused by eating too many astringent and acidic foods (vinegar, lemon juice, garlic, and pickled foods) and not enough red meat. Other blood complaints include "thin blood" that cannot nourish the body, causing a person to feel chilly; "bad blood," due to hereditary, natural, or supernatural contamination; "unclean blood" when impurities collect over the winter months and the blood carries more heat; and "clots," when the blood thickens and settles in one area, associated with menstruation or stomach and leg cramps.[35,36,37]

Tea made from the yellowroot shrub (*Xanthorhiza simplicissima*) is thought to cure stomachache and fever and used to treat diabetes. Some blacks also believe peppermint candies are helpful in diabetes. Sassafras tea or hot lemon-flavored water with honey is considered good for colds, and raw onion helps to break a fever. Turpentine sweetened with sugar reputedly cures intestinal worms when consumed orally, while a mixture of figs and honey will eliminate ringworm. Goat's milk with cabbage juice is used to cure a stomach infection. In some areas, eggs and milk may be withheld from sick children to aid in their recovery.[20,38]

Pica, the practice of eating nonnutritive substances such as clay, chalk, and laundry starch, is one of the most perplexing of all food habits practiced by African Americans, whites, and other ethnic groups. Studies have determined that pica is most often practiced by black women during pregnancy and the postpartum period, and that rates are unchanged since the 1970s (information on pica among other ethnic groups or age groups in the United States is limited). It is common in the South, where anywhere from 16 to 57 percent of pregnant African American women admit to pica. But pica is also found in other areas of the country where large populations of African Americans reside. In rural regions the substance ingested is usually clay. In urban areas, laundry starch is often the first choice, though instances of women who ate large amounts of milk of magnesia, coffee grounds, plaster, ice, and paraffin have also been reported. Many causes for pica have been postulated—a nutritional need for minerals, hunger or nausea, a desire for special treatment, and cultural tradition are the most common hypotheses. One study found pica was more common among women with limited social support. Another theory suggests it may be related to obsessive-compulsive disorder (OCD). Reasons for pica reported by women include flavor, anxiety relief, texture, and the belief that clay prevents birthmarks or that starch makes the skin of the baby lighter and helps the baby to slip out during delivery.[25,38,39,40]

Recent immigrants from Africa may hold traditional beliefs about maintaining health through a balance of proper diet, exercise, good relations with family and community, emotional well-being, and spirituality.[13] Poor diet is identified as a tangible cause of some conditions, and overweight is often valued as a sign of health. Meat consumption may be associated with longevity. Many people in Africa have limited access to biomedical care and make extensive use of botanical home remedies. In Nigeria, for example, unripe plantains and dried soursop (a tropical fruit) are two treatments used for diabetes.[40]

Contemporary Food Habits in the United States

Most researchers have noted that the food habits of African Americans today usually reflect their current socioeconomic status, geographic location, and work schedule more than their African or Southern heritage. Even in the South, many traditional foods and meal patterns have changed due to the pressures of a fast-paced society. Nevertheless, the same foods consumed by blacks and by whites are likely to have different meanings within each cultural context, and blacks often report that their food habits are uniquely African American.[70]

Adaptations of Food Habits

Ingredients and Common Foods

Food preferences do not vary greatly between blacks and whites in similar socioeconomic groups living in the same region of the United States. Comparisons do show that African Americans choose items such as pork (especially chops, bacon, and sausage), poultry, fresh fish and seafood, sugar, and non-carbonated fruit drinks far more often than the general population. African American households also purchase fewer fruits and vegetables, fewer dairy products, less cereal and baked goods, fewer snack foods (such as potato chips), and less coffee than do any other households, however.[56] Fast foods are popular, and one study determined that there are nearly 60 percent more fast-food restaurants in predominantly black neighborhoods than in mainly white neighborhoods.[42]

The popularity of soul food—a term coined in the 1960s for traditional Southern black cuisine—is notable. It is associated with fresh meats and vegetables made from scratch and thoroughly cooked. Items are preferred well spiced.[34] Many African Americans have adopted this cuisine as a symbol of ethnic solidarity, regardless of region or social class. Today soul food serves as an emblem of identity and a recognition of black history for many African Americans.

Meal Composition and Cycle

While the common foods that African Americans eat reflect geographic location and socioeconomic status, meal composition and cycle have changed more in response to work habits than to other lifestyle considerations. The traditional Southern meal pattern of the large breakfast with fried foods, followed by the large dinner with boiled foods and a hearty supper, has given way to the pressures of industrial job schedules. One seminal study showed that Southern breakfast habits were maintained for only eighteen months after migration to the North and then were replaced with a meal typically consisting of sausage and biscuits or toast.[43]

Research indicates that some African Americans no longer identify certain foods or preparation methods, such as okra and yams, and one-pot meals, as African in origin.[33] Many of the items known to be traditional fare are not eaten often, including pig's feet and chitterlings (some of these foods are also associated with the poor, and respondents may be reluctant to admit eating such items). The exception was greens, which were identified most often as a traditional African American food and were also the most popular with respondents: 78 percent eat greens at least once a month.[44]

African Americans throughout the country now eat lighter breakfasts and consume sandwiches at a noontime lunch. Dinner is served after work, and it has become the biggest meal of the day. Snacking throughout the day is still typical among most African Americans. In many households meal schedules are irregular, and family members eat when convenient. It is not unusual for snacks to replace a full meal.

Frying is still one of the most popular methods of preparing food. An increase in consumption of fried dinner items suggests that the customary method of making breakfast foods has been transferred to evening foods (which were traditionally boiled) when time constraints prevent a large morning meal.[43] Boiling and baking are second to frying in popularity.[47] African Americans use convenience foods and fast foods as income permits.

Research is limited on the food habits of more recent African immigrants. It has been noted that among Somalis, cheese, sodas, and sweetened fruit drinks are very popular. Recipes are being adapted to available ingredients. For example, wheat flour or pancake mix is now used to make anjeero.[46] A study of Sudanese Americans found that foods typically consumed at breakfast and lunch included fava or lentil beans made with feta cheese, vegetables, and sesame oil; eggs, fried liver, meat and vegetables stews, bread, salad, fresh fruit, yogurt, custard, Jell-O, and highly sweetened tea. Frying, stewing, sautéing in sesame oil (which was called "boiling"), fermenting, and grilling were the most common methods of preparation. Baking and steaming were extremely rare.[46]

Nutritional Status

The nutritional status of African Americans is difficult to fully characterize because a limited number of studies have addressed this population, and conflicting data exist. In general, however, research has shown that African Americans' nutritional intake is similar to that of the general population and varies more by socioeconomic status than by ethnicity.

Nutritional Intake

The Healthy Eating Index (HEI—developed by the USDA Center for Nutrition Policy and Promotion), which measures compliance with Food Guide Pyramid recommendations and other dietary factors, such as total fat, saturated fat, sodium and cholesterol intake, and variety, found a mean score of sixty-one for blacks, just slightly below the national population mean of sixty-three. However, a larger percentage of African Americans were determined to have poor diets compared to the total population, 21 percent to 16 percent, respectively. Low intakes of dairy products and vegetables and higher intakes of sodium were contributing factors.[48,49] African Americans have been reported to have diets high in fat—similar to that of the typical U.S. diet—associated with high meat intake, the popularity of frying, and fast food consumption.[34,49,51,52,53] The percentage of calories from animal proteins for blacks is often greater than for whites, however, due in part to a high intake of fatty meats, such as bacon and sausage.[54] Ninety percent of midlife African American women in one year-long study reported they intended to reduce their fat consumption, yet 77 percent consumed over 30 percent of the daily calories as fat, and 61 percent consumed more than 10 percent of total calories as saturated fat.[36] High intake of cholesterol has been found in some studies.[55]

Dairy foods are consumed less often by African Americans than by whites.[55] Approximately 60 to 95 percent of adult Americans of African descent are lactose intolerant. Some studies show that milk is widely disliked and avoided; others indicate milk is consumed as often by blacks as it is by whites. One trial found that lactose digestion in African American adolescent girls improved on a dairy-rich diet.[57] The necessity of consuming dairy products is debatable, and research on older women reported that dietary calcium intake was similar between blacks and whites; however, grain products, such as fortified cereals, were the primary source of the nutrient for the African American respondents.[58]

Many African Americans' diets are low in fruits, vegetables, and whole-grain products and low intakes of dietary fiber have been reported.[50,59]

Even as income increases, fresh produce is sometimes ignored in favor of increased expenditure on meat and other protein foods. However, it is noteworthy that many dietary comparisons use food frequency data without defined portion sizes. A study of rural blacks found that when portion size was explained, participants ate on average larger portions of fruits and most vegetables than standard definitions, increasing their daily intake of fruits and vegetables by a significant two-thirds serving.[60] Further, national surveys suggest that prevalence of low fruit and vegetable intake among blacks is only slightly higher than among whites and may be more typical of the general American food pattern than of ethnic variation, except among some lower-income groups with limited access to fresh produce.[61,62]

Nutrient intake deficiencies may be found among some African Americans living at or near poverty levels in the United States, especially among older study subjects. Rural African American elders consumed fewer servings of meats, fruits, and vegetables, and fats, oils, sweets, and snacks than whites in one survey.[63] The most frequent insufficiencies are of calories, iron, and calcium.[27,50,58,64] Deficiencies in vitamins D, E, B_6, folate, potassium, copper, selenium, and zinc have also been reported.[59,65,66,67] Insufficient intakes are often similar to, or only marginally lower than, those of whites living in the same location, however.[58,69] Although significant dietary differences have been reported in some studies, a review of the literature on black elders' low nutrient intake compared to white elders questioned the significance of reported deficiencies: study samples were small, and actual anthropometric, clinical, and biochemical data confirming differences were sparse.[61] In deficiencies typical of the standard American diet, it is noteworthy that use of vitamin and mineral supplements, which sometimes provides sufficiency in whites, is lower in the black population.[58,66,67]

On average, black men live 6.4 years less than white men in the United States, and black women live 4.6 years less than white women. However, the gap in life expectancy is far greater for certain subpopulations. African American men living in high-risk inner-city neighborhoods are likely to die twenty-one years earlier than Asian-American women in general. Life expectancy rates have not improved substantially since the 1980s and are not explained by ethnicity, income, or health care access alone.[71]

Morbidity and mortality rates for black mothers and their infants are also disproportionately high. Maternal deaths are over three times higher for African Americans than for whites, and infant deaths are more than two times higher.[72] Dietary factors, a large number of teenage pregnancies, and inadequate prenatal care may contribute to a higher incidence of low-birth-weight infants (13.55 percent, almost double the rate for whites).[72] Low mean daily folate intake is of particular concern because it is associated with a greater risk of both preterm delivery and low birth weight.[73] Overall, African American women are three times more likely than whites to have a preterm infant of very low birth weight (less than 1,500 grams).

Some studies report African American mothers breast-feed their infants at higher rates than white mothers, though others suggest low rates of breast feeding.[70] One study found that the primary reason low-income women do not initiate breast feeding is fear of passing dangerous things to their infants through the milk.[74] Some blacks differ somewhat from whites in regards to which solid foods they feed their infants and how soon after birth these are introduced. One-third of low-income mothers offered non-milk liquids or solids to their infants at seven to ten days, 77 percent did so by sixteen weeks, and 93 percent by sixteen weeks.[75]

Overweight is a common problem for adult Americans of African descent. Recent data estimates that as many as 51% of African American women are classified as overweight or obese and 39% of African American men as overweight or obese.[76] Adolescents and children are also at risk, especially girls.[77,78] Fat patterning has also been shown to differ between African Americans and whites. African Americans may have more upper-body and deep-fat depositions than whites.[79]

Excess weight gain may be attributed to many factors, including difference in body-size ideal, preference in body shape by members of the opposite gender,[80,81] and a more permissive attitude regarding obesity.[82,83] Though some research suggests income and education levels are inversely associated with the risk for obesity, especially among children,[78,84] one large study of adolescents found that the prevalence of overweight remained elevated or even increased with socioeconomic status among African American girls.[85] An environment that promotes high intake of fast foods and limits access to healthy items may be another factor.[78,86,87,89]

A sedentary lifestyle is also more prevalent among African Americans and is not associated with income, education, occupation, marital status, poverty levels, or other indicators of social class.[39,81,89] Data from one study found families living in low-income areas and those with higher percentages of minority residents were significantly less likely to have access to recreational facilities.[85] In addition, a desire to consume traditional African American foods and to care for others through cooking, as well as a lack of family support were barriers to weight loss cited in studies.[85,86]

Some researchers have suggested that standard anthropometric measures may be inappropriate for African Americans. One study found that when body mass index (BMI) values were compared to body fatness, black women had lower body fatness than did white women with identical BMI numbers.[90] A similar evaluation reported that when the percentage of body fat is used to measure obesity instead of BMI, rates for African American women drop dramatically, and rates for white men exceed those for black men.[91] BMI does not account for differences in fat-free mass, such as muscle and bone, which accounts for the discrepancy between measures. Weight-for-height growth charts as indicators of percentage of body fat and the use of waist-to-hip ratios in defining heart disease risk have also been found misleading in some studies.[92,93] However, a study of overweight and obesity in children did not find significant differences in rates when using national standards for assessment compared to African American specific standards, and cautioned that ethnicity-specific standards may be confounded by differences in socioeconomic status.[77]

Studies propose that African American women do not necessarily equate being overweight with being unattractive and that they are less preoccupied with dieting than are white women.[94,95] Research on disordered eating in African Americans has been limited and contradictory. Some studies suggest black women have lower rates of dieting and are protected from eating disorders due to a collective acceptance of larger body size.[96–98] Other researchers suggest low rates are due to an assumption that disordered eating does not affect blacks (hence African American girls are often excluded from studies) and that pressures to conform to societal norms regarding weight are eroding any possible preexisting cultural buffers, especially among children and adolescents.[99–102]

Concurrent with obesity is a disproportionately high rate of type 2 diabetes mellitus among African Americans, with a prevalence rate in 2008 of 14.7%.[103] There is a genetic predisposition for Type 2 diabetes mellitus but also a significant relationship with lifestyle factors such as obesity, sedentary lifestyle, and westernized dietary habits. [103–106]

Hypertension is a significant health problem for African Americans. The prevalence of high blood pressure is approximately 39% in African American men and 43% for African American women.[76,107] Hypertension has been found to be a potent risk factor for coronary heart disease (CHD) in blacks, especially women (whereas having diabetes was not predictive for CHD). The incidence rate for CHD in African American women is higher than for white women, but lower in African American men than in white men.[108] Blacks also have nearly double the age-specific death rate from strokes, also believed to be due in part to high rates of high blood pressure.[109]

Rates of iron-deficiency anemia among African Americans are higher than for whites at every age, regardless of sex or income level. This incidence remains excessive even after adjustments for differences in hemoglobin distributions are made using reference standards appropriate for blacks.[110] Pica may result in anemia among pregnant women and newborns. Hookworm can also be a cause in the rural South. Other blood disorders resulting in anemias prevalent in African Americans include alpha-thalassemia, sickle-cell disease, and glucose-6-phosphate dehydrogenase deficiency. Researchers have also suggested that undiagnosed celiac disease in blacks may underlie some cases of iron-deficiency anemia.[111]

Little has been reported on the nutritional adequacy of the traditional diet of recent Africans. Studies in Israel of Ethiopian immigrants found deficiencies in vitamin D (resulting in rickets in children), iodine (due in part to food goitrogens), and calcium. Vitamin A deficiency has led to xerophthalmia and blindness in many regions. Consumption of foods made from enset, a banana leaf plant, common in many parts of Ethiopia, is associated with esophageal cancer.[112–114] Among Sudanese immigrants in the United States, blindness due to trachoma is also common. High rates of extreme malnutrition, malaria, typhoid, hepatitis B, HIV infection, dengue fever, tuberculosis, syphilis, dental problems, diabetes, and parasitic infection have also been reported.[115–116]

Dietary changes of Ethiopians in Israel are marked. A survey of teens found that within eighteen months of arrival, only 30 percent maintained a traditional diet, 60 percent consumed a mixed diet, and 15 percent ate only Israeli foods.[117] More than half of daily calories came from snacks and fast foods, especially sweets and soft drinks. Most milk products were disliked, with the exception of hot chocolate, a favorite with the youth. Fat intake increased, while fruit and vegetable intake decreased. Though obesity is unusual, glucose intolerance is common, and the prevalence rate of type 2 diabetes increased from 0.4 percent to between 5 and 8 percent within a few years. Over 20 percent of men also developed hypertension after immigration.[118] In Australia, immigrants from Ghana experienced similar changes in diet and health status. Fat intake accounted for 33 to 35 percent of total calories, and overweight was observed in 71 percent of men and 66 percent of women. High rates of diabetes, hypertension, and dyslipidemia were reported.[119]

In Nigeria, some women believe that edema during pregnancy is an indication that the infant is male, and treatment may not be sought for the condition.[120] Studies of Ethiopian women in the United States found that most breast-fed their infants on average four months, and breast-feeding is acceptable in public. Going back to work and reduction in mother's milk were the primary reasons given for cessation.[121–122] Somali immigrants often associate fatness with health and may overfeed their children.[123] The Somali Bantus are in particularly poor health due to acute or chronic malnutrition when living in African refugee camps prior to arrival in the United States. Most have little knowledge of American foods. The prevalence of low-birth-weight infants is high, and weaning often occurs before six months due to subsequent pregnancy. Diarrheal diseases and infections are common. Post-traumatic stress syndrome is also frequently found.[124]

Counseling

Many African Americans have limited access to health care. In 2009, 19% of African Americans were without health insurance.[76] Cost, including time off from work, is often an issue. Self-reliance is highly valued and may lead to delay in seeking care or minimization of symptoms. Furthermore, an attitude that fate determines wellness may restrict medical visits. When a doctor's care is sought, it is usually for treatment of symptoms (often after home remedies have been tried) rather than for prevention of illness and health maintenance.[125] Many blacks are present-oriented, and flexible scheduling and on-time policies may be helpful.[126]

Some African American clients feel patronized by non-black providers and may choose to suffer at home rather than submit to humiliation. Others may be suspicious or hostile when working with non-black health care professionals. Such attitudes are rarely directed specifically at the health care worker; instead, they are an adaptation to what is perceived as a prejudicial society.[125,126] Some African Americans may not consider themselves active participants in their interaction with providers and may not communicate needs or questions. This is sometimes done to test the competency of the provider, who is expected to diagnose without help from the client.[127] It may also be a reflection of the belief that their health care is out of their control, up to luck or destiny.[13]

African American conversational style is fully engaged and very expressive.[13] Interjections of agreement or disagreement are frequent. Words are often spoken rhythmically and passionately. Response time is very quick. When counseling African Americans, it is helpful to be direct yet respectful. Eye contact is made while speaking, but prolonged eye contact is considered rude.[34] African Americans may avert their eyes while listening; however, some blacks may interpret rapid eye aversion as an insult. Attentive listening is more important than eye contact to many blacks. A firm handshake and smile are the expected greeting; hugging and kissing may also be included. Touching is common, and reluctance to touch may be interpreted as personal rejection.

Nonnutritive food intake during pregnancy may be missed unless information about pica is solicited during the interview. Most women who eat clay, laundry starch, or other nonfood items will willingly list the items consumed when asked directly about the habit. The nutritional effects of pica are uncertain. Possible problems include excessive weight gain (from laundry starch), aggravated hypertension (from the sodium in clay), iron-deficiency anemia, and hyperkalemia.[78] Furthermore, over-the-counter remedies for the gas and constipation that can accompany pica may be harmful during pregnancy.[128] Traditional health practices, such as using diet to cure high and low blood, may complicate some nutrition counseling. Pregnancy is sometimes considered to be a high blood state, and pregnant women will avoid red meats. Patients who confuse hypertension with high blood may eat astringent foods, which are often high in sodium, to balance the condition. Home remedies for diabetes, such as peppermint candies or yellowroot bush tea, should also be investigated. A client is unlikely to mention any use of other healers.[129] It may be useful to directly inquire whether a rural African American believes an illness is due to outside forces or witchcraft and what additional treatment is being sought.

Counseling recommendations should be action- or task-oriented. Several studies of African Americans have suggested family-oriented programs and group classes may be more successful than individual counseling, and use of community resources, such as churches, can provide additional support for nutritional change.[87,130] Culturally relevant education may include elements of spirituality, ethnic pride, group planning, and the use of peer counselors.[131–134] Diets that limit traditional African American foods are often resisted by clients, due to preference, expense, family desires, and ethnic identity.[87,135,136] One study found that blacks felt socially isolated when restricted to nontraditional foods.[137] Researchers have suggested that because many African Americans take pride in the adaptations made in their cuisine due to historical circumstances, the potential for dietary improvement associated with environmental change is high.[33]

Little information is available regarding counseling recent immigrants from Africa. Recommendations for Ethiopians include use of an interpreter from the client's community; a warm, personable communication style; a positive outlook; and disclosure of poor prognosis or terminal illness to the patient's family (preferably not the wife or mother) or friend, who will then inform the patient.[138] Ethiopians and Somali Bantus may view illness as a punishment from God or angry spirits, and they may employ herbal remedies or prefer spiritual healing.[121,123] Injections may be requested by Ethiopians instead of oral prescriptions, and Sudanese may stop taking medications as soon as symptoms

DISCUSSION STARTERS: HEALTH RISKS OF AFRICAN AMERICANS

As this chapter makes clear, African Americans today often suffer from obesity, diabetes, hypertension, infant and maternal morbidity and mortality, and depression in greater numbers than other American races. Make a list of all the possible reasons for this phenomenon. Don't limit yourself to just one or two, and look back at the diet and culture of the geographical area from which the ancestors of most modern African Americans came. In a small group, share your lists of reasons and come up with one group list. Again, your task is to list all the possible reasons that you can.

After making this group list, individually take a blank sheet of paper and cluster the items on that list. That is, look the listed items and identify reasons that you see as possibly being somehow linked. On the blank sheet of paper, rewrite the reasons on the group list but locate them as best you can near to other reasons that you as somehow related. Circle each reason and draw lines between the ones that you believe are connected.

Finally, imagine that you have been hired by a local health department in an area of the country with a large African American population to write a brief (1-page) resource handout for local health professionals on the reasons why many of their African American clients may have the above health problems. As you plan this short paper, consider whether—or not—it might have made a difference had most African slaves come from East Africa.

subside, saving them for a later recurrence of illness in themselves or among family members or friends.[116] It has been noted that Eritreans have faith in certain botanical cures, and if they are unobtainable in the United States, they may return to Africa for treatment.[121]

An in-depth interview is especially important with African American clients. Variability in diet related to region, socioeconomic conditions, and degree of ethnic identity should be considered. Information regarding country of origin may be significant: Clients from the Caribbean or Central America are more likely to identify with the foods and food habits of Latinos than with those of blacks in the United States. Religious affiliation may also be important, such as among Americans of African descent who are members of Islam or an Eastern Orthodox faith.

REVIEW QUESTIONS

1. Describe the cuisine that West African slaves brought to America and one American food recipe that has its origins in Africa. What traditional food might be served on Juneteenth?
2. Compare similarities and differences in West and East African traditional cuisines. What countries have influenced East African cuisine?
3. Name the presented symbolic foods used in the Nigerian child-naming ceremonies, and explain what they symbolize.
4. Describe three therapeutic uses of food among African Americans. What is pica, and why is it practiced?
5. For African Americans, how might diet affect the incidence and treatment of hypertension and type 2 diabetes mellitus?

REFERENCES

1. Population Reference Bureau. PRB 2010 World Population Data Sheet. 2011. Available from: http://www.prb.org/Publications/Datasheets/2010/2010wpds.aspx#. Accessed: January 25, 2011.
2. McKinnon, J.D., & Bennett, C.E. 2005. *We the people: Blacks in the United States.* Washington, DC: U.S. Census Bureau.
3. Terrazas A. African Americans in the U.S. 2009. Available from: http://www.migrationinformation. org/USfocus/display.cfm?id=719. Accessed: January 31, 2011.
4. US Census. American Community Survey Five Year Estimates. 2011. Available from: http://www.census. gov/acs. Accessed: January 25, 2011.
5. Bigelow, B.C. 2000. African Americans. *In Gale Encyclopedia of Multicultural America,* R.V. Dassanowsky & J. Lehman (Eds.). Farmington Hills, MI: Gale Group.
6. US Department of Health and Human Services. Office of Minority Health. African American Profile. Available from: http://minorityhealth.hhs.gov/templates/browse.aspx?lvl=2&lvlID=51. Accessed: January 25, 2011.
7. U.S. Census. Immigration Statistics Staff. 2011. Foreign-Born Profiles (STP-159). Available from: http://www.census.gov/population/www/socdemo/foreign/STP-159-2000tl.html. Accessed: January 25, 2011.
8. Sarkodie-Mensah, K. 2000. Nigerian Americans. In *Gale Encyclopedia of Multicultural America,* R.V. Dassanowsky & J. Lehman (Eds.). Farmington Hills, MI: Gale Group.
9. Walker, D. 2000. Ghanaian Americans. In *Gale Encyclopedia of Multicultural America,* R.V. Dassanowsky & J. Lehman (Eds.). Farmington Hills, MI: Gale Group.
10. Ockerstrom, L. 2000. Eritrean Americans. In *Gale Encyclopedia of Multicultural America,* R.V. Dassanowsky & J. Lehman (Eds.). Farmington Hills, MI: Gale Group.
11. Gruen A. 2008. *Somali Bantu Literature Review.* EthnoMed, University of Washington Harborview Medical Center. Available at: http://ethnomed.org/culture/somali-bantu/somali-bantu-literature-review/. Accessed January 29, 2011.
12. Knight, J., & Mabunda, L. 2000. South African Americans. In *Gale Encyclopedia of Multicultural America,* R.V. Dassanowsky & J. Lehman (Eds.). Farmington Hills, MI: Gale Group.
13. Yehieli, M., & Grey, M.A. 2005. *Health matters: A pocket guide for working with diverse cultures and underserved populations.* Boston: Intercultural Press.
14. U.S. Census Bureau, Census 2000 Summary File 3, Matrices P18, P19, P21, P22, P24, P36, P37, P39, P42, PCT8, PCT16, PCT17, and PCT19. Available at:
15. Rudolph, L.C. 2000. Kenyan Americans. In *Gale Encyclopedia of Multicultural America,* R.V. Dassanowsky & J. Lehman (Eds.). Farmington Hills, MI: Gale Group.
16. Sarkodie-Mensah, K. 2000. Nigerian Americans. In *Gale Encyclopedia of Multicultural America,* R.V. Dassanowsky & J. Lehman (Eds.). Farmington Hills, MI: Gale Group.
17. Mbiti, J.S. 1970. *African religions and philosophy.* Garden City, NY: Doubleday Anchor.
18. Jackson, B. 1976. The other kind of doctor: Conjure and magic in black American folk medicine. In W.D. Hand (Ed.), *American Folk Medicine.* Berkeley: University of California Press.
19. Spector, R.E., 2004. *Cultural diversity in health and illness* (6th ed.). Upper Saddle River, NJ: Pearson Education, Inc.
20. Ravenell, J.E., Johnson, W.E., Jr., & Whitaker, E.E. 2006. African-American men's perceptions of health: A focus group study. *Journal of the National Medical Association,* 98, 544–550.

21. Hsiao, A.F., Wong, M.D., Goldstein, M.S., Yu, H.J., Anderson, R.M., Brown, E.R., Becerra, L.M., & Wenger, N.S. 2006. Variation in complementary and alternative medicine (CAM) use across racial/ethnic groups and the development of ethnic-specific measures of CAM use. *Journal of Alternative and Complementary Medicine, 12,* 281–280.
22. Chester DN, Himburg, Weatherspoon LF. Spirituality of African-American women: correlations to health promoting behavious. *J Natl Black Nurses Assoc.* 2006; 17: 1–8.
23. Lecca, P.J., Quervalu, I., Nunes, J.V., & Gonzales, H.F. 1998. *Cultural competency in health, social, and human services.* New York: Garland Publishing, Inc.
24. Brandon, E. 1976. Folk medicine in French Louisiana. In W.D. Hand (Ed.), *American Folk Medicine.* Berkeley: University of California Press.
25. Glanville, C.L. 2003. People of African American Heritage. In *Transcultural Health Care* (2nd ed.), L.D. Purnell & B.J. Paulanka (Eds.). Philadelphia: FA Davis Company.
26. Boyd, E.L., Taylor, S.D., Shimp, L.A., & Semier, C.R. 2000. An assessment of home remedy use by African Americans. *Journal of the National Medical Association, 92,* 341–353.
27. Geller, S.E., & Derman, R. 2001. Knowledge, beliefs, and risk factors for osteoporosis among African-American and Hispanic-American women. *Journal of the National Medical Association, 93,* 13–21.
28. Zibart, E. 2001. *The ethnic food lovers companion: Understanding the cuisines of the world.* Birmingham: Menasha Ridge Press.
29. Kobel, P.S. 2000. Ethiopian Americans. In *Gale Encyclopedia of Multicultural America,* R.V. Dassanowsky & J. Lehman (Eds.). Farmington Hills, MI: Gale Group.
30. Heymann, T.D., Bhupulan, A., Zureikat, N.E., Bomanji, J., Drinkwater, C., Giles, P., & Murray-Lyon, I.M. 1995. Khat chewing delays gastric emptying of a semi-solid meal. *Alimentary Pharmacological Therapy, 9,* 81–83.
31. Foster, D. 2002. The global etiquette guide to Africa and the Middle East. New York: John Wiley & Sons, Inc.
32. Miller, O. 2000. Ugandan Americans. In *Gale Encyclopedia of Multicultural America,* R.V. Dassanowsky & J. Lehman (Eds.). Farmington Hills, MI: Gale Group.
33. Airhihenbuwa, C.O., & Kumanyika, S. 1996. Cultural aspects of African American eating patterns. *Ethnicity & Health, 1,* 245–261.
34. Weatherspoon L, Chester D, Kidd T. African American Food Practices. In: Goody CM, Drago L (eds). *Cultural Food Practice.* Chicago IL: American Dietetic Association. 2010.
35. Jackson, J.J. 1981. Urban black Americans. In A. Harwood (Ed.), *Ethnicity and Medical Care.* Cambridge, MA: Harvard University Press.
36. Smith, S.L., Quandt, S.A., Arcury, T.A., Wetmore, L.K., Bell, R.A., & Vitolins, M.Z. 2005. Aging and eating in the rural, southern United States: Beliefs about salt and its effect on health. *Social Science & Medicine, 62,* 189–198.
37. Hunter, J.M. 1973. Geophagy in Africa and the United States. *Geographical Review, 63,* 170–195.
38. Edwards, C.H., Johnson, A.A., Knight, E.M., Oyemade, U.J., Cole, O.J., Jones, S., Laryea, H., & Westney, L.S. 1994. Pica in an urban environment. *Journal of Nutrition, 124,* 954S–962S.
39. Rose, D., & Bodor, J.N. 2006. Household food insecurity and overweight status in young school children: Results from the Early Childhood Longitudinal Study. *Pediatrics, 117,* 464–473.
40. Nwosu, M.O. 1998. Aspects of ethnobotanical medicine in Southeast Nigeria. *The Journal of Alternative and Complementary Medicine, 4,* 305–310.
41. National Center for Health Statistics. 2005. *Health, United States, 2005 with Chartbook on Trends in the Health of Americans.* Hyattsville, MD: Author.
42. Block, J.P., Scribner, R.A., & DeSalvo, K.B. 2004. Fast food, race/ethnicity, and income: A geographic analysis. *American Journal of Preventive Medicine, 27,* 211–217.
43. Jerome, N.W. 1969. Northern urbanization and food consumption of southern born Negroes. *American Journal of Clinical Nutrition, 22,* 1667–1669.
44. Byars, D. 1996. Traditional African American foods and African Americans. *Agriculture and Human Values, 13,* 74–78.
45. Wheeler, M., & Haider, S.Q. 1979. Buying and food preparation patterns of ghetto blacks and Hispanics in Brooklyn. *Journal of the American Dietetic Association, 75,* 560–563.
46. Haq, A.S. 2003. Report on Somali diet: Common dietary beliefs and practices of Somali participants in WIC nutrition education groups. EthnoMed, University of Washington Harborview Medical Center. Online at http://ethnomed.org/clin_topics/nutrition/somali_diet_report.html
47. Elmubarak, E., Bromfield, E., & Bovell-Benjamin, A.C. 2004. Focused interviews with Sudanese Americans: Perceptions about diet, nutrition, and cancer. *Preventive Medicine, 40,* 502–509.
48. Basiotis, P.P., Carlson, A., Gerrior, S.A., Juan, W.Y., & Lino, M. 2002. *The Healthy Eating Index: 1999–2000.* Washington, DC: U.S. Department of Agriculture, Center for Nutrition Policy and Promotion. CNPP–12.
49. Furomoto-Dawson AA, Pandey D, Elliott WJ, DeLeon CF, Al-Hani AJ, Hollenberg S, Camba N, Wicklund R, Black HR. Hypertension in Women: The Women Take Heart Project. *J Clin Hypertens.* 2003; 5: 38–46.
50. Daroszewski, E.B. 2004. Dietary fat consumption, readiness to change, and ethnocultural association in midlife African American women. *Journal of Community Health Nursing, 21,* 63–75.
51. Bowman, S.A., Gortmaker, S.L., Ebbeling, C.B., Pereira, M.A., & Ludwig, D.S. 2004. Effects of fast-food consumption on energy intake and diet quality among children in a national household survey. *Pediatrics, 113,* 112–118.
52. Gans, K.M., Burkholder, G.J., Risica, P.M., & Lasater, T.M. 2003. Baseline fat-related dietary behaviors

of white, Hispanic and black participants in a cholesterol screening and education project in New England. *Journal of the American Dietetic Association, 103,* 699–706.

53. Satia, J.A., Galanko, J.A., & Siega-Riz, A.M. 2004. Eating at fast-food restaurants is associated with dietary intake, demographic, psychosocial and behavioral factors among African Americans in North Carolina. *Public Health Nutrition, 7,* 1089–1096.
54. Steyen, K., Fourie, J., Lombard, C., Katzenellbogen, J., Bourne, L., & Jooste, P. 1996. Hypertension in the black community of the Cape Peninsula, South Africa. *East African Medical Journal, 73,* 758–763.
55. Daroszewski, E.B. 2004. Dietary fat consumption, readiness to change, and ethnocultural association in midlife African American women. *Journal of Community Health Nursing, 21,* 63–75.
56. Tucker, K.L., Maras, J., Champagne, C., Connell, C., Goolsby, S., Weber, J., Zaghloul, S., Carithers, T., & Bogle, M.L. 2005. A regional food-frequency questionnaire for the US Mississippi Delta. *Public Health Nutrition, 8,* 87–96.
57. Pribila, B.A., Hertzler, S.R., Martin, B.R., Weaver, C.M., & Savaiano, D.A. 2000. Improved lactose digestion and intolerance among African-American adolescent girls fed a dairy-rich diet. *Journal of the American Dietetic Association, 100,* 524–528.
58. Mojtahedi, M.C., Plawecki, K.L., Chapman-Novakofski, K.M., McAuley, E., & Evans, E.M. 2006. Older black women differ in calcium intake source compared to age- and socioeconomic status-matched white women. *Journal of the American Dietetic Association, 106,* 1102–1107.
59. Diaz, V.A., Mainous, A.G., Koopman, R.J., & Geesey, M.E. 2005. Are ethnic differences in insulin sensitivity explained by variation in carbohydrate intake? *Diabetologia 48,* 1264–1268.
60. Campbell, M.K., Polhamus, B., McClelleand, J.W., Bennett, K., Kalsbeek, W., Coole, D., Jackson, B., & Demark-Wahnefried, W. 1996. Assessing fruit and vegetable consumption in a 5 a day study targeting rural blacks: The issue of portion size. *Journal of the American Dietetic Association, 96,* 1040–1042.
61. Fahm, E.G., & Senborn, C. 1998, Spring. Nutritional status of black elders: A review of the literature. *Journal of the Family and Consumer Sciences,* 23–27.
62. Baker, E.A., Schootman, M., Barnidge, E., & Kelly, C. 2006. The role of race and poverty in access to foods that enable individuals to adhere to dietary guidelines. *Preventing Chronic Disease, 3,* A76.
63. Zenk, S.N., Schulz, A.J., Hollis-Neely, T., Campbell, R.T., Holmes, N., Watkins, G., Nwankwo, R., & Odoms-Young, A. 2005. Fruit and vegetable intake in African Americans income and store characteristics. *American Journal of Preventive Medicine, 29,* 1–9.
64. Vitolins, M.Z., Quandt, S.A., Bell, R.A., Arcury, T.A., & Case, L.D. 1992. Quality of diets consumed by older rural adults. *Journal of Rural Health, 18,* 49–56.
65. Maitland, T.E., Gomez-Marin, O., Weddle, D.O., & Fleming, L.E. 2006. Associations of nationality and race with nutritional status during perimenopause: Implications for public health practice. *Ethnicity & Disease, 16,* 201–206.
66. Arab, L., Carriquiry, A., Steck-Scott, S., & Gaudet, M.M. 2003. Ethnic differences in the nutrient intake adequacy of premenopausal US women: Results from the Third National Health Examination Survey. *Journal of the American Dietetic Association, 103,* 1008–1014.
67. Ervin, R.B., & Kennedy-Stephenson, J. 2002. Mineral intakes of elderly adult supplement and non-supplement users in the Third National Health and Nutrition Examination Survey. *Journal of Nutrition, 132,* 3422–3427.
68. Lewis, L.B., Sloane, D.C., Nascimento, L.M., Diamant, A.L., Guinyard, J.J., Yancey, A.K., & Flynn, G. 2005. African Americans' access to healthy food options in south Los Angeles restaurants. *American Journal of Public Health, 95,* 668–673.
69. Lewis, S.M., Mayhugh, M.A., Freni, S.C., Thorn, B., Cardoso, S., Buffington, C., Jairaj, K., & Feuers, R.J. 2003. Assessment of antioxidant nutrient intake of a population of southern US African American and Caucasian women of various ages when compared to dietary reference intakes. *Journal of Nutrition, Health & Aging, 7,* 121–128.
70. Cricco-Lizza, R. 2004. Infant-feeding beliefs and experiences of Black women enrolled in WIC in the New York metropolitan area. *Quality Health Research, 14,* 1197–1210.
71. Murray, C.J.L., Kulkarni, S.C., Michaud, C., Tomijima, N., Bulzacchelli, T.J., Iandiorio, T.J., & Ezzati, M. 2006. Eight Americas: Investigating mortality disparities across races, counties, and race-counties in the United States. *PLoS Med 3,* e260, 1513–1524.
72. National Center for Health Statistics. 2005. Health, United States, 2005 with Chartbook on Trends in the Health of Americans. Hyattsville, MD: Author.
73. Scholl, T.O., Hediger, M.L., Schall, J.I., Khoo, C.S., & Fischer, R.L. 1996. Dietary and serum folate: Their influence on the outcome of pregnancy. *American Journal of Clinical Nutrition, 63,* 520–525.
74. England, L., Brenner, R., Bhaskar, B., Simons-Morton, B., Das, A., Revenis, M., Mehta, N., & Clemens, J. 2003. Breastfeeding practices in a cohort of inner-city women: The role of contraindications. *BMC Public Health,* 20, 28.
75. Bronner, Y.L., Gross, S.M., Caulfield, L., Bentley, M.E., Kessler, L., Jensen, J., Weathers, B., & Paige, D.M. 1999. Early introduction of solid foods among urban African-American participants in WIC. *Journal of the American Dietetic Association, 99,* 457–461.
76. Surgeon General. *The Surgeon General's Call To Action To Prevent and Decrease Overweight and Obesity.* Washington, DC: US Government Printing Office; 2001.
77. Dwyer, J.T., Stone, E.J., Minhua, Y., Webber, L.S., Must, A., Feldman, H.A., Nader, P.R., Perry, C.L., & Parcel, G.S. 2000. Prevalence of marked overweight and obesity in a multiethnic pediatric population: Findings from the Child and Adolescent Trial for

Cardiovascular Health (CATCH) study. *Journal of the American Dietetic Association, 100,* 1149–1156.

78. Kumanyika, S., & Grier, S. 2006. Targeting interventions for ethnic minority and low-income populations. *The Future of Children, 16,* 187–207.
79. Zillikens, M.C., & Conway, J.M. 1990. Anthropometry in blacks: Applicability of generalized skinfold equations and differences in fat patterning between blacks and whites. *American Journal of Clinical Nutrition, 52,* 45–51.
80. Becker, D.M., Yanek, L.R., Koffman, D.M., & Bronner, Y.C. 1999. Body image preferences among urban African-Americans and whites from low-income communities. *Ethnicity & Disease, 9,* 377–386.
81. Sanchez-Johnsen, L.A., Fitzgibbon, M.L., Martinovich, Z., Stolley, M.R., Dyer, A.R., & Van Horn, L. 2004. Ethnic differences in correlates of obesity between Latin-American and black women. *Obesity Research, 12,* 652–660.
82. Blixen, C.E., Singh, A., & Thacker, H. 2006. Values and beliefs about obesity and weight reduction among African American and Caucasian women. *Journal of Transcultural Nursing, 17,* 290–297.
83. Eckstein, K.C., Mikhail, L.M., Ariza, A.J., Thomson, J.S., Millard, S.C., & Binns, H.J. 2006. Parents' perceptions of their child's weight and health. *Pediatrics, 117,* 681–690.
84. Rose, D., & Bodor, J.N. 2006. Household food insecurity and overweight status in young school children: Results from the Early Childhood Longitudinal Study. *Pediatrics, 117,* 464–473.
85. Gordon-Larsen, P., Adair, L.S., & Popkin, B.M. 2003. The relationship of ethnicity, socioeconomic factors, and overweight in US adolescents. *Obesity Research, 11,* 121–129.
86. Lewis, L.B., Sloane, D.C., Nascimento, L.M., Diamant, A.L., Guinyard, J.J., Yancey, A.K., & Flynn, G. 2005. African Americans' access to healthy food options in south Los Angeles restaurants. *American Journal of Public Health, 95,* 668–673.
87. James, D.C.S. 2004. Factors influencing food choices, dietary intake, and nutrition-related attitudes among African Americans: Application of a culturally sensitive model. *Ethnicity & Health, 9,* 349–367.
88. Zenk, S.N., Schulz, A.J., Hollis-Neely, T., Campbell, R.T., Holmes, N., Watkins, G., Nwankwo, R., & Odoms-Young, A. 2005. Fruit and vegetable intake in African Americans income and store characteristics. *American Journal of Preventive Medicine, 29,* 1–9.
89. Crespo, C.J., Smit, E., Andersen, R.E., Carter-Pokras, O., & Ainsworth, B.E. 2000. Race/ethnicity, social class and their relation to physical inactivity during leisure time: Results from the Third National Health and Nutrition Examination Survey, 1988–1994. *American Journal of Preventive Medicine, 18,* 46–53.
90. Evans, E.M., Rowe, D.A., Racette, S.B., Ross, K.M., & McAuley, E. 2006. Is the current BMI obesity classification appropriate for black and white post menopausal women? *International Journal of Obesity, 30,* 837–843.
91. Cawley, J., & Burkhauser, R.V. 2006. Beyond BMI: The value of more accurate measures of fatness and obesity in social science research. National Bureau of Economic Research (NBER) at http://papers. nber. org/papers/w12291.
92. Croft, J.B., Keenan, N.L., Sheridan, D.P., Wheeler, F.C., & Speers, M.A. 1995.Waist-to-hip ratio in a biracial population: Measurement, implications, and cautions for cardiovascular disease. *Journal of the American Dietetic Association, 95,* 60–64.
93. Hoerr, S.L., Nelson, R.A., & Lohman, T.R. 1992. Discrepancies among predictors of desirable weight for black and white obese adolescent girls. *Journal of the American Dietetic Association, 92,* 450–453.
94. Kumanyika, S., Wilson, J.F., & Guilford-Davenport, M. 1993. Weight-related attitudes and behaviors of black women. *Journal of the American Dietetic Association, 93,* 416–422.
95. Stevens, J., Kumanyika, S.K., & Keill, J.E. 1994. Attitudes towards body size and dieting: Differences between elderly black and white women. *American Journal of Public Health, 84,* 1322–1325.
96. Neumark-Sztainer, D., Croll, J., Story, M., Hannan, P.J., French, S.A., & Perry, C. 2002. Ethnic/racial differences in weight-related concerns and behaviors among adolescent girls and boys: Findings from Project EAT. *Journal of Psychosomatic Research, 53,* 963–974.
97. Rhea, D.J. 1999. Eating disorder behaviors of ethnically diverse urban female adolescent athletes and non-athletes. *Journal of Adolescence, 22,* 379–388.
98. White, M.A., Kohlmaier, J.R., Varnado-Sullivan, P., & Williamson, D.A. 2003. Racial/ethnic differences in weight concerns: Protective and risk factors for the development of eating disorders and obesity among adolescent females. *Eating and Weight Disorders, 8,* 20–25.
99. Mitchell, K.S., & Mazzeo, S.E. 2004. Binge eating and psychological distress in ethnically diverse undergraduate men and women. *Eating Behaviors, 5,* 157–169.
100. Perez, M., & Joiner, T.E. 2003. Body image dissatisfaction and disordered eating in black and white women. *International Journal of Eating Disorders, 33,* 342–350.
101. Robinson, T.N., Chang, J.Y., Haydel, K.F., & Killen, J.D. 2001. Overweight concerns and body image dissatisfaction among third-grade children: The impacts of ethnicity and socioeconomic status. *Journal of Pediatrics, 138,* 181–187.
102. Williamson, L. 1998. Eating disorders and the cultural forces behind the drive for thinness: Are African American women really protected? *Social Work and Health Care, 28,* 61–73.
103. National Institute of Diabetes and Digestive and Kidney Diseases. National Diabetes Statistics, 2007. Bethesda, MD: U.S. Department of Health and Human Services, National Institutes of Health, 2008. http://www.diabetes.niddk.nih.gov/dm/pubs/statistics/.

104. Shai, I., Jiang, R., Manson, J.E., Stampfer, M.J., Willett, W.C., Colditz, G.A., & Hu, F.B. 2006. Ethnicity, obesity, and risk of type 2 diabetes in women: A 20-year follow-up study. *Diabetes Care, 29,* 1585–1590.
105. Diaz, V.A., Mainous, A.G., Koopman, R.J., & Geesey, M.E. 2005. Are ethnic differences in insulin sensitivity explained by variation in carbohydrate intake? *Diabetologia 48,* 1264–1268.
106. Banini, A.E., Allen, J.C., Allen, H.G., Boyd, L.C., & Lartey, A. 2003. Fatty acids, diet, and body indices of type II diabetic American whites and blacks and Ghanaians. *Nutrition, 19,* 722–726.
107. National Center for Health Statistics. *Health, United States, 2008.* Hyattsville, MD: National Center for Health Statistics; 2008. Available from: http://www.cdc.gov/bloodpressure/facts.htm. Accessed: April 7, 2011.
108. Jones, D.W., Chambless, L.E., Folsom, A.R., Heiss, G., Hutchinson, R.G., Sharrett, A.R., Szklo, M., & Taylor, H.A. 2002. Risk factors for coronary heart disease in African Americans. *Archives of Internal Medicine, 162,* 2565–2571.
109. Centers for Disease Control. 2005. Disparities in deaths from stroke among persons aged <75 years—United States, 2002. *MMWR, 54,* 477–481.
110. Beutler, E., & West, C. 2005. Hematologic differences between African-Americans and whites: The roles of iron deficiency and a-thalassemia on hemoglobin levels and mean corpuscular volume. *Blood, 106,* 740–745.
111. Brar, P., Lee, A.R., Lewis, S.K., Bhagat, G., & Green, P.H. 2006. Celiac disease in African-Americans. *Digestive Diseases and Sciences, 51,* 1012–1015.
112. Belachew, T., Nida, H., Getaneh, T., Woldemariam, D., & Getinet, W. 2005. Calcium deficiency and causation of rickets in Ethiopian children. East African *Medical Journal, 82,* 153–159.
113. Fogelman, Y., Rakover, Y., & Luboshitzky, R. 1995. High prevalence of vitamin D deficiency among Ethiopian women immigrants to Israel: Exacerbation during pregnancy and lactation. *Israeli Journal of Medical Sciences, 31,* 221–224.
114. Mengesha, B., & Ergete, W. 2005. Staple Ethiopian diet and cancer of the oesophagus. *East African Medical Journal, 82,* 353–356.
115. Ackerman, L.K. 1997. Health problems of refugees. *Journal of the American Board of Family Practice, 10,* 337–348.
116. Rasbridge, L.A. 2006. Sudanese. Refugee Health-Immigrant Health, Baylor University. Online at http://www3.baylor.edu/~Charles_Kemp/refugees.htm.
117. Trostler, N. 1997. Health risks of immigration: The Yemenite and Ethiopian cases in Israel. *Biomedicine and Pharmacotherapy, 51,* 352–359.
118. Bursztyn, M., & Raz, I. 1993. Blood pressure, glucose, insulin and lipids of young Ethiopian recent immigrants to Israel and in those resident for 2 years. *Journal of Hypertension, 11,* 455–459.
119. Salah, A., Amanatidis, S., & Samman, S. 2002. Cross-sectional study of diet and risk factors for metabolic diseases in a Ghanaian population of Sydney, Australia. *Asia Pacific Journal of Clinical Nutrition, 11,* 210–216.
120. Okafor, C.B. 2000. Folklore linked to pregnancy and birth in Nigeria. *Western Journal of Nursing Research, 22,* 189–202.
121. EthnoMed. 2006. Ethiopian cultural profile. EthnoMed. University of Washington, Harborview Medical Center. Online at http://ethnomed.org/ethnomed/cultures/ethiop/ethiop_cp.html.
122. Meftuh, A.B., Tapsoba, L.P., & Lamounier, J.A. 1991. Breastfeeding practices in Ethiopian women in southern California. *Indian Journal of Pediatrics, 58,* 349–356.
123. Rasbridge, L.A. 2006. Sudanese. Refugee Health-Immigrant Health, Baylor University. Online at http://www3.baylor.edu/~Charles_Kemp/refugees.htm.
124. Owens, C.W. 2004. Somali Bantu refugees. EthnoMed, University of Washington Harborview Medical Center. Online at http://ethnomed.org/cultures/somali/somali_bantu.html
125. Giger, J.N., Davidhizar, R.E., & Turner, G. 1992. Black American folk medicine health care beliefs: Implications for nursing plans of care. *The ABNF Journal, 3,* 42–46.
126. Galanti, G.A. 2004. *Caring for patients from different cultures,* 3rd ed. Philadelphia: University of Pennsylvania Press.
127. Jackson, B. 1976. The other kind of doctor: Conjure and magic in black American folk medicine. In W.D. Hand (Ed.), *American Folk Medicine.* Berkeley: University of California Press.
128. Boyle, J.S., & MacKay, M.C. 1999. Pica: Sorting it out. *Journal of Transcultural Nursing, 10,* 65–67.
129. Graham, R.E., Ahn, A.C., Davis, R.B., O'Connor, B.B., Eisenberg, D.M., & Phillips, R.S. 2005. Use of complementary and alternative medical therapies among racial and ethnic minority adults: Results from the 2002 National Health Interview Survey. *Journal of the National Medical Association, 97,* 535–545.
130. Peregrin, T. 2006. Cooking with soul: A look into faith-based wellness programs. *Journal of the American Dietetic Association, 106,* 1016–1020.
131. Fitzgibbon, M.L., Stolley, M.R., Ganschow, P., Schiffer, L., Wells, A., Simon, N., & Dyer, A. 2005. Results of a faith-based weight loss intervention for black women. *Journal of the National Medical Association, 97,* 1393–1402.
132. Kieffe, E.C., Willis, S.K., Odoms-Young, A.M., Guzman, J.R., Allen, A.J., Two Feathers, J., & Loveluck, J. 2004. Reducing disparities in diabetes among African-American and Latino residents of Detroit: The essential role of community planning focus groups. *Ethnicity & Disease, 14,* S27–S37.
133. Kreuter, M.W., Sugg-Skinner, C., Holt, C.L., Clark, E.M., Haire-Joshu, D., Fu, Q., Booker, A.C., Steger-May, K., & Bucholtz, D. 2005. Cultural tailoring for mammography and fruit and vegetable intake among low-income African-American women in urban public health centers. *Preventive Medicine, 41,* 53–62.

134. Williams, J.H., Auslander, W.F., de Groot, M., Robinson, A.D., Houston, C., & Haire-Joshu, D. 2006. Cultural relevancy of a diabetes prevention nutrition program for African American women. *Health Promotion Practice, 7,* 56–67.
135. James, D.C.S. 2004. Factors influencing food choices, dietary intake, and nutrition-related attitudes among African Americans: Application of a culturally sensitive model. *Ethnicity & Health, 9,* 349–367.
136. Patel, C., & Nicol, A. 1997. Adaptation of African-American cultural and food preferences in end-stage renal disease clients. *Advances in Renal Replacement Therapy, 4,* 30–39.
137. Horowitz, C.R., Tuzzio, L., Rojas, M., Monteith, S.A., & Sisk, J.E. 2004. How do urban African Americans and Latinos view the influence of diet on hypertension? *Journal of Health Care for the Poor and Underserved, 15,* 631–644.
138. Beyene, Y. 1992. Medical disclosure and refugees: Telling bad news to Ethiopian patients. *Western Journal of Medicine, 157,* 328–332.

Mexicans and Central Americans

CHAPTER 9

Latinos are the largest non-European ethnic group in the United States, representing 13 percent of the total population. Yet Latinos are not a single cultural group, coming from more than twenty-five Latin American nations with diverse ethnic populations. Though a majority share Spanish as a common language of origin, others speak English, French, Portuguese, or a native Indian dialect as their mother tongue.

Immigrants from Mexico and the countries of Central America bring a rich cultural history (see Figure 9.1). The Olmec culture, known for its sophisticated sculpture, existed in southeastern Mexico as early as 1200 B.C.E. The great Aztec, Mayan,w and Toltec civilizations thrived while Europe was in its dark ages. Their independent mastery of astronomy, architecture, agriculture, and art astonished later explorers. Spanish occupation of Mexico and Central America introduced new ideas and traditions, most notably Roman Catholicism; British, French, and Austrian intrusions also provided minor contributions. The foods of Mexico and Central America reflect the native Indian and European heritage of the region. This chapter examines Mexican cuisine and the food habits of Americans of Mexican descent. An overview of recent immigration from Central America and the traditional fare is also included. The following chapter reviews Latinos from the Caribbean Islands and South America.

Mexicans

Estados Unidos Mexicanos, the United Mexican States, is the northernmost Latin American country. Though Mexico seems logically a part of Central America, it is geographically a part of North America. It is more than one-fourth the size of the United States, with 756,065 square miles of territory. The varied geography includes a large central plateau surrounded by mountains except to the north. Coastal plains edge the country along the Gulf of Mexico and the Pacific. The separate Baja peninsula is found in the west, and the Yucatán peninsula juts out in the southeast. Snow-capped volcanoes, such as Orizaba, Popocatépetl, Ixtacchiuatl, and El Chichón, and frequent earthquakes also affect the landscape. The climate ranges from arid desert in the northern plains to tropical lowlands in the south.

Over 80% of Mexicans are mestizos—that is, of mixed Indian and Spanish ancestry, based on genomics, while European and indigenous ancestry are approximately even. Spanish is the official language. Only 1.5 percent of Mexicans speak a single Indian language, mostly Nahuatl.[1]

The term Latino describes people originally from Mexico, the Caribbean, and Central and South America. It suggests culture of Latin heritage, not exclusively of Spanish background. For example, Haitians and Brazilians are both Latinos but speak French and Portuguese, respectively. Hispanic is preferred by some, though there is no clear definition of this term. It can mean "born in Latin America," "ancestors born in Latin America," "Spanish surnames," or "Spanish speaking."

Cultural Perspective

History of Mexicans in the United States

Immigration Patterns. Mexican immigration patterns have changed over the years since

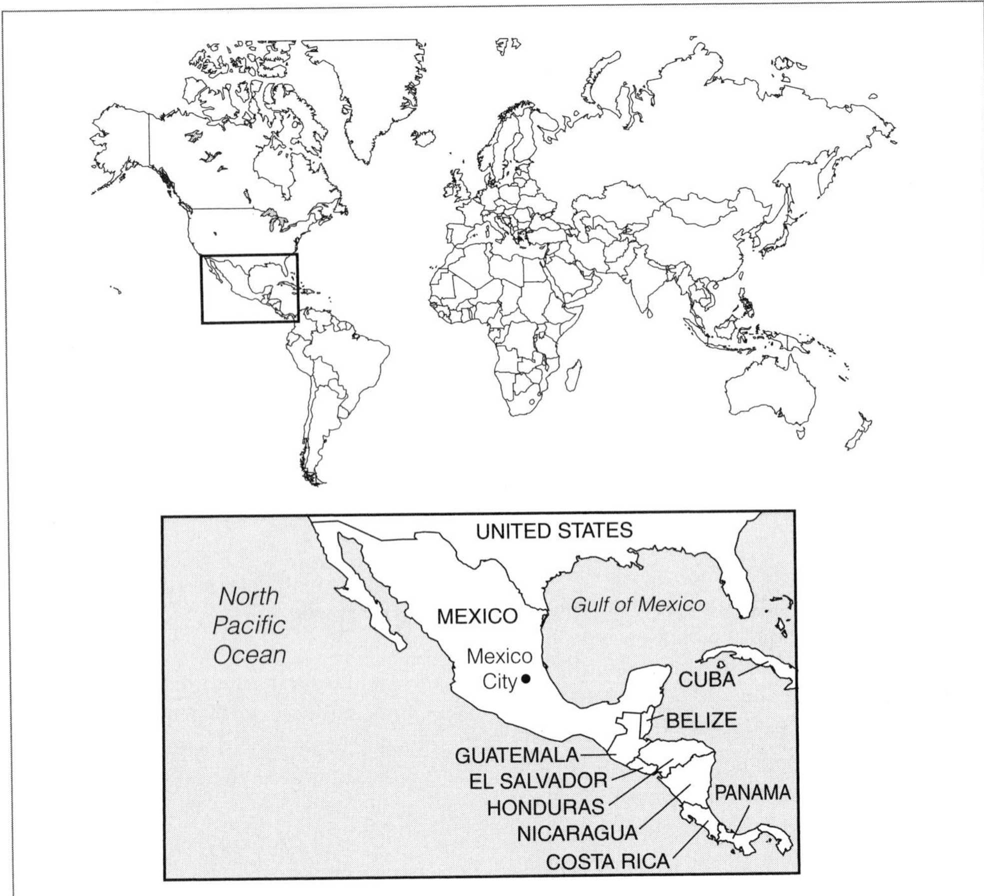

Figure 9.1
Mexico and Central America.

the Mexican-American War clearly defined the U.S.-Mexican border. Today, Mexicans living in America and their descendants can be classified in the following groups: (1) Chicanos—those who are born in the United States (from the descendants of the wealthy Mexican landowners who controlled the area from California to Texas in the eighteenth and early nineteenth centuries to the children of the most recent arrivals), as well as those who immigrated from Mexico and became U.S. citizens; (2) Braceros—those who work here legally but remain Mexican citizens; and (3) unauthorized migrants—those who enter the country illegally. It should be noted, however, that use of identifiers may vary regionally or by generation. For example, many who live in New Mexico prefer the terms Spanish American or Hispano, while in Texas, Tejano is popular. Although many younger Mexican Americans take pride in being known as a Chicano, some older Mexican Americans associate the name with unsophisticated immigrants and may be insulted by use of the term.[2]

Mexicans lived in what is now the American Southwest for hundreds of years before the United States declared its independence in 1776. Although they welcomed American settlers, they soon found themselves outnumbered, and their economic and political control of the region weakened. At the end of the Mexican-American War in 1848, the 75,000 Mexicans living in the ceded territories became U.S. citizens.

Between 1900 and 1935, it is estimated that 10 percent of the Mexican population, approximately 1 million persons, emigrated north to the United States. Then, during the Great Depression, tens of thousands of unauthorized migrants, plus those admitted legally under the 1917 contract labor laws, were repatriated and sent back to Mexico.

After the Great Depression, the need for cheap labor increased. The Bracero program was created to meet this need. Thousands of Mexicans were offered jobs in agriculture and on the railroads. Following World War II, the continued need for farm workers encouraged additional immigration, resulting in a total of over 4.5 million contracts (some individuals obtained multiple contracts over several years) issued to Mexican nationals between 1942 and 1964. In the past thirty-five years, Mexicans have been the largest single group of legal immigrants to the United States.

Current Demographics. Population estimate from 2009 list over 32 million Chicanos and Braceros in the United States (of whom over 11.5 million were born in Mexico), accounting for 59 percent of the Latino population.[3,4] The majority of Chicanos live in California and Texas. However, the settlement pattern for new immigrants has changed since the 1990s, becoming more nationwide. Destination states for Mexican Americans are now Florida, Georgia, Idaho, Minnesota, Nevada, New York, New Jersey, Oregon, and Utah. Eighty percent of the immigrants from Mexico settle in U.S. cities such as San Diego, Los Angeles, San Antonio, and Chicago—and more recently, Dallas, Houston, Las Vegas, Minneapolis, New York City, and Phoenix.[5] Many recent immigrants, legal and unauthorized, settle in the urban Latino neighborhoods called barrios.

Research on unauthorized migrants suggests that between 11 and 12 million were living in the United States in 2006, a majority of whom came from Mexico.[6] Economic pressures in Mexico have increased the number of Mexican immigrants entering the United States each year. Many are attempting to escape the life-threatening poverty affecting many of the populace.

Socioeconomic Status. Mexican Americans occupy three main socioeconomic classes. There are the migrant farm workers, who maintain a culturally isolated community; the residents of the urban barrios, who also are segregated from much of American society; and a growing number of acculturated middle-class Chicanos. Unauthorized migrants tend to move in with family members who already reside in the United States. Usually, this is in a predominantly Latino neighborhood where they can live inconspicuously among the residents while becoming familiar with the American social and economic systems. Specific data on Mexican Americans is limited because they are often grouped with data for all Hispanics.

Despite a rapidly growing Chicano middle to upper class, many Americans of Mexican descent have low socioeconomic status. Median family incomes are 30 percent below the national average, and more than 23 percent of families fall below the poverty level.[3] This is due in part to a disproportionate number of Chicanos employed in unskilled or semiskilled labor. Although only 3 percent of Mexican Americans are currently working in agricultural jobs, more than 50 percent are employed in manufacturing or service occupations. Approximately 16 percent hold professional or managerial positions. Just 55 percent of Mexican Americans graduate from high school—one study found the figure is only 16 percent for adults who were born in Mexico, but rates rise to 80 percent among third-generation Mexican Americans. Yet, this is still lower than for blacks and whites. Less schooling and fewer skills may account for up to three-quarters of the wage gap with whites.[7,8]

Research on unauthorized migrants of all nationalities indicates that over 7.8 million were employed in 2009, accounting for over 5 percent of the civilian workforce.[9] They hold 24 percent of all jobs in farming, 17 percent in cleaning, 14 percent in construction, and 12 percent in food preparation.[6]

Worldview

Chicanos, Braceros, and unauthorized migrants frequently live in culturally homogeneous communities. Many proudly maintain their ethnic identity, speaking Spanish and enjoying Mexican music and food. The concept of *la raza* (meaning "the people") was first promoted in the 1960s as a pride and solidarity movement for all persons of Latin American heritage.

When exposed to the mainstream American society, however, immigrants from Mexico can be highly adaptive. Some Chicanos are completely assimilated. They may speak no Spanish and consider themselves white or simply American. Others are successfully bicultural. Cross-cultural marriages are becoming more common, especially in the northern regions of the United States.

Religion. Approximately 75 percent of Americans of Mexican descent are Roman Catholics

The term *Chicano* came from the Aztec word for Mexicans, *Meshicano*.

The 2001 Canadian Census listed more than 36,000 residents of Mexican heritage and an additional 41,000 of Latin of unspecified nationality.

Some immigrants from Mexico seeking migrant agricultural jobs are Mexican Indians (mostly Mixtecos from Oaxaca), who speak neither English nor Spanish.

When a girl turns fifteen years old, her family hosts a quinceañera, an elaborate coming-out party with music, feasting, and dancing.

(nearly one in every three U.S. Catholics are of Hispanic heritage). Traditional religious ceremonies, such as baptism, communion, confirmation, marriage, and the novenas (nine days of prayer for the deceased) are important family events (see Chapter 4, "Food and Religion").

A strong faith in the will of God influences how many immigrants from Mexico perceive their world. Many believe they have no direct control over their own fate. Nearly all Mexican Americans who are not Catholic practice Protestant faiths. Evangelical churches are particularly popular in urban areas.

The Chicano Family. The most important social unit in the Chicano community is the family. In contrast to American majority society, the well-being of the family comes before the needs of the individual.

The Chicano father (or eldest male relative) is typically the head of the household. He is the primary decision maker and wage earner. Machismo, roughly translated as manhood, is the word for the pride and self-worth a man feels when fulfilling his obligations and duties to his family and community. In traditional Mexican culture the wife is a homemaker, the person who provides all childcare and holds the family together. Women are expected to be subservient to their husbands. In America, however, this role is changing. One-half of Mexican-American women work outside the home, are responsible for household management, and are likely to be involved in family decisions. Men rarely increase involvement in chores as women increase their hours of employment. Some women find that their new roles are occasionally in conflict with their self-concept as mother and caregiver.

Aztec medicine was a highly developed system featuring an elite group of certified practitioners with access to a zoo and a herbarium for research. It was abolished during the Spanish conquest.

Children are cherished in the Chicano family. Families are typically large among new immigrants and first-generation families, but research suggests lower rates of marriage and improvements in educational attainment and socioeconomic status for women of subsequent generations result in the birth of fewer children and smaller families.[10] Children are taught to share and to work together; sibling rivalry is minimal. When possible, an extended family is the preferred living arrangement for many Mexican Americans.

Grandparents are honored and are often involved in child care. Because of space limitations in the United States, however, many Chicano elders live in separate apartments. During periods of hardship, other relatives such as aunts, uncles, and godparents willingly accept the care of children. Girls were traditionally raised differently from boys and were kept at home to learn household skills; they were carefully chaperoned in public. Although such strict supervision diminished with successive generations born in the United States, family expectations may limit the educational and professional attainments of some young women.

Traditional Health Beliefs and Practices. Traditional health care in Mexico includes elements of Indian supernatural rituals combined with European folk medicine introduced from Spain. Beliefs and practices are closely interrelated with the culture, resulting in a health system widely shared throughout Latin America. Most Mexican Americans are familiar with the conditions specific to the culture, and many use the traditional cures.

Health is a gift from God, and illness is almost always due to outside forces (unless one is being punished by God for one's sins). An individual must endure illness as inevitable. Prayer is appropriate for all illness, and beseeching the saints for intervention through the lighting of candles on behalf of a sick person is common. Pilgrimages may be made to religious shrines, especially those devoted to the Virgin Mary or St. Francis.

Health care is traditionally sought from a hierarchy of healers.[11,12] Treatment is first discussed with *señoras* or *abuelas*—mothers, grandmothers, wives, or older female neighbors who are the health experts in each family. Home remedies, especially teas and over-the-counter remedies such as Pepto Bismol, Alka Seltzer, and Vicks® VapoRub®, are nearly always tried first before outside help is sought.[11] Laxatives and enemas are common. If a cure is not found, a *yerbero* (herbalist), *sobador* (massage therapist or occupational therapist), or a *patera* (a midwife who also specializes in the care of small children) may be consulted. Traditional herbal remedies, homeopathic cures, and amulets are available at pharmacies called *botánicas*.[14] Therapeutic items may also be sold at religious fiestas.

When an ailment is unresponsive to these cures, the services of a healer known as a *curandero* (or *curandera* if the healer is female) are sought. *Curanderos* are esteemed members of each community who see patients without an appointment and customarily do not charge for their services, though they may accept gratuities.[12] Their healing powers may be God-given at birth, learned, or received through a calling.[15] *Curanderos* are sought for a broad range of complaints, such as marital problems, infertility, alcoholism, and business failure, as well as for specific illnesses, including diabetes and cancer. They are important in diagnosing the underlying causes of a condition, which may be natural or supernatural in nature. Thus, diabetes in a client may be due to lifestyle and diet, amenable to biomedicine, or it may be due to evil spirits or witchcraft, in which case biomedicine is ineffective.[12] *Curanderos* specialize in somatic ailments and are essential to curing illnesses due to supernatural causes. In regions where witchcraft is practiced, a *curandero* can counteract the hexes or spells of a *brujo* (a person who works on behalf of the devil). Faith is crucial to the success of a *curandero*. Prayer is his or her primary treatment; the lighting of candles or the use of wood or metal effigies formed in the shape of the afflicted body part (called *milagros* or *exvoto*) may also be used. Cleansing rituals are applied in certain conditions.

Illness is believed to be due to (1) excessive emotion, (2) dislocation of organs, (3) magic, or (4) an imbalance in hot or cold; or (5) is considered an Anglo disease, such as pneumonia and appendicitis.[15] Treatment is based on the cause of the disorder.

Susto is an ailment considered due to excessive emotion—such as smoldering anger or shame—usually associated with a specific event. The response may be physical, physiological, or psychological, and symptoms include anxiety, depression or sadness, lack of appetite, paleness, shaking, headaches, bad dreams and too much sleep, and ennui.[16,14,17] A type of *susto* known as *espanto*, which occurs when an individual is so frightened by a ghost that the soul leaves the body, is the most typical form of the disorder. *Susto* is considered a serious condition, associated with other illnesses such as epilepsy and infectious diseases. Mild *susto* is sometimes treated at home with sugar or sugar water. More serious *susto*, particularly when the soul is involved, must be cured by a *curandero* and may require lengthy treatment. *Susto* is also believed to be a cause of *nervios*, a condition affecting primarily adult women (*nervios* may also cause *susto*). Symptoms include crying attacks, sleep problems, headache, trembling, sadness or hopelessness, ill-temper, lack of appetite, stomachache, feeling of choking, chills, itching, and general body ache. *Nervios* responds best to sedatives (provided by a biomedical physician), prayer, or massage. Unlike *susto*, *nervios* cannot be cured by a traditional healer. *Nervios* is often a chronic condition in response to poor diet, alcoholism, drug use, or other underlying health behaviors. Some Mexicans believe that *nervios* can cause diabetes if not cured.

Studies report that *susto* is associated with an increased risk for other illnesses and a higher mortality rate.[165]

Feeling too much rage and suffering from revenge fantasies can result in *bilis*, a condition where excess bile is thought to spill into the blood, causing symptoms such as loss of appetite, vomiting, headaches, nightmares, and inability to urinate. *Envidia* is another ailment taking the form of various illnesses (some terminal) caused by the emotion of envy among one's friends and neighbors. A person's success may be tempered by the misfortune of *envidia*.

A problem caused by the displacement of organs in infants is *caida de la mollera*, or fallen fontanel. It occurs from a fall, yanking the nipple out of a baby's mouth too quickly, or holding a baby vertically when it is too young. The fontanel appears depressed, and the palate is believed to drop, preventing the infant from feeding. The inability to suckle and a change in stools are symptoms, and serious weight loss may occur. Tight caps on the infant can help prevent the condition, and the application of salt poultices or olive oil (followed by a dip in water accompanied by prayers) may be used to treat it. The baby may be held upside down and shaken gently, the hair pulled, the fontanel sucked, or the palate pressed up with a finger or thumb to reposition the fontanel.

Mal de ojo (evil eye) is a condition with supernatural origins. Children are considered especially vulnerable to the ailment, which has flu-like symptoms, including fever and headache. It is caused when one person, usually inadvertently, casts a strong, admiring look on another person. Irrational behavior and mental disabilities are often attributed to *mal de ojo*. This condition can result in death, so prompt diagnosis and cure are imperative. A curandero is required for treatment. A cleansing ritual is performed that includes sweeping over the ill individual with

Photo by Laurie Macfee'

▲ *Traditional Latin American foods. Some typical foods include achiote, avocado, bacalao, black beans, cassava (yucca), chili peppers, cilantro, corn tortillas, jícama, papaya, plantains, pinto beans, pork, tomatillos, and tomatoes.*

an egg, then breaking the egg into a saucer. The egg is read to see whether the cure has been effective. It may be read immediately or left under the bed overnight before examination. Prayers, herb teas, and sweeping with herb bundles may also be part of the treatment. *Mal aire* (bad air or wind) may cause headaches and colds,[14] and *mal puesto* (witchcraft) accounts for certain other disorders, such as swelling, trembling, or paralytic twitching.

Empacho, a digestive ailment characterized by nausea, gas, and weakness, is widely known throughout Latin America. *Empacho* is sometimes classified as an illness due to eating too many hot or cold foods, or a hot-cold imbalance in the stomach due to emotional upset (see Therapeutic Uses of Foods section in this chapter). The direct cause is believed to be a ball or wad of food adhered to the stomach. Herb teas are administered at home, and if they are ineffective, a *curandero* is employed. Treatment consists of prayer, pinching the spine, and stomach massage to restore a proper hot-cold balance.

One researcher has shown that some Mexican dishes reflect Islamic culinary traditions (which arrived via Spain), such as aromatic nut-thickened sauces, fruit pastes, and sugar figurines (see Chapters 6 and 13).[166]

"Chile" (with an e) comes from the Nahuatl word *chilli*. Foods like the powders, sauces, and stews made with chile peppers are conventionally called "chili" (with an i).

The scientific name for the cacao tree is *Theobroma*, meaning "food of the gods."

Traditional Food Habits

Mexicans are very proud of their culinary heritage, which is a unique blend of native and European foods prepared with Indian (mostly Aztec) and Spanish cooking techniques. There are even some French and Viennese influences from the Maximilian reign. The resulting cuisine is both spicy and sophisticated.

Ingredients and Common Foods

Many people associate the cooking of Mexico with chili peppers. Although chili peppers are used frequently, not all Mexican dishes are hot and spicy. Other New World foods such as beans, cocoa (from the Aztec word for "bitter"), corn, and tomatoes add equally important flavors to the cuisine. These indigenous ingredients were the basis of Indian fare throughout Mexico before the arrival of the Spanish.

Aztec Foods. The Aztec empire had an estimated population of 25 million people at its peak in the fifteenth century. About one-quarter of the population was an elite class of nobles supported by the remaining 75 percent of the Indian slave populace. The capital city of Tenochtitlán was surrounded by lakes on which were built *chinampas*, rich agricultural fields of mud scooped from the lake bottoms. These drought-resistant fields are believed to have produced enough food to feed 180,000 people annually. The monarchy stored surplus crops to protect against famine. The Aztecs were also known for their animal husbandry and game protection laws.

Documents from early Spanish expeditions recorded in glowing terms the enormous variety of foods enjoyed by the Aztec nobility.[18] More than 1,000 dishes are described. Montezuma II reportedly ate up to thirty different items per meal, each kept warm on a pottery brazier. These items included roast turkey, quail, and duck; fish, crab, lobster, frog, turtle, newt, and insect dishes garnished with red, green, or yellow chiles; squash blossoms; and sauces of chiles, tomatoes, squash seeds, or green plums.[19] Chocolatl, a hot, unsweetened chocolate drink made from native cacao beans, was the most popular beverage.

Corn was the staple grain. Legumes, fruits, and vegetables were plentiful; turkeys and dogs were domesticated for meat; and some game was available, including deer, peccary, and rabbits. The notable deficiency of the Aztec diet was a consistent source of fat or oil, and the average Indian ate a mostly vegetarian diet of corn and beans.[18]

Spanish Contributions. The Spanish arrived in Mexico with cinnamon, garlic, onions, rice, sugar cane, wheat, and, most importantly, hogs, which added a reliable source of domesticated protein and lard to the native diet. These additions combined with indigenous ingredients produce the classic flavors and foods of Mexican cuisine, such as corn tortillas with pork filling; tomato, chile, and onion sauces or salsas; rice and beans; and pan-fried boiled beans, known as frijoles refritos, or, as they are incorrectly called in English, refried beans. The Spanish also introduced the distillation of alcohol to native Mexican fermented beverages; tequila and mescal were the result.

Staples. The cuisine of Mexico is very diverse, and many inaccessible regions have retained their native diets. Others have held on to traditional foods and food habits despite Aztec or Spanish domination. The diets of still other areas differ because of the availability of local fruits, vegetables, or meats. The majority of poor Mexicans have little variety in their diet; some subsist almost entirely on corn, beans, and squash. This divergence makes it difficult to typify Mexican foods in general (Table 9.1). Nevertheless, some foods are found, in varying forms, throughout Mexico.

Tortillas are the flat bread of Mexico. Traditionally they are made by hand. Corn kernels are heated in lime solution until the skins break and separate. The treated kernels, called *nixtamal*, are then pulverized on a stone slab (*metate*). The resulting flour, *masa harina*, is combined with water to make the tortilla dough. Small balls of the dough are patted into round, flat circles, about six to eight inches across. The tortillas are cooked on a griddle (often with a little lard) until soft or crisp, depending on the recipe.

Beans are ubiquitous in Mexican meals. They are served in some form at nearly every lunch and dinner and are frequently found at breakfast, too. They are often the filling in stuffed foods and are common in side dishes, such as simmered frijoles de olla ("out of the pot") and frijoles refritos.

One-dish meals are typical, almost always served with warm tortillas. Hearty soups or stews called caldos are favorite family dinner entrées. Casseroles known as sopas-secas, using stale tortilla pieces, rice, or macaroni, are eaten as main dishes. Stale tortillas can also be broken up and softened in a sauce to make chilaquiles, which are served as a side dish or light entrée. They can also be soaked in milk overnight, and then pureed to make a thick dough. This dough is used to prepare gordos, which are fat-fried cakes, or bolitos, which are added to soup and are similar to dumplings.

San Antonio Light Collection, UTSA's Institute of Texan Cultures, No. L-1535

▲ *Handmade tortillas, made from* **masa harina** *(a type of cornmeal) or wheat flour, are the staple bread of Mexico. In this photograph, taken in a tortilla factory, the lime-soaked corn kernels are pulverized on a stone* **metate**. *On the right, a tortilla is patted into a flat circle by hand.*

Meats are normally prepared over high heat. They are typically grilled, as in carne asada (beef strips), or fried, as in chicharrónes (fried pork rind). Slow, moist cooking (stewing, braising, etc.) may also be used. These techniques help to tenderize the tough cuts that are generally available, as does marinating, another common preparation method. Nearly all parts of the animal are used, including the variety cuts and organs. Sausage, such as spicy pork or beef chorizo, is especially popular.

Mexico is famous for its stuffed foods, such as tacos, flautas, enchiladas, tamales, quesadillas, and burritos. These are found throughout the country, with regional variations. Tacos are the Mexican equivalent of sandwiches. Tortillas, either soft or crisply fried, are filled with anything from just salsa to meat, vegetables, and

Corn is believed to have been domesticated from extinct wild varieties in southern Mexico somewhere between 8000 and 7000 B.C.E. It spread south into Central America and north into what is now the United States.

Table 9.1 Cultural Food Groups: Mexican

Group	Comments	Common Foods	Adaptations in the United States
Protein Foods			
Milk/milk products	Few dairy products are used (incidence of lactose intolerance estimated at two-thirds of the population). Dairy products are used more in northern Mexico than in other regions.	Milk (cow, goat), evaporated milk, *café con lèche*, hot chocolate; *atole*; cheese	Aged cheese is used in place of fresh cheese; more milk (usually whole) is consumed; ice cream is popular.
Meat/poultry/fish/eggs/legumes	Vegetable protein is the primary source for majority of rural and urban poor. Pork, goat, poultry are common meats. Beef is preferred in northern areas and seafood in coastal regions. Meat is usually tough, prepared by marinating, chopping, grinding (sausages are popular), or slicing thinly. It is cooked by grilling, frying, stewing, or steaming, and is usually mixed with vegetables and cereals.	*Meats:* beef, goat, pork (including *chicharrónes* and variety cuts) *Poultry:* chicken, turkey *Fish and seafood: camarónes* (shrimp), *huachinango* (red snapper), other firm-fleshed fish *Eggs:* chicken *Legumes:* black beans, chickpeas (garbanzo beans), kidney beans, pinto beans	Traditional entrées remain popular. Fewer variety cuts are used. Protein intake may decline in second-generation Mexican Americans. Beans are eaten less frequently.
Cereals/Grains	Corn and rice products are used throughout the country; wheat products are more common in the north. Principal bread is tortilla; European-style breads and rolls are also popular.	Corn (*masa harina, pozole*, tortillas); wheat (breads, rolls, *pan dulce*, pasta); rice	Wheat tortillas are used more than corn tortillas; convenience breads are used. Increased consumption of baked sweets, such as doughnuts, cake, and cookies, is noted. Increased consumption has occurred of sugared breakfast cereals.
Fruits/Vegetables	Vegetables are usually served as part of a dish, not separately. Semitropical and tropical fruits are popular in most regions (limited availability in north).	*Fruits:* bananas, *carambola, casimiroa, cherimoya*, coconut, custard apple, *granadilla* (passion fruit), *guanábana*, guava, lemons, limes, *mamey*, mangoes, melon, oranges, papaya, pineapple, strawberries, sugar cane, sweet sop, *tuna* (cactus fruit), *zapote* *Vegetables:* avocados, cactus (*nopales* or *nopalitos*), *calabaza criolla* (green pumpkin), chilechili peppers, corn, *jícama*, lettuce, onions, peas, plantains, potatoes, squashes (*chayote*, pumpkin, summer, etc.), squash blossoms, sweet potatoes, *tomatillos*, tomatoes, yams, *yuca* (cassava)	Fruit remains popular as dessert and snack item; apples and grapes are accepted after familiarization.
Additional Foods			
Seasonings	Food is often heavily spiced; 92 varieties of chiles are used. Regional sauces are typical.	Anise, *achiote* (annatto), chiles, cilantro (coriander leaves), cinnamon, cocoa, cumin, *epazote*, garlic, *hoja santa*, mace, onions, vanilla	Use of spices depends on availability.
Nuts/seeds	Seeds are often used in flavoring.	*Piñons* (pine nuts), *pepitas* (pumpkin seeds), sesame seeds	
Beverages		*Atole*, beer, coffee (*café con lèche*), hot chocolate, soft drinks, *pulque*, mescal, tequila, whiskey, wine	Noted are increased consumption of fruit juices, Kool-aid, soft drinks, and beverages with caffeine; and decrease in use of hard spirits.
Fats/oils	Traditional diet is relatively low in fat.	Butter, *manteca* (lard)	Fat intake increases, including use of mayonnaise and salad dressings.
Sweeteners	Spanish-influenced pastries, candies, and custards/puddings are popular.	Sugar, *panocha* (raw brown cane sugar)	

sauce. Flautas ("flutes") are a variation on the taco, with tortillas tightly rolled around the filling, then fried until crispy. They may be served with a red or green chile sauce or guacamole. Enchiladas are tortillas softened in lard or sauce and then filled with meat, poultry, seafood, cheese, or egg mixtures. The tortilla rolls are then baked covered with sauce. Tamales are one of the oldest Mexican foods, dating back at least to the Aztec period. Dough made with either *masa harina* or leftover pozole (hominy) is placed in corn husks (in the north) or young leaves of avocados or bananas (in the south). The leaves are folded and then baked in hot ashes or steamed over boiling water. The tamale may be plain, filled with a meat or vegetable mixture, or sweetened for a dessert (tamales dulce). After cooking, the husk or leaf is unfolded, revealing the aromatic tamale. Quesadillas are tortillas filled with a little cheese, leftover meat, sausage, or vegetable, then folded in half and heated or crisply fried. Burritos are popular in northern Mexico. They are similar to tacos, but large, thin, wheat flour tortillas—instead of corn tortillas—are folded around a filling such as beans with salsa.

© Photodisc/PhotoLink/Getty Images

A large variety of chili peppers are used in Mexican cuisine, providing complexity of flavor, heat (very mild to incendiary), and color (yellow, orange, red, green, brown, and black).

Vegetables are usually part of the main dish or served as a substantial garnish. Potatoes, greens, tomatoes, and onions are most common. Chili peppers are used extensively in seasonings and sauces, and are stuffed, as in chiles relleños and the Independence Day dish chiles en nogada, garnished with the colors of the Mexican flag—white sauce, green cilantro, and red pomegranate seeds.

Sugar cane grows well in Mexico, and sweets of all kinds are popular. Dried fruits and vegetables, candied fruits and vegetables, and sugared fruit or nut pastes are eaten alone and used in more complex desserts. The Spanish make many desserts with eggs, and some of these recipes have been adopted in Mexico. Flan, a sweetened egg custard topped with caramelized sugar, is the most common. Huevos reales is another popular dessert, made with egg yolks, sugar, sherry, cinnamon, pine nuts, and raisins. Dulce de leche is made by boiling condensed milk down until the sugars caramelize and it thickens. It is used as a spread or hardened to make candy. A similar, distinctive, Mexican specialty is cajeta, made like dulce de leche but using goat's milk flavored with cinnamon. It is eaten by itself as a pudding or used to top fresh fruit or ice cream.

The most common beverage in Mexico is coffee, which is grown in the south. Soft drinks and fresh fruit blended with water and sugar, called aguas naturales, are also favored. Adults drink milk infrequently, except in sweetened, flavored beverages such as hot chocolate with cinnamon, or café con leche (coffee with milk). The most popular alcoholic beverage in Mexico is beer. The Mexican wine industry is also developing rapidly, in part due to a 1982 ban on the import of foreign wine. Men may drink alcoholic beverages at occasions when they gather socially. In addition to tequila and mescal, whiskey is typically served at these times.

Epazote, a pungent herb with minty overtones, is added to many dishes, especially those with beans; because it is thought to reduce flatulence.

Regional Variations

Mexican Plains. The northern and central regions of Mexico, nearly half the nation, consist of mostly arid plains and high mountain valleys. The Indians who originally inhabited the area were called *Chichimecs* ("sons of the dog") by the Aztecs because of their seminomadic lifestyle. Their diet probably consisted of corn, beans, squash, greens, cactus fruit (*tuna*), and young cactus leaves (*nopales*). They also hunted small game and ate domesticated poultry, such as turkey. When the Mexican Indians mixed with

Menudo is a tripe and hominy soup believed to have curative properties, particularly for hangovers. It is a popular weekend breakfast dish.

Placement of a worm from the maguey plant in a bottle of mescal or tequila—supposedly a signature of authenticity—was actually begun as a marketing gimmick.

the Indians of the American Southwest, they adopted *piñons* (also called pine nuts or pignolis), pumpkin, and plums. Some specialties of the region, such as hominy-based stews known as pozoles and salads made with sliced, cooked *nopales*, reflect this history. Due to the limited variety of foods available, traditional preparations emphasize the natural flavors of the ingredients; and sauces, when used, feature simple seasoning.

The Spanish introduced longhorn cattle to the northern plains, as well as dairy cows and wheat. It is the only area in Mexico where beef is frequently consumed, often served as steaks or in stews. Another favorite way to prepare beef is to air-dry it in thin slices called cecinas, which are similar to American chipped beef but more assertive in flavor. Cecinas can be used in stews or soups; they can be fried or used as fillings for other foods. In Chihuahua, beef is shredded and then fried to make a snack known as móchomos. Goat and sheep are also raised. Spit-roasted kid, cabrito al pastor, is served in the Monterrey area, and birria—kid or lamb stewed in a sauce flavored with roasted chiles and vinegar—is a specialty in Guadalajara. This area is also known for *barbacoa*, a method of pit-roasting meats that are first wrapped in maguey leaves (see below). Cow cheeks (or more traditionally, the whole head) and kid are commonly prepared this way in the more northern sections, while in the more central regions lamb is preferred.[20]

In the Baja Peninsula and along the Gulf of California and Gulf of Mexico coasts, fish is important in the diet. *Huachinango*, red snapper, is a specialty, cooked in orange sauce, used as a filling in tacos, or served chilled, marinated in vinegar with onions and chiles. Shark is found on the eastern coast, shredded and layered between tortillas and beans in the dish from Campeche known as pan de cazón.[21] Shrimp and clams are other favorites. In the more inland areas, fish from freshwater lakes are available, such as *pescado blanco*, a small whitefish that is popularly served fried or added to soups.

Cheese (see Table 9.2) is more common in the north than in other parts of Mexico, and one specialty is queso flameado (called queso fundido in the Guadalajara area), which is a fondue-like dish sometimes topped with chorizo crumbles and served with fresh tortillas and salsa. Wheat products are more popular as well, particularly wheat tortillas.

A very popular dessert of northern Mexico is buñuelos, which may be made at home or purchased at street stands. Circles of sweet pastry dough are fried until they slightly puff. They are eaten fresh, sprinkled with sugar and cinnamon, or broken up and added to hot cinnamon-laced syrup. Café con leche is a common accompaniment to buñuelos.

The sap of the maguey cactus (century plant) is credited with being a reliable substitute for fresh water in the arid countryside; it is called *aguamiel* or "honey water." Tequila is probably the best-known beverage of the region, made from the distillation of fermented *aguamiel*, known as pulque. Pulque is the sour, mildly alcoholic beverage that was drunk throughout Mexico before the arrival of the Spaniards. The Spaniards, lacking grain, tried their distillation methods on pulque, producing mescal, a harsh brew. Tequila is the more refined, twice-distilled version of mescal. It is produced in the central-western state of Jalisco around the towns of Tequila and Tepatitlan from the maguey subspecies *Agave tequiliana*.

Tropical Mexico. The southern coastal areas of eastern Mexico include hot lowlands and tropical forests. Seafood and freshwater fish are prominent in the cuisine. Red snapper with a Spanish-influenced sauce of tomatoes, garlic, onions, olives, capers, and chiles, called Huachinango a la Veracruzana ("from Veracruz"), is a specialty. Arroz a la Tumbada, rice cooked and seasoned with tomatoes and garlic, topped with fresh fish, shrimp, octopus, crabs, and clams, is another favorite. It is often served in individual clay pots. Tamales and tostadas stuffed with shrimp are also popular. One unusual food enjoyed in some areas is black iguana.[22]

Tomatoes, green tomato-like tomatillos, chayote squash, onions, jícama (a sweet, crispy root), bananas and starchy plantains, carambola (star fruit), cherimoya (custard apple), guanábana (soursop), guava, mamey (a type of plum), mango, pineapple, yucca (a tuber also called cassava or manioc), and zapote (the fruit of the sapodilla tree) are just a small sampling of the produce available in this area. Avocados are also cultivated. They vary in size from two to eight inches across, in skin color from light green to black, and in flavor from bland to bitter. Their succulent, smooth flesh is added to soups, stews, and salads.

Table 9.2 Traditional Mexican Cheeses

Fresh: Unripened cheeses that do not melt, but soften when heated. Often used in stuffed foods and as a garnish on top of dishes.	
Queso blanco	Crumbly cow's milk cheese similar to a soft, dry, fresh mozzarella. Sometimes made by using lemon juice to coagulate the milk, providing a distinctive flavor. Sold commercially in blocks.
Queso fresca	A fine-grained, creamy farmer's cheese or pot cheese. Sometimes called *queso metate* when formed in a traditional stone grinder. Often made daily in homes; sold commercially in small rounds.
Queso panela	A salty, semi-soft white cheese formed in small baskets that leave an imprint. Is often served cubed or sliced with chorizo sausage or guava paste.
Ranchero seco	Drier version of *queso fresco* with much stronger flavor (similar to Romano); curds are broken apart and remolded during processing. Used grated over foods.
Requesón	Milky, ricotta-like cheese good for fillings, spreads, and desserts.
Soft: Smooth, aged cheeses that melt well when heated, often used for baked dishes.	
Queso añejo	Aged *queso fresco* that is firm and salty, similar to feta.
Queso asadero	Buttery, mild cheese that tastes like Provolone but with the slightly stringy texture of aged Mozzarella; used in *queso flameado*.
Queso Chihuahua	Mild, cheddar-like yellow cheese formed in large wheels, introduced by Mennonites in the farming communities of Chihuahua.
Queso Oaxaca	Tangy string-cheese from the city of Oaxaca. Sold in balls
Firm: Semi-hard or hard aged cheeses, often eaten thinly sliced or added for flavoring to dishes.	
Queso añejo enchilada	Strong, salty cheese made from longer aged *añejo*, coated with spicy ground chiles.
Queso Cotija	Sharp goat's milk cheese that is crumbled over beans or salads, sometimes called "Mexican Parmesan." From the Michoacan city of Cotija.
Queso Manchego	Spanish-style, full-bodied, hard cheese that is served shaved with fruit or eaten with snacks, beer, or other drinks.

They are most popular in guacamole—mashed avocado with onions, tomatoes, chili peppers or ground chiles, and cilantro, the pungent leaf of the coriander plant. Guacamole is used as a side dish, a topping, or a filling for tortillas.

More than ninety varieties of chili peppers are found in the region, varying enormously in degree of hotness (see Figure 9.2). In general, the smaller the chile the hotter it is, and each variety develops more heat as it ripens. The heat also intensifies when it is dried. Some of the more common varieties include ancho (dried, red-ripened poblano chiles), chilaca (thin, ribbed, dark green chiles which are often cut into very thin strips after roasting), chipotle (dried, smoked jalapeño chiles), de arbol (long, very thin, curved green or red chile used fresh and dried), guajillo (long, thin, dark red chile used primarily dried), habañero (lantern-shaped, yellow, orange, or dark red, added fresh to sauces but removed before serving, or roasted and chopped into dishes), pasilla (dried chilaca chiles), jalapeño (short, three-inch chiles with smooth green or red skin and blunt end, used fresh in salsas and sauces, or sliced and pickled), mulato (dried chile with very dark brown, wrinkled skin, and sweet overtones, often used to make moles—see below), pequin (tiny, oval chiles used whole in stews, added pureed in sauces, soups, salsas, or pickled as a table condiment), poblano (large, heart-shaped, black-green chiles that are often stuffed), and serrano (small, two-inch dark green or red, torpedo-shaped chiles that are often chopped to add zest to salsa and cooked dishes or sliced and pickled, known as escabeche).

Yucatán. The cuisine of the Yucatán peninsula reflects its unique history. It was isolated from the rest of the country by dense, mountainous jungles until modern times. Many of the residents are descendants of the Mayans, the early Indian dynasty of the region. Some regional favorites date from this time. For example, one popular

Avocado comes from the Nahuatl word for the fruit, *ahuactl*, meaning "testicle," which avocados resemble while hanging from the tree.

The chemical heat of chiles comes from the alkaloid capsaicin, found mostly in the fleshy ribs and seeds inside the fruit. One theory as to why chile-eating is pleasurable is that capsaicin may cause the body to release pain-killing endorphins in reaction to the irritation, creating a comfortable, gratified feeling.

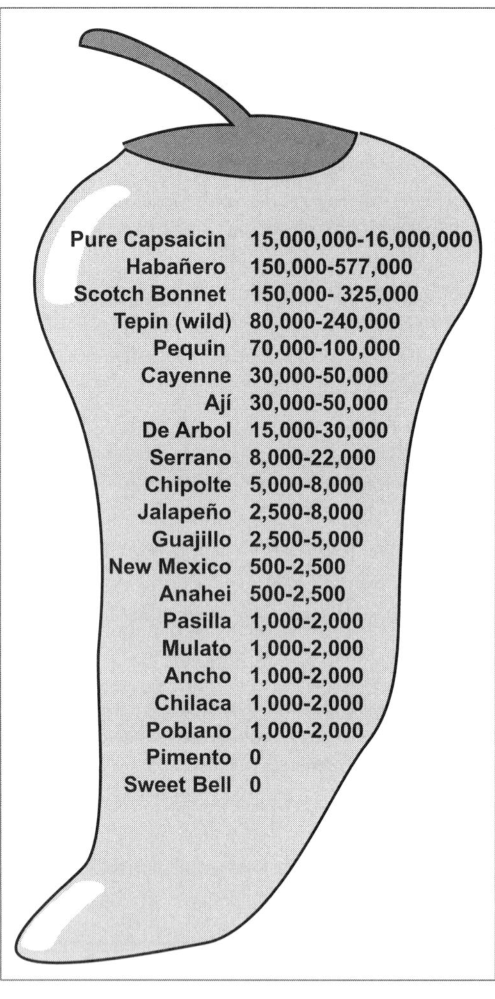

Figure 9.2
Scoville Scale of Hotness for Selected Chili Peppers: Scoville units, defined as the amount of sugar water needed to dilute chili pungency to the point it no longer burns the mouth, is one measure of chili hotness. It is a subjective scale, and hotness may vary within each variety due to seed lineage, soil, and climate, as well as taster individuality. High Pressure Liquid Chromotography (HPLC) is another measure, but does not account for differences in capsaicin release properties, which also affect chili heat.

Achiote is called annatto in the United States. Made from the seeds of a tropical tree, achiote is sometimes used to color cheddar-style cheeses, ice creams, margarine, and some baked goods.

preparation method is to steam foods wrapped in banana leaves, called píbil. Traditionally, food was cooked this way in an outdoor pit, but today it is prepared more often in a covered pot. Salbutes, small corn tortillas (some made with black bean paste so that they are black in color), are often fried until they puff, then layered with lettuce or cabbage, tomato, onion, bell pepper, and píbil-cooked meats. In some versions the black beans are stuffed inside the pocket of the puffed tortilla, in which case the treat is known as panucho.

Citrus fruits flavor some dishes of the region. For example, in the recipe for chicken píbil, the poultry is marinated in sour orange juice, garlic, and cumin before steaming. A popular sliced pork dish, poc chuc, also uses sour orange juice to tenderize the meat before grilling. Bitter lime juice is featured in a specialty of the area called sopa de lima, a chicken and vegetable soup made with bits of fried tortillas. Lime juice is also found in frijoles con puerco, a traditional dish of black beans with pork served weekly in most homes of the region. Some foods are pickled in citrus juice, including vegetables such as onions, and other items, including fish and oysters. Achiote is the other hallmark seasoning of the region, a bright red, nutty-flavored seed mixed with sour orange juice and mild spices to make a flavorful paste called recado colorado, used to coat foods before cooking.

Shrimp are a local specialty; the long coastline of the Yucatán along the Gulf of Mexico provides ample seafood. Grouper with recado baked in banana leaves is one example, and fish soups are also common. Eggs are prominent, served with tortillas, black beans, cheese, tomato sauce, and fried plantains, or wrapped in tortillas and served with a pumpkin seed sauce. Sauces of the region are often thickened with toasted squash seeds.

Southern Mexico. The foods of southern Mexico are similar to those of the Yucatán in that they are more tropical and more Indian-influenced than the foods of other regions. Cacao trees are cultivated in this area, and chocolate flavors both savory and sweet dishes. In particular, the renowned sauces of the region, moles (probably from the Aztec word for "sauce with chiles," *molli*), sometimes include unsweetened chocolate. Chiles, tomatoes, nuts, raisins, sesame seeds, avocado leaves, and seasonings are other typical ingredients. The complex, spicy sauces are the base for thick stews with added pork, beef, poultry, or game, though some recipes call for the mole to be poured over the top of the meat. In Oaxaca mole manchamanteles (meaning "stew that stains the tablecloth") is a deep-red sauce with yams, pineapple, plantains, and chicken or pork. Mole negro, another Oaxacan specialty, includes dark roasted chiles, blackened tortillas, and chicken. The best-known mole in Puebla is poblano de guajolote, a rich brown sauce served with turkey. Other variations include mole amarillo (an orange-colored sauce) and mole coloradito (a brick-red version also known as mole roja). Mole verde is yet another example, popular along the coast, made with green chiles and tomatillos. Hoja santa (Piper sanctum), an herb with a peppery, anise flavor, seasons the sauce—the large leaves of the herb are also used to wrap steamed and grilled foods in the region.

Poultry, goat, and pork are the most popular meats of the region. One favorite is pork cut into thin strips and coated with ground chiles to make cecina enchilada (a version of the northern recipe above). It is frequently served as a topping on crunchy, platter-sized baked tortillas called tlayuda, which also include layers of black beans (more popular in the south than pintos), cabbage, salsa, tesajo (thinly sliced beef), asiento (bits of pork skin fried in lard), chorizo, and cheese. Game, such as venison and quail, are eaten in some areas. One delicacy of the region is chapulines, a grasshopper found in the corn fields. They are pan-fried with chiles, garlic, salt, and lemon juice and traditionally served with beer or mescal.[23]

Hot chocolate, coffee, atole (a warm beverage of thin cornmeal and milk gruel) and horchata, a sweetened rice-based drink, are favorite beverages. Ice creams, including those made with fresh vanilla bean, and fruit ices are popular, as is chocolate flan.

Meal Composition and Cycle

Daily Patterns. In families where income is not limited, the preferred meal pattern is four to five daily meals: *desayuno* (breakfast), *almuerzo* (coffee break), *comida* or *almuerzo* (lunch), *merienda* (late afternoon snack), and *cena* (dinner). Most meals are eaten at home and served family-style. If there are too many people to sit at the table, each one is served individually from the stove.

Desayuno is a quick, early breakfast, which features pan dulce (sweet bread, pastry, or cake) or fresh fruit, served with café con lèche. Late morning is when *almuerzo* (similar to brunch) is eaten, often including tortillas, eggs, meat, beans left over from the previous night, bolillos (wheat rolls), pan dulce, and fruit. Coffee and hot chocolate are the preferred beverages.

Comida is traditionally the largest meal of the day, traditionally eaten in the early afternoon. A complete *comida* includes several courses, individually served. Customarily this would include a soup, a sopa-seca (including items that would be considered side dishes in the United States, such as seasoned rice), a main course, beans, salad, and dessert. Today, the courses may be combined or, often, fewer are served. For example, soup, a sopa-seca, or a vegetable dish may precede the entrée. When possible, an afternoon rest period (*siesta*) follows this meal. *Merienda* is a light meal of sweet rolls, cake, or cookies eaten around 6:00 p.m. Coffee, hot chocolate, or atole accompanies the sweets. *Cena*, a light supper (often leftovers), follows between 8:00 and 10:00 p.m. This meal may be skipped entirely or expanded into a substantial feast on holidays or other formal occasions. Recently, many Mexicans have adopted the American habit of eating a light lunch (sometimes called *almuerzo*) and a heavy supper, eliminating *merienda* altogether.

Snacking is frequent in urban Mexico; munching occurs from morning to midnight. Antojitos, or "little whims," include foods made with masa, such as tostadas (called chalupas in northern Mexico), fried tortillas topped with shredded lettuce, cheese, or meat. These may be eaten as a snack or served for a light supper. Nearly every block offers street-side food vendors, providing everything from fresh fruits to grilled meats. *Pastelerías* offer coffee, chocolates, and pastries. In addition, many neighborhoods feature an open-air market that also sells ready-to-eat foods. Cantinas are popular gathering places. Originally males-only drinking establishments, many now welcome women and families. Botanas, roughly translated as "cocktail foods," are small plates of items that go well with alcoholic beverages (especially beer), such as cheese, sausages, fritters, tortillas with beans or salsa, and sometimes Anglicized Mexican dishes such as guacamole with tortilla chips, or other non-Mexican items, such as hummus. The term is often used for appetizers of any kind.

Special Occasions. Sundays, family celebrations (such as weddings, baptisms, and *quinceañeras*), and holidays are typically times when more difficult preparations are served. For example, homemade tamales or moles are labor intensive and may require numerous ingredients, so these foods are often reserved for special occasions.[7] Turkey, arroz con leche (rice pudding), and pastel de tres leches, a rich sponge cake lightly soaked in three types of milk (whole or cream, condensed milk, and evaporated milk) and then topped with whipped cream, are other festive foods eaten throughout the year.

Many foods are associated with specific holidays. For example, Día de los Santos Reyes (also called Día de los Reyes Magos, or Three Kings Day) on January 6 is customarily celebrated with rosca de reyes, a raisin-studded, ring-shaped loaf

[SAMPLE MENU]

A Traditional Oaxacan Comida

***Chilaquiles* (Tortilla Casserole)[a,b]**

Chicken or Pork *Coloradito* (*Mole Rojo*)[a,b,c]

Simmered Black Beans[a,b] and Rice

Fresh Tropical Fruit or Fruit Ice/Sherbet[a,b]

Fruit Juice or Beer

[a]Martinez, Z. 1997. *The food and life of Oaxaca: Traditional recipes from Mexico's heart.* New York: Macmillan.

[b]*Oaxaca's Tourist Guide (Cultural Attractions): Recipes* at http://www.oaxaca-travel.com/guide/index.php?lang=us

[c]Suite 101.com at http://www.suite101.com/content/mole-sauce-a16706

of bread. Baked inside the bread is a figurine of the infant Jesus, and the person who receives it is obligated to give a party on *Candelaria* Day (February 2). *Candelaria* Day includes a mass, followed by games and sweets such as tamales dulce and pink-tinted atole.[24]

During Lent, capirotada is a traditional dessert. Many families have their own recipe for this holiday bread pudding, often made with honey or brown sugar, cinnamon, nuts, raisins, and cheese. Another holiday food is bread decorated with a skull and crossbones (pan de muerto) eaten on *Días de los Muertos* (Days of the Dead, November 1 and 2) as part of a large feast honoring the deceased (the first day for children, the second for adults—additional days, for deaths due to certain causes, such as accidents, are identified in some regions). Sugar paste is used to mold skulls and skeletons of the dead, known as *alfeñique* (a word of Arab origin), which are colorfully decorated with icing. Altars for the dead are set up in homes with a bowl of water to quench the thirst of the spirit, his or her favorite foods, pan de muerto, sugar skulls, fruit pastes, and dulce de calabaza (pumpkin cooked with brown sugar). Coffee, chocolate, atole, soft drinks, and preferred alcoholic beverages are offered as well.[25] After the departed soul has absorbed the essence of the meal through the aromas, the family eats the remaining food.

Christmas festivities, called *posadas*, frequently feature *piñatas*, or brightly decorated papier-mâché animals and figures that are filled with sweets. Blindfolded children take turns swinging a large stick at the hanging *piñata* until it breaks and candies fly everywhere. In some regions buñuelos are a Christmastime treat, drenched in syrup and served in pottery bowls that, when empty, are smashed on the street for good luck. On Christmas Eve a salad of fruits, nuts, and beets is served.

Role of Food and Etiquette in Mexican Society

In family-centered Mexican society, food-related activities facilitate interactions between family members and help delineate family roles. Meal planning is usually the wife's responsibility. Depending on economic status, food is prepared by the wife or by servants supervised by her, because Mexican foods can be laborious to prepare.[19] The final dishes are greatly appreciated by all who partake in the meal, and it is considered an insult not to eat everything that is served. In rural areas, food sharing is an important social activity, reflecting the Indian worldview. To reject offered food or drink is a severe breach of social conduct. Even refusal of an invitation to dine may be considered rude, though not attending an agreed upon event is often completely acceptable.[26]

No guest should begin eating at a Mexican meal until the host says "*¡Buen provecho!*"[26] Mexicans share many dining rules with Europeans. The fork remains in the left hand and the knife in the right one. No switching is done when cutting food. When not eating, the hands should remain above the table, with the wrists resting on the edge. Dishes are passed to the left. Portions are usually large, and it is all right to leave some food on your plate. Leaving the table for any reason before others are finished with the meal is impolite.

Therapeutic Uses of Food

Some Mexican Indians and rural poor practice a hot-cold system of diet and health. It is believed to have derived from the Arab system of humoral medicine brought to Mexico by the Spanish, combined with the native Indian worldview. Although it has parallels with other classification systems, such as the Asian practice of yin-yang (see Chapter 11, "East Asians," for details), the Mexican system is only applied to foods and to the prevention and treatment of illness. It does not encompass moral or social beliefs.

The Mexican hot-cold theory is based on the concept that the world's resources are limited and must remain in balance. People must stay in harmony with the environment. Hot has the connotation of strength; cold, of weakness. When the theory is applied to foods, items can be classified according to proximity to the sun, method of preparation, or how the food is thought to affect the body. Meals balanced between hot and cold foods are considered to be health promoting. Unbalanced meals may cause illness. Thus a typical *comida* in a rural village would consist of rice (hot), soup (made with hot and cold ingredients), and beans (cold).

Although the hot-cold classification of foods does vary, items generally considered hot are alcohol, aromatic beverages, beef, chiles, corn husks, oils, onions, pork, radishes, and tamales. Cold foods include citrus fruits, dairy products, most fresh vegetables, goat, and tropical fruits. Some foods, such as beans, corn products, rice products, sugary foods, and wheat products, can be classified as either hot or cold depending on how they are prepared.

Illnesses are also believed to be hot or cold and are usually treated with a diet rich in foods of the opposite classification. Examples of hot conditions include pregnancy, hypertension, diabetes, indigestion, *susto*, *bilis*, and *mal de ojo*.[11] In particular, many Mexican women increase their intake of cooling fruits, such as melons, mangoes, and bananas, and avoid hot, spicy foods and chiles during pregnancy. Some also believe that very cold foods, including cucumbers, tomatoes, and watermelon, can create a sudden imbalance. Examples of cold conditions are pneumonia, colic, and *empacho*. Sour foods are thought by some to thin the blood and are avoided by menstruating women because they are thought to increase blood flow; acidic foods may also be avoided because they are said to cause menstrual cramps (menstruation is considered a hot condition by some, and a cold condition by others).

Though the hot-cold system of food classification is practiced by small numbers of Mexicans, one study reported that Latinos are more likely than non-Latinos to consider certain foods herbal medicines, and other research suggests home remedies are common.[27,28,29] For example, chamomile is believed by many to cure colic, menstrual cramps, anxiety, insomnia, and itching eyes. Mint and anise tea are also prepared for nausea, gas, diarrhea, and colic. Garlic is chewed for yeast infections in the mouth, toothache pain, and stomach disorders; boiled peanut broth is used to cure diarrhea; boiled corn silk is taken for kidney pain; honey and water are given to infants for colic; oregano is used for fever, dry cough, asthma, and amenorrhea; and papaya is thought to help cure digestive ailments, asthma, tuberculosis, and intestinal parasites.[11,15]

Of particular interest are remedies for hypertension and diabetes. Hypertension may be treated with garlic, passion flower, or linden flowers. The leaves of the sapodilla tree, known as zapote blanco, which act as a strong sedative, are also used in a tea to lower blood pressure.[28,11] For diabetes, several botanical remedies are used. Sage tea is common, as are infusions made from tronadora root (trumpet flower) and prodigiosa leaves and flowers (bricklebush). Preparations made with Indian plantains (matarique), papayas, bitter gourds, aloe vera juice, and prickly pear cactus (both the tuna and nopales) are also popular.[30,28] Several of these remedies have been shown to have potent diuretic or hypoglycemic properties.

Early researchers were perplexed by the absence of the niacin-deficiency disease, pellagra, in Mexicans who consumed a corn-based diet. Pellagra was common in the southern United States, where corn was also a staple. It was found that when the corn kernels were prepared for *masa harina*, the alkaline lime solution used to soften them released the niacin that was bound to a protein.[167]

Some Mexicans avoid cold air and drafts after eating chilis (which are classified as hot) to avoid causing a sudden imbalance in their bodies.

Contemporary Food Habits in the United States

The foods of Mexico have significantly influenced cooking in regions of the United States bordering the nation. Four separate regional variations have been identified (see Chapter 15, "Regional Americans," on regional U.S. foods for more information).[31] The first region is Texas, where Mexican food has often been modified into completely American dishes, such as tamale pie and nachos. Other foods retain slightly more of their

In the late nineteenth and early twentieth century, chile con carne stands were common in San Antonio, offering a spicier version of the Texas stew made by Latina women in the community, who were known as the "Chile Queens." They were shut down in the 1930s due to supposed health concerns.[31]

American chefs, such as Stephen Pyles and Mark Miller, are creating gourmet fusion Mexican dishes, such as paté de foie gras stuffed tamales, and apple-red chili chutney rolled pork. This style is also now found in the upscale restaurants of Mexico City as well.[31]

Mexican heritage, such as chili con carne, which was developed just after the Mexican-American War in the 1850s. It probably began as a chile colorado (red chile stew), but was tamed by reducing the spicing and diluting the traditional meat dish with beans. Barbecued chile-spiced meat kebobs called anacuchos and capriotada with whiskey sauce known as "drunken pudding" are other examples of Tex-Mex creativity. The most commonly used chiles in Tex-Mex cooking are anchos and jalapeños, though pequins are also popular in some dishes, and a favorite seasoning is cumin. Beef is commonly ground rather than shredded for stuffed dishes.[32] The Mexican restaurant staple in the United States—known as a combo plate, featuring a selection of enchiladas, tacos, and other stuffed items served with rice and beans all at one time—is also thought to be a Texas invention.[31]

[SAMPLE MENU]

A Border Dinner

Fresh Red Salsa[a,b] and Tortilla Chips

***Chimichangas*[a,b,c]**

***Arroz a la* Mexicana (Red Rice)[a,b]**

Refried Beans

Almond Pudding[a,b]

[a]Tausend, M. 1997. *Cocina de la familia: More than 200 authentic recipes from Mexican-American home kitchens.* New York: Simon & Schuster.

[b]*Southwestern Recipes* at http://www.recipegoldmine.com/sw/sw.html

[c]*Chimichanga History and Recipe* at http://whatscookingamerica.net/History/Chimichanga.htm

The second region is New Mexico, where a single chili pepper developed for the region, known as the New Mexico chile, dominates seasoning. It is mildly pungent, used green in chile verde and red in chile colorado. Unlike the complex sauces of southern Mexico that include numerous types of ground chiles and different seasonings, these northern-Mexico influenced sauces are often made simply with ground red chili pepper, water, garlic, oregano, and salt to taste. Pork replaces beef, kid, and lamb in many dishes. The third region is Sonora, encompassing both the Mexican state and southern Arizona. As in New Mexico, milder chiles are preferred (in this case, Anaheims), and in some recipes the seasoning is so tepid that "chile" has been dropped from the name, resulting in "carne verde," for example.[31] Beef is the favored meat, and a traditional dried beef jerky, machaca, is still used, shredded, for stuffed foods. Large, finely textured wheat tortillas are a specialty, and this region may be the original home of burritos and their deep-fried version, chimichangas.

The fourth and final region is along the California border, where the fluid movement of people back and forth between the nations has resulted in fare that cannot be claimed by either nation as its own, nor as a cuisine unique to the area.[33] Numerous American fast-food franchises are established in Mexico, and on the U.S. side taco shops and Latino grocery stores offer Mexican dishes. However, most businesses provide a mix of products, even the franchise restaurants. For example, Mexicans can purchase hamburgers and fries with jalapeños on the side, and diners in the United States can order "American burritos" filled with refried beans, carne asada, and French fries.

The market for Latino foods throughout the United States has grown dramatically since the 1980s, when the fare of Texas, New Mexico, and Arizona gained national recognition. It is now a more than 7 billion dollar a year industry.[34] In the 1990s sales of salsa surpassed those of catsup for the first time, and salsa has continued to dominate the condiment market. Tortillas and tortilla chips are also selling well. At the same time that Americanized Mexican foods, such as tacos in hard shells and fajitas, are spreading throughout the country, an appreciation of authentic regional Mexican fare is also increasing.

Adaptations of Food Habits

Ingredients and Common Foods. Many Mexicans in the United States eat a diet similar to that of their homeland. Recent immigrants, those who live near the U.S.-Mexican border, and migrant workers are most likely to continue traditional food habits. In a comprehensive review of published research on the effect of acculturation on the diet of latinos in the United States, no relationship between acculturation and dietary fat intake or percent energy from fat was evident. However, the source of the fat differed depending on acculturation. Less acculturated fat sources were from consumption of whole milk and fat added during food preperation, whereas among the more acculturated Latinos, fat sources were fast food, snacks, and added fats. In addition, the less acculturated individuals consumed more fruit, rice, and beans and less less sugar and sugar-sweetened beverages than the more acculturated Latinos.[35]

Chicanos and Mexicans who are well established in the United States often become quite acculturated. An early marketing study noted that Mexican immigrants living in the southwestern United States are more likely to eat a diet with a high intake of red meats, white bread, sugared cereals, caffeine-containing beverages, and soft drinks than their socioeconomic counterparts in Mexico or white neighbors. This suggests that, rather than adopting a diet that falls somewhere between the food habits of Mexico and those of the United States, the Mexican immigrants in the survey accepted the stereotypical American consumption patterns of the 1950s through 1970s.[36] In the intervening years, other studies have offered confirmation of these data. Studies on Latinos in southern California (who would be predominantly Mexican American) and Mexican Americans living in Washington have found higher intakes of fast foods, convenience items, salty snacks, chocolate, and added fat at the table with bread and potatoes, combined with lower intakes of beans, peas, fruits, and vegetables associated with acculturation.[37–40]

A more comprehensive study has detailed these changes. In a large sample of Mexican Americans, it was found that the greatest dietary changes occurred between the first generation born in Mexico and the second generation, born in the United States—changes between the second generation and third generation were less significant. The exception was consumption of corn tortillas, which decreased 69 percent between the first and second generations, with an additional 30 percent drop observed between the second and third generations.[41,42] Conversely, intakes of some items remained relatively constant between generations, particularly those of beef, green vegetables other than salad, and some fruits (bananas, apples, oranges and orange/grapefruit juice, and cantaloupe).

Specific changes seen between the Mexican-born subjects and the first generation of U.S.-born subjects included reduced consumption of legumes, most vegetables, rice, and pasta, and increased consumption of breads, cereals, margarine, mayonnaise, butter, potato chips, and French fries.[41,42] Intake of eggs and of American-style cheese also increases with these subjects. Whole milk, which is preferred among Mexican-born subjects, is replaced with low-fat or non-fat milk by U.S.-born subjects. Soda intake is high in all generations, although the U.S.-born subjects prefer diet sodas. Coffee is popular in all groups. Alcohol intake in general increases, with beer consumption among males especially popular (increasing 51 percent).[43]

A preference for sweet or carbonated beverages usually increases in the United States. Soft drinks, Kool-aid, and juices are popular with meals and as snacks. More acculturated Mexican Americans buy many prepared and convenience foods. Baked goods are usually purchased, including tortillas (often wheat tortillas are chosen over corn), breads, pan dulce, and even special desserts like flan. Extra income is usually spent on meats, especially more expensive cuts such as steaks and pork chops, and processed meats, such as hot dogs and bologna.[44,41,33]

Though consumption decreases with acculturation, it is estimated that Mexican Americans eat more beans than any other group in the United States—thirty-four pounds per person annually, compared to six pounds per person each year for whites.[32]

Meal Composition and Cycle

Daily Patterns. Little current data on the meal patterns of Mexican migrant farm workers have been reported, but older data indicate that traditional foods are preferred for most meals. These include eggs, beans or meat, and tortillas or pan dulce for breakfast; a large lunch of beans, tortillas, and meat, or a soup or stew; and a lighter dinner of tortillas, beans or meat, and rice or potatoes. As in Mexico, vegetables tend to be served as part of a soup or stew. Fruit remains a typical snack and dessert, especially familiar varieties such as bananas, oranges, mangoes, guava, pineapple, strawberries, and melon.[45,46]

Chicanos more often adopt the American meal pattern of small breakfast, small lunch, and large dinner; and meal-skipping may occur. As tortilla consumption declines, breads and breakfast cereals have become popular with all family members, and sandwiches are a common lunch item. Meats and cheese become more prevalent at meals, beans are eaten less frequently, and vegetables are served as side dishes. Snacking has been found to increase, especially in the evenings, and is associated with education. College graduates are more likely to snack daily than are those who attended only elementary or high school.[24,47–49] Soda, Kool-aid, juice, beer, and coffee are consumed with meals or snacks throughout the day. Milk is considered a superfood for children, but adults, especially men, may consider it to be a juvenile drink. Milk is often flavored with chocolate, eggs, and bananas in a drink called licuado; mixed with coffee; or mixed with cornmeal to make atole. Some adults reject milk completely, saying that they are allergic to it (it is estimated that 50 percent of Latinos are lactose intolerant[50]).

In Mexico, the prevalence of malnutrition resulting in stunting of children varies between 10 and 35 percent depending on region.[168]

Changes in preparation methods may also occur. Recent immigrants may not know how to use the baking and broiling apparatus on an oven and may continue to fry and grill foods outdoors. Newer immigrants sometimes avoid canned and frozen foods because they do not know how to prepare them. Soup is prepared at home three times more often by Hispanics than by non-Hispanics, according to one marketing survey, but this rate drops significantly with acculturation.[51] General spending patterns, however, suggest that cooking at home is still common: Hispanics spend more than double the average for flour and 166 percent more on dried beans.[44] Single women, those with larger families to feed, and those who identify more with their Mexican heritage have been found more likely to shop at small ethnic groceries or convenience stores that often have fewer healthy food options and may charge more for items such as low-fat milk compared to whole milk.[37,52,53]

Hispanics frequent restaurants more than any other ethnic group in the United States. Among Mexican Americans' favorite types of establishments (in order of popularity) are fast food, pizza, Mexican fast food, Chinese, coffee shops, and full-service Mexican.[54] One study of Latina women in the Los Angeles area found that younger, employed women with lower incomes who had lived more years in the United States preferred fast-food restaurants, stating that distance, price, and a child-friendly environment were deciding factors in choice of establishment.[37] In 2010 it was reported that over 13 percent of Hispanic children eat at a fast food or sit down restaurant three or more times a week.[55]

Special Occasions. The Mexican custom of reserving foods requiring extensive preparation, such as tamales and enchiladas, for Sunday and holiday meals is continued in the United States.[7] Even if such dishes are only served occasionally at family celebrations, Mexican Americans reconnect with their heritage through the preparation and consumption of these traditional items.[31] In one study that included holiday practices of migrant workers,[45] no main dish preferences were found for Easter, and tamales were favored for Christmas. Turkey with mashed potatoes was the most popular Thanksgiving entrée, indicating that this American holiday was adopted along with its traditional foods.

In addition to religious holidays, two secular celebrations are significant in the Mexican-American community. The first is Mexican Independence Day on September 16, commemorating the war of liberation from Spain. Observations emphasize ethnic unity, such as mariachis and traditional clothing. Foods the color of the Mexican flag, such as white rice, green avocado, and red or green chile peppers, are eaten. Cinco de Mayo (May 5), the second secular holiday celebrated, is more widely recognized by all ethnic groups throughout the United States even though the meaning of the event (remembrance of a historic victory over France) is often forgotten amid the parades, piñatas, and Aztec dancing that typify the day.

More recently, Mexica Aztec New Year's Day has emerged as a celebration in some Mexican-American communities. It welcomes the beginning of the Aztec solar year, comprised of eighteen periods of twenty days each, plus several separate days of reflection. Four signs—"rabbit," "reed," "flint," and "house"—are used to designate each year, along with a number between one and thirteen, resulting in a fifty-two-year cycle. Each new cycle begins with an elaborate New Fire Ceremony that involves fasting and sacrifice. The annual event emphasizes indigenous Indian identity, as opposed to the Latino or Hispanic ethnicity that incorporates aspects of European culture, such as use of the Spanish language. Some Catholic Latinos disapprove of the holiday.[56]

Nutritional Status

Nutritional Intake. It can be difficult to determine health statistics on Chicanos, Braceros, and unauthorized migrants because they are often grouped with whites or other Latinos in collected data. Information from research on Latinos can be used cautiously, since Mexicans comprise a majority of the total Latino population. Nevertheless, some nutritional problems have been identified in both new and acculturated immigrants from Mexico through studies of Spanish-surnamed patients, especially those residing in Texas, the Southwest, and California.

Life expectancy for Mexican Americans is similar to that of whites in the United States despite disadvantages such as higher rates of poverty, lower levels of educational achievement, and reduced access to health care. Overall mortality rates as well as cause-specific mortality rates are lower for Mexican Americans compared to whites when controlled for gender, age, nativity, marital status, socioeconomic status, and demographic variables. (One notable exception is mortality among younger Mexican Americans, ages eighteen to forty-four, who have elevated mortality risks, mostly due to external causes of death.)[57–60] Researchers propose several reasons for the lower-than-expected mortality rates in Mexican Americans. Some report that nativity outside the United States is a factor, accounting for health-promoting lifestyle differences.[58] Others suggest the discrepancy is an artifact of changes in classification of deaths from use of Spanish surname to use of Hispanic-origin questions in vital statistics.[61] Others have described underascertainment in the National Death Index when compared with data from a large survey of Hispanic elders.[62] Another theory is that older immigrants return home to Mexico before they die and are therefore not listed in vital statistics figures, resulting in a mortality advantage due to what has been dubbed "salmon-bias" effect.[63,64]

Similar trends are seen in infant mortality statistics. The birth rate among Hispanic women is nearly twice that of white women, and the birth rate among women aged fifteen to eighteen is more than three times as high. Nearly 25 percent of Mexican-American women receive no prenatal care during the first trimester of pregnancy.[65,66] Despite these risk factors, rates for low-birth weight infants and infant mortality are lower for Mexican Americans than for the total population (see Cultural Controversy: Breaking the Mold).

In a multi-site study of Mexican mothers, it was found that over 95 percent breast fed their infants in the first week of life.[67] Over 80 percent of Mexican-American women in a national survey breast-fed, and by six months, 23 percent were still breast-feeding.[68,69] In Mexico breast-fed babies are often given other fluids, including formula, water, and sweetened herbal teas to reduce colic or cure diarrhea,[67] a practice that may continue in the United States. Babies are usually weaned from the breast to the bottle. Long-term use of the bottle or sleeping with one at night, with milk or sweetened liquids (i.e., Kool-aid, fruit juice, tea) is sometimes a problem, resulting in iron-deficiency anemia and tooth decay among toddlers (baby-bottle tooth decay).[70–72]

Mexican Americans have the highest component scores on the USDA Healthy Eating Index (HEI) of all ethnic groups in the United States, indicating good compliance with the Food Guide Pyramid recommendations, especially dietary variety and high fruit consumption, but also a high intake of sodium. Respondents born in Mexico had slightly higher scores than U.S.-born respondents.[73] Most studies agree with the HEI findings that dietary quality may decline with acculturation. Fat and cholesterol intake increases, while intake of beta-carotene, vitamin C, and fiber decreases[7,47] due primarily to increased consumption of high-fat snacks, fried foods, eggs, cheese, and milk and decreased consumption of legumes, fruits, and vegetables.[74,38,39] Researchers caution, however, that acculturation is difficult to measure, and different methods may result in different findings. A study of low-income Hispanic women in San Diego found that when acculturation was defined as number of years in the United States, fat intake did not change.

CULTURAL CONTROVERSY—Breaking the Mold: The Mexican-American Immigrant Experience

One of the assumptions regarding assimilation is that life improves continually for immigrants the longer they reside in the United States. Yet recent studies of the largest immigrant population contradict this model. Foreign-born Mexican Americans are found to be healthier overall, to eat slightly better diets, and to have lower rates of infant mortality than U.S.-born Mexican Americans with foreign-born parents and for U.S.-born Mexican Americans with U.S.-born parents—despite higher rates of poverty and less access to medical care.[169,170,171,47]

The scientific community expressed little interest about differences in immigrant health status associated with place of birth until researchers in the late 1980s discovered a startling trend. Poor, disadvantaged women who had immigrated from Mexico were giving birth to babies who were as healthy as those of white U.S. women with overall higher levels of income and education. Rates of premature births, low-birth-weight rates, and newborn death rates among the immigrant women were equal to or less than those for whites.[105,172,173] The foreign-born women also demonstrated better birth outcomes and fewer maternal disorders than Mexican-American women who were born in the United States.[178,177] One study even suggests that internal migration also improves health, showing that Mexican-American women born outside the U.S. community where they give birth have better pregnancy outcomes than women born in the community.[174] A comprehensive review of statistics on adolescents was even more revealing. The longer a subject's family had lived in the United States, the poorer the subject's health and the more likely the subject was to engage in risky behaviors, even after controlling for neighborhood, family, education, and income variables. Mexican Americans who were born in the United States with U.S. parents had significantly higher rates of health problems (including obesity, asthma, and missing school due to illness) compared to those who were born in Mexico. Health risk behaviors, determined by sexual experience, delinquency, violent behavior, and use of controlled substances, were more than double in U.S.-born Mexican-American adolescents with U.S. parents than in Mexican-American youth born in Mexico.[170] Suddenly, the assimilation model of health was in question. What accounts for such significant differences? Most hypotheses have addressed the disparities in birth outcomes, suggesting that selective migration occurs (only healthy women come to the United States), that deaths during pregnancy may be greater (thus skewing the data), or that infant deaths are underreported in the foreign-born Mexican-American community (which may include high numbers of unauthorized residents). Other theories emphasize the protective factors of the Mexican culture.[175] Research suggests that pregnant, foreign-born Mexican-American women behave in ways different from those who are born in the United States. Intake of nutrients, including protein, folate, vitamin C, iron, and zinc, is better; smoking and alcohol consumption rates are substantially lower.[176,81,177] Other factors considered important to positive pregnancy outcomes, such as adequate weight gain and prenatal care, are less likely in foreign-born Mexican Americans. Researchers suggest these negatives may be compensated for by greater community, family, and spousal support, and less accumulative acculturation stress.[161,80,171,177] Determining the reasons that place of birth is so significant will help researchers devise a new assimilation model and suggest approaches for improving the diet, pregnancy, and health outcomes for all Mexican Americans.

A study on the prevalence of iron overload disorders in Hispanics suggests the rate may be slightly higher than among whites.[179]

When defined as being born in the United States, greater consumption of convenience foods and chocolate was found. When preference for speaking English at home was considered, lower intake of beans and peas was noted. When acculturation was defined as being born in the United States and a preference for speaking English at home, a higher consumption of salty snacks, high-fat foods, lower fiber and convenience items was reported.[40,75]

Initial research indicated total protein intake may have declined with the length of stay in the United States, but recent research indicates that protein intake has leveled off in the last decade.[76,77,47] Low protein consumption, combined with low iron and vitamin C intake, sometimes results in low hemoglobin levels among young children and pregnant women.[70,78] However, excessive rates of low iron intake or low blood iron status among Americans of Mexican descent have not been confirmed.

Deficiencies of calcium and riboflavin are common, often due to low consumption of dairy products. Although the traditional Mexican diet includes good sources of vitamins A and C, thiamin, niacin, B_6, folate, phosphorus, zinc, and fiber, low intakes of these nutrients by Mexican Americans have been reported; inadequate income or lack of traditional ingredients may limit consumption of these nutrients.[79–82]

Nutritional inadequacies may contribute to other diseases. More TB cases were reported among Hispanics than any other racial/ethnic population in 2008, for five consecutive years.[83] However the rate has decreased over this period. In contrast to the problem of undernutrition is the prevalence of obesity in persons of Mexican descent—36 percent of men, and 45 percent of women in 2008.[84] Overweight in children (when defined as BMI≥95 percentile) has also been reported, often at rates equal to, or slightly higher than, those for whites, between 17.4 and 26.8 percent for girls and boys, respectively.[47,84,85] Low socioeconomic status and less leisure-time physical activity are thought to be factors in high rates of overweight.[86–88] Cultural ideal weight may be greater for some Mexicans than for Anglo-Americans, and traditionally, Latina women believed it was normal to gain weight after marriage. Extra weight may indicate health and well-being, not only for adults but also for children, and some parents may not recognize that their children are overweight.[89] However, research suggests that many Hispanic adults and children perceive themselves as being overweight and are dissatisfied with their body image.[90–92] Unhealthy dieting practices are prevalent in some Latinos, and it is believed that Hispanic girls are at high risk of developing eating disorders.[92–93]

Prevalence of type 2 diabetes in Mexican Americans is approximately 2 times that for whites;[95] however, in one study Mexican Americans were more likely than whites to be diagnosed and treated.[96] This high rate is not explained by the incidence of obesity, age, or education, but may be related to percentage of Indian heritage.[97,98] Research suggests that Hispanics have greater insulin resistance than African Americans and whites, and their lower sensitivity may be due to a higher intake of carbohydrates.[99] Type 2 Diabetes in Mexican-American adolescents is three times that of white youth.[100] Complications from diabetes, including kidney failure and diabetic retinopathy, are also more prevalent.[101,102] Overall, comorbidity in Mexican-American elders is high, including heart disease, angina, hypertension, and arthritis, and complications such as vision impairment, reduced mobility, and incontinence are burdensome.[103] Death rates from diabetes are estimated to be over 64 percent higher for Hispanics than for whites, and this difference is even greater in the counties along the Mexican border.[104] Further, diabetes mortality rates are higher for Mexican Americans than for any other Hispanic group.[105]

Mexican Americans have very high rates of metabolic syndrome, a clustering of conditions that leads to type 2 diabetes and heart disease, including insulin resistance, hypertension, and dyslipdemia. Prevalence in Mexican-American was 40.1 percent, and Mexican American women were 1.5 times more like to have met the criteria for metabolic syndrome than non-Hispanic white females.[106] Mexican Americans have similar rates of hypertension to that of whites, approximately 25 to 28 percent.[107] Data on dyslipdemia show higher prevalence in Mexican Americans than in whites and African Americans.[108] Estimated ten-year coronary heart disease mortality risk was found heterogeneous in Mexican Americans in one study. Spanish-speaking, U.S.-born men and women were at highest risk, Mexican-born men and women at lowest risk, and U.S.-born, English-speaking men and women at intermediate risk.[109] The incidence of cardiovascular disease in Mexican Americans is lower in than the incidence in whites and was was approximately 32 percent in men and 34 percent in women and with 5 percent for coronary heart disease. Two to three percent have had heart attacks or strokes.[110]

Researchers have noted the heart-healthy elements of the traditional Mexican diet.[79] Though studies on cardiovascular mortality risk relative to whites are contradictory,[111–114] and subpopulation risk varies, cardiovascular diseases are the number one cause of death in Hispanic adults.[59] A higher prevalence of gallbladder disease has been found in Mexican Americans, and a higher rate is found in those born in the United States, compared to those born in Mexico. Diet may be involved in some cases, but more recent research suggests a genetic vulnerability may be a more important factor.[115,116] Knowledge of osteoporosis and risk-reducing behavior was low among Hispanic women in Chicago.[117] Cavities are common among Americans of Mexican descent, as is gingivitis. Studies show that nearly one-third of all immigrants from Mexico never receive any dental care. Migrant workers and their children are especially at risk.[118,119]

A study was done of Tarahumara Indians in Mexico to observe the effects of a high-calorie, high-fat, low-fiber diet on a population that traditionally consumes a low-fat, high-fiber diet. After five weeks, blood cholesterol levels increased 31 percent, and triglyceride levels increased 18 percent; all subjects also gained weight.[180]

Licorice root, known as *yerba dulce* or *orozús*, may be used as a general tonic or for infections, coughs, ulcers, and menstruation problems. It can be toxic in large quantities or taken over long periods, and may potentiate the effects of hypotensive drugs.[28]

Alcohol intake was found to increase by 47 percent between Mexican Americans who were born in Mexico and the first generation of Mexican Americans born in the United States.[47] In one study, one in three Latinos admitted to drinking hazardous amounts of alcohol, and 36 percent had a lifetime diagnosis of alcohol abuse or dependence.[120] Mexican-American men drank greater quantities, less often, than white men, and Mexican-American women have low alcohol consumption rates that increase with acculturation.[121]

Counseling. Access to biomedical health care may be limited for Americans of Mexican descent, especially for Braceros and unauthorized migrants. Income may restrict doctor visits, and transportation to clinics may be unavailable. Thirty–five percent of Mexican Americans do not have health insurance.[59] Further, 43 percent of Mexican Americans report that they speak Spanish at home and that they are not proficient in English[3] Spanish-speaking clients may be uncomfortable with interviews conducted in English. Those who believe God or fate determines health may be unwilling to undertake preventive procedures.[126]

The communication style of Latinos is high-context and non-confrontational; a warm, dignified relationship is most effective and crucial in difficult health care situations. Words are chosen carefully, and silence may be used to defuse disagreement.[26] Kindness and graciousness are appreciated.[126] Touching a client with a handshake is important, although men should wait for women to extend their hands first. Eye contact varies, and some Mexican Americans consider prolonged eye contact impolite. However, it is best to maintain eye contact initiated by a client, as looking away may be thought rude. Mexican Americans often sit and stand closer together than Anglos do. Most Latinos are present-oriented and polychronic (able to do several things at once). Inflexible appointments can be problematic for some Latino clients, who may prefer walk-in clinics.[15] Latinos may be uninterested in lengthy indirect discussion of a condition and prefer a direct, action-oriented approach.

Attitudes may differ from American biomedical beliefs. For instance, it is sometimes considered inappropriate for men to acknowledge illness.[122] People who go on working despite bad health are respected. Modesty and privacy are highly valued; thus a woman may wish to be treated by a female caregiver and a man by a male caregiver.

Studies suggest that anywhere from 20 to 81 percent of Mexican Americans use folk remedies,[13,14,122] and Latinos in one survey used the services offered at their local *botánica* interchangeably with those offered by their biomedical provider.[55] Another study found that increased use of herbal remedies by Mexican-American women corresponded with fewer visits to biomedical practitioners. This association may be related to lack of medical insurance, which was also found to be a factor in use of traditional medicines.[14] Sixty-one percent of surveyed Hispanic elders in New Mexico reported drinking herbal teas to maintain health, alleviate stress, and cure minor ailments.[29] The most commonly consumed infusions included peppermint (called both yerba buena and poleo), chamomile (manzanilla), lavender (alhucema), and osha (related to parsley). However, it is important not to assume adherence to certain folk beliefs, such as the hot-cold classification of foods and illness. Though prevalent among some people in Mexico, these traditional practices may be limited in the United States (even among healers), and younger, urban clients may be offended by use of such "backward" theories.[7,12,14] Traditional healers, such as *curanderos*, are consulted by anywhere from 4 to 21 percent of the Mexican-American population.[123,27,124,125] Those who consult *curanderos* often believe that healers are most effective for symptoms caused by folk illnesses. In one small sample, nearly 26 percent of Mexican-American women had personally used the services of a *curandero*; however, even larger numbers (almost 39 percent) had seen a *sabadores (*manual therapists).[14] However, it should be noted that use of traditional healers does not prevent a client from also seeking biomedical care concurrently.[126] Although traditional practices are most common in poor, rural regions, most Mexican Americans are knowledgeable about folk conditions: Studies in Texas report that *mal de ojo* had been diagnosed and treated in 63 to 70 percent of Mexican-American homes surveyed, *susto* was known in 37 to 62 percent, and *empacho* in 27 to 48 percent. *Mal de ojo* was known in 70 percent of homes, and *caída de mollera* in 34 percent.[124,125,12] Another study showed

71 percent of homes had a family member who had suffered from *nervios*, and that 46 percent had experienced it personally.[16] Mexican-American women in southern California also demonstrated familiarity with these illnesses.[14] Sometimes multiple etiologies are integrated. For example, some Mexican Americans believe that diabetes, which is an Anglo disease, is caused by eating a diet high in fat and sugar, but also due to experiencing strong emotions or chronic *susto*.[30,127]

Most traditional health beliefs and practices among Mexican Americans support the emotional well-being of a client and do not interfere with therapy. Many researchers have suggested that folk conditions provide an important release valve in Latino cultures, especially for men who are expected to endure pain. Disorders due to outside causes are not blamed on an individual, and the resulting irrational behavior or lethargy is excused.[15]

Several potentially harmful situations are noteworthy. Suboptimal medication use has been noted, especially in diabetes treatment where intake is inconsistent, and in situations where prescription drugs are mixed with home remedies.[128,129] For example, diabetes and hypertension may be treated with botanical remedies in addition to prescribed oral medications, risking excessive hypoglycemic and hypotensive activity, respectively. Clients consult friends and neighbors about effective treatments and are unlikely to disclose home remedies to their physicians.[130] Digestive complaints such as *empacho* are sometimes treated with toxic lead- or mercury-based medications, such as *greta*, *azarcón*, and *asogue*.[131] The condition of *caída de la mollera* in infants has been associated by some health practitioners with severe diarrhea and dehydration, resulting in the depressed fontanel.[131] Failure to Thrive Syndrome may also be a concern. Providers should be aware of these possibilities when presented with this disorder. In some regions, a tea made from the psychoactive wormwood (the toxic ingredient formerly found in the alcoholic beverage absinthe) is used for diarrhea. Though not widely used, *vibora de cascabel* (dried rattlesnake powder) can be a source of salmonella and botulism. Finally, babies may be given home remedies made with honey, a known cause of infant botulism. In one study, Latinos were the ethnic group least likely to discuss the use of alternative and complementary therapies with their biomedical care providers.[132]

Family participation in health care is common, and members should be consulted in both making a diagnosis and prescribing treatment. They may have specific ideas about the cause of an illness and the best approach for a cure; their confidence and cooperation can help ensure client compliance. One study found that family involvement in serious choices about issues such as life support is more important to Mexican Americans than patient autonomy.[133] Dietary changes may affect family members and social interactions; thus, gaining family support has also been suggested to increase compliance.[134,135]

An older study of Mexican-American families living on the Texas–Mexico border found that the husband traditionally exercised control of the food budget and food purchases.[136] The wife did the actual meal planning, shopping, and preparation. Women identified strongly with their food-related tasks within the family structure. Because their self-concept and status in the family and community are related to their abilities as a cook and homemaker, nutrition intervention and advice may be perceived as an accusation of inadequacy. More recent research shows that Mexican-American women who are more acculturated are likely to have a shared meal decision-making style. However, they face greater barriers to making healthy eating changes due to resistance from family members, and they are more likely to eat at fast-food restaurants, eat more saturated fat, and make fewer efforts to increase fiber intake than are women who maintain traditional meal preparation roles.[7]

Children are also an important influence on food habits in some households. Those raised in the United States may be the only English-speaking members of the family and may be responsible for translating in the market. These children have been found to prefer foods that they have seen advertised on television. The adoption of new foods is influenced by the presence of bilingual children in the family.[46] Researchers studied newly immigrated Latinos in the San Francisco area and found that the importance of the family unit can be used to motivate changes in food habits.[137] Adults unwilling to make changes that would benefit their own health may make those same changes to improve the well-being of their children.

As with all clients, an in-depth interview is crucial in effective nutrition counseling. Experts in the health care of Latinos recommend that

Queso fresco, traditionally made from raw milk, is responsible for more food-borne illness than any other cheese in the United States.[6]

Latinos listen to radio programming in greater numbers than do Anglos. An evaluation of a Spanish radio nutrition course found it improved both nutrition-related knowledge and selected practices in both rural and urban sample populations.[181]

health professionals who work often with Latinos learn Spanish. Familiarity with Spanish medical terminology is the minimum proficiency needed for meaningful communication. Further, interventions should be tailored to account for differences in acculturation.[37]

Central Americans

The seven nations of Belize, Guatemala, El Salvador, Honduras, Nicaragua, Costa Rica, and Panama make up Central America, an isthmus connecting North America to South America. The eastern coastal region edges the Caribbean Sea. An 800-mile chain of active volcanoes and mountains, beginning at the Mexican border in the north and continuing with only one break into central Panama in the south, forms the temperate backbone of the region. Central America is similar to the rest of Latin America in history of foreign intervention and heterogeneous culture.

Cultural Perspective

History of Central Americans in the United States

Immigration Patterns. Central American immigrants to the United States have arrived in two distinct waves. Early records are inexact because separate statistics on Central Americans were not kept by the U.S. Census Bureau until the 1960s and will be reported in the 2010 census and not listed with South Americans. Until the early 1980s, immigrants to the United States were of two groups. The first were well-educated professional men who arrived in search of employment opportunities. The second were women, who often outnumbered the men two to one, coming in search of temporary domestic jobs. These Central American immigrants were largely urban residents and settled mostly in New York, Los Angeles, San Francisco, Miami, and Chicago, where they blended into existing Latino communities.

The second major wave began in the late 1970s and early 1980s, with the exodus of refugees from the brutal civil wars in El Salvador, Guatemala, and Nicaragua. Millions of residents are estimated to have been displaced in these countries, about one-third of whom have emigrated. Many moved to Mexico, and a substantial number have continued on to the United States. They are known as the "foot people" because many have literally walked to the United States.[138] Less is known about this group, except that they are often younger, poorer, and less educated than the previous immigrants from the region.

Current Demographics and Socioeconomic Status. Just over 5 percent of U.S. Latinos are from the seven nations of Central America, a majority of whom arrived since 1980.[3] Exact figures are unknown, however, because it is believed that many Central Americans may enter the United States illegally at the border with Mexico and are undistinguished from unauthorized Mexicans. Most of these recent immigrants have settled in California, Texas, and Florida. The largest populations are the 1.7 million Salvadorans, 1.1 million Guatemalans, 490,000 Hondurans, and 366,000 Nicaraguans. It is believed that there may be an equal number of unauthorized migrants from these countries living in the United States as well.[139,140] Immigration from Costa Rica, Belize, and Panama is minimal, with fewer than 350,000 altogether from these other Central American nations listed in the 2005 U.S. Census estimates.[141]

Nearly one in every four Central Americans is born in the United States. Central American immigrants, even those from the first wave, are slow to naturalize, and only 20 percent of Central Americans in the United States have obtained citizenship.[3] Those who are not refugees often return to Central America for visits and maintain active contact with their homeland. Those who are have sought asylum, and those who are unauthorized migrants, cannot visit their homelands without fear of persecution or difficulties in returning to the United States.

Many immigrants live in neighborhoods where other Central Americans reside, especially in Los Angeles and Miami. Smaller numbers locate in Houston; Chicago; Washington, DC; and New York City. Central Americans are a very heterogeneous population, however, and it is a mistake to assume similar settlement patterns for each group. Identity is sometimes more related to race and class than to country of origin. The majority of immigrants before the 1980s were white professionals; more recent refugees are predominantly Ladino (mixed Spanish and Indian heritage) or Indian *campesinos* (peasants). Even among recent immigrants, differences in associations are found. Hispanic Guatemalans, for example, may assimilate into the broader Latino community, while Mayan Guatemalans, some of whom do not speak Spanish, often establish ethnic enclaves.[142] Significant numbers of black Central Americans have immigrated from Belize and Panama. Little is known about these specific groups, though many black Panamanians choose New York City as their home.[143]

Some socioeconomic data on Central Americans are combined with statistics on South Americans. Information on the first wave of Central American immigrants suggests that they are a middle-class population with income and education levels well above many other Latino groups. American-born children of these first immigrants graduate from high school in numbers greater than whites.

These figures do not reflect the large numbers of recent immigrants employed as migrant farm workers, gardeners, domestic cleaners, dishwashers, and food-service workers, as well as those placed in other low-skill jobs. According to U.S. Census data on foreign-born Central Americans, recent Salvadoran immigrants often find work in the construction, manufacturing, and hospitality industries. Just 34 percent have graduated from high school, and 41 percent have less than a ninth-grade education. Families falling below the poverty line comprise 20 percent of the Salvadoran population in the United States.[88] Guatemalan immigrants share a very similar socioeconomic profile. Honduran immigrants have slightly higher levels of educational attainment, but greater numbers of families live in poverty. Foreign-born Nicaraguans also have slightly higher levels of education, and more work in retail and in the education, health, and social service fields. Their poverty levels are also lower. Immigrants from Belize and Costa Rica have similar educational attainment (about 70 percent have graduated from high school), and both groups often work in the education, health, and social service sectors. The poverty rate for Belize families is 16 percent, and for Costa Rican families it is 14 percent. The socioeconomic status of foreign-born Panamanians, who as a group have been in the United States for a longer period of time, is significantly improved over that of other Central Americans. Eighty-two percent have completed high school, 22 percent have college degrees, two-thirds work in management, professional, or sales and office jobs, and poverty levels are below the U.S. average.

Mexican School/Bridgeman Art Library/Getty Images

◀ *A Mayan chocolate container.*

Information on unauthorized residents is scanty. It is thought that they often face difficulties in obtaining employment and education opportunities.[139] In addition, disposable family income may be impacted by money sent to support relatives still living in the homeland.

Many Guatemalan Americans prefer to be called Chapines. The term was originally a derogatory term for residents of Guatemala City but has new meaning in the United States, reflecting ethnic pride.

Worldview

The large numbers of recent immigrants from Central America suggest that ethnic identity is preserved by many new residents in the United States. For example, Salvadorans often establish highly insular neighborhoods within the larger Latino community, where an immigrant can live and conduct business exclusively with other Salvadorans.[139] Guatemalans are a more diverse population of immigrants, and it is the Mayan communities that are most likely to keep traditional beliefs and practices.[142] In contrast, Nicaraguans are dispersed among other Latinos and are adapting more to the pan-Latino community rather than retaining their own heritage exclusively.[140]

One unique Central American immigrant population in the United States is the Garifuna, who are black Caribs from the eastern, coastal areas of Belize and Honduras. Traditionally, men traveled great distances for work, leaving the women to farm and raise children. Today, their way of life is threatened by coastal development for tourism and lack of government support. Garifuna men often come to the United States for employment, sending much of their earnings home in an effort to save their culture.[145]

Religion. Most Central Americans are Roman Catholic. Some Guatemalans observe Catholic practices while adhering to Mayan religious beliefs; participation in native religions declines in the United States because they are usually dependent on sacred locations in Guatemala. Evangelical and fundamentalist denominations, such as the Pentecostal Church, have attracted many Central Americans after they arrive in the United States. Small storefront congregations that involve active participation and those churches that offer traditional Central American social activities in addition to worship have been especially successful.

Family. Central Americans highly value family and extended kinship. It has been noted that some apartment buildings in Latino neighborhoods are rented entirely to several families from the same village in Central America. The father is traditionally the undisputed head of the household and provider for the family. Many Costa Ricans define the term *family* as having both a father and mother in the home.[146] However, some studies suggest this dominance is changing, and shared decision making between men and women is becoming more common.[147,148] Children are often carefully controlled, especially daughters.

The roles of the men and women often change even further in the United States, where women are sometimes more easily employed, and husbands must take on some domestic responsibilities. Family disintegration has taken place in some refugee camps, where overcrowding and unemployment led to intergenerational conflict prior to immigration to the United States.[149] In other situations, family members were forced to immigrate separately. Some married outside the Central American community for immigration benefits; others found that when their families were reunited, children had become more independent.[139]

Mayan creation myths describe how humans were improved over time. They were first made of mud, then wood, and finally perfected when their flesh was made from corn dough.

Traditional Health Beliefs and Practices. A good diet, especially consumption of fruits and vegetables, fresh air, and regular hours, are thought necessary to preserve health by many Central Americans.[150] Exercise is considered important by some Guatemalans and Panamanians, though the concept of structured exercise is unfamiliar to some Central Americans.[151,147,148] Salvadorans believe that being too thin can cause sickness, and Americans are considered at risk for ill health because they are so thin.

Some Central Americans view health as a balance between the spiritual and social worlds. For most, health is a gift of God, and prayer is often used to restore harmony during illness.[147,152] Some Nicaraguans believe in witchcraft, practiced by *brujos* or *brujas* who can assume the shapes of animals and have the power to cure illness. Guatemalans consider outside forces to be the cause of some illnesses, which include diseases sent by Satan to punish unbelievers and sickness due to witchcraft. Traditional healers include *curanderos* and *sabadores*, as well as *jeberos* (herbalists) and *espiritistas* who treat witchcraft with prayer. Naturalist doctors (sometimes called naturopaths) who work in association with naturalist shops that provide botanical remedies may be used by some Guatemalans.[153] Priests may also be sought to help with prayers for health.

A balance of hot and cold is also necessary to health and can be disrupted by sudden exposure to extremes in temperature or strong emotions. In addition to *susto* and *mal de ojo,* other folk conditions include *bilis* and *cólera*, which in extreme cases are associated with anger and general distress and precipitate stroke. In a survey of Hispanic immigrants (with a large sampling of Central Americans) regarding beliefs about hypertension, respondents reported that *cólera* and *susto* may lead to high blood pressure, as may living at too high an altitude or having too much blood.[154,149]

In the culturally diverse region of Nicaragua's east coast, more than 200 plants with traditional medicinal uses have been identified. One study reported rural ethnic groups in Nicaragua were found to use traditional healing practices more often than urban residents of mixed heritage.[155] Another survey, however, found that more than three-quarters of respondents in an urban barrio used herbal remedies.[154]

Guatemalans believe that strength is maintained through the quantity and quality of a person's blood. In urban regions of Guatemala, researchers have noticed the emergence of new categories of food items, such as strong or health-promoting foods, perhaps due to the influence of modern health promotion concepts.[149]

Over-the-counter remedies, such as analgesics and cough suppressants, are commonly used by Guatemalan Americans, although they are considered weak by Guatemalan standards. Medications (including antibiotics) and herbs, such as chamomile, are sometimes brought to immigrant families by new arrivals from Guatemala.

Traditional Food Habits

Ingredients and Common Foods

Central American cuisine offers many of the foods common throughout Latin America. The native Indian dishes remain prominent in the highland areas, Spanish influences are found in the lowland regions, and the cooking of the multicultural eastern coast shares many similarities with Caribbean Islander fare. The northern nations have foods similar to those of southern Mexico; the southern countries have been more greatly influenced by European and African cuisines.

Staples. Early Mayan records indicate that the foundation of their diet was corn and beans, supplemented with squash, tomatoes, chiles, tropical fruit, cocoa, and some game. Indian foods were particularly important in the development of Guatemalan cuisine but gradually become less significant in the south of Central America. Rice, introduced by the Spanish, has become a staple in most regions. (See the cultural food groups listed in Table 9.3.)

Beans are eaten daily. Black beans are especially popular in Guatemala, while red beans are common in other nations. Beans are served simmered with spices (called frijoles sancochadas in El Salvador), pureed, or fried and are often paired with rice. In Nicaragua, red beans and rice fried with onions are called gallo pinto ("painted rooster") due to the colors of the dish.

Corn is eaten mostly as tortillas. Enchiladas in Central America are open-faced sandwiches similar to Mexican tostadas. They typically feature meat covered with pickled vegetables such as cabbage, beets, and carrots. They are known as mixtas in Guatemala, and here the tortilla is spread first with guacamole, then topped with a sausage and pickled cabbage. In El Salvador, a stuffed specialty is called pupusas. A thick tortilla is filled with chicharrónes, cheese, or black beans and then completed with another tortilla; the edges are sealed and the pupusa is then fried. They are traditionally served with pickled cabbage. Tamales are also common, often stuffed with poultry or pork. They are called nactamal in Nicaragua, where the dough is flavored with sour orange juice, and the filling includes meat, potatoes, rice, tomatoes, onions, sweet peppers, and mint. Black tamales are served on special occasions in Guatemala, stuffed with a mixture of chicken, chocolate, spices, prunes, and raisins. Empanadas, small turnovers made with wheat flour dough and filled with a savory meat mixture, are popular.

French bread, introduced from Mexico, is eaten regularly in the form of small rolls in Honduras and Guatemala. In El Salvador, French bread is used with native turkey and pickled vegetables to make sandwiches. Coconut bread is a specialty on the Caribbean seacoast. Rice is often fried before boiling, cooked with coconut milk, or, in Costa Rica, served as pancakes.

Soups and stews are popular throughout Central America, often including fruit or fruit juices. Beef, plantains, and cassava in coconut milk, spicy beef stew, beef in sour orange juice, pork and white bean stew, chicken cooked in fruit wine; mondongo (Nicaraguan tripe soup), sopa de hombre ("a man's soup") made with seafood, and plantains in coconut milk are a few specialties. In Guatemala the stews of meat and poultry, such as pepián and jocon, are thickened with toasted squash seeds. Meat, poultry, and fish are frequently roasted as well.

Fruits and vegetables are numerous. Although bananas, coconut, plantains, yucca (cassava), tomatoes, sweet peppers, cabbage, chayote squash (known as huisquil in Guatemala), mangoes (considered an aphrodisiac in Guatemala),

The Mayan word for corn, *wah*, also means "food."

In Guatemala, refried black beans (fríjoles volteados) are fondly called "Guatemalan caviar."

Table 9.3 Cultural Food Groups: Central Americans

Group	Comments	Common Foods	Adaptations in the United States
Protein Foods			
Milk/milk products	Milk is not widely consumed as a beverage, but evaporated milk and cream are popular in some regions.	Milk (evaporated), cream; cheese (aged and fresh—crumbly farmer's cheese type)	Milk and hard cheese may be disliked by Guatemalans, but increased intake reported for Salvadorans.
Meat/poultry/fish/eggs/legumes	Legumes are important in the cuisine and are often served with rice. All types of meat/poultry are eaten, but pork is popular throughout the region. Eggs are commonly served. Fish and shellfish are consumed in the coastal regions. Sea turtle eggs are popular.	*Meat:* beef, iguana, lizards, pork (all parts, including knuckles, tripe, and skin), venison *Poultry:* chicken, duck, turkey *Fish and shellfish:* clams, conch, flounder, mackerel, mussels, sea snail, shark, shrimp, sole, tarpon, turtle *Eggs:* poultry, turtle *Legumes:* beans—black, chickpeas, fava, kidney, red, white	Bean dishes remain popular.
Cereals/Grains	Rice and corn are the predominant grains of the region. Wheat flour breads are common.	Corn (tamales, tortillas), rice, wheat (bread, rolls)	Tortillas may be replaced by breads.
Fruits/Vegetables	Tropical fruits are abundant. Some temperate fruits such as grapes and apples are also available. Salads and pickled vegetables are popular.	*Fruits:* apples, bananas, breadfruit, *cherimoya*, coconut, custard apple, grapes as well as raisins, guava, *mameys*, mangoes, *nances*, oranges (sweet and sour types), papaya, passion fruit, *pejihaye*, pineapples, prunes, sour-sop, sweetsop, tamarind, tangerines, *zapote* (*sapodilla*) *Vegetables:* asparagus, avocados, beets, cabbage, *calabaza* (green pumpkin), carrots, cauliflower, *chayote*, chile peppers, corn, cucumbers, eggplant, green beans, hearts of palm, leeks, lettuce, *loroco* flowers, onions, *pacaya* buds (palm flowers), peas, plantains, potatoes, pumpkin (*ayote*), spinach, sweet peppers, *tomatillos*, tomatoes, watercress, yams, *yuca* (cassava), yucca flowers (*izote*)	Increased intake of potato chips has been reported. Increased consumption of vegetable salads.
Additional Foods			
Seasonings	Cilantro (fresh coriander) and *epazote* are important herbs. Sour orange juice gives a tang to some food; coconut milk flavors others. *Achiote* is used to color foods orange.	*Achiote* (annatto), chile peppers, cilantro, cinnamon, cloves, cocoa, *epazote*, garlic, onions, mint, nutmeg, thyme, vanilla, Worcestershire sauce	
Nuts/seeds		Palm tree nuts, *pepitoria* (toasted squash seeds)	
Beverages	Hot chocolate and coffee, grown in the region, are favorite hot beverages. *Refrescas*, cold drinks, are made with tropical fruit flavors. *Boj*, *chicha*, and *venado* are locally made alcoholic beverages.	Coffee, chocolate, tropical fruit drinks, alcoholic beverages (rum, beer, and fermented or distilled fruit, sugar cane, and grain drinks)	Increased intake of soft drinks.
Fats/oils	Lard is the most commonly used fat.	Butter, lard, vegetable oils, shortening	Lard and shortening use may decrease; vegetable oils and mayonnaise may increase.
Sweeteners	Honey and sugar are used as sweeteners.	Honey, sugar, sugar syrup	Increased intake of candy is noted.

oranges, and avocado predominate, cabbage, cauliflower, carrots, beets, radishes, green beans, lettuce, spinach, pumpkin, breadfruit, passion fruit (granadilla), pineapples, mameys, and nances (similar to yellow cherries) are also common. Flowers from yucca, palms (pacaya buds), and loroco (Fernandia pandurata) are eaten as vegetables throughout the region.[165] Starchy fruit from the peach palm (pejibaye) and spiny palm (coyoles) are especially popular in Costa Rica and Honduras. Onions and garlic flavor many dishes. Salads and pickled vegetables are common as appetizers, as side dishes, and on sandwiches.

Coffee, grown throughout the region, is a popular drink, usually consumed heavily sweetened. Hot chocolate is another favorite, and fresh milk, if consumed, may be sweetened with added sugar. Refrescas, cold beverages, are made in tropical fruit flavors, such as mango and pineapple. Tiste, a Nicaraguan favorite, is made with roasted corn, cocoa powder, sugar, cold water, and cracked ice. Beer is widely available. Fermented beverages such as boj (from sugar cane) and chicha (a wine made from fruit or grain, fortified with rum) are consumed. Venado is a common distilled drink made from sugar cane. Sweets, such as the praline-like candy called nogada, sweetened baked plantains, ices made with fruit syrups, custards, rice puddings, and cakes or fritters flavored with coconut or rum, are eaten as snacks and for dessert.

Regional Variations. Although many foods of Central America are similar, they are often flavored with local ingredients for a unique taste. Coconut milk flavors many dishes in Belize and Honduras; seafood specialties include conch and sea turtle. The foods of El Salvador are often fried and feature many indigenous flavors including corn, beans, tomatoes, chiles, and turkey. Achiote is common in mild seasoned Guatemalan fare. The juice of sour oranges is mixed with sweet peppers or mint in many Nicaraguan recipes. Costa Ricans prefer foods simmered with herbs and seasonings such as cilantro, thyme, oregano, onion, garlic, and pimento; rice is also frequently consumed. Panamanian fare is more international in flavor; one specialty is sancocho, a stew of pork, beef, ham, sausage, tomato, potato, squash, and plantains.

Meal Composition and Cycle

Daily Patterns. As in other Latin American regions, beans and corn are the cornerstones of the daily diet, eaten at every meal by the poor. Rice is also common. Queso blanco (a fresh cheese) or meat is added whenever resources permit. Dinner in wealthier areas usually includes soup, meat or poultry (sometimes fish), tortillas or bread, and substantial garnishes such as avocado salad, fried plantains, and pickled vegetables. Appetizers, such as slivers of broiled beef, bites of meat- or cheese-filled pastry, and soft-boiled turtle eggs, are eaten in some urban regions before dinner; dessert may also be served, typically including custards, ice creams, cakes, or fritters.

Special Occasions. Celebrations in Central America are focused on Catholic religious days. Christmas, Easter and Lent, saints' days (including All Saints' Day), and even Sundays may mean a change in fare. Special dishes include the cheese-flavored batter bread called quesadilla that is served in El Salvador on Sundays; sopa de rosquillas, a soup made with ring-shaped corn dumplings traditionally eaten on the Fridays of Lent in Nicaragua; gallina rellena Navidena, a Nicaraguan Christmas dish of chicken stuffed with papaya, chayote squash, capers, raisins, olives, onions, and tomatoes; and plantains served in chocolate sauce during Semana Santa (the Holy Week before Easter) in Guatemala.

Though iguana is eaten throughout Central America and parts of Mexico, South America, and the Caribbean, it is especially popular with Indians in Nicaragua.

Chocolate was so prized in Mayan culture that cacao beans were used as currency.

In Guatemala, eggs poached and served with a seasoned broth are used to treat hangovers.

Tropical fruit from Latin America.

Courtesy of the Florida Division of Tourism

Chicken in tomato sauce (guisado), chicken served with cornmeal porridge, or stews thickened with masa harina are Indian specialties eaten at ceremonial occasions. In some areas the stews are provided by the village headman to serve the community. In Guatemala All Saints' Day is celebrated with a unique salad called fiambre. These enormous salads involve a family social event at which as many as fifty friends and relatives share the creation.[156] They feature vegetables (e.g., green beans, peas, carrots, cauliflower, beets, radishes, cabbage) mixed with chicken, beef, pork, and sausages and then artfully garnished with salami, mortadella, cheese, asparagus, pacaya buds, and hard-boiled eggs. The dressing is either a vinaigrette or a sweet-and-sour sauce.

Etiquette

Dining customs in Central America are often similar to those of Mexico. For example, no guest should begin eating at a meal until the host says "*¡Buen provecho!*"[26] Most Central Americans eat European-style, with the fork remaining in the left hand and the knife in the right one. However, certain more Americanized groups, such as some Nicaraguans, may eat in the American fashion of using the fork in the right hand, switching to the left when cutting food. When not eating, the hands should remain above the table, with the wrists resting on the edge. Bread or tortillas may be served (typically without butter) and should be placed on the side of the plate. It is acceptable to scoop up small bits of food with pieces of tortilla. Dishes are passed to the left. Diners are expected to clean their plates, so taking small portions is appropriate. Asking for seconds is considered a compliment.

[SAMPLE MENU]

A Guatemalan Dinner

Chicken *Jocon*[a,b] or Pollo en Pipian

(Chicken in tomato-pumpkin seed sauce)[b]

Rice

***Frijoles Volteados* (Refried Black Beans)[a,b]**

Radish Salad,[a,b]

Plátanos al Horno (Baked sweet plantains) or Coconut Candy[a]

Hot Chocolate or Coffee

[a]Marks, C. 2004. *False tongues and Sunday bread: A Guatemalan and Mayan cookbook.* Takoma Park, MD: Takoma Books.

[b]*Guatemalan Cuisine & Recipes* at http://www.whats4eats.com/central-america/guatemala-cuisine

Therapeutic Uses of Foods

Some Central Americans follow the hot-cold theory of health and illness, and some also go by the need to balance wet-dry. Guatemalan Americans commonly believe that diarrhea is caused by hot weather and can be alleviated by consuming cold drinks, such as Kool-aid or Gatorade.[149] However, ice cubes may be avoided during hot weather. Panamanians may avoid cold foods when sick, but in one sample, none applied hot-cold principles to daily meals.[147,148] Guatemalans also appeared not to balance hot and cold foods; however, it has been suggested that the practice is so enculturated that it is done without conscious effort.[148] Fatty foods and highly spiced dishes may also be avoided by both Guatemalans and Panamanians when ill.

Herbal remedies are popular throughout Central America, especially teas. Studies of Guatemalans and Panamanians found that the teas were consumed to maintain health, and even more often, to cure minor illness.[147,148] Examples include teas such as manzanilla (chamomile) for improving circulation, menstrual cramps, and flu or colds; banana leaf and *hierbabuena* (mint) for good digestion and regularity; and lemon for general health. *Rosa de jamaica* (hibiscus) was used for respiratory illness, diarrhea, and urinary tract infections, while papaya-leaf tea was considered good for gastritis and as a laxative. Lime, fig leaf, and grapefruit teas were consumed for anxiety and alleviation of stress. Notably, avocado, garlic, ginseng, and valleriana were mentioned as remedies for hypertension and diabetes among Panamanians. Coca leaves, the source of cocaine, are also reported to be used medicinally in some areas.[155]

Contemporary Food Habits in the United States

Adaptations of Food Habits

There is scant information on the Central American diet in the United States. Low rates of assimilation among many Central American immigrants are assumed to result in preservation of traditional food habits. Most Central American ingredients are available in the Latino communities where they settle. One study found that more than half of Honduran women living in New Orleans continued to consume a diet very similar to what they ate in their homeland.[157] Rice, beans, fruit juices, tortillas, cheese, bananas or plantains, beef, and eggs were the items eaten most often. Few new foods were added by a small number of women, and only kiwi fruit, plums, canned vegetables, and olive oil were used by more than 10 percent of the sample. Prepared items, especially hamburgers (eaten by 30 percent of respondents), fried chicken, pizza, and regional dishes, such as jambalaya and Cajun foods (13 percent for each dish), were other new items consumed by a few of the Honduran women. Some also reported baking more foods, frying less, and using more vegetable oil instead of lard or coconut oil in cooking. A meal was defined as having courses, including meat of some sort, and requiring the diner to sit down. Though the women reported skipping meals, this sometimes meant that they ate a sandwich for lunch, which was not considered a meal.

Salvadoran refugees report that, in general, the quality of their diet has declined since arriving in the United States. They state that in El Salvador more foods were made at home from fresh ingredients; they believe that in the United States more processed items and junk foods are eaten, and some nutritious foods are too costly to consume.[151] However, findings from another study of Salvadorans found that there were some beneficial dietary changes after immigration to the United States. Though intake of high-sugar and high-fat foods such as jams or jellies, soft drinks, ice cream, mayonnaise, and vegetable oil increased, consumption of lard, shortening, and fatty meats, including chicharrones and sausage, decreased. Bean dishes, such as frijoles sancochados, remain popular, although other traditional items including pupusas, tamales, and plantain empanadas were eaten significantly less often. Although milk, fruit juice, and fresh salad intake increased, some Salvadorans sampled consumed inadequate servings of dairy products, fruits, and vegetables.[158] A recently published study found that fruit and vegetable intake among all subgroups of Hispanic subgroups in the California, including Central Americans, were higher when compared to non-Hispanic ethnic groups.

Health workers in Florida report that Guatemalan refugees believe that if a food is tasty and does not cause stomach discomfort, it must be good to eat. High intake of candy, soft drinks, and potato chips has been noted. Milk, which is often not well tolerated, may be avoided. WIC (Supplemental Food Program for Women, Infants, and Children) nutritionists found that some food supplements, including milk and cheese, are disliked because of their taste or texture and are sometimes discarded.[149]

Nutritional Status

Nutritional Intake. Limited data on the nutritional status of Central American immigrants have been published. Those who arrive after spending time in refugee camps may suffer high rates of malnutrition resulting in diseases such as beriberi, pellagra, scurvy, and vitamin A deficiency problems, especially in children younger than the age of five. Infectious diseases often follow; tuberculosis and parasites are common.[151] Endemic infections may cause problems as well. Chagas' heart disease, resulting from infections with Trypanosoma cruzi (found in most of Central America), presents symptoms similar to other coronary artery conditions. U.S. outbreaks of cyclosporiasis due to contaminated raspberries imported from Guatemala in 1996 and 1997 suggest another source of infection. Rates of sickle-cell anemia were found to be high (5.7 percent) among mostly Central American adolescents in Los Angeles, and the disease appears to be associated with this population independent of African heritage.[159]

Lactose intolerance may be prevalent among Central Americans.

Illness is a sign of weakness among some Guatemalans, and a person may be stigmatized because he or she is unable to fulfill responsibilities.

Infant mortality rates for Central Americans in the United States are below the average for whites.[160] Low-birth-weight infants were not found to be a problem among Central Americans in a Chicago study. Even those at significant personal or environmental risk (i.e., living in low income, urban neighborhoods) showed no excessive low birth weight.[161] Researchers report that Guatemalans consider breast feeding healthy for infants but impractical. Breast feeding often is used as supplementation to formula and solid foods for the first two to three years of a child's life.[149]

A study of Salvadoran-American youth aged six to eighteen years in Washington, DC, found a prevalence of overweight double that of the national average and 1.7 times higher than for Mexican-American children in national surveys. Thirty-eight percent were overweight (BMI ≥ 95 percentile), and another 22 percent were at risk for overweight. Overweight in this sample was associated with elevated blood pressure, body fat percentage over 30 percent, and early puberty.[162]

An occupational hazard for many Central Americans employed as U.S. farm workers is pesticide or herbicide poisoning. Exposure occurs when labor codes are unenforced or through worker mishandling of the dangerous products.

Counseling. Access to biomedical heath care can be especially difficult for Central Americans. Many are economically and linguistically isolated within their communities; others are unauthorized residents avoiding detection by authorities. Failure to utilize health care is also thought to be due to cost: a high percentage of young adults of Central/South American origins were uninsured in one study, with the highest rates (73 percent) among those who were not U.S. citizens.[163] Further, nearly 57 percent of Central Americans living in the United States report that they do not speak English well.[3]

Most immigrants from Central America are present oriented, though many are also polychronic, viewing time more as a circle than a straight line.[26] They typically view health from

NEW AMERICAN PERSPECTIVE

MARGARET K. WARD, MS, RD, LD/N

I have worked with Hispanic clients, primarily Mexican Americans, since 1990 when I returned to the Western states after living eight years in the Midwest. In my experience many of the clients/patients tend to cook without recipes. The less acculturated client/patient tends to use many more "basic" or less processed foods. They may or may not have been influenced by American food practices to the extent that they use less lard *(manteca)* and more oil (though the vegetable oils chosen are not necessarily the "best/healthiest" choices). I would say it's dependent upon the length of time the person has been in the United States and their level of acculturation. If they've been here a while and are learning English, they tend to acquire more of the Western food culture. Also, the number and age of children can have an effect. Families with older, school-aged children tend to acculturate more because of the influences of school and interaction with "American" children.

My advice to new health care professionals working with this community is to acquire Spanish-language skills as soon as possible. I began my learning in the clinic; then in desperation I took two semesters of Spanish in the community college to acquire the grammatical background that the clinic wasn't providing. Facility in the language is critical to communication. I'm a teacher of nutrition; if I can't communicate with my student, I'm not able to do my job very well. Having good Spanish skills made working with clients on modifying the diet much easier, and most Mexican foods can be modified to meet nutritional needs.

Don't be afraid to make mistakes with Spanish-language speakers. I have found that my Hispanic patients have almost always been extremely forgiving of my horrible Spanish. They seem to appreciate any attempt that one makes to communicate with them in their language. The worst that will happen if you do make an error is that you'll both laugh. I have several Spanish error stories. One took place in the Women, Infant and Children (WIC) clinic when a post-partum patient returned for a follow-up hemoglobin determination, which was low again the second time. We reviewed the foods to include in her diet to increase her hemoglobin, and I again encouraged her to continue her prenatal vitamin use. At the end of my spiel, I told her that if she took her vitamins and ate well that she would *"sentarse bien."* Unfortunately, I told her that she would sit well, rather than feel well (which is *sentirse bien*).

day to day, believing that they have no control over the future. The concept of scheduled appointments may be unfamiliar, and there is little interest in arriving on time. Touching is used to communicate feelings. Men usually embrace close friends, and women are likely to hug all acquaintances. Salvadorans, Guatemalans, and Nicaraguans prefer a light handshake that is lingered over. Men should wait for a woman to extend her hand before shaking, however. Eye contact may be direct when speaking, but it is often downcast when listening politely. Salvadorans use their hands expressively, but it is considered impolite to point with the fingers or the feet. Most Central Americans have a different sense of personal space than do Anglos and prefer to sit and stand closer than is comfortable for many whites, but backing away may be seen as an insult.

Central Americans are high-context communicators, using a calm, measured voice and emotional restraint, except among family and friends. A respectful, yet warm and caring speaking style is most effective.[26]

Culturally based descriptions of symptoms were found among Guatemalans in Florida.[149] "Weak heart" referred to palpitations or dizziness; "weak stomach" meant indigestion; and "weak nervous system" was applied to headaches or insomnia. Taking blood samples was very anxiety provoking because of the Guatemalan belief in the need for strong, ample blood. Anemia was associated with weak blood, to be cured by "eating iron." Guatemalans also identify *susto* and *nervios* as illnesses (see Traditional Health Beliefs and Practices in previous Mexican section) found more often in women than men. If not treated, these conditions are thought to cause diabetes.[16] Guatemalans believe injections are the most effective treatment for illness—more potent than pills. However, one study found that some Guatemalans fear the "chemicals" used in biomedicine.[153] Treatment is successful if symptoms are alleviated.

A study of Salvadorans found that the most important source of emotional support came from family and friends. Single immigrant men preferred living with other Salvadorans when possible.[151] Post-Traumatic Stress Disorder is prevalent among refugees from El Salvador and other Central American nations; it is especially acute among those who are here illegally and suffer continuous anxiety regarding the possibility of deportation.[164]

Central Americans sometimes assimilate into other Latino communities. Cross-cultural exchanges of health beliefs and practices may occur in some areas, and practitioners should be aware of possible Mexican or Caribbean Islander concepts adopted by Central Americans. The in-depth interview is crucial in counseling Central Americans because so little data about food habits, health practices, and nutritional status are available. In addition, information from family members or community experts may be needed.

DISCUSSION STARTERS: COMIC BOOKS AS NUTRITION EDUCATION

In a story on helping migrant workers in the United States improve their diets, Evelyn Theiss, a medical writer for the Cleveland, Ohio newspaper, *The Plain Dealer* (http://www.cleveland.com/healthfit/index.ssf/2010/07/comic_book_helps_families_in_m.html), tells how Jill Kilanowski, an assistant professor at Case Western Reserve University, developed a comic book for migrant worker mothers and their young children. Kilanowski explains, "The mothers told me they wanted reading materials with primary colors to use as a teaching tool for their small children, as well as a story line and pictures." It turns out that comic books are very popular in Mexican and Mexican-American cultures. The major focus of this comic book story is on limiting portion sizes, because Mexican-American immigrants often suffer from obesity, diabetes, and hypertension.

Imagine that you have been hired to write another comic book for Mexican-American migrant workers on making some change to their dietary or health habits, in order to improve their health. Since limiting portion sizes has been done, what other change might you focus on? Outline the storyline, if you can. Consider a comic book for Central American immigrants. Would the same themes and storylines work for them? In small groups, share your themes and storylines, getting feedback from your group members, and then revise your ideas for your comic books.

REVIEW QUESTIONS

1. Compare and contrast the staple foods of Mexico's different regions.
2. Describe the hot-cold system of diet and health practiced traditionally by Mexicans.
3. List two regional U.S. foods that are modifications of Mexican recipes. First, describe the possible original dish, and then explain how it is modified.
4. Which countries make up Central America? Roughly, what are the demographics of immigrants in the United States from Central America?
5. Compare the traditional health beliefs and practices of Mexicans and Central Americans.
6. Describe the food staples of Central America.
7. What are the most common health problems of Mexicans and Central Americans and their decedents living in the United States? How may acculturation to the American diet contribute to these problems?

REFERENCES

1. Silva-Zolezzi I., Hidalgo-Miranda A, Estrada-Gil J, Fernandez-Lopez JC, Uribe-Figueroa L, Contreras A, Balam-Ortiz E, del Bosque-Plata L, Velazquez Fernandez D, Lara C, Goya R, Hernandez-Lemus E, Davila C, Barrientos E, March S, Jimenez-Sanchez G. Analysis of genomic diversity in Mexican Mestizo populations to develop genomic medicine in Mexico. *Proc Natl Acad Sci U S A.* 2009 May 26; 106(21): 861–16.
2. Englekirk, A., & Marin, M. 2000. Mexican Americans. In *Gale Encyclopedia of Multicultural America,* R.V. Dassanowsky & J. Lehman (Eds.). Farmington Hills, MI: Gale Group.
3. Ramirez, R.R. 2004. *We the People: Hispanics in the United States.* Washington, DC: U.S. Census Bureau.
4. Mexican Immigrants: How Many Come? How Many Leave? By Jeffrey Passel and D'Vera Cohn accessed 2/11/2011; http://pewhispanic.org/reports/report.php?ReportID=112
5. Durand, J., Massey, D.S., & Charvet, F. 2000. The changing geography of Mexican immigration to the United States: 1910–1996. *Social Science Quarterly, 81,* 1–15.
6. Passel, J. S. 2006. *The size and characteristics of the unauthorized migrant population in the U.S.* Washington, DC: Pew Hispanic Center.
7. Algert, S. J., Brzezinski, E., & Ellison, T. H. 1998. *Mexican American food practices, customs, and holidays. Ethnic and Regional Food Practices.* Chicago: American Dietetic Association/American Diabetes Association.
8. Grogger, J., & Trejo, S.J. 2002. *Falling behind or moving up? The intergenerational progress of Mexican Americans.* San Francisco: Public Policy Institute of California.
9. U.S. Unauthorized Immigration Flows Are Down Sharply Since Mid-Decade. Jeffrey S. Passel, D'Vera Cohn. Pew Hispanic Center September 1, 2010, 2/13/2011http://pewhispanic.org/files/reports/126.pdf Accessed.
10. Hill, L. E., & Johnson, H. P. 2002. *Understanding the future of Californians' fertility: The role of immigrants.* San Francisco: The Public Policy Institute of California.
11. Neff, N. 1998. Folk Medicine in Hispanics in the Southwestern United States. Online at http://www.rice.edu/projects/HispanicHealth/Courses/mod7/mod7.html Accessed 2/13/2011.
12. Trotter, R. T., & Chavira, J. A. 1997. *Curanderismo: Mexican American Folk Healing.* Athens, GA: University of Georgia Press.
13. Mikhail, B. I. 1994. Hispanic mothers' beliefs and practices regarding selected children's health problems. *Western Journal of Nursing Research, 16,* 623–638.
14. Lopez, R. A. 2005. Use of alternative folk medicine by Mexican American women. *Journal of Immigrant Health, 7,* 23–31.
15. Spector, R. E., 2004. *Cultural diversity in health and illness* (6th ed.). Upper Saddle River, NJ: Pearson Education, Inc.
16. Baer, R.D., Weller, S.C., De Alba Garcia, J.G., Glazer, M., Trotter, R., Pachter, L., & Klein, R.E. 2003. A cross-cultural approach to the study of the folk illness nervios. *Culture, Medicine, and Psychiatry, 27,* 315–337.
17. Zoucha, R., & Purnell, L.D. 2003. People of Mexican heritage. *In Transcultural Health Care: A Culturally Competent Approach,* 2nd ed., L.D. Purnell & B.J. Paulanka (Eds.). Philadelphia: EA Davis Company.
18. Laudan, R. 1999. Chiles, chocolate, and race in New Spain: Glancing backward to Spain or looking forward to Mexico? *Eighteenth Century Life, 23,* 59–70.
19. Leonard, J.N. 1968. *Latin American cooking.* New York: Time-Life Books.
20. Peterson, J., & Peterson, D. 1998. *Eat smart in Mexico: How to decipher the menu, know the market foods, and embark on a tasting adventure.* Madison, WI: Gingko Press, Inc.
21. Kennedy, D. 1990. *Mexican regional cooking.* New York: HarperCollins.
22. Kennedy, D. 1998. *My Mexico: A culinary odyssey with more than 300 recipes.* New York: Clarkson Potter.
23. Menzel, P., & D'Aluisio, F. 1998. *Man eating bugs: The art and science of eating insects.* Berkeley: Ten Speed Press.
24. Romero-Gwynn, E., & Gwynn, D. 1994. Food and dietary adaptation among Hispanics in the United States. In T. Weaver, N. Kanellos, & C. Esteva-Fabregat (Eds.), *Handbook of Hispanic Cultures in the United States: Anthropology.* Houston: Arte Publico Press.
25. Carmichael, E., & Sayer, C. 2005. Feasting with dead souls. In *The Taste Culture Reader,* C. Korsmeyer (Ed.). Gordonsville, VA: Berg Publishers.
26. Foster, D. 2002. *The global etiquette guide to Mexico and Latin America.* New York: John Wiley & Sons.
27. Bharucha, D.X., Morling, B.A., & Niesenbaum, R.A. 2003. Use and definition of herbal medicines differ by ethnicity. *Annals of Pharmacotherapy, 37,* 1409–1413.

28. Davidow, J. 1999. *Infusions of healing: A treasury of Mexican American herbal remedies.* New York: Simon & Schuster.
29. Dole, E.J., Rhyne, R.I., Zeilmann, C.A., Skipper, B.J., McCabe, M.L., & Low Dog, T. 2000. The influence of ethnicity on use of herbal remedies in elderly Hispanic and non-Hispanic whites. *Journal of the American Pharmaceutical Association, 40,* 359–365.
30. Coronado, G.D., Thompson, B., Tejada, S., & Godina, R. 2004. Attitudes and beliefs among Mexican-Americans about type 2 diabetes. *Journal of Health Care for the Poor and Underserved, 15,* 576–588.
31. Pilcher, J.M. 2001. Tex-Mex, Cal-Mex, New Mex, or whose Mex? Notes on the historical geography of Southwestern cuisine. *Journal of the Southwest, 43,* 659–680.
32. Heise, D. 2002. *Hispanic American influence on the U.S. food industry.* Washington, DC: USDA Agricultural Research Service.
33. Ryan, R.W. 2003. Is it border cuisine or merely a case of NAFTA indigestion? *Journal for the Study of Food and Society, 6,* 21–30.
34. *Hispanic Food and Beverages in the U.S.: Market and Consumer Trends in Latino Cuisine,* 4th Edition July 2010 Publisher: Packaged Facts
35. Guadalupe X. Ayala,; Barbara Baquero; Sylvia Klinger, A Systematic Review of the Relationship between Acculturation and Diet among Latinos in the United States: *Implications for Future Research J Am Diet Assoc.* 2008; 108: 1330–1344.
36. Wallendorf, M., & Reilly, M.D. 1983. Ethnic migration, assimilation, and consumption. *Journal of Consumer Research, 10,* 292–302.
37. Ayala, G.X., Mueller, K., Lopez-Madurga, E., Campbell, N.R., & Elder, J.P. 2005. Restaurant and food shopping selections among Latino women in Southern California. *Journal of the American Dietetic Association, 105,* 38–45.
38. Gregory-Mercado, K.Y., Staten, L.K., Ranger-Moore, J., Thomson, C.A., Will, J.C., Ford, E.S., Guillen, J., Larkey, L.K., Giuliano, A.R., & Marshall, J. 2006. Fruit and vegetable consumption of older Mexican-American women is associated with their acculturation level. *Ethnicity & Disease,* 16, 89–95.
39. Neuhouser, M.L., Thompson, B., Coronado, G.D., & Solomon, C.C. 2004. Higher fat intake and lower fruit and vegetables intakes are associated with greater acculturation among Mexicans living in Washington State. *Journal of the American Dietetic Association, 104,* 51–57.
40. Norman, S., Castro, C., Albright, C., & King, A. 2004. Comparing acculturation models in evaluating dietary habits among low-income Hispanic women. *Ethnicity & Health, 14,* 399–404.
41. Monroe, K.R., Hankin, J.H., Malcolm, C.P., Henderson, B.E., Stram, D.O., Park, S., Nomura, A.M.Y., Wilkens, L.R., & Kolonel, L.N. 2003. Correlation of dietary intake and colorectal cancer incidence among Mexican-American migrants: The Multiethnic Cohort Study. *Nutrition and Cancer, 45,* 133–147.
42. Ayala GX, Baquero B, Klinger S. A systematic review of the relationship between acculturation and diet among Latinos in the United States: implications for future research. *J Am Diet Assoc.* 2008; 108: 1330–44.
43. Barquera S, Hernandez-Barrera L, Tolentino ML, Espinosa J, Ng SW, Rivera JA, Popkin BM. Energy intake from beverages is increasing among Mexican adolescents and adults. *J Nutr.* 2008; 138: 2454–61.
44. New Strategist Publications. 2005. *Who's buying groceries* (3rd ed.). Ithaca, NY: Author.
45. Bruhn, C.M., & Pangborn, R.M. 1971. Food habits of migrant workers in California. *Journal of the American Dietetic Association, 59,* 347–355.
46. Dewey, K.G., Strode, M.A., & Ruiz Fitch, Y. 1984. Dietary change among migrant and nonmigrant Mexican-American families in northern California. *Ecology of Food and Nutrition, 14,* 11–24.
47. Dwyer, J.T., Stone, E.J., Minhua, Y., Webber, L.S., Must, A., Feldman, H.A., Nader, P.R., Perry, C.L., & Parcel, G.S. 2000. Prevalence of marked overweight and obesity in a multiethnic pediatric population: Findings from the Child and Adolescent Trial for Cardiovascular Health (CATCH) study. *Journal of the American Dietetic Association, 100,* 1149–1156.
48. Moreno, C., Alvarado, M., Balcazar, H., Lane, C., Newman, E., Ortiz, G., & Forrest, M. 1997. Heart disease education and prevention program targeting immigrant Latinos: Using focus group responses to develop effective interventions. *Journal of Community Health,* 22, 435–450.
49. Romero-Gwynn, E., Gwynn, D., Grivetti, L., McDonald, R., Stanford, G., Turner, B., West, E., & Williamson, E. 1993. Dietary acculturation among Latinos of Mexican descent. *Nutrition Today, 28,* 6–12.
50. Jackson KA, Savaiano DA. Lactose maldigestion, calcium intake and osteoporosis in African, Asian, and Hispanic Americans. *J Am Coll Nutr.* 2001; 20(suppl 2): 198S–207S.
51. NPD Group. 2005. *At the table with Hispanic families across America.* Port Washington, NY: Author.
52. Wechsler, H., Basch, C.E., Zybert, P., Lantigua, R., & Shea, S. 1995. The availability of low-fat milk in an inner-city Latino community: Implications for nutrition education. *American Journal of Public Health, 85,* 1690–1692.
53. Lisabeth LD, Sánchez BN, Escobar J, Hughes R, Meurer WJ, Zuniga B, Garcia N, Brown DL, Morgenstern LB. Health Place. 2010 May; 16(3): 598–605. Epub 2010 Feb 1.*The food environment in an urban Mexican American community.*
54. Elder, J., Sallis, J.F., Zive, M.M., Hoy, P., McKenzie, T.L., Nader, P.R., & Berry, C.C. 1999. Factors affecting selection of restaurants by Anglo- and Mexican-American families. *Journal of the American Dietetic Association, 99,* 856–857.
55. Alicia Moag-Stahlberg, MS, RD, LD. The American Dietetic Association Foundation's 2010 Family Nutrition and Physical Activity Survey January, 2011 http://www.eatright.org/foundation/fnpa/.
56. Normand, V. 2005. Montezuma's revenge. Metro, July 13–19. Accessed 2/14/2011 http://www.metroactive.com/papers/metro/07.13.05/aztecs-0528.html

57. Hummer, R.A., Rogers, R.G., Amir, S.H., Forbes, D., & Frisbie, W.P. 2000. Adult mortality differentials among Hispanic subgroups and nonHispanic whites. *Social Science Quarterly, 81,* 859–876.
58. Hummer, R.A., Rogers, R.G., Nam, C.B., & LeClere, F.B. 1999. Race/ethnicity, nativity, and U.S. adult mortality. *Social Science Quarterly, 80,* 1083–1118.
59. National Center for Health Statistics. 2005. *Health, United States, 2005 with Chartbook on Trends in the Health of Americans.* Hyattsville, MD: Author.
60. Xu JQ, Kochanek KD, Murphy SL, Tejada-Vera B. *Deaths: Final data for 2007.* National vital statistics reports; vol 58 no 19. Hyattsville, MD: National Center for Health Statistics. 2010.
61. Smith, D.P., & Bradshaw, B.S. 2006. Rethinking the Hispanic paradox: Death rates and life expectancy for US non-Hispanic white and Hispanic populations. *American Journal of Public Health, 96,* 1686–1692.
62. Patel, K.V., Eschbach, K., Ray, L.A., & Markides, K.S. 2004. Evaluation of mortality data for older Mexican Americans: Implications for the Hispanic paradox. *American Journal of Epidemiology, 159,* 707–715.
63. Markides, K.S., & Eschbach, K. 2005. Aging, migration, and mortality: Current status of research on the Hispanic paradox. *Journals of Gerontology.* Series B, Psychological Sciences and Social Sciences, 60, 68–75.
64. Palloni, A., & Arias, E. 2004. Paradox lost: Explaining the Hispanic adult mortality advantage. *Demography, 41,* 385–415.
65. National Center for Health Statistics. 2005. *Health, United States, 2005 with Chartbook on Trends in the Health of Americans.* Hyattsville, MD: Author.
66. Mathews TJ, MacDorman MF. *Infant mortality statistics from the 2006.* National vital statistics reports; vol 58 no 17. Hyattsville, MD: National Center for Health Statistics. 2010.
67. Mennella, J.A., Turnbull, B., Ziegler, P.J., & Martinez, H. 2005. Infant feeding practices and early flavor experiences in Mexican infants: An intracultural study. *Journal of the American Dietetic Association, 105,* 908–915.
68. Margaret M. McDowell, M.P.H., R.D.; Chia-Yih Wang, Ph.D.; and Jocelyn Kennedy-Stephenson, M.S. *Breastfeeding in the United States: Findings from the National Health and Nutrition Examination Survey,* 1999–2006 NCHS Data Brief Number 5, April 2008.
69. Rachel Tolbert Kimbro, Scott M. Lynch and Sara McLanahan. The Influence of Acculturation on Breastfeeding Initiation and Duration for Mexican-Americans. *Population Research and Policy Review* Volume 27, Number 2, 183–199.
70. Brotanek JM, Schroer D, Valentyn L, Tomany-Korman S, Flores G. Reasons for prolonged bottle-feeding and iron deficiency among Mexican-American toddlers: an ethnographic study. *Acad Pediatr.* 2009 Jan–Feb; 9(1): 17–25.
71. Brotanek, J.M., Halterman, J.S., Auinger, P., Flores, G., & Weitzman, M. 2005. Iron deficiency, prolonged bottle-feeding, and racial/ethnic disparities in young children. *Archives of Pediatric and Adolescent Medicine, 159,* 1038–1042.
72. Huntingdon, N.L., Kim, I.J., & Hughes, C.V. 2002. Caries-risk factors for Hispanic children affected by early childhood caries. *Pediatric Dentistry, 24,* 536–542.
73. Ervin RB. Healthy Eating Index scores among adults, 60 years of age and over, by sociodemographic and health characteristics: United States, 1999–2002. Advance data from vital and health statistics; no 395. Hyattsville, MD: National Center for Health Statistics. 2008
74. Evans, A.E., Sawyer-Morse, M.K., & Betsinger, A. 2000. Fruit and vegetable consumption among Mexican-American college students. *Journal of the American Dietetic Association, 100,* 1399–1402.
75. Montez JK, Eschbach K. Country of birth and language are uniquely associated with intakes of fat, fiber, and fruits and vegetables among Mexican-American women in the United States. *J Am Diet Assoc.* 2008 Mar; 108(3): 473–80.
76. Guendelman, S., & Abrams, B. 1995. Dietary intake among Mexican-American women: Generational differences and a comparison with white non-Hispanic women. *American Journal of Public Health, 85,* 20–25.
77. Wright JD, Wang C-Y. *Trends in intake of energy and macronutrients in adults from 1999–2000 through 2007–2008.* NCHS data brief, no 49. Hyattsville, MD: National Center for Health Statistics. 2010.
78. Ramakrishnan, U., Frith-Terhune, A., Cogswell, M., & Kettel Khan, L. 2002. Dietary intake does not account for differences in low-iron stores among Mexican American and non-Hispanic white women: Third National Health and Nutrition Examination Survey, 1988–1994. *Journal of Nutrition, 132,* 996–1001.
79. Dixon, L.B., Sundquist, J., & Winkleby, M. 2000. Differences in energy, nutrient, and food intakes in a US sample of Mexican-American women and men: Findings from the Third National Health and Nutrition Examination Survey, 1988–1994. *American Journal of Epidemiology, 152,* 548–557.
80. Hampl, J.S., Taylor, C.A., & Johnston, C.S. 2004. Vitamin C deficiency and depletion in the United States: The Third National Health and Nutrition Examination Survey, 1988 to 1994. *American Journal of Public Health, 94,* 870–875.
81. Harley, K., Eskenazi, B., & Block, G. 2005. The association of time in the US and diet during pregnancy in low-income women of Mexican descent. *Paediatric and Perinatal Epidemiology, 19,* 125–134.
82. Zive, M.M., Taras, H.L., Broyles, S.L., & Frank-Spohrer, G.C. 1995. Vitamin and mineral intakes of Anglo-American and Mexican-American preschoolers. *Journal of the American Dietetic Association, 95,* 329–335.
83. Trends in Tuberculosis Incidence—United States, 2008 March 20, 2009 / 58(10);249-253. Morbidity and Mortality Weekly Report, Centers for Disease Control and Prevention.
84. Cynthia L. Ogden, Ph.D., and Margaret D. Carroll, M.S.P.H., Prevalence of Overweight, Obesity, and Extreme Obesity Among Adults: United States, Trends 1976–1980 Through 2007–2008. Division of Health and Nutrition Examination Surveys, acccessed

2/16/2011. http://www.cdc.gov/NCHS/data/hestat/obesity_adult_07_08/obesity_adult_07_08.pdf

85. Freedman, D.S., Khan, L.K., Serdula, M.K., Ogden, C.L., & Dietz, W.H. 2006. Racial and ethnic differences in secular trends for childhood BMI, weight, and height. *Obesity* (Silver Spring, MD), *14,* 301–308.
86. Ayala, G.X., Elder, J.P., Campbell, N.R., Slymen, D.J., Roy, N., Engleberg, M., & Ganiats, T. 2004. Correlates of body mass index and waist-to-hip ratio among Mexican women in the United States: Implications for intervention development. *Womens Health Issues, 14,* 155–164.
87. Crespo, C.J., Smit, E., Andersen, R.E., Carter-Pokras, O., & Ainsworth, B.E. 2000. Race/ethnicity, social class and their relation to physical inactivity during leisure time: Results from the Third National Health and Nutrition Examination Survey, 1988–1994. *American Journal of Preventive Medicine, 18,* 46–53.
88. Gordon-Larsen, P., Adair, L.S., & Popkin, B.M. 2003. The relationship of ethnicity, socioeconomic factors, and overweight in US adolescents. *Obesity Research, 11,* 121–129.
89. Eckstein, K.C., Mikhail, L.M., Ariza, A.J., Thomson, J.S., Millard, S.C., & Binns, H.J. 2006. Parents' perceptions of their child's weight and health. *Pediatrics, 11,* 681–690.
90. Robinson, T.N., Chang, J.Y., Haydel, K.F., & Killen, J.D. 2001. Overweight concerns and body image dissatisfaction among third-grade children: The impacts of ethnicity and socioeconomic status. *Journal of Pediatrics, 138,* 181–187.
91. Sanchez-Johnsen, L.A., Fitzgibbon, M.L., Martinovich, Z., Stolley, M.R., Dyer, A.R., & Van Horn, L. 2004. Ethnic differences in correlates of obesity between Latin-American and black women. *Obesity Research, 12,* 652–660.
92. Talamayan, K.S., Springer, A.E., Kelder, S.H., Gorospe, E.C., & Joye, K.A. 2006. Prevalence of overweight and weight control behaviors among normal weight adolescents in the United States. *Scientific World Journal, 26,* 365–373.
93. Byrd, T.L., Balcazar, H., & Hummer, R.A. 2001. Acculturation and breast feeding intention and practice in Hispanic women on the US-Mexico border. *Ethnicity & Disease, 11,* 72–79.
94. Regan, P.C., & Cachelin, F.M. 2006. Binge eating and purging in a multi-ethnic community sample. *International Journal of Eating Disorders, 39,* 523–526.
95. Cowie, C.C., Rust, K.F., Byrd-Holt, D.D., Eberhardt, M.S., Flegal, K.M., Engeigau, M.M., Saydah, S.H., Williams, D.E., Geiss, L.S., & Gregg, E.W. 2006. Prevalence of diabetes and impaired fasting glucose in adults in the U.S. population: National Health and Nutrition Examinations Survey 1999–2002. *Diabetes Care, 29,* 1263–1268.
96. Hertz, R.P., Unger, A.N., & Ferrario, C.M. 2006. Diabetes, hypertension, and dyslipidemia in Mexican Americans and non-Hispanic whites. *American Journal of Preventive Medicine, 30,* 103–110.
97. Burke, J.P., Williams, K., Gaskill, S.P., Hazuda, H.P., Haffner, S.M., & Stern, M.P. 1999. Rapid rise in the incidence of type 2 diabetes from 1987 to 1996: Results from the San Antonio Heart Study. *Archives of Internal Medicine, 159,* 1450–1456.
98. Haffner, S.M., Hazuda, H.P., Mitchell, B.D., Patterson, J.K., & Stern, M.P. 1991. Increased incidence of Type II diabetes mellitus in Mexican Americans. *Diabetes Care, 14,* 102–108.
99. Diaz, V.A., Mainous, A.G., Koopman, R.J., & Geesey, M.E. 2005. Are differences in insulin sensitivity explained by variation in carbohydrate intake? *Diabetologia, 48,* 1264–1268.
100. 2011 National Diabetes Fact Sheet, Centers for Disease Control and Prevention accessed 2/15/2011 available at: www.cdc.gov/diabetes/pubs/figuretext11.htm#fig4
101. Harris, M.I. 1998. Diabetes in America: Epidemiology and scope of the problem. *Diabetes Care, 21,* C11–C14.
102. Pugh, J.A. 1996. Diabetic nephropathy and end-stage renal disease in Mexican Americans. *Blood Purification, 14,* 286–292.
103. Black, S.A. 1999. Increased health burden associated with comorbid depression in older diabetic Mexican Americans. Results from the Hispanic Established Population for the Epidemiologic Study of the Elderly survey. *Diabetes Care, 22,* 56–64.
104. Centers for Disease Control. 2006. Diabetes death rate for Hispanics compared to *nonHispanic* whites—United States versus counties along the U.S.-Mexico border, 2000–2002. MMWR,
105. Smith, C.A., & Barnett, E. 2005. Diabetes-related mortality among Mexican Americans, Puerto Ricans, and Cuban Americans in the United States. *Revista Panamericana de Salud Pública, 18,* 381–387.
106. Ervin RB. Prevalence of metabolic syndrome among adults 20 years of age and over, by sex, age, race and ethnicity, and body mass index: United States, 2003–2006. National health statistics reports; no 13. Hyattsville, MD: National Center for Health Statistics. 2009.
107. Statistics. *Health, United States, 2008.* Hyattsville, MD: National Center for Health Statistics; 2008.
108. Statistical Fact Sheet — Risk Factors 2010 Update High Blood Cholesterol and Other Lipids — Statistics. Accessed 2/15/2011 - available at http://www.americanheart.org/downloadable/heart/1261004785748FS13CH10.pdf
109. Sundquist, J., & Winkleby, M.A. 1999. Cardiovascular risk factors in Mexican American adults: A transcultural analysis of NHANES III, 1988–1994. *American Journal of Public Health, 89,* 723–730.
110. American Heart Assn Heart Dosease and Stroke Statistics 2010 update at a glance. Accessed 2/15/2011 - Available at http://www.americanheart.org/downloadable/heart/1265665152970DS-3241%20HeartStrokeUpdate_2010.pdf
111. Hunt, K.J., Resendez, R.G., Williams, K., Haffner, S.M., Stern, M.P., & Hazuda, H.P. 2003. All-cause and cardiovascular mortality among Mexican-American and non-Hispanic white older participants in the

San Antonio Heart Study—evidence against the "Hispanic paradox." *American Journal of Epidemiology, 158,* 1048–1057.

112. National Center for Health Statistics. 2005. *Health, United States, 2005 with Chartbook on Trends in the Health of Americans.* Hyattsville, MD: Author.
113. Nguyen, H.T., & Stack, A.G. 2006. Ethnic disparities in cardiovascular risk factors and coronary disease prevalence among individuals with chronic kidney disease: Findings from the Third National Health and Nutrition Examination Survey. *Journal of the American Society of Nephrologists, 17,* 1716–1723.
114. Pandey, D.K., Labarthe, D.R., Goff, D.C., Chan, W., & Nichaman, M.Z. 2001. Community-wide coronary heart disease mortality in Mexican Americans equals or exceeds that in non-Hispanic whites: The Corpus Christi Heart Project. *American Journal of Medicine, 110,* 81–87.
115. Puppala, S., Dodd, G.D., Fowler, S., Arya, R., Schneider, J., Farook, V.S., Granato, R., Dyer, T.D., Almasy, L., Jenkinson, C.P., Diehl, A.K., Stern, M.P., Blangero, J., & Duggirala, R. 2006. A genomewide search finds major susceptibility loci for gallbladder disease on chromosome 1 in Mexican Americans. *American Journal of Human Genetics, 78,* 377–392.
116. Tseng, M., Deveillis, R.F., Maurer, K.R., Khare, M., Kohimeier, L., Everhart, J.E., & Sandler, R.S. 2000. Food intake patterns and gallbladder disease in Mexican Americans. *Public Health Nutrition, 3,* 233–243.
117. Geller, S.E., & Derman, R. 2001. Knowledge, beliefs, and risk factors for osteoporosis among African-American and Hispanic women. *Journal of the National Medical Association, 93,* 13–21.
118. Lukes, S.M., & Simon, B. 2005. Dental decay in southern Illinois migrant and seasonal farmworkers: An analysis of clinical data. *Journal of Rural Health, 21,* 254–258.
119. Nurko, C., Aponte-Merced, L., Bradley, E.L., & Fox, L. 1998. Dental caries prevalence and dental health care of Mexican American workers' children. *ASDC Journal of Dentistry in Children, 65,* 65–72.
120. Saitz, R., Lepore, M.F., Sullivan, L.M., Amaro, H., & Samet, J.H. 1999. Alcohol abuse and dependence in Latinos living in the United States: Validation of the CAGE (4M) questions. *Archives of Internal Medicine, 159,* 718–724.
121. Fryar CD, Hirsch R, Porter KS, et al. Smoking and alcohol behaviors reported by adults, United States, 1999–2002. Advance data from vital and health statistics; no 378. Hyattsville, MD: National Center for Health Statistics. 2006.
122. Sobralske, M.C. 2006. Health care seeking among Mexican American men. *Journal of Transcultural Nursing, 17,* 129–138.
123. Hsiao, A.F., Wong, M.D., Goldstein, M.S., Yu, H.J., Andersen, R.M., Brown, E.R., Becerra, L.M., & Wenger, N.S. 2006. Variations in complementary and alternative medicine (CAM) use across racial/ethnic groups and the development of ethnic-specific measures of CAM use. *Journal of Alternative and Complementary Medicine, 12,* 281–290.
124. Risser, A.L., & Mazur, L.J. 1995. Use of folk remedies in a Hispanic population. *Archives of Pediatric and Adolescent Medicine, 149,* 978–981.
125. Tafur MM, Crowe TK, Torres E. A review of curanderismo and healing practices among Mexicans and Mexican Americans. *Occup Ther Int. 2009;* 16(1): 82–8.
126. Yehieli, M., & Grey, M.A. 2005. *Health matters: A pocket guide for working with diverse cultures and underserved populations.* Boston: Intercultural Press.
127. Poss, J., & Jezewski, M.A. 2002. The role and meaning of susto in Mexican Americans' explanatory model of type 2 diabetes. *Medical Anthropology Quarterly, 16,* 360–377.
128. Espino, D.V., Bazaldua, O.V., Palmer, R.F., Mouton, C.P., Parchman, M.L., Miles, T.P., & Markides, K. 2006. *Journals of Gerontology.* Series A, Biological Sciences and Medical Sciences, 61, 170–175.
129. Kuo, Y.F., Raji, M.A., Markides, K.S., Ray, L.A., Espino, D.V., & Goodwin, J.S. 2003. Inconsistent use of diabetes complications, and mortality in older Mexican Americans over a 7-year period: Data from the Hispanic established population for the epidemiologic study of the elderly. *Diabetes Care, 26,* 3054–3060.
130. Poss, J., Jezewski, M.A., & Stuart, A.G. 2003. Home remedies for type 2 diabetes used by Mexican Americans in El Paso, Texas. *Clinical Nursing Research, 12,* 304–323.
131. Galanti, G.A. 2004. *Caring for patients from different cultures,* 3rd ed. Philadelphia: University of Pennsylvania Press.
132. Graham, R.E., Ahn, A.C., Davis, R.B., O'Conner, B.B., Eisenber, D.M., & Phillips, R.S. 2005. Use of complementary and alternative medical therapies among racial and ethnic minority adults: Results from the 2002 National Health Interview Survey. *Journal of the National Medical Association, 97,* 535–545.
133. Blackhall, L.J., Murphy, S.T., Frank, G., Michel, V., & Azen, S. 1995. Ethnicity and attitudes toward patient autonomy. *Journal of the American Medical Association, 274,* 820–825.
134. Horowitz, C.R., Tuzzio, L., Rojas, M., Monteith, S.A., & Sisk, J.E. 2004. How do urban African Americans and Latinos view the influence of diet on hypertension? *Journal of Health Care for the Poor and Underserved, 15,* 631–644.
135. Wen, L.K., Shepherd, M.D., & Parchman, M.L. 2004. Family support, diet, and exercise among older Mexican Americans with type 2 diabetes. *Diabetes Education, 30,* 980–993.
136. Yetley, E.A., Yetley, M.J., & Aguirre, B. 1981. Family role structure and food related roles in Mexican-American families. *Journal of Nutrition Education, 13* (Suppl. 1), S96–S101.
137. Ikeda, J., & Gonzales, M.M. 1986. *Food habits of Hispanics newly immigrated to the San Francisco Bay area* (abstract). Oakland: California Dietetic Association Annual Meeting.
138. Melville, M.B. 1985. Salvadorans and Guatemalans. In D.W. Haines (Ed.), Refugees in the United States. Westport, CT: Greenwood.

139. Mumford, J. 2000. Salvadoran Americans. In *Gale Encyclopedia of Multicultural America,* R.V. Dassanowsky & J. Lehman (Eds.). Farmington Hills, MI: Gale Group.
140. Smagula, S. 2000. Nicaraguan Americans. In *Gale Encyclopedia of Multicultural America,* R.V. Dassanowsky & J. Lehman (Eds.). Farmington Hills, MI: Gale Group.
141. Hispanic or Latino Origin by Specific Origin-Universe: Total Population. 2009. American Community Survey. Accessed 2/15/2011. http://factfinder.census.gov/servlet/DTTable?_bm=y&-geo_id=01000US&-ds_name=ACS_2009_1YR_G00_&-_lang=en&-redoLog=true&-mt_name=ACS_2009_1YR_G2000_B03001&-format=&-CONTEXT=dt
142. Hong, M. 2000. Guatemalan Americans. In *Gale Encyclopedia of Multicultural America,* R.V. Dassanowsky & J. Lehman (Eds.). Farmington Hills, MI: Gale Group.
143. Dean, R.S. 2000. Panamanian Americans. In *Gale Encyclopedia of Multicultural America,* R.V. Dassanowsky & J. Lehman (Eds.). Farmington Hills, MI: Gale Group.
144. U.S. Census. Immigration Statistics Staff. 2004. Foreign-Born Profiles (STP-159). Accessed 2/15/2011 at http://www.census.gov/population/www/socdemo/foreign/STP-159-2000tl.html
145. Maxwell, W. 2000. Honduran Americans. In *Gale Encyclopedia of Multicultural America,* R.V. Dassanowsky & J. Lehman (Eds.). Farmington Hills, MI: Gale Group.
146. Chase, C.S. 2000. Costa Rican Americans. In *Gale Encyclopedia of Multicultural America,* R.V. Dassanowsky & J. Lehman (Eds.). Farmington Hills, MI: Gale Group.
147. Purnell, L. 1999. Panamanians' practices for health promotion and the meaning of respect afforded them by health care providers. *Journal of Transcultural Nursing, 10,* 331–339.
148. Purnell, L. 2001. Guatemalans' practices for health promotion and the meaning of respect afforded them by health care providers. *Journal of Transcultural Nursing, 12,* 40–47.
149. Miralles, M.A. 1989. *A matter of life and death: Health-seeking behavior of Guatemalan refugees in south Florida.* New York: AMS Press.
150. Leonard, J.N. 1968. *Latin American cooking.* New York: Time-Life Books.
151. Boyle, J.S. 1991, April/May. Transcultural nursing care of Central American refugees. *National Student Nurses Association, Inc./Imprint,* 73–77.
152. Rutherford, M.S., & Roux, G.M. 2002. Health beliefs and practices in rural El Salvador: An ethnographic study. *Journal of Cultural Diversity, 9,* 3–11.
153. Zapata, J. 1999. The use of folk healing and healers by six Latinos living in New England: A preliminary study. *Journal of Transcultural Nursing,* 136–142.
154. Ailinger, R.L., Molloy, S., Zamora, L., & Benavides, C. 2004. Herbal remedies in a Nicaraguan barrio. *Journal of Transcultural Nursing, 15,* 278–282.
155. Barrett, B. 1995. Ethnomedical interactions: Health and identity on Nicaragua's Atlantic coast. *Social Science and Medicine, 40,* 1611–1621.
156. Marks, C. 2004. *False tongues and Sunday bread: A Guatemalan and Mayan cookbook.* Takoma Park, MD: Takoma Books.
157. Edmonds, V.M. 2005. The nutritional patterns of recently immigrated Honduran women. *Journal of Transcultural Nursing, 16,* 226–235.
158. Romieu, I., Hernandez-Avila, M., Rivera, J.A., Ruel, M.T., & Parra, S. 1997. Dietary studies in countries experiencing a health transition: Mexico and Central America. *American Journal of Clinical Nutrition, 65,* 1159S–1165S.
159. Hamdallah, M., & Bhatia, A.J. 1995. Prevalence of sickle-cell trait in USA adolescents of Central American origin. *Lancet, 346,* 707–708.
160. Mathews TJ, MacDorman MF. *Infant mortality statistics from the 2006 period linked birth/infant death data set.* National vital statistics reports; vol 58 no 17. Hyattsville, MD: National Center for Health Statistics. 2010.
161. Prevalence of low birth weight among Hispanic infants with United States–born and foreign-born mothers: The effect of urban poverty. *American Journal of Epidemiology, 139,* 184–192.
162. Mirza, N.M., Kadow, K., Palmer, M., Solano, H., Rosche, C., & Yanovski, J.A. 2004. Prevalence of overweight among inner city Hispanic-American children and adolescents. *Obesity Research, 12,* 1298–1310.
163. Callahan, S.T., Hickson, G.B., & Cooper, W.O. 2006. Health care access of Hispanic young adults in the United States. *Journal of Adolescent Health, 39,* 627–633.
164. Molesky, J. 1986. Pathology of Central America refugees. *Migration World, 14,* 19–23.
165. Bladholm, L. 2001. *Latin & Caribbean grocery stores demystified.* Los Angeles: Renaissance Books.
166. Laudan, R. 2004. The Mexican's kitchen's Islamic connection. *Saudi Aramco World, 55,* 32–39.
167. Looker, A.C., Loria, C.M., Carroll, M.D., McDowell, M.A., & Johnson, C.L. 1993. Calcium intakes of Mexican Americans, Cubans, Puerto Ricans, non-Hispanic whites, and non-Hispanic blacks in the United States. *Journal of the American Dietetic Association, 93,* 1274–1279.
168. Romieu, I., Hernandez-Avila, M., Rivera, J.A., Ruel, M.T., & Parra, S. 1997. Dietary studies in countries experiencing a health transition: Mexico and Central America. *American Journal of Clinical Nutrition, 65,* 1159S–1165S.
169. Basiotis, P.P., Carlson, A., Gerrior, S.A., Juan, W.Y., & Lino, M. 2002. *The Healthy Eating Index: 1999–2000.* Washington, DC: U.S. Department of Agriculture, Center for Nutrition Policy and Promotion. CNPP–12.
170. Harris, K.M. 2000. The health status and risk behaviors of adolescents in immigrant families. *In Children of Immigrants: Health Adjustment and Public Assistance,* D.J. Hernandez (Ed.). Washington, DC: National Research Council.

171. Landale, N.S., Oropesa, R.S., & Gorman, B.K. 1999. Immigration and infant health: Birth outcomes of immigrant and native-born women. In D.J. Hernandez (Ed.), *Children of Immigrants: Health, Adjustment, and Public Assistance.* Washington, DC: National Academy Press.
172. Franzini, L., & Fernandez-Esquer, M.E. 2004. Socioeconomic, cultural, and personal influences on health outcomes in low-income Mexican-origin individuals in Texas. *Social Science & Medicine, 59,* 1629–1646.
173. Gould, J.B., Madan, A., Qin, C., & Chavez, G. 2003. Perinatal outcomes in two dissimilar immigrant populations in the United States. *Pediatrics, 111,* e676–682.
174. Wingate, M.S., & Alexander, G.R. 2006. The healthy migrant theory: Variations in pregnancy outcomes among US-born migrants. *Social Science & Medicine, 62,* 491–498.
175. Guendelman, S., Thornton, D., Gould, J., & Hosang, N. 2006. Mexican women in California: Differentials in maternal morbidity between foreign and US-born populations. *Paediatrics and Perinatal Epidemiology, 20,* 471–481.
176. Acevedo, M.C. 2000. The role of acculturation in explaining ethnic differences in the prenatal health-risk behaviors, mental health, and parenting beliefs of Mexican-American and European American at risk women. *Child Abuse and Neglect, 24,* 111–127.
177. Page, R.L. 2004. Positive pregnancy outcomes in Mexican immigrants: What can we learn? *Journal of Obstetrical and Gynecological Nursing, 33,* 783–790.
178. Jenny, A.M., Schoendorf, K.C., & Parker, J.D. 2001. The association between community context and mortality among Mexican-American infants. *Ethnicity & Disease, 11,* 722–731.
179. Guendelman, S., Thornton, D., Gould, J., & Hosang, N. 2006. Mexican women in California: Differentials in maternal morbidity between foreign and US-born populations. *Paediatrics and Perinatal Epidemiology, 20,* 471–481.
180. McMurray, M.P., Ceriqueira, M.T., Conner, S.L., & Conner, W.E. 1991. Changes in lipid and lipoprotein levels and body weight in Tarahumara Indians after consumption of an affluent diet. *New England Journal of Medicine, 325,* 1704–1708.
181. Wright, J., Romero-Gwynn, E., Cotter, A., Powell, C., Garrett, C., Grajales-Hall, M., Parnell, S., Stanford, G., Turner, B., Wrightman, N., & Williamson, E. 1996. Radio is effective in teaching nutrition to Latino families. *California Agriculture, 50,* 14–18.

Caribbean Islanders and South Americans

CHAPTER 10

Latinos from the Caribbean Islands and South America often seem more different from one another than similar. Their homelands vary from the tropics of the islands and northern Brazil to the highland plains of Argentina and the snow-topped mountains of Peru. Their ethnic backgrounds include native Indian, Spanish, Portuguese, French, British, Danish, Dutch, African, Asian Indian, Chinese, Italian, German, and Japanese. And though Roman Catholicism is practiced by a majority, many others follow Protestant faiths, Judaism, and numerous indigenous Afro-European religions including voodoo, santeria, and candomblé.

One commonality between Caribbean Islanders and South Americans is a variety of regional fares with few national cuisines. Dishes typically combine native ingredients with foods introduced from Europe, Africa, and Asia, with a broad preference for strong, spicy flavors. This chapter reviews Caribbean Islanders and their fare, focusing on Puerto Ricans, Cubans, and Dominicans. A summary of South Americans is also presented (see Figure 10.1). Other Latinos are covered in Chapter 9, "Mexicans and Central Americans."

Caribbean Islanders

More than 1,000 tropical islands in the Caribbean stretch from Florida to Venezuela. They include the Bahamas, the Greater Antilles (Jamaica, Cuba, Hispaniola, and Puerto Rico), and the Lesser Antilles. The largest island is Cuba, and the smallest islands are barely more than exposed rocks. Most were claimed at one time by Spain, Britain, France, the Netherlands, Denmark, or the United States, but now include the independent nations of Antigua/Barbuda, the Bahamas, Barbados, Cuba, Dominica, Dominican Republic, Grenada, Haiti, Jamaica, St. Christopher/Nevis, St. Lucia, St. Vincent/Grenadines, and Trinidad and Tobago, as well as the U.S. territory of Puerto Rico. Many islands, such as the Virgin Islands (U.S.) and Martinique (France), are still under foreign control.

The islands are uniformly scenic. The tropical warmth and torrential rains provide the ideal climate for a lush plant cover that includes numerous indigenous fruits and vegetables. Later immigrants found the region suitable for imported crops. The Caribbean Islands share a history of domination by foreign powers and political turmoil. Native Indians, Europeans, blacks from Africa, and Asians from China and India have intermarried over the centuries to produce an extremely diverse population.

When Columbus landed in the Bahamas in 1492, he believed he had discovered a new route to Indonesia and called the native people "Indians." The Caribbean Islands later became known as the West Indies.

Cultural Perspective

History of Caribbean Islanders in the United States

Immigration Patterns. It is estimated that the Latino population in the United States was more than 41 million people as of 2005, and Caribbean

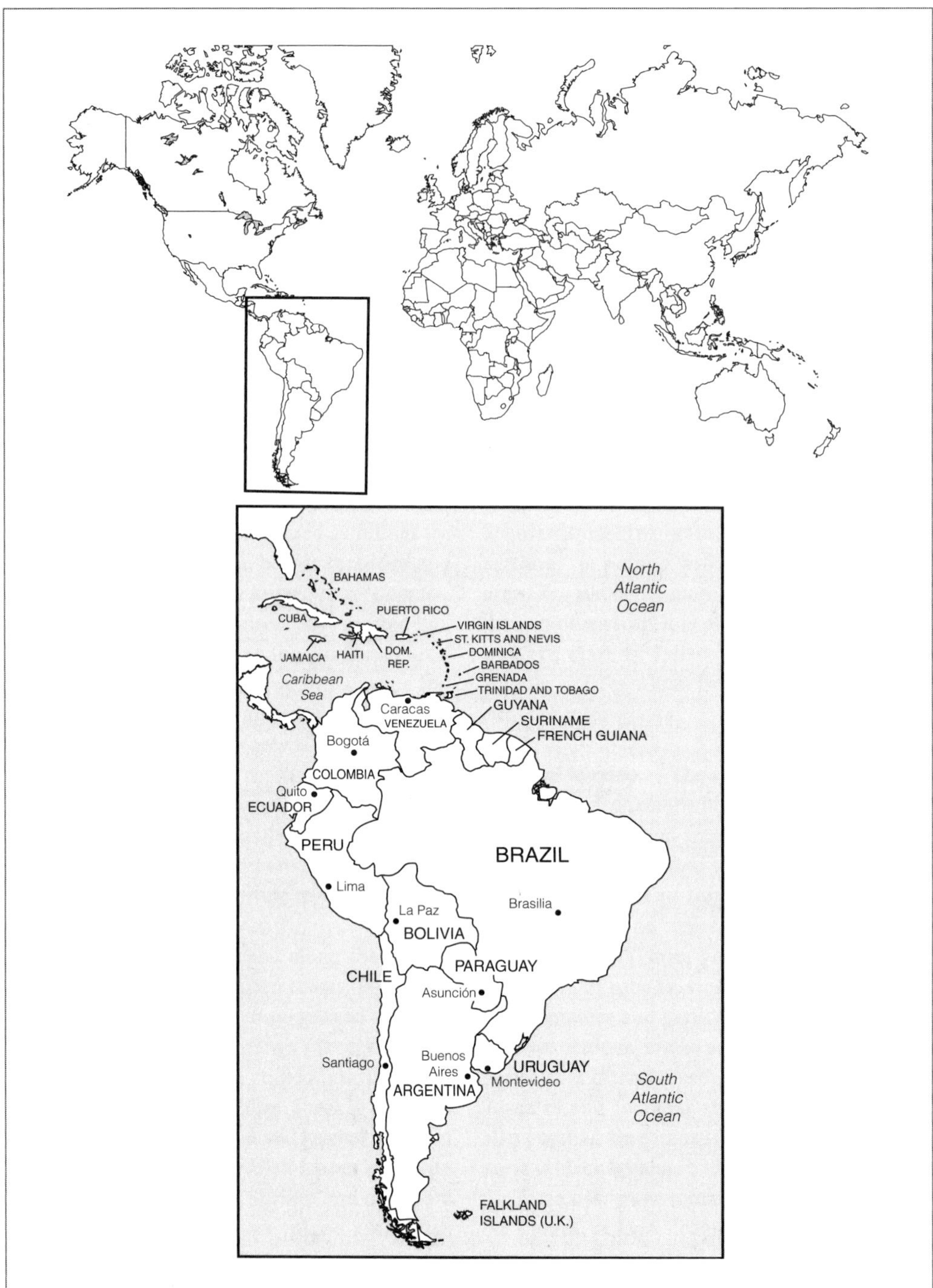

Figure 10.1
Caribbean Islands and South America.

Islanders accounted for just over 20 percent of that total. Residents from Puerto Rico constituted the largest group, followed by those from Cuba and the Dominican Republic. In addition, there were small groups of immigrants from other Caribbean nations, most significantly from Jamaica and Haiti.

Puerto Ricans. Puerto Ricans differ from most other people who come to the United States in that they are technically not immigrants. They come to the mainland as U.S. citizens and are free to travel to and from Puerto Rico without restriction. Over half the population of Puerto Rico resides on the mainland of the United States, and the number of Puerto Ricans who live in New York City is almost double the number living in San Juan, the largest city in Puerto Rico.

Small numbers of political exiles from Puerto Rico arrived in America in the 1800s, but most returned home when Puerto Rico became a U.S. possession. Others arrived when unemployment increased in the depressed agricultural economy of the island during the 1920s and 1930s. It was after World War II that the largest number of Puerto Ricans moved to the mainland. Unlike other immigrants, the Puerto Rican population in the United States is in continual flux. Many Puerto Ricans live alternately between the mainland and the island, depending on economic conditions.

Cubans. Cubans have immigrated to the United States since the early 19th century. In the early years, the majority were those who found economic conditions disadvantageous or who were politically out of favor with the current government.

The majority of Cubans came to the United States after Fidel Castro overthrew the dictatorship of Fulgencio Batista in 1959. In the three years following the revolution, more than 150,000 Cubans arrived in America. Most of these were families from the upper socioeconomic group fleeing the restraints of communism. Commercial air travel between Cuba and the United States was suspended after the Cuban missile crisis in 1962. Airlifts of immigrants from 1965 to 1973 increased the total number of Cubans in the United States to nearly 700,000. Due to the political differences between the two countries, most Cubans have not been subject to the usual immigration quotas.

Immigration from Cuba slowed with the end of the airlifts. In 1980, another large group of 110,000 Cubans arrived in Florida in private boats (the Mariel boatlift) seeking asylum. Unlike earlier immigrants, these recent arrivals, called marielitos, were mostly poor, unskilled laborers. They were often single and black. Today, a trickle of exiles continues to come to the United States. Some arrive by boat and others go through legal immigration channels from a neutral third country.

Dominicans. Information on early immigrants from the Dominican Republic is limited because prior to 1990, Dominicans were counted within the broader Hispanic category in the U.S. Census data (in 1990 "Dominican" was a write-in category; in 2000 it became a check-off category).[1] Four immigration groups have been identified.[2] The first was during the Trujillo era (1930–1960), when political dissidents came to the United States to escape the regime of President Rafael L. Trujillo. The second group came during the post-Trujillo era (1961–1981), when improved social and economic conditions slowed emigration to a trickle. The third is the flotilla group (1982–1986); this comprises Dominicans who left the country by boat, seeking escape from oppressive poverty, hoping for a new start in the United States. The fourth group includes those who have left since the early 1980s, many of whom are urban Dominicans better educated than those they leave behind, seeking employment opportunity. Some Dominicans enter the U.S. mainland after making Puerto Rico their home. The numbers of unauthorized Dominicans, or those who return to the island, are unknown.

Current Demographics

Puerto Ricans. The number of Puerto Ricans living on the U.S. mainland reached over 4 million in the 2008 census estimates.[3] Nearly 25 percent of Puerto Ricans make New York State their home, with an estimated 900,000 living in New York City.[4] In the 1930s, Puerto Ricans began moving into East Harlem, which became known both as El Barrio and as "Spanish Harlem." Over 800,000 Puerto Ricans now reside in New York City, and because Puerto Ricans are a young population, they now constitute nearly one-quarter of the students in the city's public schools. Boston, Chicago, Philadelphia, Newark, Cleveland,

Terms of identity vary for Puerto Ricans, including Puerto Rican American, *Borrinqueño* or *Boricua* (used by those who prefer the native Taíno Indian name for the island), and "Nuyorican" (used by second-generation Puerto Ricans living in New York City).

Puerto Ricans living in Puerto Rico are not countedin the U.S. Census totals. If all Puerto Ricans were included, they would constitute nearly 20 percent of all Latinos in the United States.

Puerto Ricans living inPuerto Rico are not countedin the U.S. Census totals. Ifall Puerto Ricans wereincluded, they would consti-tute nearly 20 percent of allLatinos in the United States.

▼ *Cuban cafeteria, Miami, Florida.*

Courtesy of the Metro Dade Department of Tourism

Figure 10.2
Top Cities for Hispanic Immigration.
Source: Adapted from the U.S. Census Bureau, http://minorityhealth.hhs.gov/templates/browse.aspx?lvl=2&lvlID=55

Miami, San Francisco, and Los Angeles also have substantial Puerto Rican communities.

Cubans. In 2008 census estimates, Americans of Cuban descent totaled 1.6 million.[5] Many live in the Miami area, which is sometimes called "Little Havana," where the climate is similar to that of their homeland. The Cuban population in this region is the largest in the United States. Cubans living in Los Angeles; Chicago; New York City; and Jersey City and Newark, New Jersey, often choose to move to Miami after gaining job skills. Those Cubans who remain in other cities are most likely to be employed in technical or professional occupations. Nearly all Cubans who live in the United States are urban dwellers (Figure 10.2).

Dominicans. As of 2008, it is estimated that approximately 700,000 individuals from the Dominican Republic are either legal permanent residents or are eligible to become naturalized citizens of the United States.[6] The majority lived in the urban areas of New York, New Jersey, Massachusetts, and Florida. It has been estimated that as many as 300,000 unauthorized migrants also reside in those regions.[8]

Other Caribbean Islanders. Steady immigration to the United States during the 1990s had increased numbers of other Caribbean Islanders in 2005 to over 840,000 Jamaicans and nearly 700,000 residents from Haiti.[7] Jamaicans have settled in the cities of the Northeast and South, including New York City, Philadelphia, Baltimore, Atlanta, Miami, Ft. Lauderdale, and Orlando. Haitians often choose to live in New York City, Miami, Chicago, New Orleans, Los Angeles, and Boston.

Socioeconomic Status. Immigrants from the Caribbean vary in both economic and educational attainment. Among Latinos, Puerto Rican Americans have the highest rate of unemployment (in 2000, it was more than double the national average), and Cuban Americans have the lowest. More than 25 percent of Puerto Ricans live in poverty, over three times the rate for non-Hispanics. However, it should be noted that second-generation mainland Puerto Ricans living in regions outside New York may have a different socioeconomic profile, with significantly higher rates of college graduation and white-collar employment.[4] Though nearly half of Cuban Americans are employed in professional, technical, managerial, sales, or administrative support positions, and annual family income is high relative to other Latino groups, earnings are below the national average, and 15 percent of Cuban Americans live below poverty level.[7]

Large numbers of Dominican professionals have immigrated to the United States; however, most obtain lower-paying jobs in manufacturing and service industries, including restaurant work and home care.[10] Dominicans have the lowest median family income of all Latinos in the United States, and more than 27 percent live in poverty.[7] Though some Jamaicans are limited by their lack of skills, a majority hold professional, management, technical, or sales jobs. Haitian immigrants have usually come from poor segments of their society and have entered migrant labor and service positions. It is estimated that nearly one-third of Haitians in Miami are unemployed.[11]

Educational rates for first-generation immigrants from the Caribbean are very similar to those of other Latinos. Figures from 2000 show that for adults, both Puerto Ricans and Cuban Americans graduate from high school at a rate somewhat below that of the general population (63 percent); however, rates of high school attendance are lower for those who were born in Puerto Rico or Cuba. In contrast, Cubans graduate from college at rates close to those of the general population, and nearly double those of Puerto Ricans.[7] Cuban Americans are also more likely than any other Latino group to pay for education; nearly one-half have attended private schools. Just over half of adult Dominican Americans have graduated from high school, with rates slightly lower among those who were foreign born.[7,12] Nearly three-quarters of adult Jamaican Americans who are foreign born are high school graduates.[12] Dominicans in New York have demonstrated their commitment to education through political involvement, such as running for control of local school boards.[1] Statistics of school attainment of Haitians in the United States are usually combined with those of other Latinos and are not generally available for individual groups. Among foreign-born Haitian adults, 62 percent have graduated from high school.[12]

Many factors influence socioeconomic differences among Latino groups from the Caribbean. Puerto Ricans living on the mainland are free to travel between the United States and their homeland, and the frequent changes in residence may hamper socioeconomic improvement. Many Puerto Ricans have chosen to reside in New York City, where as a new ethnic minority they took over the decayed neighborhoods previously occupied by African Americans. In general, their low education levels and undeveloped job skills translate into unemployment and low-level employment.

Furthermore, some Puerto Ricans, Dominicans, Jamaicans, and Haitians face biases similar to that experienced by African Americans because many are of African heritage. Racial distinctions are not as significant in the Caribbean Islands, and more overt discrimination is a new challenge for some immigrants.[13] In contrast, Cubans in the United States are mostly political refugees who emigrated out of necessity, not choice. There are a disproportionate number of Cubans in the United States from the upper socioeconomic levels, and although many lost all their material goods when they emigrated, they brought upper-class values, including the importance of educational and financial success. Instead of displacing any ethnic minority in Florida, they immigrated in such numbers that they immediately became the dominant ethnic group. This is not to say that there are no well-educated, wealthy Puerto Ricans living on the mainland or no poor Cuban Americans. However, circumstances surrounding their immigration have influenced the general socioeconomic status of both groups.

Worldview

Ethnic identity is strongly maintained in the Puerto Rican, Cuban, and Dominican communities in the United States. Puerto Ricans on the mainland continue close ties with the island, frequently returning to visit family and friends. Cubans believe it is important to retain their heritage because they cannot return. Dominicans often consider their stay in the United States temporary and may resist acculturation in order to maintain their identity. In all three groups, Spanish may be spoken exclusively in the home and used frequently to conduct business within the community.[1,4,14] In contrast, Caribbean Islanders who suffer dysfunctions due to the immigration process, and those who are unauthorized residents, often try to blend into existing Latino communities or assimilate into mixed ethnic neighborhoods.[11,13]

Religion. A majority of Caribbean Islanders are Roman Catholics. The role of the Catholic Church has been less important in the Caribbean than in other regions of Latin America and is less significant to immigrants in the United States. For example, the number of Puerto Ricans living on the mainland who are Catholic has declined since the 1960s, approximating 70 percent today,[4] and it is estimated that only 64 percent of Cuban Americans born in the United States are followers of the faith.[1] However, higher percentages of Catholics may be found in other Caribbean Islander groups who have been in the United States for shorter periods of time. For instance, about 95 percent of Dominicans are adherents to Roman Catholicism.[14]

A number of other religions are practiced in the islands, including Protestantism and Judaism. For example, a majority of Jamaicans, who at one time lived under British rule, belong to Protestant congregations, such as the Church of God and Seventh-Day Adventists. Folk religions are found as well. The best known of these is voodoo, a unique combination of West African tribal rituals with Catholic beliefs and local customs. Saint Patrick is associated with the African snake deity Damballah, for example, and St. Christopher is identified with Bacoso, the god responsible for infectious illness. Certain rites, such as repeating the Hail Mary, making the sign of the cross, and baptism, are practiced in conjunction with ancestor worship, drums, and African dancing. Worship is family based, and there is no central leadership or organization of activities. Typically ceremonies are conducted for annual events such as Christmas and the harvest and for funerals.

Voodoo originated in Haiti, although very similar Afro-Catholic cults are found on the other islands. In Cuba and Puerto Rico, these are called *santería*. Many followers of voodoo or *santería* are also members of Christian faiths and do not believe there is any contradiction in practicing both religions simultaneously.[9] Rastafari is another Afro-Caribbean faith indigenous to Jamaica. Rastas practice a natural, simple lifestyle typified by bare feet, loose clothing, dreadlocks, and sacramental use of marijuana. It is also considered a political movement due to Rastafari opposition to traditional government and support for repatriation of blacks to Africa.

Family. The Puerto Rican family is based on the concept of *compadrazgo*, which means co-parenting. Grandparents, aunts and uncles, cousins, and godparents are all considered part of the immediate family, responsible for the care of children.[4] Men are the heads of households as well as being in charge of community matters. The oldest boys in the family are expected to help with supervision of younger siblings, particularly daughters. Women maintain the home. Men are expected to be aggressive; women are traditionally reserved. As in most Latino cultures, age is respected and elders are honored. Younger children are taught to defer even to older children. Duty to family is extremely important.[15]

Traditional Cuban families are also patriarchal and extend to include relatives. Godparents are significant in childrearing. Children are deferential to elders and well chaperoned in public.

Caribbean Islander families often change in the United States. Women, who work in greater numbers than men, and who may make a higher income, often gain greater authority within the home, and Caribbean Islander children gain greater autonomy in the United States. Economic pressures, American values of individualism and equality, and intergenerational stress are often cited as responsible for nontraditional adaptations. Studies of early immigrants have shown that one-third of Dominicans, for example, lived in nuclear family groupings in the United States even though only 1 percent did so in the Dominican Republic. Dominican women in the United States also had fewer children than those on the island.[1] Another dramatic change is in the number of households headed by women, which totaled nearly 34 percent of all Dominican families in 2000 (three times the national average).[7] It is thought that in some cases men who were unable to support their families have deserted them, often leaving uneducated and unskilled women to manage alone.[1] Among Jamaican Americans, parents often separate when the wife or husband comes first to the United States to obtain employment and establish residency. Long periods apart often result in divorce or abandonment, and single-parent households result. Children may lapse in school attendance, and gang activity among adolescents is of concern in the community.[13]

Among the rural population of Haiti, common-law marriage is frequent, and it is acceptable for a man to maintain several different households as long as he supports each wife and their children. This *placaj* system is believed to be a remnant of the polygamous societies found in parts of West Africa.[16] Gender roles are inflexible, with men responsible for farming and providing for the family, and women in charge of the household budget, marketing, and child care. Haitians have maintained more traditional families than some other Caribbean groups living in the United States. Typically, men still head the family, though Haitian-American women often insist on a greater role in making decisions than is customary in Haiti. Haitian children are still expected to obey their parents, bring honor to their family, and to reside at home until marriage. Haitian Americans establish a close network with other Haitian immigrants and keep in touch with family and friends remaining in Haiti. Those who are unable to return for political reasons may sponsor voodoo ceremonies on their behalf.[11]

© Francoise de Mulder/Corbis

▲ *Serious conditions, such as sickness due to supernatural causes, may require the cures associated with the healing practice of* santeria.

Traditional Health Beliefs and Practices. Many Caribbean Islanders hold health beliefs similar to those of other Latin American cultures. For example, Puerto Ricans, Cubans, Dominicans, and Haitians often believe that illness is a punishment from God, or that fate determines life and death. Prayer, the lighting of candles to saints, and the laying on of hands are important ways of maintaining health and curing disease.[17,18,19,20] Many Dominicans employ the *promesa*, a promise or obligation to be performed by a supplicant in exchange for maintenance or restoration of health.[14] In addition, some believe that all individuals have a guardian angel who protects them from evil. Some also believe illness can be caused by evil spirits or the devil, particularly Dominicans and Haitians.

The conditions of *empacho, susto, nervios,* and *mal de ojo* are known by many Puerto Ricans, but not by all.[17] Of special note is *nervios*, which takes several forms.[21,22] Someone who has experienced trauma as a child may become a nervous person for life, *ser nervioso,* with crying bouts, headaches, stomach maladies, and a tendency toward violence in men. This condition can be tempered by use of herbal teas and talking with family members, religious advisors, or mental health professionals. *Padecer de los nervios* is a mental illness associated with depression that develops in adults. It is treated with the help of psychologists or psychiatrists. *Ataques de nervios,* also known as *ataques*, is a hysterical reaction to stressful events. It may include acute breathing difficulties, frenzy, or the sudden onset of illness. *Nervios* is a problem found more often in women than men, and it is associated with a weak character.[23] In general, *nervios* is helped by prayer, massage, sedatives, and herbal teas. Physicians and mental health specialists are also useful. Other folk conditions reported by Puerto Ricans include *pasmo*, a type of paralysis due to an imbalance of hot and cold, and *fatique*, acute breathing difficulties. *Pasmo* is cured through folk remedies, whereas *fatique* responds to emergency care provided by a physician.

Haitians are especially concerned with the flow of blood, considered essential in the balance of hot-cold categories. Many blood irregularities are recognized, classified as hot, cold, weak, thin, thick, dirty, and yellow.[24] *Febles* occurs when there is insufficient blood or anemia due to poor diet.

The condition is characterized by general weakness and is cured by eating items such as liver, red meat, pigeon meat, cow's feet, or leafy green vegetables. *Sezisman* is a disruption of normal blood flow, due to sudden emotional trauma or chronic ill-treatment by others. It can cause vision loss, headaches, high blood pressure, or stroke. It is treated with relaxation, cold compresses, sipping cool water, or drinking coffee mixed with rum.[25] Haitian women may be encouraged to eat red fruits and vegetables (such as beets or pomegranates) to strengthen their blood.[16]

Gaz (gas) is another common condition for some Haitians. Gas may settle between the ears, causing headache, in the stomach, causing indigestion, or in other parts of the body, where it causes pain. Eating leftovers (especially beans) is one cause of *gaz*.[25] A nursing mother may undergo a thickening of her milk, which causes headaches or depression in the woman and impetigo in her baby. *Move san* is a more serious condition in which a nursing mother experiences fright or negative emotions, causing her milk to spoil, resulting in diarrhea and failure-to-thrive syndrome in her infant.[16] Some Haitians believe *mal dyok* (evil eye) can also cause illness.

The traditional healing practices common in the Caribbean are more closely related to African beliefs than the Arab-Spanish humoral system used in hot-cold applications. Mild conditions are treated through an informal system of older women (mothers, grandmothers, or neighbors) who are knowledgeable about the use of teas, herbs, amulets, and charms.[26] A study of Dominican healers found that the women learned their skills from relatives or through spiritual guidance.[27] When at home in the Dominican Republic, they treated only clients they knew personally from within their communities. In the United States, the healers expanded their practice to include strangers and clients of all ethnicities. They typically accepted a wide variety of physical and psychological cases. When possible, the healers preferred to consult with the entire family in order to gain full support for recommended treatment.

More serious conditions, such as those due to supernatural causes, particularly witchcraft, require the cures associated with voodoo and *santería*. Voodoo priests (*hougans* or *bokors*) or priestesses (*mambos*) or spiritualist healers, known as *espiritos* and *santeros*, intervene with the saints on behalf of a bewitched person (practices related to those of the American South; see Chapter 8, "Africans"). *Santeros* specialize in soul possession and mental disorders. Dreams may play an important role in health care because they are a connection with the supernatural world. Ancestors provide instructions to an individual regarding health behaviors through dreams. *Brujos* (witches) and *curenderos* (healers) (see Chapter 9) may also be sought for medical care.[18]

Dominicans often believe the best way to treat illness is to use a traditional healer who will consult with Catholic saints about which home remedies are appropriate for the symptoms.[14] This allows Dominicans to address the spiritual and emotional aspects of physical problems as well as seek symptom relief. Many Haitians recognize two types of illness. The first is natural illness, due to a poor diet, blood conditions, bone displacement, or cold drafts and other environmental factors. These can be cured by home remedies or visits to biomedical practitioners. The second is supernatural illness, due to angry spirits. These can only be treated by a voodoo *manger mort* (feast for the dead) ceremony.[25] Cubans often consider *santería* a link to their past and may use a biomedical provider for relief of physical symptoms but employ a *santero* to help them restore balance or counteract the circumstances that led to their illnesses.[20]

Good hygiene, especially daily bathing, is done to promote health among Puerto Ricans. Some Puerto Ricans, Dominicans, and other Caribbean Islanders practice a modified version of the hot-cold classification system in their diet, and some use it for categorizing illness (see Therapeutic Uses of Foods in this chapter).[19,27] Dominican healers may consider junk food, red meat, lack of exercise, emotional distress, contact with negative people, and environmental stresses as some contributing factors in health problems.[27] Haitians consider eating well, cleanliness, and regular sleep essential to health. Laxatives or enemas may be used to refresh the bowel, remove impurities, and prevent acne in children.[25]

Most Caribbean Islanders use herbal teas, and over-the-counter medications are often used to relieve symptoms—Dominicans may also take baths with herbs or flowers. Many Americans of Caribbean Island heritage are also likely to use home remedies or visit a *bontánicas* (herbal pharmacy) or a *bodega* (small market) to purchase cures (including antibiotics obtained without prescription)[28] as a first step in treating symptoms.[20] One study notes that some Dominicans refer to home remedies of all types as *zumos*, a word that, strictly translated, means "juices."[29]

Traditional Food Habits

Ingredients and Common Foods

Caribbean food habits are remarkably similar for an area influenced by so many other cultures. The indigenous Indians, the Spanish, French, British, Dutch, Danes, Africans, Asian Indians, and Chinese have all had an impact on the cuisine. The basic diet is similar throughout the region, with regional variations found on each island. In recent years, tourism has helped spread specialties from one nation to another in order to meet visitor expectations about what dishes are available, and the global economy has furthered the development of a pan-Caribbean cuisine.[30,31,32]

Indigenous Foods. Columbus likened the West Indies to paradise on Earth. The islands are naturally laden with fresh fruits and vegetables originally from Central or South America, including the staple cassava (two varieties of tuber, bitter and sweet, also known as manioc and yuca; tapioca is a starch product of manioc), acerola (Barbados cherry, a small, sour fruit with exceptionally high vitamin C content), avocados, bananas and plantains, some varieties of beans, calabaza (a type of pumpkin), cashew apples (fruit of the cashew nut), cocoa, coconuts, corn, guavas, malanga (a mild yam-like tuber sometimes called cocoyam, yautia, tannier, or tannia), mammee apples (a small green fruit with flesh reminiscent of apricots), papayas (sometimes called pawpaws), pineapple, sapodilla or naseberry (a small fruit with aromatic flesh that has a gritty texture similar to pears), soursop (a fruit with a cotton-like consistency), several types of squash (including chayote, called chocho or christophene in the islands), sweet potatoes, and tomatoes. Fish and small birds are also plentiful.

As in other Latin American areas, chili peppers grow profusely in the West Indies. Extremely hot varieties are favored, including Scotch bonnet and bird peppers (also called tepins—see Chapter 9, for more information on chiles). The native cuisine makes frequent use of these for flavoring, especially in pepper sauces, such as coui, a mixture of cassava juice and chiles. Other native seasoning includes allspice, recao (Eryngium foetidum, a pungent herb also known as culantro, long cilantro, or shandon beni), and annatto (achiote).

Because of the abundance of fresh fruits and vegetables year-round, traditionally there was little need for preparation or preservation of foods. Consequently, cooking techniques were underdeveloped in the native populations. Cassava was baked most often in a kind of bread, made from pressed, dried, grated cassava that was fried in a flat loaf. Fish and game were either covered with mud and baked in a pit or grilled over an open fire.

Cassava contains hydrocyanic acid, which is toxic in large amounts. The acid must be leached out and the tuber cooked before it can be eaten safely.

Foreign Influence. The Europeans who settled in the Caribbean were impressed with the abundant supply of native fruits and vegetables. Yet they longed for the accustomed tastes of home. The Spanish brought cattle, goats, hogs, and sheep to the islands, in addition to introducing rice. Plants introduced for trade by the Europeans included breadfruit, coffee, limes, mangoes, oranges (both sweet and sour varieties), and spices such as ginger, nutmeg, and mace. The African slaves brought in to work the sugarcane fields cultivated akee (a mild, apple-sized fruit), yams, okra, and taro (also called eddo or dasheen; both the roots and the leaves are eaten). The demand for Asian ingredients by later immigrants resulted in the introduction of soybean products, Asian greens, lentils, and tamarind to the Caribbean.

Nearly all parts of the akee fruit contain hypoglycins, which can cause fatal hypoglycemia. Most akee products are banned in the United States.

Staples. Legumes are eaten throughout the Caribbean, most often in the dish "rice and peas." Rice with red (kidney) beans is popular in Puerto Rico and is also found in the Dominican Republic and Jamaica (where the dish is nicknamed "coat of arms"). In Cuba, black beans with rice are preferred, called Moros y Cristianos ("Moors and Christians," a reference to Spanish history). In Haiti, black-eyed peas (a type of cowpea from Africa) are combined with the rice. The legumes in all countries are prepared similarly, flavored with lard and salt. Onions, sweet peppers, and tomatoes or coconut milk are added in some variations. Other popular legumes include pigeon peas, popularly known as gungo (originally from Africa, often cooked with rice, also), lentils (from India), chickpeas (also known as garbanzo beans, introduced from Europe), and bodi beans (another variety of cowpea eaten as a green bean, also called Chinese long beans).

Examples of other foods common throughout the West Indies are found in the Cultural Food Groups list (Table 10.1). They include native Indian foods such as cassava bread, chili sauces, and pepper pot (a meat stew made with the boiled juice of the cassava, called cassarep). Tamales and pasteles are steamed cornmeal, cassava, or plantain dough packets with savory (such as meat, seafood, or cheese) or sweet (including coconut or guava) fillings. European-influenced items popular in many Caribbean countries include escabeche (fried, marinated fish, seafood, or poultry), asapao (a thick rice soup with chicken, pork, or seafood, often garnished with Parmesan cheese in Puerto Rico and slices of avocado or fried plantains in the Dominican Republic), morcillas (a type of blood sausage), flaky pastry turnovers with meat, poultry, seafood, or fruit fillings, and fried corn cakes (known as surrulitos in Puerto Rico). Foods from Africa found throughout the region include callaloo (a dish of taro or malanga greens cooked with okra), dried salt cod fritters (called bacalaitas in Puerto Rico, these cakes have a different name on nearly every island, from *arcat de marue* to "stamp and go"), foofoo (okra and plantain), and coocoo (cornmeal-okra bread). Dishes from India and Asia are also common on many islands, although they are better known in the areas where cheap labor was most needed: the French-, British-, and Dutch-dominated islands (few Asians immigrated to Puerto Rico, Cuba, or the Dominican Republic). Curried dishes, called kerry in the Dutch-influenced islands and colombo on the French-influenced islands, and variations of pilaf are considered Caribbean foods. Chinese cuisine is also popular and Chinese-owned restaurants are omnipresent.

The most popular beverage in the Caribbean is coffee. It is often mixed with milk and is consumed at meals, as a snack, and even as dessert, flavored with orange rind, cinnamon, whipped cream, coconut cream, or rum. Some of the most expensive coffee in the world is produced in the Blue Mountains of Jamaica, where the cool, moderately rainy climate is ideal for coffee cultivation. Most of the rich beans are exported to England and Italy, although small amounts can be found in the United States.

The most important beverage in the Caribbean, at least historically, is the spirit distilled from fermented molasses—rum. This alcoholic drink is believed to have originated on the island of Barbados in the early 1600s as a by-product of sugarcane processing. Molasses is the liquid that remains after the syrup from the sugarcane has been crystallized to make sugar. It is fermented, naturally or with the addition of yeast, and then distilled to make a clear, high-proof alcoholic beverage. Rum can be bottled immediately or aged in oak casks from a few months to twenty-five years. Caramel is added to achieve the desired color. Nearly every island produces its own variety of rum.

The molasses produced in the West Indies was crucial to the development of the region during the seventeenth and eighteenth centuries. The Caribbean Islands were one corner of the infamous slave triangle, formed when molasses was shipped to New England for distillation into rum, rum was shipped to Africa and exchanged for slaves, and slaves were shipped to the West Indies to work in the sugarcane fields.

[SAMPLE MENU]

A Puerto Rican Lunch

***Escabeche* (*Escovitched* Fish)**[a,b]

***Surrulitos*[c] or Caribbean Johnnycake**[a]

***Arroz con Pollo* (Peppery Chicken and Rice)**[a,b,c]

***Habichuelas Guisado* (Stewed Beans)**[a,c]

***Plántanos en Alimbar* (Candied Plantains/Baked Bananas)**[a,c]

Fruit Juice, Beer, and Coffee with Milk

[a]DeMers, J. 1997. *Caribbean cooking*. New York: HPBooks.

[b]*Cocina Criollo* at http://www.hechoenpuertorico.org/comida

[c]*The Boricua Kitchen* at http://www.elboricua.com/recipes.html

Table 10.1 Cultural Food Groups: Caribbean Islands

Group	Comments	Common Foods	Adaptations in the United States
Protein Foods			
Milk/milk products	Few dairy products are used; incidence of lactose intolerance is assumed to be high. Infants are given whole, evaporated, or condensed milk.	Cow's milk (fresh, condensed, evaporated), *café con leche, café latte*; aged cheeses	More milk and cheese are consumed.
Meat/poultry/fish/eggs/legumes	Traditional diet is high in vegetable pro-tein, especially rice and legumes; red and kidney beans are used by Puerto Ricans, black beans by Cubans. Pork and beef are used more in Spanish-influenced countries. Dried salt cod is preferred over fresh fish; some seafood specialties. Eggs are a common protein source, especially among the poor. Entrées are often fried in lard or olive oil.	*Meats*: beef, pork (including intestines, organs, variety cuts), goat *Poultry*: chicken, turkey *Fish and shellfish*: *bacalao* (dried salt cod), barracuda, bonito, butterfish, crab, dolphin fish (*dorado*), flying fish, gar, grouper, grunts, land crabs, mackerel, mullets, *ostiones* (tree oysters), porgie, salmon, snapper, tarpon, turtle, tuna *Eggs*: chicken *Legumes*: black beans, black-eyed peas, chick-peas (garbanzo beans), kidney beans, lima beans, peas, red beans, soybeans	More beef and poultry are eaten as income increases, though pork intake may decline. Less fresh fish is consumed. Traditional entrées remain popular.
Cereals/Grains	Breads of other countries are well accepted. Fried breads are popular.	Cassava bread; cornmeal (fried breads, *surrulitos*, puddings); oatmeal; rice (short-grain); wheat (Asian Indian breads, European breads, pasta)	Short-grain rice is still preferred. More wheat breads are eaten.
Fruits/Vegetables	Starchy fruits and vegetables are eaten daily; leafy vegetables are consumed infrequently. Great diversity of tropical fruits is available, eaten mostly as snacks or dessert. Lime juice is used to "cook" (a marinating method called escabeche or ceviche) meats and fish.	*Fruits*: acerola cherries, akee, avocados, bananas and plantains, breadfruit, *caimito* (star apple), cashew apple, *cherimoya*, citron, coconut, cocoplum, custard apple, gooseberries, *granadilla* (passion fruit), grapefruit, guava, *guanábana* (soursop), jackfruit, kumquats, lemons, limes, *mamey*, mangoes, oranges, papayas, pineapple, pomegranates, raisins, *sapodilla*, sugar cane, sweetsop, tamarind *Vegetables*: *arracacha*, arrowroot, black-eyed peas, broccoli, cabbage, *calabaza* (green pumpkin), *callaloo* (malanga or taro leaves), cassava (*yuca*, manioc), chiles, corn, cucum-bers, eggplant, green beans, lettuce, *malan-gas*, okra, onions, palm hearts, peppers, potatoes, radishes, spinach, squashes (chay- ote, summer, and winter), sweet potatoes, taro (*eddo, dasheen*), tomatoes, yams	Temperate fruits are substituted for tropical fruits when latter are unavailable. More fresh fruit is eaten. Starchy fruits and vegetables are still frequently consumed. Low intake of leafy vegetables is often continued.
Additional Foods			
Seasonings	Aromatic, piquant sauces are often used to flavor foods. Very hot chiles popular in some regions.	Anise, annatto, bay leaf, chiles, chives, cilantro (coriander leaves), cinnamon, *coui* (chiles mixed with cassava juice), garlic, mace, nutmeg, onions, parsley, *pimento* (allspice), *recao* (culantro), scallions, thyme	
Beverages	Teas of all sorts are common and are often thought to have therapeutic value. Rum is especially popular and is often added for flavoring to foods and beverages.	Beer, coffee (*café con leche*), teas, soft drinks, milk, rum, Irish moss (seaweed extract), sorrel	Fruit juice and soft drink consumption may increase.
Fats/oils		Butter in French-influenced countries; coconut oil; *ghee* (Asian Indian clarified butter); lard in Spanish-influenced countries; olive oil	
Sweeteners		Sugar cane products, such as raw and unrefined sugar and molasses	

Juices made from tropical fruits such as lime, otaheite apple (also called ambarella, originally from Polynesia), pineapple, roselle (also known as "Jamaican sorrel," brought from Africa), soursop, and tamarind are common. Ginger often spices the juice mixtures, and coconut milk or condensed milk may also be added.

Regional Variations. Despite the similarities in foods throughout the Caribbean, some regional differences are notable. Same-named dishes prepared on one island may not taste the same on another island due to variations in ingredients and seasoning. For example, butter is the preferred cooking fat in French-influenced countries, whereas lard is more popular in Spanish-influenced nations. Coconut oil is common in Jamaica. In British-influenced countries, dishes often include scallions, parsley or cilantro, and thyme. On French-influenced islands roux (flour blended with butter or oil, then cooked until browned) is used to thicken stews and sauces, and sauce chien ("dog sauce") is a popular fresh condiment served with pork, chicken, and seafood, made with olive oil and lime juice seasoned with ginger, garlic, scallions, parsley, chiles, allspice, and thyme. A similar preparation known as sauce ti-malice is found in Haiti. Spanish-influenced islands use more piquant seasonings with less heat, including a greater use of tomatoes, onions, annatto, and sweet bell peppers.

▼ *A typical Cuban meal includes savory picadillo (right), black beans and rice, bread, and flan (a Spanish-style custard) for dessert.*

Courtesy of the Florida News Bureau

Each island is also known for its specialties. In addition to rice and red beans, Puerto Rican fare is notable for its use of distinctive flavorings, such as alcaparrado, a pickle mix of capers, olives, and pimento, and recaito, an aromatic blend of recao, onions, garlic, and bell peppers. Sofrito, an all-purpose sauce that is the foundation for many Puerto Rican dishes, combines alcaparrado and recaito with tomatoes. All ingredients are then fried in lard colored with annatto seed until a thick paste is formed. Some foods are seasoned with adobo, a mixture of lemon, garlic, salt, pepper, and other spices. Ajilimojili sauce is a puree of bell peppers, garlic, olive oil, and lemon juice. Sazón, a commercial spice blend that is primarily MSG (monosodium glutamate, used to enhance flavors), is a popular seasoning as well.[15,31]

Starchy foods have a central role in Puerto Rican cuisine, traditionally consumed at nearly every meal as a side dish or in soups and stews. They are known as viandas and include bland-tasting, white or creamy colored roots, tubers, and fruits that must be cooked, such as cassava, malanga, potatoes, sweet potatoes (white or yellow are preferred), yams, celery root, breadfruit (and breadfruit seeds), under-ripe bananas, and plantains.[15,31] One especially popular preparation is mofongo, fried and mashed plantains flavored with either pork cracklings or bacon. Calabaza and carrots are considered vegetables because their starch content is lower and they are a significant source of vitamins; ripe bananas are considered a fruit that is eaten raw.

Pork is a favorite meat in Puerto Rico, especially roast pork adobo. It is also used frequently for added flavor in the form of salt pork, ham, cracklings, or bacon. Beef and goat are also consumed. One very popular stew is sancocho, which includes beef short ribs, calabaza, malanga, yams, and corn. Chicken is very popular, frequently prepared with rice as arroz con pollo, which is usually served with stewed beans (known as habichuelas guisada), or in asopao. Land crabs and ostiones, a type of oyster that grows on the roots of mangrove trees,[33] are eaten, as is some seafood, including shrimp, lobster, and conch, often prepared as soups or stews. Fresh fish is not consumed frequently (though a few dishes, such as escabeche, are popular), but dried salt cod, called bacalao, is used in many dishes. It is soaked and drained before use to remove some of the salt and then added to numerous dishes including serenata, a mixture of cod and potatoes. Variety meats are featured in several national dishes, such as mondongo (tripe soup), lengua relleno (stuffed tongue), rinones guisados (calf kidneys), and sesos empanados (calf brains).[30]

Fritturas, or finger foods, are also a specialty, consumed as snacks or appetizers or added to meals. They include simple fritters (e.g., banana, squash, or bacalaitos); alcapurrias (starchy vegetable dough stuffed with spicy beef, pork rind, poultry, or seafood and then fried); piñones (plantain strips wrapped around sausage, poultry, or seafood fillings and fried); pastilillos (fried meat or cheese turnovers); and cuchifritos (deep-fried chitterlings or variety meats). Empanadillas, small baked turnovers typically filled with ham, beef, lobster, conch, or cheese, are also a favorite. Sweets including cakes, pastries, puddings, and cookies are popular for desserts and snacks. One specialty is tembleque, the Puerto Rican version of Spanish-style flan. Flans are also flavored with chocolate, coconut, pineapple, pumpkin, or rum. Candied ripe plantains and baked bananas are a common fruit-based sweet in Puerto Rico, also found throughout the Caribbean.

Nearly 70 percent of food in Puerto Rico is imported from the mainland.[15] American dishes are common, especially among younger diners. Pizza, canned spaghetti, hot dogs, canned soups, and cold cereals have become favorites.[15]

Cuba is noted for the prominent use of black beans in its cuisine. In addition to black beans and rice, spicy black bean soup is very popular. As in Puerto Rico, viandas are standard fare, especially in the more rural eastern sections of the island where Indian heritage is prominent. Examples include foofoo (cassava balls) and tostones (plantain slices that have been pressed to make them larger and thinner, then fried in olive oil). Meats and viandas are often served with mojito, a sauce of olive oil, juice from limes or sour oranges, onions, and garlic.

The western parts of the island are more urban and cosmopolitan, especially around Havana, where Spanish and Asian culinary influences are evident. Picadillo is a type of beef hash flavored with the traditional Spanish ingredients featured in alcaparrado (the same mix of olives, raisins, and capers used in Puerto Rico), as well as Caribbean tomatoes and chili peppers. Picadillo is served with fried plantains or boiled rice, or topped with fried eggs. Other Spanish-influenced beef dishes are ropa vieja ("old clothes"), spicy beef strips cooked until they begin to shred, and brazo gitano, a cassava dough pastry filled with corned beef. Roast pork is popular, and eggs are often prepared as Spanish-style potato omelets. In addition, the rice and beans are usually served separately in this region. Asian ingredients are less prominent, but notable. Chicharrónes de pollo is prepared with small pieces of chicken marinated in lime juice and soy sauce, breaded, and then fried in lard. Another example is arroz salteado, a fried rice dish made with eggs, shrimp, and vegetables cooked in olive oil and seasoned with soy sauce. Fish are eaten in western coastal areas, and one specialty is grilled or stewed crocodile.[30,31] Fruit pastes, such as those made from guava, are typical desserts, sometimes served with a slice of salty cheese. Spanish-style egg desserts are also found, especially custards, flans, and puddings. Turrones, a nougat candy made with peanuts, is a Cuban favorite.

Stews are a specialty in the Dominican Republic. Examples include pollo guisado (chicken with bell peppers, tomatoes, onions, and olives, seasoned with oregano), mondongo (similar to the Puerto Rican tripe soup), and stews made with fish or seafood, such as shrimp, conch, or herring. Best known is the Dominican version of sancocho, made with several kinds of meats (including pork, chicken, beef, and Spanish-style longaniza pork sausage), plus numerous starchy vegetables cooked in sour orange juice. On special occasions additional types of meats (e.g., goat, ham) are added to make sancocho prieto. Stews are often served with rice and red beans, and cassava bread.

Locrio is another Dominican favorite—a rice dish that has its origins in Spanish paella (see Chapter 6, "Northern and Southern Europeans," for more information), but differs in that only a single item, such as chicken, shrimp, or sardines, distinguishes each version. Other common dishes include chicharrónes de pollo (prepared like the Cuban recipe), rice with chicken and pigeon peas, and mangu (mashed plantains topped with olive oil–fried onions). Salads are especially popular in the Dominican Republic. A few feature lettuce and tomatoes, but, more often, cooked vegetables such as okra, potatoes, chayote squash, or cabbage are cooled and dressed with oil and vinegar. Avocado and hearts of palm (a specialty of the island) are featured in other versions. Habichuelas con dulce is a unique Dominican dish served as a side dish or as a dessert, combining red beans, coconut milk, evaporated milk, whole milk, sugar, and butter.[14] Desserts include fruit compotes, Spanish-style flan, plantains or guavas with caramel sauce, coconut biscuits, or sweet potato or squash puddings.

Jamaica specialties include akee and salt cod, curried goat, bammies, a type of cassava bread, mackerel rundown, cooked in coconut milk with vegetables, and jerked foods (see Exploring Global Cuisine box for more information). Haiti is known for its banana-stuffed chicken dish called poulet rôti à la créole and barbecued goat with chili peppers (kabrit boukannen ak bon piman). Griot is another popular dish made with pork that is first marinated in seasoned sour orange juice, then boiled, and then fried. Patties, a curried meat turnover, are a Haitian specialty now served throughout the Caribbean. Common Haitian side dishes include cornmeal mush and diri a djon djon (also called riz noir, or black rice, this is rice cooked in a broth made by boiling dried mushrooms native to the island called djon djon—the mushrooms themselves are not consumed). Curaçao is famous for its orange-flavored liqueur of the same name, and, in Dominica, crapaud, or "mountain chicken," a large, tasty frog, is considered a delicacy. In Barbados, many more unusual seafood dishes are popular, including those made with flying fish, green turtles, and sea urchins.

[SAMPLE MENU]

A Caribbean Sampler

Patties (Haiti, Jamaica)[a,b]

Fritters-Black-eyed Pea, Salt-Cod, or Conch (Pan-Island)[a,b,c]

Callaloo Soup (Pan-Island)[a,b]

***Puerco Asado* (Cuban Pork Roast)[a,b,c]**

***Mangú* (Dominican Republic)[a,d]**

Black Cake/Rum Cake (Pan-Island)

[a]DeMers, J. 1997. *Caribbean cooking*. New York: HPBooks.

[b]*Caribbean Recipes* at http://www.recipezaar.com/recipes/caribbean

[c]*Cuban Recipes* at http://www.recipehound.com/Recipes/cuba.html

[d]*Aunt Clara's Kitchen Dominican Cooking* at http://www.dominicancooking.com/dominican-recipes/

Meal Composition and Cycle

The most typical aspect of a Caribbean meal is its emphasis on starchy vegetables with some meat, poultry, or fish served with rice and beans. Breads of all sorts are now common in many areas. Meats are frequently fried or grilled. Sometimes before meat is added to mixed dishes it is cooked first with sugar to caramelize it (a technique thought to have been brought by Africans).[30,31] Soups and stews are also popular. Soups are sometimes served in two courses—the strained broth first, followed by the cooked meats and vegetables. Leafy vegetables are sometimes ingredients in soups, stews, and stuffed foods, and only served uncooked as part of the lettuce and tomato salads found in many regions, including Puerto Rico, Cuba, and the Dominican Republic. Fruits are eaten infrequently in many areas but are found fresh in some desserts and as snacks.

Ethnic heritage and social class determine which dishes are served.[30] A poor native Indian may eat mostly cassava, tomatoes, and chiles with a bit of salted fish at every meal. An Asian Indian may serve typically Asian-Indian meals adapted to Caribbean ingredients, such as a curried dish garnished with coconut, fried plantains, and pineapple. Most menus, however, consist of a multicultural mix, such as European blood sausage and accra, West African–style fritters made from the meal of soybeans or black-eyed peas. More meats and foreign dishes are consumed by wealthy Caribbean islanders and

EXPLORING GLOBAL CUISINE—Specialty Cooking of Jamaica

The nearly half-million Americans of Jamaican ancestry have had significant impact on U.S. pop culture. Calypso, reggae, the *Rastafari* religion, and dreadlocks are among the many cultural additions. In cuisine, two regional specialties have piqued American interest: jerk and *i-tal*.

Jerk is believed to be related to the dried meat called jerky. Legend is that the technique was created by escaped African slaves known as Maroons (from the Spanish word for "untamed"—*cimarron*) who spiced and smoked wild pig meat to preserve it. Today, the word *jerk* is used to identify the wet spice mixture used as a barbecue seasoning. It includes allspice, black pepper, cinnamon, ginger, nutmeg, thyme, scallions, and extremely hot Scotch bonnet chile peppers—some recipes also call for garlic, onions, ground coriander, bay leaves, brown sugar, or other seasonings. The spices are moistened with a little oil, lime juice, or soy sauce to make a paste. Traditionally, the meat is rubbed with the jerk blend and marinated for several hours. It is then grilled in a pit over Jamaican pimento (allspice) wood, covered with banana leaves, typically one to four hours, depending on the meat. Though pork and chicken are found at every street jerk stand in Jamaica, more recently the cooking technique has been applied to turkey, fish, seafood, and even vegetables. Jerk pork is used to make jerk sausage in some parts of Jamaica. For a complete meal, rice and peas, cassava bread, or cornsticks accompany the meat.[67,120]

I-tal, meaning "vital," is the Rasta way of life. Applied to food it emphasizes simple, unprocessed vegetarian fare. Fruit, vegetables, and grains are permitted, while pork, red meat, salt, and artificial additives are prohibited. Some Rastas will eat chicken or fish (but shun bottom feeders such as shrimp and lobster, scaleless fish such as shark, and any fish more than twelve inches long). In general, milk, coffee, soft drinks, and alcohol are not consumed. *I-tal* foods are ideally eaten raw or cooked over a fire (microwave ovens are avoided by many Rastas), prepared and served using pots, dishes, and utensils made from natural products, such as wood or earthenware. A woman is not allowed to prepare food for others when she is menstruating. Typical *i-tal* dishes include rice and peas, cassava bread, baked yams, vegetable stews, cornmeal porridge, sautéed plantains, and freshly squeezed juices. Thyme, cinnamon, allspice, coconut, and reputedly marijuana are used to flavor foods.[82,120]

some reportedly visit the United States weekly to shop for groceries.[14,30] American fast foods have become popular throughout the region with rich and poor alike.

Daily Patterns. Meal patterns vary somewhat throughout the region. Three meals each day, with lunch the largest meal, is typical in most regions. In Haiti, however, two meals a day is not uncommon.

In Puerto Rico, the traditionally large lunch and smaller dinner are gradually changing to a dining schedule similar to that on the mainland, especially in urban areas. Toast and coffee are a common breakfast, though eggs are popular as well, often served as a Spanish-style omelet. Lunch and dinner menus may be similar, starting with soup (such as black bean soup or chicken with rice soup), followed by a stew served with rice and beans, fried plantains, and chayote squash. Quick lunches, such as fast food fare, may replace the full meal. Dessert is usually eaten daily, following whichever meal is largest, lunch or dinner. Bread puddings with rum sauce are favored. Soda, fruit juice, beer, or rum accompany the meal. Restaurants are widely available in the cities, serving traditional Puerto Rican cuisine, as well as international fare, such as Spanish, Italian, and Japanese. Snacking is prevalent, particularly on fried items, such as bacalaitos, surrulitos, and cuchifritos.

Toast and coffee is a customary breakfast in Cuba, often followed by a midmorning coffee break with pastries or cakes. Lunch and dinner menus are similar, with meat, poultry, or fish (if available) served with fried plantains, rice and black beans, and often cassava. Custards and puddings (bread or rice) are typical desserts. Coffee is served after the meal. Lunch is typically the largest meal of the day, even in urban areas, and dinner is often leisurely, and may include beer, rum, or wine. Snacking on fruit, fruit juices, batidas (fruit juice blended with milk and ice), or ice cream is frequent.

In the Dominican Republic, breakfast may be just bread and coffee, but more often it is larger, including eggs, cheese, and salami or longaniza sausage (scrambled together, or each fried separately), fried or mashed plantains, and espresso or hot chocolate. Lunch is usually the biggest meal of the day. Traditionally served between noon and 2:00 p.m., it is known as La Bandera Dominicana ("the Dominican flag"), a plate with rice and beans, a meat or chicken dish, and salad—incorporating colors similar to those found on the national banner.[14] Plantains or other starchy vegetables may also accompany lunch. Dessert always follows, and espresso or sweetened coffee with milk ends the meal. American meal patterns are influencing many Dominicans, and abbreviated lunches are becoming more common, including only a main dish, dessert, and coffee. When lunch is the main meal of the day, dinner is light, consisting of scrambled eggs or soup, and fried plantains or cassava bread. But when lunch is light, dinner is more substantial, similar to the traditional lunch.

Jamaicans often include fish at breakfast, including sardines, mackerel, herring, or salt cod. Other common items are eggs, fried plantains, cornmeal porridge, and bammies. On the weekends, liver with bananas is a breakfast specialty. Lunches and dinners are similar to those of other Caribbean Islanders, including soups, rice, and peas with added beef, chicken, or curried goat, pork stews, fish dishes, and tossed salad or sweet potatoes on the side.[31,34]

Special Occasions. The early European dominance in the West Indies resulted in an emphasis on Christian holidays. Christmas is important, especially in the Spanish-influenced islands that are predominantly Catholic. In Puerto Rico pasteles are prepared to celebrate Christmas. Similar to Mexican tamales, pasteles are a savory meat mixture surrounded by cornmeal or mashed plantains, wrapped in plantain leaves, and steamed. Carolers traditionally stop at houses late at night to request hot pasteles from the occupants. Christmas Eve, or *Noche Buena*, includes Mass and a feast with lechón asado (spit-roasted pig), morcillas, rice with pigeon peas, coquito (rum and coconut milk), and special desserts such as rice pudding and coconut custard. In the Dominican Republic a whole-roasted pig is also customary, served with rice and peas, and a salad. Cubans associate pasteles (large turnovers) or pastelitos (smaller turnovers) with the holidays as well, but make them with a dough that is similar to French puff-pastry, stuffed with spicy meat or cheese fillings, or sweet fillings, such as guava, mango, or coconut.[31]

Other holidays reflect the multicultural history of the islands. *Carnival* is celebrated in some Caribbean countries such as Trinidad and Tobago and is similar to Mardi Gras in the United States. The pre-Lenten festivities feature parades of dancing celebrants; many are elaborately costumed as traditional European or African figures. Food booths that line the parade route provide a day-and-night supply of carnival treats. Fried Asian Indian fritters are particularly popular.

Examples of nonreligious events include the day-long birthday open house on Curaçao for friends, relatives, and acquaintances. Thanksgiving is observed in Puerto Rico. Turkey, stuffed with a Spanish-style meat filling, is the main course. Rum cake (also known as black cake) is a fruitcake specialty of the Caribbean, especially in Jamaica, where it is served at weddings, Christmas, and other special occasions. Dominican cake, a citrus-flavored cake with a cooked pineapple filling, topped with a caramelized sugar meringue, is popular in that nation for all holidays and events.

Traditionally, Sunday meals emphasize fresh meats when available, especially beef or pork roasts. In the Dominican Republic lunch on Sunday is very large and may last into the early evening. On many islands Sundays are also times when picnics are enjoyed, called *día del campo* ("field day") in Spanish-speaking nations. One dish often served is carne fiambre, a selection of cold-cuts served with pickles, olives, and green salad.[30,35]

Etiquette. In Puerto Rico forks and knives are held European style—the fork in the left hand and the knife in the right hand with no switching hands for cutting food. This pattern is also the norm in Cuba, though in the Dominican Republic both European-style and American-style use of utensils is accepted. In these nations, dishes are passed to the left, and when not eating, hands should be kept visible, with the wrists resting on the edge of the table. In Puerto Rico it is impolite to start eating until a host says, "*Bon appetite!*" whereas in Cuba and the Dominican Republic it is rude to eat before the host says, "*¡Buen provecho!*"[35]

Food may be in short supply in some parts of the Caribbean, and respectful behavior is expected when eating. For example, in Cuba vegetables and fruits should not be consumed with the hands. In Puerto Rico, food should not be wasted, and one should not take more than one can eat.

Therapeutic Uses of Food

Some Caribbean Islanders adhere to a hot-cold classification system of diet and health similar to that found in Mexico (see Chapter 9, on Mexican therapeutic uses of food). In addition to the categories of hot and cold, Puerto Ricans add cool. Imbalances in hot and cold—for example, sitting in the shade of a tree after being out in the sun—can cause illness even years after the imbalance has occurred.[36] Haitians believe that women are warmer than men and that a person cools as he or she ages.[24] Among Caribbean Islanders, it is mostly Puerto Ricans, Dominicans, and Haitians who follow dietary and disease hot-cold classifications, and only small numbers are strict adherents.[31]

The hot-cold theory of foods practiced in the Caribbean sometimes includes not only the category of cool foods, but also those considered heavy or light. A balance of hot-cold elements is attempted at meals, and heavy foods, such as starches, are consumed during the day, whereas light foods, such as soup, are eaten in the evening. Although the specific classification of items varies from person to person, one guideline for Puerto Ricans indicates bananas, coconuts, and most vegetables are cold; chiles, garlic, chocolate, coffee, evaporated milk and infant formula, and alcoholic beverages are hot.[37] Cool foods include fruit, chicken, bacalao, whole milk, honey, onions, peas, and wheat. Excessive intake of cool or cold foods can make a cold condition, such as a cough, develop into a chronic illness, such as asthma.

Pregnancy, defined as a hot condition by most Puerto Ricans, is a time when a hot-cold balance is practiced carefully, and hot foods are avoided. When infants suffer from hot ailments, including diarrhea or rash, infant formula may be replaced with whole milk, or cooling ingredients such as barley water, mannitol, or magnesium carbonate may be added to the formula. High-calorie tonics (eggnogs and malts are popular types) are taken by some Puerto Ricans to stimulate the appetite and provide strength or energy. These are considered especially appropriate for pale children and for pregnant or postpartum women.

Research with Dominican Americans suggests the use of hot-cold classifications for numerous conditions. Examples cited are excessive cold causing asthma and fibroids, whereas perimenopausal hot flashes are a hot problem.[27,29] Home remedies given to children for asthma include warming foods, such as oils (whale, cod liver, almond, and castor), honey or royal jelly (bee-larva food), onion, garlic, oregano, lemon, and aloe vera juice. Beets combined with molasses are used by traditional Dominican healers to treat fibroids (and may be used by some Caribbean Islanders to lower blood pressure or treat arthritis and ulcers).[34,38] Research on Dominican mothers suggests that nutritional practices during lactation may sometimes include avoidance of certain protein foods and increased intake of fluids such as malt beer, milk, orange juice, chocolate milk, and noodle soup.[39] Formula may be withheld from sick infants and tea provided instead.

Haitians apply the hot-cold theory, including the heavy-light categories, to a broader number of conditions impacting health. A person's life cycle, a woman's reproductive cycle, the climate, and the time of day are categorized, and they must be balanced to maintain health.[16] For example, heavy foods should be eaten in the morning and light foods in the evening. Environmental forces (such as wind, or seeing a lightning strike) and social interactions can disrupt equilibrium and result in illness. Women and their newborn infants may spend the first month after birth in seclusion to avoid excessive chilling.

Therapeutic use of food is not limited to balancing hot-cold conditions. Some Haitians believe that certain illnesses in infants can be caused if a nursing mother's milk is too thick or too thin. Further, if a woman is frightened while breast feeding, her milk goes to her head, causing a headache in her and diarrhea in the baby. *Gaz*, another condition, causes pain in the shoulders, back, legs, or appendix, and headaches, stomachaches, or anemia. Foods such as corn or a tea made from garlic, cloves, and mint are home remedies for *gaz*.[11]

Other Caribbean Islanders, including Cubans, do not generally subscribe to the hot-cold theory but often use food-based home remedies.[34,20] One study of Hispanics in the Miami area reported 75 percent had used herbal cures during the previous twelve months. Cubans reportedly use grapefruit and garlic for hypertension, chayote to calm nerves, and beets to treat anemia and flu. Star anise tea is consumed to relieve intestinal pain and flatulence in adults and colic in infants. Other teas used for stomach aches include those made with aloe vera or spearmint. Teas with cinnamon, sour orange, or honey and lemon are used for colds and coughs. Cinnamon tea is also believed useful for menstrual cramps. Gastrointestinal parasites are treated with pumpkin seed tea. Linden leaf, also popular, is used for anxiety.[20,34]

Some non-Hispanic Caribbean Islanders believe cassava helps prevent heart disease and cancer. Plantains are also used to decrease the risk of heart disease, as well as for treating hypertension, ulcers, and constipation. Teas are used for many ailments, including lemon-grass tea for fever, and ginger tea for indigestion and flatulence (ginger may also be added to rum for diabetes). Cerasse tea, made from Asian bitter melon (which has hypotensive properties), may be consumed to lower blood sugar levels, and wild sage (*ma Bbzou*) tea is also used to treat diabetes.[31,34,40]

Contemporary Food Habits in the United States

Adaptations of Food Habits

Traditional food habits are easily maintained in the self-sustaining immigrant communities of Spanish Harlem in New York City and Little Havana in Miami. Ingredients for Caribbean cuisine are readily available through Puerto Rican and Cuban American markets. Cubans, for example, may consider drinking strong coffee a way to maintain their ethnic identity; by comparison, Americans drink weak coffee. Changes do occur, however, as immigrants settle into culturally mixed communities and children grow up as Americans.

Little current data on Caribbean Islander food habits in the United States are available. One older study compared the diet of three groups of women from Puerto Rico: (1) those living in New York (forward migrants), (2) those who had lived on the mainland but later returned to the island (return migrants), and (3) those who never lived on the mainland (nonmigrants). It was found that nonmigrants and return migrants ate more starchy vegetables, sugar, and sweetened foods than did forward migrants. Forward migrants ate a greater variety of foods, including more beef, eggs, bread, fresh fruit, and leafy green vegetables. Puerto Rican women who had lived on the mainland quickly reverted to their traditional food habits when they returned to the island.[41]

Recent research on Afro-Caribbean immigrants (mostly Jamaicans) in Great Britain found that traditional items such as fish (boiled, baked, or fried), chicken (fried, roast, or curried), homemade soups, rice and peas, plain rice, and boiled potatoes were consumed by respondents several times each week. However, some foods, including patties, salt cod fritters, akee and salt cod, calaloo, breadfruit, and cassava were consumed infrequently. Few Western foods were popular: 83 percent ate hamburgers less than once a month, and similar numbers reported rarely eating pizza, pasta, butter, and margarine.[42]

Ingredients and Common Foods. Research on the food habits of Caribbean immigrants in the United States is limited. It is thought that rice, beans, starchy vegetables, sofrito, and bacalao remain the basis of the daily diet of many Puerto Ricans who live on the mainland. Poultry is used when possible, and egg intake decreases. Because of Cuban Americans' greater discretionary income, their diet usually includes additional foods, such as more pork and beef. Recent poorer immigrants from Cuba, including the marielitos, are more restricted in what foods they purchase, however, and may follow a diet that is closer to the subsistence-level Puerto Rican regimen.

According to marketing studies, Caribbean Islanders accept some American foods, especially convenience items, and purchase frozen and dehydrated products when they can afford to do so. The proportion of meat in the diet often increases on the mainland, as does the consumption of milk (and other dairy foods) and soft drinks.[31,43,44] Intake of leafy vegetables continues to be low. Local fruits often replace the tropical fruits of Puerto Rico.

Data regarding Dominicans in the United States indicate that only small changes have occurred in consumption patterns. Protein and fat intake has increased slightly, mostly due to eating more meat, while carbohydrate intake has decreased. Dominican women reported that their diet was more varied and abundant than in their homeland.[44]

A study of Hispanics over the age of fifty-five in Massachusetts, including Dominicans, Puerto Ricans, and other Latino groups, made dietary comparisons among less-acculturated Hispanics, more-acculturated Hispanics, and non-Hispanic whites. It was found that rice was the major contributor of energy for both groups of Hispanics, compared to bread for non-Hispanic whites. But Hispanics who had lived in the United States for at least twenty years had macronutrient profiles closer to non-Hispanic whites than to less-acculturated Hispanics, with a lower consumption of complex carbohydrates and an increased consumption of simple sugars.[45]

Meal Composition and Cycle. The meals of Puerto Ricans residing on the mainland are similar to those of people on the island, with a few changes. A light breakfast of bread and coffee may be followed by a light lunch of rice and beans or a starchy vegetable, with or without bacalao. Often this traditional midday meal becomes a sandwich and soft drink, however. A late dinner consists of rice, beans, starchy vegetable, meat if available, or soup. Salad is included in some homes. Many researchers have reported an increase in the amount of snacking between meals, mostly on high-calorie foods with little nutritional value.[31,44]

Dominicans have started eating lighter lunches.[44] Interviews with Haitians living in New York City suggest that some traditional dietary practices are discontinued; for example, the main meal is eaten in the evening instead of at noon. Although some Haitians adhere to hot-cold classifications of food, they may differ from those used in Haiti.[24]

One older study reported that many low-income Latina women living in New York did not plan menus far in advance and that this hampered their ability to add variety to their diets.[45] It was found that food shopping serves as a social occasion for many of the women and is one of the few opportunities they have to get out of the house; thus they may go to the grocery store more often than is really necessary. The investigators reported that nearly half of the women questioned preferred to fry main dishes. Boiling was the second choice, baking third. Broiling food was a distant fourth choice. The researchers noted that in many low-income households the oven or broiler element may not work, restricting food preparation methods to frying and boiling.

Special Occasions. It is assumed that many Caribbean Islander holiday food traditions are retained after immigration to the United States. Several events have been added to the annual calendar, however, often featuring traditional foods and music of the region. One of the largest is the West Indian Carnival that has been held annually for over sixty-five years in New York City. The carnival celebrates the cultures of the Caribbean, featuring an enormous parade, music competitions, and street vendors selling items such as curried goat and Jamaican jerk barbecue. In June, cities with large Puerto Rican populations often host Puerto Rican Day parades. The Dominican Day Parade is held every August in New York City. Major reggae music festivals with ample Caribbean food are held throughout the United States on February 6—reggae artist Bob Marley's birthday.

In Haiti protein foods are served first to the father in the household; leftovers go to the wife and children. This pattern is believed to continue in Haitian-American homes.

Nutritional Status

Nutritional Intake. There is limited information on the nutritional status of Caribbean-American immigrants to the United States. The few studies available suggest several health trends in these immigrants that have important nutritional implications. Health disparities for Hispanics compared to the general population have been reported, including lower rates of preventative care (such as inoculations and screenings) health care insurance coverage, and higher rates of risk factors.[46] The most recent data indicate that Hispanics have the highest rate of uninsured health care among any ethnic group in the U.S.[50,91] More specifically, disparities among different Caribbean Islander groups have been identified. In general, Puerto Ricans have the worst health indicators and Cubans have the best.[47] Differences among Puerto Ricans have also been reported between those living on the mainland and those living in Puerto Rico, with those living on the mainland experiencing more physical illness and having less access to health care.[48] Low socioeconomic and education levels are often associated with some disparities.[47,49,50,51] In a study of elders finding chronic health conditions, such as disability and diabetes, more prevalent in Puerto Ricans than in whites living in the same neighborhoods, the researchers hypothesized that physiological responses to life stress may be mediated by nutritional status, particularly intake of B vitamins and antioxidants, and that this may account for some of the difference in Puerto Ricans.[52]

Mortality data suggest that although Puerto Rican and Cuban men have lower overall rates compared to whites, younger men die in disproportionately higher numbers (often due to preventable causes). Among Puerto Ricans, those who are born in Puerto Rico have lower mortality rates than those born on the mainland.[53]

Nearly 10 percent of Puerto Rican infants born on the mainland are of low birth weight, and over 13 percent are born preterm. These factors contribute to a high infant mortality rate. Nationally, infant mortality rates for Puerto Ricans living on the mainland are 40 percent higher than for whites, and one earlier study of Puerto Ricans in New York reported an infant mortality rate 70 percent higher than that of the total population. Risk factors including poverty, young maternal age, low educational attainment, and inadequate prenatal care are positively correlated with these figures.[54] The numbers for low birth weight, preterm delivery, and infant mortality are even higher in Puerto Rico; however, one study found recent arrivals from the island had lower infant mortality rates than Puerto Ricans who had lived for an extended period on the mainland.[55] High rates (8.2%) of low-birth-weight infants have also been reported in the Haitian American community, associated with hypertension and preeclampsia.[56] In contrast, the infant mortality rate of Cuban American babies is well below the national average.

Recent data on breast feeding practices are limited. One anecdotal report on Puerto Rican women states that breast feeding is common.[15] However, earlier studies found breast-feeding infrequent among Puerto Rican, Cuban, and Haitian women in the United States, and in a more recent study, overweight Hispanic women in New York were found less likely to initiate and more likely to discontinue breast-feeding than lower-weight women. This finding is significant when obesity rates in this population are considered (see below).[57] Those few who started breast-feeding often switched to bottle feeding after two to four weeks. Whole milk, condensed milk, and evaporated milk were frequently fed to infants, as were juices. Solid food typically was introduced at a young age.

When the traditional Caribbean diet is limited because of low income, it inevitably results in low intake of many vitamins and minerals. The emphasis on carbohydrates and vegetable protein, with a low consumption of leafy vegetables and often fruit, provides inadequate intake of calories, vitamins A and C, iron, and calcium. An older study indicated that the only foods consumed by at least one-half of recent Cuban immigrants were eggs, rice, bread, legumes, lard and oils, sugar, and crackers.[58] Of the same group, 79 to 100 percent reported never eating leafy green vegetables or fresh fruits. Dietary variety and micronutrient content were found low in Puerto Ricans and Dominicans living in Massachusetts.[45] Other broader studies of Hispanics in New York and Boston suggest higher rates of fruit and vegetable intake, but still below national and state norms.[59] Deficiencies of B_1, B_{12}, folate, and sulfur amino acids have been reported.[60] One study of Puerto Rican and Dominican elders found that the high prevalence of B_{12} deficiency was due to insufficient intake, and that supplementation or frequent consumption of fortified cereals reduced the risk of insufficiency.[61] Low iron intake among African-Caribbean Islanders has been reported in Great Britain.[62] Anthropometric measurements and physical observation suggested that 20 percent of the children under fifteen showed signs of malnutrition, 37 percent of the men and 17 percent of the women had adipose tissue measurements consistent with adult marasmus, and 12 percent of the immigrants suffered from anemia.[63,64]

Many native Puerto Ricans suffer from parasitic diseases, including dysentery, malaria, hookworm, filariasis, and schistosomiasis.[65] A study of children from Cuba found 19 percent tested positive for intestinal parasites, and suggested rates may be much higher.[66] Further, 23 percent of these children had elevated blood lead levels. Recent immigrants settled in Massachusetts, Minnesota, and Rhode Island have also been identified to have higher rate of elevated blood lead levels.[67,68,69] These conditions can contribute to general poor health and nutritional deficiencies among poorer immigrants.

In contrast to the malnutrition evident in some areas of Puerto Rico and Cuba, some studies suggest rates of overweight and obesity are higher than national or state averages in some Hispanic populations. Data from 2008 indicate that hispanics have a 21 percent higher rate than whites.[70] (Figure 10.3) Among Hispanic elementary school students in New York City, who are assumed to be primarily of Puerto Rican and Dominican heritage, 31 percent were obese (BMI ≥95 percentile) compared to 20 percent of students overall,[71] and Hispanic preschool children enrolled in Women, Infants, and Children (WIC) programs in New York City were twice as likely as black children to be overweight.[72]

Research on Puerto Rican and Dominican elders found that obesity (as measured by BMI and central waist circumference) was associated with a traditional diet based on rice and beans and poultry and oil and that this diet was more prevalent among less acculturated subjects.[73] In contrast, studies of Puerto Rican women in Connecticut report factors associated with acculturation resulted in an approximately 54 percent increase in obesity.[74,75] Calorie and fat intake is reportedly high in many Caribbean Islander groups.[76,51] Further, larger portion size has been noted in some research on Caribbean Islanders.[103,51,77] Low levels of physical activity and cultural norms regarding weight and health may be significant factors of being overweight among Caribbean Islanders.[59] Puerto Ricans, Cubans, and Haitians often associate well-being with being *gordita*, or a little fat. This is particularly true for children, even when a thinner body ideal is desired by mothers.[78] Thinness is thought by some Caribbean Islanders to be indicative of poor health due to emotional or psychological conditions.[24,25]

Research has established a genetic contribution to the development of the clustering of health characteristics (including obesity/waist circumference, insulin resistance, hypertension, and dyslipidemia) known as metabolic syndrome in Caribbean-Hispanic families.[79] Persons with metabolic syndrome are at increased risk for type 2 diabetes and cardiovascular disease. Health statistics in 2009 estimate that 11.8 percent of Hispanics had diagnosed diabetes. Among Hispanics, rates were 7.6 percent for both Cubans and for Central and South Americans, 13.3 percent for Mexican Americans, and 13.8 percent for Puerto Ricans.[80]

Studies indicate that the prevalence of type 2 diabetes mellitus is two to three times higher among Puerto Ricans and Dominicans than among whites; Cuban Americans develop the condition at rates slightly higher than whites.[31,80,81,82] African-Caribbean women also developed impaired glucose metabolism at rates nearly double those for white women after gestational diabetes in one study (50% and 28%, respectively).[83] Rates of renal failure due to diabetes among hispanics are approximately 1.7 times higher than in whites. Death rates associated with diabetes are also 1.5 times higher in Hispanics.[84]

The prevalence of hypertension is slightly lower in Hispanics than whites.[85] Rates of mortality due to hypertension, heart disease, and stroke, however, vary between groups. Puerto Ricans have the highest rates of all Latinos, approximately 13 percent above whites. Higher rates of diabetes, which is a risk factor for high blood pressure, may be one reason for the discrepancy. In comparison, Cuban Americans have the lowest rates, 39 percent below those of whites. Deaths from hypertension are higher for men than for women in all Hispanics. It has been noted that a high prevalence of hypertension and intercranial atherosclerosis, as well as high rates of noncompliance regarding medications, is a contributing factor in strokes among black Caribbean Islanders in Miami.[86] Although Hispanics overall experience cardiovascular disease at rates lower than the national average, there is some evidence that it may be higher than average among some subpopulations, such as in Hispanic women living in New York City. Heart disease is still the leading cause of death among Hispanics.[59]

Chronic liver disease and cirrhosis are ranked as the seventh leading cause of death among Hispanics. Men are more likely to drink heavily than women. Among Puerto Rican men, 17 percent reported drinking one or more ounces of alcohol each day; 9 percent of Cuban men have similar drinking habits.[87] Dental health is problematic for Hispanics. Data indicates that Hispanic preschool

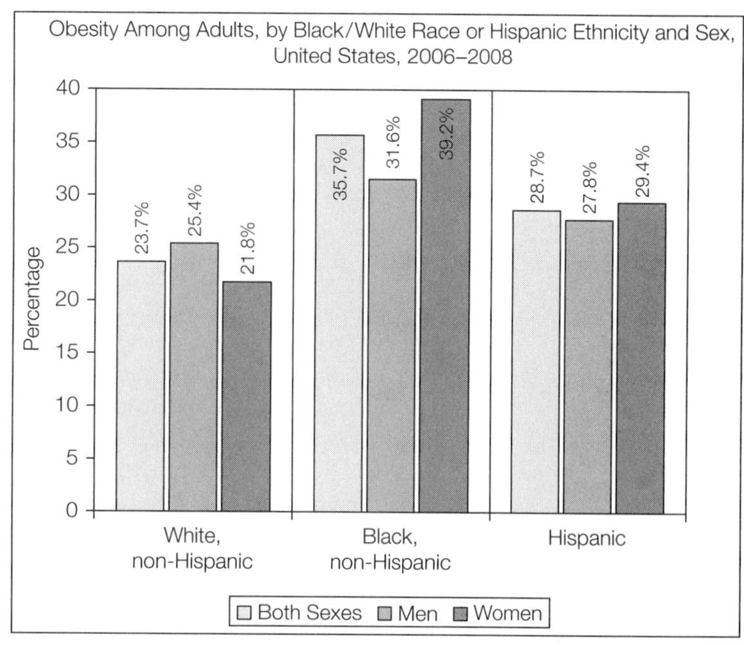

Figure 10.3
Obesity Rates in Hispanics.

The rates of type 1 diabetes in Puerto Rican children on the island are the highest of all children living in the Americas.

Lactose intolerance is thought to be a problem among many Caribbean Islanders, although estimated incidence has not been reported.

children have a high rate of dental caries. Hispanic adults have a larger proportion of untreated caries than whites.[88,89]

Counseling. Counseling Americans of Caribbean descent is similar to counseling other Latino clients and patients (see Chapter 9) Language problems may make interviews difficult, health beliefs may differ from those of the provider, and access to medical care may be limited. For example, census figures show 93 percent of Dominicans speak Spanish at home, and 54 percent are not proficient in English.[7] The idea that God or fate determines the outcome of illness may also interfere with seeking health care. For example, Puerto Ricans perceive diabetes as a chronic condition and believe that complications of the disease are unavoidable.[21] One study found Puerto Ricans in New York engaged in preventive behaviors, such as getting vaccinations or cholesterol testing, at rates below state averages.[48] Illness may be viewed as a sign of personal weakness among Haitians, revealed only to family members for as long as possible. In addition, some Haitians may believe that supernatural illness worsens if biomedical treatment is used, and they may delay seeking care until very ill.[25,90] Lack of health insurance coverage varies among Caribbean Islander groups, but it is important to remember that Hispanics have the highest uninsured rates of any racial or ethnic group within the United States. Approximately 20 percent of Puerto Ricans, 22.8 percent of Cubans, and 32.3 percent of other Hispanic or Latino groups are uninsured.[92]

Caribbean Islanders use an expressive conversational style. Respect and politeness are practiced, but it is not considered rude to interrupt a speaker. Periods of silence are uncommon.[35] Shaking hands in greeting and in leaving is customary, firmly between men, and less vigorously with women. Touching is very common, especially between members of the same sex, who may hug freely. Even if touching within a group is not direct, passing items (such as a piece of paper) from practitioner to client (or to every client in a classroom situation) increases the personal connection.[15] Direct eye contact is expected throughout the Caribbean, with some variations. Looking away suggests disrespect or dishonesty among Cubans; Dominicans maintain eye contact depending on the situation, and men are more likely to look directly at one another than are women; Haitians may avert their eyes from authority figures.

Puerto Ricans are open about physical and emotional complaints, although they may hesitate to ask questions because this might be interpreted as disrespectful. Adequate time and consideration of symptoms are necessary for both diagnosis and the client-provider relationship. High degrees of modesty are found in both women and men, who may prefer health providers of the same gender. In contrast, Haitians may expect a quick physical examination (with a stethoscope) and a fast, accurate diagnosis. Lack of a prescription to address a complaint may be seen as incompetence. Haitians may also assume that men are physicians and women are nurses.[25]

Puerto Ricans and Cubans are typically present oriented, and may have a relativistic view of time. Flexible appointments and relaxed attitudes regarding timeliness can help facilitate interactions with these Caribbean Islanders. It has been suggested that time limitations should be explained at the outset of the appointment so clients know what to expect.[19,20,35]

A few traditional Puerto Rican hot-cold practices may be problematic. Pregnant women may avoid iron supplements, which are classified as hot. Infant formula diluted with mannitol or magnesium carbonate to cool it (see Therapeutic Uses of Foods section) may cause diarrhea in infants. A mother may cool a hot medication, such as vitamins or aspirin, by providing a cool beverage with it, possibly fruit juice or milk of magnesia.[37] Puerto Ricans also may interpret high blood pressure as being too much blood or thick blood; conversely, low blood pressure means weak blood or anemia. Hypertension may be treated with herbal teas.[93] Some Puerto Ricans believe that ulcers lead to cancer.

A study of Long Island Cuban Americans showed that 20 percent acknowledged some degree of belief in *santería.* Some Dominicans believe that biomedicine is not responsive to their spiritual and emotional needs and that traditional healers provide the treatments needed to address all aspects of illness.[14] Estimates of *espirito* and *santero* use vary considerably, from 1 to 23 percent.[20,21] Many people who use traditional folk-healing practices and practitioners may be unwilling to admit to these folk practices or use them only in time of stress.[94]

Biomedical prescriptions may also present difficulties. Many medications available only by prescription in the United States are readily available over the counter or through the black market in Latin American countries. Overuse, inappropriate use, and addiction can occur. For example, one study found that New York Latinos did not understand that antibiotics are useful only for certain infectious conditions, and they were obtained without prescription to treat pain. A study regarding Metamizole (dypyrone), a toxic nonsteroidal anti-inflammatory drug banned in the United States, found that 28 percent of Hispanics surveyed in Miami had purchased it in their homelands, and 13 percent had used it in the previous year.[95,96] Another study found that medications for chronic conditions were only used by Dominicans during a medical crisis. As soon as symptoms subsided, the prescription therapies were replaced with home remedies. Dominicans often express concern about taking "chemicals" and worry that biomedical prescriptions cause addiction and serious side effects.[27,29] In addition, sharing prescription medications with family members or friends is common among Americans of Cuban, Dominican, and Haitian heritage.[20,25,29] Some Haitians reportedly have difficulty with dietary compliance.[24]

Research on a diverse group of Latinos in New York reported that physicians were the primary source of health care information, followed by television. Educational programs with Puerto Ricans may be more successful if a respected member of the community is enlisted in support of the project, and women often respond better if men in the family approve of their participation.[15] Caribbean Islanders are high context communicators, and the relationship and process are as important as the results.[35]

A Caribbean Islander client may be of African, Asian, European, or Indian descent; however, the island of origin is equally important in determining ethnic identity and diet. A black from Puerto Rico, for example, is more likely to eat Latino foods than African dishes or soul food. Early researchers working with low-income Latinas and black women in New York recommended that significant attention be placed on the client's socioeconomic status, a suggestion still valid today.[97] Nutrition therapy for a poor Puerto Rican immigrant is likely to require a different approach than that for a wealthy Cuban immigrant within the context of the basic Caribbean cuisine. An in-depth interview should be conducted with each client of Caribbean descent to establish country of origin and ethnic identity. The client's degree of acculturation, socioeconomic status, use of traditional heath practices, and personal food preferences also should be determined.

NEW AMERICAN PERSPECTIVES—Puerto Rican

PAULA VELAZQUEZ

I was born and raised in Puerto Rico and came to the United States when I was thirteen years old. In Puerto Rico I grew up on a farm, and my diet included rice and beans and products that we grew, such as fresh green bananas, sweet potatoes, and plantains. When my parents went to town to shop, that was when we ate meat and bread. Once a year we had pork for Christmas and turkey for Thanksgiving because these animals were raised by my parents for the holidays. My sister still owns the farm, and I try to get back for visits.

Here in the United States I live with my aunt and uncle, and their diet is high in meat, vegetables, and "healthy" foods. They are in their eighties and still very healthy. The food that made the biggest impression on me when I came to the U.S. was mashed potatoes. I had never seen or eaten mashed potatoes, and it took me a while before I would even try them. The second thing was having meat every day—that was really a treat. However, my favorite foods are bread and then the vegetables. I like all kinds of seafood. I will eat seafood rather than meat. Meat I could give up but not bread and vegetables.

South Americans

South America is a vast land that features the rugged ridge of the Andes Mountains stretching from north to south. Highland plains, tropical rainforests, temperate valleys, and desert dunes extend from where the mountain peaks slope toward the coastal edges of the continent. Extremes in terrain and climate limit agriculture in many areas. The continent contains twelve independent nations: Argentina, Brazil, Bolivia, Chile, Colombia, Ecuador, Guyana, Paraguay, Peru, Suriname, Uruguay, and Venezuela. In addition, France retains control of the territory called French Guyana, and Great Britain claims the Falkland Islands.

Numerous native Indian groups populated the continent prior to settlement by the Europeans. Although the Spanish were the first to arrive, significant numbers of Portuguese, Italians, and

Germans also settled in South America. Forced labor from West Africa introduced blacks to the continent, followed by Asian-Indian workers after slavery was outlawed. Indian and mixed Indian-European populations live in the tropical highlands, Creoles (descendants of the Europeans) have concentrated in the southern, temperate regions of the continent, and parts of northeastern Brazil are populated primarily by blacks and mulatto black Europeans. In more recent times, Japanese immigration to South America has become notable.

Cultural Perspective

History of South Americans in the United States

Immigration Patterns. The first documented South Americans in the United States were Chileans who came to California to participate in the Gold Rush. It is believed that several thousand worked in the mines. Approximately half returned to Chile, and those who remained quickly intermarried and were absorbed into the general population. Prior to the 1960s, all South American immigrants were counted as Other Hispanics in the U.S. Census. Specific figures regarding the individual nations before this time are uncertain but are thought to be minimal. Most South American immigration has occurred in the past twenty years during periods of land reform, economic hardship, or political repression. Jobs and educational opportunities are the primary attractions for the majority of immigrants, although significant numbers of political exiles are from Argentina and Chile.

Current Demographics. South Americans accounted for only 5 percent of Latinos in the United States in 2005. Yet according to U.S. Census estimates, many South American populations have almost doubled since 2000. There were approximately 730,500 Americans of Colombian ancestry, 436,000 from Ecuador, 412,000 from Peru, 204,000 from Guyana, 303,500 from Brazil, 186,000 from Argentina, 102,000 from Chile, 164,000 from Venezuela, 65,500 from Bolivia, 60,000 from Uruguay, 15,000 from Paraguay, and 74,500 listed as "Other South American."

According to immigration statistics, 102,878 visas were awarded to individuals from South America in 2009.[98] Immigration statistics also estimate that there are approximately 170,000 Ecuadorians and 150,000 Brazilians in the United States as unauthorized immigrants in 2009.[99]

Most South Americans settle in the Northeast, especially New York and New Jersey. In New York City, Colombians, Ecuadorans, and Peruvians have established ethnic enclaves in Queens, and "Little Brazils" are found in both Queens and Manhattan. Miami and Los Angeles also host large South American populations from most nations. In addition, Brazilians are found in Pennsylvania and Washington, DC; Chileans have settled in Texas; Colombians have clustered in Stamford, Connecticut, the urban areas of Illinois, and California; and Peruvian neighborhoods have developed in Houston, Chicago, and Washington, DC.

Most South American immigrants are proud of their heritage and differentiate themselves from Americans of Mexican, Caribbean, or Central American background. Brazilians in particular resent being mistaken as Hispanics who speak Spanish (Portuguese is their official language). Some South Americans suffer from discrimination directed toward Mexican Americans or Latinos in general. Others, who are mostly of European heritage, are not recognized as Latinos but may continue the prejudices between some South Americans. For example, there may be lingering hostilities between Bolivians and Chileans. Second- and third-generation South Americans often leave homogeneous neighborhoods and relocate into mixed communities.

Socioeconomic Status. Few data are available on the socioeconomic status of Americans of South American descent who have lived in the United States for extended periods. Most data are related to foreign-born immigrants who have come in the past twenty-five years. Regardless of arrival date, a majority of immigrants from South America come to the United States in search of employment opportunities. Many are well-educated professionals; however, they sometimes find that their credentials are not accepted after arrival, forcing them to accept positions in the sales, service, trade, and labor fields, such as restaurant work, construction, child care, or textile and garment industry jobs. It is believed that many second- and third-generation Americans of South American ancestry obtain advanced education and work in professional occupations.

Foreign-born Columbians arrive in the United States with overall education levels slightly lower than the U.S. average, though some are professionals with college degrees who hope to find employment commensurate with their skills.[100] Many Columbians have been successful as entrepreneurs,[108] and others have found employment in management, the education, health, and social services fields, the service industry, in manufacturing, and in sales. However, slightly higher numbers of Columbians live in poverty than the U.S. average. Most Ecuadorans who come to the United States are from the working class,[101] and nearly 40 percent of foreign-born immigrants do not have a high school education.[100] Many work in the service industry and manufacturing, and the percentage of families who live in poverty is above the national average. Yet, some Ecuadorans have found success founding restaurants, travel offices, and other services. Family poverty rates are more than 20 percent above the U.S. average.

Although over 40 percent of Peruvians hold agricultural, fishing, and farming jobs in their homeland,[102] nearly three-quarters of foreign-born immigrants have found employment in management and professional occupations, in the service industry, and in sales or office work.[100] Those from Guyana have also obtained jobs in those fields, with over 30 percent employed in sales and office work. Nursing is a preferred field for some Afro-Guyanese women.[103] Family poverty rates are much lower than the national average for both groups.[100]

Foreign-born Brazilians arrive with high levels of education, and they have occupational success as a group, taking jobs in management, technical fields, the service industry, sales, and as manufacturing operators or fabricators.[100] However, census data suggest that there are some immigrants who are not doing as well. In 2000, almost one-third of Brazilians reported working in the service industry, and there were more Brazilian families living in poverty than the national average.

Census data show that foreign-born Argentineans, Bolivians, and Chileans in the United States have also arrived with high levels of educational attainment.[100] Though some professionals must sometimes accept jobs below their skill levels,[104–106] family poverty rates are very low in these populations.[100] Over 40 percent of foreign-born Venezuelans have obtained a college degree. Forty percent work in professional or management careers,[100] particularly in the oil industry, banking, and in media, such as publishing, radio, and television.[107] Despite this success, the percentage of families living in poverty is near the national average. Foreign-born Paraguayans who live in urban areas typically find jobs in the service industry. Women often work in hotel housekeeping. Some men have found employment in agriculture, settling in California or Kansas.[108] Uruguayans who come to the United States are either middle-class professionals or blue-collar workers who held jobs in their homeland as skilled or unskilled laborers.[109] Thirty percent are employed in management and professional careers, and larger numbers work in sales, office jobs, and the service industry.

Worldview

Religion. South Americans are overwhelmingly Roman Catholic, a legacy of the European conquest. In most nations approximately 90 to 95 percent of the population are members of the Catholic Church, and its influence is seen in many South American institutions. The constitution of Argentina offers protections for Catholicism, and in Ecuador political leadership is established through sponsorship of local fiestas in honor of the saints. The Catholic faith is taught in Peruvian public schools.

In some regions, the practice of Catholicism is often blended with other belief systems. In Peru, Incan gods may be included in Catholic rites, for example, and in Venezuela the Cult of Maria Lionza mixes indigenous, Catholic, and African practices. Religious syncretism is greatest in Brazil. Spiritism, which was imported from France originally, combines Christian precepts with scientific principals. Popular with the upper middle class of the country, adherents communicate with the dead through spiritual mediums. Umbanda is very common in rural areas and among the urban poor, combining several Afro-Brazilian faiths with spiritism and the idea of Christian charity. Candomblé is probably the best known of the mixed religions, an Afro-Brazilian faith founded by blacks in the Bahia region that is now practiced nationwide by followers of all ethnicities.[110] African-derived beliefs dealing with earthly matters such as health and wealth are combined with Catholic cosmology. Yoruba deities called *orixá* or *orisha* are venerated with rites of worship that include animal sacrifice, feasting, and dancing. Over twenty *orixás* are recognized in Brazil and most are correlated with a Catholic entity: Oxalá, god of creation, with Jesus

Christ; Exú, the messenger god, with the devil; Ogun, god of war and iron-craft, with St. Anthony or St. George; Oxoosi, god of affluence, with St. Sebastian; Omolu, god of plagues and illness, with St. Lazarus; and Oxum, the fertility goddess (also called the goddess of vanity), with the Virgin Mary. Each *orixá* is associated with certain personality traits, day of the week, color, plants, animals, foods, and drinks; and each person has an *orixá* owner of his or her head who influences individual temperament and behavior.

Protestant missionaries were active in South America during the twentieth century and were especially successful in Guyana, where a majority of Guyanese attend the Episcopal church, and in Ecuador, where in some regions as many as 40 percent practice some Protestant faith. In Brazil the Baptist, Pentecostal, Seventh-Day Adventist, and Universalist denominations are most popular. Chilean Protestants are usually members of the Pentecostal or Seventh-Day Adventist churches, though those of German ancestry often follow the Lutheran or Baptist faiths. Small numbers of Jews and Buddhists are also found in South America, in total less than 1 percent of the population together.

Most South Americans who emigrate to the United States are Roman Catholic and very involved with their local parish. In some areas, particularly the Bronx in New York City, tension between parishioners of Irish or Italian descent and South American immigrants caused the South Americans to leave the traditional Roman Catholic faith. Colombians, for instance, formed a church based on charismatic Catholicism led by a Colombian priest. A majority of Guyanese Americans belong to an Episcopal church headed by a Guyanese priest and often also send their children to church-run schools with Guyanese teachers. In addition, Some Guyanese Americans frequent "Unity Centers," which function as community centers and promote spirituality but are not affiliated with any organized religion.[103] As many as one-third of Ecuadoran Americans belong to Protestant denominations.[101] How many South Americans who come to the United States and practice blended religions is unknown.

Family. Family life is important in all South American societies. In Argentina, Spanish and Italian traditions have shaped family structure. The extended family usually gathers together at least once a week and on holidays as well. Grandparents are involved in most family decisions, and children often stay at home until marriage. In Brazil, extended family members typically live close to one another, and daily visits are common. Relatives mentor children through rites of passage such as confirmation, graduation, the start of a career, and marriage.

The father is the head of the household in Chilean homes, but the mother makes all decisions regarding the family. In Colombia the father holds all authority, and children are taught to obey their parents. Ecuadoran families follow two models: Spanish-influenced families are ruled by the father, who has few responsibilities to the home other than financial support; in Indian-influenced families, the father and mother share more power and household responsibilities. In Peru, extended families typically include godparents, who sponsor baptisms and provide both social and economic assistance. Families are predominantly patriarchal, though more so in the Spanish-speaking upper and middle classes than in poor, rural Indian homes. In contrast to most of South America, the Venezuelan family has changed rapidly in the past decades due to increased national prosperity. Much of the population has relocated to urban centers. Many families have declined in size, and the extended family is less common.

In many areas of South America, it is unacceptable for women to work outside the home. Even those with a profession traditionally stay at home after marriage. Among some Indian groups, however, women contribute to the well-being of the family through farm work, and in the urban areas of Venezuela many women have outside jobs but remain responsible for household chores. In Chile, women are usually involved in local social and political issues.

Most South Americans prefer to immigrate as family groups, although financial pressures often demand that a single family member become established in the United States before the rest of the family follows. Individual immigrants commonly move to neighborhoods where relatives, godparents, or friends have settled. They depend on these contacts for housing and support. This system of mutual assistance is maintained after the immediate family arrives, bringing more relatives into the extended family. Colombians and Ecuadorans often broaden their relationships beyond national boundaries to form strong bonds with other Latinos.

Many families suffer from the stresses of American informality and freedom. Men lose some authority over wives and children, and women find it difficult to adjust to working outside the home. Furthermore, many upper- and middle-class women, who had paid help with the housework in South America, must learn to balance a job with responsibility for running a home.

Traditional Health Beliefs and Practices. There are few data available on how Americans of South American heritage maintain health or how they approach illness. Brazilians often attribute bad health to liver problems or an imbalance between hot and cold, such as drinking a glass of cold water on a hot day or taking a cool shower after eating a hot meal. Many South Americans self-diagnose or seek health advice from their mothers or friends. They then visit a pharmacist where they can purchase many medications, such as antibiotics, by the pill.

Most Brazilians associate faith with health. Catholics may believe in fate and seek intervention from patron saints when ill. Spiritists employ homeopathy, exorcism, past-lives therapy, acupuncture, yoga therapy, and chromotherapy to cure sickness.[111] Followers of candomblé believe that health is maintained by achieving balance between the earthly and spiritual spheres. The *pai-de-santo* or *babalorixá* (high priest) or the *mäe-de-santo* or *ialorixá* (high priestess) may be hired to read the oracle of a personal *orixá*, for example, so that an individual can improve his or her relationship with the deity.[110] Harmonious relations with one's *orixá* can maximize *axé* (vital force). Spiritual equilibrium is maintained by observing the preferences and prohibitions of one's *orixá*, including certain food and beverages, colors, therapeutic herbs, beaded necklaces, and other limitations. The priest or priestess also serves as the local *curendiero*, diagnosing physical and spiritual problems, prescribing healing herbal baths or botanicals, and manipulating occult forces. In Ecuador, either a healer, called a *curandero*, or a witch doctor, called a *brujo*, treats many illnesses in small villages. In Peru, urban residents typically obtain biomedical health care, but in rural regions home remedies and ritual magic are often preferred.[102]

Herbal teas are a favorite remedy throughout most of South America, where street stands and small markets called *yerbeterías* sell medicinal botanicals for home use.[112] Numerous plants, many unfamiliar in the United States, are used therapeutically.[40] Soursop leaves are used to treat diabetes, and the seeds of the guaraná are thought to relieve fatigue and help with weight loss. Retained urine is treated with avocado leaves, and papaya leaves are considered useful in getting rid of intestinal worms. Rue is taken for uterine pain and as an abortive, and black nightshade is used for coughs. *Pau d' arco,* the bark of a tree native to Brazil, is widely used to treat rheumatism, diabetes, venereal diseases, yeast infections, enlarged prostate, and several cancers.

Traditional Food Habits

Ingredients and Common Foods

Staples. The cooking of South America is similar to that of other Latin American regions in that it combines some native ingredients and preparation techniques with the foods of colonial Europeans. The diet is largely corn based and spiced with chili pepper (see Table 10.2).Tomatoes are common, and in tropical areas cassava (called *yuca*) is a popular tuber. Pumpkins, bananas, and plantains are consumed often. Beef, rice, onions, and olive oil, introduced by the Spanish and the Portuguese, are eaten regularly. Tropical fruits, such as those found in the Caribbean (see above), are plentiful in many regions. However, South American fare also features a number of ingredients used infrequently in the dishes of other Latin American areas. Potatoes, which were first cultivated by the Incas on mountain terraces, are particularly important in the highlands of Peru and Ecuador. Sweet potatoes (the orange-fleshed root vegetable usually called yams in the United States) are also native to the region. A white *root* similar to a mild carrot, known as *apio* or *arracacha*, is found in Colombia, Peru, and Venezuela; oca (a tuber similar to the potato in appearance but with leaves like clover) and *yacón* (an elongated tuber that has the taste and texture of a sweet turnip) are commonly eaten raw and cooked in Bolivia, Brazil, Colombia, Ecuador, and Peru; and the tuber known as *ahipa* (called jicama in the United States and Mexico) is native to the Amazon River basin.

Beans, a foundation food in many Latin American regions, are common in most South American countries yet not eaten at every meal. Other legumes and nuts, such as peanuts and cashews, are used often in dishes. Indigenous meats, including llama, deer, rabbit, wild pig, capybara,

Table 10.2 Cultural Food Groups: South Americans

Group	Comments	Common Foods	Adaptations in the United States
Protein Foods			
Milk/milk products	Milk is not usually consumed as a beverage but used in fruit-based drinks and added to coffee. Many milk-based desserts are enjoyed.	Cow's, goat's milk; evaporated milk; fresh and aged cheeses	Available cheeses are sometimes substituted for unavailable traditional cheeses.
Meat/poultry/fish/eggs/legumes	Beef is a foundation of the diet in parts of Argentina, Brazil, Paraguay, and Uruguay. Some game meats are consumed. Fish and seafood are significant in coastal regions, popular as *ceviche* in Ecuador and Peru. Beans are commonly consumed.	*Meat:* alligator, armadillo, beef (including variety cuts), capybara, frog, goat, guinea pig (*cuy*), iguana, llama, mutton, pork, rabbit, tapir *Poultry:* chicken, duck, turkey *Fish and shellfish:* abalone, bass, catfish, cod (including dried salt cod), crab, eel, haddock, lobster, oysters, scallops, shrimp, squid, trout, tuna *Eggs:* chicken, quail, turtle *Legumes:* beans (black, cranberry, kidney), black-eyed peas	Less acceptable meats such as guinea pig may no longer be eaten.
Cereals/Grains	*Cuzcuz*, made from cornmeal, is prepared in parts of Brazil; *arepa*, cornmeal bread, is staple in some areas. Pasta is popular in Argentina, Paraguay, and Uruguay. Rice and corn puddings are a favorite.	Amaranth, corn, rice, quinoa, wheat	
Fruits/Vegetables	Tropical and temperate fruits are plentiful and popular, added to savory and sweet dishes. Fruit compotes and fruit pastes are enjoyed. Potatoes are a staple in the Andes. Cassava flour and meal are common in many areas; tapioca is used in desserts.	*Fruits: Abiu, acerola*, apples, banana/plantains, cashew apple (*cajú*), *caimito, casimiroa*, cherimoya, custard apple, *feijoa*, guava, grapes, jackfruit, *jabitocaba*, lemons, limes, *lulo (naranjillo)*, mammea, mango, melon, olives, oranges (sweet and sour), palm fruits, papaya, passion fruit, peaches, pineapple, *pitango*, quince, raisins, roseapple, *sapote*, soursop, sweetsop, strawberries, sugar cane *Vegetables: abipa (jicama), arracacha (apio)*, avocado, *calabaza* (green pumpkin), cassava (*mandioca; yuca*), green peppers, hearts of palm, kale, okra, *oca*, onions, *roselle*, squash (chayote, winter), sweet potatoes, tomatoes, *yacón*, yams	
Additional Foods			
Seasonings	Toasted cassava meal, *farinha*, is sprinkled over foods in Brazil. Spicy hot foods are preferred in many areas; salsas are common.	*Achiote*, allspice, chiles (*aji*, malgueta, pimento), cilantro, cinnamon, citrus juices (lemon, lime, and sour orange), garlic, ginger root, oregano, paprika, parsley, scallions, thyme, vinegar	
Nuts/seeds	Coconut and coconut milk are added to numerous dishes. Peanut sauces flavored with chiles are common in the Andes.	Brazil nuts, cashews, coconut, peanuts, pumpkin seeds	
Beverages	Coffee is often served concentrated, then diluted with evaporated milk or water. *Maté* is more popular than coffee or tea in parts of Argentina, Brazil, and Paraguay.	*Batidas* (tropical fruit juices, sometimes made with alcoholic beverages), coffee, *guaraná*, soft drinks, sugarcane juice, tea, *yerba maté* and alcoholic beverages: beer, *cachaça* (sugarcane brandy), *pisco* (grape brandy), *chicha* (distilled corn liquor), wine	
Fats/oils	Dendê oil flavors and colors many dishes in the Bahia region of Brazil.	Dendê (palm) oil, olive oil, butter	Vegetable or peanut oil is substituted for dendê oil.
Sweeteners		Sugar cane, brown sugar, honey	

tapir, and cuy (guinea pigs who are raised for consumption), are consumed in some areas. Fish, such as anchovies and tuna, and shellfish, particularly shrimp, crab, spiny lobster, oysters, clams, giant sea urchins (evisos), and giant abalone (locos), are significant foods in the extensive coastal regions. Iguana is consumed occasionally, and alligator is a specialty in some areas.

A favorite way to prepare meats in South America is grilling. Traditionally, sides of beef, whole lambs, hogs, and kids (young goats) are hung over smoldering wood to slowly cook for hours in a method called asado. Today, a grill is used more often. Steaks and marinated kebobs (which often include organ meats) are favorites. Another, even older cooking tradition is to steam foods in a pit oven. In Peru, this method is called *a pachamanca* and typically includes a young pig or goat with guinea pigs, chickens, tamales, potatoes, and corn tucked around layers of hot stones and aromatic leaves and herbs.[113] In Chile, a curanto is closer to an elaborate coastal clambake, including shellfish, suckling pig, sausages, potato patties, peas, and beans layered with seaweed.

Stuffed foods are also common, including pastry turnovers filled with savory meat, fish, or cheese fillings. They are called empanadas in Argentina, which specializes in a turnover with a flaky, Spanish-style dough enriched with indigenous ingredients such as mashed potatoes, cassava, or corn. The turnovers are usually baked, but sometimes they are fried. Fillings are as many as there are cooks. Chopped meat, olives, raisins, and onions are popular. In Chile, the turnovers may be filled with abalone, and in Brazil, where they are known as empadinhas, a spicy shrimp mixture is traditional. In Bolivia, where they are called salteñas, the turnovers are filled with cheese. Tamale-like steamed packets of dough-wrapped fillings are also popular throughout South America. In Peru, chapanas are made with cassava dough, while in Ecuador bollos are formed around cooked chicken meat with plantain dough. In Brazil, a freshly grated corn kernel dough is mixed with coconut and cassava (and no filling) to prepare pamonhas. Ground cornmeal dough flavored with annatto and tomatoes is preferred for Venezuelan hallacas. A favorite in Argentina, Bolivia, Brazil, Chile, and Ecuador are humitas, which feature fresh kernel or ground cornmeal dough wrapped around a variety of savory or sweet meat, fish, or vegetable fillings.

Regional Variations. National differences exist, although there are few clearly distinctive divisions in South American fare. Several countries share similar dishes, and only a few nations have well-developed regional cuisines.

Peru and Ecuador. The cooking of Peru and Ecuador is divided into the highland fare of the Andes and the lowland dishes of the tropical coastal regions. The cuisine of the mountain areas is among the most unique in South America, preserving many ingredients and dishes of the Inca Indians. Potatoes are eaten at nearly every meal and often for snacks. Over one hundred varieties

© Yoshio Tomii/SuperStock

▲ *A market in Pisaq, Peru.*

[SAMPLE MENU]

An Ecuadoran Dinner

Cebiche de Pescado **(Fish *Ceviche*)[a,b]**

Locro **(Potato Soup)[a,c]**

Humitas **(Fresh Corn Tamales)[a,b]**

Chucula **(Plantain and Milk)[a] or Juice**

[a]Kijac, M.B. 2003. *The South American table*. Boston: Harvard Common Press.

[b]*Ecuadoran Recipes* at http://www.galapagosonline.com/predeparture/Food/Recipes.htm

[c]*Ecuadoran Cuisine & Recipes* at http://www.whats4eats.com/4rec_ecuad.html

are cultivated. Ocopa, boiled potatoes topped with cheese sauce and chile peppers or peanuts, is a typical dish in Peru. In Ecuador fried potato and cheese patties, called llapingachos, and potato cheese soup served with slices of avocado, known as locro, are common. Traditionally, the tubers are preserved by freezing in the cold night air and then drying in the hot daytime sun. Papa seca are boiled first and then dried until the potatoes are rocklike chunks that must be rehydrated before consumption; chuño are not cooked before drying and are often ground into a fine potato starch. Corn is also grown in the mountains. Some varieties have kernels the size of small strawberries that when prepared as hominy are known as mote and are a popular snack item. Bananas and plantains are cooked as savory chips and made into flour for breads and pastries.

The foods of Peru and Ecuador are preferred picante and feature abundant use of chili peppers in both the highlands and along the coast. Salsa de ají, a combination of fresh chopped chile, onion, and salt, is served as a condiment at most meals. Orange- or yellow-hued dishes are favored; along the coast, annatto colors foods, and in the Peruvian highlands an herb known as palillo is used. Charqui, dried strips of llama meat, is a specialty of the Andes. Anticuchos, chunks of beef heart marinated in vinegar with chiles and cilantro, then skewered and grilled, are a spicy Peruvian favorite also from the Andes. Rabbit dishes are also popular in the region. Along the coast, seafood dominates the diet. The region is famous for its ceviches (also spelled *cebiche*), a method of preparing fresh fish, shrimp, scallops, or crab by marinating small raw chunks in citrus juice. The acidity of the juice cooks the fish and turns it opaque. At many beaches, cevicherias offer the dish as a snack or light meal with beer. Chopped onion, tomato, avocado, and cilantro are often added. In Peru, ceviche is typically garnished with sliced sweet potato. Chucula is a thick plantain and milk beverage flavored with cinnamon popular along the coastal regions of Ecuador.[114] Pisco, a grape brandy that originated in Peru, is a national favorite, often mixed with orange juice to make the refreshing drink called yugeno.

Argentina, Chile, Bolivia, Uruguay, and Paraguay. Hearty, ample fare with an emphasis on beef exemplifies the cooking of these southern nations. Argentina is a major beef-producing region, and its people eat more beef per capita than in any other country worldwide. The temperate weather permits the cultivation of numerous fruits and vegetables, notably strawberries, grapes, and Jerusalem artichokes (known as topinambur in Chile). The cooking of Argentina, Chile, Paraguay, and Uruguay has been influenced more by their immigrant populations than by the numerous small Indian groups native to the area. The Spanish introduced cattle, and the Italians brought pasta. Smaller numbers of Germans, Hungarians, and other central Europeans have added their foods as well.

The national dish of Argentina is matambre, which means "to kill hunger." A special cut of flank steak is seasoned with herbs, then traditionally rolled pinwheel fashion around a filling of spinach, whole hard-boiled eggs, and whole or sliced carrots, and then tied with a string and poached in broth or baked. Matambre can be served as a main course, but it is often chilled first and offered as a cold appetizer. Grilled steaks are particularly popular in Argentina and surrounding nations. In Paraguay, steaks are typically served with sopa Paraguay, a cornmeal and cheese bread. In Uruguay, beef is eaten nearly as often as in Argentina, although mutton and lamb are also common.

Robust soups and stews are everyday fare. In Bolivia, beef stew is made with carrots, onions, hominy, and chuño. The stews of Argentina often pair meat with fruits as well as vegetables, such as carbonada criolla (beef cooked with squash, corn, and peaches) or carbonada en zapallo (veal stew cooked in a pumpkin). In Paraguay, soups reveal European inspiration, such as bori-bori, beef with cornmeal and cheese dumplings, and so' o-yosopy, beef soup with bell peppers, tomatoes, and vermicelli or rice, topped with Parmesan cheese. Fish soups and stews are popular in Chile, which has an extensive coastline and plentiful seafood. A specialty is clam or abalone chowder with beans (chupe de loco) and congrio (an elongated, firm-fleshed fish that looks a little like an eel) cooked with potatoes, onions, garlic, and white wine.

National favorites include pasta (e.g., spaghetti, ravioli, and lasagna), which is served in many homes on Sundays in Argentina. It is considered lucky to eat it on the twenty-ninth of every month as well. In Chile, beans are especially popular, and seafood is eaten regularly. Wines from the temperate midlands of the country are considered some of the best on the continent. Pisco is consumed in both Bolivia and Chile, where it is mixed with lemon juice, sugar, and egg whites to make a pisco sour. In Bolivia, legs from the giant frogs found in the Andean Lake Titicaca are a specialty, and chicha, a distilled corn liquor, is popular.

Although coffee is consumed throughout the area, another caffeinated beverage is equally popular in some regions. Called maté, it is an infusion made from the leaves of a plant (*Ilex paraguariensis*) in the holly family native to Paraguay. Served hot or chilled, maté is consumed nearly every afternoon with small snacks in Paraguay and in parts of Argentina. The dried, powdered leaves are called yerba and are traditionally mixed in a gourd with boiling water. A special metal straw is inserted to drink the brew.

Colombia and Venezuela. The fare found in Colombia and Venezuela is colonial Spanish in character, cooked with olive oil, cream, or cheese and flavored with ground cumin, annatto, parsley, cilantro and chopped onions, tomatoes, and garlic. Yet native tastes are still evident. Guascas, or huascas (Galinsoga parvilora Lineo), an herb native to Colombia, provides a flavor similar to boiled peanuts in soups and stews. Hot chili-pepper sauces are served on the side of most dishes. Tropical fruits and vegetables, including avocados, bananas and plantains, naranjillo (a small, orange fruit related to tomatoes used for its tart juice), pineapple, and coconut milk or cream are other common regional ingredients.

In Colombia, Bogatá chicken stew (made with chicken, two types of potatoes, and cream) and sancocho (a boiled dinner traditionally made with beef brisket or other roast, and ample starchy vegetables such as potatoes, sweet potatoes, plantains, or cassava) are typical Spanish-influenced dishes. Examples in Venezuela include ropa vieja ("old clothes"), shredded flank steak served in a sauce made with tomatoes, onions, and olive oil; and pabellón caraqueño, flank steak served on rice with black beans, topped with fried eggs, and garnished with fried plantain chips. Dishes with more indigenous flavors include arepa, the staple cornmeal bread of Venezuela that is formed into one-inch-thick patties and cooked on a griddle (it is sometimes stuffed with meat or cheese before it is fried), cachapas, tender cornmeal crepes, and mashed black beans, known as caviar criollo or "native caviar." Tropical fruits, such as guavas and pineapple, are often sweetened and dried to make favorite snacks of fruit leathers and fruit pastes.

Guyana. Guyana has a cuisine widely influenced by its proximity to the Caribbean, as well as by the many immigrants from throughout the world who have called Latin America home. For example, one national favorite is pepper pot, a stew made with a variety of meats and onions and flavored with the Caribbean cassava-based sauce cassareep (see Chapter 9). Other common dishes similar to those in the Caribbean include salt-fish cakes, blood pudding, coocoo (cornmeal and okra bread), cookup rice (rice with black-eyed peas or split peas), bammies, and ginger beer. Caribbean desserts are common, such as the dense fruitcake known as black cake, and konkee, a tamale made from sweetened cornmeal, coconut milk, and raisins wrapped in banana leaves, then boiled. African influence is found in foofoo, a pounded plantain paste like African fufu, and stews made with fish or meat, plantains, onions, and okra (see Chapter 8). Dumplings, called metamgee, are made from starchy vegetables and are often added to stews. Asian foods include Indian curries, roti (flat bread), the use of dal (a type of legume; see Chapter 14, "South Asians"), and Chinese noodle dishes. One national specialty is Portuguese garlic pork, which is marinated in vinegar, then fried. The country is famous for Demerara sugar, a very rich, brown-colored, and crumbly raw cane sugar named for a region in Guyana. It is the source of Demerara rum, a Guyanese specialty.

Brazil. The cooking of Brazil is very different from that of other South American countries due to Portuguese and African influences. The Portuguese arrived in the sixteenth century, looking for land on which to cultivate sugar cane. They contributed dried salt cod and linguiça to the diet, stews known as cozidos made with many different meats and vegetables (known as cocido in Portugal), and a variety of exceptionally sweet desserts based on sugar and egg yolks, such as caramel custards and corn (canjica) or rice (pirão de arroz) puddings flavored with coconut. African slaves put to work on the sugar plantations brought foods unknown in nearby countries, such as dendê oil (a type of palm oil) and okra. Spicy dishes were preferred. In West Africa, malagueta peppercorn, a small, hot grain, was used to season foods; in Brazil, Africans adopted a very small, mouth-searing chile pepper indigenous to the area and also called it malagueta. It is typically minced and added to dendê oil, often with dried shrimp and grated ginger root, to make a hot sauce.

[SAMPLE MENU]

A Brazilian Celebration

***Feijoada Completa* (Black Beans with Meats)[a,b,c]**

***Farofa* (Toasted Manioc Meal)[a,b,c]**

Braised Collard Greens[a,b,c]

Brazilian Rice[a,b]

Coconut Bread Pudding[a] or *Torte de Banana* (Banana Pie)[b]

***Capirinhas,*[a,b,c] Beer or Juice**

[a]Kijac, M.B. 2003. *The South American table*. Boston: Harvard Common Press.

[b]*Maria's Cookbook* at http://www.maria-brazil.org

[c]*Cook Brazil* at http://www.cookbrazil.com

Although Indian, Portuguese, and African tastes and textures have influenced cooking throughout Brazil, nowhere are they more prominent than in the state of Bahia. Known as Afro-Brazilian fare, or cozinha baiana, this cuisine is famous for fritters made from dried shrimp, dried salt cod, yams, black-eyed peas, mashed beans, peanuts, and ripe plantains fried in dendê oil. Vatapá, another specialty, is a paste made with smoked dried shrimp, peanuts, cashews, coconut milk, and malagueta chiles. It is used as a filling for black-bean fritters called aracanjá and sometimes served with rice as an entree.

The national dish of Brazil is feijoda completa, which originated in Rio. Black beans cooked with smoked meats and sausages are served with rice, sliced oranges, boiled greens, and a hot sauce mixed with lemon or lime juice. Toasted cassava meal, called farinha, is sprinkled over the top like Parmesan cheese. Farinha is served with most dishes and is often mixed with butter and other ingredients, such as bits of meat, pumpkin, plantains, or coconut milk to create crunchy side dishes called farofa. Rice or cornmeal porridge, called pirão, is another type of side dish. Middle Easterners who immigrated to the southeastern areas of Brazil brought the concept of couscous to the country and adapted the dish to native ingredients. Cuzcuz paulista is prepared with cornmeal in a cuzcuzeiro, which looks like a colander on legs that is inserted over a pot of boiling water to steam. The basket of the cuzcuzeiro is first lined with seafood or poultry and vegetables, which flavors the cornmeal as it cooks and looks decorative when the cuzcuz cake is inverted.

In the far south, the cuisine has been influenced by the foods of Argentina. Grilled meats are a favorite in Brazil, especially in the south, home of the frontiersmen known as gauchos, who herded cattle on the grassland plains. Sides of beef were traditionally staked at the edges of a bonfire for slow cooking in a method called *churrasco*. The popularity of the outdoor barbecue led to *churrascaria rodizio*, restaurants located in cities throughout the nation that specialize in spit-roasted beef, pork, lamb, and sausages brought to the table on large skewers and carved to taste. Specialties include picanha (rump roast) and beef heart. Assorted side dishes such as salads, potatoes, condiments, and desserts round out the meal. Brazilians in the South also drink maté, which they call chimmarão. Coffee, rum, and beer are common beverages in Brazil, but several other drinks are also popular. Guaraná is a delicious, stimulating carbonated soft drink made from the seeds of the native guaraná fruit, which contain caffeine. Cachaça (called aguardiente in other South American nations) is an alcoholic beverage, often compared to brandy, distilled from sugar cane. It is used to make batidas, a refreshing punch with fruit juice, or caipirinhas, mixed with a little lime juice, sugar and mint, then consumed over ice.

Meal Composition and Cycle

Daily Pattern. Three meals a day are traditional among middle-class and affluent South Americans, with an afternoon snack often added. The poor, especially those in rural areas, are often limited to an early breakfast with a large dinner around 6:00 p.m.[112]

For those who can afford more than two meals daily, breakfast is typically light, often bread or a roll with jam and a cup of coffee, served black or with milk. A more complete meal features fresh fruit or pastries and occasionally ham or cheese. Lunch is usually the main meal, consumed in a leisurely manner with family or friends. Appetizers such as fritters, humitas, or empanadas may start the meal, followed by a meat or seafood stew or a grilled meat dish. Side dishes may include rice, beans, farofa, fried potatoes, and greens such as kale. Salads, typically featuring cooked vegetables, are popular in some areas, including Brazil, and are served with the meal. Dessert, most often flan or another sweet custard or pudding, is usually served. In Argentina, the time spent relaxing and socializing after lunch is called *la sombremes* and sometimes includes a nap. Dinner is traditionally lighter, sometimes just cold cuts, a seafood salad, or a serving of soup or stew, and usually eaten around 9:00 each evening, often continuing past midnight. Beer, wine, fruit juice, and soft drinks are beverages commonly consumed at meals.

An afternoon break is typical in much of South America: coffee is typically consumed in Argentina, Colombia, Ecuador, and Brazil; tea is served in the late afternoon in Chile and Uruguay; and maté is popular in parts of Argentina, Paraguay, Uruguay, and Brazil. Snacks eaten with the beverage are often fruit, cachapas or arepa, sandwiches, or a pastry. Street vendors offering coffee, fruit juice, and snacks throughout the day are common in urban areas. Unlike wealthier South Americans, the poor often skip lunch and eat a large dinner. The meal may consist of soup or a serving of stew with a side dish of potatoes, plantains, cassava, corn, or rice and beans, depending on the region.

Special Occasions. Catholic traditions have influenced many South American holidays. A rich Christmas Eve dinner is traditional in most nations, often with a roast, such as lechón (suckling pig) in Brazil and cuy or lechón in Ecuador. Italian specialties including torrone and panettone are Christmas items in Argentina, where Epiphany is another significant religious holiday (see Chapter 6).

Easter is important in many homes, and Carnival (*Carnaval*) festivities featuring dancing, parties, and traditional fare are popular in Brazil, Ecuador, Peru, and Uruguay. Americans from these countries sometimes celebrate with parties during the three days before Lent. During the time of Lent, animals associated with water habits, such as alligators, armadillos, capybaras, iguanas, and turtles, were traditionally classified as fish; thus the Catholic Church permitted their consumption on meatless days.[112] These game meats are still considered Lent specialties in some regions.

Saint John's Day is a favorite in Brazil, featuring foods made with corn and pumpkin, and it is also celebrated by the citizens of Otavalo, Ecuador with all-night feasting and dancing. In Peru, All Soul's Day on November 2 includes gifts of food and family picnics at the gravesites of deceased kin. Also significant for many Americans of South American descent are the independence days observed in various nations. Brazilian Americans commemorate their independence on September 7 with daylong festivities in Boston, New York, and Newark, New Jersey. Americans from Chile sponsor traditional food and craft booths for fairs to celebrate their Independence on September 18. Colombian Americans consume tamales, empanadas, arepas, and other specialties on their independence day, July 20. In Ecuador, the *primer grito* ("first cry" of independence) is held on August 10 and is officially marked as Ecuador Day in New York City. Independence Day in Peru is July 28. The Day of Tradition is popular in the Argentinian American community, with customary foods, folk music, and equestrian displays by men dressed as gauchos.

Etiquette. Traditionally, women prepared meals and served them to men, who consumed their food first. Women would eat after the men finished. This custom is continued in many rural, and even some urban, homes in South America today, especially among families of Indian ancestry.[112]

European-style dining is common in most of South America. The fork is kept in the left hand, and the knife in the right, with no switching for cutting food. Bread is often served without butter and placed on the side of the plate. It is the only food that should be eaten with the hands. All other items, including fruit, require cutlery. Salads, however, should not be cut. Instead the lettuce should be folded into bite-sized packets with the fork. In Brazil, even sandwiches are eaten with a knife and fork. All items are passed to the left. The hands should remain above the table when a person is not eating, with wrists resting on the table edge. In Colombia, it is impolite to start a meal until the host says, "¡*Buen provecho*!" In Bolivia, it is an insult to pour wine with your left hand or to hold the bottle at the base when pouring, which is interpreted to mean you dislike the person for whom the glass is intended.[35]

Therapeutic Uses of Foods. *Candomblé orixás* are associated with certain foods, and followers honor their deity by eating those items. Examples include white corn, white beans, rice, porridge, yams, and water with Oxalá; rice, black beans, black-eyed peas, and roasted corn with Omolu; black beans with Ogun; farofa made with dendê oil, black beans, honey, steak with onions, and chachaça with Exú; tapioca, pudding, cooked corn, and a ginger-flavored drink called aluá with Oxoosi; and pudding, banana, ximxim (chicken stew), and champagne with Oxum. A hot-cold system of medicine, most likely introduced by European immigrants, has also been adapted by *candomblé* healers, treating hot conditions associated with hot *orixás* with cool prescriptions associated with cool *orixás*. Classification is inconsistent, however, and cold conditions are rarely treated with hot remedies.[110]

Some South Americans may adhere to more general hot-cold classifications not associated with *candomblé*.[112] Foods that are hot in temperature, or irritating to the stomach, may be avoided during fevers, for example. Conditions such as menstruation, pregnancy, and lactation also require specific foods. In addition, some people believe that certain foods should be eaten at specific times of day, such as fruit, which is considered wholesome in the morning but harmful in the evening. Some South Americans avoid combinations of some foods, such as eating acidic fruits at the same time as drinking milk.

Contemporary Food Habits in the United States

Adaptations of Food Habits

Very little has been reported on the adapted food habits of South Americans living in the United States. Many continue cooking their favorites from home, although recipes are often adapted to accommodate U.S. ingredients or to improve acceptability (e.g., cuy is not often prepared). Substitutions for unavailable ingredients, such as feta cheese for fresh farmer's cheese or peanut oil for dênde oil, are common. Sometimes the fact that certain customary ingredients are unobtainable makes other dishes that can be prepared traditionally more popular in the United States than these dishes are in their countries of origin. For instance, llapingachos is probably eaten more often by Peruvian Americans than by Peruvians.[115] Among Chileans, many find it difficult to adapt to typical American schedules with a work day that begins earlier than in Chile (difficult after a late dinner) and has a short lunch period precluding a leisurely meal. The diets of poorer Chileans may improve in the United States where food is less expensive.[106]

Nutritional Status

Nutritional Intake. There are minimal data on the nutritional status of Americans of South American descent. Parasitic infection, iron-deficiency anemia, and protein-calorie malnutrition are common in many rural areas of South America and in some crowded urban neighborhoods as well. Chronic Chagas' disease involving the esophagus and colon is endemic in some regions and may be a risk factor in cardiovascular disease.

Studies in South America reveal trends that may be applicable to the population in the United States, particularly recent immigrants. Overweight and obesity is prevalent in some regions, with reported rates for BMI $\geq$25 of 46 percent in one Colombian adult sample, 52 percent in a study of Ecuadoran elders, and 33 to 45 percent in Brazilian studies.[116–119] Reports on obesity show rates of BMI >30 of 12 percent in a study of Brazilian adults, 31 percent of one adult sample in Paraguay, and 64 percent in postmenopausal women in Argentina.[119,120,121] An increase in overweight and obesity in Bolivian women has been reported.[122] Among Brazilians, traditional diets low in calorie density and high in fiber were associated with lower BMI, whereas more Westernized diets with foods high in added fats (especially butter, margarine, and fried snacks) and sugars (particularly soft drinks) were associated with a higher BMI. In the research on Ecuadoran elders, many were deficient in vitamins and minerals despite high rates of overweight, including vitamins B_{12}, D, and folate, as well as iron and zinc.

Concerns regarding the development of metabolic syndrome (the clustering of symptoms related to heart disease and diabetes) have been published. Hypertension is common in many populations: 50 percent in Colombian elders, 39 percent in postmenopausal Ecuadoran women, 42 and 46 percent respectively in Guyanese women and men over fifty, 33 percent in postmenopausal Argentinean women, and over 25 percent in Paraguayan elders.[116,123–126] Type 2 diabetes rates are in the 4 to 8 percent range. Excessive central body fat was found in one group of overweight Brazilian subjects.[117] A study in Ecuador reported 41 percent of postmenopausal women had metabolic syndrome.[124] Research in Venezuela estimated that one-third of the adult population in the region had metabolic syndrome associated with dyslipidemia. Among men, rates varied widely by ethnicity and were highest in those of mixed heritage (37 percent), followed by whites and blacks. The lowest rates (17 percent) were found in the Indian group.[127]

Counseling. Access to health care may be limited for some Americans of South American heritage. In some countries, such as Chile, preventive health practices are uncommon and care may not be sought except in emergency situations.[106] One study that grouped Central and South American Latino women found that 36 percent of those women had no health insurance, and 22 percent did not have a usual biomedical clinic or physician for care.[128] Unauthorized South Americans living in the United States illegally typically have no medical insurance and may avoid contact with any government agency. Nearly 90 percent of South Americans do not speak English at home, preferring to maintain their native languages, including Spanish, Portuguese, or various Indian languages. Guyanese Americans may speak a Creole language

Chileans commonly use both their paternal and maternal surnames.

DISCUSSION STARTERS: UNDERSTANDING CULTURAL DIFFERENCES

Chapter 10 covers many cultures. Some share traits; others do not. Cultural differences can be subtle. It's easy to stereotype people from unfamiliar cultures, but doing so can cause poor outcomes for U.S. health care professionals trying to counsel immigrant patients. Below are examples of two matrixes, intended to help us differentiate among immigrants from various Caribbean Island and South American cultures. Your task is to create two tables using the following models, each with two rows: Caribbean Islands and South America. The first table lists health issues, and the second lists cultural issues for the two groups. Place a plus sign in the squares where the features are present for the immigrants in each row.

In the first matrix, identify which of the immigrant groups have been generally recognized as suffering from which diseases. This is an example of how you can organize your information

	Obesity	Hypertension	Diabetes	Liver disease	Parasitic
List the country in this column					

Counseling patients from different cultures is not just complicated by the fact that immigrants from different cultures may tend to suffer from different diseases but also by the fact that different cultures tend to behave differently. In the second matrix, identify which problematic features are associated with which immigrant groups.

	Language Problems	No health Insurance	Fatalism[1]	Modest reserved behavior	Direct eye Contact
List the country in this column					

In small groups, share your completed matrixes. Within your group, come to a consensus on your identifications and discuss the implications this information might have for training health professionals. Are there categories shared by several immigrant groups? Do the matrixes suggest some general guidelines for addressing the dietary needs of immigrants from the Caribbean Islands and South America?

[1]By *fatalism*, we mean the idea that God or fate determines a person's health.

that combines African dialects with English, and some Venezuelans use Spanglish, a mixture of Spanish with English. Of those who do not speak English at home, over 47 percent reported they do not speak English "very well."[7]

South Americans tend toward formality in their interactions with others. Conversations are often reserved, with little emotional expression. However, Venezuelans are noteworthy for their directness, and Brazilians are typically restrained with strangers but very animated with family and friends.[35] Most South Americans prefer to sit and stand closer than is usual in the United States. Direct eye contact is common throughout South America, except in Colombia, where eye contact may be avoided with authority figures or elders, or in embarrassing situations as a sign of subordination. Most South Americans are present oriented and polychronic, which may result in relaxed attitudes toward appointment and treatment schedules. Generally speaking, immediate interventions are valued more than either preventive or long-term care. In Argentina, a patient is often protected from a negative prognosis.

The role of traditional health beliefs and practices is largely undocumented, though it is assumed that as in other Latin American nations, only small numbers of people adhere completely to any single folk system. Homeopathic remedies and over-the-counter antibiotics sent from family members still in South America may be obtained, and faith healers may be sought by some immigrants.[32]

One study of women with type 2 diabetes living in Southeastern Brazil reported that many found dietary restrictions burdensome. Some were concerned with loss of autonomy, and others mentioned cravings for sweets. Some felt sadness when following the diet. Many mentioned that they had no symptoms of diabetes, and therefore believed dietary restrictions were unnecessary. Compliance was poor in this group.[129]

An in-depth client interview should be used to determine South American country of origin, preferred language, and length of stay in the United States. Client's socioeconomic status, degree of acculturation, and use of traditional health practices will be significant factors in care.[32]

REVIEW QUESTIONS

1. Choose one Caribbean country and summarize the worldview of its immigrants living in the United States. Include an example of the use of the hot-cold system for cause or treatment of an illness. Describe the types of traditional healers used in this region.
2. Select one indigenous food found in the Caribbean. Describe its taste and use in recipes from the region. Next, select a foreign food that was brought to the region and still commonly consumed—provide a recipe. Which foods are now the staples of the diet? Describe a holiday meal in one Caribbean country plus the specialties of the island.
3. What health problems have become common for people from the Caribbean living in the United States? If you were a nutritionist, how would you modify the diet in the treatment of these disorders?

REFERENCES

1. Buffington, S.T. 2000. Dominican Americans. In *Gale Encyclopedia of Multicultural America,* R.V. Dassanowsky & J. Lehman (Eds.). Farmington Hills, MI: Gale Group.
2. Torres-Saillant, S., & Hernandez, R. 1998. *The Dominican Americans.* Westport, CT: Greenwood Press.
3. U.S. Census Bureau. 2005–2009 American Community Survey 5-Year Estimates. S0506. Selected Characteristics of the Foreign-Born Population by Region of Birth.
4. Green, D. 2000. Puerto Rican Americans. In *Gale Encyclopedia of Multicultural America, R.V.* Dassanowsky & J. Lehman (Eds.). Farmington Hills, MI: Gale Group.
5. U.S. Census. Selected Population Profile in the United States Population Group: Cuban Data Set: 2007 American Community Survey 1-Year Estimates.
6. U.S. Census. Selected Population Profile in the United States Population Group: Dominican Data Set: 2007 American Community Survey 1-Year Estimates.
7. Ramirez, R.R. 2004. *We the People: Hispanics in the United States.* Washington, DC: U.S. Census Bureau.
8. Hoefer, Michael, Nancy Rytina, and Bryan C. Baker, 2009. "Estimates of the Unauthorized Immigrant Population Residing in the United States: January 2009." Office of Immigration Statistics, Policy Directorate, U.S. Department of Homeland Security. http://www.dhs.gov/xlibrary/assets/statistics/publications/ois_ill_pe_ 2009.pdf.
9. Monger, Randall, 2010. "U.S. Legal Permanent Residents: 2009." Office of Immigration Statistics, Policy Directorate, U.S. Department of Homeland Security. http://www.dhs.gov/xlibrary/assets/statistics/publications/lpr_fr_2009.pdf.
10. Lewis Mumford Center for Comparative Urban and Regional Research. 2001. *The new Latinos: Who they are, where they are.* Albany, NY: Author.
11. Unaeze, F.E., & Perrin, R.E. 2000. Haitian Americans. In *Gale Encyclopedia of Multicultural America,* R.V. Dassanowsky & J. Lehman (Eds.). Farmington Hills, MI: Gale Group.

12. U.S. Census. Immigration Statistics Staff. 2004. Foreign-Born Profiles (STP-159). Online at http://www.census.gov/population/www/socdemo/forign/STP-159-2000tl.html
13. Murrell, N.S. 2000. Jamaican Americans. In *Gale Encyclopedia of Multicultural America,* R.V. Dassanowsky & J. Lehman (Eds.). Farmington Hills, MI: Gale Group.
14. Lopez-De Fede, A., & Haeussler-Fiore, D. 2002. *An introduction to the culture of the Dominican Republic for rehabilitation service providers. CIRRIE Monograph Series,* J. Stone (Ed.). Buffalo, NY: Center for International Rehabilitation Research Information & Exchange (CIRRIE).
15. Dahl, M. 2004. Working with Puerto Rican clients. *Health Care Food & Nutrition Focus, 21,* 10–12.
16. Miller, N.L. 2000. Haitian ethnomedical systems and biomedical practitioners: Directions for clinicians. *Journal of Transcultural Nursing, 11,* 204–211.
17. Freidenberg, J., Mulvihill, M., & Caraballo, L.R. 1993. From ethnology to survey: Some methodological issues in research on health seeking in east Harlem. *Human Organization, 52,* 151–161.
18. Anderson NLR, Andrews M, Bent KN, Douglas MK, Elhammoumi CV, Keenan C, Kemppainen JK, Lipson JG, Martin CT, Mattson S. Chapter 5. Culturally based health and tllness beliefs and practices across the life span. In: Douglas MK, Pacquiao DF. (eds). 2010; Core curriculum in transcultural nursing and health care. *Journal of Transcultural Nursing, 21*(Suppl 1).
19. Juarbe, T.C. 2003. People of Puerto Rican heritage. In *Transcultural Health Care,* 2nd ed. L.D. Purnell & B.J. Paulanka (Eds.). Philadelphia: FA Davis Company.
20. Purnell, L.D. 2003. People of Cuban heritage. In *Transcultural Health Care,* 2nd ed. L.D. Purnell & B.J. Paulanka (Eds.). Philadelphia: FA Davis Company.
21. Caban, A., & Walker, E.A. 2006. A systematic review of research on culturally relevant issues for Hispanics with diabetes. *Diabetes Educator, 32,* 584–600.
22. Guarnaccia, P.J., Lewis-Fernandez, R., & Marano, M.R. 2003. Toward a Puerto Rican popular nosology: Nervios and ataque de nervios. *Culture, Medicine and Psychiarty, 27,* 339–366.
23. Baer, R.D., Weller, S.C., Garcia, de Alba, Garcia, J., Glazier, M., Trotter, R., Pachter, L., & Klein, R.E. 2003. *Culture, Medicine and Psychiatry, 27,* 315–337.
24. Laguerre, M.S. 1981. Haitian Americans. In A. Harwood (Ed.), *Ethnicity and Medical Care.* Cambridge, MA: Harvard University Press.
25. Colin, J.M., & Paperwalla, G. 2003. People of Haitian heritage. In *Transcultural Health Care,* 2nd ed. L.D. Purnell & B.J. Paulanka (Eds.). Philadelphia: FA Davis Company.
26. Zapata, J., & Shippee-Rice, R. 1999. The use of folk healing and healers by six Latinos living in New England: A preliminary study. *Journal of Transcultural Nursing, 10,* 136–142.
27. Reiff, M., O'Conner, B., Kronenberg, F., Balick, M., & Lohr, P. 2003. Ethnomedicine in the urban environment: Dominican healers in New York City. *Human Organization, 62,* 12–26.
28. Larson, E.L., Dilone, J., Garcia, M., & Smolowitz, J. 2006. Factors which influence Latino community members to self-prescribe antibiotics. *Nursing Research, 55,* 94–102.
29. Bearison, D.J., Minian, N., & Granowetter, L. 2002. Medical management of asthma and folk medicine in a Hispanic community. *Journal of Pediatric Psychology, 27,* 385–392.
30. Houston, L.M. 2005. *Food culture in the Caribbean.* Westport, CN: Greenwood Press.
31. Drago L. Carribean Hispanic food practices. In: Goody CM, Drago L. (Eds). *Introduction: Cultural competence and nutrition counseling.* In: Cultural Food Practices. Chicago IL: American Dietetic Association, 2010.
32. Pereira RF, Kurtz C. South American food practices. In: Goody CM, Drago L. (Eds). *Introduction: Cultural competence and nutrition counseling.* In: Cultural Food Practices. Chicago IL: American Dietetic Association, 2010.
33. Wolf, L. 1970. *The cooking of the Caribbean Islands.* New York: Time-Life Books.
34. Drago, L. 2005. Are you Caribbean savvy? Ethnically correct nutrition counseling tips. *Newsflash: Diabetes Care and Education, 26,* 4–9.
35. Foster, D. 2002. *The global etiquette guide to Mexico and Latin America.* New York: John Wiley & Sons.
36. Spector, R.E. 2004. *Cultural diversity in health and illness,* 6th ed. Upper Saddle River, NJ: Pearson Education, Inc.
37. Harwood, A. 1981. *Ethnicity and medical care.* Cambridge, MA: Harvard University Press.
38. Fugh-Berman, A., Balick, M.J., Kronenberg, F., Oroski, A.L., O'Conner, B., Reiff, M., Roble, M., Lohr, P., Brosi, B.J., & Lee, R. 2004. Treatment of fibroids: The use of beets (Beta vulgaris) and molasses (Saccharum officinarum) as an herbal therapy by Dominican healers in New York City. *Journal of Ethnopharmacology, 92,* 337–339.
39. Sanjur, D. 1995. Hispanic foodways, nutrition, and health. Boston: Allyn & Bacon.
40. Ortiz, B.I., & Clauson, K.A. 2006. Use of herbs and herbal products by Hispanics in south Florida. *Journal of the American Pharmaceutical Association, 46,* 161–167.
41. Immink, M.D.C., Sanjur, D., & Burgos, M. 1983. Nutritional consequences of U.S. migration patterns among Puerto Rican women. *Ecology of Food and Nutrition, 13,* 139–148.
42. Sharma, S., Cade, J., Landman, J., & Cruickshank, J.K. 2002. Assessing the diet of the British African-Caribbean population: Frequency of consumption of foods and food portion sizes. *International Journal of Food Sciences and Nutrition, 53,* 439–444.
43. Chávez, N., Sha, L., Persky, V., Langenberg, P., & Pestano-Binghay, E. 1994. Effect of length of stay on food group intake in Mexican and Puerto Rican women. *Journal of Nutrition Education, 26,* 79–86.
44. Sanjur, D. 1995. *Hispanic foodways, nutrition, and health.* Boston: Allyn & Bacon.
45. Bermudez, O.I., Falcon, L.M., & Tucker K.L. 2000. Intake and food sources of macronutrients among older Hispanic adults: Association with ethnicity, acculturation, and length of residence in the United States.

Journal of the American Dietetic Association, 100, 665–673.
46. Centers for Disease Control. 2011. Fact Sheet: CDC Health Disparities and Inequalities Report—US 2011. Available from: http://www.cdc.gov/minorityhealth/reports/CHDIR11/FactSheet.pdf. Accessed: March 5, 2011.
47. Hajat, A., Lucas, J.B., & Kington, R. 2000. Health outcomes among Hispanic subgroups: Data from the National Health Interview Survey, 1992–1995. *Advance Data, 25,* 1–14.
48. Hosler, A.S., & Melnik, T.A. 2005. Population-based assessment of diabetes care and self-management among Puerto Rican adults in New York City. *Diabetes Educator, 31,* 418–426.
49. Centers for Disease Control. *Eliminate disparities in diabetes.* Available from: http://www.cdc.gov/omhd/AMH/factsheets/diabetes.htm. Accessed: March 5, 2011.
50. USDHHS. Office of Minority Health. Hispanic/Latino Profile. Available from: http://minorityhealth.hhs.gov/templates/browse.aspx?lvl=2&lvlid=54. Accessed: March 5, 2011.
51. Borrell, L.N., Dallo, F.J., & White, K. 2006. Education and diabetes in a racially and ethnically diverse population. *American Journal of Public Health, 96,* 1637–1642.
52. Tucker, K.L. 2005. Stress and nutrition in relation to excess development of chronic disease in Puerto Rican adults living in the Northeastern USA. *Journal of Medical Investigation, 52,* 252–258.
53. Hummer, R.A., Rogers, R.G., Amir, S.H., Forbes, D., & Frisbie, W.P. 2000. Adult mortality differentials among Hispanic subgroups and non-Hispanic whites. *Social Science Quarterly, 81,* 459–476.
54. Centers for Disease Control. CDC 2010. *Infant Mortality Statistics from the 2006 Period Linked Birth/Infant Death Data Set.* National Vital Statistics Reports. Table 2. Available from: http://www.cdc.gov/nchs/data/nvsr/nvsr58/nvsr58_17.pdf. Accessed: March 5, 2011.
55. Landale, N.S., Gorman, B.K., & Oropesa, R.S. 2006. Selective migration and infant mortality among Puerto Ricans. *Maternal and Child Health Journal, 10,* 351–360.
56. Odel, C.D., Kotelchuck, M., Chetty, V.K., Fowler, J., Stubblefield, P.G., Orejuela, M., & Jack, B.W. 2006. Maternal hypertension as a risk factor for low birth-weight infants: Comparison of Haitian and African American women. *Maternal and Child Health Journal, 10,* 39–46.
57. Kugyelka, J.G., Rasmussen, K.M., & Frongillo, E.A. 2004. Maternal obesity is negatively associated with breastfeeding success among Hispanic but not black women. Journal of Nutrition, 134, 1746–1753.
58. Gordon, A.M. 1982. Nutritional status of Cuban refugees: A field study on the health and nutrition of refugees processed at Opa-Locka, Florida. Americans *Journal of Clinical Nutrition, 35,* 582–590.
59. Liao, Y., Tucker, P., Okoro, C.A., Giles, W.H., Mokdad, A.H., & Harris, V.B. 2004. REACH 2010 surveillance for health status in minority communities—United States, 2001–2002. MMWR, 53, 1–36.
60. Ackerman, L.K. 1997. Health problems of refugees. *Journal of the American Board of Family Practice, 10,* 337–348.
61. Kwan, L.L., Bermudez, O.I., & Tucker, K.L. Low vitamin B_{12} intake and status are more prevalent in Hispanic older adults of Caribbean origin than in neighborhood-matched non-Hispanic whites. *Journal of Nutrition, 2002: 132,* 2059–2064.
62. Vyas, A., Greenhalgh, A., Cade, J., Sanghera, B., Riste, L., Sharma, S., & Cruickshank, K. 2003. Nutrient intake of an adult Pakistani, European and African-Caribbean community in inner city Britain. *Journal of Human Nutrition and Dietetics, 16,* 327–337.
63. Shoham J, Duffield A. Proceedings of the World Health Organization/UNICEF/World Food Programme/United Nations High Commissioner for Refugees Consultation on the management of moderate malnutrition in children under 5 years of age. *Food Nutr Bull.* 2009 Sep; 30(3 Suppl): S464–74.
64. Broach JP, McNamara M, Harrison K. Ambulatory care by disaster responders in the tent camps of Port-au-Prince, Haiti, January 2010. *Disaster Med Public Health Prep.* 2010 Jun; 4(2): 116–21.
65. Entzel, P.P., Fleming, L.E., Trepka, M.J., & Dominick, S. 2003. The health status of newly arrived refugee children in Miami-Dade County, Florida. *American Journal of Public Health, 93,* 286–287.
66. Carter-Pokras, O., Pirkle, J., Chavez, G., & Gunter, E. 1990. Blood lead levels of 4–11-year-old Mexican American, Puerto Rican, and Cuban children. *Public Health Reports, 104,* 388–393.
67. Eisenberg KW, van Wijngaarden E, Fisher SG, Korfmacher KS, Campbell JR, Fernandez ID, Cochran J, Geltman PL. Blood lead levels of refugee children resettled in Massachusetts, 2000 to 2007. *Am J Public Health. 2011*; 101(1): 48–54.
68. Proue M, Jones-Webb R, Oberg C. Blood lead screening among newly arrived refugees in Minnesota. *Minn Med. 2010*; 93: 42–6.
69. Hebbar S, Vanderslice R, Simon P, Vallejo ML. Blood levels in refugee children in Rhode Island. Med *Health R I. 2010*; 93: 254–5.
70. Centers for Disease Control. Overweight and Obesity. Available at: http://www.cdc.gov/obesity/data/trends.html. Accessed: March 5, 2011.
71. Thorpe, L.E., List, D.G., Marx, T., May, L., Helgerson, SD, Frieden, T.R. 2004. Childhood obesity in New York City elementary school students. *American Journal of Public Health, 94,* 1496–1500.
72. Nelson, J.A., Carpenter, K., & Chiasson, M.A. 2006. Diet, activity, and overweight among preschool-age children enrolled in Special Supplemental Nutrition Program for Women, Infants, and Children (WIC). *Preventing Chronic Disease, 3,* A 49.
73. Lin, H., Bermudez, O.I., & Tucker, K.L. 2003. Dietary patterns of Hispanic elders are associated acculturation and obesity. *Journal of Nutrition, 133,* 3651–3657.
74. Fitzgerald, N., Himmelgreen, D., Damio, G., Segura-Perez, S., Peng, Y.K., & Perez-Escamilla, R. 2006. Acculturation, socioeconomic status, obesity and lifestyle factors among low-income Puerto Rican women in Connecticut, U.S., 1998–1999. *Pan American Journal of Health, 19,* 306–313.

75. Himmelgreen, D.A., Perez-Escamilla, R., Martinez, D., Bretnall, A., Eells, B., Peng, Y., & Bermudez, A. 2004. The longer you stay, the bigger you get: Length of time and language use in the U.S. are associated with obesity in Puerto Rican women. *American Journal of Physical Anthropology, 125,* 90–96.
76. Gans, K.M., Burkholder, G.J., Upegui, D.I., Risica, P.M., Lasater, T.M., & Fortunet, R. 2002. Comparison of baseline fat-related eating behaviors of Puerto Rican, Dominican, Colombian, and Guatemalan participants who joined a cholesterol education project. *Journal of Nutrition Education and Behavior, 34,* 202–210.
77. Sharma, S., Cade, J., Landman, J., & Cruickshank, J.K. 2002. Assessing the diet of the British African-Caribbean population: Frequency of consumption of foods and food portion sizes. *International Journal of Food Sciences and Nutrition, 53,* 439–444.
78. Contento, I.R., Basch, C., & Zybert, P. 2003. Body image, weight, and food choices of Latina women and their young children. *Journal of Nutrition Education, 35,* 236–248.
79. Lin, H.F., Boden-Albala, B., Juo, S.H., Park, N., Rundek, T., & Sacco, R.L. 2005. Heritabilities of the metabolic syndrome and its components in the Northern Manhattan Family Study. *Diabetologia, 48,* 2006–2012.
80. Centers for Disease Control. 2011 Diabetes Fact Sheet. Available from: http://www.cdc.gov/diabetes/pubs/estimates11.htm#4. Accessed: March 5, 2011.
81. Cleghorn, G.D., Nguyen, M., Roberts, B., Duran, G., Tellez, T., & Alecon, M. 2004. Practice-based interventions to improve health care for Latinos with diabetes. *Ethnicity & Disease, 14,* S117–S121.
82. Ho, G.Y., Qian, H., Kim, M.Y., Melnik, T.A., Tucker, K.L., Jimenez-Velazquez, I.Z., Kaplan, R.C., Lee-Rey, E.T., Stein, D.T., Rivera, W., & Rohan, T.E. 2006. Pan *American Journal of Public Health, 19,* 331–339.
83. Kousta, E., Efstathiadou, Z., Lawrence, N.J., Jeffs, J.A., Godsland, I.H., Barrett, S.C., Dore, C.J., Penny, A., Anyaoku, V., Millauer, B.A., Cela, E., Robinson, S., McCarthy, M.I., & Johnston, D.G. 2006. The impact of ethnicity on glucose regulation and the metabolic syndrome following gestational diabetes. *Diabetologia, 49,* 36–40.
84. CDC 2010. National Diabetes Surveillance System. Rate of initiation of treatment for end-stage renal disease related to diabetes per 100,000 diabetic population (2006). Available from: http://www.cdc.gov/diabetes/statistics/esrd/fig5.htm. Accessed: March 5, 2011.
85. USDHHS. Office of Minority Health. Heart Disease and Hispanics. Available from: http://minorityhealth.hhs.gov/templates/content.aspx?lvl=3&lvlID=6&ID=3325. Accessed: March 5, 2011.
86. Koch, S., Pabon, D., Rabinstein, A.A., Chirinos, J., Romano, J.G., & Forteza, A. 2005. Stroke etiology among Haitians in Miami. *Neuroepidemiology, 25,* 192–195.
87. USDHHS. Office of Minority Health. Liver Disease and Hispanics. Available from: http://minorityhealth.hhs.gov/templates/content.aspx?ID=6207. Accessed: March 5, 2011.
88. Ismail, A.I., & Szpunar, S.M. 1990. The prevalence of total tooth loss, dental caries, and periodontal disease among Mexican Americans, Cuban Americans, and Puerto Ricans: Findings from HHANES 1982–1984. *American Journal of Public Health, 80* (Suppl.), 66–70.
89. USDHHS. Office of Minority Health. Racial and Ethnic Specific Oral Health Data 2010. Available from: http://minorityhealth.hhs.gov/templates/browse.aspx?lvl=3&lvlid=209. Accessed: March 5, 2011.
90. Saint-Jean, G., & Crandall, L.A. 2005. Sources and barriers to health care for Haitian immigrants in Miami-Dade County, Florida. *Journal of Health Care for the Poor and Underserved, 16,* 29–41.
91. Freeman, G., & Lethbridge-Cejku, M. 2006. Access to health care among Hispanic or Latino women: United States, 2000–2002. *Advance Data, 20,* 1–25.
92. USDHHS. Office of Minority Health. Hispanic/Latino profile. Available from: http://minorityhealth.hhs.gov/templates/browse.aspx?lvl=2&lvlid=54. Accessed: March 5, 2011.
93. Vergara, C., Martin, A.M., Wang, R., & Horowitz, S. 2004. Awareness about factors that affect management of hypertension in Puerto Rican patients. *Connecticut Medicine, 68,* 269–276.
94. Pasquali, E.A. 1994. Santeria. *Journal of Holistic Nursing, 12,* 380–390.
95. Garcia, S., Canoniero, M., Lopez, G., & Soriano, A.O. 2006. Metamizole use among Hispanics in Miami: Report of a survey conducted in a primary care setting. *Southern Medical Journal, 99,* 924–926.
96. Larson, E.L., Dilone, J., Garcia, M., & Smolowitz, J. 2006. Factors which influence Latino community members to self-prescribe antibiotics. *Nursing Research, 55,* 94–102.
97. Haider, S.Q., & Wheeler, M. 1979. Nutritive intake of black and Hispanic mothers in a Brooklyn ghetto. *Journal of the American Dietetic Association, 75,* 670–673.
98. Monger R. U.S. Legal Permanent Residents: 2009. Department of Immigration Statistics. Available from: http://www.dhs.gov/xlibrary/assets/statistics/publications/lpr_fr_2009.pdf. Accessed: March 5, 2011.
99. Hoefer M, Rytina N, Baker BC. Estimates of the Unauthorized Immigrant Population Residing in the United States: January 2009. Department of Immigration Statistics. Available from: http://www.dhs.gov/xlibrary/assets/statistics/publications/ois_ill_pe_2009.pdf. Accessed: March 5, 2011.
100. U.S. Census. Immigration Statistics Staff. 2004. Foreign-Born Profiles (STP-159). Online at http://www.census.gov/population/www/socdemo/forign/STP-159-2000tl.html.
101. Mumford, J. 2000. Ecuadoran Americans. In *Gale Encyclopedia of Multicultural America,* R.V. Dassanowsky & J. Lehman (Eds.). Farmington Hills, MI: Gale Group.
102. Packel, J. 2000. Peruvian Americans. In *Gale Encyclopedia of Multicultural America,* R.V. Dassanowsky & J. Lehman (Eds.). Farmington Hills, MI: Gale Group.

103. McLeod, J.A. 2000. Guyanese Americans. In *Gale Encyclopedia of Multicultural America,* R.V. Dassanowsky & J. Lehman (Eds.). Farmington Hills, MI: Gale Group.
104. Rodriguez, J. 2000. Argentinean Americans. In *Gale Encyclopedia of Multicultural America,* R.V. Dassanowsky & J. Lehman (Eds.). Farmington Hills, MI: Gale Group.
105. Eigo, T. 2000. Bolivian Americans. In *Gale Encyclopedia of Multicultural America,* R.V. Dassanowsky & J. Lehman (Eds.). Farmington Hills, MI: Gale Group.
106. Burson, P.J. 2000. Chilean Americans. In *Gale Encyclopedia of Multicultural America,* R.V. Dassanowsky & J. Lehman (Eds.). Farmington Hills, MI: Gale Group.
107. Walker, D. 2000. Venezuelan Americans. In *Gale Encyclopedia of Multicultural America,* R.V. Dassanowsky & J. Lehman (Eds.). Farmington Hills, MI: Gale Group.
108. Miller, O. 2000. Paraguayan Americans. In *Gale Encyclopedia of Multicultural America,* R.V. Dassanowsky & J. Lehman (Eds.). Farmington Hills, MI: Gale Group.
109. Spear, J.E. 2000. Uruguayan Americans. In *Gale Encyclopedia of Multicultural America,* R.V. Dassanowsky & J. Lehman (Eds.). Farmington Hills, MI: Gale Group.
110. Voeks, R.A. 1997. *Sacred leaves of Candomblé: African magic, medicine, and religion in Brazil.* Austin: University of Texas Press.
111. Jefferson, A.W. 2000. Brazilian Americans. In *Gale Encyclopedia of Multicultural America,* R.V. Dassanowsky & J. Lehman (Eds.). Farmington Hills, MI: Gale Group.
112. Lovera, J.R. 2005. *Food culture in South America.* Westport, CN: Greenwood Press.
113. Leonard, J.N. 1968. *Latin American cooking.* New York: Time-Life Books.
114. Kijac, M.B. 2003. *The South American table.* Boston: Harvard Common Press.
115. Novas, H., & Silva, R. 1997. *Latin American cooking in the U.S.A.* New York: Knopf.
116. Bautista, L.E., Orostegui, M., Vera, M., Prada, G.E., Orozco, L.C., & Herran, O.F. 2006. Prevalence and impact of cardiovascular risk factors in Bucaramanga, Colombia: Results from the Countrywide Integrated Noncommunicable Disease Intervention Programme (CINDI/CARMEN) baseline survey. *European Journal of Cardiovascular Prevention and Rehabilitation, 13,* 769–775.
117. Ramos de Marins, V.M., Varnier Almeida, R.M., Pereira, R.A., & Barros, M.B. 2001. Factors associated with overweight and central body fat in the city of Rio de Janeiro: Results of a two-stage random sampling survey. *Public Health, 115,* 236–242.
118. Sempertegui, F., Estrella, B., Elmieh, N., Jordan, M., Ahmed, T., Rodriguez, A., Tucker, K.L., Hamer, D.H., Reeves, P.G., & Meydani, S.N. 2006. Nutritional, immunological, and health status of the elderly population living in poor neighbourhoods of Quito, Ecuador. *British Journal of Nutrition, 96,* 845–853.
119. Sichieri, R. 2002. Dietary patterns and their associations with obesity in the Brazilian city of Rio de Janeiro. *Obesity Research, 10,* 42–48.
120. de Sereday, M.S., Gonzalez, C., Giorgini, D., De Loredo, L., Braguinsky, J., Cobenas, C., Libman, C., & Tesone, C. 2004. Prevelance of diabetes, obesity, hypertension and hyperlipidemia in the central area of Argentina. *Diabetes & Metabolism, 30,* 3335–3339.
121. Jimenez, J.T., Palacios, M., Canete, F., Barriocanal, L.A., Medina, U., Figueredo, R., Martinez, S., de Melgarejo, M.V., Weik, S., Kiefer, R., Alberti, K.G., & Moreno-Azorero, R. 1998. Prevalence of diabetes mellitus and associated cardiovascular risk factors in an adult urban population in Paraguay. *Diabetic Medicine, 15,* 334–338.
122. Perez-Cueto, F.J., & Kolsteren, P.W. 2004. Changes in nutritional status of Bolivian women 1994–1998: Demographic and social predictors. *European Journal of Clinical Nutrition, 58,* 660–666.
123. Etchegoyen, G.S., Ortiz, D., Goya, R.G., Sala, C., Panzica, E., Sevillano, A., & Dron, N. 1995. Assessment of cardiovascular risk factors in menopausal Argentinean women. *Gerontology, 41,* 166–172.
124. Hidalgo, L.A., Chedraui, P.A., Morocho, N., Alvarado, M., Chavez, D., & Huc, A. 2006. The metabolic syndrome among menopausal women in Ecuador. *Gynecological Endocrinology, 22,* 447–454.
125. Inamo, J., Lang, T., Atallah, A., Inamo, A., Larabi, L., Chatellier, G., & de Guademaris, R. 2005. Prevalence and therapeutic control of hypertension in French Caribbean regions. *Journal of Hypertension, 23,* 1341–1346.
126. Ramirez, M.O., Pino, C.T., Furiasse, L.V., Lee, A.J., & Fowkes, F.G. 1995. Paraguayan National Blood Pressure Study: Prevalence of hypertension in the general population. *Journal of Human Hypertension, 9,* 891–897.
127. Florez, H., Silva, E., Fernandez, V., Ryder, E., Sulbaran, T., Campos, G., Calmon, G., Clavel, E., Castill-Florez, S., & Goldberg, R. 2005. Prevalence and risk factors associated with metabolic syndrome and dyslipdemia in white, black, Amerindian and mixed Hispanics in Zulia State, Venezuela. *Diabetes Research and Clinical Practice, 69,* 63–77.
128. Freeman, G., & Lethbridge-Cejku, M. 2006. Access to health care among Hispanic or Latino women: United States, 2000–2002. *Advance Data, 20,* 1–25.
129. Peres, D.S., Franco, L.J., & dos Santos, M.A. 2006. Eating behavior among type 2 diabetes women. *Revista de Saúde Pública, 40,* 310–317.

CHAPTER 11

East Asians

Asia is one of the world's largest continents, stretching from the Ural Mountains and Suez Canal in the East and the Arctic Circle in the North, to the tropical peninsulas of India and Southeast Asia. It encompasses almost one-third of the world land mass and nearly two-thirds of the global population. Asia is divided into the regions of East Asia, Southeast Asia, and South Asia. Though the continent has historically included parts of Russia and several nations of the former Soviet Union (sometimes known as Central Asia) and the Middle East (sometimes called West Asia, or Asia Minor), the people of these countries are culturally distinct from the rest of Asia and are covered in Chapters 7, "Central Europeans," and 13, "People of the Balkans and the Middle East," respectively.

East Asia is defined as China (the People's Republic of China), Taiwan (Republic of China), Japan, the Democratic People's Republic of Korea (North Korea), the Republic of Korea (South Korea), and the Mongolian People's Republic (see Figure 11.1). Immigrants from these nations, particularly China and Japan, have been coming to the United States since the 1800s. Many settled on the West Coast, where the majority of their descendants still live. In recent years, large numbers from throughout the region have arrived in the United States; many are refugees from political oppression, whereas others seek education and employment opportunities. This chapter introduces the peoples and cuisines of China, Japan, and Korea. Southeast Asians are discussed in Chapter 12, "Southeast Asians and Pacific Islanders," and South Asians are considered in Chapter 14, "South Asians."

Chinese

Chinese civilization is more than 4,000 years old and has made numerous significant contributions in agriculture, the arts, religion and philosophy, and warfare. Silk and embroidered brocade cloth, intricate jade sculpture, Chinese porcelain and lacquerware, book printing, Confucianism, Taoism, and gunpowder are just a few examples. The name China, meaning "middle kingdom," or center of the world, is probably derived from a ruling dynasty of the third century B.C.E.

China's landscape is dominated by the valleys of two great rivers, the Huang (Yellow) River in the North and the Chang Jiang (Yangtze) in the South. The climate is monsoonal, with most of China's rainfall occurring in the spring and summer months. The northern plain through which the Huang River flows is agriculturally very fertile. The area is cold, and sometimes the severe winter results in a growing season of only four to six months. In the South the Chang Jiang River starts in Tibet, traverses the southern provinces, and eventually empties into the China Sea near the city of Shanghai. South of the mouth of the Chang Jiang delta is a rugged and mountainous coastline off which are located the islands of Hong Kong and Taiwan. The southern provinces

Marco Polo, the famous European traveler who went to China in the late thirteenth century, is said to have brought Chinese noodles to Italy. The Italians, however, were undoubtedly making pasta long before the times of Marco Polo—noodles are thought to have been developed independently in each region.

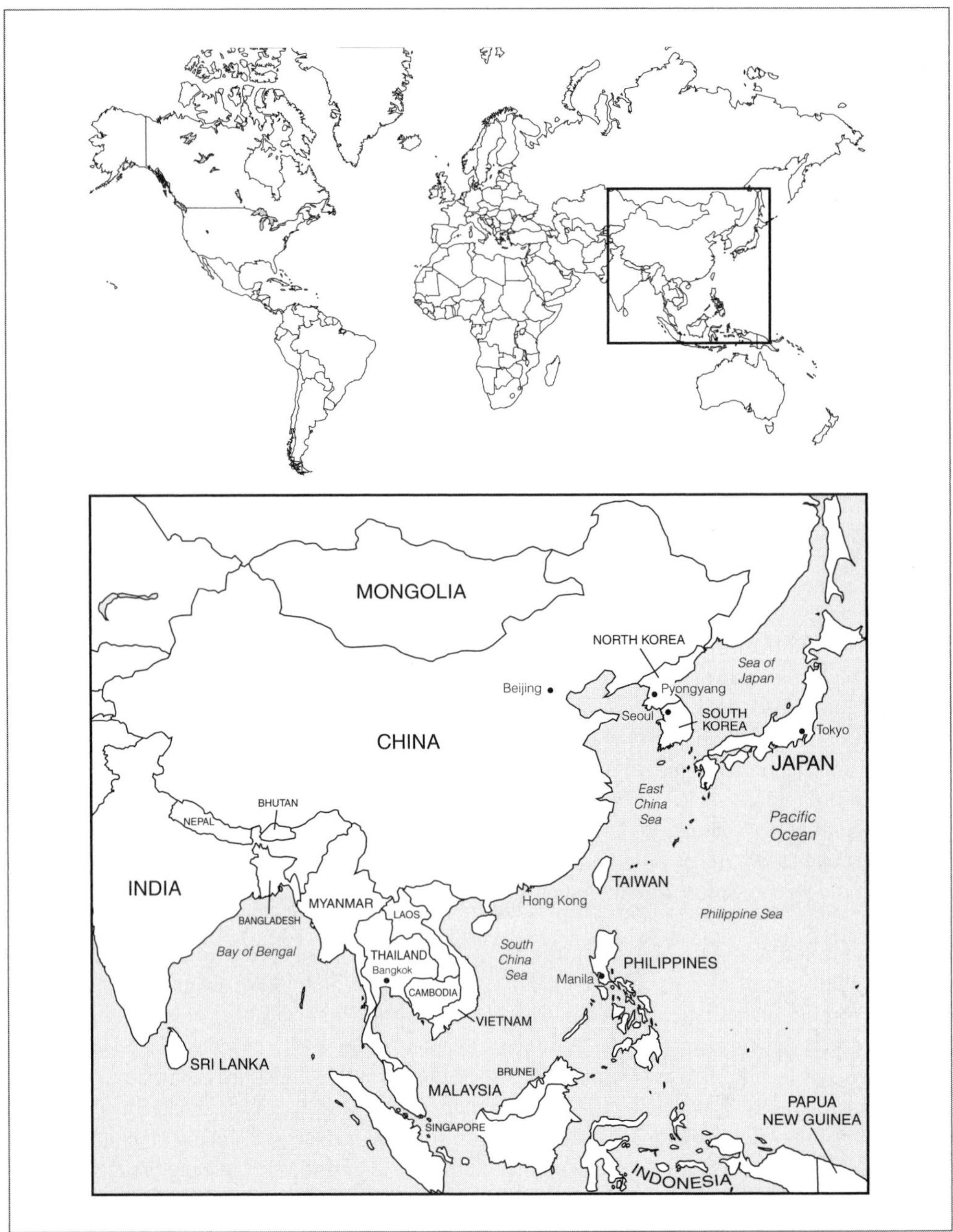

Figure 11.1
China, Japan, and Korea.

Chow-chow is a vegetable relish made with mustard brought to the United States by Chinese railroad workers. Its name may derive from the Chinese word for "mixed," ***cha*****.**

are warmer and wetter and have a longer growing season (six to nine months) than in the North.

The population of China is estimated to exceeds 1.3 billion people, more than four times as large as the population of the United States. The Chinese have a heterogeneous society with numerous ethnic and racial groups. The Chinese language is equally diverse, with many dialects, some of which are incomprehensible to people of other Chinese regions.

Cultural Perspective

History of Chinese in the United States

Immigration Patterns. The first major surge in Chinese immigration to the United States occurred in the early 1850s when the Chinese joined in the gold rush to California; many Chinese still refer to America as the "Land of the Golden Mountain." As mining became less lucrative, the Chinese opened their own businesses, such as laundries and restaurants, but also found employment in other occupations. The Central Pacific Railroad, which joined the Union Pacific as the first cross-country line, was built primarily by 10,000 Chinese workers.

By 1870, there were 63,000 Chinese, mostly male, in the United States, nearly all on the West Coast. Another 120,000 Chinese are estimated to have entered the United States during the following decade. Racial discrimination against Asians increased as their numbers swelled. By 1880 Chinese immigration slowed to a trickle due to exclusion laws directed against Asians. The Chinese also immigrated to Hawaii, and when the islands were annexed by the United States in 1898, approximately 25,000 Chinese were living there.

Most early Chinese immigrants were from the southeastern Guangdong province of China, formerly referred to as Canton. Most were young men with no intention of staying—they came to make their fortune and then return to China and their families. Many married before coming to the United States, and more than half returned to China. By the 1920s the Chinese population in the United States had dropped to 1870 levels.

In each city where the Chinese settled, they usually lived within a small geographic area known as "Chinatown." Large Chinatowns evolved in San Francisco, New York, Boston, Chicago, Philadelphia, Los Angeles, and Oakland, California. These neighborhoods offered protection against a sometimes hostile social and economic environment; assimilation was not an option. Conditions were often crowded and unusual in the predominance of men, but were tolerated with the expectation of eventual return to China. It was not until 1943 that Chinese could become naturalized U.S. citizens.

Current Demographics and Socioeconomic Status. When the exclusion laws were repealed in 1943, people from many Asian countries once again entered the United States. Chinese immigrants who have arrived since World War II are usually not from Guangdong. They are urban dwellers from other regions and are generally better educated than earlier immigrants.

Political instability in the People's Republic of China has led to a recent exodus of Chinese. Following the pro-democracy uprising in Tiananmen Square in 1989, U.S. immigration laws were changed with the Chinese Student Protection Act of 1992. More than 41,000 Chinese residents were granted visas and are eligible for citizenship under the provision. Furthermore, the return of Hong Kong to mainland China and the uncertainty of Taiwan's future have led to increased immigration from these islands.

In 2007 the U.S. Census estimated that 3.3 million Chinese (including Taiwanese) are living in the United States, representing nearly a quarter of all Asians in the nation. Approximately one-third are foreign-born.[1] Nearly one third of the Chinese population in the United States have arrived since 2000. More than half reside in California and New York. Large numbers of Chinese Americans are found today in Hawaii, Texas, New Jersey, Massachusetts, Illinois, Washington, and Maryland.

As of 2000, estimates are that there are over 342,000 Taiwanese living in the United States.[1] Taiwanese Americans have settled in areas where professional employment opportunities, particularly in the sciences, are available. Sizable populations are found in the Los Angeles area (which includes the community of "Little Taipai"), Houston; San Jose; New York City; Atlanta; Baltimore; Boston; Chicago; Cincinnati; San Diego; Washington, DC; and New Jersey.

Four types of Chinese American households have been identified.[2] First are the sojourners, mostly men born in the early twentieth century, who came to the United States with the intent of returning after accumulating some wealth. Political changes in China have stranded this group in urban Chinatowns throughout America. The second type are sojourners who were successful in bringing their wives to the United States. These elderly first-generation couples still live in Chinatown neighborhoods, although their children have often moved from the area. The third type of household is established by new immigrant families. Often one member will arrive to establish residency, and then family members will follow. These immigrants usually settle with other Asians, often in Chinatowns, where the parents speak Chinese and children learn English in school. Although many families include two wage earners or combine the incomes of extended members, they may remain poor due to low salaries, large families, and supporting relatives in the homeland. Chinese American households of the fourth type include acculturated suburban

The first Chinese restaurants in the United States opened in San Francisco in 1849. Over 40,000 Chinese restaurants are in operation today.[172]

The number of Chinese American women in the United States did not equal men until the 1970s.

families. These families include both new immigrants who are white-collar professionals and those Americans of Chinese descent who have lived in the United States for several generations. Both parents may be college educated, often with specialization in engineering and the sciences. Unlike other immigrants of previous generations, these affluent Chinese families live in areas with university or defense research facilities, including the metropolitan areas of the Silicon Valley (California), Houston, Seattle, Pittsburgh, San Diego, and Dallas.

Chinese Americans value education, and there are disproportionately large numbers of them (48 percent according to 2000 census figures) holding college and graduate degrees.[1] High levels of educational attainment often translate into well-compensated professional employment. Over half work in management, professional, or related occupations, while others have successfully opened their own businesses, particularly small markets, dry cleaning, and restaurants.[3]

Americans of Chinese descent are so well known for their successful transition in the United States that a stereotype has developed, obscuring what is in reality a divided population.[3] Though many Chinese Americans are in the upper and middle classes, immigrants who came before the 1950s were often poorly educated, and many found work in sweatshops. Even those who had college degrees were sometimes unable to find jobs suitable to their skills due to discrimination. In addition, small numbers of recent immigrants have also found work in low-paying service jobs or manufacturing. Estimates indicate 28 percent of foreign-born Chinese in the United States do not have a high school diploma.[4]

A large percentage of Taiwanese Americans have immigrated to the United States since 1980, mostly seeking professional opportunities. Two-thirds of the foreign-born population have obtained college degrees, and 64 percent work in management and professional jobs, with another 22 percent employed in office work and sales. Family poverty rates are very low.[4] More recently, some immigrants, called *taikongren* (literally meaning "astronauts") have chosen transnational business careers, living in the United States while their families remain in Taiwan, traveling frequently between the two nations.[5]

Early immigrants and their descendants, as well as many current immigrants from Hong Kong, speak Cantonese, a dialect that typically is difficult to understand by Chinese immigrants who speak Mandarin.

2001 census data report over one million Canadians of Chinese heritage.

Confucius also developed protocols for cooking and eating that were practiced in China until the nineteenth century.[17]

Worldview

Religion. Most Americans of Chinese descent are not affiliated with a specific church. Religious practices are often eclectic, a combination of ancestor worship, Confucianism, Taoism, and Buddhism. Many early Chinese immigrants were not formally schooled in any religion; instead, beliefs and practices were passed orally from generation to generation. Spirituality is integrated into family and community life. Daily living includes avoiding any actions that might offend the gods, nature, or ancestors.

Early Religion. The ancient faith of China was probably a mixture of ancestor worship and respect for the forces of nature and the heavenly bodies. The supreme power was either *Tien* (Heaven) or *Shang Ti* (the Supreme Ruler or the Ruler Above). One gained favor with the spirits by the correct performance of ceremonies. These beliefs and practices were later incorporated into subsequent Chinese religions.

Ceremonies for the dead are a prominent Chinese religious practice. The dead are supposed to depend on the living for the conditions of their existence after death. In turn, the dead can influence the lives of the living.

Confucianism. Confucius was a sage, one of many who gave order to Chinese society by defining the ways in which people should live and work together. Confucianism incorporated the ceremonies of earlier religions, with the following cornerstones:

1. Fatherly love and filial piety in the son (i.e., children are expected to obey their parents and adults are expected to take care of their children).
2. Tolerance in the eldest brother and humility in the younger.
3. Proper behavior by the husband and submission by the wife.
4. Respect for one's elders and compassion in adults.
5. Allegiance to rulers and benevolence by leaders.

Inherent in these relationships is the ideal of social reciprocity, which means that one should treat others as one would like to be treated. To enhance harmony in the family and in society as a whole, one must exercise self-restraint. An individual must never lose face—meaning a person's favorable name and position in society—because that would defame the whole family. Many of these values influence Chinese behavior today.

Taoism. The Taoist, like the Confucianist, believes that heaven and humanity function in unison and can achieve harmony, but under Taoism people are subordinate to nature's way. There is a fundamental duality within the universe of interacting, opposite principles or forces—the *yang* (masculine, positive, bright, steadfast, warm, hard, and dry; sometimes referred to as *shen*) and the *yin* (feminine, negative, dark, cold, wet, mysterious, and secret; also called kwei) (Figure 11.2). Everything in nature contains both yin and *yang*, and a balanced unity between them is necessary for harmony. This balance occurs when Tao, the way of nature, is allowed to take its course unimpeded by human willfulness. Taoism advocates a simple life, communion with nature, and the avoidance of extremes.

Buddhism. Buddhism grew during the T'ang dynasty but then suffered a slow decline. The Mahayana sect dominated in China, blending with traditional Chinese beliefs and resulting in a unique Chinese form of Buddhism. Ten schools of Buddhism flourished in China at one time, but only four were left by the twentieth century. The two dominant schools in China are *Ch'an* (Zen Buddhism in Japan) and Pure Land (see Chapter 2, "Traditional Health Beliefs and Practices," for more information).

Chinese American Spirituality. Both Catholic and Protestant churches were established in the early Chinatown neighborhoods, usually organized by the Chinese dialect spoken in the area. Few first-generation Chinese Americans joined Christian religions, but converts were found in subsequent generations. Others maintain aspects of Buddhism, Taoism, or spirit and ancestor worship in their daily lives, keeping small altars at home in which to offer respect and perform the rites that will preserve good relations with the gods and bring good fortune.

Only a minority of Taiwanese belong to Protestant faiths in their homeland; however, Baptist, Presbyterian, and several smaller evangelical churches have found Taiwanese followers in the United States. Services are conducted in Mandarin or Taiwanese dialects, and the church serves as a social network for the immigrant community. Buddhism has also gained adherents in recent years.[5]

Figure 11.2
Yin-yang symbol. This symbol represents the fundamental duality of the universe and the balance between the forces of *yang* (light) and *yin* (dark). Each force has a little of the other in it (indicated by the dot of the opposite color).

Family. Confucian teachings about correct relationships are still important for many Chinese American families, even if they have become Christians. Chinese American families are usually patriarchal. Women are traditionally taught to be unassuming and yielding. They live by the formula of "thrice obeying": Young girls are submissive to their fathers, wives are subordinate to their husbands, and mothers obey their sons. Children are expected to be quiet, acquiescent, and deferential to their elders.

Harmony in the family is the ideal, so children are taught not to fight or cry. Showing emotion is discouraged. Chinese parents may be very strict, and children are commanded to honor the family. Many of these ideas conflict with American ideals of equal rights and freedom of speech and may lead to intergenerational conflict in the Chinese American home.

Traditional Health Beliefs and Practices. Chinese medicine includes a complex humoral system of professional practice by physicians, known as Traditional Chinese Medicine (TCM), as well as correlated folk remedies used by laypersons at home. Health beliefs and practices have developed over generations, incorporating Confucian, Taoist, and Buddhist concepts regarding the interdependencies of man and nature and the need for balance and moderation in life (Figure 11.3).

Professional TCM follows texts prepared between approximately 2500 B.C.E. and the third century B.C.E., outlining the dynamic equilibrium of forces necessary for health. These include the five elements, or five evolving phases, of fire, earth, metal, water, and wood, each of which may become unbalanced, much as fire consumes wood or wood (as a tree) absorbs the earth. These elements correspond with five organs: the heart, spleen, lungs, kidneys, and gallbladder, respectively. Associations with secretions (perspiration, saliva, mucous, spit, and tears); the seasons (summer, late summer, autumn, winter, and spring); colors (red, yellow, white, blue, and green); tastes (bitter, sweet, pungent, salty, and sour); and directions (south, center, west, north, and east); as well as times of day, odors, sounds, and emotions may also occur.[6–8]

Most Chinese believe in *fengsui*, the way in which a home should be situated and its furnishings arranged to promote optimal flow of energy and personal well-being.

Photo by Kathy Sucher

▶ *Botanical remedies are usually combined in formulary mixtures in traditional Chinese medicine.*

Chinese physicians were traditionally paid for their services when the client was healthy. Payment stopped if the client became ill.

Jade charms are worn to keep children safe and to bestow health, fertility, long life, power, and wisdom on adults.

This system was further elaborated somewhere between the third and sixth centuries by the adoption of Buddhist principles of hot and cold humoral medicine, which were congruent with the Taoist system of *yin* and *yang*.[9,10] The concept of harmony was refined to include a balance of these opposites; illness develops when disequilibrium occurs. Organs such as the liver, heart, spleen, kidneys, and lungs are yin, as is the outside and the front of the body. The gallbladder, stomach, intestines, and bladder are *yang*, as well as the body surface and the back. Outside forces, such as the seasons, are also defined as yin (winter/spring) and *yang* (summer/fall) and illnesses associated with these times may fall into corresponding categories.

Symptoms of disease usually reflect an imbalance between *yin* and *yang*. When there is an excess of *yang*, acne, rashes, conjunctivitis, hemorrhoids, constipation, diarrhea, coughing, sore throat, ear infections, fever, or hypertension may occur. Anemia, colds, flu, frequent urination, nausea, shortness of breath, weakness, and weight loss suggest that an excess of *yin* is the problem. Also associated with *yin* and *yang* is the condition of the blood. Weak blood (*yin*) may develop during growth or pregnancy, postpartum, and in old age. Treatment includes *yang* therapies, particularly the intake of herbs and certain foods.

The vital force of life is *qi* and is equated with energy, air, and breath. *Qi* flows along twelve defined meridians in the body, and some conditions are related to the disruption of *qi* or to excessive *qi*. Other types of energy that must be balanced for health include *jing*, sexual or primordial energy, and *sheng*, spiritual energy or the essence of consciousness.[6,10]

Other lesser forces that may influence health include wind (including natural drafts and those resulting from fans, air conditioners, exposure, or a symptom that rhymes with the Chinese word for wind); poison, which is somewhat related to the Western concept of allergies; and fright, a condition in children that includes listlessness, anorexia, low fever, and crying. Fright is believed to occur when the soul becomes scattered and is mostly limited to Chinese from Taiwan and Hong Kong.[2]

A major difference between Chinese medicine and U.S. biomedicine is the idea that the body and mind are unified, governed by the heart. There is no English word to describe the concept. Emotions are often somaticized, meaning that feelings are related to specific conditions. More than 500 symptoms corresponding to emotions have been identified, each characteristic of one or more organs. For example, *tou yun* (or *tou hun*) is vertigo, the most common complaint made by Chinese patients worldwide. Dizziness or a confused state of mind indicates significant imbalance and serious illness. It is a nonspecific condition thought to originate from anger or anxiety manifested in liver, heart, or kidney dysfunction (if the patient is a young man, too much sexual intercourse or masturbation may be believed to be the cause). Liver disorders develop from suppressed hostility. Anger is discouraged in Chinese culture and may accumulate in the liver, causing it to expand and attack other organs. This diagnosis is common for many gastrointestinal complaints. Generalized stomachaches are believed to be due to eating bitterness in life, often including an inadequate diet when one is young. Anxiety, nervousness, or the stronger emotion of fear results in heart palpitations.[11] Most Chinese prefer this integrated approach to health.[12]

Many Chinese maintain health through a properly balanced diet, moderation in activities and sleep, and avoidance of sudden imbalance caused by forces such as wind. *Qi* must flow freely and blood must be strengthened through nourishment. For instance, if an individual is imprudent and celebrates too much by eating excessive *yang* foods (see the section Therapeutic Uses of Foods), indigestion or a hangover may occur. Eating *yin* foods, which are often bland, is the remedy.[9] One study of Chinese Americans found that 94 percent used home remedies to treat illness.[12] When home cures are ineffective, advice from a TCM physician may be sought. Diagnosis is made through taking an extensive history, and examination of the client, particularly palpitation of pulses and evaluation of the tongue. Through this process a medical pattern is detected, in contrast to determining a specific disease or condition based on symptoms or laboratory testing. It is the medical pattern that determines the appropriate intervention, not the illness.

Treatment for nearly all illness involves the restoration of harmony. Therapy may emphasize dietary and lifestyle changes, or attempt to balance the organs so that emotional balance results. Nearly every visit to the doctor results in a botanical remedy, and most medicinal herbs are only available through prescription. For instance, ginseng may be used to fortify *qi*, and antelope horn can help cool too much *yang* in the liver.[10] Formulary mixtures of five to ten substances are common. Most TCM remedies are prepared as decoctions, taken in a single dose. The client owns the prescription and can reuse it when symptoms occur or share it with family and friends.

Acupuncture is another traditional Chinese treatment. It involves the use of nine types of exceptionally thin metal needles inserted at various points on the body where the *qi* meridians surface. Meridians are considered yin or *yang* and correspond with specific organs. The needles are placed to facilitate a balanced flow of *qi*, restoring harmony to the afflicted organ, mostly for symptoms of excess *yang*. Acupuncture may be performed by a TCM physician or by a specialist. Another, less common, therapy is moxibustion. Small bundles of dried wormwood are heated and carefully applied to certain meridians, usually to balance a *yin* condition. Moxibustion is particularly used during labor and delivery. Massage or therapeutic exercise is another traditional therapy, found more often in China and rarely in the United States.

TCM physicians practicing in the United States do not have specified training and competency varies. Word-of-mouth recommendations are common within the Chinese community. TCM practitioners may use first aid on injuries or broken bones, prescribe herbs, perform acupuncture or moxibustion, or they may diagnose the condition and provide a recommended course of therapy by a specialist in one of these practices. Asians concerned with humoral conditions use TCM (often concurrently with biomedical therapies), and in recent years these practitioners have attracted a growing multi-ethnic clientele. One study reported that 32 percent of Chinese immigrants had returned to China or Taiwan during the previous two years for treatment of a condition.[12]

Traditional Food Habits

The Chinese eat a wide variety of foods and avoid very few. This may have developed out of necessity, as China has long been plagued with recurrent famine caused by too much or too little rainfall. Chinese cuisine largely reflects the food habits and preferences of the Han people, the largest ethnic group in China, but not to the exclusion of other ethnic groups' cuisines. For example, Beijing has a large Muslim population whose restaurants serve lamb, kid, horse meat, and donkey, but no pork. Foreigners have also introduced ingredients that have been incorporated into local cuisines. Some foods now common in China, but not indigenous, are watermelon, tomatoes, bananas, peanuts, and chile peppers.

Ingredients and Common Foods

Staples. Traditional Chinese foods are listed in Table 11.1. In China, numerous fruits, vegetables, and protein items are consumed but few dairy

Figure 11.3

Traditional Chinese Medicine is a humoral system that incorporates Confucian, Taoist, and Buddhist concepts regarding the interdependencies of man and nature and the need for balance and moderation in life.

Source: New Yorker Cartoon Bank, www.cartoonbank.com; image ID 121130, originally published in The New Yorker on July 25, 2005.

"Possible side effects include disharmony, lack of oneness, and blurred third-eye vision."

© The New Yorker Collection 2005 Alex Gregory from cartoonbank.com. All Rights Reserved.

Table 11.1 Cultural Food Groups: Chinese

Group	Comments	Common Foods	Adaptations in the United States
Protein Foods			
Milk/milk products	Dairy products are not routinely used in China. Many Chinese are lactose intolerant. Traditional alternative sources of calcium are tofu, calcium-fortified soy milk, small bones in fish and poultry, and dishes in which bones have been dissolved.	Cow's milk, buffalo milk	Many Chinese consume dairy products, especially milk and ice cream, some cheese. Some alternative sources of calcium may no longer be used.
Meat/poultry/fish/eggs/legumes	Few protein-rich foods are not eaten. Beef and pork are usually cut into bite-size pieces before cooking. Fish is preferred fresh and is often prepared whole and divided into portions at the table. Preservation by salting and drying is common. Shrimp and legumes are made into pastes.	*Meat*: beef and lamb (brains, heart, kidneys, liver, tongue, tripe, oxtails); pork (bacon, ham, roasts, pig's feet, sausage, ears); game meats (e.g., bear, moose) *Poultry*: chicken, duck, quail, rice birds, squab *Fish*: bluegill, carp, catfish, cod, dace, fish tripe, herring, king fish, mandarin fish, minnow, mullet, perch, red snapper, river bass, salmon, sea bass, sea bream, sea perch, shad, sole, sturgeon, tuna *Eggs*: chicken, duck, quail, fresh and preserved *Shellfish and other seafood*: abalone, clams, conch, crab, jellyfish, lobster, mussels, oysters, periwinkles, prawns, sea cucumbers (sea slugs), shark's fin, shrimp, squid, turtle, *wawa* fish (salamander) *Legumes*: broad beans, cowpeas, horse beans, mung beans, red beans, red kidney beans, split peas, soybeans, white beans, bean paste	More meat and poultry are consumed, though some traditional protein sources are still popular
Cereals/Grains	Wheat is the staple grain in the North, long-grain rice in the South. Fan (cereal or grain) is the primary item of the meal; ts'ai (vegetables and meat or seafood) makes it tastier. Rice is washed before cooking.	Buckwheat, corn, millet, rice, sorghum, wheat	Chinese Americans eat less *fan* and more *ts'ai*. The primary staple remains rice, but more wheat bread is eaten.
Fruits/Vegetables	Many non-Asian fruits and vegetables are popular. Potatoes, however, are not well accepted. Vegetables are usually cut into bite-size pieces before cooking. Slightly unripe fruit is often served as a dessert.	*Fruits*: apples, bananas, custard apples, coconut, dates, dragon eyes (longan), figs, grapes, kumquats, lily seed, lime, litchi, mango, muskmelon, oranges, papaya, passion fruit, peaches, persimmons, pineapples, plums (fresh and preserved), pomegranates, pomelos, tangerines, watermelon.	More temperate fruits are consumed.
	Both fresh fruits and vegetables preferred; seasonal variation dictates the type of produce used. Many vegetables are pickled or preserved. Fruits are often dried or preserved.	*Vegetables*: amaranth, asparagus, bamboo shoots, banana squash, bean sprouts, bitter melon, cassava (tapioca), cauliflower, celery, cabbage (*bok choy* and *napa*), chile peppers, Chinese broccoli (*gai lan*), Chinese long beans, Chinese mustard (*gai choy*), chrysanthemum greens, cucumbers, eggplant, flat beans, fuzzy melon, garlic, ginger root, green peppers, kohlrabi, leeks, lettuce, lily blossoms, lily root, lotus root and stems, luffa, dried and fresh mushrooms (black, button, cloud ear, wood ear, enoki, straw, oyster, monkey's head), mustard root, okra, olives, onions (yellow, scallions, shallots), parsnip, peas, potato, pumpkin, seaweed (agar), snow peas, spinach, taro, tea	More raw vegetables/salads are eaten. Data on overall consumption trends are contradictory.

Table 11.1 Cultural Food Groups: Chinese (*continued*)

Group	Comments	Common Foods	Adaptations in the United States
		melon, tomatoes, turnips, water chestnuts, watercress, wax beans, water convolvus, winter melon, yams, yam beans	
Additional Foods			
Seasonings	Complex, sophisticated seasoning combinations common. Various tastes appreciated, such as the moldy flavor of lily flower buds. Spice and herb preferences distinguish regional cuisines.	Anise, bird's nest, chili sauce, Chinese parsley (cilantro), cinnamon, cloves, cumin, curry powder, five-spice powder (anise, star anise, clove, cinnamon or cassia, Sichuan pepper), fennel, fish sauce, garlic, ginger, golden needles (lily flowers), green onions, hot mustard, mace, monosodium glutamate (MSG), mustard seed, nutmeg, oyster sauce, parsley, pastes (*hoisin*, sweet flour, brown bean, Sichuan hot beans, sesame seed, shrimp), pepper (black, chile, red, and Sichuan), red dates, sesame seeds (black and white), soy sauce (light and dark), star anise, tangerine skin, tumeric, vinegar	Many Chinese restaurants use MSG, but it is not usually used in the home.
Nuts/seeds	Nuts and seeds are popular snacks and may be colored or flavored.	Almonds, apricot kernels, areca nuts, cashews, chestnuts, ginkgo nuts, peanuts, walnuts; sesame seeds, watermelon seeds	
Beverages	In northern China the beverage accompanying the meal often is soup. In the South, it is tea. Alcoholic drinks, usually called wines, are rarely made from grapes. They are either beers or distilled spirits made from starches or fruit.	Beer, distilled alcoholic spirits, soup broth, tea	
Fats/oils	Traditionally lard was used if affordable. In recent years, soy, peanut, or corn oil is more common.	Bacon fat, butter, lard, corn oil, peanut oil, sesame oil, soybean oil, suet	Fat intake increases with consumption of fast foods and snacks
Sweeteners	Sugar not used in large quantities; many desserts made with bean pastes.	Honey, maltose syrup, table sugar (brown and white)	Sugar consumption has increased due to increased intake of soft drinks, candy, cakes, and pastries.

products, whether fresh or fermented, are eaten. Grains are the foundation of the diet. As recently as 1985, 75 percent of grain crops were grown by rural farmers for home consumption. In recent decades, however, productivity has improved, and a move away from self-production of foods has occurred, especially in the amount of food consumed away from home. This has resulted in an overall reduction of grain product consumption, although it is still a staple.[13]

Rice is essential in the cuisine of southern China, believed to have been introduced to the region from India in the first century B.C.E. It is so common that people in southern China greet each other by asking, "Have you had rice today?" There are approximately 2,500 different forms of rice, but the Chinese prefer a polished, white, long-grain variety that is not sticky and remains firm after cooking. Short, sticky, glutinous rice is used occasionally, mainly in sweet dishes. Although it is usually steamed, rice can also be made into a porridge called *congee*, eaten for breakfast or as a late-night snack, with vegetables, meat, or fish added for flavor. *Congee* is also fed to people who are ill. Rice flour is used to make rice sticks, which can be boiled or fried in hot oil.

Wheat is also common throughout China, although it is used more often in the North than the South. It is popular as noodles, thin wrappers, dumplings, pancakes, and steamed bread. Noodles are popular in soups, or pan-fried, then topped with meats and vegetables that have been

The custom of throwing rice at newlyweds is believed to come from China, where rice is a symbol of fertility.

Congee may be eaten at any meal in Hong Kong, where a family version topped with lobster is popular. In Taiwan, congee is also consumed throughout the day. The dish is known in Chinese American restaurants as "sizzling rice soup."

Bird's nest soup is often served at special occasions. It is made from the cleansed nests of swifts from the South China Sea. The flavor is bland, but the dish is very expensive and is reputed to be an aphrodisiac.

stir-fried separately. Thin, square wheat-flour wrappers are used to make steamed or fried egg rolls with a meat, vegetable, or mixed filling and wontons (in which the wrapper is folded over the filling), served either fried or in soup. Spring rolls, similar to egg rolls, are made with very thin, round, wheat-flour wrappers. Dumplings can be small steamed bundles made with wontons filled with bits of shrimp, crab, and vegetables (called *sui mai*) or more substantial, bread-like versions, filled with spiced pork, minced beef, or sweetened bean paste, then baked, steamed, or pan-fried. Buckwheat is grown in the North and commonly made into noodles.

The Chinese eat a variety of animal protein foods. Pork, mutton, chicken, and duck are common in many regions (see Regional Variations below). Fish and seafood of all kinds are specialties. Eggs are frequently consumed. They are sometimes cooked as thin omelets in which to wrap foods or to add to mixed dishes. However, eggs are most often prepared cooked, then salted and brined, and are particularly popular in southern China.[14] Thousand-year-old eggs (also called hundred-year-old, century, and pine flower eggs) are duck or chicken eggs cured for three months in a lime, ash, and salt mixture. The whites become black and gelatinous; the yolks turn greenish. In Taiwan a similar specialty called iron eggs is common. Chicken, pigeon, or quail eggs are cooked repeatedly in soy sauce, tea, or other flavored liquids until they shrink and become very chewy. They are eaten for breakfast and snacks.[15] More unusual items include snakes, frogs, turtles, sea cucumbers (also known as sea slugs, shell-less echinoderms related to starfish and sea urchins), and seahorses. The Chinese also raise many kinds of insects for consumption, such as scorpions, which are prepared fried or in soups.[16]

In China soybeans are known as the poor man's cow, as they are made into products resembling milk and cheese. Soybeans are transformed into an amazing array of food products that are indispensable in Chinese cooking (see Table 11.2). Other beans are also popular, made into pastes, flour, or even thin, transparent noodles known in the United States as cellophane noodles or bean threads.

Chinese cuisine makes extensive use of vegetables. Many are those known in other regions of the world such as asparagus, broccoli, cabbage, cauliflower, eggplant, green beans, mushrooms, onions, peas, potatoes, radish, and squash. Chinese varieties may differ, however. For instance, leafy bok choy and wrinkled napa cabbage are preferred over European types; long beans, small purple eggplant, *gai lan* (Chinese broccoli, also called Chinese kale), gai choi (Chinese mustard), and the large white icicle radish are featured in many dishes. One popular squash variety is called a winter melon (when immature it is called fuzzy melon); it is pale green and mild in flavor. Mushrooms of all types, including black mushrooms (a Japanese native also called *shiitake*), the tiny enoki, grayish oyster mushrooms, straw mushrooms, and dried kinds such as cloud (or wood) ears flavor numerous dishes. Lily buds, snow peas (pea pods), bamboo shoots, chrysanthemum greens, water chestnuts, bitter melon, water convolvulus, and lotus root are other, more distinctively Asian, vegetables found in Chinese cuisine.

The Chinese eat fresh fruit infrequently, occasionally for a snack or for dessert, and it is preferred slightly unripe or even salted. Chinese dates (jujubes), persimmons, pomegranates, and tangerines are favorites. A few fruits are typically preserved in syrups, such as pungent kumquats, yellow-orange loquats, longans (dragon eyes), and litchis, a tropical fruit with creamy, jellylike flesh.

Traditionally people cooked with lard if they could afford it. In recent years, soy, peanut, or corn oil is more common. Until recently sugar

▼ *Traditional foods of China. Some typical foods include bitter melon, bok choy, Chinese eggplant, ginger root, long beans, lotus root, mushrooms, oyster sauce, pork, long-grain rice, shrimp, soy sauce, and water chestnuts.*

Photo by Laurie Macfee

Table 11.2 Common Chinese Soy Bean Products

Soy sauce	Cooked soybeans that are first fermented and then processed into sauce. The southern Chinese prefer light-colored soy sauce in some dishes over the darker, more opaque kind used in Japanese and some regional Chinese cooking.
Soy milk	Prepared with soaked soybeans that are first pureed, then filtered, and then boiled to produce a white, milk-like drink.
Doufu (tofu)	Made by boiling soy milk and then adding gypsum, which causes it to curdle. The excess liquid is pressed from the bean curd, producing a soft or firm, bland, custard-like product. *Doufu* can be purchased fresh, frozen, smoked, dried, sweetened, or in sheets to make wrapped dishes.
Fuyu (*sufu*)	Sometimes called Chinese cheese—bean curd is fermented in brine and 100-proof liquor. The aroma is tangy but the flavor is mild, except when chile peppers are added to the process.
Black bean sauce	Cooked fermented soybeans preserved with salt and ginger. Black beans are usually added as a flavoring in dishes.
Brown bean sauce	Similar to black bean sauce, but made with yellow soy beans.
Sweet bean sauce	Similar to soy sauce but with reduced spicing and added sugar. It is common in northern Chinese cooking.
Hoisin sauce	Thick, brownish-red sweet-and-sour sauce that combines fermented soybeans, flour, sugar, water, spices, and garlic with chiles—often used in southern Chinese cuisine.
Oyster sauce	Thick brown sauce prepared from oysters, soybeans, and brine that is also used in southern Chinese fare
Chili bean paste	Very hot, thick paste made from brown bean sauce spiced with mashed chile peppers and vinegar. A favorite in Sichuan cuisine.

was not used in large quantities; many desserts were made with sweetened bean pastes.

Hot soup or tea is the usual beverage accompanying a meal. Tea, used in China for more than 2,000 years, was first cultivated in the Chang Jiang valley and later introduced to Western Europe in the seventeenth century. There are three general types of tea: green, black (red), and oolong (black dragon). Green tea is the dried, tender leaves of the tea plant. It brews a yellow, slightly astringent drink. Black tea is toasted, fermented black-colored leaves; it makes a reddish drink. Black tea is commonly drunk in Europe and America. Oolong tea is made from partially fermented leaves and is a Taiwanese specialty. Some teas in China are flavored with fruits or flowers, such as black tea with litchis, or orange blossoms, or oolong with jasmine. Other infusions of fruits or flowers are called tea, including longan and chrysanthemum. Pearl tea, or bubble tea, made with chewy, pea-size balls of tapioca that are sucked up through large straws, was also created in Taiwan and has become popular with Chinese youth worldwide.

Chinese alcoholic drinks are often called wines, but they are not usually made from grapes.[17] They are typically distilled alcohols made from grains or from fruit, such as plums. A few examples are bamboo leaf-green (95 proof), *fen* (made from rice, 130 proof), *hua diao* (yellow rice wine), *mou tai* (made from sorghum, 110 proof), and red rice wine. Beer is also very popular.

Most Chinese food is cooked, and very little raw food, except fruit, is eaten. Cooked foods may be eaten cold. Common cooking methods maximize the limited fuel available and include stir-frying, steaming, deep-fat frying, simmering, and roasting. In stir-frying, foods are cut into uniform, bite-size pieces and quickly cooked in a wok (a hemispherical shell of iron or steel) in which oil has been heated. The wok is placed over a gas burner or in a metal ring placed over an electric burner. Food can also be steamed in the wok. Bamboo containers, perforated on the bottom, are stacked in a wok containing boiling water and fitted with a domed cover. Roasted food is usually bought from a commercial shop, not prepared in the home.

The Chinese usually strive to obtain the freshest ingredients for their meals, and in most American Chinatowns it is common to find markets that sell live animals and fish. However, because of seasonal availability and geographic distances, many Chinese foods are preserved by drying or pickling.

In some regions, two longans or *litchis* are placed under the pillows of newlyweds because their names sound like the words "to have children quickly."[25]

In China, a marriage is consummated when the new in-laws officially accept a cup of tea from the bride. If the groom dies before the wedding, the bride-to-be is said to have "spilled her tea."

Unlike northern or southern Chinese cooking, Taiwanese fare makes ample use of herbs such as basil and parsley. Vodka is a popular alcoholic beverage; the most popular brand is named for the infamous Mongolian ruler, Ghengis Khan.[9]

Dumplings shaped like animals, such as birds or frogs, are specialties in Xian. A dim sum banquet may feature several dozen varieties.

Regional Variations. China is usually divided into five culinary regions characterized by flavor or into two areas (northern and southern) based on climate and the availability of foodstuffs. In recent years, however, regional differences have diminished due to increased global influences, particularly television.[18]

Northern. This area includes the Shandong and Honan regions of Chinese cooking. The Shandong area (Beijing is sometimes included in this area, sometimes considered a third division of northern cooking) is famous for Peking duck and mu shu pork, both of which are eaten wrapped in Mandarin wheat pancakes topped with *hoisin* sauce. Honan, south of Beijing, is known for its sweet-and-sour freshwater fish, made from whole carp caught in the Huang River. Much of the North is bordered by Mongolia, whose people eat mainly mutton.

Grilling or barbecuing is a common way of preparing meat in this area. One specialty is the Mongolian hot pot, featuring sliced meats and vegetables cooked at the table in a pot of broth simmering over a charcoal brazier. The food is eaten first, and the broth is consumed as a beverage afterward. (See also the box on Mongolian Fare.)

Northern China has a cool climate, limiting the amount and type of food produced. Traditionally, foods were often preserved, resulting in a preference for salty flavors. In general, its staples are millet, sorghum, and soybeans. Winter vegetables such as cabbage, turnips, and onions are common. A delicacy from this area is braised bear paw. Hot, clear soup is the beverage that usually accompanies a meal.

Southern. Southern China is divided into three culinary areas: Sichuan-Hunan, Yunnan, and Cantonese (with Fukien and Hakka regional specialties). Sichuan-Hunan (some English translations still use the term *Szechuan* or *Szechwan*), which is an inland region, features fare distinguished by the use of chiles, garlic, and the Sichuan pepper fagara. Typical dishes include hot and sour soup, camphor and tea-smoked duck, and an oily walnut paste and sugar dessert that may be related to the nut halvah of the Middle East. Yunnan cooking is distinctive in its use of dairy products, such as yogurt, fried milk curd, and cheese. Dishes are often hot and spicy, and some of the best ham and head cheese in China are found in this area.

Cantonese cooking is probably the most familiar to Americans, because the majority of Chinese restaurants in the United States serve Cantonese-style food. It is characterized by stir-fried dishes, seafood (fresh and dried or salted), delicate thickened sauces, and the use of vegetable oil instead of lard. The Cantonese are known for *dim sum* ("small bites," such as *sui mai*, pork ribs, meatballs, and other tidbits) served with tea. The staple foods of the South are rice and soybeans. As in the North, a variety of vegetables from the cabbage family are used, as well as garlic, melon,

EXPLORING GLOBAL CUISINE—Mongolian Fare

The Mongolians once ruled an empire that stretched from China to Europe. In more recent times, it has been colonized by Russia and China. Today, it is an independent nation reestablishing its cultural identity through shared language, customs, and cuisine.

Historically, Mongolians consumed red foods (meat) and white foods (dairy), and this tradition continues today with the addition of some grain products. Meats, especially mutton, goat, and beef, are favorites (camel meat is eaten when available, though it is banned in some areas). Meat is enjoyed barbecued on a grill or over charcoal in a specially designed hot pot that sits on the table. It is also added to soups, stuffed into pancakes, and served on sesame seed buns.[173]

Dairy foods are numerous, prepared from cow, sheep, goat, or camel's milk. There are three types of butter (liquid, yellow, and white), a type of milk *doufu*, sour milk (similar to yogurt), milk leather (made from the film skimmed off boiled milk and air-dried), and fresh cheese. Milk is added to tea, called Mongolian tea, sometimes with a little salt or fried millet.[174] Cheese is mixed with sugar and flour and then baked, to make a dessert known as milk pie. *Kumys*, a wine distilled from fermented milk (traditionally mare's milk), is a Mongolian specialty and consumed at many occasions.

Millet is the staple grain in Mongolia. It is cooked like a porridge or roasted until it pops like popcorn. Flour is made from millet, buckwheat, or wheat and cooked as fried pancakes or steamed flat breads. Tea is consumed at every meal and with snacks. Three meals a day are typical, consumed with the fingers. Special occasions include Lunar New Years and the *Naadam* festival, a three-day event featuring wrestling, archery, and horse races.

onions, peas, green beans, squashes, and a range of root-like crops, such as taro, water chestnuts, and lotus root. Southern cooking uses mushrooms of many types to enhance the flavor of the foods and takes advantage of an abundance of fruits and nuts. Fish, both fresh and saltwater, are popular. Also important are poultry and eggs. Pork is the preferred meat. Tea is the beverage served with meals.

Along the coast, Fukien provincial fare includes numerous seafood dishes and clear broths. Paper-wrapped foods and egg rolls are thought to have originated there. In the city of Shanghai, chefs specialize in new food creations and elaborate garnishes. Red foods are also a specialty due to the use of red wine paste (a sediment remaining after the fermentation of rice wine) on pork, on poultry, in soups, and even in dumpling dough.[19]

A southern regional specialty is Hakka cuisine, sometimes called the soul food of southern China. The Hakkas fled to the South in the fourth century B.C.E. when the Mongols invaded the North. They remained an insular ethnic group, preserving their traditional language, dress, and foods. Their fare is hearty and robust, featuring dishes made with red rice wine and pungent seasonings, cooked for a lengthy time, often in clay pots. Salt-baked chicken, greens simmered with pork fat, and meat-stuffed doufu are examples.

[SAMPLE MENU]

Cantonese Dim Sum

Spring Rolls[a,b] or Fried Wontons[a,b]

Har gau (shrimp dumplings)[a,b]

***Sui mai* (pork dumplings)[a,b]**

Char siu bao (steamed BBQ buns)[a,b]

Egg Custard Tartlets[a,b]

Jasmine Tea or Chrysanthemum Tea

[a]Simonds, N. 1994. *Classic Chinese cuisine*. Boston: Houghton Mifflin Co.

[b]*About: Chinese Cuisine* at http://chinesefood.about.com/od/diningout/p/dim_sum.htm

Meal Composition and Cycle

Daily Patterns. The Chinese customarily eat three meals per day, plus numerous snacks. Breakfast often includes the hot rice or millet porridge, congee, which in southern China may be seasoned with small amounts of meat or fish. In northern China hot steamed bread, deep-fried crullers, dumplings, or noodles are served for breakfast. In Taiwan both southern- and northern-style breakfasts are popular. In urban areas lunch is a smaller version of dinner, including soup, a rice or wheat dish, vegetables, and fish or meat. Sliced fruit may be offered at the end of the meal.

Although the Chinese are receptive to all types of food, the composition of a meal is governed by specific rules—a balance between *yin* and *yang* foods and the proper amounts of fan and cai. *Fan* includes all foods made from grains, such as steamed rice, noodles, porridge, pancakes, or dumplings, which are served in a separate bowl to each diner. *Cai* includes cooked meats and vegetables, which are shared from bowls set in the center of the table. *Fan* is the primary item in a meal; cai only helps people eat the grain by making the meal more tasty. A meal is not complete unless it contains *fan*, but it does not have to contain *cai*. At a banquet the opposite is true. An elaborate meal must contain *cai*, but the *fan* is usually an afterthought and may not be eaten.

Street stalls and tea houses provide snacks and small meals when away from home. Although restaurants were traditionally uncommon in rural areas, today they are found throughout the nation. The Chinese all-you-can-eat buffet, which originated in the United States, is now found in some regions. Other American restaurant ideas that have made their way to China include fast-food franchises and food courts.[14]

Etiquette. The traditional eating utensils are chopsticks and a porcelain spoon used for soup.

EXPLORING GLOBAL CUISINE—Tibetan Fare

The Chinese-controlled province of Tibet has a unique fare due to the isolation provided by its locale in the Himalayan Mountains. The foundation of the diet is *zampa*, a toasted flour produced from barley or buckwheat. It is traditionally mixed with the butter obtained from yak, cow, or sheep milk (called "crispy oil"), sugar, milk or cream, and sometimes tea to make flattened balls consumed with tea or soup. The *zampa* can also be used to make momos, a Tibetan dumpling filled with meat. Yak and mutton are common, but most Tibetans who are Buddhist do not eat pork, poultry, or fish.[25] Dairy products are also prevalent. Butter-tea, made by churning crispy oil, milk, and salt with brewed tea, is consumed throughout the day. Sour milk, milk solids preserved from the crispy oil process, and the milk film skimmed from boiled milk and then dried are all consumed. Cabbage, radishes, onions, garlic, leeks, and potatoes are available. Wine, made from barley or buckwheat, is served at special occasions.

Chopsticks were likely invented as an extension of the fingers. They are made from bamboo, ivory, or plastic. Chopsticks are used in most countries that have been influenced by China, including Japan (where the chopsticks are shorter and have rounded rather than squared sides and more pointed tips) and Korea (where the chopsticks are typically made of metal, the same length as the Japanese type, but flatter). Chopsticks are found frequently in Vietnam (the Chinese type), though forks, spoons, and fingers are also commonly used. Other Southeast Asian cultures use chopsticks only occasionally, mostly for rice or noodles.

Teacups are always made out of porcelain, as are rice bowls. Few foods are eaten with the fingers, though that is changing somewhat in China today, where it is sometimes acceptable to pick up wrapped or stuffed items, such as dumplings.[20] All courses of a meal are traditionally served at once. Each place setting includes a bowl of rice or noodles, and each diner then takes what is desired from the communal serving plates. At the meal all diners should take equal amounts of the *cai* dishes, and younger diners wait to eat until their elders have started; it is rude to reject food. It is also considered bad manners to eat rice or noodles with the bowl resting on the table; instead, it should be raised to the mouth. It is rude to pick at your food or to lick your chopsticks. Laying your chopsticks across the top of the rice bowl or dropping them brings bad luck. It is also improper to stick chopsticks straight up in a rice bowl because in some areas this symbolizes an offering to the dead. Any bones or other debris should be placed on the small plate at each place setting, or on the table next to the rice bowl.

Proper Chinese behavior at the table was first outlined over 4,000 years ago, and many practices remain unchanged.[14] Rules include not making noises while eating (except when consuming soup, when slurping facilitates cooling the soup and expresses pleasure), not grabbing food, not eating quickly, not putting food back on the communal plate after tasting it, and not picking one's teeth. Beverages, such as tea, should be served to others at the table before pouring for one's self, and the cups should not be filled to the brim. Both hands are used to offer a cup of tea, and the cup should be taken with both hands as well. Wine and other alcoholic drinks should not be consumed alone, and when the toast *gambei* ("bottoms up") is made, everyone at the table drains his or her glass.

Though strict rules regarding dining behavior are observed in China, it is not uncommon to play games at the table during a meal.[14] This is especially true at banquets, with guests joining and leaving the game as they please. Multiple conversations may take place at once, and interruptions are frequent. It is considered polite to compliment the host throughout the meal on the deliciousness of the food and on his or her good taste and wisdom.[21]

Special Occasions. Traditionally, the Chinese week did not include a day of rest. Consequently, there were numerous feasts to break up the continuous workdays. Chinese festival days do not fall on the same day each year because their calendar is lunar. Celebrations are traditionally *yang* occasions because heat symbolizes activity, noise, and excitement in China.[9] *Yang* foods, such as meats, fried dishes, and alcoholic beverages, are featured at festive banquets (see Therapeutic Uses of Foods). Most Chinese homes are small and unsuited to entertaining, so special meals with guests are generally held at restaurants.[14]

The most important festival is New Year's, which can fall any time from the end of January to the end of February. Traditionally the New Year was a time to settle old debts and to honor ancestors, parents, and elders. The New Year holiday season begins on the evening of the twenty-third day of the last lunar month of the year. At that time the Kitchen God, whose picture hangs in the kitchen and who sees and hears everything in the house, flies upward to make his annual report on the family to the Jade Emperor. To ensure that his report will be good, the family smears his lips with honey or sweet rice before they burn his picture. A new picture of the Kitchen God is placed in the kitchen on New Year's Eve. Food preparation must be completed on New Year's Eve, as knives cannot be used on the first day of the year because they might "cut" luck. Deep-fried dumplings, made from glutinous rice and filled with sweetmeats, and steamed turnip and rice flour puddings, are usually included in the New Year's Day meal.

During the New Year festivities, only good omens are permitted and unlucky-sounding words are not uttered. Foods that sound like lucky words, such as tangerine (good fortune), fish (surplus), chicken (good fortune), chestnuts (profit), and *doufu* (*fu* means "riches"), are eaten. Friends and relatives visit each other during the first ten days of the new year, and good wishes, presents, and food are exchanged. Children receive money in small red envelopes. Traditionally the Feast of Lanterns, the fifteenth day of the first month, ends the New Year's season and is marked by the dragon dancing in the streets and exploding firecrackers to scare away evil spirits.

Ch'ing Ming, the chief spring festival, falls 106 days after the winter solstice. Families customarily go to the cemetery and tend the graves of their relatives. Food is symbolically fed to the dead and then later eaten by the family. Sweets and alcoholic beverages are popular offerings. *Duan wu*, the Dragon Boat Festival, is held on the fifteenth day of the fifth month to commemorate the drowning death of a famous third-century B.C.E. poet. A boat race and special dumplings are traditional. The Moon Festival occurs at the end of September on a full moon (the fifteenth day of the eighth lunar month). Because the moon is a *yin* symbol, this festival was traditionally for women, but today it also symbolizes the togetherness of the family. It is sometimes called the harvest festival or moon's birthday. Large round cakes filled with spices, nuts, fruit, or red bean paste, called moon cakes, are typically eaten during this event.

Therapeutic Uses of Food

Most Chinese believe eating the proper balance of *yin* and *yang* foods is necessary to assure physical and emotional harmony and to strengthen the body against disease (see Chapter 1, "Food and Culture"). Extra care should be taken with children's diets because they are more susceptible to imbalance.

Hot foods generally include those high in calories, cooked in oil, and irritating to the mouth and those that are red, orange, or yellow in color. Examples include most meats, eggs, chile peppers, tomatoes, eggplant, persimmons, pomegranates, onions, leeks, garlic, ginger, and alcoholic beverages. Cold foods are often low in calories, raw or boiled/steamed, soothing, and green or white in color. Many vegetables and fruits are considered cold items, as are some legumes. Pork, duck, crab, clams, shrimp, snake meat, and honey also are classified as cold in some regions. Staples, such as boiled rice and noodles, and other commonly eaten foods, such as soy sauce and red or black tea, are typically placed in a third, neutral category.[9,22,14] Some food preparations can make foods hotter or colder by the infusion or removal of heat. Foods classified as *yin* or *yang* vary from region to region. Acculturated Chinese Americans may be uncertain about some categorizations and thus identify many foods as neutral.

Typically, hot foods are eaten in the winter by menstruating women and for fatigue. Pregnancy is considered a cold condition, and birth is a dangerously cooling experience. Postpartum women often remain indoors and eat hot foods, such as chicken fried in sesame oil and pig's feet simmered in vinegar, for four to six weeks after delivery.[23] This period is known as *tso yueh-tzu*, "doing the month." In addition to eating warming items, raw and cooling foods are avoided, as is contact with cold air, wind, and water (bathing in hot water with ginger in it is permitted after a few days). Other conditions caused by too much *yin* and that respond to eating more *yang* foods include colds, flu, nausea, anemia, frequent urination, shortness of breath, weakness, and unexplained weight loss. It is also believed that as a person grows older, the body cools off and more hot foods should be eaten.

Conditions due to excessive *yang* that improve with an increase in *yin* food intake include constipation, diarrhea, hemorrhoids, coughing, sore throat, fever, skin problems, conjunctivitis, earaches, and hypertension. Cool foods are consumed in the summer, for dry lips, and to relieve irritability.

In addition to *yin* and *yang*, some foods are believed to affect the blood or promote wound

According to the Chinese, a child is one year old at birth and becomes two years old after the New Year.

The New Year's dragon dance and firecrackers are thought to inhibit the *yin* element and promote the *yang* forces. Red, the color of *yang*, is used throughout the New Year's season.

In 1718, a Jesuit missionary in Quebec discovered an American species of ginseng that is nearly identical to the Chinese variety. Growing demand in China led many Americans, including Daniel Boone, to hunt the root for export.

Eating crab and persimmons together is one food taboo maintained by some Chinese Americans elders because these foods represent extreme hot and extreme cold and are considered to be poisonous if mixed.

healing and are labeled pu, or bo, meaning "strengthening." This classification is separate from the concept of *yin* and *yang* but often used in conjunction with it; most strengthening foods are also categorized as hot. The *yin* condition of weak blood (most associated with pregnancy, postpartum, and surgery) is treated with specific hot items such as protein-rich soups made with chicken, pork liver, eggs, pig's feet, or oxtail. Other health-promoting foods identified by Chinese Americans include royal jelly (made from honey), bee pollen, *lin chih* (edible fungus), rattlesnake meat, dog meat, roasted beetles, barley juice, garlic, *dong gwai* (angelica, a celery-like herb), fruit juice, and milk. However, it is believed that too many of certain *yang* items can cause the blood to thin, and these foods are avoided for conditions such as hypertension.[9,22,8,24]

Ginseng is the best-known health-promoting Chinese food. It is made from an herb (genus *Panax*) found in Asia and the Americas. The root is boiled until only a sediment remains, then powdered for use in teas and broths. Ginseng reputedly cures cancer, rheumatism, diabetes, sexual dysfunctions, and complaints associated with aging. It is most often used as a restorative tonic. Taro root is also thought to have therapeutic properties, such as improving eyesight, curing vaginal discharge, reducing weakness, and promoting multiple births; it will also bring good luck if eaten on the fourth day of the first lunar month.[25] Bitter orange is used to alleviate bloating and constipation. Guava, which has some hypoglycemic properties, is used for diabetes.[26] Other popular remedies include deer antlers, rhinoceros horns, and pulverized sea horses.[27] The concept "like cures like" (sympathy healing) is seen in many food cures for specific illnesses.[28] Walnuts (which resemble brains) are eaten as a remedy for headaches in Hong Kong and to increase intelligence in China.[14] Red jujubes may be consumed for strengthening blood, soups made with bones are used for treating broken bones, and male genital organs from sea otters, deer, or other animals are eaten to cure impotence. Chinese foods and herbs, such as such as "bird nest" and "glucose drink"[29] are also used in the infant weaning diet. The weaning diet is considered the semi-solid food that is added to a child's diet of formula or breast milk to increase appetite, balance the *yin* and *yang* system, restore *qi*, or treat diarrhea.[30] Base on some traditional beliefs, orange skin is believed to enhance the taste and flavor of the soups and contributes to the *ying/yang* balance. Pork bone is believed to add calcium into the soup and is needed for growth. Moreover, some Chinese believes that alligator meats will benefit the respiratory system.[31]

Some food taboos have been noted during pregnancy. Soy sauce may be avoided to prevent dark skin, and iron supplements may not be taken because they are thought to harden the baby's bones and make birth difficult. Shellfish may also be shunned for the same reason.[32]

Contemporary Food Habits in the United States

Adaptations of Food Habits

Generally speaking, changes in the eating habits of Chinese Americans correlate with increased length of stay in the United States, particularly in subsequent generations. Dinner often remains the most traditionally Chinese meal, whereas breakfast, lunch, and snacks tend to become more Americanized. Younger persons are also more likely than their elders to accept U.S. fare.

Ingredients and Common Foods. Most Americans of Chinese descent regularly consume several Chinese foods, such as rice, pork, seafood, soup broth, soybean products, cooked vegetables, tea, and fruit.[33,34] One preliminary study suggests that the majority (88 percent) of foreign-born immigrants prefer Chinese fare at home, although younger respondents (ages twenty to thirty-four) expressed preference for American foods.[24] Meat and poultry intake increases, while some traditional protein items like pig's liver and bone marrow soup often remain popular. Greater consumption of protein foods, in addition to increased intake of fast foods, soft drinks, candy, and pastries, results in higher fat and sugar intake among more acculturated Chinese Americans.

The impact of acculturation on fruit and vegetable intake is less clear. Traditional fruits and vegetables may be replaced by more commonly available American items, such as potatoes, lettuce, apples, peaches, and watermelon.[35,24] Some studies have found that greater fruit and vegetable intake is associated with acculturation, education level, and income.[36,37] Among families, however, data indicate that pressure to maintain a traditional diet by elders living at home results in a higher intake of fruits and vegetables among all members.[38]

Even though milk is not a familiar item in the typical Chinese diet, several studies suggest that milk is consumed by nearly half to three-quarters of Chinese Americans in the United States.[37,39,24] Cheese, yogurt, and ice cream have also been found to be well accepted.

One study found that dietary variety increased after immigration to the United States,[36] and another noted that U.S.-born Chinese women have a more varied diet than Chinese American women who were foreign born.[40] Respondents ate more breads, cereals, dairy foods, meats, vegetables, and ethnic items, such as Italian and Mexican foods. In another study acculturation was significantly associated with improved dietary variety but with lower dietary moderation.[41]

Meal Composition. Skipping meals and increased snacking have been reported in Asian students and in Chinese American and Chinese Canadian women.[36,42,35] Surveys of Chinese in North America suggest that traditional foods are the choice of older, less acculturated adults, and their preferences sometimes influence household meals.[33,43,38] Lunches and dinners may consist mainly of Chinese-style foods, while breakfasts are more variable. Many Chinese Americans attempt to balance hot and cold items in their diets. Other studies suggest that the use of *yin* and *yang* in the diet may diminish over time and that Chinese Americans may practice some aspects of it but without knowledge as to why certain food combinations are preferred.[43,8,14] Stir-frying, simmering, and steaming remain the favored cooking methods.[24,44] In one study, more acculturated respondents found cooking Chinese meals was inconvenient.[38]

Americans of Chinese descent usually celebrate the major Chinese holidays of New Year's and the Moon Festival with traditional foods. Chinese American Christians sometimes combine the spring festival of *Ch'ing Ming* with Easter festivities. In addition, some Chinese Americans recognize the founding of the People's Republic of China (mainland China) on October 1 (on the solar calendar) or the establishment of the Republic of China (Taiwan) on October 10 with cultural performances and banquets.

Nutritional Status

Nutritional Intake. The traditional Chinese diet is low in fat and dairy products and high in complex carbohydrates and sodium. As length of stay and the number of generations living in the United States increase, the diet becomes more like the majority American diet—higher in fat, protein, sugar, and cholesterol, and lower in complex carbohydrates. Research on women in the United States and Canada reported even less acculturated respondents consumed milk, ate cheese, ate fast foods, and snacked regularly,[37] and another study of women including foreign-born Chinese Americans, U.S.-born Chinese Americans, and white Americans revealed that all three groups consumed more than recommended levels of fat in their diets, suggesting that some changes in food consumption may occur very quickly.[39] The U.S.-born cohort also demonstrated high levels of nutrition knowledge, and their diet contained a higher concentration of nutrients than either the foreign-born Chinese Americans or the white Americans.[40,41]

Some Americans of Chinese descent continue to avoid fresh dairy products because of lactose intolerance, which may be found in as many as 75 percent of Asians. Low calcium intake has been reported in some samples.[45,39] Alternative calcium sources are bean curd, soy milk—if fortified with calcium—and soups or condiments made with vinegar in which bones have been partially dissolved. However, as noted previously, many Chinese Americans do consume milk, cheese, and yogurt, as well as leafy green vegetables, and calcium deficiency should not be presumed. Low vitamin A and C intake has been observed in some Americans of Chinese descent, but iron intake is satisfactory, perhaps due in part to the use of iron-containing cooking tools, such as woks.[44]

A study of Asian college students found fast foods and sweet/salty snacks very popular. Intake of fats, sweets, dairy products, and fruit increased, while intake of meats and vegetables decreased.[42]

Obesity and overweight are found to be low among Chinese Americans.[46,47,48,49] In one national survey Chinese were found to have very low median body mass indexes (BMIs) when compared to the general U.S. population. However, median BMI and proportion obese went up significantly for U.S.-born subjects when compared to foreign-born subjects and with acculturation.[46,50] Some Chinese Americans feel pressured to overeat due to traditional Chinese eating behavior[51]. In a study of Chinese American children, 33 percent were reported overweight

(BMI ≥ 85th percentile); however, the mother's degree of acculturation was found inversely associated with risk.[52,53]

Concerns that overweight and obesity may become problematic in this population as demographics change over time are as yet unconfirmed. Research on anthropometric measures indicates that BMI and waste cicumferences underestimate obesity in Chinese Americans.[54,55] Chinese heritage was found to modify waist circumference measurements and metabolic risk factors.[56] Calculated energy requirements may differ as well. Predictive equations for basal metabolic rate (BMR) and for resting energy expenditure (REE) are found to overestimate BMR and REE in adult Chinese Americans. There is concern in some Asian nations about the increasing incidence of eating disorders in young women.[57,58,59]

Studies of type 2 diabetes in Asians show unadjusted rates lower (6.2 percent) than in whites. Yet studies indicate that when adjusted for BMI, Asians are 60 to 74 percent more likely than whites to develop the condition.[47,60] Further, data suggest that weight gain associated with incidence of type 2 diabetes was particularly detrimental in Asians—each five-kilogram gain increased risk by 84 percent, nearly double the increase found in any other ethnic group in the study. Data on diabetes specific to Chinese Americans are sparse, but these pan-Asian studies suggest type 2 diabetes may become a significant health issue if overweight and obesity rates grow.

It is generally assumed that many Chinese eat a diet high in sodium, which may contribute to high blood pressure. Hypertension rates among Chinese Americans are lower than for whites but 17 percent of adult Chinese Americans have hypertension.[61] One study found Chinese subjects are 30 percent more likely than whites to have high blood pressure when adjusted for age, BMI, prevalence of diabetes, and smoking.[62] Another study found that when compared to whites, Chinese Americans who suffered from stroke had higher risk profiles, including history of hypertension, history of diabetes, and higher levels of blood lipids and glucose.[63,64] Hypertension is considered a *yang* condition and is often treated by the consumption of *yin* foods.[65]

A study of acculturation and diet in Chinese American women found that Chinese-language newspapers and friends were primary sources of nutrition information.[35]

The percentage of Chinese Americans with heart disease is 5.6 percent. [61] However, cardiovascular disease rates in China increased 60 percent between 1993 and 2003, paralleling increased rates of overweight and obesity, diabetes, and hypertension.[66] Prevalence in the United States may also increase with these changes in first-generation immigrants, as well as with possible changes in subsequent generations. Cancer risk, especially for colorectal and breast cancers, has been found to increase with length of stay in the United States, and cancer is now the leading cause of death in Chinese Americans.[67,68] Inadequate preventative screenings and dietary changes, including lower intake of protective foods (e.g., soybean products) and higher intake of saturated fats, are thought to be factors. Rates of liver cancer among men and cervical cancer among women are significantly higher than among whites.[52,69,70,71]

Infant mortality rates for Chinese Americans are very low.[72] Most studies report low-birthweight rates among Chinese Americans similar to those of whites. However, some researchers suggest that the definition of low birth weight may be inappropriate for Chinese infants who are typically smaller than the U.S. average.[73] reast-feeding is reportedly common in China,[27,74] and 65 percent of Chinese women in Australia breast-fed their infants in one study.[75]

High rates of tuberculosis, parasitic infection, and hepatitis B have been found in many recent Asian-American immigrants.[76] Limited information regarding prevalence among Chinese immigrants specifically has been reported. Clonorchiasis, a liver fluke infection of the biliary passage or pancreatic ducts, has been identified in 25 percent of immigrants from Hong Kong and a smaller number of those from China.[77] Also noted are high rates of certain inherited conditions, including thalassemias and glucose-6-phosphate dehydrogenase deficiency.[78]

Studies do not typically differentiate Chinese American immigrant groups. Compared to immigrants from the mainland, recent Taiwanese Americans are more likely to experience conditions due to overconsumption than underconsumption. Obesity has increased dramatically in Taiwan during the past several decades; impaired fasting glucose occurs at rates three times that of the general U.S. adult population, and age-adjusted prevalence of type 2 diabetes (9.2%) is nearly double.[79,80] However, in the past 10 years the incidence of obesity has dramatically increased in mainland China and that 20 perecent of of the overweight and obese individuals in the world are Chinese.[81]

Counseling. Americans of Chinese descent accept personal responsibility for their health; keeping healthy is considered an obligation to family and society. However, biomedical health care is underutilized. Language barriers are thought to be one issue. Only 15 percent of Chinese Americans speak English at home, and nearly 50 percent report that they speak English less than "very well."[1] Low income, long work hours, and inconvenient locations are other reasons believed to limit access. Lack of health care insurance is another factor, which may be due to expense, or because purchasing insurance is sometimes seen as inviting death.[12,78,82] Mistrust of biomedicine, especially its possible side effects, is also found.[12,83] The concept of preventive checkups is unfamiliar to many Chinese Americans.[84,85]

Perhaps because Chinese Americans frequently believe that hospitals are where a person goes to die, hospitalization rates of Chinese Americans are lower than for any other ethnic group in America. Blood tests, thought to permanently diminish the blood supply, are of particular concern to some Chinese clients, who may avoid all biomedical health care for this reason.

Some Chinese Americans favor biomedical providers of Chinese heritage, citing common language, mutual sympathy, and flexible appointments (many Chinese are polychronic).[12] Effective treatment is the primary concern for most clients, however. Preferred communication style is formal and includes unrushed dialogue (focusing on time is considered offensive), detailed explanation of the origins and symptoms of any condition in understandable terms, simple treatment, and a positive outlook. Chinese patients consider it important to maintain hope. When possible, terminal illness should be discussed first with family members to determine how and when a client is informed. Medical confidentiality is not widely practiced.[2]

The Chinese have a quiet conversational approach, especially with strangers and acquaintances. Some speakers may pause during conversation as a sign of thoughtfulness. Interruptions should always be avoided. Many Chinese avoid confrontation and may initially say "yes" to questions that require a positive or negative response. Asking questions can be interpreted as disrespect, a sign that the person speaking is being unclear. Surprise or discomfort may be expressed by quickly and noisily sucking in air.[86] If a person is offended, he or she may become very direct and even loud in expressing anger. Conversations between friends or family members are often animated.[21] Eye contact is made briefly during introductions, but indirect eye contact is standard. Direct eye contact may be interpreted as confrontational.

Elder or less acculturated Americans of Chinese descent may show deference to authority by means of acceptance and submission. In the hospital setting, patients are often silent rather than voicing complaints; providers should not necessarily accept this as compliance but should actively seek information about patient satisfaction. Emotional displays are considered immature, but most Chinese patients are willing to discuss feelings in conjunction with somatic symptoms (see the section Traditional Health Beliefs and Practices).[78]

The traditional Chinese greeting is a nod or a slight bow from the waist while holding palms together near the chest, often without a smile. Traditionally, surnames come first, followed by given names, and women do not customarily take their husband's last name when married, so surnames may differ within the same family. Use of any titles demonstrates respect. Touching between strangers and acquaintances is uncommon.[82] Even handshaking may be inappropriate (wait for the extended hand, especially with women), except for westernized Chinese Americans and people from Hong Kong. Hugging, kissing, and back-patting should be avoided. Good posture is expected, and slouching or putting one's feet on a desk is considered rude. Personal space is typically farther apart than in Western cultures.[21]

Chinese American women may be very modest, especially regarding touching. Traditionally,

In Asia nurses perform only medical procedures. Family members stay at the hospital to provide feeding, bathing, and general care for the patient.

In 2004 the U.S. Food and Drug Administration banned the Chinese herb *ma huang* (ephedra), a methamphetamine-like stimulant with serious side effects such as heart attack, stroke, seizures, and psychosis. In a subsequent lawsuit, a court ruling partially overturned the ban, permitting small amounts to be included in diet preparations.

Chinese women were never touched by male health care providers (today, more than 90 percent of obstetricians and gynecologists in China are women). Symptoms would be discussed by pointing to an alabaster figurine.[42] If a male must do an examination, a formal, polite attitude, explanations of all procedures, and avoidance of tension-relieving jokes or comments will help the client feel more at ease.[2] Furthermore, within the family, sons receive more concern and attention over minor symptoms than daughters. Women, consequently, may believe that their complaints do not warrant care.

Researchers are unsure how many Chinese Americans use Traditional Chinese Medicine.[87,88,89] The majority are believed to first self-diagnose and self-prescribe at home before seeking outside care, although the reasons why certain foods or medications are consumed for an illness (particularly the complementary use of *yin* and *yang*) are often lost through acculturation.[43,8] Biomedicine is completely accepted by many Chinese Americans; one unpublished study found that 88 percent of foreign-born Chinese American subjects preferred biomedical care for the treatment of illness.[24] Others consider biomedicine best in the treatment of acute symptoms, and TCM best for chronic conditions.[12] For example, a client with type 2 diabetes may consult a biomedical physician regarding symptoms but, when he or she finds that no immediate cure is offered, may seek a TCM doctor to restore balance to the body and treat the actual cause of the disease.[90]

Practitioners of TCM are often consulted concurrently with biomedical care in an effort to maximize the chances of a cure. Few conflicts in therapies have been identified, although the active agents in most herbal medicines remain unidentified in biomedical terms. It is possible that a formulary preparation might be additive with a prescribed medication, producing excessive response, such as taking guava juice with hypoglycemics. Conversely, a TCM product taken for a different condition might counteract a drug therapy. For example, bitter orange peel (which contains synephrine, a chemical similar to stimulant ephedrine) may be taken for constipation, thus reducing the effectiveness of a client's hypotensive prescription. Providers should encourage traditional practices if desired by a client, but ask for information regarding herbal medicines consumed. A prescription from a TCM doctor remains in possession of the client, who may reuse it if symptoms reoccur or who may share it with family and friends. Some prescriptions are passed along from generation to generation; others are obtained directly from China. Occasionally, a client may present multiple burn marks from moxibustion, and the client's use of this treatment should be determined before presumption of abuse.

A Chinese client expects that the provider will perform few tests and ask a limited number of questions during an examination. Recommendations on diet, relaxation, and sleep are desired as an integral part of treatment. Long-term therapy intended to cure a disease is preferred over short-term surgical solutions or invasive treatment, even at the expense of pain or discomfort from symptoms. Most Chinese clients are resolved to die at home (many actually return to China), and their wishes should be accommodated.

Few compliance problems have been noted. One difficulty that sometimes arises is the issue of lengthy or continuous medication. Many Chinese are accustomed to single-dose Chinese remedies and may discontinue a prescription if directions are not thoroughly explained. Many researchers have remarked on the difficulty of eliminating high-salt items (e.g., soy sauce) from Chinese fare and recommend reduction as a goal in cases where a low-sodium diet is required.[91] Dairy products may be accepted when clients become familiar with the foods.[39] One study of nursing home residents found a significant dislike of western foods among Chinese elders.[92]

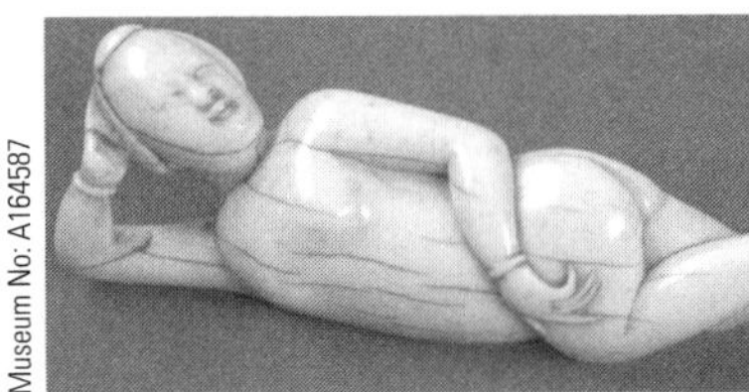

© Wellcome Library, London, Museum No: A164587

▲ *Traditionally, Chinese women were never touched by their male health care providers. Symptoms would be discussed by pointing to an alabaster figurine, like the one shown.*

An in-depth interview should tactfully be conducted to identify which traditional practices, if any, are followed by a client. Even when concepts such as *yin* and *yang* are considered primitive or rustic by some Chinese Americans, they may still adhere to certain food combinations learned at home without knowing why. Birthplace and acculturation may significantly impact dietary intake of Chinese Americans, and individual preferences should be determined.

Japanese

The multi-island nation of Japan is off the coast of East Asia and has approximately the same latitude and range of climate as the East Coast of the United States. The capital of Japan is Tokyo, located on the island of Honshu. Today, Japan is a prosperous country of over 127 million people that has overcome the limitations of a mountainous geography, a rugged coastline, and few mineral resources. Perhaps Japan's greatest natural resource is the sea, which provides one of the richest fishing grounds in the world.

Cultural Perspective

History of Japanese in the United States

Immigration Patterns. Significant Japanese immigration to the United States occurred after 1890 during the Meiji era. The immigrants were mostly young men with four to six years of education from the rural southern provinces of Japan. Most came for economic opportunities and many eventually returned to Japan. They settled primarily in Hawaii and on the West Coast of the United States and often worked in agriculture, on the railroads, and in canneries. Like the Chinese before them, Japanese immigrants opened small businesses, such as hotels and restaurants, to provide services for their countrymen. In contrast to the Chinese, many Japanese became farmers, ran plant nurseries, and were employed as gardeners. The Japanese prospered within their ethnic communities. Most Japanese women came to the United States as picture brides: Their marriages were arranged by professional matchmakers, and they were married by proxy in Japan. They did not usually meet their husbands until they disembarked from the ship in the United States. Among Japanese Americans, first-generation immigrants born in Japan are called *Issei*, second-generation Japanese Americans born in the United States are known as *Nisei*, and the third and fourth generations are known as *Sansei* and *Yonsei*, respectively.

Discrimination against the Japanese was commonplace. The Issei were classified as aliens who were ineligible to become naturalized U.S. citizens, and in 1913 land ownership became illegal in California. Although Japanese bought land in the names of their children, who were Americans by birth, the amount of land owned and leased by Japanese was reduced by half in the 1920s. In 1907 the Japanese government had informally agreed to limit the number of emigrants, and in 1924 the Japanese Exclusion Act halted Japanese immigration completely.

World War II heightened the prejudice against the Japanese on the West Coast. After Japan attacked Pearl Harbor, all West Coast Japanese, even if they were U.S. citizens, were evacuated to war relocation camps, and many remained there for the duration of the war. Most lost or sold their businesses as a result of internment. Nevertheless, many *Nisei* volunteered for combat duty and fought in Europe.

▼ *Japantown, San Francisco.*

Kevin Fleming/Encyclopedia/Corbis

Current Demographics and Socioeconomic Status. After the war, most Japanese Americans resettled on the West Coast, and the most discriminatory laws were repealed or ruled unconstitutional. The successful postwar recovery of Japan resulted in reduced emigration to the United States, usually far below the quota allotted under current immigration laws. According to 2007 U.S. Census estimates, 864,000 Japanese Americans live in the United States, a majority of whom reside in California or Hawaii. Many West Coast cities have a section of town called "Little Tokyo" or "Japantown," and a small number of older Japanese still live in these homogeneous neighborhoods. Most Japantowns contain Japanese American–owned restaurants, markets, and other small businesses, as well as Chinese churches or Buddhist temples.

More than 95 percent of Japanese Americans live in culturally mixed urban and suburban areas. Americans of Japanese descent are unique for a nonwhite ethnic group in the high rate of citizenship, degree of assimilation, and economic mobility they have experienced.[1] Over 90 percent of all Japanese Americans have graduated from high school, and of the third-generation Sansei, many of whom were born in the relocation camps, 72 percent have attended college and most hold professional jobs. In 2007, median family income was 30 percent above the national median, and fewer than 5 percent of families lived in poverty. Some Japanese Americans have noted that few Asians are found in top private and public sector positions, however, and believe that they suffer from the stereotype of being too unassertive for upper management.[93]

Over 85,000 residents of Japanese ancestry were living in Canada in 2001.

Heikegani **crabs are believed to be reincarnations of drowned warriors due to the face-like markings on their shells.**

Japanese physicians were polled to discover why they began prescribing traditional herbal remedies; a majority said they were influenced by pharmaceutical company salespeople and advertising.

Worldview

Religion. Early Japanese immigrants usually joined a Buddhist temple (Pure Land sect) or a Christian church after arriving in America. The church frequently provided employment and an opportunity to learn English. Today, it is thought that there are more Japanese Americans who belong to Protestant faiths than who follow Buddhism.[1]

Shintoism, the indigenous religion of Japan, does not have a formal organization, but its beliefs are a fundamental part of Japanese culture. The *Shinto* view is that humans are inherently good. Evil is caused by pollution or filthiness—physical as well as spiritual; goodness is associated with purity. Evil can be removed through ritual purification. *Shinto* deities, called *kami*, represent any form of existence (human, animal, plant, or geologic) that evokes a sense of awe. *Kami* are worshipped at their shrines as ritual expression of veneration and thankfulness. Prayers are also said for divine favors and blessings, as well as for avoidance of misfortunes and accidents.

Family. Until World War II the structure of the Japanese American family had its roots in Japan and was similar to that of the Chinese due to the strong influence of Confucianism. In addition, the rigid pattern of conduct that evolved in Japan during the sixteenth century resulted in the following practices among the *Issei* and their descendants:

1. *Koko.* Filial piety defines the relationship between parents and children, between siblings, and between individuals and their community and rulers. (See the Chinese religion section on Confucianism in the first part of this chapter for further explanation.) One outcome is that the *Issei* expect their children to care for them in their old age.
2. *Gaman.* Most Japanese believe it is virtuous to suppress emotions. The practice of self-control is paramount.
3. *Haji.* Individuals should not bring shame on themselves, their families, or their communities. This Japanese cultural concept exerts strong social control.
4. *Enryo.* There is no equivalent word in English, but many Japanese believe it is important to be polite and to show respect, deference, self-effacement, humility, and hesitation. Thus, some older or less-acculturated Americans of Japanese descent are neither aggressive nor assertive.

Japanese clan or village affiliation has traditionally been much weaker than in China, and Japanese immigrants arrived in the United States prepared to raise nuclear families similar to those in white America. Even when picture brides were trapped in unhappy marriages, they persevered on behalf of their children. Most *Issei* women worked alongside their husbands to support the family financially.

The internment of Japanese Americans during World War II brought further changes in family structure and accelerated acculturation into mainstream society after the war. In the camps very low wages were paid and the pay was the same for everyone; thus the father could no longer be the primary wage earner. The camps were run democratically, but positions of authority could be held only by American citizens, so the younger generation held these more prestigious jobs. The camps also allowed the Japanese to work in a wider range of fields than those available to them on the outside. After their internment, the *Nisei* no longer had to follow the few occupations of their parents, and many chose to follow different paths.

Sansei couples generally form dual-career households. Nearly 50 percent marry outside their ethnic group. The societal problems prevalent in majority American homes, such as spousal abuse, have surfaced among Japanese Americans as well,[93] and it is not known if the family values that have thrust Americans of Japanese descent into educational and financial success will continue in the fully assimilated fourth-generation of *Yonsei*.

Traditional Health Beliefs and Practices. Early Japanese health beliefs involved *Shinto* concepts of purity and pollution. Health was maintained through cleanliness and avoidance of contaminating substances such as blood, skin infections, and corpses.[94] Botanical remedies were used, particularly purgatives, in the prevention and treatment of disease.

When Buddhism was introduced in the sixth century, the concept of harmony was applied within the context of Japanese culture to mean a person's relationship with nature, family, and society. Imbalance resulting from poor diet, insufficient sleep, lack of exercise, or conflict with family or society disrupts the proper flow of energy within the body, leading to illness. Chinese practices such as acupuncture, moxibustion, and massage were accepted as ways to restore the energy flow along the meridians of the body (see the section Traditional Health Beliefs and Practices in the section about Chinese people in this chapter). The application of *yin* and *yang* in health and diet was limited in Japan.

The more complex herbal medications of China were brought to Japan as *kanpo*. However, the numerous plants, animals, and minerals necessary for *kanpo* were not widely available on the islands, so its use was confined to the elite, urban aristocracy until recent times. Practitioners of the profession were called *kanpo-i* and underwent rigorous training.

Kanpo-i approached each case individually, reviewing symptoms carefully and in detail before determining the best combination of therapies and medications for the specific patient. Diagnosis was an art that recognized that symptoms may present differently in every consultation.[95] Western biomedicine was introduced to Japan in the sixteenth century with the arrival of the Portuguese. It was widely embraced; Japanese *kanpo-i* were required to retrain if they wished to continue working as doctors. The majority of Japanese Americans migrated to the United States during the time when *kanpo* was rarely practiced, and they were often unfamiliar with its therapies.

Since 1960 Japan has been in the middle of a *kanpo* boom and is now approved reimbursement under the Japanese health insurance policy.[95] Concerns about the side effects of biomedical therapeutics and a growing interest in holistic and herbal healing have prompted the resurgence.

Kanpo-i take a generalized approach, using natural medications with broad effects to stimulate the immune system. Some herbs also have known bacteriostatic action or anti-inflammatory properties. Small doses of the medications are taken for lengthy periods of time to promote gradual improvement. Physicians trained in biomedicine are also prescribing kanpo for many clients (though without the extensive education of *kanpo-i*); mass production of herbal medications by pharmaceutical companies began in the 1970s.

Stress-induced illness is of particular concern in Japan. Work-related fatigue and symptoms of anxiety and depression have risen dramatically in the past decade.[96] An estimated 10,000 men die annually from *koroshi* (literally "death from overwork," but referring to suicide). The healing industry in Japan is an estimated $30 billion-a-year business.[97] Stress-reducing therapies, called

Photo by Laurie Macfee

▲ *Traditional foods of Japan. Some typical foods include daikon, dried sardines, Japanese eggplant, Japanese pickles, nori (seaweed), red beans, shiitake mushrooms, short-grain rice, shrimp, soy sauce, and wheat noodles.*

iyashi, include herbs, teas, and ten-minute massage parlors. One aquarium provides overnight accommodations in its tank rooms, where visitors can sleep to relaxing music in the company of swaying jellyfish. Spas offer specialized soaking alternatives, such as bathing in coffee, green tea, red wine, or sake, to rejuvenate and energize clients. Researchers report such activities result in reduced levels of stress hormones as measured in saliva tests.[98] Napping on the job and at school has also gained some acceptance following studies on how short rests can improve productivity.[99,100]

Traditional Food Habits

Japanese ingredients, as well as cooking and eating utensils, are very similar to those of the Chinese, due to China's strong influence on Japan. Yet Japanese food preparation and presentation are unique. The Japanese reverence for harmony within the body and community and with nature has resulted in a cuisine offering numerous preparation methods for a limited number of foods. Each item is to be seen, tasted, and relished. The Japanese also place an emphasis on the appearance of the meal so that the visual appeal reflects balance among the foods and the environment. For example, a summer meal may be served on glass dishes so that the meal looks cooler, while a September meal may include the autumn colors of reds and golds.

Sea urchin roe, called *uni*, is thought to enhance male sexual potency in Japan, where wholesalers pay up to $100 per pound for it. California imposed strict urchin harvesting laws to prevent extinction along the coast. Maine suffers from similar overharvesting.

Ingredients and Common Foods

Japan's mountainous terrain and limited arable land have contributed historically to a less-than-abundant food supply. Even today much of Japan's food supply is imported.

Staples and Regional Variations. The basic foods of the Japanese diet are found in the Cultural Food Groups list (Table 11.3). Several key ingredients were adopted from China, including rice, soybeans, and tea. Rice or *gohan* (the word for "cooked rice," and also for "meal") is the main staple eaten with almost every meal. In contrast to the Chinese, the Japanese prefer a short-grain rice that contains more starch and is stickier after cooking.

Rice mixed with rice vinegar, called *su*, is used in *sushi*, one of the most popular Japanese specialties in both Japan and abroad. *Sushi* rice is formed with fish and seafood to make decorative, bite-sized mounds served with soy sauce for dipping. Types of *sushi* include *Nigirisushi*, which features rice topped with items such as sliced raw fish or squid (called *sashimi*), cooked octopus, crab or shrimp, omelet strips, or roe of salmon (*ikura*), sea urchin (*uni*), or flying fish (*tobikko*), sometimes wrapped in a strip of seaweed; *Makisushi*, a roll of sushi rice, often including cucumber (*kappamaki*), tuna, mushrooms, or other fillings, then wrapped in a sheet of seaweed and sliced into individual pieces; and *Chirashisushi*, with the topping ingredients literally scattered over a large mound of rice and eaten with chopsticks.

Rice is also made into noodles, although those made from wheat (known as *udon*, *somen*, and *ramen*) or buckwheat (*soba*) are more commonly consumed. Other noodles made from less familiar starches, such as *kudzu*, are also eaten.

Soybean products are an important component of Japanese cuisine. *Tofu* (bean curd), soy sauce (*shoyu*), and fermented bean paste (*miso*) are just a few. *Miso* comes in numerous varieties. Those made with the addition of rice are most popular; however, miso mixed with barley is found in western regions, and plain *miso* with just soybeans and salt is favored in a few central areas.[101] Red *miso* (*akamiso*) is very salty and is used most often. White *miso* (*shiromiso*) is sweeter, often used in cooked salads. Specialty *misos*, with added vegetables such as *kombu* or *daikon*, or seasoned with *shiso*, are also available.

Table 11.3 Cultural Food Groups: Japanese

Group	Comments	Common Foods	Adaptations in the United States
Protein Foods			
Milk/milk products	Japanese cooking does not utilize significant amounts of dairy products. Many Japanese are lactose intolerant. Soybean products, seaweed, and small bony fish are alternative calcium sources.	Milk, butter, ice cream	First-generation Japanese Americans drink little milk and eat few dairy products. Subsequent generations eat more dairy foods.
Meat/poultry/fish/eggs/legumes	Soybean products and a wide variety of fish and shellfish (fresh, frozen, dried, smoked) are the primary protein sources in the Japanese diet. Fish and shellfish often eaten raw. Chicken is used more often than beef; price is the limiting factor in meat consumption.	*Meat*: beef, deer, lamb, pork, rabbit, veal *Poultry*: capon, chicken, duck, goose, partridge, pheasant, quail, thrush, turkey *Fish*: blowfish, bonito, bream, carp, cod, cuttlefish, eel, flounder, herring, mackerel, porgy, octopus, red snapper, salmon, sardines, shark, sillago, snipefish, squid, swordfish, trout, tuna, turbot, yellowtail, whale *Shellfish*: abalone, *ayu*, clams, crab, earshell, lobster, mussels, oysters, sea urchin roe (*uni*), scallops, shrimp, snails *Legumes*: *adzuki*, black beans, lima beans, red beans, soybeans	Dried fish and fish cakes are available in the United States, but some varieties of fresh fish are not. Japanese Americans eat more poultry and meat than fish.
Cereals/Grains	Short-grain rice is the primary staple of the diet and is eaten with every meal. Wheat is often eaten in the form of noodles, such as *ramen*, *somen*, and *udon*.	Wheat, rice, buckwheat, millet	Rice is still an important staple in the diet and usually eaten at dinner.
Fruits/Vegetables	Fresh fruits and vegetables are the most desirable; usually eaten only in season. Many fruits and vegetables are preserved, dried, or pickled.	*Fruits*: apples, apricots, bananas, cherries, dates, figs, grapes, grapefruits (*yuzu*), kumquats, lemons, limes, loquats, melons, oranges, peaches, pears, pear apples, persimmons, plums (fresh and pickled), pineapples, strawberries, *mikan* (tangerine) *Vegetables*: artichokes, asparagus, bamboo shoots, beans, bean sprouts, broccoli, brussels sprouts, beets, burdock root (*gobo*), cabbage (several varieties), carrots, chickweed, chrysanthemum greens, eggplant (long, slender variety), ferns, ginger, ginger sprouts and flowers (*myoga*), and pickled ginger (*beni shoga*), green onions, green peppers, gourd (*kanpyo*, dried gourd shavings), kudzu, leeks, lettuce, lotus root, *mizuna*, mushrooms (*shiitake, matsutake, nameko*), okra, onions, peas, potatoes, pumpkins, radishes, rhubarb, seaweed, snow peas, *shiso*, sorrel, spinach, squash (including *kabocha*), sweet potatoes, taro, tomatoes, turnips, watercress, yams	Fewer fruits and vegetables are eaten; freshness is less critical.
Additional Foods			
Seasonings	Sugar, *shoyu*, and vinegar are a basic seasoning mixture. *Shoyu* and *mirin* can vary in strength; amounts used will vary according to taste.	Alum, anise, bean paste (*miso*), caraway, chives, *dashi*, fish paste, garlic, ginger, mint, *mirin*, MSG, mustard, red pepper, sake, seaweed, sesame seeds, *shiso*, *shoyu*, sugar, thyme, vinegar (rice), *wasabi* (green, horseradish-like condiment)	
Nuts/seeds		Chestnut, gingko nuts, peanuts, walnuts; poppy (black and white), sesame seeds	
Beverages	Green tea is the preferred beverage with meals; coffee or black tea is drunk with western-style foods. Sake or beer is often served with dinner.	Carbonated beverages, beer, coffee, gin, tea (black and green), sake, scotch	Japanese Americans drink less tea and more milk, coffee, and carbonated beverages.
Fats/oils	The traditional Japanese diet is low in fat and cholesterol.	Butter, cottonseed oil, olive oil, peanut oil, sesame seed oil, vegetable oil	Japanese Americans consume more fats and oils because of increased use of western foods and cooking methods.
Sweeteners		Honey, sugar	Increased use of sugar; sweet desserts are noted.

When a family moves to a new home, they give *soba* noodles to the neighbors on either side and across the street as a gesture of friendship.

***Kaiseki ryo¯ri* meals have recently become trendy in both Japan and the United States, costing up to $100 per person at restaurants featuring the ceremonial menu.**

Sugar, *shoyu*, and vinegar are a basic seasoning mixture for foods. *Teriyaki* sauce ("shining broil") made from soy sauce and *mirin*, a sweet rice wine, is another common flavoring for foods. *Shoyu* and *mirin* can vary in strength, and the amounts used depend on personal taste. In addition to soy beans, small, red *adzuki* beans are significant in Japanese cuisine, most often made into sweetened red bean paste and a popular red bean jelly dessert (similar to gelatin) called *yo-kan*.

Green tea is served with most meals. Tea was originally used in a devotional ceremony in Zen Buddhism. The ritual was raised to a fine art by Japanese tea masters, and as a result, they also set the standards for behavior in Japanese society. Today, the tea ceremony and the accompanying food (*kaiseki ryo¯ri*) remain a cultural ideal that reflects the search for harmony with nature and within one's self. The meal features six small courses balancing the tastes of sweet, sour, pungent, bitter, and salty. The tea used for the ceremony is not the common leaf tea usually used for meals, but rather a blend of ground, dry tea or a tea powder. Hot water is added to the tea, and the mixture is whipped together using a handmade whisk, resulting in a frothy green drink. Diners demonstrate their sophistication and sensitivity through deliberate eating of each course after expressions of appreciation for the presentation.

Soybean products and a wide variety of fish and shellfish (fresh, dried, or smoked) are the primary protein sources. Fresh fish and shellfish are often eaten raw. Beef, pork, and poultry are also popular but very expensive. One specialty is *Kobe* (or *Tajima*) beef, from a Japanese breed of cattle that is fed beer as an appetite stimulant and regularly massaged to relieve stress. It often costs more than one hundred dollars per pound. Pork is a favorite as cutlets. Chicken, which is often served as *teriyaki* sauce-glazed skewers, may also be very thinly sliced and served raw like *sashimi*.[102] Only small amounts of meat, poultry, or fish are added to the vegetables in traditional Japanese recipes. Japanese fare does not use many dairy foods.

Fresh fruits and vegetables are the most desirable and are eaten only when in season. As in China, many Asian and European varieties are consumed (see the section Staples in the Chinese part of the chapter). A few favorites include herbs and greens such as chrysanthemum greens, *mizuna* (potherb), and shiso (*perilla*, a member of the mint family—the red variety is used to color pickled foods); many tubers, including *gobo* (burdock root, which is shaved and leached in water to remove bitterness), sweet potatoes, a small variety of taro, and yams; and others such as *daikon* (a white radish similar to the Chinese radish, but longer, up to twelve inches in length), *edamame* (young soybean pods boiled in salt water, then popped open for a snack, often with beer), *kabocha* (winter squash), *shiitake* mushrooms, and the winter tangerine known as *mikan*. Pickled vegetables are available year round and are eaten extensively. Fresh fruit is a traditional dessert.

The Japanese use large amounts of seaweed and algae in their cooking for seasoning, as a wrapping, or in salads and soups. There are many types. *Nori* is a paper-thin sheet of algae that is rolled around sushi. *Kombu* is an essential ingredient in *dashi*, or soup stock made from dried bonito fish and seaweed. *Misoshiru* is a popular soup made with *dashi* and *miso* (either red or white miso can be used). *Wakame* and *hijiki* are used primarily in soups and salads. *Aonoriko* is powdered green seaweed used as a seasoning agent.

Japanese dishes are classified by the way the food is prepared (see Table 11.4). *Tempura* is an example of an *agemono* dish. Adapted from a dish introduced in the sixteenth century by the Catholic Portuguese for religious fast days, it consists of shrimp and sliced vegetables—such as eggplant, carrots, sweet potato, lotus root, and green beans—lightly battered and deep-fried. *Katsu* is another *agemono* dish of deep-fried breaded pork cutlets or fish filets. *Sukiyaki* is a simmered beef dish usually prepared at the table. The name means "broiled on the blade of the plow" and probably dates back to ancient times.[103] The current version of *sukiyaki* is mislabeled, however, because it is a *nimono*-style, not a *yakimono*-style food. *Shabu shabu,* a *nabemono* dish of small pieces of beef and vegetables cooked in broth at the table, is similar to a Mongolian hot pot (see Chinese Regional Specialties above). After the meat and vegetables are cooked and eaten, the broth is ladled into bowls and consumed. *Teriyaki* is one type of grilled *yakimono* dish, as is yakitori (grilled, marinated chicken skewers). *Teppanyaki* is a Japanese term for stovetop grilling. The style

Table 11.4 Selected Japanese Cooking Styles

Suimono	Clear soups, such as *dashi* or *misoshiru*
Yakimono	Broiled or grilled food (often marinated), such as *teriyaki* or *yakitori*
Nimono	Foods (usually a single item) simmered in seasoned water or broth, such as fish in sake-flavored broth, served hot or at room temperature
Mushimono	Steamed foods, such as *chawanmushi*
Agemono	Deep-fried foods, such as deep-fried tofu or *katsu*, usually served with a dipping sauce
Aemono	Fresh or cooked mixed foods tossed with thick sauces, such as salad with *miso* dressing
Sunomono	Mixed salads with vinegared dressing, such as crab and cucumber with rice vinegar and soy sauce dipping sauce
Chameshi	Rice cooked with other ingredients, such as chicken, fish, vegetables (especially mushrooms)—one specialty is rice with red adzuki beans, served for celebrations
Men rui	Noodle dishes, served hot or cold (plain or topped with fish or vegetables) with dipping sauce, or in a broth with items such as meats, seafood, tofu, and vegetables
Nabemono	Foods that are cooked at the table and one-pot dishes (usually a type of *nimono* dish), usually hearty combinations, such as *sukiyaki* and *shabu shabu*

familiar to U.S. diners was invented to take advantage of the tourist trade. Beef, chicken, shrimp, and vegetables are cooked on a hot grill in the center of a large table, then served with *ponzu*, a soy sauce and citrus juice mixture. *Chawanmushi*, a savory egg custard in which meats and vegetables are cooked, is a typical steamed *mushimono* dish.

Seafood, fish, fruits, and vegetables that are pickled in a mixture of *miso*, soy sauce, vinegar, and the residue from sake production are known as *tsukemono*, and they are served at nearly every meal.

Japanese foods are usually cut into small pieces if the item is not naturally easy to eat with chopsticks, and dishes are frequently modified for children, as it is believed that adult recipes are too spicy for them. Cooking style varies from region to region in Japan. Kyoto is known for its vegetarian specialties, Osaka and Tokyo are known for their seafood, and Nagasaki's cooking has been greatly influenced by the Chinese.

Meal Composition and Cycle

Daily Pattern. Traditionally the Japanese eat three meals a day, plus a snack called *oyatsu*. Simple meals, such as breakfast and lunch, are often *ichiju—issei*, meaning "soup with one side." For example, breakfast usually starts with a salty sour plum (*umeboshi*), followed by rice garnished with *nori*, soup, and pickled vegetables. A side dish such as an egg or fried fish is served with the rice. A *nabemono* can replace the side dish, which often happens at lunchtime. The meal is typically simple and often consists only of rice topped with leftovers from the previous night. Sometimes hot tea or dashi is added to the rice mixture. A bowl of noodles cooked or served with meats, poultry, or fish and vegetables is a popular alternative to leftovers. One such dish is *oyakodon*, which means "parent and child on rice," a mixture of boiled chicken and scrambled eggs on a bed of rice.

Dinner is usually *ichiju sansei*, meaning "soup and three sides," including rice, soup, and tsukemono, and three dishes: a raw or vinegared fish, a simmered dish, and a grilled or fried dish.[101] Pink pickled ginger (beni shoga) garnishes many meals, and soy sauce is usually available. The pungent, green, horseradish-like condiment called *wasabi* may also be offered, and it is mixed in small individual bowls with soy sauce to taste.

The Japanese tend not to serve meals by courses. Instead, all the dishes are presented at the same time in individual portions, each food in its own bowl or plate. The soup, however, is sometimes served last or near the end of the meal, and tsukemono may be placed on a communal plate for diners to add according to their personal preference. Traditionally, desserts were

A small grill in Japan is called a *hibachi*, meaning "fire bowl."

Sansai ryo¯ri **is a style of cooking with fresh wild herbs and vegetables such as goosefoot, mugwort, nettles, ferns, and bracken. It is considered the essence of spring.**

Myoga **ginger is prized in Japan for its tender shoots in the spring and its flower buds in the fall.**

Kombu **sounds like the word for "happiness," and it is often presented as a hostess gift by guests.**

The Japanese use 25 billion sets of disposable wood chopsticks in restaurants annually. Nearly all are made in China, which has recently restricted production to preserve forests, prompting skyrocketing prices and fears of shortages in Japan.

not common in Japan; meals usually ended with fruit.

In addition, the Japanese often eat a boxed meal called *bento*. A pleasing assortment of at least ten items is packaged attractively for consumption at school, picnics, or even between acts at the theater. Some restaurants specialize in *bento* meals.

Snacks include several kinds of sweets, rice crackers, or fruit. Traditional Japanese confections include *mochi gashi* (rice cakes with sweet red bean paste), *manju* (dumplings), and *yo-kan*. Green tea is served after all meals except when western-style food is eaten; then coffee or black tea is served. Beer or sake (rice wine, usually served warm) may be served with dinner.

Eating out is common. Numerous small restaurants specialize in certain preparations, such as *sushi, yakitori,* or noodle dishes. Restaurant windows often display their menu items with plastic replicas of their dishes.

Etiquette. The Japanese, like the Chinese, eat with chopsticks and follow many of the same customs regarding their use (see the section Etiquette in the Chinese part of the chapter); the rice bowl is not held as closely to the mouth, however. Soups are consumed directly from the bowl; the only dish eaten with a spoon is *chawanmushi*. Slurping soups and noodles is permitted and may be seen as a sign of appreciation. Tea should always be silently sipped.[21]

Traditionally the Japanese eat their meals at low tables, in a kneeling position with the heels tucked under the buttocks. In less formal situations men may sit cross-legged and women with their legs tucked to the side. Shoes are removed first. Dishes on the left are picked up with the right hand, and dishes to the right are lifted with the left hand. It is impolite to serve sake, beer, or tea to oneself. Each diner is obliged to fill his or her neighbor's glass whenever it is half-empty.

Guests are usually entertained at restaurants, where the host chooses the menu in advance. The meal may include frequent toasts, particularly *kampai* ("bottoms up"). Games may be played at the table after the meal. *Karaoke* singing is common, and guests are expected to good-naturedly participate.[21]

▼ *The* **o sonae mochi** *is traditional in Japanese Buddhist homes at New Year's, symbolizing prosperity and happiness in the future.*

© Gary Conner/PhotoEdit, Inc.

Special Occasions. In Japan there are numerous festivals associated with the harvesting of specific crops or with local Shinto shrines or Buddhist temples. The most important and largest celebration in Japan is the New Year celebration. The Japanese share many holiday traditions with the Chinese. Homes are cleaned thoroughly, and all debts are settled before the New Year; food is also prepared ahead so that no knives or cooking will interfere with the seven-day event. The Japanese celebrate New Year's Day on January 1. The New Year's foods consist of ten to twenty meticulously prepared dishes served in a special set of nesting boxes. Each dish symbolizes a specific value, such as happiness, prosperity, wealth, long life, wisdom, and diligence. For example, fish eggs represent fertility, mashed sweet potatoes and chestnuts protect against bad spirits, and black beans represent a willingness to keep healthy through hard work and sweat.

An important New Year's food is mochi, a rice cake made by pounding hot, steamed rice into a sticky dough. Traditionally a Buddhist *o sonae mochi* is set up in many homes. A large rice cake represents the foundation of the older generation and is placed on the bottom, and a smaller rice cake symbolizing the younger generation is placed atop it, followed by a tangerine indicating generations to come. The *o sonae mochi* is as meaningful to the Japanese as a Christmas tree is to Americans, preserving good fortune and happiness for future generations. Another special food is *ozoni*, a soup cooked with *mochi*, vegetables, fish cakes, and chicken or eggs. A special rice wine called *otoso* is consumed to preserve health in the coming year. Japanese Buddhist temples usually hold an *Obon* festival in the second or third week of July to appreciate the living, honor the dead, and comfort the bereaved. Food and dancing are a traditional part of the holiday. Certain birthdays are considered either hazardous or auspicious in Japanese culture. When a man turns forty-two or a woman becomes thirty-three, special festivities are held to prevent misfortune. Age sixty-one marks the beginning of second childhood, and a person dons a red cap for this honor. At age seventy-seven a person puts on a long red overcoat, and at the most propitious birthday of all, age eighty-eight, the celebrant may begin wearing both the hat and the coat.

Therapeutic Uses of Food

Although the use of *yin* and *yang* is not as prevalent among the Japanese as it is among the Chinese, there are many beliefs about the harmful or beneficial effects of specific foods and food combinations. Traditionally certain food pairs, such as eel and pickled plums, watermelon and crab, or cherries and milk, are thought to cause illness.

Pickled plums and hot tea, which are customarily eaten for breakfast, are believed to prevent constipation. Both pickled plums and rice porridge, called okayu, are thought to be easily digested and well tolerated during recovery from sickness.

Contemporary Food Habits in the United States

Adaptations of Food Habits

Little recent information on the food habits of Japanese Americans has been reported, but available data suggest the adoption of a westernized diet continues with each subsequent generation in the United States. It is thought that when acculturated Japanese Americans eat a typical American diet, they may eat more rice and use more soy sauce than non-Asians. Traditional foods are still prepared for special occasions. A westernized diet is increasingly followed even in Japan. Bread and butter are becoming staples, and consumption of meat, milk, and eggs is increasing.

A detailed study of *Nisei* and *Sansei* (mothers of Japanese immigrants and their daughters) reported several trends in dietary acculturation.[104] Compared to the predominantly Japanese diet consumed by the *Issei* who immigrated to the United States, *Nisei* women in the study continued to eat rice daily but added other starches, such as breads and cereals. Traditional protein sources, including seafood, tofu, and eggs, were often replaced with more meats and dairy items. Condiments such as soy sauce and *miso* were consumed less often, while butter and margarine intake increased substantially. When compared to their mothers, the *Sansei* daughters consumed less rice and more pasta, ate less dried fish and more cheese, consumed fewer traditional preserved or pickled foods and more fresh fruits and vegetables (particularly potatoes), and used less butter and margarine. The *Sansei* women also ate fewer Japanese-style sweets and more salty snacks. Green tea consumption declined with the *Sansei* daughters, and soft drink consumption increased. Notably, meal consumption became more irregular with the *Sansei*, and eating out and use of take-out foods were significantly more common.

Historically, rice was often in short supply in Japan, so eating the round rice cakes, mochi, at New Year's represented wealth and prosperity.

[SAMPLE MENU]

A Japanese Family Dinner

***Misoshiru* (*Miso* Soup)[a,b]**

***Sashimi*[a,b]**

***Tempura*[a] or Yellowtail *Teriyaki*[b]**

Pickled Cucumber[a,b]

Steamed Rice, Pickled Ginger, *Wasabi*, and Soy Sauce

Sake (Rice Wine) or Green Tea

[a]Fukushima, S. 2001. *Japanese home cooking*. Boston: Periplus.

[b]*Yasuko-san's Home Cooking* at http://www.nsknet.or.jp/~tomi-yasu/index_e.html

Nutritional Status

Nutritional Intake. The traditional Japanese diet is high in carbohydrates and very low in fat and cholesterol. Most cooking fats are polyunsaturated, and butter is rarely used. Japanese Americans, however, consume a more typically American diet, and this change may contribute to increased incidence of several diseases. According to classic epidemiological studies, mainland Japanese Americans have a higher risk of developing colonrectal cancer and heart disease than the Japanese in Hawaii, and those in Hawaii have a higher risk than the Japanese in Japan.[105,106] More detailed evaluation has shown Japanese Americans have more rapid atherosclerosis progression than do Japanese men and women.[107] It has been postulated that the increase is caused by diet because it is correlated to a higher intake of cholesterol and animal fat and a lower intake of dietary fiber and a lower intake of fish oil (omega 3 fatty acids).[108] Other cancers, such as those of the breast andprostate, have also increased in Japanese Americans as their stay in the United States lengthens.

Changes in diet have also been implicated in the high rates of type 2 diabetes found among Japanese American men.[109,110,111] Among *Nisei* men in one study, the rates for diabetes were twice that for similarly aged white men living in the same region of the United States and four times that for similarly aged men in Japan.[112] Data showed that the Japanese American men were consuming carbohydrate, protein, and fat proportions similar to the overall American diet, but with fewer total calories. (Though the rate of overweight and obesity has been found to be higher in U.S.-born Japanese Americans than in foreign-born Japanese Americans, total rates are low.)[46] However, insulin resistance and metabolic syndrome was greater in the Japanese American men when compared to Japanese subjects, and increased intra-abdominal fat deposits were found.[113,114] Among women, intra-abdominal fat increased with menopause.[109] Intra-abdominal fat deposition was found predictive of type 2 diabetes in Japanese Americans independent of total adiposity, family history, and other risk factors.[115,116] A genetic predisposition for diabetes combined with increased fat consumption, especially animal fats, may account for the disproportionately high rates.

A small percentage of Asians lack the ability to metabolize alcohol well. This inherited condition causes immediate skin flushing (reddening) and may even result in heart palpitations when alcohol is consumed. This reaction may contribute to the number of abstainers in the population.

Earlier studies suggest Japanese American elders may have a low intake of calcium because of limited consumption of dairy products.[45] The Japanese have a high incidence of lactose intolerance. Although seaweed, tofu, and small bony fish contain calcium, they may not be eaten in adequate amounts to provide sufficient intake. Prevalence of osteoporosis may be higher than among whites (see Dairy Foods, BMD, and Osteoporosis). Calorie consumption and meat intake have also been found to decline with age in Japanese Americans.

The traditional Japanese diet tends to be high in salt from soy sauce, *dashi*, *miso*, monosodium glutamate (MSG), dried preserved fish, and pickled vegetables. Rates of conditions sometimes linked to high-sodium diets, such as hypertension, stroke, and stomach cancer, are extremely high in Japan but have been dropping as the Japanese adopt western fare.

Infant mortality rates for Japanese Americans are lower than those for the general population.[72] However, a study comparing pregnancy outcomes in Japanese Americans noted that U.S.-born mothers are significantly more likely to have low-birth-weight infants than are foreign-born mothers.[117,118]

A comparison of Asian (Chinese, Japanese, and Korean) alcohol consumption habits found that Americans of Japanese heritage had the most permissive attitude toward drinking, particularly among women. Japanese American men had high rates of heavy drinking (nearly 30%) and the fewest abstainers (16%). Japanese American women showed similar trends, with the highest rates of heavy drinking (almost 12%) and the lowest number of abstainers (27%). Although the number of women engaged in chronic heavy drinking is lower than for white American women, the rates for men were comparable. Friends who drink and social occasions where drinking was expected were significant influences on consumption among men.[119] This study confirms previous work indicating that alcohol consumption may be more frequent than previously assumed among Japanese Americans, although behavior problems related to drinking have not been widely reported.[120]

Counseling. Japanese American values such as placing the family before the individual, preserving harmony with society, and respecting and

CULTURAL CONTROVERSY—Dairy Foods, BMD, and Osteoporosis

Osteoporosis, which means "porous bone," affects 10 million women and 2 million men in the United States. Another 16 million have low bone mineral density (BMD), which may put them at risk of developing the disease. Osteoporosis is characterized by reduced height, a stooped spinal deformity, and over 1.5 million bone fractures annually, most often of the spine, the hip, and the wrist. The causes of osteoporosis are not completely understood. Contributing factors include ethnicity, family history, low calcium intake, insufficient weight-bearing exercise, smoking, high alcohol consumption, and low levels of estrogen in women and testosterone in men.[175]

White women have long been considered at highest risk for osteoporosis. In particular, thin white women have been thought most vulnerable because higher body mass is related to better BMD. However, recent findings in a study of postmenopausal U.S. women cast doubt on the assumption that osteoporosis is a white woman's disease.[176] Data from the National Osteoporosis Risk Assessment Initiative indicated that almost 12 percent of Native American women, about 10 percent of Asians and Hispanics, 7 percent of whites, and 4 percent of blacks have osteoporosis. At every age, blacks had the highest BMD, and Asians had the lowest. However, when adjusted for weight, age, and other risk factors, the researchers report the relative risk for fracture was highest in whites and Hispanics, followed by Native Americans, blacks, and Asians. Differences in fracture rates among Asians are also seen. Chinese Americans have the lowest incidence of hip fracture compared to whites, followed by Korean Americans, then Japanese Americans.[177,178] Protective factors that lower the risk for fracture in Asians may be diminishing, however. Hip fracture rates are rising dramatically in Japan, Hong Kong, and among the Chinese in Singapore.[179,180,181,182]

Dietary recommendations regarding osteoporosis have traditionally emphasized a high intake of calcium-rich foods. Dairy foods are considered good sources because in addition to calcium they also contain vitamin D, which enhances absorption of the mineral. Although many Asians do not eat dairy foods, it has been thought that they obtained adequate calcium from eating small fish with bones (e.g., sardines), mineral-rich fish sauces, and ample dark green leafy vegetables. Some scientists also suggest that soybean intake may be protective, slowing bone mineral loss after menopause.[183,184] Yet, if prevalence for low BMD and osteoporosis among Asians is higher than previously calculated, and if fractures are increasing, do nondairy foods provide adequate calcium intake? Or, if the intake of traditional calcium-rich foods declines with acculturation in the United States (and westernization worldwide), are dairy foods needed to provide the calcium no longer being consumed?[185]

More research on the bioavailability of calcium in different foods and the role of diet in the development of BMD, osteoporosis, and fractures is needed to determine whether dairy food recommendations are sensible for all Americans or simply ethnocentric.

caring for elders may have a positive impact on health.[93] Illness may be regarded as both a symptom of an unbalanced life as well as an impediment to fulfilling personal obligations. Japanese Americans were found less accepting of pain behaviors than whites in one study.[121]

Formality and politeness are essential conversational elements in Japan. Addressing Japanese elders or Japanese American *Issei*, and some *Nisei*, by their first names is insulting. *Sansei* and *Yonsei* are often more informal.[94] Emotional displays are avoided, especially of anger. The Japanese are nonconfrontational and may be reluctant to say "no" even when the answer to a question is negative. Waving a hand in front of the face with the palm outward indicates "I'm unsure" or "I don't know." Conversational style is typically indirect, and frequent pauses, up to several minutes, are common. It is usually best to remain silent during such pauses. Direct eye contact is disrespectful—glancing around or downcast eyes are expected. Smiling can indicate pleasure but is also used to hide displeasure. Sucking in air through the teeth can also be a sign of discomfort or anger.

The Japanese are a non-touching culture, and they often stand or sit farther apart than do most Americans. Their communication style is extremely high-context, and the slightest gesture may have meaning (see Chapter 3, "Intercultural Communication," for more details). Broad hand or body gestures may be misconstrued. Though touching between strangers or acquaintances is infrequent, most Japanese Americans are comfortable with a light handshake. The traditional greeting is a bow from the waist with palms against thighs. The lower the bow and the longer it is held, the more respect is shown for the status of the other person.[21] Slouching and putting one's feet on a desk are signs of disinterest.

Japanese Americans often believe that the health care provider is a knowledgeable authority figure who will meet their needs without assistance. Most Americans of Japanese descent expect to be directed in their health care, yet are insulted if they are ordered to do anything that they feel requires only an explanation. Criticism of a client's health habits can lead to embarrassment and loss of effective communication. Concrete, structured approaches based on information gathered through an unhurried, in-depth interview that determines degree of acculturation and personal preferences are most effective.

Koreans

Korea is called *Choson* by Koreans, meaning "Land of Morning Calm."

The status of Koreans in the United States during World War II was unique. Many remained technically Japanese citizens and as such were declared enemy aliens (although none were interred). Yet many Koreans had been involved for years in anti-Japanese activities. They wore "I am Korean" buttons to distinguish themselves from Japanese residents.

There were approximately 101,000 Canadians of Korean ancestry in the 2001 census.

The economic success of South Korea has lured thousands of second-generation Korean Americans to immigrate there, where they have taken positions in marketing, public relations, and the entertainment industry.

The mountainous peninsula that forms Korea is suspended geographically and culturally between China and Japan. Korea has historically been caught in the middle of both Chinese and Japanese expansionism yet has maintained a homogeneous population with an independent, distinctive character. Little land is arable, and the climate fluctuates between cold, snowy winters and hot, monsoonal summers that limit agriculture significantly. The peninsula is currently divided into two nations. The Democratic People's Republic of Korea (North Korea), with the capital city of Pyongyang, has a communist government. The Republic of Korea (South Korea) is a democracy supported by the United States. Seoul is the country's capital. Both nations desire the reunification of Korea through political and military domination of the other.

Cultural Perspective

History of Koreans in the United States

Immigration Patterns. Early Korean immigration to the United States was severely restricted by the isolationist policies of Korea. A small number of Koreans arrived before 1900, most of whom were Protestants seeking to escape discrimination and further their education in America.

Between 1903 and 1905 Christian missionaries recruited more than 7,000 Korean men, women, and children to work in Hawaiian sugarcane fields. In 1905, when Korea was under Japanese rule, overseas emigration of Koreans was barred, and in 1907 the United States entered a gentlemen's agreement with Japan limiting both Japanese and Korean immigration. During the next seventeen years, only picture brides and oppressed political activists were permitted entry. In 1924 the Japanese Exclusion Act was applied to Koreans as well, preventing all immigration. Most early Korean immigrants worked as field hands or in domestic service; many first- and second-generation Koreans living in urban areas were barred from professional jobs and established small businesses such as vegetable stands and restaurants.[122]

Between 1959 and 1971, nearly all Korean immigrants were the wives and children of U.S. soldiers who fought in the Korean War. Following relaxation of U.S. immigration laws in 1965, the numbers of Korean immigrants increased, including many college-educated, middle-class professionals and their families.

Current Demographics. The Korean immigrant population has increased dramatically in recent years. According to 2008 census estimates, over 1.3 million Koreans are living in the United States, over half of whom have arrived since 1980. More than 50 percent of Koreans in America were born in Korea, and 50 percent of those have become citizens.[123] Large numbers of immigrants to the United States relocated first from North Korea to South Korea, seeking greater freedom. Koreans coming to America hope to find economic opportunity and to avoid any North Korean–South Korean conflict that may arise in the future. Large numbers of Koreans have settled in California, particularly Los Angeles. Other states with significant populations include New York, Illinois, New Jersey, Texas, Washington, Virginia, Pennsylvania, Maryland, and Hawaii. Difficulties in adjustment have led Korean Americans to form expatriate communities in many areas.

Socioeconomic Status. Many Koreans must accept temporary or permanent nonprofessional employment due to language difficulties or licensing restrictions.[122] Families working together toward the success of a small business and the purchase of a home are common. Korean American descendants of the 1903 to 1905 immigrants to Hawaii are securely middle class. Contributing factors include high achievement in education and professionalism; quick mastery of English (faster than Japanese and Chinese immigrants of the same period); and a greater willingness to give up Korean traditions, perhaps due to their experience with nonconformity in Korea as Christians in a Confucian and Buddhist society.[124] Almost 50 percent have college degrees, and large numbers work in management and professional positions. Family median incomes were slightly above the U.S. average, and poverty rates were above the national percentage. Some Koreans who do not find the economic prosperity they seek in the United States return to Korea.[123]

A notable conflict persists in the immigrant Korean community. Korean Americans may not accept recent immigrants who demand immediate social standing based on prior status in Korea. Recent professional immigrants are sometimes contemptuous of Korean war brides, who are considered uneducated and lower class. The large numbers of new arrivals may serve to coalesce the Korean community because they share similar experiences in acculturation and are a significant percentage of the growing Korean American population.

NEW AMERICAN PERSPECTIVES—Korean

MICHAEL HAN, Foodservice Management

I am a Korean American born in South Korea, and my parents moved to the United States in 1979 when I was eight months old. Members of my father's family came over first and then we followed.

We pretty much ate Korean meals at home. Most meals consisted of short-grain (sticky) rice, kimchi, with a soup/stew. The soup contains meat (including Spam) and vegetables. My father really likes meat, so we probably had more than other Korean families, and really unusual is that he added butter as well. Now he adds olive oil instead because his cholesterol is too high. Once in a while we would have Korean barbecue, but that was for special occasions or as a treat. Dessert was mostly fruit.

When I was young I wanted to eat more American food. For a while my mom would make pancakes for breakfast on the weekend, and when she stopped, I was really upset. Now I eat American food for breakfast and lunch but not dinner, and I usually have rice once a day. My aunt has taken over making kimchi for the family, but I don't live that close to her, so I only get it once in a while. When I do have it—I eat it all the time, so my supply only lasts a week.

In general, it would be hard to get a Korean in the United States to modify the amount and type of rice they eat, and I don't think they would ever give up pickled, spicy vegetables like kimchi.

Worldview

Religion. In South Korea Buddhism and Confucianism are the majority religions. Approximately 28 percent of South Koreans are Christian, and smaller numbers adhere to shamanism (belief in natural and ancestral spirits) and the national Korean religion *Chundo Kyo*, formerly known as *Tonghak* (a mixture of Confucian, Taoist, and Buddhist concepts). In North Korea, all religious beliefs other than the national ideology of Marxism and self-reliance are suppressed.

The first Korean immigrants at the turn of the century were Christians attracted to the United States as a religious homeland. Many suffered discrimination in the largely Confucian/Buddhist society of Korea. Although specific figures are unknown, it is believed that many recent immigrants to the United States are also Christian. Over 2,000 Korean Protestant churches, some with as many as 5,000 members, serve the community.[122] A study of elder Korean immigrant women in Washington, DC, indicated that 85 percent were Christian (Baptist, Methodist, Presbyterian, and Catholic), 10 percent practiced ancestor worship (veneration of ancestors often coexists with other religious practices), and 5 percent were Buddhist.[125] Other research suggests that larger numbers of Buddhists have immigrated.

Family. Hundreds of years of Confucianism in Korea have significantly influenced family structure regardless of current religious affiliation. Family is highly valued, and loyalty to one's immediate and extended family is more important than individual wants or needs. Generational ties are more important than those of marriage, and parents are especially close to children. Korean Americans often invite family members in Korea to join them in the United States, and extended families are not uncommon.

In Korea, a male is always the head of the family; if a father is unable to fulfill that role, the eldest son (even if still a child) takes on that responsibility. Birth sequence orders life's events within the family, and older male children are traditionally awarded privileges, such as advanced education, that younger children and daughters are denied. The role of women is to take care of the house and care for children. Parenting tends to be authoritative, and any child over the age of five years exhibiting inappropriate behavior brings disrespect upon the entire family.[126] Elders are esteemed and cared for in Korea. The two major birthday celebrations in Korean culture occur at age one and when an individual reaches age sixty, meaning that the person has survived a full five repeats of the twelve-year cycle of life and attained old age. The opinions of elders are respected, and after age sixty a person is allowed to relax and enjoy life.

Many changes occur in the Korean family after immigration to the United States. For example, the marriage bond often becomes more important than obligations to one's parents. Few elders maintain traditional arrangements of living with an eldest son's family. Some live with unmarried children or with married daughters, or live alone. Many elders feel that they are now a burden to their children and that old age is a negative experience instead of a privilege.[125] Male dominance has declined with increased participation of women in the workplace, yet most married women assume total responsibility for home and their job and often are not paid for their work if employed in a family-owned business. Divorce rates are reportedly high, and intergenerational conflict increases with length of time in the United States.[122]

Traditional Health Beliefs and Practices. Traditional Korean concepts related health to happiness, to the ability to live life fully, to function without impairment, and to not be a burden on one's children.[125] A good appetite is a significant indicator of health.

The Korean system of health and illness is closely related to Chinese precepts (see the section Health Beliefs and Practices in the Chinese part of the chapter). The proper balance of *um* (*yin*) and *yang* must occur to maintain health, influenced by the relationships of the five evolutive elements (fire, water, wood, metal, and earth) and *ki* (vital energy). Too much or too little of these forces results in illness. For example, cold, damp, heat, or wind can enter a body through the pores and then interfere with *ki* and weaken *yang*. Symptoms of the imbalance include indigestion, arthritis, and asthma. Other disruptive causes such as physical exhaustion, eating too much or too little food, and spiritual intervention (by ancestors or supernatural deities) may also result in disease.[127]

Both digestion and circulation are prominent in the maintenance of health, because energy is absorbed into the body through the stomach when food is mixed with air or one of the forces, and the blood distributes this vital energy. Good-quality food is restorative, but too much food can block *ki*, resulting in cold hands and feet, cold sweats, or even fainting.[125] A few Koreans attribute diabetes to eating too much rich food, such as meat, sugar, or honey, and getting too little exercise. Blood conditions that interfere with the distribution of vital energy include a lack of blood; a drying or hardening of the blood (typical in old age, causing indigestion, neuralgia, and body aches); and bad blood, caused by a sudden fright, which can result in chronic pain.

Korean-specific folk illnesses often include somatic complaints that are an expression of psychological distress.[128] Excessive emotions such as joy, sadness, depression, worry, anger, fright, and fear are believed to result in certain physical conditions. *Hwabyung*, attributed most often to anger, victimization, or stress, is associated with poor appetite, indigestion, stomach pain, chest pain, shortness of breath, weight gain, and high blood pressure among other symptoms. A study of middle-age Korean women found that nearly 5 percent suffered from *hwabyung*, with rates higher in women of lower socioeconomic status, those who lived in rural areas, those who were separated or divorced, and those who smoked or drank alcohol.[129,130] *Han*, which causes a painful lump in the throat, occurs when a person suffers disappointments and regrets, such as guilt over the neglect of one's children, parents, or spouse. *Shinggyongshaeyak*, resulting from stress (especially from oversensitivity and lack of happy interactions with family and friends), can cause insomnia, weight loss, and nervous collapse. Traditional cures include use of a shaman or spiritual mediator (*mansin* or *mudang*) to determine whether the cause of an illness is due to disharmony with one's ancestors or natural and supernatural forces.[127] Sacred therapeutic rituals to rectify such spiritual disruptions are conducted with the patient, the family, and sometimes the community.

Hanyak is the traditional approach to natural cures in Korea. It is typically practiced by a *hanui*. When a client visits a *hanui*, he or she obtains a medical history, observes how the patient looks, listens to the quality of the voice, and takes the patient's pulse. More than twenty-four pulse conditions are defined, from floating to sunken, and smooth, vacant, or accelerated.[131] *Hanyak* medications are classified according to their plant, animal, or mineral source, and mixed in ways to balance um (*yin*), *yang*, and *ki*. Other physical therapies to restore harmony and vital energy, such as acupuncture, moxibustion, cupping, and sweating (see Chapter 2), may also be applied. In the United States, the *hanui* may use some biomedical procedures in conjunction with traditional practices. It has been reported that some *hanui* take blood pressures and body temperatures. Some even offer the convenience of taking *hanyak* prescriptions in pill form so that bitter-tasting broths or teas are avoided.[131]

Many professional Korean health practices have been popularized and may be used as home remedies. In the United States, where access to traditional healers may be limited in some regions, the mother or grandmother in the home often takes responsibility for administering these cures. Some Koreans believe that a person's fate is determined by the forces of *um* and *yang* at the moment of birth. Christian Koreans may believe strongly in faith healing and in fate as determined by God.

Traditional Food Habits

Ingredients and Common Foods

Korean cuisine is neither Chinese nor Japanese, although it has been influenced by both styles of cooking. It is a distinctly hearty Asian fare that is highly seasoned and instantaneously recognizable as Korean by its flavors and colors. Sweet, sour, bitter, hot, and salty tastes are combined in all meals, and foods are often seasoned before and after cooking. Five colors—white, red, black, green, and yellow—are also important considerations in the preparation and presentation of dishes.

Staples. Korean cooking is based on grains flavored with spicy vegetable and meat, poultry, or fish side dishes. Korean staples are listed in Table 11.5. Rice is the foundation of the Korean diet. Rice cooking is an important skill in Korea; the rice must be neither underdone nor overcooked and mushy. Short-grain varieties are usually preferred, both a regular and glutinous (sticky) type, the latter most often in sweets. Long-grain rice, called Vietnamese rice in Korea, is available but not common. Millet and barley are used, most often as extenders for rice.

Noodles are also an important staple and are made from wheat, buckwheat, mung beans, and from the starch of sweet potatoes and the kudzu plant (a ubiquitous vine).[132] The buckwheat variety is often used in cold dishes. Fritters, dumplings, and pancakes flavored with scallions, chile pepper, and sometimes fish or meat placed directly in the batter before cooking are other popular grain dishes.[133]

Vegetables are served at every meal. Chinese cabbage (both *bok choy* and napa), European cabbage, and a long white radish (similar but not identical to the Japanese *daikon* radish) are eaten most often. Eggplant, cucumbers, perilla (a mint-family green also called *shiso*), chrysanthemum greens, bean sprouts, sweet potatoes, and winter melon are also very popular. Vegetables are added to soups and braised dishes and are often served individually as hot or cold side dishes. Pickled, fermented vegetables are included at every meal, usually in the form of kimchi, which comes in many types, based on seasonal availability of produce. A common version of kimchi is made with shredded Chinese cabbage and white radish, heavily seasoned with garlic, onions, and chile peppers, then fermented. Cucumber, eggplant, turnip, and even fruits or fish are sometimes added. Some recipes are mild, but most are very hot. Seaweed is eaten as a vegetable, including kelp and laver (called *kim*). *Kim* is brushed with sesame oil, salted, and toasted to make a condiment. The seaweed called *wakame* in Japan is used in Korean soups.[134]

Fruits are eaten mostly fresh. Crisp, juicy Asian pears (known as apple pears in the United States) are very popular; apples, cherries, jujubes (red dates), plums, melons, grapes, tangerines, and persimmons are also common.

Fish and shellfish are eaten throughout Korea. Fresh fish dishes are preferred near the coast or in river regions, but dried or salted fish is more common in the inland areas. *Saewujeot*, a Korean fermented fish sauce, is made from tiny shrimp-like crustaceans. *Saewujeot* flavors many dishes. Beef and beef variety cuts are especially popular in Korea.[102] Cubes, thin slices, or small ribs of marinated beef are barbecued or grilled at the

In order to make an accurate diagnosis, a hanui must determine a client's character through questions, such as "Do you like meats or vegetables?" "What season is your favorite?" "Do you worry much?" or "Are you stubborn?"[131]

The preparation of kimchi every autumn is a special family event. In the past a family's wealth was demonstrated by the ingredients in their kimchi, with rare vegetables and fruits used by the most affluent. Today, Koreans consume over 70 pounds per capita annually.

One of the most popular black market commodities in Korea is the diced pork and gelatin product Spam.

It is believed that a Portuguese Catholic priest introduced chile peppers into Korea in the sixteenth century. They were quickly adopted and have become ubiquitous in the cuisine.

Table 11.5 Cultural Food Groups: Koreans

Group	Comments	Common Foods	Adaptations in the United States
Protein Foods			
Milk/milk products	Milk and milk products are generally not consumed or used in cooking.		Milk, yogurt, and cheese product consumption increases.
Meat/poultry/fish/eggs/legumes	Beef and beef variety cuts are especially popular. Barbecuing is a popular method for cooking meat. Fish and shellfish, either fresh, dried, or salted, are eaten throughout Korea. Soybean products are added to many dishes.	*Meat*: beef, variety meats (heart, kidney, liver), oxtail, pork *Fish and shellfish*: abalone, clams, codfish, crab, cuttlefish, jellyfish, lobster, mackerel, mullet, octopus, oysters, perch, scallops, sea cucumber, shad, shrimp, squid, whiting *Poultry and small birds*: chicken, pheasant *Eggs*: hen *Legumes*: *adzuki*, lima beans, mung beans, red beans, soybeans	Beef, pork, and poultry consumption rises. Fish is eaten frequently. Many still consume tofu several times a week. Younger Korean Americans may consume more pork products.
Cereals/Grains	Rice is the most important component of the Korean diet. Noodles made from wheat, mung bean, or buckwheat flours are an important staple.	Barley, buckwheat, millet, rice (short-grain glutinous), wheat	Rice consumption declines, but it is still eaten every day. Breads, cereals, and pasta popular with well-acculturated Korean Americans.
Fruits/Vegetables	A wide variety of fruits are consumed. Vegetables are often pickled and are eaten at every meal.	*Fruits*: apples, Asian pears, cherries, dates (jujubes, red date), grapes, melons, oranges, pears, persimmons, plums, pumpkin, tangerines *Vegetables*: bamboo shoots, bean sprouts, beets, cabbage (Chinese, European), celery, chives, chrysanthemum leaves, cucumber, eggplant, fern, green beans, green onion, green pepper, leaf lettuce, leeks, lotus root, mushrooms, onion, peas, perilla (*shiso*), potato, seaweed (*kim*), spinach, sweet potato, turnips, water chestnut, watercress, white radish	Increased intake of fruits and vegetables is noted. The majority eat *kimchi* daily.
Additional Foods			
Seasonings	Sweet, sour, bitter, hot, and salty tastes are combined in all meals. Dishes are often seasoned during and after cooking.	Chile peppers (*kochujang*—fermented chile paste), Chinese parsley (cilantro), chrysanthemum greens, cinnamon, garlic, ginger root, green onions, MSG, hot mustard, red pepper sauce, pine nuts, rice wine, *saewujeot* (fermented fish sauce), sesame seed oil, sesame seeds, soy sauce, sugar, vinegar, sea salt	
Nuts/seeds		Chestnuts, gingko nuts, hazelnuts, peanuts, pine nuts, pistachios, sesame seeds, walnuts	
Beverages	Herbal teas are popular, as well as rice tea. They are commonly served after the meal. Soup or barley water is used as a beverage with the meal.	Barley water, beer, coffee, fruit drinks, green tea, honey water, jasmine tea, magnolia flower drink, rice tea, rice water, rice wine, *soju* (sweet potato vodka), spiced teas (ginseng, cinnamon, ginger), wines made from other ingredients	Hot barley water is still the preferred beverage and is served after the meal. Increased consumption of soft drinks.
Fats/oils	Animal fat is rarely used.	Sesame oil, vegetable oils	
Sweeteners	Sweets are made for snacks and special occasions.	Honey, sugar	Cookies and other sweets popular with Korean American youth.

table over a small charcoal brazier or gas grill. Bulgogi, grilled strips of beef flavored with garlic, onions, soy sauce, and sesame oil, is best known. Another Korean specialty is the fire pot (*sinsullo*), similar to the Mongolian hot pot, featuring beef or liver, cooked egg strips, sliced vegetables (e.g., mushrooms, carrots, bamboo shoots, onions), and nuts that are cooked in a seasoned broth heated over charcoal. After the morsels of food have been eaten, the broth is served as a soup. Chicken and poultry are not especially popular in Korea. Soybean products, however, are common, including soy sauce and soy paste. Bean curd, made from soybeans (*tobu*) or mung beans (*cheong-po*), is a favorite. Mung beans, *adzuki* beans, and other legumes are steamed and added to many savory and sweet dishes. Nuts, such as pine nuts, chestnuts, and peanuts, and toasted, crushed sesame seeds are frequent additions as ingredients or garnish.

Seasonings are the soul of Korean cooking. Garlic, ginger root, black pepper, chile peppers, scallions, and toasted sesame in the form of oil or crushed seeds flavor nearly all dishes. Ginseng is added to some soups.[134] Prepared condiments, such as soy sauce, *toenjang* (fermented soy bean paste), fish sauce (*saewujeot*), and hot mustard are also frequently added to foods. *Kochujang*, a fermented jam-like chile paste, is prepared by each family on March 3 and then traditionally stored for use throughout the year in black pottery crocks. Marinades and dipping sauces are common.

Soup or a thin barley water is used as a beverage. Herbal teas are very popular; ginseng tea flavored with cinnamon is a favorite. Ginger, cinnamon, or citron can also be used separately to make a spice tea. A common drink is rice tea, made by pouring warm water over toasted, ground rice or by simmering water in the pot in which rice was cooked. On special occasions, wine might be served. Wines are made from rice and other grains; some include various flower blossoms or ginseng as flavorings. Beer is also well liked. Milk and other dairy products are generally not consumed or used in cooking.

Meal Composition and Cycle

Daily Patterns. Three small meals, with frequent snacking throughout the day, are typical in Korea. Breakfast was traditionally the main meal in Korea, but today is more likely to be something light, caught on the run. Soup is almost always served at breakfast, along with rice (usually as gruel). Eggs, meat or fish, or vegetables may top the meal. Kimchi and dipping sauces are the usual accompaniments.

Lunch is typically noodles served with broth of beef, chicken, or fish and garnished with shellfish, meat, or vegetables. Supper is more similar to breakfast, but with steamed rice. In many modern homes, it has become the main meal of the day. Snacks are widely available from street vendors, including grilled and steamed tidbits of all types. Sweets, such as rice cookies and cakes, or dried fruit (especially persimmons) are also popular snacks.

Rice is considered the main dish of each meal. Everything else is served as an accompaniment to the rice and is called panch'an. For dinner, at least one meat or fish dish is included, if affordable, and two or three vegetables are usually served. Kimchi is always offered. Soup is very popular and is served at most meals. Individual bowls of rice and soup are served to each diner, and panch'an dishes are served on trays in the center of the table for communal eating. Wine may be served before the meal with appetizers such as batter-fried vegetables, seasoned tobu, pickled seafood, meatballs, or steamed dumplings. Dessert is seldom eaten, although fresh fruit sometimes concludes the meal. Hot barley water or rice tea is served with the meal.

Michael Freeman/Documentary/Corbis

▲ *Korean street vendors shredding daikon for the preparation of the spicy vegetable pickle kimchi.*

At Korean weddings, sweetened dates rolled in sesame seeds are tossed at the bride by her in-laws to ensure health, prosperity, and numerous children.

© Gary Conner/Photo Edit, Inc.

▲ ***Korean Americans celebrate Sol, or the New Year, on the first full moon.***

Drinking is a social ritual, practiced mostly by men. Distilled beverages like soju, a sweet potato vodka, are consumed with snacks like spicy squid or chile peppers stuffed with beef.

A special category of foods, called *anju*, are considered an alternative to a full meal. Analogous to *tapas* in Spain (see Chapter 6, "Northern and Southern Europeans"), they are small dishes especially well suited to eating while socializing and drinking.[102] Examples include scallion-flavored pancakes usually accompanied by Korean rice wine, mukhuli, Japanese-style sashimi, quail eggs, bits of pork, raw crab in chili paste, or dumplings. These items are often served as appetizers in Korean American restaurants.

Etiquette. Chopsticks and soup spoons are the only eating utensils used in Korea. In the past, seating was around a low, rectangular table on the floor, and young women were assigned the awkward corner seats (today most tables are set with chairs). Traditionally, elders are served first, and children are served last at the meal. It is considered polite to fill the soy sauce dish of the people one is sitting beside. Food is always passed with the right hand, and a communal beverage may be passed for all to share.

Dog meat soup is eaten by some men to enhance their strength and physical prowess.

***Boyak* is often given as a gift, particularly to one's parents, to promote long life.**

Special Occasions. Korean cooking was historically divided into everyday fare and cuisine for royalty. The traditions of palace cooking and food presentation, including the use of numerous ingredients in elaborate dishes, are seen today in meals for special occasions. At a meal celebrating a birthday or holiday, or one shared with guests, more dishes are served and both wine and dessert are offered. For special occasions, Koreans offer a thick drink of persimmons or dates, nuts, and spices, or a beverage made with molasses and magnolia served with small, edible flowers floating on top.

Both Koreans and Korean Americans celebrate several holidays throughout the year. The first is New Year's, called *Sol*, a three-day event at which traditional dress is worn and the elders in the family are honored. Festivities include feasts, games, and flying kites. On the first full moon, in a tradition reflecting ancient religious rites, torches are lighted, and firecrackers are set off to frighten evil spirits away. Shampoo Day (*Yadu Nal*) on June 15 is when families bathe in streams to ward off fevers. Thanksgiving (*Chusok*) is a fall harvest festival; duk, steamed rice cakes filled with chestnuts, dates, red beans, or other items, are associated with the holiday.

A special ceremony is observed on a child's first birthday. He or she is dressed in traditional clothing and placed among stacks of rice cakes, cookies, and fruit. Family and friends offer the child objects symbolizing various professions, such as a pen for writing or a coin for finance, and the first one accepted is thought to predict his or her future career.[122]

Therapeutic Uses of Food

Many Koreans follow the *um* and *yang* food classification system. Little has been reported about specific foods, although Koreans are believed to adhere to categorizations similar to other Asians: *um* (cold) foods include mung beans, winter melon, cucumber, and most other vegetables and fruits; meats (e.g., beef, mutton, goat, dog), chile peppers, garlic, and ginger are considered *yang* (hot) foods.

Preparation of healthy, tasty food is an important way that Korean women show affection for family and friends.[133,125] Good appetite is considered a sign of good health. Foods that are believed to be health promoting include bean paste soup, beef turnip soup, rice with grains and beans, broiled seaweed, kimchi, and ginseng tea. Ginseng products are often used to promote health and stamina and to alleviate tiredness. One very popular tonic, called *boyak*, combines ginseng and deer horn or bear gallbladder.[131]

More than one-half of Korean American respondents in one survey reported using ginseng, with older women more likely to use such products than men or younger women.[135] Other home remedies commonly used include ginger tea, yoojacha (hot citrus beverage), bean sprout soup, and lemon with honey in hot water. Restorative herbal medicines, vitamin supplements, meat soups, bone marrow soup, and samgyetang (game-hen soup) also were mentioned by subjects as useful when feeling weak.

A study of pregnant Korean American women suggests that certain foods, such as seaweed soup, beef, and rice, are thought to build strength during this difficult time.[136] Food taboos during pregnancy often involve the concept of "like causes like." For instance, eating blemished fruit may result in a baby with skin problems. Other such items mentioned are chicken, duck, rabbit, goat, crab, sparrow, pork, twin chestnuts, and spicy foods. Although most food avoidance was attributed to personal preference and availability, many of the women acknowledged familiarity with the traditional beliefs. Korean women traditionally consumed seaweed soup, *miyuk kook*, three times a day for seven weeks after the birth of a child to restore their strength.

[SAMPLE MENU]

Dinner in Korea

Soybean Sprout Soup[a,b]

***Bulgogi* (Korean Barbecue Beef)[a,b]**

Seasoned *Tobu* (Bean Curd),[a] Chrysanthemum Leaf Salad[a]

Seasoned Eggplant[a]

Steamed Rice, *Kimchi*

Apple Pear

Barley Water or Ginseng Tea

[a]Kwak, J. 1998. *Dok Suni: Recipes from my mother's Korean kitchen*. New York: St. Martin's Press.

[b]*Korean Recipes* at http://koreanrecipes.org/

Contemporary Food Habits in the United States

Adaptations of Food Habits

Ingredients and Common Foods. A survey of Korean Americans in the San Francisco area showed that many traditional Korean food habits continue after immigration to the United States. Nearly all respondents ate rice at one meal every day, and two-thirds ate kimchi daily. Beef and beef variety cuts were consumed regularly, and fish was eaten at least once a week. Sesame oil/seeds and vegetable oils were used more often than butter or mayonnaise. Soy sauce, soybean paste, kochujang, and garlic were the most popular condiments. More than 40 percent of Korean Americans surveyed consumed tobu several times a week. Pork and pork products were not commonly eaten, although younger respondents reported more frequent use of these foods than older subjects. Length of stay in the United States had little impact on diet.[137] Among Korean American adults in Chicago, only one-third reported increased intake of beef, dairy products, bread, coffee, and soda (combined with decreased consumption of fish and rice), and a majority continued to follow Korean dietary patterns.[138]

A national study of mostly first-generation Korean Americans found similar trends. Regular consumption continued of some traditional dishes and ingredients, such as rice, kimchi, garlic, scallions, Korean soup, sesame oil, Korean stew, soybean paste, and kochujang. Frequently consumed American foods included oranges, low-fat milk, bagels, tomatoes, and bread. Regularly eaten foods common to both cultures included onions, coffee, apples, eggs, beef, carrots, lettuce, fish, and tea. The researchers found that most respondents had access to Korean items, but as participation in American social life increased, acceptance of American foods increased. Persons who had someone familiar with Korean fare to cook for them were more likely to continue eating traditional dishes than were those who cooked for themselves.[139]

In contrast, one study of Korean Americans, in New York found that while less acculturated adults consumed a relatively traditional Korean diet, well-acculturated adults consumed significantly more bread, cereals, spaghetti, ham, green salad, corn, chocolate, candies, and diet soft drinks.[140] Another study reported that well-acculturated Korean American mothers were less likely to prepare Korean dishes at home and were more likely to dine out.[141] Research on Korean, Korean American and American adolescents found that Korean American teens adopted a diet in-between that of traditional Korean and American food patterns, eating less rice and kimchi, but more cookies and other sweets and soft drinks.[142]

Although milk and dairy products are not consumed in Korea, these foods are often well accepted in the United States. One study indicated that more than one-half of the subjects reported drinking milk or eating yogurt or cheese one or more times each week.[137] Another study of pregnant Korean American women reported a similar finding that 56 percent of respondents drank milk daily.[136] Korean American adolescent boys were found to consume milk at 22 percent of meals, and teen girls drank it at 17 percent of meals.[142]

It should be noted that many changes in the traditional diet have been observed in Korea. A study of adolescents found only 30 percent consumed a diet dependent on rice and kimchi, while 70 percent ate a modified diet with added bread, noodles, cookies, pizza, and hamburgers.[143] Recent immigrants from Korea may prefer a more westernized diet than earlier immigrants, even when length of stay is shorter.

Meal Composition and Cycle. Little has been reported regarding Korean American meal composition and cycle. It is assumed that because Korean meal and snacking patterns are similar to those in the United States, three meals a day remain common. One study found American foods were most popular at breakfast and lunch, but traditional Korean dishes were favored for dinner.[144] Elderly Korean Americans are more likely to consume a Korean meal pattern for most of their meals.[145] Survey data suggest that hot barley water served after meals is still the preferred beverage of Korean Americans.[137] Acculturation and/or length stay has been found to increase the frequency of eating out for Korean Americans.[146,147]

Korean Americans observe traditional Korean holidays (see Special Occasions in this section) as well as other events, such as Buddha's birthday (April 8), Korean Memorial Day (June 6), South Korean Constitution Day (July 17), and Korean National Foundation Day (October 3). Fathers are honored on June 15. In addition, Christian Americans of Korean descent celebrate the major religious holidays.

Nutritional Status

Nutritional Intake. Few health studies focusing on Korean Americans have been published. Most research on Asian Americans combines heterogeneous Asian and sometimes Pacific Islander populations. One study found that Korean Americans maintained a relatively traditional diet and that even with the addition of some American foods, 60 percent of calories came from carbohydrates and only 16 percent from fat.[138] Higher intakes of vitamins A and C, beta-carotene, niacin, and fiber are associated with a more traditional diet. High sodium intake is also common when Korean foods are preferred.[140,138,148,142] Korean Americans who adopt more American foods are reported to consume more calories and have higher intakes of total fat, saturated fat, cholesterol, B_1, vitamin E, folate, iron, and zinc. Calcium, which is often deficient in the Korean diet, is increased with the use of dairy products. A survey of Korean Americans conducted in a midwestern city found that those with the worst eating habits were younger, currently not married, less educated, and acculturated to American society.[149]

Age-adjusted mortality rates for Korean Americans are lower than for the general population, although infant mortality rates are somewhat higher than those for whites.[73,150,72,151] The leading cause of death among Koreans is stomach cancer; higher incidence of stomach, liver, and esophageal cancers have been found among Korean Americans as compared to whites.[69,152,153] However, self reported incidence of digestive diseases decreased with increase length of residence in the U.S but increased with more servings of rice/rice dishes.[154] Factors associated with a risk of cancers in the Korean diet include consumption of soybean paste and kimchi, high sodium intake, and possibly toxins resulting from the food fermentation process.

Aflatoxins are also sometimes found in Korean soy sauce.[155] In addition, the rate of hepatitis B (responsible for most liver cancer) is very high in Korean Americans.

Information on obesity in Korean Americans is limited. A study of Korean American women found that the proportion of overweight and obesity was only 9 percent in foreign-born subjects, compared to 31 percent in U.S.-born subjects.[148] Data suggest that length of stay in the United States is positively correlated with being overweight. A survey of Korean Americans in California determined 38 percent were overweight, and 8 percent were obese when Asian-specific criteria were used, and men were more likely than women to be overweight or obese.[156] A more recent nationwide study found similar rates of being overweight (28 percent in men and 6 percent in women).[157] It has also been reported that many Korean Americans do not exercise regularly.[158,159] In general, Asians and Pacific Islanders in the United States develop type 2 diabetes at rates significantly higher than those of the white population. One study in Hawaii found that Korean Americans had a prevalence rate twice that of whites, and self-reported rates in a sample of middle-age and older Koreans in Chicago were almost three times higher.[160]

Korean Americans appear to be at risk for hypertension. Though some studies report high blood pressure rates in Korean Americans below the U.S. average of 24 percent, others have found rates 25 percent higher.[160,161,162,163] Notably, national rates reported in Korea show hypertension affecting 25 to 45 percent of the population. Family history is the major risk factor, though gender (Korean American men have higher rates than do women), higher levels of education and acculturation, lack of exercise, and obesity may also be significant. No correlation with diet (including alcohol, spicy food, or salt intake) has been noted.[164] Research suggests Korean Americans diagnosed with high blood pressure are less likely to take hypertension medications or to follow their physician's advice on losing weight or lowering sodium intake.[165,162] Incidence of death from heart disease also has been found to be very low.

Koreans have higher alcohol consumption rates than some other Asian groups. Older data on alcohol consumption patterns in Korean Americans suggest that 25 percent of Korean American men can be characterized as heavy drinkers, whereas 44.5 percent abstain completely; less than 1 percent of Korean American women are heavy drinkers, and 75 percent are abstainers. This prevalence rate of alcohol abuse among men is equivalent to that of the general U.S. population.[119] Having friends who drink was a major factor in predicting drinking behavior, confirmed in another study indicating that, for 77 percent of Korean Americans, drinking occurred at social occasions.[120]

Counseling. Korean American clients are similar to many Asians in that language barriers may interfere with counseling. Proficient translators and cultural interpreters are often crucial to effective communication. Studies suggest that compared to other Asian American groups, Korean Americans have lower rates of health insurance coverage. As many as one in every three may be uninsured, reducing use of biomedical care.[166,161,167] Cultural attitudes may also interfere with health care. A client may be ashamed of needing help and fear being a burden to other family members during illness. Many Korean Americans avoid screening exams.[126] The stresses of acculturation may be especially severe in the Korean American community due to the large number of recent immigrants.

Koreans use a quiet, nonassertive approach to conversation. Emotional expression over pleasant topics may be animated, but during confrontation, emotional displays are avoided.[21] Loud talking or laughing is considered impolite (although some Koreans may laugh excessively when embarrassed). A measured, indirect approach to topics is appreciated. Koreans may be hesitant to say "no" or to disagree with a statement. Instead, tipping the head back and sucking air through the teeth are often used to signal dissension. Direct eye contact is expected, and it is used to demonstrate attentiveness and sincerity. Koreans use few hand gestures when talking. Touching is uncommon, except for a light, introductory handshake between men. Hugging, kissing, and back patting should be avoided, as should crossing one's legs or putting feet on the furniture. Rising whenever an elder enters the room and touching the palm of the left hand to the elbow of the right arm when shaking hands

Korean Americans have high rates of glucose-6-phosphate dehydrogenase deficiency disease.

or passing an item to an elder is considered respectful.[86,168]

The family is responsible for all its members and is usually involved in health care. In one study it was found that most Korean Americans believed a patient should not be told of a terminal illness, and the family should make any life support decisions, although younger, better educated, and wealthier Korean Americans were more open to patient autonomy.[169] It also is helpful to determine who is head of the household (often the father, eldest son, or other male member). Some Koreans find biomedical health systems very unsatisfactory. They become frustrated when a physician relies on lab tests to determine illness, especially if they are told there is no reason for a specific complaint, or conversely, that some problem exists even though they have no symptoms. They also may disagree with the cause of a complaint; some somatic symptoms may result from emotional distress, yet mental illness is highly stigmatized. Furthermore, they may expect physical treatments (i.e., acupuncture or cupping) from the provider and are likely to want a permanent cure.[125] Inconvenient hours and the need for an appointment may also discourage the use of biomedical health care.

As with other Asians who use moxibustion or cupping, it is important for health care providers to determine the cause of burns and bruises on clients before assuming abuse has occurred.

In the United States clients often administer home remedies and consult shamans or hanui before (or while) they seek advice from biomedical health care providers.[161] A study of Korean Americans in Los Angeles revealed that more acculturated and better educated individuals were more likely to use traditional health practices than those of lower socioeconomic status, perhaps due to greater affordability of multiple providers.[155] There is concern regarding the safety of some traditional products and overmedication due to herbal remedies used simultaneously with prescriptive medications.[127] For example, clamshell powder (*haigefen*) may cause stomach and muscle pain or fatigue from lead toxicity.[126] Ginseng may act as an antihypertensive and multiply the effect of other antihypertensive drugs.[155] Korean Americans must sometimes choose between treatments when traditional and biomedical health practitioners work at cross purposes; professionals of both systems have been known to advise clients against using the services of the other.[131,170]

Religious affiliation varies in Korean American clients and may have a significant impact on health and nutritional care. A recent study found that the influence of messages from religious leaders and congregants that discouraged excessive eating or encouraged exercise was associated with fewer overweight or obese in Korean American congregants.[171] An in-depth interview should be used to ascertain religious beliefs, degree of acculturation, traditional health practices used, and personal dietary preferences.

DISCUSSION STARTERS: EXAMINING THE DIET OF ASIAN-AMERICANS

In small groups of 3–4, compare and contrast the diet and culture of Chinese Americans, Japanese Americans, and Korean Americans, with each group focusing on a different aspect of the diet and culture of these groups:

Group A: The food habits and the typical eating etiquette and meal composition of these three immigrant groups

Group B: Issues involved in counseling these immigrant groups on diet and health

Group C: Attitudes within each immigrant group toward diet, health, and medical treatment, notably attitudes toward traditional home culture medical treatment and U.S. biomedicine

Group D: Amount of obesity, diabetes, hypertension, and other diseases within each immigrant group

Within your group, try to come to a consensus on what findings to report to the rest of the class. Before breaking up, assign a number to each group member: A1, A2, A3, A4; B1, B2, and so forth. Form new groups with all the 1's in a group; all the 2's in another group; all the 3's another group; and so on. In your new group, report the findings of your previous group, and, as a group, discuss the relationship of traditional attitudes toward diet and health and changes in diet and health due to immigration to the United States.

REVIEW QUESTIONS

1. List the countries for each region of Asia: East Asia, Southeast Asia, and South Asia.
2. Visit a Chinese restaurant or just look at the menu. Can you tell which region of China the recipes represent? How? Why do so many Chinese restaurants in the U.S. cook Cantonese style (Guangdong)?
3. What are the basic tenets of Confucianism, Taoism, and Buddhism? How might these religions influence Asian food culture?
4. Describe what is meant by Traditional Chinese Medicines, Kanpo, and Hanyak. How might these be integrated into the Chinese, Japanese, and Korean meaning of a balanced diet and life?
5. What are the staples of the Asian diet? Describe some common foods derived from soybeans. What are different types of tea from Asia—describe how they differ, including their tastes.
6. How did the Chinese influence the cuisine and worldview of the Koreans and the Japanese? Provide two example of their influence, one each in Korea and Japan.

REFERENCES

1. US Census Bureau. *2008 American Community Survey.* Accessed from Steven Ruggles, Matthew Sobek, Trent Alexander, et al., Integrated Public Use Microdata Series: Version 3.0. Minneapolis, MN: Minnesota Population Center, 2004.
2. Gould-Martin, K., & Ngin, C. 1981. Chinese Americans. In A. Harwood (Ed.), *Ethnicity and Medical Care.* Cambridge, MA: Harvard University Press.
3. Wang, L.L.C. 2000. Chinese Americans. In *Gale Encyclopedia of Multicultural America,* R.V. Dassanowsky & J. Lehman (Eds.). Farmington Hills, MI: Gale Group.
4. U.S. Census. Immigration Statistics Staff. 2004. *Foreign-Born Profiles* (STP-159). Online at http://www.census.gov/population/www/socdemo/foreign/STP-159-2000tl.html
5. Jones, J. S. 2000. Taiwanese Americans. In *Gale Encyclopedia of Multicultural America,* R.V. Dassanowsky & J. Lehman (Eds.). Farmington Hills, MI: Gale Group.
6. Flaws, B., & Sionneau, P. 2001. *The treatment of modern Western medical diseases with Chinese medicine: A textbook and clinical manual.* Boulder, CO: Blue Poppy Press.
7. Goldberg, B. 2002. *Alternative medicine: The definitive guide,* 2nd ed. Berkeley, CA: Celestial Arts.
8. Ludman, E. K., Newman, J. M., & Lynn, L. L. 1989. Blood-building foods in contemporary Chinese populations. *Journal of the American Dietetic Association, 89,* 1122–1124.
9. Anderson, E. N. 1987. Why is humoral medicine so popular? *Social Science and Medicine, 25,* 331–337.
10. Molony, D. 1998. *The American Association of Oriental Medicine's complete guide to Chinese herbal medicine.* New York: Berkley.
11. Ots, T. 1990. The angry liver, the anxious heart and the melancholy spleen: The phenomenology of perceptions in Chinese culture. *Culture, Medicine, and Psychiatry, 14,* 21–58.
12. Ma, G. X. 1999. Between two worlds: The use of traditional and Western health services by Chinese immigrants. *Journal of Community Health, 24,* 421–437.
13. Gale, F., Tang, P., Xianhong, B., & Xu, H. 2005. Economic Research Report, Number 8: Commercialization of Food Consumption in Rural China. Electronic Report from the Economic Research Service, US Department of Agriculture. Online at http://www.ers.usda.gov/publications/ERR8
14. Newman, J. M. 2004. *Food Culture in China.* Westport, CN: Greenwood Press.
15. Newman, J. M. 2006. Iron eggs: Taiwanese snacks. *Flavor & Fortune, 13*(1), 5.
16. Menzel, P., & D'Alusio, F. 1998. *Man eating bugs: The art and science of eating insects.* Berkeley, CA: Ten Speed Press.
17. Hahn, E. 1968. *The cooking of China.* New York: Time-Life Books, Inc.
18. Newman, J. M. 1998a. China: Transformations of its cuisine, a prelude to understanding its people. *Journal for the Study of Food and Society, 2,* 5–6.
19. Newman, J. M. 1999a. Fujian, the province and its foods. *Flavor & Fortune, 6*(2), 13, 20.
20. Newman, J. M. 2005. Wraps, Chinese style. *Flavor & Fortune, 12*(2), 24.
21. Foster, D. 2000. *The global etiquette guide to Asia.* New York: John Wiley & Sons.
22. Ludman, E. K., & Newman, J. M. 1984. *Yin* and *yang* in the health-related practices of three Chinese groups. *Journal of Nutrition Education, 16:* 3–5.
23. Pillsbury, B. L. K. 1978. "Doing the month": Confinement and convalescence of Chinese women after childbirth. *Social Science & Medicine, 12,* 11–12.
24. Sun, H. P. 1996. *Dietary habits, health beliefs, and food practices of Chinese Americans living in the San Francisco Bay Area.* Unpublished master's thesis, California State University, San Jose.
25. Newman, J. M. 1998b. Chinese ingredients: Both usual and unusual. In J. M. Powers (Ed.), *From Cathay to Canada: Chinese Cuisine in Transition.* Willowdale, Canada: Ontario Historical Society.
26. Cheng, J. T., & Yang, R. S. 1983. Hypoglycemic effect of guava juice in mice and human subjects. *American Journal of Clinical Nutrition, 11,* 74–76.
27. Spector, R. E., 2004. *Cultural diversity in health and illness* (6th ed.). Upper Saddle River, NJ: Pearson Education, Inc.
28. Koo, L. C. 1984. The use of food to treat and prevent disease in Chinese culture. *Social Science and Medicine, 18,* 757–766.
29. Sit, C., David, L., & Yeung, L. (2001). The growth and feeding patterns of 9 to 12 month old Chinese Canadian infants. *Nutrition Research, 21,* 505–516.
30. S. Liu, K. Sucher, L. McProud. Use of Chinese Herbs for Chinese Infant Feeding in the San Francisco Bay Area *Journal of the American Dietetic Association Volume 105,* Issue 8, Supplement, Page 23, August 2005.
31. Yun, W. (1996). *A growing kid.* Hong Kong, China: Sun Ya publication
32. Campbell, T., & Chang, B. 1981. Health care of the Chinese in America. In G. Henderson & M. Primeaux

(Eds.), *Transcultural health care.* Menlo Park, CA: Addison-Wesley.

33. Chau, P., & Lee, H. S. 1987. *Dietary habits, health beliefs, and related food practices of female Chinese elderly.* Master's project, California State University, San Jose.
34. Tseng, M., & Hernandez, T. 2005. Comparison of intakes of US Chinese women based on food frequency and 24-hour recall data. *Journal of the American Dietetic Association, 105,* 1145–1148.
35. Satia, J. A., Patterson, R. E., Taylor, V. M., Cheney, C. L., Thornton, S., Chitnarong, K., & Kristal, A.R. 2000. Use of qualitative methods to study diet, acculturation, and health in Chinese American women. *Journal of the American Dietetic Association, 100,* 934–940.
36. Lv, N., & Cason, K.L. 2004. Dietary pattern change and acculturation of Chinese Americans in Pennsylvania. *Journal of the American Dietetic Association, 104,* 771–778.
37. Satia, J. A., Patterson, R. E., Kristal, A. R., Hisiop, T. G., Yasui, Y., & Taylor, V. M. 2001. Development of scales to measure dietary acculturation among Chinese Americans and Chinese-Canadians. *Journal of the American Dietetic Association, 101,* 548–553.
38. Satia-Abouta, J., Patterson, R. E., Kristal, A. R., Teh, C., & Tu, S. P. 2002. Psychosocial predictors of diet and acculturation in Chinese American and Chinese Canadian women. *Ethnicity & Health, 7,* 21–39.
39. Schultz, J.D., Spindler, A. A., & Josephson, R. V. 1994. Diet and acculturation in Chinese women. *Journal of Nutrition Education, 26,* 266–272.
40. Spindler, A. A., & Schultz, J. D. 1996. Comparison of dietary variety and ethnic food consumption among Chinese, Chinese American, and white women. *Journal of Agriculture and Human Values, 131,* 64–73.
41. Amy Liu, Zekarias Berhane, & Marilyn Tseng. Improved Dietary Variety and Adequacy but Lower Dietary Moderation with Acculturation in Chinese Women in the United States. *Journal of the American Dietetic Association* Volume 110, Issue 3 , Pages 457–462, March 2010.
42. Pan, Y. L., Dixon, Z., Himburg, S., & Huffman, F. 1999. Asian students change their eating patterns after living in the United States. *Journal of the American Dietetic Association, 99,* 54–57.
43. Chau, P., Lee, H. S., Tseng, R., & Downes, N.J. 1990. Dietary habits, health beliefs, and food practices of elderly Chinese women. *Journal of the American Dietetic Association, 90,* 579–580.
44. Zhou, Y. D., & Britten, H. C. 1994. Increased iron content of some Chinese foods due to cooking in steel woks. Journal of the American Dietetic Association, 94, 1153–1154.
45. Kim, K. K., Yu, E. S., Liu, W. T., Kim, J., & Kohrs, M. B. 1993. Nutritional status of Chinese-, Korean-, and Japanese American elderly. *Journal of the American Dietetic Association, 93,* 1416–1422.
46. Lauderdale, D. S., & Rathouz, P. J. 2002. Body mass index in a US national sample of Asian Americans: Effects of nativity, years since immigration and socioeconomic status. *International Journal of Obesity, 24,* 1188–1194.
47. McNeely, M. J., & Boyko, E. J. 2004. Type 2 diabetes prevalence in Asian Americans. *Diabetes Care, 27,* 66–69.
48. Popkin, B. M., & Udry, J. R. 1998. Adolescent obesity increases significantly in second and third generation U.S. immigrants: The National Longitudinal Study of Adolescent Health. *Journal of Nutrition, 128,* 701–706.
49. Barnes P. M., Adams P. F., Powell-Griner E. *Health characteristics of the Asian adult population: United States, 2004–2006.* Advance data from vital and health statistics; no 394. Hyattsville, MD: National Center for Health Statistics. 2008.
50. Body Weight and Length of Residence in the US Among Chinese American.s Ming-Chin Yeh, Marianne Fahs, Donna Shelley, Rajeev Yerneni, Nina S. Parikh, Dee Burton. *J Immigrant Minority Health* (2009).
51. Liou, Doreen; Bauer, Kathleen. Obesity Perceptions among Chinese Americans: The Interface of Traditional Chinese and American Values Food, Culture and Society: An *International Journal of Multidisciplinary Research,* Volume 13, Number 3, September 2010 , pp. 351–369 (19).
52. Chen, M.S. 2005. Cancer health disparities among Asian Americans. *Cancer, 104,* 2895–2902.
53. Jyu-Lin Chen. Household Income, *Maternal Acculturation, Maternal Education Level and Health Behaviors of Chinese American Children and Mothers Journal of Immigrant and Minority Health* Volume 11, Number 3, 198–204 (2009).
54. Chenli Xu, Xiaolin Yang, Shuyu Zu, Shaomei Han, Zhengguo Zhang, Guangjin Zhu Association between Serum Lipids, Blood Pressure, and Simple Anthropometric Measures in an Adult Chinese Population. *Archives of Medical Research* Volume 39, Issue 6, Pages 610–617 (2008).
55. Pamela L. Lutsey*, Mark A. Pereira, Alain G. Bertoni, Namratha R. Kandula and David R. Jacobs. Interactions Between Race/Ethnicity and Anthropometry in Risk of Incident Diabetes: The Multi-Ethnic Study of Atherosclerosis. *Am. J. Epidemiol.* (2010) 172 (2): 197–204.
56. Lear, S. A., Chen, M. M., Frohlich, J. J., & Birmingham, C. L. 2002. The relationship between waist circumference and metabolic risk factors: Cohorts of European and Chinese descent. *Metabolism, 51,* 1427–1432.
57. Case, K. O., Brahler, C. J., & Heiss, C. 1997. Resting energy expenditures in Asian women measured by indirect calorimetry are lower than expenditures calculated from prediction equations. *Journal of the American Dietetic Association, 97,* 1288–1292.
58. Liu, H. Y., Lu, Y. F., & Chen, W. J. 1995. Predictive equations for basal metabolic rate in Chinese adults: A cross-validation study. *Journal of the American Dietetic Association, 95,* 1403–1408.
59. Wong, W., Tang, N.L.S., Lau, T.K., & Wong, T.W. 2000. A new recommendation for maternal weight gain in Chinese women. *Journal of the American Dietetic Association, 100,* 791–796.
60. Shai, I., Jiang, R., Manson, J. E., Stampfer, M.J., Willett, W.C., Colditz, G.A., & Hu, F.B. 2006. Ethnicity, obesity, and risk of type 2 diabetes in women: A 20-year follow-up study. *Diabetes Care, 29,* 1585–1590.
61. Barnes P. M., Adams P. F., Powell-Griner E. *Health characteristics of the Asian adult population: United States, 2004–2006.* Advance data from vital and

health statistics; no 394. Hyattsville, MD: National Center for Health Statistics. 2008.

62. Kramer, H., Han, C., Post, W., Goff, D., Diez-Roux, A., Cooper, R., & Jinagouda, S., & Shea, S. 2004.Racial/ethnic differences in hypertension treatment and control in the multi-ethnic study of atherosclerosis (MESA). *American Journal of Hypertension, 17,* 963–970.
63. Fang, J., Foo, S. H., Jeng, J.S., Yip, P. K., & Alderman, M.H. 2004. Clinical characteristics of stroke among Chinese in New York City. *Ethnicity & Disease, 14,* 378–383.
64. Asian Americans have greater prevalence of metabolic syndrome despite lower body mass index. L P Palaniappan, E C Wong, J J Shin, S P Fortmann and D S Lauderdale. *International Journal of Obesity,* (3 August 2010).
65. Newman, J.M. 2000a. Chinese meals. In H.L. Meiselman (Ed.), *Dimensions of the Meal: The Science, Culture, Business, and Art of Eating.* Gaithersburg, MD: Aspen.
66. Wang, Y., Mi, J., Shan, X.Y., Wang, Q.J., & Ge, K.Y. 2007. Is China facing an obesity epidemic and the consequences? The trends in obesity and chronic disease in China. *International Journal of Obesity, 31,* 177–188.
67. Disparities in Breast Cancer Survival Among Asian Women by Ethnicity and Immigrant Status: A Population-Based Study. Scarlett Lin Gomez, PhD, Christina A. Clarke, PhD, Sarah J. Shema, MS, Ellen T. Chang, ScD, Theresa H. M. Keegan, PhD and Sally L. Glaser, PhD May 2010, Vol 100, No. 5 | *American Journal of Public Health* 861–869.
68. Miller BA, Chu KC, Hankey BF et al. (2008). Cancer incidence and mortality patterns among specific Asian and Pacific Islander populations in the US. *Cancer Causes Control 19*(3): 227–256.
69. Jenkins, C.N.H., & Kagawa-Singer, M. 1994. Cancer. In N.W.S. Zane, D.T. Takeuchi, & K. N.J. Young (Eds.), *Confronting Critical Health Issues of Asian and Pacific Islander Americans.* Thousand Oaks, CA: SAGE.
70. Wong, S.T., Gildengorin, G., Nguyen, T., & Mock, J. 2005. Disparities in colorectal cancer screening rates among Asian Americans and non-Latino whites. *Cancer, 104,* 2940–2947.
71. Cervical cancer incidence among 6 Asian ethnic groups in the United States, 1996 through 2004 Sophia S. Wang. J. Daniel Carreon, Scarlett L. Gomez, Susan S. Devesa. *Cancer,* Volume 116, Issue 4, pages 949–956, 15 February 2010.
72. Infant Mortality Statistics from the 2005 Period Linked Birth/Infant Death Data Set by T.J. Mathews, M.S., and Marian F. MacDorman, Ph.D., Division of Vital Statistics National Vital Statistics Reports. Volume 57, Number 2.
73. Gardner, W.E. 1994. Mortality. In N.W.S. Zane, D.T. Takeuchi, & K.N.J. Young (Eds.), *Confronting Critical Health Issues of Asian and Pacific Islander Americans.* Thousand Oaks, CA: Sage.
74. Fenglian Xu Liqian Qiu Colin W Binns and Xiaoxian Liu Breastfeeding in China: a review. *International Breastfeeding Journal* 2009, 4: 6.
75. Diong, S., Johnson, M., & Langdon, R. 2000. Breastfeeding and Chinese mothers living in Australia. *Breastfeeding Review, 8,* 17–23.
76. U.S. Department of Health and Human Services (HHS), Office of Minority Health Resource Center (OMHRC), Asian American / Pacific Islander Profile http://minorityhealth.hhs.gov/templates/browse.aspx?lvl=2&lvlid=53 Accessed 2/4/11.
77. Hann, R.S. 1994a. Parasitic infections. In N.W.S. Zane, D.T. Takeuchi, K.N.J. Young (Eds.), *Confronting Critical Health Issues of Asian and Pacific Islander Americans.* Thousand Oaks, CA: Sage.
78. Wang, Y. 2003. People of Chinese heritage. In *Transcultural Health Care* (2nd ed.), L.D. Purnell & B.J. Paulanka (Eds.). Philadelphia: FA Davis Company.
79. Chen, K.T., Chen, C.J., Gregg, E.W., Williamson, D.F., & Yarayan, K.M. 1999. High prevalence of impaired fasting glucose and type 2 diabetes mellitus in Penghu Islets, Taiwan: Evidence of a rapidly emerging epidemic? *Diabetes Research and Clinical Practice, 44,* 59–69.
80. Gao, M., Ikeda, K., Hattori, H., Miura, A., Nara, Y., & Yamori, Y. 1999. Cardiovascular risk factors emerging in Chinese populations undergoing urbanization. *Hypertension Research, 22,* 209–215.
81. Ma G, Li Y, Wu Y, Zhai F, Cui Z, Hu X, et al. . The prevalence of body overweight and obesity and its changes among Chinese people during 1992 to 2002. *Chin J Prev Med 2005*; 39: 311–5.
82. Yehieli, M., & Grey, M.A. 2005. *Health matters: A pocket guide for working with diverse cultures and underserved populations.* Boston, MA: Intercultural Press.
83. Ma, G.X. 2000. Barriers to the use of health services by Chinese Americans. *Journal of Allied Health, 29,* 64–70.
84. Choe, J.H., Tu, S.P., Lim, J.M., Burke, N.J., Acorda, E., & Taylor, V.M. 2006. "Heat in their intestine": Colorectal cancer prevention beliefs among older Chinese Americans. *Ethnicity & Disease, 16,* 248–254.
85. Liang, W., Yuan, E., Mandelblatt, J.S., & Pasick, R.J. 2004. How do older Chinese women view health and cancer screening? Results from focus groups and implications for interventions. *Ethnicity & Disease, 9,* 283–304.
86. Axtell, R.E. 1991. *Gestures: The do's and taboos of body language around the world.* New York: Wiley.
87. Erika A Muse Cultural competency and Chinese Medicine: Immigrant Chinese beliefs of utilization and plurality in health seeking behaviors and health care coverage. *International Journal of Transdisciplinary Research,* Vol. 2, No.1, 2007.
88. Jing Wang, Judith Tabolt Matthews Chronic Disease Self-Management: Views Among Older Adults of Chinese Descent. *Geriatr Nurs 2010; 31:* 86–94.
89. Christine Wade, Maria T. Chao and Fredi Kronenberg Medical Pluralism of Chinese Women Living in the United States. *Journal of Immigrant and Minority Health,* Volume 9, Number 4, 255–267 (2007).
90. Cultural and Family Challenges to Managing Type 2 Diabetes in Immigrant Chinese Americans. Catherine A. Chesla, Kevin M. Chun, and Christine M.L. Kwan. *Diabetes Care* October 2009 vol. 32 no. 10.
91. Chew, T. 1983. Sodium values of Chinese condiments and their use in sodium-restricted diets. *Journal of the American Dietetic Association, 82,* 397–401.
92. Chan, J., & Kyser-Jones, J. 2005. The experience of dying for Chinese nursing home residents: Cultural considerations. *Journal of Gerontological Nursing, 31,* 26–32.

93. Easton, S.E., & Ellington, L. 2000. Japanese Americans. In *Gale Encyclopedia of Multicultural America,* R.V. Dassanowsky & J. Lehman (Eds.). Farmington Hills, MI: Gale Group.
94. Hashizume, S., & Takano, J. 1983. Nursing care of Japanese American patients. In M.S. Orque, B. Bloch, & L.S.A. Monrroy (Eds.), *Ethnic Nursing Care: A Multicultural Approach.* St. Louis: Mosby.
95. Lock, M. 1990. Rationalization of Japanese herbal medication: The hegemony of orchestrated pluralism. *Human Organization, 49,* 41–47.
96. Pollack, R., & Kuo, I. 2004. Advances in the Treatment of Anxiety Disorders—Transcultural Issues. Online at http://www.medscape.com/viewarticle/47156
97. Faiola, A. 2006. Sleeping with the fish for inner peace. *San Jose Mercury News,* June 6, 9A.
98. Toda, M., Morimoto, K., Nagasawa, S., & Kitamura, K. 2006. Change in salivary physiological stress markers by spa bathing. *Biomedical Research, 27,* 11–14.
99. Faiola, A. 2006. Nation of workaholics sleeps on the job. *Washington Post,* June 21, A01.
100. Takeyama, H., Kubo, T., & Itani, T. 2005. The nighttime nap strategies for improving night shift work in workplace. *Industrial Health, 43,* 24–29.
101. Hosking, R. 1996. *A dictionary of Japanese food: Ingredients and culture.* Rutland, VT: Tuttle.
102. Zibart, E. 2001. *The ethnic food lover's companion: Understanding the cuisines of the world.* Birmingham, AL: Menasha Ridge Press.
103. Steinberg, R. 1969. *The cooking of Japan.* New York: Time-Life Books, Inc.
104. Kudo, Y., Falciglia, G.A., & Couch, S.C. 2000. Evolution of meal patterns and food choices of Japanese American females born in the United States. *European Journal of Clinical Nutrition, 54,* 665–670.
105. Marmot, M.G., & Syme, S.L. 1976. Acculturation and coronary heart disease in Japanese Americans. *American Journal of Epidemiology, 104,* 225–247.
106. Wenkam, N.S., & Wolff, R.J. 1970. A half century of changing food habits among Japanese in Hawaii. *Journal of the American Dietetic Association, 57,* 29–32.
107. Watanabe, H., Yamane, K., Fujikawa, R., Okubo, M., Egusa, G., & Kohno, N. 2003. Westernization of lifestyle markedly increases carotid intimamedia wall thickness (IMT) in Japanese people. *Atherosclerosis, 166,* 67–72.
108. Sekikawa A, Curb JD, Ueshima H, El-Saed A, Kadowaki T, Abbott RD, Evans RW, Rodriguez BL, Okamura T, Sutton-Tyrrell K, Nakamura Y, Masaki K, Edmundowicz D, Kashiwagi A, Willcox BJ, Takamiya T, Mitsunami K, Seto TB, Murata K, White RL, Kuller LH; ERA JUMP 77. (Electron-Beam Tomography, Risk Factor Assessment Among Japanese and U.S. Men in the Post-World War II Birth Cohort) Study Group. Marine-Derived n-3 Fatty Acids and Atherosclerosis in Japanese, Japanese-American, and White Men—A Cross-Sectional Study. *J Am Coll Cardiol, 2008;* 52: 417–424,
109. Fujimoto, W. Y., Bergstrom, R. W., Newell-Morris, L. and Leonetti, D. L. (1989), Nature and nurture in the etiology of type 2 diabetes mellitus in Japanese Americans. *Diabetes/Metabolism Reviews, 5*: 607–625.
110. Nakanishi, S., Okubo, M., Yoneda, M., Jitsuiki, K., Yamane, K., & Kohno, N. 2004. A comparison between Japanese Americans living in Hawaii and Los Angeles and native Japanese: The impact of lifestyle westernization on diabetes mellitus. *Biomedicine & Pharmacotherapy, 58,* 571–577.
111. Brandon L Pierce, Melissa A Austin, Paul K Crane, Barbara M Retzlaff, Brian Fish, Carolyn M Hutter, Donna L Leonetti, and Wilfred Y Fujimoto. Measuring dietary acculturation in Japanese Americans with the use of confirmatory factor analysis of food-frequency data. *Am J Clin Nutr 86* (2007) 496–503.
112. Tsunehara, C.H., Leonetti, D.L., & Fujimoto, W.Y. 1990. Diet of second generation Japanese American men with and without noninsulin-dependent diabetes. *American Journal of Clinical Nutrition, 52,* 731–738.
113. Bergstrom, R.W., Newell-Morris, L.L., Leonetti, D.L., Shuman, W.P., Wahl, P.W., & Fujimoto, W.Y. 1990. Association of elevated fasting C-peptide level and increased intra-abdominal fat distribution with development of NIDDM in Japanese American men. *Diabetes, 39,* 104–111.
114. Masayasu Yoneda, Kiminori Yamane, Kuniaki Jitsuiki, Shuhei Nakanishi, Nozomu Kamei, Hiroshi Watanabe, Nobuoki Kohno. Prevalence of metabolic syndrome compared between native Japanese and Japanese Americans. *Diabetes Research and Clinical Practice,* March 2008 (Vol. 79, Issue 3, Pages 518–522).
115. Pribila, B.A., Hertzler, S.R., Martin, B.R., Weaver, C.M., & Savaiano, D.A. 2000. Improved lactose digestion and intolerance among African-American adolescent girls fed a dairy-rich diet. Journal of the *American Dietetic Association, 100,* 524–528.
116. Christine G. Lee, Wilfred Y. Fujimoto, John D. Brunzell, Steven E. Kahn, Marguerite J. McNeely, Donna L. Leonetti, Edward J. Boyko. Intra-abdominal fat accumulation is greatest at younger ages in Japanese American adults. *Diabetes research and clinical practice 89* (2010) 58–64.
117. Alexander, G.R., Mor, J.M., Kogan, M.D., Leland, N.L., & Kieffer, E. 1996. Pregnancy outcomes of US born and foreign-born Japanese *Americans. American Journal of Public Health, 86,* 820–824.
118. Cheng Qin and Jeffrey B. Gould Maternal Nativity Status and Birth Outcomes in Asian Immigrants. *J Immigrant Minority Health* (2010) 12: 798–805.
119. Chi, I., Lubben, J.E., & Kitano, H.H.L. 1989. Differences in drinking behavior among three Asian-American groups. *Journal of Studies in Alcohol, 50,* 15–23.
120. Zane, N.W.S., & Kim, J.H. 1994. Substance use and abuse. In N.W.S. Zane, D.T. Takeuchi, & K.N.J. Young (Eds.), *Confronting Critical Health Issues of Asian and Pacific Islander Americans.* Thousand Oaks, CA: Sage.
121. Hobara, M. 2005. Beliefs about appropriate pain behavior: Cross-cultural and sex differences between Japanese and Euro-Americans. *European Journal of Pain, 9,* 389–393.
122. Nash, A. 2000. Korean Americans. In *Gale Encyclopedia of Multicultural America,* R.V. Dassanowsky & J. Lehman (Eds.). Farmington Hills, MI: Gale Group.
123. Selected Population Profile in the United States Population Group: Korean alone or in any combination. Data Set: 2008 American Community Survey 1-Year Estimates Survey: American Community Survey, U.S.Census.

124. Kim, H.C. 1980. Koreans. In S. Thernstrom, A. Orlov, & O. Handlin (Eds.), *Harvard Encyclopedia of American Ethnic Groups.* Cambridge, MA: Belknap Press.
125. Pang, K.Y.C. 1991. *Korean elderly women in America: Everyday life, health, and illness.* New York: AMS Press.
126. Purnell, L.D., & Kim, S. 2003. People of Korean heritage. In *Transcultural Health Care* (2nd ed.). L.D. Purnell & B.J. Paulanka (Eds.). Philadelphia: FA Davis Company.
127. Chin, S.Y. 1992. This, that, and the other: Managing illness in a first-generation Korean American family. *Western Journal of Medicine, 157,* 305–309.
128. Pang, K.Y.C. 1994. Understanding depression among elderly Korean immigrants through their folk illnesses. *Medical Anthropology Quarterly, 8,* 209–216.
129. Park, Y.J., Kim, H.S., Kang, H.C., & Kim, J.W. 2001. A survey of hwa-byung in middle-age Korean women. *Journal of Transcultural Nursing, 12,* 115–112.
130. Park, Y.J., Kim, H.S., Schwartz-Barcott, D., & Kim, J.W. 2002. The conceptual structure of hwa-byung in middle-aged Korean women. *Health Care for Women International, 23,* 389–397.
131. Pang, K.Y.C. 1989. The practice of traditional Korean medicine in Washington, DC. *Social Science and Medicine, 28,* 875–884.
132. Shin-Hepinstall, H.S. 2001. *Growing up in a Korean kitchen.* Berkeley, CA: Ten Speed Press.
133. Marks, C., & Kim, M. 1993. *The Korean kitchen.* San Francisco: Chronicle.
134. Kwak, J. 1998. *Dok Suni: Recipes from my mother's Korean kitchen.* New York: St. Martin's Press.
135. Yom, M.S., Gordon, B.H.J., & Sucher, K.P. 1995. Korean dietary habits and health beliefs in the San Francisco Bay area. *Journal of the American Dietetic Association, 95* (Suppl.), A–98.
136. Ludman, E.K., Kang, K.J., & Lynn, L.L. 1992. Food beliefs and diets of pregnant Korean American women. *Journal of the American Dietetic Association, 92,* 1519–1520.
137. Gordon, B.H.J., Kang, M.S.Y., Cho, P., & Sucher, K.P. 2000. Dietary habits and health beliefs of Korean Americans in the San Francisco Bay Area. *Journal of the American Dietetic Association, 100,* 1198–1201.
138. Kim, K.K., Yu, E.S., Chen, E.H., Cross, N., Kim, J., & Brintnall, R.A. 2000. National health status of Korean Americans: Implications for cancer risk. Oncology *Nursing Forum, 27,* 1573–1583.
139. Lee, S.K., Sobel, J., & Frongillo, E.A. 1999b. Acculturation, food consumption, and diet-related factors in Korean Americans. *Journal of Nutrition Education, 31,* 321–330.
140. Kim, J., & Chan, M.M. 2004. Acculturation and dietary habits of Korean Americans. *British Journal of Nutrition, 91,* 469–478.
141. Park, S.Y., Paik, N.Y., Skinner, J.D., Ok, S.W., & Spindler, A.A. 2003. Mothers' acculturation and eating behaviors of Korean American families in California. *Journal of Nutrition Education, 35,* 142–147.
142. Park, S.Y., Paik, N.Y., Skinner, J.D., Spindler, A.A., & Park, H.R. 2004. Nutrient intake of Korean American, Korean, and American adolescents. *Journal of the American Dietetic Association, 104,* 242–245.
143. Song, Y., Joung, H., Engelhart, K., Yoo, S.Y., & Paik, H.Y. 2005. Traditional v. modified dietary patterns and their influence on adolescents' nutritional profile. *British Journal of Nutrition, 93,* 943–949.
144. Lee, S.K., Sobel, J., & Frongillo, E.A. 1999a. Acculturation and dietary practices among Korean Americans. *Journal of the American Dietetic Association, 99,* 1084–1089.
145. Young Hee Lee, Jongeun Lee, Miyong T. Kim, Hae-Ra Han, ,In-Depth Assessment of the Nutritional Status of Korean. *American Elderly Geriatric Nursing, 30* (5) 304-311 (2009).
146. Soo-Kyung Lee Acculturation, meal frequency, eating-out, and body weight in Korean Americans. *Nutr Res Pract.* 2008 Winter; 2(4): 269–274.
147. LRajagopal, T Zheng, J Kang, J Y Lee. Influence of acculturation on dining-out behavior of Koreans living in the United States—an exploratory study. Journal of Foodservice. 20: 6, pp 321–329, (2009).
148. Park, S.Y., Murphy, S.P., Sharma, S., & Kolonel, L.N. 2005. Dietary intakes and health-related behaviours of Korean American women born in the USA and Korea: The Multiethnic Cohort Study. *Public Health Nutrition, 8,* 904–911.
149. Shin CN, Lach H.Nutritional Issues of Korean Americans. Clin Nurs Res. 2010 Dec 15. [Epub ahead of print].
150. Lauderdale, D.S., & Kestenbaum, B. 2002. Mortality rates of elderly Asian American populations based on Medicare and Social Security data. *Demography, 39,* 529–540.
151. Laurence C. Baker, PhD; Christopher C. Afendulis, PhD; Amitabh Chandra, PhD; Shannon McConville, MPP; Ciaran S. Phibbs, PhD; Elena Fuentes-Afflick, MD, MPH. Differences in Neonatal Mortality Among Whites and Asian American Subgroups Evidence From California. *Arch Pediatr Adolesc Med. 2007;* 161(1): 69–76.
152. Kim, K.E. 2003. Gastric cancer in Korean Americans. *Korean and Korean American Studies Bulletin, 13,* 84–90.
153. McCracken M, Olsen M, Chen MS Jr, Jemal A, Thun M, Cokkinides V, Deapen D, Ward E. Cancer incidence, mortality, and associated risk factors among Asian Americans of Chinese, Filipino, Vietnamese, Korean, and Japanese ethnicities. *CA Cancer J Clin.* 2007 Nov–Dec; 57(6): 380.
154. Yang EJ, Chung HK, Kim WY, Bianchi L, Song WO. Chronic diseases and dietary changes in relation to Korean Americans' length of residence in the United States. *J Am Diet Assoc.* 2007 Jun; 107(6): 942–50.
155. Sawyers, J.E., & Eaton, L. 1992. Gastric cancer in the Korean American: Cultural implications. *Oncology Nursing Forum, 19,* 619–623.
156. Cho, J., & Juon, H.S. 2006. Assessing overweight and obesity risk among Korean Americans in California using World Health Organization body mass index criteria for Asians. *Preventing Chronic Disease, 3,* A79.
157. Soo-Kyung Lee Acculturation, meal frequency, eating-out, and body weight in Korean Americans. *Nutr Res Pract.* 2008 Winter; 2(4): 269–274.
158. Crews, D.E. 1994. Obesity and diabetes. In N.W.S. Zane, D.T. Takeuchi, & K.N.J. Young (Eds.),

Confronting Critical Health Issues of Asian and Pacific Islander Americans. Thousand Oaks, CA: Sage.

159. Jiwon Choi, Joellen Wilbur and Mi Ja Kim. Patterns of leisure time and non-leisure time physical activity of Korean immigrant women. *Health Care Wom. 32:* 2 (2011)pp. 140–153 (14).
160. Cross, N.A., Kim, K.K., Yu, E.S.H., Chen, E.H., & Kim, J. 2002. Assessment of the diet quality of middle-aged and older adult Korean Americans living in Chicago. *Journal of the American Dietetic Association, 102,* 552–554.
161. Kim, M., Han, H.R., Kim, K.B., & Duong, D.N. 2002. The use of traditional and Western medicine among Korean American elderly. *Journal of Community Health, 27,* 109–120.
162. Kim, M.T., Kim, K.B., Juon, H.S., & Hill, M.N. 2000. Prevalence and factors associated with high blood pressure in Korean Americans. *Ethnicity & Disease, 10,* 364–374.
163. Yang EJ, Chung HK, Kim WY, Bianchi L, Song WO. Chronic diseases and dietary changes in relation to Korean Americans' length of residence in the United States. *J Am Diet Assoc.* 2007 Jun; 107(6): 942–50.
164. Tamir, A., & Cachola, S. 1994. Hypertension and other cardiovascular risk factors. In N.W.S. Zane, D.T. Takeuchi, and K.N.J. Young (Eds.), *Confronting Critical Health Issues of Asian and Pacific Islander Americans.* Thousand Oaks, CA: Sage.
165. Kim, M.J., Ahn, Y.H., Chon, C., Bowen, P., & Khan, S. 2005. Health disparities in lifestyle choices among hypertensive Korean Americans. *Biological Research for Nursing, 7,* 67–74.
166. Hill, L., Hofstetter, C.R., Hovell, M., Lee, J., Irvin, V., & Zakarian, J. 2006. Korean's use of medical services in Seoul, Korea and California. *Journal of Immigration and Minority Health, 8,* 273–280.
167. Sohn, L. 2004. The health and health status of older Korean Americans at the 100-year anniversary of Korean immigrants. *Journal of Cross Cultural Gerontology, 19,* 203–219.
168. Morrison, T., Conaway, W.A., & Borden, G.A. 1994. *Kiss, bow, or shake hands: How to do business in sixty countries.* Holbrook, MA: Adams.
169. Blackhall, L.J., Murphy, S.T., Frank, G., Michel, V., & Azen, S. 1995. Ethnicity and attitudes toward patient autonomy. *Journal of the American Medical Association, 274,* 820–825.
170. Jeong Hee Kang; Hae-Ra Han;Kim B. Kim,; Miyong T. Kim,. Barriers to care and control of high blood pressure in Korean American elderly. *Ethn Dis. 2006; 16:* 145–151.
171. John W. Ayers, C. Richard Hofstetter, Veronica L. Irvin, Yoonju Song, Hae-Ryun Park, Hee-Yong Paik, Melbourne F. Hovel. Can Religion Help Prevent Obesity? Religious Messages and the Prevalence of Being Overweight or Obese Among Korean Women in California. *Journal for the scientific study of religion 2010 49:* 3, pp. 536–549.
172. Newman, J.M. 2006. Chinese restaurants beginning. *Flavor & Fortune, 13*(2), 17–18.
173. Newman, J.M. 2000b. Mongolians and their cuisine. *Flavor & Fortune, 7*(1), 9–10, 24.
174. Ang, C. 2000. Tibetan food and beverages. *Flavor & Fortune, 6*(3), 21.
175. National Institute of Health (NIH) Osteoporosis and Related Bone Diseases—National Resource Center. 2000. Available online: http://www.niams.nih.gov/Health_Info/Bone/Osteoporosis/overview.asp
176. Barrett-Connor, E., Siris, E.S., Wehren, L.E., Miller, P.D., Abbott, T.A., Berger, M.L., Santora, A.C., & Sherwood, L.M. 2005. Osteoporosis and fracture risk in women of different ethnic groups. *Journal of Bone Mineral Research, 20,* 185–194.
177. Lauderdale, D.S., Jacobsen, S.J., Furner, S.E., Levy, P.S., Brody, J.A., & Goldberg, J. 1997. Hip fracture incidence among elderly Asian-American populations. *American Journal of Epidemiology, 146,* 502–509.
178. Ross, P.D., Norimatsu, H., Davis, J.W., Yano, K., Wasnich, R.D., Fujiwara, S., Hosod, Y., & Melton, L.J. 1991. A comparison of hip fracture incidence among native Japanese, Japanese Americans, and American Caucasians. *American Journal of Epidemiology, 133,* 801–809.
179. Koh, L.K., Saw, S.M., Lee, J.J., Leong, K.H., & Lee, J. 2001. Hip fracture incidence rates in Singapore 1991–1999. Osteoporosis International, 12, 311–318.
180. Lau, E.M. 1996. The epidemiology of hip fracture in Asia: An update. *Osteoporosis International, 6* (Suppl. 3), 19–23.
181. Orimo, H., Hashimoto, T., Sakata, K., Yoshimura, N., Suzuki, T., & Hosoi, T. 2000. Trends in the incidence of hip fracture in Japan, 1987–1997: The third nationwide survey. *Journal of Bone Mineral Metabolism, 18,* 126–131.
182. D. K. Dhanwal,1 C. Cooper,1, 2 and E.M. Dennison1. Geographic Variation in Osteoporotic Hip Fracture Incidence: The Growing Importance of Asian Influences in Coming Decades. *Journal of Osteoporosis (2010) 2:* 6 pp 1–5.
183. Harrison, E., Adjei, A., Ameho, C., Yamamoto, S., & Kono, S. 1998. The effect of soybean protein on bone loss in a rat model of postmenopausal osteoporosis. *Journal of Nutritional Science and Vitaminology, 44,* 257–268.
184. Messina, M., & Messina, V. 2000. Soyfoods, soybean isoflavones, and bone health: A brief overview. *Journal of Renal Nutrition, 10,* 63–68.
185. Anne McTiernan, Jean Wactawski-Wende,LieLing Wu, Rebecca J Rodabaugh,Nelson B Watts,Frances Tylavsky, Ruth Freeman, Susan Hendrix, and Rebecca Jackson. Low-fat, increased fruit, vegetable, and grain dietary pattern, fractures, and bone mineral density: the Women's Health Initiative Dietary Modification Trial. *Am J Clin Nutr* June 2009 vol. 89 no. 6 1864–1876.

Southeast Asians and Pacific Islanders

CHAPTER 12

Southeast Asians and Pacific Islanders live in similar tropical regions and may share some common ancestors, yet their cultures have diverged markedly over the centuries. The countries of Southeast Asia have developed under hundreds of years of Chinese domination. Spanish expansionism in the Philippines and French occupation in Vietnam were also significant. In contrast, Asian and European contact in the Pacific Islands was limited until the eighteenth century, and subsequent foreign influence almost completely overwhelmed the traditional indigenous societies.

Southeast Asians and Pacific Islanders are some of the newest, youngest, and fastest-growing populations in the United States. More than 2 million mainland Southeast Asians and Filipinos have arrived since 1975; the number of Pacific Islanders has increased by 50 percent since 1990. This section discusses the cultures and cuisines of the Southeast Asians who have immigrated in substantial numbers to the United States—Filipinos, Vietnamese, Cambodians, and Laotians—as well as those Pacific Islander groups with significant American populations—native Hawaiians, Samoans, Guamanians, and Tongans.

Southeast Asians

Cultural Perspective

History of Southeast Asians in the United States

Immigration Patterns. Most Filipinos immigrate to the United States for educational and economic opportunities. In contrast, the majority of mainland Southeast Asians who have come to the United States have arrived since the 1970s as refugees from the political conflicts of the region.

The Philippines. Substantial immigration from the Philippines to the United States started in 1898 after the country became a U.S. territory. Approximately 113,000 young male Filipinos traveled to the Hawaiian Islands between 1909 and 1930 to work in the sugarcane fields. Many of these immigrants later moved to the U.S. mainland. These early immigrants were considered U.S. nationals and carried U.S. passports, yet they were not allowed to become citizens or own land. Most were uneducated peasant laborers from the island of Illocos. Because they were not permitted to bring their wives or families, social and political clubs replaced the family as the primary social structure.

In 1924 the immigration of Filipinos slowed as a result of Asian exclusion laws (see Chapter 11, "East Asians," for more information). After World War II, it became legal for Filipinos to become U.S. citizens, and the number of immigrants increased. Significant numbers of Filipinos arrived in the United States after 1965 when the U.S. immigration laws were changed, and by 1980 more than 350,000 had emigrated to mostly urban areas. Two-thirds of the Filipinos who arrived after World War II qualified for entrance as professional or technical workers. Yet discrimination against Asians often forced Filipinos into

The only Filipinos allowed to apply for U.S. citizenship prior to 1946 were those who had enlisted in the U.S. Navy, Naval Auxiliary, or Marine Corps during World War II; had served at least three years; and had an honorable discharge.

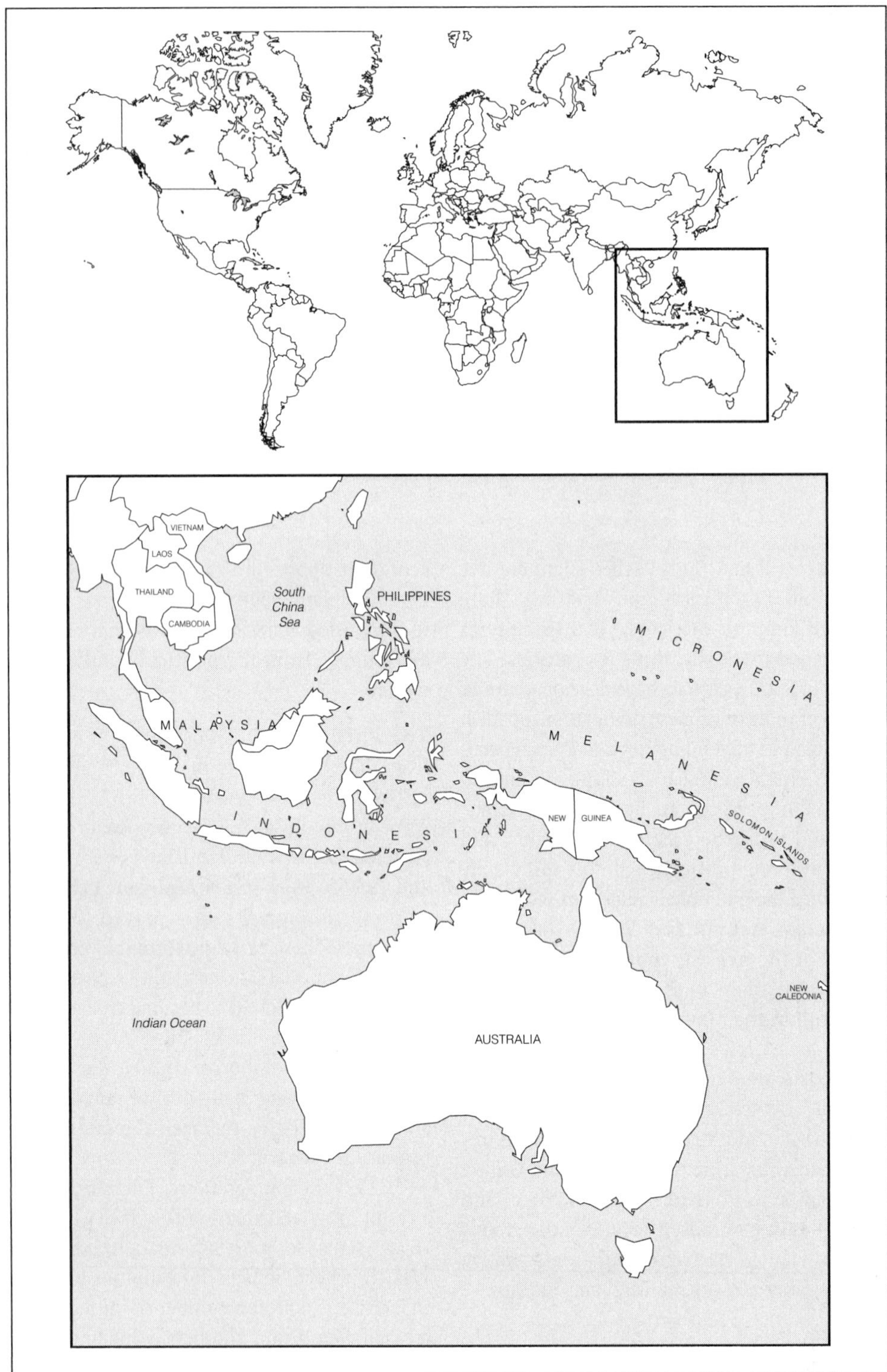

Figure 12.1
Southeast Asia and the Pacific Islands.

low-paying jobs. "Little Manilas" formed in many California cities, and similar homogeneous neighborhoods were found in Chicago, New York City, and Washington, DC, where some Filipinos opened small service businesses to meet the needs of their community.

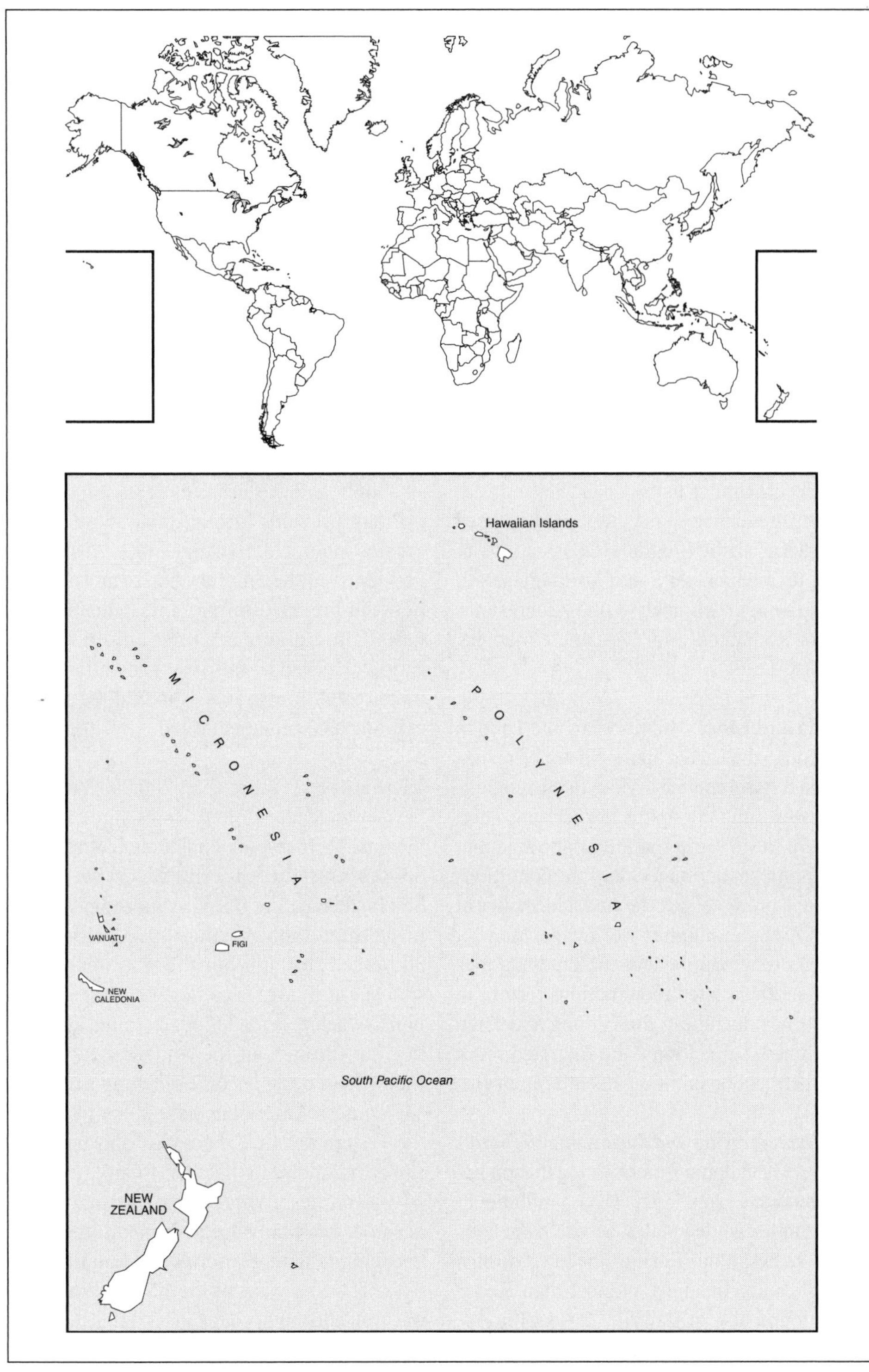

Figure 12.1
Southeast Asia and the Pacific Islands (continued).

Vietnam. Vietnamese immigration to the United States is characterized by three distinct waves. The first occurred when South Vietnam fell to the North in 1975, and 60,000 Vietnamese left the country with the assistance of the United States. Another 70,000 managed to flee on their own. Most of these refugees had been employed by the United States or were members of the upper classes. They immigrated in intact family groups and were able to bring their property with them. From 1975 to 1977 another wave of Vietnamese left for political or economic reasons, often escaping by sea. Many of these boat people left their families and what little money they had to find freedom. The third phase of immigration started in 1978, when increasing numbers of ethnic Chinese living in Vietnam fled the country, again by boat. This wave of immigration was accelerated by the Chinese invasion of northern Vietnam in 1979. This second group of boat people left with no financial resources, and many lost family members escaping in unseaworthy vessels that were easy prey for pirates. After being rescued, the boat people often lived for several months, even years, in refugee camps in various Southeast Asian countries before coming to the United States. In addition, the United States and Vietnam developed the Orderly Departure Program (ODP) in 1979 to bring imprisoned former South Vietnamese soldiers and the approximately 8,000 Amerasians, children of U.S. fathers and Vietnamese mothers, to the United States.

Cambodia and Laos. Cambodian and Laotian immigration to the United States did not begin until the United States granted asylum to the residents of the refugee camps along the border with Thailand in 1976 to 1979. International concern about the large numbers of refugees and the conditions in the camps prompted accelerated admittance of Southeast Asian immigrants worldwide. Many of the refugees were members of tribal populations who lived in the isolated mountainous regions of Southeast Asia, including the Hmong and Mien (ethnic Chinese populations who migrated south to escape persecution in the eighteenth century).

Current Demographics and Socioeconomic Status

Filipino. The Filipino American population has more than doubled since 1980. Over 3 million Filipinos live in the United States, according to 2007 census estimates. Almost half reside in California, with significant populations of more than 30,000 Filipinos found also in Hawaii, Florida, Illinois, New York, New Jersey, Texas, and Washington.[1] It should be noted that Filipino Americans are a very heterogeneous group who come from different cultural groups within the Philippines and often speak a variety of native languages.[2,1]

There were over 327,000 Canadians of Filipino heritage in the 2001 census.

During the 1920s and 1930s, Filipinos were often barred from restaurants, swimming pools (and other recreational facilities), and movie theaters because of their dark skin and heavy accents.

Over 50 percent of Filipino Americans are foreign-born, citizenship rates are higher than in other Asian groups. Educational attainment is also high. Nearly 44 percent of Filipino Americans hold college degrees.[1] Family median incomes are 34 percent above the national average, combined with poverty rates approximately half that of the general U.S. population, both due to higher earnings and, sometimes, to pooled income of several adults living in each household.

Some Filipinos find that their education or professional experience is not recognized in the United States, forcing them to accept blue-collar jobs when they first arrive. Over time, many professionals do obtain licensing or accreditation and work as physicians, nurses, or lawyers. Nearly 40 percent are employed in professional or management careers.[1] Others are employed in the sales and service industries, or in construction, and this group is economically better off than previous immigrants, many of whom are now poor, elderly single men. A schism has developed in the Filipino-American community between pre-1965 immigrants, who believe the newer immigrants are ungrateful for the advances achieved by the older generation, and the more recent immigrants, who find the older generation naïve and uneducated.[1]

Vietnamese. More than 700,000 Vietnamese have entered the United States since 1975, and the total Vietnamese population, including those born in America, was estimated at over 1.6 million in 2006 by the U.S. Census, indicating it has more than doubled since 1990, with roughly 50 percent foreign born. Some were initially sponsored by American agencies or organizations, which provided food, clothing, and shelter in cities throughout the United States until the Vietnamese could become self-supporting. Many Vietnamese Americans have since relocated to the western and Gulf states, probably because the climate is similar to that of Vietnam.

Vietnamese Americans live primarily in urban areas. Many are young and live in large households with grandparents and other relatives. The first wave of Vietnamese immigrants was well educated, could speak English, and had held white-collar jobs in Vietnam. Many of them had to accept blue-collar jobs initially and had difficulties supporting their extended families. The boat people were even less prepared for life in the United States because they often had no English language skills, were illiterate, had less job training, and did not have support from an extended family.

Language acquisition has been important in the Vietnamese community,[3] and many immigrants have made the decision to assimilate as quickly as possible, even changing their Asian names, such as Nguyen, to something more Anglicized, such as Newman. However, over 80 percent do not speak English at home. Education is highly valued, and a child's academic achievement is considered a reflection on the whole family. High school dropout rates are very low among Vietnamese Americans, and almost half attend college, compared to 40 percent of white Americans. However, when all Vietnamese Americans are counted, including the adults who immigrated later in life, 27 percent have less than a high school education, and slightly lower than U.S. average numbers have college degrees.[4]

Vietnamese have high rates of employment, over 50 percent have professional or service related empolyment. Those who obtain college degrees often prefer technical professions, such as engineering. Median family income is slightly lower than the U.S. average, and poverty rates are higher.

Cambodian. The 2000 U.S. Census identified approximately 178,000 Americans of Cambodian descent, but it is thought that this figure may underestimate actual numbers due to a low response rate by Cambodians unfamiliar with American society in the census count.[5] Approximately 50 percent are foreign born.[4] Nearly half of Cambodian Americans reside in California, with the greatest populations found in Long Beach and Stockton. Another large Cambodian community has developed in Lowell, Massachusetts. Texas, Pennsylvania, Virginia, New York, Minnesota, and Illinois also have significant numbers of immigrants.

Adjustment to American life has been difficult for Cambodians, most of whom came from rural regions where they worked as farmers. They have a high unemployment rate (although the probability of employment increases with length of stay), and those who work have found employment mostly in service jobs, manufacturing, transportation, and manual labor. Median family income is 30 percent below the national average, and over 30 percent of Cambodians live below the poverty level. Low levels of education have limited economic success for many Cambodian Americans. More than half of adult men and women have less than a high school education.[6] Low dropout rates among younger Americans of Cambodian descent may contribute to better economic outcomes in the future.[5]

Laotian. About 193,000 nontribal Laotians live in the United States, according to 2005 U.S. Census figures. Most Laotians have settled in the California cities of Fresno, San Diego, Sacramento, and Stockton. Large populations are also found in Amarillo and Denton, Texas; Minneapolis, Minnesota; and Seattle, Washington. Another 183,000 Hmong and approximately 25,000 Mien have been identified. Refugee resettlement services dispersed the Hmong, irrespective of family clan associations, in fifty-three American cities. Secondary migration of Hmong Americans has reunited clan groups, mostly in the suburban and rural communities of Minnesota (St. Paul has the largest single population, over 25,000) and California's Central Valley—Fresno, Merced, Sacramento, Stockton, Chico, Modesto, and Visalia. It is believed that nearly all Mien live in four states: California, Oregon, Washington, and Alaska.

Poor fluency in English and low education levels have hindered economic achievement for Laotian Americans. Only 64 percent of adults have a high school degree or higher.[7] Current Laotian students are generally believed to do well in school, but higher-than-average high school dropout rates and lower-than-average college attendance rates have been noted.[8] Many Laotians hold lower-paying jobs in transportation and as machine operators, fabricators, and laborers, yet it is estimated that only 14 percent live in poverty.

Nearly 50 percent of Americans of Hmong descent have less than a high school education, and most Hmong are employed in blue-collar jobs or manual labor.[9,7] Some enterprising Hmong women have been successful in marketing the traditional Hmong needlecraft known as *paj ntaub*. The colorful wall hangings, pillows, and bed coverings have become popular among collectors of native arts. Median family income is among the lowest of all immigrant groups—over 35 percent below the national norm—and poverty rates of more than 33 percent are more than three times the U.S. average.

In 2001 the Canadian Census reported approximately 187,000 residents of Southeast Asian heritage, 151,000 of whom were Vietnamese.

There are five major tribes among the Hmong—the White Hmong, the Black Hmong, the Blue Hmong, the Red Hmong, and the Flowery Hmong—named according to legend by the colors of clothing they were forced to wear by the ancient Chinese for identification purposes. Most Hmong refugees in the United States come from the White and Blue tribes.

Over 100,000 Thais have immigrated to the United States since the 1970s; most have settled in Los Angeles, New York, and Texas. They are known for their rapid acculturation.[170]

Worldview

Religion. Many Southeast Asians hold beliefs dating back to the ancient religions prevalent before the introduction of Buddhism, Catholicism, and Islam. Most believe in spirits and ghosts, especially of ancestors, who have the ability to act as guardians against misfortune or cause harm and suffering. Ideas about spirit intervention have often been incorporated into Eastern and Western religious practices or survive as significant superstitions.

Filipino. The majority of Americans of Filipino descent are Roman Catholics, although it is estimated that as many as 5 percent are Muslim. Religion significantly affects the worldview of Filipinos, especially elders. Many hold that those who lead a good life on Earth will be rewarded with life after death. Human misfortunes come from violating the will of God. One should accept one's fate because supernatural forces control the world. Time and providence will ultimately solve all problems.

Vietnamese. Nearly 70 percent of Vietnamese Americans are Buddhists, and 30 percent are Roman Catholic. Small numbers of Protestants also are found. Those who are Buddhists follow the Mahayana sect, influenced particularly by the Chinese school of Ch'an Buddhism (Zen Buddhism in Japan) called Tien in Vietnam. Buddhists believe that their present life reflects their past lives and also predetermines their and their descendants' future lives. They consider themselves as part of a greater force in the universe (see Chapter 4, "Food and Religion," for more information about Buddhism).

Hmong ancestors are fed at every festive occasion with a little pork and rice placed in the center of the feast table.

In Thailand, miniature Thai temples are built and posted on pedestals next to a home to shelter the family ancestors. Such spirit houses are believed to prevent the ghosts from moving in with the family and causing trouble.

Cambodian. The predominant religion in Cambodia is Theravada Buddhism, which places greater emphasis on a person's efforts to reach spiritual perfection than does the Mahayana sect, which employs the help of deities. Merit through good deeds, the participation in religious ritual, and the support of monks and temples is critical to one's progress through reincarnation. Although some Cambodian Americans have converted to Christian faiths, most still practice Buddhism, often in temples established in apartments or homes.

Laotian. Almost all Laotians are also Theravada Buddhists. Making merit for Laotians includes the expectation that every man will devote some time in his life to living as a practicing monk, either before marriage or in his old age. In the United States, men find it difficult to fulfill their obligation to the faith. Women may become nuns for periods in their lives as well, especially if widowed. Most Laotians in the United States worship at Buddhist temples alongside Cambodians or Thais who share their religious practices.

About half the Hmong population in the United States is now Christian as a result of French and American missionary work in Laos and conversions after arrival. Baptists, Presbyterians, Mormons, Jehovah's Witnesses, and members of the Church of Christ have actively recruited the Hmong. The other half of Hmong Americans practice animism, shamanism, and ancestor worship. They believe that the world is divided into two spheres—that which is visible, containing humans, nature, and material objects, and that which is invisible, containing spirits. The shaman acts as an intermediary between the two; some spirits, such as those of ancestors, are available to those who aren't shamans. Women are generally responsible for making contact with these more accessible spirits.

It has been suggested that many Hmong became Protestants because they believed it was necessary to gain entry to the United States or in deference to their church sponsors. Some try to combine ancestor worship with their new faiths so as not to offend the spirits.[10] Small numbers of other Hmong belong to a modern Hmong religion called Chao Fa ("Lord of the Sky"), started by a prophet in the 1960s who encouraged the Hmong to break with both Laotian and Western ways.

Family. Southeast Asians share a high esteem for the family, respect for elders, and interdependence among family members. Behavior that would bring shame to the family's honor is avoided, as is direct expression of conflict. Social acceptance and smooth interpersonal relationships are emphasized.

Filipino. The Filipino family is highly structured. At the center is the extended family, containing all paternal and maternal relatives. Kinship is extended to friends, neighbors, and fellow workers through the system of compadrazgo. Lifelong relationships are initiated through shared Roman Catholic rituals, particularly the selection of godparents and baptism of new babies. Community obligations created through this system include shared food, labor, and financial resources.

The first Filipino immigrants were single men hoping to make their fortune and return to the Philippines. They developed surrogate family systems of fellow workers who lived together, with the eldest man serving as patriarch. More recent immigrants have come as whole families and have been able to maintain much of their social organization. For example, as many as 200 sponsors may be appointed in a Filipino-American baptism, although individual commitment may be less than if it were held in the Philippines.[1]

Filipino children are adored by the family and are typically indulged until the age of six. At that time, socialization through negative feedback (e.g., shame) begins. Children are taught to be obedient and respectful, to contain their emotions and avoid all conflict, and to be quiet and shy. Politeness is emphasized.[11] People must avoid shaming themselves or their families; discord is minimized by the use of euphemisms and by sending go-betweens in sensitive situations. The prestige of a family may be measured by how well children adhere to traditional values. Though Filipino parents often support their children in advanced education and professional careers, women may be discouraged from attending schools where they would be beyond family supervision.[12]

Vietnamese. The extended Vietnamese family has been modified in the United States to adapt to American norms. A nuclear family is more typical, though somewhat larger than the average family in the United States. Close relatives are encouraged to move into homes next door or in the same neighborhood.

Family values are in transition. The father is traditionally the undisputed head of the household, but patriarchy is diminishing as women attain higher levels of education and professional achievement. American-style dating has become common. Although most Vietnamese marry within their ethnic group, women are more likely than men to wed non-Vietnamese mates. Divorce is uncommon.[3]

The level of intergenerational conflict is reportedly high.[3] Children are often the first to learn English and acculturate more easily than their parents, causing value conflicts and loss of respect for elders. The role of elders is also changing. Old age is valued in Southeast Asia, but in the United States older relatives are often physically isolated from their peers and even younger family members and may be linguistically isolated within the larger community.

Cambodian. Large extended families are common in Cambodia; children are considered treasures. Cambodians are notable in that their traditional kinship system was bilateral, emphasizing both the paternal and maternal lines. The family was primarily a matriarchy until the 1930s, when French influence strengthened the authority of the father. Today, men are responsible for providing for their families, while women make all decisions regarding the family budget.

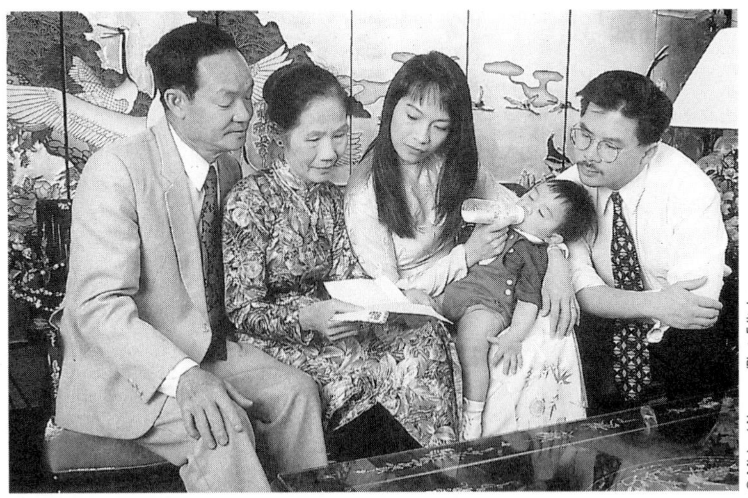

© Michael Newman/PhotoEdit, Inc.

▲ *The traditional Vietnamese extended family home is less common in the United States, but relatives often live near each other.*

It has been difficult for Cambodian Americans to retain their traditional family structure in the United States. A large percentage of Cambodian homes are headed by single women, due most likely to the large numbers of men killed in recent years during conflicts in Cambodia. Furthermore, Cambodian American women are permitted formal education, and their financial contributions are needed to help support the whole family. Differences between immigrants who have lived most of their lives in Cambodia and the children of these immigrants raised in the United States (with no memory of Cambodia) can be enormous in terms of language, acculturation, and values. It has been noted that many Cambodian teenagers suffer from identity problems.[5]

A Cambodian proverb states, "If you don't take your wife's advice, you'll have no rice the next year."

When Cambodians wed, a Buddhist monk cuts a lock of hair from both the bride and groom to mix together symbolically in a bowl.

Laotian. Most families are agriculturally based in Laos. Extended members live and work together in the fields to support the whole family. In Laos men represent the family in village affairs, and women run the home. Great significance is given to the site of the family's house; it is believed that as long as the site exists, the family will exist.

Extended families are still important to Laotian Americans due to dependency on relatives for social and economic support. Although nuclear families have become the norm, extended members tend to live nearby each other. Women have attained nearly equal status with men in the United States, and it is not unusual for Laotian men to share responsibility for household chores. Laotians have a notably low divorce rate.[8]

Hmong American families are among the youngest and largest in the United States: more than half of all Hmong are under the age of 18.[6] Typical family size is about six members; the Asian average is 3.7, and the white American average is just 3.1. These large families are usually nuclear, reflecting numbers of children rather than relatives. Extended family members are often located nearby, and families frequently congregate with other families from the same traditional clan from Laos, sometimes pooling resources.[13]

Men remain the heads of households in the United States. Women are traditionally held in high regard in their roles as mothers, yet it has been noted that fathers are assuming a much larger role in child care in the United States.[13] Children are the heart of the home, and much of family life revolves around them. At age five, however, children are expected to behave as adults.

As in some other Southeast Asian groups, a generation gap has developed between recently arrived immigrants and their westernized children. Teenage runaways have become increasingly common among the Hmong.[9] Some Hmong customs have come in direct conflict with U.S. laws. Polygyny, marriage to more than one wife, is unusual but not illegal in Laos (by custom, Hmong men marry the widows of their deceased brothers and support their nieces and nephews). Clan leaders might wed several women to establish political connections, and having several wives was indicative of wealth. The practice that has received the most press in the United States, however, is the kidnapping of young women and enforced marriage. Traditionally, girls married between the ages of fourteen and eighteen after a bridal price (paid by the groom) was agreed to by both families. If no agreement was reached, the couple could elope, and a mediator would help to settle the differences. If the bride was unwilling, the groom could kidnap the girl, and the marriage would be recognized after some payment to the bride's family was arranged. Men who have attempted this in the United States have been charged with abduction and sexual assault, often by the young women involved. Many Americans of Hmong descent believe that it is best to wait until a woman is in her late teens or early twenties to wed; arranged marriages are still common.

Hmong marriages are traditionally formalized at a two-day feast featuring a roasted pig.

Conversion to Christianity has split many Hmong families when some members change faiths and others retain their traditional beliefs. Conflict over marriage traditions is especially prevalent.

Traditional Health Beliefs and Practices. Southeast Asian health concepts typically combine facets of multiple belief systems. Indigenous ideas about the origins of illness center on the supernatural world, particularly the intervention of malevolent spirits or the ghosts of angry ancestors. Chinese medical practices involving yin and yang or the five evolutive elements (see Chapter 11) are considerations in some areas of mainland Southeast Asia, while the Mexican hot-cold theory is more prevalent in the Philippines (see Chapter 9, "Mexicans and Central Americans").

Religious precepts regarding rewards for making merit, or performing good deeds, and punishment for violating God's will are also involved in health maintenance. In the most general terms, keeping healthy requires personal harmony with the supernatural world, nature, society, and family fulfilled through one's obligations to one's ancestors, one's religion, and one's kin and community. Illness is usually defined by its cause, not its symptoms.

Most Filipinos adhere to the concept of bahala na, meaning that life is controlled by the will of God and by supernatural forces. If a person behaves properly, shows consideration of others and sensitivity in relationships, fulfills debts and obligations, shows gratitude, and avoids shame, he or she is rewarded with health in this life and eternal life after death.[11] Many Filipinos believe that illness is a punishment for transgressions against God. Religious medals are worn for protection from evil.

Spanish control of the islands administered through Mexico led to the adoption of some aspects of humoral medicine in the Philippines, including not only the hot-cold theory, but also the condition commonly known as wind or air (*mal aire* in Mexico; see Chapter 9, Mexican Traditional Health Beliefs and Practices section).

Supernatural illnesses in the Philippines are most often due to the unhappy ghosts of one's ancestors, although witchcraft, or the powers of animal spirits, may also be involved. *Usog* or *tuyaw* occurs when a person transmits illness through the power of the evil eye or the use of hands, fingers, words, or even physical proximity.[14] Undesirable traits or conditions can be transferred magically through contact with a person or object. A pregnant woman will try to gaze upon beautiful objects or people and avoid looking at a person with a deformity to prevent a similar occurrence in her fetus. Some believe if a pregnant woman craves dark-skinned fruit, her infant will have a dark complexion.[12,15] Some Filipino Americans who believe in supernatural causes of disease do not think that these forces apply in the United States, because ghosts and spirits cannot cross the ocean, nor can they survive in the noisy cities where many Filipinos now live.

In the Philippines health is maintained through the balance (timbering) of natural and supernatural elements.[11,2,14,15] A person is thought to be predisposed to certain illnesses, and the timing of external events contributes to the development of disease. Unbalanced conditions, such as working too much, overeating, excessive drinking, inadequate diet or sleep, unhygienic conditions, infections, accidents, emotional stress (especially fright or anxiety), or loss of self-esteem, may increase a person's vulnerability, as do factors such as the season and the weather.

Three practices are used to produce balance: heating, protection, and flushing. It is widely believed that a warm body is needed to prevent illness. Heating means that a person balances hot and cold—whether through exposure to the elements or through eating the right proportions of foods classified as hot or cold—so that warmth is maintained and overheating is avoided. For example, cold or cooling foods, such as orange juice, are not consumed first thing in the morning. Bathing, often twice per day, is also used to maintain warmth in the body. Any imbalance, whether too hot or too cold, is believed to cause illness by reducing blood flow, causing loss of appetite, and lowering of the body's ability to fight off sickness. For example, a nursing mother who becomes overheated by too much sun or from exposure to a hot kitchen may find that her milk has become rancid, producing colic or diarrhea in the baby.[16] Specifics on the application of hot and cold classifications and treatment vary tremendously from person to person.

Protection safeguards the body from natural and supernatural forces. For instance, a layer of fat is needed to protect the body from external cooling. Wind is of special concern. It may cause disease directly through drafts or be absorbed through the pores or any wounds. Wind that is too cold or too hot affects the blood, causing increased or decreased circulation, resulting in a general malaise and increased susceptibility to illness. A postpartum woman avoids bathing for nine to forty days after birth of the baby to prevent wind from entering her vagina; a newborn's umbilicus is bound to keep wind from entering that opening; and coconut oil may be rubbed into the skin to block the pores. Whooping cough and mental illnesses are two of the more serious conditions that can be caused by wind.

Balance is also important in other areas as well. Any rapid change is avoided, such as going from the activity of heavy exercise to the inactivity of showering (thus a short rest period is needed in-between). Emotional restraint is maintained when possible,[2] because strong emotion may cause certain symptoms of illness. For instance, some Filipinos feel that excessive anger or envy is a hot condition and great fright or joy is a cold condition. Somaticized complaints are common.[17,12]

Flushing is used to cleanse the body of impurities or evil forces through perspiration, flatulence, vomiting, or menstrual blood.[15] Vinegar mixed with water, salt, and chile peppers is one example of a flushing treatment, taken to stimulate sweating.

Several types of traditional healers are common in the Philippines: midwives, masseurs, curers (who diagnoses through evaluation of the pulse), arbularyos (herbalists), and shamans, who cure supernaturally caused illness through use of folk remedies. In urban regions, where belief in ghosts and spirits is not as prevalent, faith healers are gaining in popularity. Faith healers do not diagnose illness but cure it through prayer, anointing with oil, and the laying on of hands, which transmits a sacred healing energy to the patient.[14,12,15]

For the Vietnamese, health is related to personal destiny. How one behaved in past lives and the number of good deeds performed by one's ancestors determine one's experiences in this life. Current behavior, such as pleasing good spirits and avoiding evil spirits, can also impact health. Similar to the Filipinos, pregnant Vietnamese women may avoid funerals or ugly objects or leaving their homes at the times malevolent spirits are active (noon and 5:00 p.m.). The use of divination, through fortune-telling, astrology, or physiognomy (the shape of the body, especially the head, as it correlates to the mind), is popular for predicting how a person might expect his or her life to proceed and what interventions might be needed to prevent certain negative experiences.[18,19]

Traditionally, the Vietnamese believe that the human body is sustained by three separate souls: one that encompasses the life force, one that represents intelligence, and one that embodies emotions. In addition, nine vital spirits provide assistance to the souls. Soul loss can be an important and life-threatening reason for illness. Typically, strong feelings, especially fright, can cause the soul to leave the body.[20]

Among the ethnic Chinese-Vietnamese immigrants, many are Traditional Chinese Medicine practitioners.

The Chinese medical system is commonly used by ethnic Chinese Vietnamese and by some other Vietnamese as well. Maintaining a balance of *yin* and *yang*, especially through diet and the treatment of disease, is a primary consideration in health. Like the Filipinos, wind (or air) is sometimes seen as a cause of illness. Some Hmong are also concerned with hot, cold, and wind as well. Hmong women are reportedly thought to be in a cold condition immediately following birth (having lost hot blood) and must avoid cold drinks, cold drafts and wind, and sexual intercourse to reestablish balance.[21]

Laotians often wear copper or silver bracelets or colored strings around their wrists, necks, and ankles to keep their souls from leaving. White string is used by families; red or black strings are tied on by shamans during ritual ceremonies.

Cambodians, Laotians, and Hmong are also concerned with spiritual intervention in health. Laotians identify thirty-two spirits that oversee the thirty-two organs of the body.[8] The Hmong recognize the world of the invisible, where the spirit of every animal, tree, and rock resides, amid the souls of the living, ancestor spirits, caretaker spirits, and evil spirits. Ancestor spirits require special consideration because if they become angry, they may leave their progenies or fail to protect them from evil. The Laotians have elaborate rituals called baci, mostly performed at all special occasions by older men who have been monks, that bind the spirits to their possessor. Among the Hmong, the loss of one's soul, usually due to strong emotional distress, is the single most important cause of illness. It generally results in malaise and weight loss, leading to more serious disease. Related to soul loss is the condition called *ceeb*, or fright illness. *Ceeb* typically occurs in children (although it can happen to adults as well) if they are in an accident, chased by a dog, startled by a noise, or plunged into cold water. The soul becomes disconnected; the blood cools down and slows, resulting in a chilling effect that begins in the extremities and can progress to the vital organs.[22]

In parts of Southeast Asia, opium is traditionally grown in home herb gardens for use as a pain killer.[13]

Unique to Southeast Asians in the United States is the unexplained condition known as sudden unexpected nocturnal death syndrome (SUNDS), when a seemingly healthy person dies in his or her sleep. It is especially prevalent among Cambodians, Laotians, and Hmong, although it may occur in other immigrants from the mainland as well. It was a leading cause of death among Hmong men ages twenty-five to forty who were the earliest immigrants to the United States.[10] Although biomedical hypotheses have been proposed to account for the fatal syndrome, such as heart irregularities or sleep apnea, none has been proven. It is now believed the phenomenon known as sleep paralysis is involved.[23,24] Some researchers believe that death is caused within the cultural context of the nightmare experience. Specifically, the nightmare spirit, *dab tsog*, enters the room at night and the victim "wakes" to the sensation of the spirit sitting on his or her chest; he or she is unable to move and is terrified. Although many immigrants report having experienced nightmares in Southeast Asia, the attack by the spirit does not usually result in death. Cultural disruptions are believed to have intensified the episodes. A Hmong man in the United States is often unable to perform his obligations to ancestor spirits and may also have difficulties in fulfilling his role as breadwinner and head of household. Guilt and depression create increased vulnerability to fatal nightmare experiences. Post-traumatic stress disorder, panic attacks, exposure to chemical warfare agents, or blood electrolyte imbalances may be other risk factors.[24,20]

Traditional healers are typically specialized practitioners among mainland Southeast Asians. They may provide services for broken bones, skin infections, or objects stuck in the throat. Hmong herbalists (*kws tshuaj*) treat natural disorders, such as menstrual problems, impotence, infertility, stomach disorders, and diarrhea, with teas and poultices.[21] Hmong shamans treat patients for spiritual disorders at great personal risk due to interaction with the spirit world.[25] They heal conditions such as mental illness, hypertension, diabetes, breathing difficulties, and fainting. In many cases, they often deal with lost souls. Magic healers are not spiritually chosen for their profession but may interface with the spirit world to treat injuries and stones (such as those found in kidneys, or those placed in bodies by evil spirits). Monks may lead religious rituals. Among most Southeast Asians, minor illnesses may be treated by anyone with healing experience, typically a grandmother or mother in the home. The family takes responsibility for the illness of an individual and will usually exhaust all remedies available within the house before seeking outside help.[13]

Botanical remedies are very popular with many mainland Southeast Asians living in the United States. Cambodians, Laotians, and Hmong sometimes maintain herb gardens for easy access to therapeutic ingredients. Some immigrants frequent Chinese herbalists or will buy imported products from Asia.[26] Herbs and other substances (e.g., rhinoceros horn, black chicken dung, dead turtle, or chicken parts) are prepared as teas, broths, steam inhalants, or balms.[27,13] Physical therapies may include massage; cupping (a heated cup or a cup with a small amount of burning paper is placed over a certain spot on the body until the fire goes out, leaving a round red spot on the skin); moxibustion (burning a small bundle of herbs on the skin or using a lit cigarette); and coining (rubbing a coin or spoon dipped in tiger balm or eucalyptus ointment across the skin with pressure), scratching, or pinching affected areas. In most cases the therapy is used to release any bad wind or excess heat and to restore balance to the body.[28,13]

Religious rituals are also used to intervene on behalf of an ill person. Hmong soul callers perform the "Mandate of Life" ceremony to return a lost soul to its host body, and the Mien appeal to ancestor spirits to protect family members and assist in healing. These rituals sometimes include animal offerings. A butchered animal, typically a chicken, pig, or occasionally a cow, is purchased from a packinghouse prior to the ceremony; then it is cooked and consumed after the rite as part of a feast. Its soul is offered in exchange for the victim's missing soul.[21,13] In Vietnam small shrines are sometimes constructed to appease ancestor spirits or the souls of premature infants who have died and still wander the Earth. Offerings may also be made to the Goddess Quang Am for good health. Among Catholic Vietnamese and Filipinos, appeals are made to the Virgin Mary; group prayer has assumed significance for many Protestant Southeast Asians.

It has been noted that Christian Hmong often avoid the use of shamans, soul callers, and other traditional practitioners, depending on the clergy and the power of prayer to promote physical and spiritual healing. Herbs, however, may still be used at home.[21,25]

Traditional Food Habits

The cuisines of Southeast Asia have many ingredients in common, but food preparation methods and meal patterns reflect the foreign cultures that have influenced each nation. For example, the Vietnamese often serve cream-filled French pastries for dessert, whereas Filipinos frequently have Spanish-style custard flan. As in China and Japan, the staple foods are rice (primarily long-grain), soybean products, and tea. A meal is not considered complete unless rice is included. Instead of soy sauce, however, Southeast Asians often season their food with strongly flavored fermented fish sauces and fish pastes.

Mien rituals involving ancestor spirits require a genealogical record of the family going back ten generations.

Ingredients and Common Foods: Staples and Regional Variations

Filipino. Filipino fare has blended Malaysian, Polynesian, Spanish, and Chinese influences into a distinctive cuisine. There are three principles in Filipino cooking: First, never cook any food by itself; second, fry with garlic in olive oil or lard; and third, foods should have a sour-cool-salty taste.[29] For example, adobo, one of the most popular Filipino preparations, combines marinated chicken, pork, and sometimes fish or shellfish, that is then fried with garlic in lard and then braised in soy sauce, vinegar, garlic, chili peppers, bay leaf, and peppercorn with whatever vegetables are on hand, such as plantains, potatoes, greens, or bamboo shoots.[30] Filipinos traditionally used a clay pot for cooking but now use a large wok called a *kawali*, especially for frying. They tend to leave the food in longer than the Chinese do, to allow it to absorb more fat. The common foods of the Philippines are listed in Table 12.1.

▼ ***Traditional foods of Southeast Asia and the Pacific Islands. Some typical foods include coconut, dried anchovies, dried mango, French bread, lemon grass, lime,*** **nuoc mam*****, pineapple, pork, rice, rice paper, rice sticks, taro root, and water chestnuts.***

Photo by Laurie Macfee

Table 12.1 Cultural Food Groups: Filipino

Group	Comments	Common Foods	Adaptations in the United States
Protein Foods			
Milk/milk products	Filipinos make one of the few native cheeses in Asia, from *carabao* (water buffalo) milk. U.S. influence has resulted in the availability of many western dairy products. Many Filipinos may be lactose intolerant. In desserts, coconut milk is frequently used in place of cow's milk.	Evaporated milk (cow, goat), white cheese (*carabao*)	Consumption of milk and other dairy products has increased.
Meat/poultry/fish/eggs/legumes	Protein intake is often dependent on income.	*Meat:* beef, *carabao*, goat, pork, monkey, variety meats (liver, kidney, stomach, tripe), rabbits *Poultry and small birds:* chicken, duck, pigeon, sparrow *Fish and shellfish:* anchovies, bonita, carp, catfish, crab, crawfish, cuttlefish, *dilis*, mackerel, milkfish, mussels, prawns, rock oyster, salt cod, salmon, sardines, sea bass, sea urchins, shrimp, sole, squid, swordfish, tilapia, tuna *Eggs:* chicken, fish *Legumes:* black beans, black-eyed peas, chickpeas, lentils, lima beans, mung beans, red beans, soybeans, white kidney beans, winged beans	Consumption of fish has decreased; intake of meat, poultry, and eggs has increased.
Cereals/Grains	Rice is the main staple and is usually eaten at every meal.	Corn, oatmeal, rice (long- and short-grain, flour, noodles), wheat flour (bread and noodles)	Rice is not usually eaten at breakfast but is eaten at least once per day.
Fruits/Vegetables	Vegetables are often consumed in mixed stews, stir-fries, and soups. Braised vegetables may be consumed as entree or side dish. Pickled fruits and vegetables are very popular.	*Fruits:* apples, avocados, banana blossoms, bananas (100 varieties), breadfruit, *calamansi* (lime), citrus fruit, coconut, durian, grapes, guava, jackfruit, Java plum, litchi, mangoes, melons, papaya, pears, persimmons (*chicos*), pineapples, plums, pomegranates, pomelo, rambutan, rhubarb, star fruit, strawberries, sugar cane, tamarind, watermelon *Vegetables:* amaranth, bamboo shoots, bean sprouts, beets, bitter melon, burdock root, cabbage, carrots, cashew nut leaves, cassava, cauliflower, celery, Chinese celery, drumstick plant (sili leaves), eggplant, endive, garlic, green beans, green papaya, green peppers, hearts of palm, hyacinth bean, *kamis*, leaf fern, leeks, lettuce, long green beans, mushrooms, nettles, okra, onions, parsley, pigeon peas, potatoes, pumpkins, purslane, radish, safflower, snow peas, spinach, sponge gourd, squash blossoms, winter and summer squashes, sugar palm shoot, swamp cabbage, sweet potatoes, taro leaves and roots, tomatoes, turnips, water chestnuts, watercress, yams	More green vegetables are consumed. More raw vegetables/salads are eaten.
Additional Foods			
Seasonings	Food is spicy, but the variety of spices used is limited. Regional cooking is differentiated in part by seasoning preferences.	*Atchuete* (annatto), *bagoong, baggong-alamang*, chile peppers, garlic, lemon grass, *patis*, seaweed, soy sauce, turmeric, vinegar	
Nuts/seeds		Betel nuts, cashews, *kaong* (palm seeds), peanuts, pili nuts	
Beverages		Soy milk, cocoa, coconut juice, coffee with milk, tea	Chocolate milk is substituted for soy milk. Soft drinks are popular.
Fats/oils	Traditional diet is considerably higher in fat than are other Asian cultures.	Coconut oil, lard, vegetable oil	
Sweeteners		Brown and white sugar, coconut, honey	

Rice is the foundation of the diet, and the long-grain variety accompanies the meal. It is typically steamed or fried (the preferred method of serving leftover rice). Garlic fried rice is a favorite, topped with bits of meat, sausages, and a fried egg. Vinegar, additional garlic, and a spicy vegetable/fruit pickle called atchara are added to taste. A common bread, pan de sal, is made from rice flour. Noodles are also used extensively. Pancit is a popular dish made with rice, wheat, or mung bean noodles mixed with cooked chicken, ham, shrimp, or pork in a soy and garlic-flavored sauce. Short-grain, glutinous rice is used for sweet desserts such as puto, a fluffy cake made from rice, sugar, and sometimes coconut milk.

The amount of meat, poultry, or fish a family eats depends on economic status. Pork, chicken, and fish are popular, added as available to mixed dishes such as sinigang, a soup of fish or meat cooked in water with sour fruits, tomatoes, and vegetables; puchero, a beef, chicken, sweet potato, tomato, and garbanzo bean stew with an eggplant sauce; gulay, fried fish with vegetables; and lumpia, the Filipino version of egg rolls, stuffed with pork, chili peppers, and vegetables like hearts of palm. Other traditional dishes popular for special occasions include chicken relleno, a whole chicken stuffed with boiled eggs, pork, sausage, and spices; paella, a Spanish recipe for saffron-flavored rice typically topped with chicken, sausage, pork, seafood, tomatoes, and peas; and lechon, a whole roasted pig.

Filipinos use all parts of the animal in their cooking; in addition to the pork meat, for example, the pig variety cuts might show up in various soups or mixed stews, such as dinuguan, consisting of diced pork, chicken, or entrails cooked in pig's blood and seasoned with vinegar and hot green chile peppers or sausage such as garlicky longaniza. The skin is commonly fried to make sitsaron (similar to the Mexican chicharrónes or U.S. cracklings), which are eaten as snacks or pulverized to top noodle dishes.

Due to the U.S. influence in the Philippines, many western dairy products are available, but cow's milk is used infrequently. Evaporated milk is a common ingredient in leche flan, a custard, and in halo-halo, a parfait-like dessert consisting of shaved ice, coconut milk, mung beans, purple yam pudding (ube), boiled palm seeds (kaong), corn kernels, pineapple jelly, and other ingredients. Halo-halo can be bought premixed with just the shaved ice needed for completion. Rural Filipinos use carabao (water buffalo) milk to make one of the few native cheeses in Asia, kesong puti. Carabao milk is also popular in desserts, such as ice cream, flan, and pastille candies.

Photo by Kathy Sucher

▲ *Two traditional Filipino dishes—lumpia (similar to an egg roll) and pancit (noodles cooked with meat or shrimp in a soy- and garlic-flavored sauce).*

A common seasoning, used instead of salt and found throughout Southeast Asia, is fermented fish paste or sauce. In the Philippines, the powerful paste is called *bagoong* and tastes somewhat like anchovies, although it can be made from a variety of fish. A similar paste made of shrimp is known as bagoong-alamang. Patis is a translucent amber fish sauce. To obtain the popular sour-cool taste, palm vinegar, or a paste made from either the cucumber-like vegetable called kamis or the pulp of the tamarind pod, may be used. Kinilaw, a Filipino specialty, uses sour ingredients to marinate and pickle raw foods, including fruits and vegetables, but also meats, organs, and seafood. Bagoong, patis, lime (calamansi) wedges, and vinegar flavored with chilies are frequently placed on the table so that each diner may add saltiness or sourness to taste.

In Filipino culture, sticky, glutinous rice cakes symbolize the cohesiveness of the family.

Among the more unusual Filipino specialties is balut, eaten occasionally as snacks. These partially developed duck eggs are soft-boiled and sold warm by street vendors. Salt and a little vinegar are added to the embryonic birds before they are popped whole into the mouth.

A principal food in many Pacific Islands is the coconut, and it is widely used in Filipino cooking. In addition, copra (dried coconut kernels used for oil extraction) is an important export crop. It takes approximately one year for a coconut to mature, but if picked at six months, the soft, jellylike coconut meat can be eaten with a spoon and is a popular delicacy. The coconut plant provides several food products, including beverages, cooking liquids, and even a vegetable. The sweet, clear liquid found in young coconuts is the juice or water. It is consumed fresh but is not used in

Durian is an acquired taste. Its odor, which has been likened to rotting onions mixed with gasoline, is so strong that some apartment buildings in Asia ban the fruit.

In rural regions, raw pork is heavily salted and then stored in jars for many months until it ripens. Called itog, small amounts are added to other dishes to enhance their flavor.

cooking. Coconut cream, which is used for cooking along with coconut milk, is the first liquid extracted from grated, mature coconut meat. After the cream is removed, coconut milk is made by adding water to the meat and then squeezing the mixture. Coconut milk is used primarily in special dishes. Coconut palm blossom sap can be fermented to produce a strong alcoholic drink called tuba, which, when distilled, is known as lambanog. Hearts of palm, sometimes called palmetto cabbage, is the firm, greenish inner core of the tree; it is used as a vegetable. Bananas, *durian* (a large, strong-smelling, sweet fruit with a creamy texture), jackfruit, mango, papaya, and pineapples are also popular.

Regional cooking styles are divided into four regions: Luzon (the largest group of islands, also home to the nation's capital, Manila), Bicolandia, the Viscayan Islands, and Mindanao.[31] Luzon is made up of various ethnic groups, and the cuisine has been strongly influenced by the Spanish. Ocean fish, such as prawns, milkfish (bangus), and halibut, as well as the ample use of anchovy sauce and shrimp paste, are preferred in the northern areas. Foods are typically boiled or steamed. Saluyot ("okra leaves"—not related to okra), a spinach-like green with a slippery texture when cooked, and drumstick plant leaves, called sili, are especially popular in the North.

Rice is grown in the central region of Luzon, known for its freshwater fish and richly sauced dishes flavored with onions and garlic. One delicacy is rellenong manok, a deboned chicken stuffed with sausage, vegetables, and ground pork mixed with raisins and spices, topped with a tangy red sauce. Stir-frying is the most common cooking technique. Coconut products and tropical fruits predominate. Sweetened rice dishes such as suman, a snack made from rice (cassava or plantains can be used) steamed in banana leaves or corn husks, are a specialty.

Bicolandia is an ethnically homogeneous region that came in contact with both Malaysian and Polynesian cooking. Foods are spicy hot with chili peppers, balanced by copious use of coconut milk and cream. Taro leaves cooked in coconut milk with ginger and chiles are one example of the unique blend of foods found in this area. Viscayan Islands fare also reflects its heritage—abundant use of seafood (including a distinctive fermented shrimp paste called guinamos) and seaweed, as well as many dessert specialties, such as candies and pastries developed due to the sugarcane plantations in the area. The Mindanao region was heavily influenced by the Indonesians and Malaysians. The many ethnic groups living there are predominantly Muslim, so little pork is consumed (see Chapter 4). Sauces made from peanuts and chiles are popular, as are curries and other spicy dishes, such as piarun (fish spiced with chiles) and tiola sapi (boiled beef curry).

[SAMPLE MENU]

A Filipino Lunch

***Lumpia* (Egg Rolls)[a,b,c]**

***Sinigang* (Pork in Sour Broth)[a,c]**

Chicken *Adobo*[a,b,c]

Steamed Rice

***Pansit Canton* (Sautéed Rice Noodles)[a,c]**

***Halo-halo* (Shaved Ice Dessert)[a,b] or *Puto* (Steamed Rice Cakes)[a,b]**

[a]Gelle, G.G. 1997. *Filipino cuisine: Recipes from the Islands.* Santa Fe: Red Crane Books.

[b]*Filipino Recipe* at www.filipinorecipe.com

[c]*Filipino Recipes* at http://www.filipinofoodrecipes.net/

Vietnamese, Cambodian, and Laotian. Ingredients are similar in all the mainland Southeast Asian countries, but recipes and meal patterns vary. Indigenous fish and seafood, tropical fruits and vegetables, and glutinous rice were the foundation of the native diets. The Chinese introduced long-grain rice, soy products, stir-frying, hot pots, fried pastries, and chopsticks to areas they ruled. French regional occupation popularized such items as French bread, meat pâtés, asparagus, potatoes, pastries, and strong coffee.

EXPLORING GLOBAL CUISINE—The Cooking of Malaysia, Singapore, and Indonesia

Malaysia, which includes a western section contiguous with Thailand and an eastern section on the island of Borneo, extends south into the gap between Southeast Asia, the Philippines, and Australia (see Figure 12.1). At its tip is the independent city-nation of Singapore. Indonesia, comprised of over 10,000 islands (mostly uninhabited), arches eastward from Malaysia and includes Bali, Borneo (now known as Kalimantan), Java, New Guinea, and Sumatra. The region lies along the equator and contains a majority of the world's tropical rainforests. The fertile land is conducive to the cultivation of the herbs and spices brought by Asian, Middle Eastern, and European traders, including chile peppers, cinnamon, cloves, cumin, ginger, nutmeg, and pepper. Parts of Indonesia are still known as the Spice Islands.

The cuisines of Malaysia, Singapore, and Indonesia have been greatly influenced by the diversity of their populations: native Malays, Chinese, Asian Indians, Pakistanis, Arabs, Thais, Eurasians, Melaka Portuguese (Malaysian Portuguese), and Peranakan (southern Chinese and Malaysian or Indonesian—the women are called Nyonya and men are known as Baba).[171] Further, the numerous religious practices of the region—Islam, Hinduism, Buddhism, Christianity, and Judaism—have played a role in the fare. Today, Malaysia and Indonesia are predominantly Muslim nations (with the exception of Bali, which is mostly Hindu), while Singapore is primarily Christian.

Rice, both long-grain and glutinous, is the foundation of the diet. It is often steamed, but it is also popular fried, prepared as sticky rice balls, and especially as noodles. Noodles are typically stir-fried, added to soups, or topped with mixed vegetables, fish, or meat. They are eaten at nearly all meals, and often for snacks. One Indonesian favorite is nasi goring, Chinese fried rice topped with a European-introduced fried egg. Steamed rice in Malaysia is often served with both an Indian-influenced curried meat or fish and a Chinese stir-fried vegetable. Fish is very common (often fried) and eaten by all groups except some vegetarian Buddhists (who prefer tofu- or tempeh-based dishes). Beef and poultry are popular but costly in much of Malaysia and Indonesia, while pork is uncommon in majority Muslim areas. The exception is Singapore, a wealthier nation, which includes abundant meat and egg dishes in its fare. Temperate vegetables introduced from the Middle East and Europe, such as tomatoes, eggplants, and potatoes, are usually added to soups and rice dishes instead of being served separately. Salads are popular, however; an example is Indonesian gado gado, a mixture of cooked vegetables (including cabbage, green beans, and carrots) dressed with a peanut sauce. Tropical fruits are eaten at nearly every meal, fresh, preserved, in baked goods and puddings, and deep-fried as fritters.

Coconut flavors many foods, and seasonings are used liberally, including chiles, fresh coriander, ginger, lemongrass, pandanus leaf, pepper, and turmeric. Lemons, limes, unripe mangoes, tamarind, or vinegar is usually added for a sour taste. The most distinctive cooking in the region is Nyonya fare, found especially in Singapore, which combines Chinese preparations (often pork-based) with Malaysian seasonings, particularly coconut, turmeric, and lemongrass.

All courses are served at once in Malaysian, Singaporean, and Indonesian meals, and the dishes are categorized by preparation technique, not ingredients.[84] For example, sambals are fried dishes seasoned with chiles (or the name for just a chili dipping sauce); satays are delicate, grilled kebabs (of Middle Eastern origin) served with spicy dipping sauces; croquettes (from the Dutch influence) are fried rice, meat, vegetable; or fruit fritters; and sayur are soupy dishes with ample sauce for dipping rice balls. In most areas forks and spoons are used (knives are rarely available). The Chinese often use chopsticks, and Asian Indians frequently employ their right hand to scoop up food.[171] One tradition is universal in the region: street vendors. Meals and snacks are usually available around the clock at food stalls and small eateries, and al fresco dining is a daily event.

Indian and Malaysian influence is seen in the curries and coconut milk-flavored dishes of eastern and southern Southeast Asia. The common foods of mainland Southeast Asians are listed in Table 12.2.

Rice, both long- and short-grain, is the staple of the diet. Rice products, such as noodles, paper, and flour, are used extensively. In Vietnam, rice paper is used as egg roll or wonton wrappers. In the dish cha gio, the moistened paper is wrapped around a variety of meats, fish, vegetables, and herbs and then deep-fried. Often the rice paper is filled with meat, fresh herbs, and vegetables at the table. Dried rice noodles (sticks) are called *pho*, which is also the name of the popular noodle-based soup in Vietnam. In Laos, the sticky, glutinous short-grain rice is more prevalent than long-grain types (traditionally formed into small balls to use as scoops for other foods), and the very thin Chinese-style rice noodles are common. Wheat is used to make French bread, noodles, and some pastries.

Fried noodles topped with meats and vegetables are a favorite. Fish and shellfish are the predominant protein food on the mainland. Even landlocked Laos depends on freshwater

Table 12.2 Cultural Food Groups: Mainland Southeast Asian

Group	Comments	Common Foods	Adaptations in the United States
Protein Foods			
Milk/milk products	Most Southeast Asians do not drink milk and may be lactose intolerant. Sweetened condensed milk is used in coffee; whipping cream is used in pastries.	Sweetened condensed milk, whipping cream	It is expected that younger Southeast Asians will increase their use of dairy products. Ice cream is popular; milk and cheese are often disliked.
Meat/poultry/fish/eggs/legumes	The traditional Southeast Asian diet is low in protein. Fish, poultry, and pork are common; most parts of the animal are used (brains, heart, lungs, spleen).	*Meat:* beef, lamb, pork, goat, venison; variety meats of all types *Poultry and small birds:* chicken, duck, quail, pigeon, sparrow, doves *Eggs:* chicken, duck (both embryonic and unfertilized), fish *Fish and shellfish:* almost all varieties of fresh- and saltwater seafood, fresh and dried *Legumes:* chickpeas, lentils, mung beans (black and red), soybeans and soybean products (tempeh, tofu, soy milk), winged beans	Meat, lamb, and eggs are eaten more; fish, shellfish, and duck are eaten less because of price.
Cereals/Grains	Rice is the staple grain and is usually eaten with every meal. French bread is commonly eaten.	Rice (long- and short-grain, sticks, noodles), wheat (French bread, cakes, pastries)	Intake of baked goods increases.
Fruits/Vegetables	Hearty garnishes of fresh vegetables are commonly added to dishes. The Vietnamese eat a considerable amount of fruit and vegetables, fresh and cooked. Fruit is often eaten for dessert or as a snack.	*Fruits:* apples, bananas, cantaloupe, coconut, custard apple, dates, durian, figs, grapefruit, guava, jackfruit, jujube, lemon, lime, litchi, longans, mandarin orange, mango, orange, papaya, peach, pear, persimmon, pineapple, plum, pomegranates, pomelo, raisins, rambutan, roselle, sapodilla, star fruit, soursop, strawberries, tamarind, watermelon *Vegetables:* amaranth, arrowroot, artichokes, asparagus, bamboo shoots, banana leaves and flowers, betel leaves, beans (yard-long and string), bitter melon, breadfruit, broccoli (Chinese and domestic), cabbage (domestic, Chinese, savoy, napa), calabash, carrot, cassava (tapioca), cauliflower, celery (domestic and Chinese), chayote squash, Chinese chard, Chinese radish (*daikon*), chrysanthemum, corn, cucumber, eggplant (domestic and Thai), leeks, lotus root, luffa, matrimony vine, mushrooms (many varieties), mustard (Chinese greens), okra (domestic, lady finger), peas, peppers, potato, pumpkin (flowers, leaves), spinach, squash, sweet potatoes (tubers, leaves), taro (root, stalk, leaf, shoots), tomatoes, turnips, water lily greens, water chestnuts, water convolvulus, was gourd, yams	Use of fruits and vegetables is dependent on availability and price. It is expected that use of fruits and vegetables will decline. Fresh vegetables and herbs are sometimes grown in backyard gardens.
Additional Foods			
Seasonings	Fermented fish sauce, as well as soy sauce, is often used. Fresh herbs are very popular garnishes in Vietnamese dishes; typical Cambodian fare is delicately seasoned; Thai dishes are frequently very hot and spicy, with several types of curry and chile peppers especially popular.	Allspice, alum, basil, black pepper, borax, cayenne pepper, chile pepper, chives, cinnamon, coconut milk, fresh coriander, curry powder, fennel, galanga, garlic, ginger, kaffir lime leaves, lemon grass, lemon juice, lily flowers, lotus seed, mint, MSG, *nuoc mam* (and other fermented fish sauces and pastes), paprika, saffron, star anise, tamarind juice, vinegar	
Nuts/seeds		Almonds, betel nuts, cashews, chestnuts, macadamia nuts, peanuts, pili nuts, walnuts; locust seeds, lotus seeds, pumpkin seeds, sesame seeds, watermelon seeds	Peanut butter is often disliked.
Beverages	Beverages are usually drunk after the meal or with snacks or desserts.	Coffee, tea, sweetened soybean milk, a wide variety of fruit and bean drinks, hot water, hot soup, beer	Carbonated drinks have increased in use.
Fats/oils		Bacon, butter, lard, margarine, peanut oil, vegetable oil	The Vietnamese have increased their use of butter and margarine.
Sweeteners	Sweets are luxury foods.	Cane sugar, candy	The use of sweetened products has risen in the United States.

varieties. Fish, shrimp, and squid are often preserved through salting and drying. Poultry is widely available, and pork or goat is eaten in wealthier areas. Beef is used occasionally. Religious prohibitions often influence which meats are consumed. Like other Asians, the people of mainland Southeast Asia do not use appreciable amounts of dairy products. However, soy milk is a common beverage. Soy products, particularly a chewier version of tofu (soybean curd) called *tempeh*, are common.

Mainland Southeast Asians frequently consume vegetables, cooked in stir-fries and stews or uncooked in salads and pickles. Especially noteworthy are the many shredded vegetables and unripe fruits, such as cabbage, papaya, carrots, cucumber, radishes, jicama, and bean sprouts topped with fish, poultry, meat, or peanuts and spicy hot dressing. One example is goi go, a Vietnamese specialty featuring cabbage and chicken. Greens and leaves are often used to wrap foods, such as collard greens for Cambodian steamed fish and la lot (betel leaves) used for Vietnamese spring rolls stuffed with minced beef. Further, fresh herbs and spices, including basil, coriander leaves, chile peppers, galangal (similar to young ginger root), garlic, ginger, kaffir lime leaves, lemon grass, and mint, are often added to foods as they are served, providing distinctive flavors and color to many dishes. Both Laotian laap, a spicy ground meat or fish dish (traditionally prepared uncooked), and Vietnamese grilled lemon grass beef, bo nuong xa, are served with a substantial garnish of basil, mint, and coriander leaves. Due to the strong influence of the French, the Vietnamese also frequently eat asparagus, green beans (haricots), and potatoes; though the French had less impact on the cuisine of surrounding nations, subsequent Vietnamese rule has popularized these vegetables in other regions as well.

Tropical fruits are available, although in some areas bananas and plantains are the only fruit widely consumed. Banana leaves are used to wrap rice, vegetables, and meats for steaming in both Cambodia and Laos. Pineapple, papaya, limes, mangoes, and mangosteens are common, as are soursop, star fruit, guavas, custard apples, durian, jackfruit, and tamarind (a pod with very tart pulp). Oranges, lemons, melon, and sugarcane are also popular.

In Vietnam, foods are customarily seasoned with a salty sauce made from fermented fish called nuoc mam. It can be transformed into a hot sauce, nuoc cham, with the addition of chiles, vinegar, sugar, garlic, and citrus fruit juice. In Cambodia the fermented fish sauce is called tuk-trey; a stronger fish paste is also used, known as prahoc. The Laotian version of fish sauce is nam pa; pa dek is the fermented fish paste.

Tea is the preferred beverage throughout mainland Southeast Asia. In Vietnam, it is served before and after meals but not during the meal. Tea is often blended with flowers such as rose petals, jasmine blossoms, chrysanthemums, and lotus blossoms (which are especially popular). Coffee is popular in French-influenced areas, usually served with large amounts of sweetened condensed milk added to it. Broth is traditionally consumed at meals, and in poorer, rural regions, such as where the Hmong live, it is the only beverage besides water that is commonly available. In wealthier areas, men may drink beer, and women and children consume soft drinks during meals. Soybean drinks and fruit drinks are common; rice wine or whiskey is served at special occasions.

Regional variations are prominent, especially in Vietnam.[32] The Chinese influence is stronger in the North than in other regions. Hot pots,

© PhotoLink/Photodisc/Getty Images

◀ *Shoppers choose from a wide selection of tropical fruits and vegetables in a Southeast Asian street market.*

In Vietnam, *com*, meaning "cooked rice," is the same word used for "food."

Vietnamese Buddhists eat soybean products on the first, fifteenth, and last day of the lunar month, when meat is prohibited.

Furr, a soup containing pork, noodles, garlic, and hemp (marijuana) leaves, is a Laotian specialty.

"Without fish sauce or salt, life is nothing," according to a Vietnamese saying.

Numerous small wild animals are consumed in Southeast Asia. In 2005 a new species of rodent, related to the porcupine, was discovered by scientists for sale in a Laos market. Called *kha-nyou,* it is roasted whole, then eaten by crunching up the small bones, and spitting out the larger ones.[172]

stir-fried foods, and chao (rice gruel similar to congee) are especially popular. Soups are a specialty, particularly pho bo ha noi, a delicate broth to which rice noodles, sliced beef, bean sprouts, herbs, and other seasonings are added immediately before serving. Mein go is a chicken noodle soup served in a similar manner. Other favorites include stuffed tofu, bun cha (grilled pork over noodles), and snails (stir-fried, simmered in beer, or minced with garlic). The central region is known for sophisticated gastronomy. Presentation is emphasized and seasonal cooking reigns. Specialties include a sauce similar to nuoc mam made from shrimp called mam tom, shrimp pâté grilled on sugar cane, spicy pork sausages, sweet soups, vermicelli soups, and both sweet and salty rice cakes. The climate of the South is tropical. Cooking is simpler and seasoning is stronger; curries and spicy Indonesian-style peanut sauces are favorites. Coconut milk and caramel flavor many dishes. Clay pot cooking is common. One specialty is tidbits of grilled meats, fresh vegetables, and fruits such as guava, mango, green papaya, pineapple, or starfruit wrapped in a lettuce leaf, according to personal preference, that can be dipped in salty or spicy sauces. Sweets are more popular in the South than in other areas.

[SAMPLE MENU]

A Vietnamese Dinner

Asparagus and Crabmeat Soup[a,b]

Braised Bean Curd[a,b] or Grilled Beef with Lemongrass[a,b]

Stir-Fried Vegetables

Steamed Rice

Fruit Juice or Iced Coffee with Evaporated Milk

[a]Routhier, N. 1999. *The foods of Vietnam*. New York: Stewart, Tabori & Chang.

[b]*Vietnamese Recipes & Cuisines* at www.vietnamese-recipes.com

Khmer cooking of Cambodia features northern Indian, Malaysian, and Chinese elements. Although French cooking is much admired, it has never been integrated into the Khmer kitchen.[33] Aromatic seasonings are preferred, particularly the paste known as kroeung, made fresh for each dish from pulverized herbs and spices such as galangal, garlic, kaffir lime leaves, lemon grass, shallots, and turmeric. A touch of spice is achieved with chili peppers, particularly in curried dishes, though it is usually moderated by the use of coconut milk. Sweet ingredients, such as ripe fruit or sugar, are often included as a contrast to the sour flavor provided by vinegar, lime juice, or tamarind. Salty fish sauce or soy sauce is always added, as are bitter herbs for balance. Amok, fish in coconut milk steamed in a banana or collard leaf, is a national favorite as is num banh choc, a rice noodle and fish soup. In some areas, dishes featuring wild foods such as land crabs, snakes, and locusts are found.

Laotians prefer glutinous rice over long-grain types. Added vegetables and fish make up the basic diet, with eggs, poultry, and beef included as affordable. In rural areas game such as deer, squirrels, ducks, quail, lizards, frogs, snakes, and grasshoppers are common. Meats are frequently stewed or grilled, though a salty beef jerky prepared with *nam pa* is a specialty. Coconut cream or milk, *nam pa* and *pa dek,* lime juice, fresh coriander leaves, garlic, lemon grass, and mint are typical seasonings. Hot chili peppers add heat to most foods, though the extent of their use varies regionally. Spicy salads consisting of fresh vegetables or shredded immature papaya topped with lime juice, palm sugar, and chile pepper dressing are popular. Chinese and French influence via Vietnam is seen in some areas where French bread, croissants, spring rolls, and a soup similar to pho are popular foods.

Hmong fare traditionally differs from Laotian cooking and shares some similarities with Vietnamese cuisine. Long-grain rice is favored and stir-frying, steaming, and roasting are common preparation methods. Although rice and vegetables are the foundation of the diet, families sometimes raise chickens, ducks, pigeons, and pigs. These foods are supplemented with wild game and fish, crabs, and snails. Seasonings, however, are similar to Laotian, though the Hmong use soy sauce in addition to fish sauces. Hmong who were forced from the mountains to the lowlands during Southeast Asian conflicts have added many Laotian foods to their meals.

Meal Composition and Cycle

Daily Pattern

Filipino. The traditional meal pattern in the Philippines is three meals a day. Breakfast is garlic fried rice with eggs or broiled fish, sausage, or meat, plus coffee or hot chocolate; bread may be substituted for rice. Especially popular are sweet, cheesy rolls called ensaymada. Lunch and dinner are similar in size and composition. Both are often large meals, characteristically including soup, rice, a crispy or chewy dish (such as fried fish), a salty dish (meat or poultry cooked in fish sauce or soy sauce), a sour dish (flavored predominantly with vinegar or tamarind), a noodle dish, and often, an adobo dish. Fresh fruit or dessert concludes the meal. If the meal features mostly Spanish-style items, the courses are served consecutively. If the meal features more Filipino-style dishes, all courses are served together, including dessert.[30]

In addition to meals, two snacks, called meriendas, are consumed in the midmorning and late afternoon. Meriendas may be small or may consist of substantial amounts of food, such as fritters, pastries, fruits, ensaymadas, lumpia, or almost anything else except rice, which is served only at meals.

Vietnamese and Other Mainland Southeast Asians. Mainland Southeast Asians eat two or three meals a day with the number of meals and the amount of food consumed often based on income. Snacking is uncommon. Southeast Asians do not usually associate particular foods with breakfast, lunch, or dinner. For example, soups are especially popular and are often consumed with every meal.

In Vietnam, a traditional breakfast is large and may consist of soup with rice noodles topped with meat or poultry; a boiled egg with meat and pickled vegetables on French bread; chao with bits of leftover meat and vegetables; steamed rice cakes or Chinese-style crullers; or glutinous rice or boiled sweet potatoes with sugar, coconut, and chopped roasted peanuts. A strong cup of coffee may accompany the meal. Lunch and dinner typically include rice, fish or meat, a vegetable dish, and a broth with vegetables or meat. Fresh vegetables and pickled garnishes are served with the meal. All items are served at once, and individual diners place whatever foods they wish over their portion of rice and flavor it as desired with nuoc mam and other condiments. French bread with meat or shrimp pâté may be substituted for a lunch or dinner meal. In late afternoon, tea or coffee may be enjoyed with a sweet custard, pastry, candy, or piece of fruit.

Cambodians also eat family style. Soups are often served for breakfast and accompany the main course at nearly every other meal. Steamed or fried rice or rice noodles are the centerpiece of lunch and dinner, accompanied by grilled or steamed freshwater fish and seafood, and less frequently poultry, pork, or beef. Fresh salads are common. Fruit is often eaten as dessert, though very sweet rice or corn dishes are also popular. Tea and coffee with condensed milk are usually consumed with the meal, and fruit juices, soft drinks, and beer may be available. In Laos glutinous rice, fish, poultry or meat, soup, and a cooked vegetable dish or fresh salads make up most meals. Chili pepper paste is the standard condiment. Tea, coffee with condensed milk, and rice wine or rice whiskey round out the menu.

Turo-turo are fast-food stands in the Philippines, specializing in rice bowls topped with foods according to the customer's preference. Other hot items are also available.

Etiquette. Filipinos generally dine at tables equipped with lazy Susan turntables so that dishes are accessible to everyone. Traditionally, no one starts eating until the eldest male at the table begins. Many Filipinos use a western style of dining with forks, knives, and spoons. Others employ just forks and spoons. The spoon is used to hold the food down while the fork is used to pull bits away. The food is then pushed onto the spoon with the fork and eaten. Chopsticks may be used for Chinese dishes. In some rural areas, fingers are still more commonly used.[30] In such cases, only the right hand is used for dining. Small mounds of rice are rolled between the index finger, middle finger, and thumb to form a ball that is dipped into a sauce, then pressed into a bit of meat or poultry and popped whole into the mouth. It is considered rude to take the last bits of food from the central platter.[34]

In Vietnam it is polite to wait for the eldest person to be served and then, after everyone else is served, to ask him or her if it is okay to eat. It is a breach of good manners to refuse any offer of food, yet when served, only small amounts of any single dish should be taken. If sufficient amounts remain, seconds will be offered.[34] Throughout mainland Southeast Asia, an empty plate or cup indicates that the diner is still hungry or thirsty. Leaving a small amount of food or beverage signals satiety.

Traditional dining in Vietnam is done on a low table with family gathered round, sitting cross-legged on mats. Both hands customarily rest on the table while dining, and conversation is limited. In contrast, dinner is a time for socializing in Laos. The food is served on a low rattan tray, and women gather on one side, the men on the other side. Each diner eats from the dishes as desired.

It is said that you can trace the penetration of Chinese rule in Vietnam by the areas in which chopsticks are common, compared to those where hands are still most frequently used to dine.[29]

A variety of utensils are used to eat in Southeast Asia mainland nations. Chopsticks are used for most dishes in most of Vietnam, though spoons and fingers are considered appropriate for certain foods and in some areas. Rural Laotians often eat with their fingers, using balls of sticky rice to scoop up fish and meats and vegetables and sauce; however, spoons are used as needed. In urban areas forks are now common. Hmong typically employ forks and spoons, and Cambodians use spoons, chopsticks, or fingers dependent on the food.

Special Occasions

Hospitality is very important to the Filipinos, and food gifts to express love or appreciation are common.

Filipino. In the predominantly Catholic Philippines, religious festivals and saints' days are numerous (see Chapter 4). On all special occasions, it is customary to serve plenty of food buffet-style with a roasted pig (lechon) as the centerpiece. The Filipinos claim to have the longest Christmas season in the world, from the first Sunday of Advent in late November or early December to January 6. The midnight Mass celebrated on Christmas Eve is usually followed by the traditional *media noche,* a midnight supper of fiesta foods such as roast ham, sweet potatoes, banana flower salad, niaga—a dish made of boiled meat, onions, and vegetables whose name means "good life"—and hot chocolate. Other specialties eaten during the Christmas season are puto bumbong, a rice flour delicacy cooked in a whistling bamboo kettle, and bibingka, a glutinous rice cake cooked in a clay pan topped with salted egg slices, kesong puti cheese, and a bit of coconut.

A midnight Mass is also held on New Year's Eve, but many Filipinos attend parties to celebrate the holiday instead. Again, a midnight supper consisting of fiesta foods is traditional. There is also a superstition that eating seven grapes in succession as the clock strikes midnight will bring good luck in the coming year. For birthdays, pancit is eaten to ensure a long life.

There are numerous Filipino practices and customs associated with Easter, beginning with observances on Ash Wednesday. Late on Easter Eve, young children are awakened to partake of special meat dishes, such as adobo and dinuguan, in the belief that if they do not do so, they will become deaf. In May, fiestas honoring the Virgin Mary often include family feasts.

Vietnamese and Other Mainland Southeast Asians. Of all Vietnamese holidays, *Tet*, the New Year's celebration, is the most important. *Tet* is observed at the end of the lunar year (end of January or beginning of February) just after the rice harvest. In Vietnam, the first *Tet* ritual is an observance at the family gravesites. Offerings of cake, chicken, tea, rice, and alcohol, as well as money, are made at the graves, and then the family picnics on the offerings.

The second ritual, held on the twenty-third day of the twelfth lunar month, is to celebrate the departure of the Spirit of the Hearth, Ong Tao. He is represented by three stones on which the cooking pots are placed and is honored by a small altar. Like the Chinese Kitchen God, Ong Tao returns to the celestial realm each year and reports on the family's behavior. After the family makes an offering to symbolize his departure, they share a feast including glutinous rice cakes and a very sweet soybean soup. One week later the family celebrates Ong Tao's return to their hearth. The following day is the first day of *Tet*. Guests (especially those with favorable names, such as Tho, meaning "longevity") are entertained with tea, rice alcohol, red-dyed watermelon seeds, candied fruits, and vegetables.

Special dishes prepared for the week-long celebration include banh chung, glutinous rice cakes filled with meat and beans and boiled in banana leaves, squid soup, stir-fried young seasonal vegetables, pork with lotus root, and sometimes a special shark fin soup.

Many Vietnamese, including those in the United States, celebrate the Buddhist holiday called *Trung Nguyen*, or Wandering Souls Day. It occurs in the middle of the seventh lunar month and is celebrated with a large banquet prepared in honor of the lost souls of ancestors.[3] Traditionally, the Vietnamese did not commemorate birthdays but rather honored their ancestors on the anniversary of their death with a special celebration and meal. In the United States, it is now more common to celebrate birthdays.

The largest holiday of the year in Cambodia is also the New Year's day celebration, *Chaul Chnam,* which begins on April 13 and lasts for three days. Prayers and special foods like fried coconut and fried bananas rolled in coconut are offered to the New Year Angel, who descends with either blessings or ill will. The Water Festival, held in November after the seasonal rains have ended, features colorful floats in local rivers.

Most Laotian holidays are religious in origin and are celebrated at local temples. *Pha Vet,* which occurs in the fourth lunar month, commemorates the life of Buddha. *Boon Bang Fay,* held in the sixth lunar month, also honors the Buddha with a fireworks display. Among the Hmong and other Laotians, the New Year's celebration is a major event. It begins with the first crow of a rooster on the first day of the new moon in the twelfth lunar month, usually in December. The highlight of the festivities is the world renewal ritual, which involves an elder who chants while holding a live chicken. He circles a tree three times clockwise to remove the accumulated evil of the previous year and then circles the tree three more times counterclockwise to invoke good fortune in the upcoming year. The bad luck collects in the blood of the chicken, which is traditionally taken to a remote location and slaughtered. Customarily considered a good time to meet future wives and husbands, New Year's was the one time each year when Hmong from different clans celebrated together.

Therapeutic Uses of Food

Filipino. When the Spanish came to the Philippines, they introduced the Mexican hot-cold theory of health and diet (see Chapter 9). Foods are classified as being hot or cold based on their innate qualities or their effect on the body, not on their spiciness or temperature. Although the classification of certain foods varies regionally, avocados, alcoholic beverages, coconuts, nuts, legumes, spices, chili peppers, and fatty meats are generally considered hot items; tropical fruits, vegetables, milk and dairy foods, eggs, fish, and lean or inexpensive meats are regarded as cold.[35,17] A balance is attempted at meals between hot and cold elements. The reason Filipino dishes contain so many ingredients may be to ensure this balance.

Some illnesses are characterized as hot or cold and are treated with foods of the opposite category. Diarrhea and fevers are hot; colds and chills are cold. Other food beliefs are based on sympathetic qualities ("like causes like"); for instance, pregnant women may avoid dark foods to prevent their babies' skin from being too dark. Sometimes the meaning behind a therapeutic food use is more obscure; horseradish leaves and broth seasoned with ginger are believed to promote milk production in nursing mothers, and fish heads and onions are considered brain food by some Filipinos. Honey, as well as certain herbs such as thyme, marjoram, and chamomile, is used to treat diabetes. Licorice root is considered a general tonic, especially beneficial during times of stress. Some elderly Filipinos have adopted the Asian Indian practice of chewing areca nuts (also called betel nuts), which is believed to prevent tooth decay, although it leaves permanent stains.

The Department of Health in the Philippines has approved several herbal remedies as safe and effective, including *ampalaya* (bitter melon, prepared as a side dish or as a juice) for diabetes, *bawang* (garlic) to lower blood cholesterol levels and reduce blood pressure, *ulasaming bato* (pepperomia, which is eaten as a salad or brewed into tea) for arthritis and gout, and *sambong* (an indigenous herb) as a diuretic.[36]

Vietnamese and Other Mainland Southeast Asians. Many Vietnamese follow the Chinese yin-yang theory of health and diet (see Chapter 11, Chinese Therapeutic Uses of Foods section). Yin is known as *âm* and yang is called *duong*. As in the hot-cold system, classification is based on intrinsic characteristics rather than temperature or spiciness. Examples of *duong* (hot) foods are red meat, unripe fruit, ginger, garlic, coffee, and alcoholic beverages. *Âm* (cold) items include noodles, bananas, oranges, gelatin, and ice cream.[37,19] Some foods, such as rice, pork, eggs, chicken broth, teas, and sweets, are classified as neutral.[38] Not only must a balance be maintained within a meal, but extremes are also avoided during certain conditions, such as pregnancy. As with Filipinos and other Asians, illnesses are defined as *âm* and *duong* and are

A feast is held by the Hmong following the birth of a child. Included are two chickens representing the parents, a boiled egg signifying the child, and a small lit candle symbolic of the ancestor spirits whose blessing and protection are sought.

Licorice root, which contains glycyrrhizin, can cause fluid retention and increase blood pressure if consumed in large amounts.

sometimes caused by eating too many *âm* or too many *duong* foods. During pregnancy, which is duong, hot foods are avoided, and equilibrium is restored by eating foods of the opposite type; during the postpartum period, which is *âm*, cold foods are avoided. Yin and yang concepts are less prevalent among Cambodians and Laotians, although some hot and cold beliefs exist regarding specific foods and certain conditions.

The Chinese medical system details other influential elements in health, including the five flavors of sour, bitter, sweet, pungent, and salty; these tastes are harmonized in many Vietnamese dishes. Vietnamese believe that ingestion of specific organ meats will benefit the like internal organs. For example, consumption of liver will produce a stronger liver. Some Vietnamese believe that eating gelatinous tiger bones (produced by prolonged cooking) will make them strong. Concurrently, some foods may be injurious because they resemble certain disorders. Pregnant women may refuse to eat ginger because the multi-lobed root is thought to cause polydactyly (too many digits) in babies.

Though overall consumption of fish by Vietnamese Americans is thought to be lower than in Vietnam, data show they have the highest intakes of recent Asian immigrants, particularly of shellfish.[52]

The therapeutic value of some foods is unrelated to yin or yang or how they look. Some Vietnamese eat chile peppers to get rid of worms, or noodles with roasted rice paper and shrimp sauce for curing the flu. Cambodians may drink water with bitter melon for fevers. Vietnamese women may consume large amounts of salty foods during pregnancy,[39] and mothers may avoid feeding chicken or duck to their babies to prevent them from becoming deaf or mute.[40] Hmong women eat a diet of rice, chicken broth, black pepper, and herbs for a month after giving birth,[41] and some clans have specific taboos against eating certain foods, such as heart.[42]

Contemporary Food Habits in the United States

Adaptations of Food Habits

Filipino. Little current information on the food habits of Filipino Americans has been reported. Most are able to obtain traditional foodstuffs without much difficulty, although some of the familiar tropical fresh fruits and vegetables are not available. Older data suggest that most Filipinos still eat rice every day but not with every meal, and their diets tend to contain a greater variety of foods, especially more milk, green vegetables, meat, and sweets than they did in the Philippines.[43,44] Meriendas are not eaten as often as in the Philippines.

Filipinos born in the United States frequently consume a typically American diet. Breakfast consists of cereal, toast, eggs or meat, juice, and coffee; sandwiches, salads, and sodas are common at lunch; and dinner is usually a meat or fish dish served with rice or potatoes, followed by dessert. Traditional Filipino items may appear at some meals, such as eating longaniza sausage at breakfast or eating halo-halo (sometimes topped with vanilla ice cream) for dessert.[45]

Vietnamese and Other Mainland Southeast Asians. A study conducted in Washington, DC, found that 30 percent of the Vietnamese households surveyed had changed their eating habits since coming to the United States.[46] Although most continued to eat rice at least once a day, they ate more bread or instant noodles at lunch and more cereal at breakfast. Respondents also reported consuming more meat and poultry and less fish and shellfish than in Vietnam, mainly because of cost. Pork and pork products were still preferred to beef. They also reported consuming fewer bananas and more oranges, fruit juices, and soft drinks. These findings are similar to those reported in a survey conducted in the 1970s among Vietnamese living in northern Florida.[47]

More than 90 percent of Vietnamese-American adolescents in another study were found to prefer their native diet, although a majority listed items such as steak and ice cream as being among their favorite foods. Soft drinks and milk were well liked; cheese and peanut butter were strongly disliked. Only a small percentage of the teens snacked regularly.[48] A more recent study on the same population reported that Vietnamese-American high school students in Massachusetts consumed more fruits and vegetables than did other ethnic groups, and fewer dairy products. Over 28 percent ate at least five fruits or vegetables each day; however, only 8.5 percent consumed the recommended number of dairy servings. Notably, fruit and vegetable consumption increased with degree of acculturation, but decreased with age.[49]

A survey of Cambodian and Hmong families indicated some similar trends. Although traditional items were preferred by the adults, both American and native foods were acceptable to

EXPLORING GLOBAL CUISINE—Thai Fare

There are approximately sixteen times more Filipino Americans and ten times more Vietnamese Americans than there are Thai Americans in the United States. Yet Thai cooking is more familiar to the general population than either Filipino or Vietnamese fare. Thai restaurants have introduced the distinctive cuisine in many parts of the United States, even where few Thai Americans live, and dozens of cookbooks have further popularized the cuisine.

The country of Thailand is located on the southern end of the archipelago that is Southeast Asia. The hot, monsoonal climate is ideal for rice cultivation. Long-grain rice is the foundation of the diet, though short-grain glutinous rice is used for snacks and desserts and is preferred in the regional Issan cuisine of the Northeast (similar to Cambodian fare, also known for its culinary use of insects).[173] Noodles made of rice, wheat, or mung beans are also common. Both tropical and temperate fruits and vegetables are prominent in the cuisine. Seafood from the lengthy coast, especially shrimp, is popular. Dried herring-like fish (which are sometimes smoked as well) are often flaked into rice for added flavor. Beef, chicken, and pork are common. Duck is a favorite.

Thai food differs from that of its Southeast Asian neighbors because of its flavors. It is one of the hottest cuisines in the world, with lavish use of chili peppers. Several varieties of basil, fresh coriander leaves and root, galangal, garlic, ginger root, kaffir lime leaves, lemon grass, mint, and tamarind are typical seasonings. In addition, curried dishes are eaten daily. There are three types of curry sauces: yellow, which are smooth, mild, Indian-like sauces that include spices such as cardamom and turmeric; red, which are chunkier, hotter, and typically include ample fresh red chiles and coconut milk; and green, which are prepared with fresh green chiles whose heat is excruciating for all but the most experienced palates. Fermented fish products, such as nam pla (similar to the Vietnamese sauce called nuoc mam) and kapi (a paste made from fish or shrimp), are added to most dishes. Nam prik, a sauce that combines nam pla or kapi with other ingredients like garlic, ipeppers, shallots, lime juice, tamarind, palm sugar, and peanuts, complements dishes such as yam (fresh vegetables rolled up into a leafy package and dipped into the nam prik), salads (nam prik is the dressing), noodle dishes, dumplings, fried or grilled foods, and highly spiced raw pork called nam. Noodle dishes are usually eaten for breakfast and lunch. Phad Thai—stir-fried noodles cooked with bits of meat, seafood, and vegetables bound with eggs, then topped with peanuts and nam prik—is an example. Sweets such as coconut custards and fruit jellies are preferred snacks.

Thai cooking began as a court cuisine, and this heritage is most obvious in the evening meal.[29] Dinner often includes appetizers, such as deep-fried chicken wings stuffed with ground pork and shrimp or pastries shaped like delicate flowers. The main meal traditionally features steamed rice, soup, a curried dish, a fried dish, and a salad of raw vegetables and grilled poultry, beef, or seafood. Mee krob, a volcanic-looking mound of stir-fried noodles and meats or seafood cooked with sugar until caramelized, is a favorite addition. Various nam prik accompany the dishes. All dishes are served at the same time. Fingers and spoons are the usual implements, and forks are available for pushing food into the spoon. The meal usually concludes with elaborately carved fruits.

the children. Most-liked items among the adults included steak, oranges, candy, and soft drinks, all of which are prestige foods in Cambodia and Laos.[50] Least-liked items included cheese, chocolate milk, and milk.[51] A detailed study of poor Hmong immigrants in California revealed that the majority of adults (52 percent) consumed two meals each day of rice, greens, and meats. Pork was the preferred meat, although chicken, turkey, fish, and eggs were also eaten. The adults were mostly unfamiliar with baked products, such as bread or cookies, and most strongly disliked both milk and cheese. It was found that many Hmong grow their own vegetables and seasonings (especially varieties difficult to obtain from grocery stores) in backyard gardens. Most children ate three meals, often including a free lunch at school. Snacking was uncommon, although more prevalent among children.[41] In a study limited to fish consumption, it was reported that Laotians and Mien were most likely to harvest fish and seafood locally, and that Mien (23 percent) and Hmong (90 percent) frequently consumed the entire fish, including head, bones, eggs, and organs.[52]

Food purchasing and preparation as well as meal patterns are also changing. Southeast Asian-American women report that men frequently help with shopping or cooking. Vietnamese, Cambodian, and Laotian adolescents often are involved in food purchases, and surveys indicate as many as 60 percent of girls and 35 percent of boys have total responsibility for fixing dinner each evening. Southeast Asian women living in the United States are more likely to have a job or to attend adult education classes than in their homeland, relinquishing some household

Some young Hmong women avoid eating gizzards because they are believed to toughen the placenta and make birth difficult.

A survey of Thai Americans found that the number of meals per day decreased (breakfast was frequently skipped) and that the consumption of American foods increased while the intake of many Thai foods decreased.[174]

David Weintraub/Stock, Boston, Inc.

Vietnamese restaurants, especially those featuring the noodle soups called pho, have become popular in many communities where Vietnamese immigrants have settled.

responsibilities to other family members. Further, many families report a significant decline in eating meals together.[41,48,51,53,50,54]

Nutritional Status

Nutritional Intake

Filipino. The traditional Filipino diet is higher in total fat, saturated fat, and cholesterol than most Asian diets. Urban Filipinos living in the United States tend to have even higher intakes of these dietary components. Filipino Americans have high rates of hypertension and serum cholesterol levels equal to those of white Americans. Filipinos may have a genetic inability to process large amounts of sodium in their diet.[55] Alcohol consumption among Filipinos is also positively associated with increased blood pressure; however, low rates of drinking have been reported.[56,57]

Certain Filipino dishes are very high in purines, a concern for patients with gout; dinu-guan, for example, often includes pork liver, kidney, heart, and small intestine.

High rates of B_{12} deficiency and iron-deficiency anemia have also been reported in Thai vegetarians.[175]

Life expectancy rates for Filipino Americans are higher than for the general population.[58] Twenty-seven percent of Filipinos have reported being told that they had hypertension.[59] Heart disease is the leading cause of death in mortality statistics for Filipinos, and cerebrovascular diseases are third. Cancer risk is low among Filipinos, although incidence reportedly shifts from that in the Philippines to that in the United States with length of stay,[60] and risk for malignancies of the lip, oral cavity, pharynx, liver, and thyroid are above the national average.[58,61] While relative risk is low, it is noteworthy that survival rates among Filipino Americans who do develop cancer are lower than for white Americans.[62,63]

Infant mortality rates are slightly below those for the general population.[64] However, when compared to white, Chinese American, and Japanese American women, Filipino Americans have relatively high rates of preeclampsia, gestational diabetes, low-birth-weight infants, and preterm delivery (before 37 weeks).[65,66,67] When compared to other Asian groups, Filipino neonates are also at increased risk for death from infection, and post-neonatal infants from respiratory distress syndrome.[68]

Median body mass indexes (BMIs) in men and women are close to those of whites in the United States, and while rates of overweight and obesity are higher than in most other Asian groups.[69,59] One study of adolescents found that approximately 22 percent of boys and 13 percent of girls exceeded the 85 greater than or equal to percentile for BMI, rates lower than for most other Asians and for all other ethnic groups.[70]

Filipinos develop type 2 diabetes mellitus (9 percent) 59 than most other Asians. Data on Filipino American women in San Diego (ages 50–69 years) found a risk for type 2 diabetes of 36 percent, compared to 9 percent for whites, and the prevalence rates for adults in Texas were 16 percent for Filipino Americans compared to just over 6 percent for the U.S. total population.[71,72] Filipina women are at increased risk for developing gestational diabetes during pregnancy.[73] Filipina women in the United States have been found to have larger waist circumferences and a higher percentage of visceral adipose tissue than white women despite lower rates of overweight and obesity, suggesting more research is needed on metabolic syndrome (which is also higher in Filipinas) in nonobese populations.[74,71,75] Rates for hyperuricemia (resulting in gouty arthritis) and glucose-6-phosphate dehydrogenase deficiency (causing anemia unrelated to iron intake) are also higher than for white Americans. It should also be noted that alpha thalassemia (hemoglobin H disease) is also prevalent among Filipinos and results in a hypochromic microcytic anemia, especially during an infection or when oxidant drugs are taken.

Although Filipinos consume more milk and cheese than other Asians, many are lactose intolerant[11] and calcium intake may therefore be limited. Dried fish, fish sauce, and fish paste may provide calcium, but amounts vary depending on the source and quality of the product. One older study found some Filipino Americans have poor intakes of calcium and vitamin A.[43] Some Filipinos may be at risk for calcium deficiency, particularly newer immigrant women during pregnancy and postpartum.[12] Areca nuts may be chewed by older men, resulting in stained teeth.

Vietnamese and Other Mainland Southeast Asians. Food intake data suggest that the calcium intake of mainland Southeast Asians is low, although this observation does not account for fish sauces and other traditional foods that may contain sufficient calcium.[76] An analysis of broth made with acidified bones reported that one tablespoon provided nearly as much calcium as one-half of a cup of milk.[77] Vietnamese have been reported to have high rates of lactose intolerance.

Riboflavin, magnesium, and zinc consumption was found to be less than 80 percent of the RDA in adults among the Hmong.[41] Deficiencies during pregnancy include riboflavin; vitamins B_6, D, and E; folacin; calcium; phosphorus; potassium; and magnesium.[78] Iron intake may also be marginal,[41] particularly among children.[79] Anemia rates among refugees have been found to vary from 6 to 37 percent,[80,81,82] although genetic traits such as a high prevalence of hemoglobin E trait thalassemia syndromes may also cause anemia unrelated to iron intake.

Certain conditions common among recent immigrants from Southeast Asia may compromise their nutritional status. These include tuberculosis, usually the inactive form (approximately 4 percent of refugees are denied entry to the United States due to active cases—between 50 and 60 percent of immigrants in the United States test tuberculin positive);[83] intestinal parasites, which can contribute to anemias, fatigue, and weight loss; malaria; liver, and renal disease, whose contributing factor is the presence of hepatitis B surface antigens (chronic HBV rates have been estimated at between 5 and 15 percent);[83,84,85] and dental problems caused by chronic malnutrition and, in the United States, excessive consumption of sweets. Thalassemia, a genetic form of anemia, is more common in Southeast Asians and is not due to an iron deficiency.[86] There is concern that the persistence of continued parasitic infection suggests ongoing transmission; asymptomatic infection is common, and poor compliance with treatment may account for the excess cases.[87] Malnutrition contributes to the prevalence of short stature in recent immigrants.[81]

Life expectancy is higher for Vietnamese Americans than for the general population.[58] The leading cause of death is cancer, followed by heart diseases, and then pulmonary diseases. Malignancies of the lung, liver, prostate, colon, and stomach are most common in men, while those of the cervix (five times the rate for whites in the U.S.), breast, lung, colon, and stomach are most common among women.[88] This mortality profile is different from that of the general population and that of some other Asians, in which heart disease is the leading cause of death (see Filipinos above). One researcher suggests the excess rates may be due, in part, to registry misclassification of Vietnamese Americans.[89]

Findings suggest that obesity rates are currently very low among Vietnamese Americans. A study comparing the BMIs of different Asian subpopulations found Vietnamese had the lowest median; however, the data also suggest overweight increases with length of stay in the United States.[69] When recent Vietnamese immigrants were compared with Australians, BMI was found to be substantially less than in whites, but waist-to-hip ratios were higher. Essential hypertension is associated with higher BMI and insulin resistance in Vietnamese—even when higher BMI levels fall below the threshold of overweight in whites.[90,91] Small studies on a Vietnamese population in Mississippi reported 44 percent of subjects were hypertensive, 58 percent had high total blood cholesterol (often with high HDL levels), and 35 percent had high triglyceride levels.[92,93]

Scant data are available on the Cambodian-American and Laotian-American populations. Death rates from California show these populations may be at risk for several conditions.[94] Those born in the United States are more likely to be overweight.[53] The diabetes mortality rate for Cambodians was over 45 percent above the state average, and for Laotians it was almost 38 percent higher. The disparity in stroke deaths was even more startling: nearly double the average rate for Cambodians, and almost 65 percent higher in Laotians. Though the statistical samplings are small, the figures suggest these conditions may be of concern. It should be noted that Hmong and Mien may be subpopulations within the groups, since only nationality was noted.

Preliminary data suggest rates of obesity, diabetes, hypertension, cardiovascular diseases, and kidney failure are increasing significantly among Hmong.[21] High ratios of waist-hip measurements predicted high blood glucose in one study of Hmong in Wisconsin. Low adherence to medication schedules for the control of hypertension has also been noted.[95] Hmong are also reported to have higher rates of liver, stomach, cervical, and nasopharyngeal cancer, as well as elevated rates of leukemia and non-Hodgkin's lymphoma.[96,93] Hmong may choose to have no treatment for cancer if it involves the removal of body parts, and the mortality rate was found to be three to four times higher in Hmong women than in other Asians, Pacific Islanders, or whites.[97] Cultural factors that are potential barriers to treatment are belief in the spiritual etiology of diseases, patriarchal values, modesty, and mistrust of the Western medical system.[98,54]

Some Southeast-Asian women avoid taking supplements during pregnancy because they fear the baby will grow too big for delivery.

The Mien in California have been found to be at high risk for trichinosis infection due to the use of raw pork in dishes such as laap.

Recent immigrants have been reported to have a high incidence of low-birth-weight infants as a result of poor maternal weight gain during pregnancy. Traditionally, Southeast Asians gained less weight during pregnancy in order to have an easier birth and overeating was discouraged.[37,65] One study found that the birth weights of babies of Southeast Asians with longer residence are close to the U.S. average; these groups are believed to have benefited from medical and nutritional care.[99,100] Infant mortality rates for most Southeast Asians are reportedly low, however, regardless of low birth weight and other potentially adverse risk factors such as nutritional deficiencies, inadequate use of prenatal care, and low socioeconomic status. Hmong, Laotian, and Cambodian infant mortality rates may be the exception. Hmong infant mortality rates were reported to be slightly elevated in comparison to other Asian groups,[78] and death statistics from California found that Laotian infant mortality rates (in which Hmong may be counted) were nearly double the state average.[94] In a study of California Asian groups, Cambodian and Laotian women had adverse maternal risk profiles and higher death rates than whites for neonates, post-natal infants, and infants.[68]

Nearly all babies are breast-fed in mainland Southeast Asia for periods of about one year. Studies on infant feeding practices of Americans of Southeast Asian descent reveal a dramatic decline; breast-feeding is reduced to only 9 to 26 percent of postpartum women. Work, schooling, physical discomfort, embarrassment, and the ready availability of formula through hospitals were cited as reasons for not breast-feeding. Some women mentioned that they believed formula was nutritionally superior to breast milk.[101,102,103,104] Southeast Asian women generally introduce solid foods later (at about 8 months) than do whites (5 months) or African Americans (about 4 months). However, many Southeast Asians associate overweight with health and may try to overfeed babies.

Southeast Asians typically calculate age on a lunar calendar, often starting with being one year old at birth. Reported age may differ as much as two years from Western chronological age, which can distort the use of standardized growth curves. Some Vietnamese parents may claim, however, that their children are younger than they are to enroll them in lower school grades; this allows the children to catch up in their schooling.

Counseling

Filipino. Americans of Filipino descent may accept illness as fate, tolerating symptoms until the severity forces them to seek care. Relatives, neighbors, and traditional healers may be consulted before obtaining biomedical service. Language barriers may be significant. For example, some Filipino languages do not recognize gender, and there may be some confusion with pronouns. In 2000, over 24 percent of Filipinos reported they did not speak English well; nevertheless, the assumption that an interpreter is needed may offend those who have mastered the language.[6]

The communication style of many Filipino Americans is very high context,[12] and expression is formal and polite. Confrontation is avoided, and all attempts to maintain harmony will be made, including the use of silent pauses and laughter to hide embarrassment.[34] Raising one's voice or losing emotional control is considered rude and immature. Positive expression, no matter the situation, is expected.

Filipino elders should not be addressed by their first names, as this is disrespectful. Health care practitioners are often considered to be authority figures, so responses to questions may be deferential; Filipinos will avoid voicing disagreement. Many Filipinos avoid situations in which self-esteem may be lost, and thus health care providers should be sensitive in discussing certain subjects, such as socioeconomic background. Modesty may make other topics uncomfortable to discuss as well, including sexuality (handled best by a provider of the same gender as the client) and "shameful" conditions such as tuberculosis or mental illness.

Soft handshaking is the common greeting, although an eyebrow flash (quick lifting of the eyebrows) may be used between acquaintances.[105] Avoid any other touching, and keep hands exposed at the side of the body, not in pockets. Direct eye contact between peers of the same gender may occur, but in general quick contact and aversion is more common, particularly when addressing someone in a position of higher authority. Further, direct eye contact between men and women may be interpreted as an expression of sexual interest or aggression. Filipinos may expect quick results from their health providers and will switch to other healers if they feel that progress is too slow.[14]

Unknown numbers of traditional healers are used by some Filipino Americans. An older study in Los Angeles reported that most respondents, independent of education level, still adhered to many traditional beliefs about the cause of illness, including unbalanced conditions such as eating too much or eating the wrong combination of foods, working too hard, or being punished for one's sins against God. Immigrants from rural regions were more familiar with traditional medical practices than immigrants from urban areas, who were more likely to rely on over-the-counter therapeutics.[14]

Due to the strong family orientation, relatives play a significant role in a Filipino client's treatment and recovery. For most effective treatment, the provider should discuss diet modifications with family members as well as with the patient. Compliance may be motivated by desire to fulfill familial obligations and participate in social life. [106] The in-depth interview should be used to determine the patient's degree of acculturation, use of traditional medical practices, and personal food habits.

Vietnamese and Other Mainland Southeast Asians. Southeast Asian access to health care may be limited by lack of health insurance. Data are limited, but it is estimated that 27 percent of those living in California are uninsured. Among Vietnamese, 42 percent of those who have no health insurance report that they have no usual source of health care, compared to only 10 percent of Vietnamese with health insurance.[107] One study of recent Vietnamese immigrants found that demographic variables such as poverty and marital status were the most influential factors in health care access.[108] Language barriers may also exist. Large numbers of Southeast Asians do not speak English well: Vietnamese (62 percent), Hmong (57 percent), Cambodians (54 percent), and Laotians (53 percent).[6] In one study Vietnamese Americans with limited English proficiency considered quality translation services essential to care, expressing a preference for professional, gender-concordant interpreters instead of the use of family members.[109]

Similar to Filipinos, some Southeast Asians believe that illness is in the hands of God, spirits, or fate. For example, a study of college students found that over 38 percent of Hmong participants believe developing osteoporosis is due to chance or luck (though Vietnamese students were more likely to attribute the condition to diet).[110] Furthermore, some Southeast Asian Americans may deny discomfort and pain until it becomes intolerable, or until all home remedies prove ineffective. Many Southeast Asians also philosophically regard quality of life to be more important than length of life, believing that personal illness or suffering will diminish in the next reincarnation. As a result, a client may be very ill before deciding to go to a clinic or hospital. Traditionally, prevention of disease occurs primarily through harmonious living; most Southeast Asians have little experience with medical checkups or treating a condition when no symptoms are present.[21,108,111]

Trust is a significant issue in Southeast-Asian health care. Experiences with medical personnel in refugee camps have left many immigrants suspicious of biomedicine in general. Many Americans of Southeast Asian descent believe that western practitioners do not understand their medical needs and are disrespectful of their traditional practices; many are fearful of invasive laboratory tests, especially the taking of

Some Filipinos believe that fat is a protection against becoming too cold and losing vital energy; thus, being overweight is preferred to being too thin.

Some Filipino elders prefer their food soft and warm and will reject beverages with ice.[176]

Traditional therapies such as coining or moxibustion may leave marks on the skin; abuse should not necessarily be presumed.

Hmong may prefer unseasoned foods when hospitalized and may desire water that is boiled before drinking.[42]

Some Southeast Asians may not distinguish between fruit juices and fruit-flavored beverages in food recalls.

blood because this may upset the body's balance. Surgery may be avoided and even autopsies denied due to fears about the relationship between the body and soul. Privacy issues may also be of concern. For example, the need to completely undress or the use of hospital gowns, breast and pelvic exams, and discussions about family planning should be postponed until a client-provider relationship has been established, preferably between a provider of the same gender as the client.[21,27,111] Acceptance of invasive procedures, such as having blood taken and pelvic exams, was reportedly increased in one study of pregnant Hmong women by the use of a videotape on prenatal care narrated in the Hmong language.[112]

Southeast-Asian clients desire a full description of their disorders and therapies, and interpreter fluent in a client's dialect and culture may be essential to communication. For example, one study found that the concept of chronic illness did not exist in Hmong healing practices, and there were no words or explanations for conditions such as hypertension and diabetes. There was confusion among respondents between curing and controlling an illness.[113] Differences in medical concepts and technologies require

A very polite, unhurried, and reserved conversational style is appreciated by most Southeast Asians. Excited, informal, or frank speech may be considered rude.[111,42] Many Hmong prefer an attitude of caring and respect, showing warmth through smiling and using a positive approach—negative statements and outcomes should be avoided.[114,21] The Vietnamese place a high value on social harmony; both Confucian and Buddhist belief systems encourage modesty. The clinician should be aware that, in general, Southeast-Asian clients will be agreeable to avoid disharmony or to please the questioner. When angry or embarrassed, Southeast-Asian Americans may laugh to mask their emotion. Proper posture and appearance are important. In addition, certain nonverbal forms of communication should be carefully observed. The head is considered sacred, and it is extremely offensive to pat or even touch the head of an adult or child without permission. The feet are the lowest part of the body, and thus it is impolite to point with the foot or show the bottoms of one's shoes. It is also rude to snap one's fingers or signal by using an upturned index finger, as this is how dogs are called. Respect is shown by giving a small bow of the head when greeting elders and by using both hands to present any item to the client.

Numerous studies have noted that refugees from mainland Southeast Asia are at special risk for mental health problems, due to the horrors of war, difficulties in escape, lengthy camp confinement, and the extreme cultural differences between their homeland and the United States. Posttraumatic stress disorder is common. One study of the Hmong suggests, however, that levels of depression, anxiety, hostility, and other symptoms of adjustment problems may gradually resolve with length of stay.[115] Adherence to traditional health beliefs varies, often according to whether new religious faiths have been adopted; Christian churches strongly discourage ancestor and spirit worship.[22] Some studies suggest that the majority of immigrants continue

NEW AMERICAN PERSPECTIVES—Vietnamese

HAN LE, Student

I was born in Vietnam but came to the United States when I was about a year old. My parents left Vietnam in 1976. My grandparents got us and about thirty-five other people out on a boat. We first went to a refugee camp in Malaysia, and after a few months we were able to settle in Southern California.

At home we ate a traditional diet, and I still do. Rice and noodles are staple foods. Breakfast is a French bageuette with pâté and ham (there is a lot of pork in the diet). Lunch is soup or a bowl of rice with one entrée, usually a leftover from dinner. There are always three things for dinner: (1) always soup, (2) a stewed meat entrée, and (3) vegetable stir fry. Fruit is usually served for dessert. Snacks are commonly fruit or sticky rice. Western foods that I eat now are dairy products such as milk, cheese, and yogurt; cereal for breakfast; and more sweets for dessert. Vietnamese have always eaten a lot of sugar but not as desserts. I lost all my baby teeth due to cavities because I was given so much candy. In Vietnam, only children from wealthy families got candy, so it was a real treat here.

When we first came over, our neighbors gave my parents cereal to try, and they liked it. At that time my parents hardly spoke or read English, and there were very few Vietnamese markets. Because the cereal box had pictures of cats and dogs on it, when they went to the grocery store they bought a box of something with pictures of cats or dogs on it, but it turned out to be pet food. They ate the whole box, but they did put sugar on it because it wasn't sweet enough. This was not an uncommon problem when the Vietnamese first immigrated to the United States. A friend was enrolled in WIC while she was pregnant and had vouchers to buy WIC food at the market. She recognized the word "cheese" and she bought cream cheese and ate one package every day during her pregnancy.

When counseling Vietnamese, it is important to remember that our meal is not complete unless it contains rice or noodles. Therefore you should not recommend eliminating these elements of the diet but rather just decrease their amount.

certain practices, such as the use of botanical home remedies and coining (see the section Traditional Health Beliefs and Practices), for many years after arrival[116,27] although the costs and inconvenience of some traditional cures, including difficulties in obtaining animals and herbs, the disintegration of clan ties, and the scarcity of shamans present barriers in some communities.[10] Southeast Asians frequently develop a medical pluralism, accepting those theories and therapeutics most congruent with their acculturation experiences.

Adherence to traditional health beliefs and practices, such as hot and cold theories and the use of herbal remedies, is reportedly high among many Southeast Asians.[28,21,108] It is not unusual for Southeast Asians to consult both biomedical practitioners and healers for relief of symptoms. The Vietnamese usually characterize U.S. biomedicine as yang (hot) and traditional Vietnamese medicine as yin (cold). Biomedicine is seen as fast-acting, temporary, and likely to have side effects. It is useful for emergency situations. Vietnamese medicine is considered slow-acting, gentle, without side effects, a permanent cure, or useful in prevention. Some immigrants believe it is important to counteract the yang impact of biomedical therapies with yin herbal remedies.[20] It has been reported that some Vietnamese Americans routinely reduce the dose or duration of prescription medications because they are considered too strong or appropriate for larger people, or because they have side effects. Medication is frequently ceased when symptoms are alleviated.[116,20] Similar practices have been found in the Hmong.[13]

Little is known about the active agents in Southeast-Asian herbal remedies, and some may be of concern. High levels of lead have been found in *paylooh*, an orange powder ingested for rashes or fever. A case study found that the use of slang nut (the source of strychnine) was not toxic as applied in Cambodia, but is poisonous when used in the United States without the leaves it is traditionally wrapped in.[21,117]

Hospital nurses report that numerous conflicts have developed around traditional Vietnamese birthing practices.[20] Because the postpartum period is defined as yin (cold), women avoid exposure to cold and wind (see the section Traditional Health Beliefs and Practices). Clients may refuse to get out of bed soon after giving birth and refuse to take a shower or wash their hair. *Âm* foods, such as cold beverages and vegetables, may be refused even if prepared in a culturally sensitive manner. Women may consume alcoholic beverages because they are duong, and alcohol is also thought to cleanse the reproductive system and promote milk production.[37] Hmong women also consider the postpartum period a cold condition.

An in-depth interview is critical in order to determine the patient's country of origin (a patient may be offended if grouped with all other Southeast Asians), length of time in the United States, and any immediate health problems. Degree of acculturation and personal food preferences should also be noted.

Pacific Islanders

Pacific Islanders are the peoples inhabiting some of the 10,000 islands of Oceania. Polynesia, Micronesia, and Melanesia are the three areas that make up the Pacific region. Polynesia includes the major islands and island groups of Hawaii, American Samoa, Western Samoa, Tonga, Easter Island, and Tahiti and the Society Islands. The 2,000 small islands of Micronesia include Guam, Kiribati, Nauru, the Marshall and Northern Mariana Islands, Palau, and the Federated States of Micronesia. Although the boundaries of Melanesia are not exact, it includes the nations of Fiji, Papua New Guinea, Vanuatu, the Solomon Islands, and the French dependency of New Caledonia.

The Native Hawaiian Government Reorganization Act of 2005 proposes that Native Hawaiians have the same legal status as American Indians and Alaska Natives, including a governing body that can negotiate with federal and state authorities over land deposition. It needs legislative approval to take effect.

Although the area is geographically similar, consisting of mostly small, tropical coral or volcanic islands, the Pacific Islanders are a racially and culturally diverse population. European, American, and Japanese influences have been extensive. Today there are greater numbers of some Pacific Islander groups living in the United States than in their native homelands.

Cultural Perspective

History of Pacific Islanders in the United States

Immigration Patterns. Since the first settlement of the island Oceania, the migration of population groups has been very fluid among Polynesia, Micronesia, and Melanesia. As conditions on one island grew too crowded, colonization of surrounding islands occurred. That

©Bettmann/CORBIS

▲ *A rendering of the* Resolution, *the ship used by James Cook to explore Polynesia. Many Pacific Islander cultural traditions disappeared after European discovery of the islands in the 18th century.*

trend continues, with economic opportunity the primary motivation for Pacific Islander immigration to the United States and other nations.

Hawaiians. When British explorer James Cook first arrived in Hawaii, approximately 300,000 Native Hawaiians were living on the islands. European diseases introduced by the explorers and missionaries decimated the population, and by 1910 only a little more than 38,500 persons of Hawaiian ancestry remained. A high rate of intermarriage has resulted in a population with few Hawaiians of full native heritage. A large number of Hawaiians have migrated to the mainland United States, but the census figures do not reflect such interstate relocations.

Samoans. In 1951 the Pago Pago naval base that employed many Samoans moved to Hawaii, and many Samoans followed. Increasing population pressures and a deteriorating economy encouraged further immigration. Once in Hawaii, Samoans sometimes move to the mainland United States in search of broader job availability and for wider educational opportunities for their children. A chain of immigration between Hawaii and the West Coast has been created, with Samoans established in the mainland helping extended family members to settle nearby.

Guamanians. Following attainment of citizenship in 1950, many Guamanians enlisted in the U.S. armed forces seeking better employment. Some moved to Hawaii and the West Coast of the U.S. mainland. By the early 1970s, approximately 12,000 Guamanians had immigrated, mostly Chamorros (native Micronesians). Major populations have settled in San Diego, Los Angeles, San Francisco, the Seattle area, and Hawaii.

Tongans. The immigration of Tongans to the United States did not begin until population pressures in the 1960s decreased economic opportunities in Tonga. Under the strict hereditary social structure, only the eldest son in a family may inherit land, leaving younger men with little economic mobility. Unlike most other Pacific Islanders, Tongans usually immigrate directly to the U.S. mainland rather than settling first in Hawaii.

Current Demographics and Socioeconomic Status

Hawaiians. More than 151,000 U.S. citizens were self-identified as Native Hawaiians in the 2005 U.S. Census estimate, representing the largest percentage (38 percent) of American Pacific Islanders, and approximately 260,000 are Native Hawaiian in combination with other races. Over half live in Hawaii: 20 percent of the total state population has some Native Hawaiian heritage. The remaining Hawaiians live primarily on the West Coast, mostly in California. Scant socioeconomic data are available, but reports indicate that some Hawaiians occupy the lowest economic strata in the state of Hawaii, along with other Pacific Islander immigrants, living mostly in rural and semirural regions.[118] Hawaiian heritage is less of a handicap on the mainland, where Hawaiians have a cultural advantage over other Pacific Islanders (through extensive exposure to U.S. society) and often enter the middle class. Overall, median family income is higher than the national average, and poverty rates are higher than the total U.S. figure. Eighty percent have a high school degree or higher. College attendance rates for adults are higher than the U.S. average, although graduation rates are much lower.[119]

Samoans. Probably more Samoans are living outside Samoa than in American and Western Samoa combined. The 2005 U.S. Census figures showed over 56,000 people of Samoan heritage alone lived in the United States, most of whom were born in America. Over 98,000 Samoans total (alone and in combination with other ethnicity) are U.S. residents.[120] The largest groups are found in Honolulu, Los Angeles, the San Francisco Bay area, and Salt Lake City. Smaller numbers have settled in Laie, Hawaii; Oakland, California; and Independence, Missouri. Religion is a factor for some immigrants: Samoan Mormons tend to migrate to Mormon centers such as Salt Lake City when they move to the United States. Although most Samoan immigrants to the mainland are from American Samoa, approximately 15 to 20 percent come from the independent nation of Western Samoa.

Though nearly one-third of Samoans in the United States work in office and sales jobs, and 18 percent hold management and professional positions, another 30 percent of Samoans in the United States are employed in unskilled or semiskilled labor such as assembly line jobs, construction, janitorial or maintenance jobs, or as security guards.[120] Low pay, large immediate families, and responsibilities for extended family members in Samoa often translate into a lower standard of living.[121] Median family income is among the lowest of all Pacific Islanders and is below the national average. Approximately 15 percent of Samoans live in poverty.[120] Eighty-four percent of Samoan adults have a high school education or higher, though college attendance rates are slightly higher than the national average, completion rate rate is low.

Guamanians. The 2005 U.S. Census statistics indicate over 76,000 Guamanians are residing in the United States. Many are Chamorros, although self-identification figures are unclear on the exact percentage. The Chamorros often become part of Pacific Islander communities (with Samoan and Tongan immigrants). Large populations of Guamanians are found in Hawaii, California, Washington State, and Washington, D.C.[122]

Over one-quarter of Guamanians work in professional or management positions, and nearly one-third are employed in office and sales jobs.[123] Similar to Native Hawaiians, median family income is just below the U.S. average, and poverty rates are higher than the national average. Nearly 50 percent of Guamanian adults have attended college, though graduation rates are well below the national average. Additionally, 22 percent of adults have not graduated from high school. Some Guamanians, particularly Chomorros, may suffer the same discrimination and economic hardships as some other Pacific Islander groups.

Tongans. Nearly 27,000 Tongans live in the United States according to U.S. 2000 Census figures. Many are Mormons and are aided in their immigration by their church, and most move to communities of other Pacific Islanders in San Francisco, Los Angeles, Dallas, Fort Worth, and in Salt Lake City. A small community of Tongans has also located in Hawaii.[124]

Educational attainment in the U.S. Tongan community is low—nearly 35 percent of adults have not graduated from high school. Tongans are more likely than most other Pacific Islanders to work in the construction, maintenance, and extraction jobs. Many also report employment in the production and transportation fields.[125] Though median family income is close to the national norm, sharing earnings with relatives in Tonga is expected,[126] and nearly 20 percent of Tongans live in poverty.

Worldview

Religion. Pacific Islanders follow a wide variety of religions, often according to which missionary groups were active in their homeland. Hawaiians practice mostly Protestantism, Buddhism, or Shintoism. Samoans are largely Methodist, Catholic, Mormon, and Anglican. Chomorros are primarily Catholic, and Tongans in the United States are mostly Mormon. Religion is often prominent in the lives of Pacific Islanders, and ministers usually are held in high esteem. In Samoa, nightly readings from the Bible are common in most homes, and prayers are offered at every meal.[121]

Family. Although most native religions were abandoned, many concepts central to Pacific Islander culture remained within the structure of families. For example, on many islands social rank and power were established by birth order within the extended kinship system or clan, and even within families, younger siblings deferred to their older brothers and sisters. Elders were respected. The senior male in the group, whether in the village or in a family, managed all group matters. In Hawaii women and men were segregated under the *kapu* system, each with specific roles and responsibilities. Extended families were the foundation of society, and children were raised usually by grandparents, aunts and uncles, and even remote kin rather than by just the parents. Household composition was flexible, and all members were obligated to support the extended

Nearly 10,000 Fijians live in the United States, according to 2000 U.S. Census data, primarily in California. Most are of Asian-Indian descent, were originally indentured laborers in Fiji, and follow the Islamic or Hindu faith. Though overall educational attainment is low, median family income approaches the national average, and poverty rates are significantly lower than the norm.[125]

The 2000 U.S. Census reports nearly 6,000 residents from the Marshall Islands, most of whom are foreign-born. Many have extremely low educational attainment, median family income nearly 50 percent below the national average, and poverty rates over 38 percent.[125]

Certain fish in Hawaii were reserved for royalty, and women rarely consumed fish, although they were permitted to eat shellfish.

In Samoa, a serious offense by an individual can be ameliorated by an *ifoga* (literally "a lowering"). The extended family positions themselves in front of the victim's home and remains there until invited in and forgiven. They formally apologize through the presentation of gifts and cash.

family, resulting in substantial redistribution of resources. Generosity and sharing were highly valued. Any social transgressions committed by the individual were the responsibility of the whole family. In Samoa, if the violation was severe, the family could be disinherited from their land and stripped of any social title.

Some of these same practices are continued by Pacific Islanders today. Unlike many immigrant groups, they typically maintain extended families in the United States. Responsibility for child rearing is shared among family members (children may move freely between homes), and household chores are assigned according to age and gender. The oldest man (or occasionally the oldest woman) in the home assumes control, collecting everyone's paycheck (or weekly contribution) and providing for the household needs. The good of the whole family is considered before the benefit to the individual, and most Pacific Islanders are guided by their desire to avoid bringing shame on their family. In Guam this interdependence is known as *inafa'maolek* and extends not only to family and community, but to national affiliation as well.[122]

The stresses of acculturation in the United States usually occur due to moving from a society in which there is little anonymity (with individual behavior reflecting on the whole family or village) to a society in which Pacific Islanders are often marginalized or even invisible if misidentified as Asians or Filipinos. Most Pacific Islanders maintain close contact with their homeland and fulfill their obligation to family by sending financial support. Trips back to the islands for political and social events are common.

Traditional Health Beliefs and Practices. Religion and medicine were closely linked in Pacific Islander culture, and the loss of many traditional health beliefs and practices occurred with the adoption of nonnative faiths. Although folk healers specializing in herbs, massage, or religious and/or spiritual intervention work in Pacific Islander communities both in Hawaii and on the U.S. mainland, their current use is not well documented. Traditional practitioners are identified with a broad range of Eastern and Western religious affiliations and offer a wide spectrum of services. Their clientele are also extremely diverse, often crossing ethnic or religious lines to seek effective care.[127]

Many Native Hawaiians believe lokahi, harmony between individuals, nature, and the gods, is essential to good health. Hawaiian healers typically practice massage (*ho'olomilomi*—used for conditions such as childbirth, asthma, congestion, bronchitis, inflammation, and rheumatism), herbal medicine (*la'au lapa'au*), and/or conflict resolution (*ho'ononpono,* meaning "make things right"—used to remove emotional obstacles to healing). Meditation, deep breathing, and Chinese Traditional Medicine may also be employed. Healing is initiated at the end of each session with spiritual blessings.[128,129,130]

It is reported that more than 300 Hawaiian botanicals and animal- or mineral-based cures were available traditionally, with over fifty-eight remedies for respiratory problems alone.[131,118] Today, only thirty plants are estimated to be used regularly. Examples include aloe vera for burns, hypertension, diabetes, and cancer; plantain leaves to reduce blood sugar levels in diabetes; *polokai* (black nightshade) for asthma, coughs, and colds; and wild ginger for gastrointestinal problems, ulcers, and asthma. Native Hawaiians also practice home remedies. For example, drinking seawater followed by fresh water is believed to be a general tonic.

Samoans believe that health is maintained through a good diet, cleanliness, and harmony in interpersonal relationships. An individual is at high risk of illness if he or she does not fulfill family obligations or support village life. The concept of balance is essential: disruptions in interpersonal relationships, working too hard, sleeping too little, or eating the wrong foods can cause dislocation of the *to'ala* (the center of one's being, located just beneath the navel) to another part of the body, where it induces pain, poor appetite, or other symptoms.[132,133] Treatment typically requires the restoration of balance. A family may get together to openly air disputes so that harmony can be reestablished, or massage by an elder may be used to gently coax the *to'ala* back into position. A traditional Samoan healer may be consulted to cure certain folk illnesses, particularly those due to supernatural causes, such as spirit possession by malevolent ghosts or the actions of ancestor spirits angered by a person's conduct. One such condition is *musu*, a mental illness in young men and women typified by extreme withdrawal. Healers may be herbalists, masseuses, bone setters, midwives, *taulasea* (a general practitioner familiar with Samoan sicknesses), power healers (who provide spiritual interventions), or diviners (specializing in the determination of why an illness has occurred). Massage, herbal remedies, and communication with the supernatural elements are the usual therapies.[121,134]

Traditional Food Habits

The cooking of the Pacific Islands developed without benefit of metal pots, pans, and utensils, and many foods were eaten raw. The indigenous cuisine was probably based on breadfruit, taro, cassava, yams, and perhaps pigs and poultry. Fruits were also widely available, although those often associated with the Pacific region, such as coconuts and bananas, were not introduced from Indonesia until approximately 1000 C.E. Other items, including sugar cane and pineapple, were brought by European plantation owners.

Ingredients and Common Foods: Staples

Starchy vegetables are the mainstay of the traditional Pacific Islander diet (Table 12.3). These include the root vegetable taro, which is a little denser and more glutinous than the white potato; breadfruit, with a fluffy, bread-like interior; cassava; and yams. These foods were often cooked and then pounded into a paste. In Hawaii, taro root paste eaten fresh, or partially fermented, is called poi (a word that originally referred to the pounding method). When food was scarce, the Native Hawaiians survived on the purplish-colored *poi*, sometimes with a little seaweed or fish added to it. Although taro root is also a staple in Samoa, it is usually boiled but not pounded. Arrowroot is used to thicken puddings and other dishes. The Europeans introduced wheat, and bread is eaten in some areas; for instance, Portuguese sweet bread is known as Hawaiian bread in Hawaii. Asian settlers popularized both short- and long-grain rice, as well as noodles.

More than forty varieties of seaweed are consumed. Cooked greens, including the leaves of the taro root, yam, ti plant, and sweet potatoes, are very popular. One specialty is to wrap foods in ti or taro leaves, then steam the packets for several hours. The musky flavor of the leaves permeates the entire dish, called laulau in Hawaii.[29] Another specialty known as lu'au combines chopped taro leaves with chicken or octopus and coconut milk.[135]

© Isabelle Rozenbaum/PhotoAlto Agency RF Collections/Getty Images

▲ *Taro root is a staple in the Pacific Islands, served boiled or pounded into a paste called poi. The leaves are also consumed.*

Fish and seafood are abundant in the Pacific Islands, and in some regions they were eaten at every meal. Mullet is one of the most popular fishes, but many others, including mahimahi ("dolphin fish," not related to the mammal), salmon, shark, tuna, and sardines are also consumed. A tremendous variety of shellfish is available, such as clams, crabs, lobster, scallops, shrimp, crawfish, and sea urchins, as well as many local species. Eel, octopus, squid, and sea cucumbers are also eaten. Although some fish and seafood was stewed or roasted, some was also eaten uncooked, marinated in lemon or lime juice, which turns the fish opaque much the same as cooking it. A popular Hawaiian specialty is lomi-lomi, made with marinated chunks of salmon, tomatoes, and onions, served with or without poi as an appetizer. A similar Samoan dish, called oka, is also made with chunks of raw fish marinated in a mixture of lemon juice and coconut cream.

Pork is the most commonly eaten meat, especially for ceremonial occasions. Traditionally it was cooked in a pit called an *imu* in Hawaii, a *hima'a* in Samoa, and an *umu* in Tonga. A fire was built over the stones lining the pit, and when the coals were hot, layers of banana leaves or palm fronds were added. The pig and other foods, such as breadfruit and yams, were placed on the leaves, then covered with more leaves and sealed with dirt. In some cases water was poured over the rocks just before the pit was closed, steaming the foods instead of baking them. The pit was left sealed for hours until the food was completely cooked.

Hysterical behavior, especially by women, was called "ghost sickness" by the Tongans and attributed to possession by the spirit of a deceased female ancestor. Another traditional illness occurs in infants when the fontanel does not close (*mavae ua*).

Traditionally, Samoan men obtained full-body tattoos, in part to help them empathize with the prolonged pain of childbirth suffered by Samoan women.

Fish hooks are symbolic of good luck in Hawaii.

Table 12.3 Cultural Food Groups: Pacific Islanders

Group	Comments	Common Foods	Adaptations in the United States
Protein Foods			
Milk/milk products	Milk and other dairy products are uncommon. Many Pacific Islanders are lactose intolerant.		Increased intake of milk has occurred.
Meat/poultry/fish/eggs/legumes	Pork is the most commonly eaten meat. Soybean products are used by Asian residents. Winged beans are a popular legume on some islands.	*Meat:* beef, pork *Poultry and small birds:* chicken, duck, squab, turkey *Eggs:* chicken, duck *Fish and shellfish:* *ahi*, clams, crabs, crawfish, eel, lobster, *mahimahi*, mullet, octopus, salmon, sardines, scallops, sea cucumber, sea urchin, shark, shrimp, swordfish, tuna, turtle, whale *Legumes:* beans (long, navy, soy, sword, winged), cowpeas, lentils, pigeon peas	Dietary changes often occur before immigration. Many are dependent on imported foods such as canned meats and fish.
Cereals/Grains	Europeans introduced wheat bread, and Asians brought rice and noodles.	Rice, wheat	Increased intake of bread and rice is noted.
Fruits/Vegetables	Starchy vegetables are the mainstay of the diet. They were often cooked and pounded into a paste. More than 40 varieties of seaweed are eaten. Cooked greens are popular. Fruits are an important ingredient. Immature coconuts are considered a delicacy. Arrowroot is used to thicken puddings and other dishes.	*Fruits:* acerola cherry, apples, apricot, avocado, banana, breadfruit, citrus fruits, coconut, guava, jackfruit, kumquat, litchis, loquat, mango, melons, papaya, passion fruit, peach, pear, pineapple, plum, prune, soursop, strawberry, tamarind *Vegetables:* arrowroot, bitter melon, burdock root, cabbage, carrot, cassava, cauliflower, *daikon*, eggplant, ferns, green beans, green pepper, horseradish, jute, kohlrabi, leeks, lettuce, lotus root, mustard greens, green onions, parsley, peas, seaweed, spinach, squashes, sweet potato, taro, *ti* plant, tomato, water chestnuts, yams	The traditional starchy vegetables have decreased in use and may only be consumed at special occasions.
Additional Foods			
Seasonings	Food is not highly seasoned but often flavored with lime or lemon juice and coconut milk or cream.	Curry powder, garlic, ginger, mint, paprika, pepper, salt, scallions or green onions, seaweed, soy sauce, tamarind	
Nuts/seeds	Nuts are a core ingredient.	Candlenuts (*kukui*), litchi, macadamia nuts, peanuts	
Beverages	Coconuts provide juice for drinking and sap for fermentation.	Cocoa, coconut drinks, coffee, fruit juice, *kava* (alcoholic beverage made from pepper plant), tea	Increased consumption of sweetened fruit beverages and soft drinks has occurred.
Fats/oils	Coconut oil and lard are the preferred fats.	Butter, coconut oil or cream, lard, vegetable oil and shortening, sesame oil	Use of vegetable oils and mayonnaise has increased.
Sweeteners		Sugar	

Chicken is widely available, as are eggs. Limited grazing land kept beef from becoming a frequently eaten item. Milk and other dairy products are also uncommon. Soybean products are used by Asian residents, and winged beans are a popular legume on some islands.

Fruits and nuts are important ingredients in Pacific Islander cuisine. Bananas, candlenuts (kukui nuts), citrus fruits, coconuts, pineapples, guavas, litchis, jackfruit, mangoes, melons, papayas, passion fruit, and *vi* (ambarella) are a few of the widely available varieties. Fruits are eaten fresh or added to dishes such as Samoan papaya and coconut cream soup (supo 'esi) and deep-fried dumplings filled with pineapple or bananas (pani keki). Coconuts provide juice for drinking, sap for fermentation, and milk or cream used in numerous stewed dishes (coconut milk can also be made into foods resembling cheese and buttermilk). Immature coconuts are considered a delicacy throughout the Pacific. Haupia, a traditional gelatin-like Hawaiian dessert, is made from coconut milk sweetened with sugar.

Traditional Pacific Islander fare was not highly seasoned. The flavors of lime or lemon juice, coconut milk or cream, and salt predominate, with occasional use of ginger, garlic, tamarind, and scallions or onions. Coconut oil and lard are the preferred fats, providing a distinctive taste to many dishes. Foreign spices, such as Asian Indian curry blends, and sauces, such as soy sauce, have been incorporated into some dishes.

Meal Composition and Cycle

Daily Patterns. Traditional meals included poi or boiled taro root, breadfruit, or green bananas; fish or pork; and greens or seaweed. In Samoa, Guam, and Tonga, most dishes are cooked in coconut milk or cream. Although the evening supper was generally the largest meal, little distinction was made between the foods served at the two or three daily meals. When food was pit-cooked, amounts suitable for two or three days at a time were prepared. Fresh fruit was eaten as snacks. Beverages made from coconut juice or sap were common. The Asians introduced various teas, including those made from lemon grass and orange leaves. In Samoa a drink made from ground cacao beans mixed with water, called koko samoa, is traditional. Kava, a bland, mildly intoxicating beverage, remains a popular drink in many regions. It is made from the chewed or ground root of the native pepper plant, mixed with water in a stone bowl. It reputedly tastes a little like dirt or licorice.

Etiquette. Most Pacific Islanders consider hospitality an honor, and outsiders are usually exempt from traditional manners. In Samoa, for example, it is considered rude to eat in front of someone without sharing. When eating a meal, respect should be shown for the food because it represents the host's generosity; this includes not talking during the meal.[121] Most hosts will not eat until a guest is satisfied. As in many societies, it is impolite to refuse food, although a guest is not obligated to eat every morsel served. At some Samoan celebrations, for example, a large box of food may be placed on the table for each guest. The diner samples items from the box until full, then brings the remaining food home to share with his or her extended family.[136]

© Porterfield/Chickering/Photo Researchers, Inc.

▲ *Pork was traditionally the most commonly eaten meat, particularly for ceremonial occasions. The pig and other foods were cooked in a stone-lined pit over coals.*

Special Occasions. Throughout the Pacific Islands special events were commemorated with feasting, often including pit-roasted foods. In Hawaii weddings, childbirth, completion of a canoe or house, or a prolific harvest or abundant catch was celebrated with a luau, featuring a whole pig, poultry, fish, and vegetables cooked in the imu. In Samoa the feast is sometimes preceded by a kava ceremony, in which the beverage is distributed ritualistically to guests who are expected to drain the cup in one gulp. Traditionally kava was offered as a gesture of hospitality and for occasions such as the ordination of a new chief. In Tonga, where *umu*-cooked food accompanies celebrations such as the commemoration of a royal birthday, special coconut frond stretchers are woven for transporting the massive amounts of food prepared for the occasion.

Holidays celebrated in the Pacific Islands are usually those associated with religious affiliation. In addition, Hawaiians also honor Prince Kuhio on March 26 and the Kamehameha Dynasty on June 11. Samoans feast nearly every Sunday, and almost all denominations celebrate White Sunday on the third Sunday of October, venerating children. After a special service featuring religious recitations by children, a festive meal is served, and children are waited on by the adults in their extended family.[121] Guamanians celebrate their Liberation Day on July 21 with parades, fireworks, and feasting.[122] King Tuafa'ahau Tuopu IV's birthday is a national holiday in Tonga.[124]

Hawaii is a leader in aquaculture. As early as 1778, Captain Cook reported 360 fish farms on the island of Kauai, producing an estimated 2 million pounds of fish annually.

The enzyme papain, extracted from papayas, is used in some meat tenderizers. In Hawaii, papaya juice is applied to jellyfish stings to reduce the pain.

Red salt, made from salt mixed with iron-rich earth, is a Hawaiian specialty, traditionally reserved for important feasts.[135]

When drinking kava, it is considered polite to drip a few drops onto the ground and say "*manuia*" as a blessing before quaffing

[SAMPLE MENU]

A Native Hawaiian Dinner

***Lomi-Lomi* (Marinated Salmon)[a,b,c]**

Chicken *Lu'au* (Chicken with Taro/Spinach Leaves)[a,c]

***Poi* (fermented taro root)[a] or Sweet Potatoes[c]**

***Haupia* (Coconut Dessert)[a,b,c] or Guava Cake[a,b,c]**

Fruit Juice or Coffee

[a]Laudan, R. 1996. *The food of Paradise: Exploring Hawaii's culinary heritage.* Honolulu: University of Hawai'i Press.

[b]Local Kine Recipes at www.hawaii.edu/recipes

[c]Luau Foods and Recipes at http://alohaworld.com/ono/

Role of Food

Food holds particular importance within most Pacific Island cultures. Sharing food is a way of demonstrating generosity and support for family and village. It is also a way of expressing prosperity or social standing. Many events are celebrated with feasting, and food is eaten to excess as part of the ceremony in some regions. Traditionally, gender roles were defined by food interactions. Boys and girls were often raised similarly until the age of eight or nine, at which time they were separated for training in food procurement (farming and fishing) or food preparation (cooking and food storage). Throughout the Pacific Islands, gifts of food are given often. Because the gifts are given without expectation of reciprocity, it is a serious affront to reject any item presented.

Therapeutic Uses of Foods

Numerous botanicals are used by Native Hawaiians as cures. Most notable is the pepper plant (*Piper methusticum*—related to the betel vine), which is used to make kava or awa. It is sometimes used medicinally as an analgesic or narcotic.[137] In diluted form, it is used as a sedative and given to teething infants. Noni (Indian Mulberry) is consumed as a juice for anorexia, renal problems, urinary tract infections, hypertension, diabetes, musculoskeletal pain, to boost the immune system, and to prevent cancer.[129] Turmeric is consumed as a blood cleanser, and breadfruit is made into a tea to treat high blood sugar and elevated blood pressure.

Contemporary Food Habits in the United States

Adaptations of Food Habits

Ingredients and Common Foods. In Hawaii, today's population often eats fare that combines elements of Eastern and Western foods, typified by the plate lunch: rice heaped with one or two types of meat, poultry, or fish (such as teriyaki beef, spaghetti and meat sauce, fried mahi-mahi, pit-cooked pork, curry, or Spam) covered with gravy, served with a scoop of macaroni or potato salad, and eaten with chopsticks. Soy sauce is the most common condiment.[135] Ethnic cuisines and typical American dishes are also widely consumed. Snacking is prevalent and may include noodles topped with meat (saimin), steamed pork rolls (manapua) or a large ball of glutinous rice with a bit of pickled daikon or teriyaki tuna tucked in the middle (musubi). Crack seed is especially popular, consisting of dried, salted, and sugared fruits (e.g., guava, sliced ginger, lemon peel, mango, plum) that provide a sweet-and-sour flavor experience. Some are also seasoned with spices, such as anise. The seed kernels (in whole preserved fruit) are traditionally cracked open with the teeth, adding extra flavor and giving the descriptive name to the treats.[135]

The dietary changes made by other Pacific Islanders often begin before immigration to the mainland United States. Most Pacific Islanders in their homelands are highly reliant on imported foods, particularly processed items such as canned meats and fish, cooking oil, mayonnaise, cookies, breakfast cereals, and soft drinks. One study of Samoa indicated that 80 percent of all foods found in one market were from the United States, New Zealand, or Australia.[138] Consumption of native foods, such as taro and coconut, increases on Sundays.[139] In Guam, a high intake of meats, fried foods, foods cooked in coconut milk, white rice, and sweetened beverages and a low intake of produce have been noted.[140] Nutrient-rich native foods, such as fish, yams, papayas, and mangoes, were rarely consumed by Guamanian children in another study.[140] Traditional starches, such as taro root and cassava, are often reserved for special-occasion feasts.[141]

Little has been reported on Pacific Islander food habits in the United States. A study comparing the intakes of Western Samoans (living in a poorer, less westernized culture) to those of American Samoans (living in a wealthier, more westernized culture) found that the Western Samoans ate a diet higher in total fat due to a reliance on coconut cream compared to a diet higher in protein, cholesterol, and salt due to higher consumption of processed foods eaten by American Samoans.[142]

Older studies on Samoans who have moved to Hawaii show diets with greater variety of foods; traditional foods only contribute minimally to daily intake, and items such as rice, bread, sugar, beef, canned fish, milk, soft drinks, and sweetened fruit beverages make up most of the diet.[139] Available health statistics indicate that a diet substantially higher in fat and simple carbohydrates and lower in fruits, vegetables, and fiber has been adopted by many immigrants.

Meal Composition and Cycle. Three meals each day are common for most Pacific Islanders. Breakfast is most often cereal with coffee. More traditional meals may be eaten for lunch and dinner; a few Hawaiians still eat poi once or twice a day. Fruit appears more often as part of the meal rather than as a snack. It is believed that Sunday feasting among Samoans is still prevalent.

Nutritional Status

Nutritional Intake. Serious nutritional deficiencies are uncommon in most Pacific Islander diets. A high calorie intake generally guarantees nutritional sufficiency. However, research on Samoans in Hawaii indicated that fewer calories were consumed than in Samoa and that riboflavin, calcium, and iron intake was low.[139] Further, a study of children in Guam found that three-quarters consumed fruits and vegetables less than once daily.[143] In a survey of Native Hawaiians and Pacific Islanders, adults consume approximately one fruit or vegetable daily on average.[144] Rice, meat, powdered fruit drinks, milk, and fortified cereals were the primary sources of vitamins and minerals in the diets of Guamanian children in other research. Low intakes of calcium, vitamin E, and folate were identified.[140]

Hawaiians and Samoans born in 1992 had lower than the U.S. average life expectancy, while Guamanians were expected to live longer than average.[58] Hawaiian data indicate that life expectancy for Hawaiians is lower than that for whites and Asians living in the state.[60] Heart disease is the leading cause of death for Samoan Americans and Native Hawaiians, with rates estimated to be 44 percent higher than in the general population.[58,145] Comparative data specific to California found deaths from coronary heart disease were over 2.5 times the average for the state population among some Pacific Islanders—Samoans had rates 1.5 times higher, Guamanians 1.3 times higher, and Native Hawaiians 1.1 times higher.[94] Deaths from strokes were also notably higher. High total cholesterol levels are also noted among Hawaiians and Samoans. The few studies on diet suggest that nearly all fat in the traditional Pacific Islander diet is saturated (36 percent of calories from coconut fats alone) and that saturated fats are also predominant in acculturated diets.[139] Hypertension is slightly higher among Native Hawaiians than among whites in Hawaii, and in one study of low-income Pacific Islanders in the state, rates were five times the reported average for whites.[146] In California, 37 percent of Native Hawaiians and Picific Islanders reported they had been told that they had hypertension.[144] Native Hawaiians show a higher incidence rate for cancers of the breast, cervix, uterus, esophagus, larynx, lung, pancreas, stomach, and multiple myeloma than whites. Survival rates are also generally lower.[63] Cancer mortality rates for Chomorros in Guam suggest this group suffers disproportionately from cancers of the mouth and pharynx, lung and bronchus, colon-rectum-anus, breast, and prostate when compared to U.S. averages.[147]

Infant mortality data suggest that Native Hawaiians have rates above the national average in all categories.[64] California data also report significantly higher rates of infant deaths among Samoans and Pacific Islanders, though for Native Hawaiians and Guamanians, infant deaths were nearly half those of the total population despite

Areca nuts are used throughout the Pacific Islands, and in Hawaii they are considered a stimulant. Guamanians are especially fond of chewing them. The practice is traditionally passed from grandparent to grandchild and is common at social occasions. Chomorrans chew them after meals.[122]

One study of children in Guam found that 26 percent ate at least three meals each week at fast-food restaurants, and 53 percent consumed at least two cans of sweetened soda daily.[143]

Research on Asian, mixed ethnicity, and white adolescent girls in Hawaii found that all groups could benefit from increased intake of fruits, vegetables, and calcium-rich foods.[177]

higher rates of low-birth-weight neonates.[94] In a study comparing perinatal outcomes in U.S. Asian and Pacific Islander subgroups, Pacific Islanders were found to be at increased risk for gestational diabetes, and 25 percent of births exceeded 4,000 grams (high birth weights are also reported in American Samoa, as are rapid weight gains in infants throughout the Samoan islands).[67]

Research on infant feeding practices in Hawaii suggests that the percentage of Hawaiians in Hawaii breast feeding their babies is more than 50 percent. Another approximately 30 percent mix breast feeding with a bottle. Native Hawaiian women were more likely to introduce solid foods to their infants before four months of age than whites, Filipinos, or Japanese women in Hawaii; sweetened beverages and baby food desserts were common items included in the infants' diets. Tooth decay due to excessive consumption of sweetened beverages in a bottle is three times the U.S. average among children five years old and younger in Hawaii.[148]

Chronic disease rates associated with westernization are of primary concern. Obesity rates among Pacific Islanders are some of the highest in the world, regardless of where they live. Native Hawaiian and Samoan adults show average body mass indices that exceed those used to define obesity in the general American population.[149,150,144,151] A study of Samoans living in Western Samoa and Hawaii showed large weight gains with increasing modernization. In rural Samoa the prevalence for being overweight was 33 percent for men and 46 percent for women; in Hawaii the prevalence was 75 percent for men and 80 percent for women. Sixty-six percent of Native Hawaiians are thought to be overweight or obese, and a study on Asian Americans reported 74 percent of Pacific Islanders were in that category.[141,152,146] In a study of Hawaiian preschool children, 27 percent of Samoan two to four year olds were overweight (BMI ≥95th percentile), a rate double that of any other ethnic group.[153,154] Native Hawiian children are more susceptible to acanthosis nigricans, a hyperpigmentation disorder associated with obesity and type 2 diabetes.[155]

Obesity may be caused directly by overeating (within the context of family and church activities) combined with inadequate physical exercise, and indirectly because overweight was an aesthetic preference in the traditional cultures. Heredity may be a factor in the rate of obesity in Pacific Islanders, or it may be due in part to the change in the types and amount of carbohydrates consumed.[156] A small study of Native Hawaiians who ate as much of traditional starchy vegetables and greens as they desired supplemented with small amounts of fish and poultry reduced total calorie intake, lost weight, and lowered their blood pressure and total serum cholesterol. Those who were diabetic reduced or eliminated their insulin requirements.[157] Lower levels of physical activity among Pacific Islanders may be another factor.[158,143,159] Additionally, researchers have found that when Pacific Islanders are compared to whites with the same percentage of body fat, the Pacific Islanders have higher BMIs, suggesting greater nonfat density. BMI cutoffs for overweight and obesity that are standardized for whites may need to be modified.[160] Many Pacific Islanders value a larger body size. One study in Hawaii found that while Pacific Islanders believe whites and Europeans should be slim, they also think that Pacific Islanders should be overweight or obese.[146] Though Samoans living in westernized New Zealand identified a slim body as ideal, those who were above normal weight did not consider themselves to be obese and were positive about their body size and health.[161] Research in Micronesia reported that many mothers of overweight children associate thinness with illness.[162]

Risk for type 2 diabetes mellitus is also high for Pacific Islanders.[152,144] Prevalence estimates are the highest for Native Hawaiians (two to six times the 5 to 7 percent prevalence among whites) and from 9 to 16 percent for Samoans.[150,163,164] Health statistics from California and Hawaii suggest mortality rates from diabetes are 1.6 to 3.7 times higher for Pacific Islanders than for the general population.[94,150] Studies have found, however, that Polynesians and Pacific Islanders are not intrinsically insulin resistant, and that most risk for diabetes is accounted for by high rates of overweight and obesity.[152,164,146] Central body adiposity is also predictive.[165] It is postulated that Samoans may be especially susceptible to the kidney damage associated with hypertension because end-stage renal failure is a common cause of death in American Samoan diabetes patients.

Counseling. Concepts regarding the role of the individual in health care may be an issue for Pacific Islander American clients. Biomedicine

EXPLORING GLOBAL CUISINE—Australian and New Zealand Fare

The large island continent of Australia and the two-island nation of New Zealand are thought to have been initially populated by peoples from Southeast Asia. Centuries later, Polynesians traveled to New Zealand, intermingling with the existing native peoples to become the Maoris. Despite a shared heritage, there are only a few similarities between the cooking of Southeast Asia and the Pacific Islands and the fare of Australia and New Zealand. British colonization of the countries in the eighteenth and nineteenth centuries overwhelmed the original inhabitants and their culinary traditions.

Meat is the mainstay of the Australian and New Zealand diets. Beef is most popular in Australia, especially steak, served with fried eggs or stuffed with oysters (carpetbagger steak). Lamb is favored in New Zealand (and also well liked in Australia), typically roasted and served with mint sauce or barbecued. Other meats and poultry are uncommon in both nations, though some wild game, particularly venison, boar, duck, and pheasant, is found in New Zealand. Fish and seafood, however, are eaten often, including shrimp, oysters, scallops, spiny lobster, and crawfish (called yabbies in Australia). Most meats and fish are simply prepared, by roasting, pan-frying, braising or poaching, and grilling. Traditionally, seasoning was very limited, but today, an international array of herbs and spices are used. For example, in New Zealand, Asian Indian lamb curries and Greek gyros are popular. Minced and ground meats appear in numerous dishes. In Australia, shepherd's pie (ground lamb topped with mashed potatoes) and hamburgers topped with fried eggs and beet slices are common. Battered, fried sausages are a favorite snack in New Zealand. In both nations individual steak, sausage, bacon, and egg, or fish pies (consumed with tomato sauce or ketchup) are popular.

Numerous fruits and vegetables are grown in the mild climate. Potatoes, beets, peas, carrots, and corn are typical accompaniments to meals, though more exotic produce, such as eggplants, feijoas, kiwifruit (Chinese gooseberries), pineapple, and tamarillos are also available. Bread rounds out the meal. Sweets are eaten daily. Scones are common, as are biscuits (cookies) such as lamingtons (chocolate coconut) and ANZAC biscuits (oatmeal cookies provided to the Australian and New Zealand Army Corps during the world wars). Puddings and custards are especially popular. Both countries claim pavlova, a meringue and fruit dish topped with whipped cream, as a national dessert. Three meals each day are the norm. In Australia, an afternoon break for tea or beer is customary, and in New Zealand many people break for morning and afternoon tea. Australia is noteworthy for its beers and hearty wines, beverages that New Zealand is also successfully producing.

The Aboriginal foods of Australia, called bush tucker, include kangaroo, wombat, emu, duck, fish, shrimp, snakes, and lizards, witchetty grubs, and wild plants such as yams, onions, wattle seeds, and quandong (a peach-like fruit). Most foods were prepared simply over a fire, in the ashes, or boiled. Foods introduced by Polynesians formed the foundation of the Maori diet in New Zealand, such as kumara (sweet potatoes), taro, and ti plants. Native greens and fruits were also available. Fish, seafood, birds, and sea mammals provided protein (there were no indigenous land mammals in the islands). The Polynesian influence is also seen in the use of the *hangi*, a pit-roasting method similar to imu used for ceremonial occasions.

presumes that better health depends on behavior changes made by a client. Pacific Islanders view the role of the individual as interdependent on the group and may not take responsibility for personal health.[141] Family members may expect to be involved in the care of sick relatives, and decisions regarding treatment are often made through consultation with the entire family.[166] Because health is defined as a harmonious balance of the physical, social, natural, and spiritual worlds, many Pacific Islanders prefer a comprehensive approach to care, which may involve religious rituals, including prayer.

Samoan Americans generally seek care for symptomatic relief and are typically uninterested in long-term approaches to disease prevention or management.[134] Low compliance rates in the treatment of hypertension and diabetes in Pacific Islanders have been noted.[144] A study on Native Hawaiian health needs found that cultural differences were a primary reason for underuse of services.[167] Young adults defined health in terms of participating in their responsibilites, which include working and caring for their families.[168] Further, some Pacific Islanders believe that illness is the will of God and may delay obtaining care until symptoms are advanced.

Language difficulties or linguistic isolation may also occur. According to the 2000 U.S. Census, nearly one-third of Tongan Americans and nearly 20 percent of Samoan Americans do not speak English well. Rates for Guamanians are close to 15 percent.[125] Samoans expect exceptional politeness in interactions; showing

Many Pacific Islanders are lactose-intolerant; it is estimated that 50 percent of Samoan adults do not tolerate milk.[178]

In Guam, an extremely high prevalence of both amyothrophic lateral sclerosis (called *letigo*) and Parkinsonism dementia (known as *bodig*) has been reported. Incidence has decreased significantly in recent years, however, suggesting environmental causes may be to blame.[179]

DISCUSSION STARTERS: WORKING WITH DIFFERENCES IN DIET AND CULTURE

In small groups of 3–4, compare and contrast the diet and culture of Filipino, Vietnamese, Cambodian, and Laotian Americans, with each group focusing on a different aspect of the diet and culture of these groups:

Group A:	The food habits and the typical eating etiquette and meal composition of these four immigrant groups
Group B:	Issues involved in counseling these immigrant groups on diet and health
Group C:	Attitudes within each immigrant group toward diet, health, and medical treatment, notably attitudes toward traditional home culture medical treatment and U.S. biomedicine
Group D:	Amount of obesity, diabetes, hypertension, and other diseases within each immigrant group

Within your group, try to come to a consensus on what findings to report to the rest of the class. Before breaking up, assign a number to each group member: A1, A2, A3, A4; B1, B2, and so forth. Form new groups with all the 1's in a group; all the 2's in another group; all the 3's another group; and so on. In your new group, report the findings of your previous group, and, as a group, discuss the relationship of traditional attitudes toward diet and health and changes in diet and health due to immigration to the United States.

Estimates indicate that over 40 percent of immigrants from the Marshall Islands do not speak English well.[125]

irritation, anger, or other hostile emotion is considered rude and a sign of weakness.[121] A reserved conversational style and the desire to maintain harmony may result in some Pacific Islanders enduring pain rather than expressing it.[166] Judgmental or accusatory attitudes regarding lifestyle (especially weight gain) may cause Pacific Islanders to avoid further counseling.

When entering a room, Samoans walk around to greet each person with a smile and handshake. Eye contact is expected. Both Samoans and Tongans are concerned that all participants in a conversation be at an equal level; for example, everyone should be sitting on mats or in chairs. It is offensive to stand while speaking to someone who is sitting. When seated, legs should be crossed or kept close to the body; extended legs or pointing one's feet at a person is considered poor manners.[121]

One study reported that weight-loss interventions based on traditional Hawaiian foods were well accepted by Native Hawaiians, particularly those who emphasized cultural values and provided group support. However, most participants found it difficult to adhere to the diet long-term, citing difficulties in obtaining fresh produce and the pressures of an obesogenic environment.[169] A study in Micronesia found significant cultural conflict in homes where mothers attempted to restrict the food intake of their overweight children. Food is associated with love, and grandparents who perceive grandchildren as too thin often accuse mothers of inadequate care.[162]

Samoan women are traditionally treated to a rich coconut drink called vaisalo after childbirth.

An in-depth interview should be used to determine if a client has any traditional health beliefs or practices regarding a specific condition and if religious affiliation is a factor. Due to the paucity of research, coworkers or a client's family members may provide significant information regarding particular Pacific Islander groups.

REVIEW QUESTIONS

1. Choose one typical Filipino dish and describe it. Explain how it conforms to the principles of Filipino cooking. Select one or two ingredients and discuss whether they are due to an influence from another culture and why that might have happened.
2. Describe the traditional health and dietary beliefs for the prevention and treatment of disease in one Southeast Asian immigrant group. How does the concept of "balance" for maintenance of health fit within Southeast Asian health beliefs?
3. Pick one type of traditional healer used in Southeast Asian countries. Describe this type of healer's practice and research whether they currently practice in the United States.
4. List the indigenous foods of the Pacific Islands. Pick two from your list, describe how they might be prepared today, and discuss whether they are considered to have any special dietary or health properties.
5. Processed meat products, such as Spam, are common in the Philippines and Pacific Islands. What is Spam, where does it come from, and how did it become prevalent in the Southeast Asian and Pacific Island diet? Include a recipe using Spam.

REFERENCES

1. Selected Population Profile in the United States Population Group: Filipino alone or in any combination Data Set: 2007 American Community Survey 1- year Estimates Survey: American Community Survey http://factfinder.census.gov/servlet/IPTable?_bm=y&-context=ip&-reg=ACS_2007_1YR_G00_S0201:038;ACS_2007_1YR_G00_S0201PR:038;ACS_2007_1YR_G00_S0201T:038;ACS_2007_1YR_G00_S0201TPR:038&-qr_name=ACS_2007_1YR_G00_S0201&-qr_name=ACS_2007_1YR_G00_S0201PR&-qr_name=ACS_2007_1YR_G00_S0201T&-qr_name=ACS_2007_1YR_G00_S0201TPR&-ds_name=ACS_2007_1YR_G00_&-tree_id=306&-redoLog=false&-geo_id=01000US&-geo_id=NBSP&-search_results=16000US3651000&-format=&-_lang=en
2. Becker, G. 2003. Cultural expressions of bodily awareness among chronically ill Filipino Americans. *Annals of Family Medicine, 1,* 113–118.
3. Bankston, C. L. 2000. Vietnamese Americans. In *Gale Encyclopedia of Multicultural America,* R.V. Dassanowsky & J. Lehman (Eds.). Farmington Hills, MI: Gale Group.
4. Selected Population Profile in the United States Population Group: Vietnamese alone or in any combination Data Set: 2006 American Community Survey: American Community Survey U.S. Census. http://factfinder.census.gov/servlet/IPTable?_bm=y&-geo_id=01000US&-qr_name=ACS_2006_EST_G00_S0201&-qr_name=ACS_2006_EST_G00_S0201PR&-qr_name=ACS_2006_EST_G00_S0201T&-qr_name=ACS_2006_EST_G00_S0201TPR&-ds_name=ACS_2006_EST_G00_&-reg=ACS_2006_EST_G00_S0201:048;ACS_2006_EST_G00_S0201PR:048;ACS_2006_EST_G00_S0201T:048;ACS_2006_EST_G00_S0201TPR:048&-_lang=en&-redoLog=false&-format=
5. Bankston, C.L. 2000. Cambodian Americans. In *Gale Encyclopedia of Multicultural America,* R.V. Dassanowsky & J. Lehman (Eds.). Farmington Hills, MI: Gale Group.
6. Reeves, T.J., & Bennett, C.E. 2004. *We the People: Asians in the United States.* Washington, DC: U.S. Census Bureau.
7. Selected Population Profile in the United States Population Group: Laotian alone Data Set: 2005 American Community Survey: American Community Survey U.S. Census.
8. Bankston, C.L. 2000. Laotian Americans. In Gale Encyclopedia of Multicultural America, R.V. Dassanowsky & J. Lehman (Eds.). Farmington Hills, MI: Gale Group.
9. Bankston, C.L. 2000. Hmong Americans. In *Gale Encyclopedia of Multicultural America,* R.V. Dassanowsky & J. Lehman (Eds.). Farmington Hills, MI: Gale Group.
10. Adler, S.R. 1995. Refuge stress and folk belief: Hmong sudden deaths. *Social Science and Medicine, 40,* 1623–1629.
11. Anderson, J.N. 1983. Health and illness in Filipino immigrants. *Western Journal of Medicine, 139,* 811–819.
12. Pacquiao, D.F. 2003. People of Filipino heritage. In *Transcultural Health Care,* 2nd ed. L.D. Purnell & B.J. Paulanka (Eds.). Philadelphia: FA Davis Company.
13. Nuttall, P., & Flores, F.C. 1997. Hmong healing practices used for common childhood illnesses. *Pediatric Nursing, 23,* 247–214.
14. Montepio, S.N. 1986–1987. Folk medicine in the Filipino American experience. *Amerasia Journal, 13,* 151–162.
15. Vance, A.R., & Davidhizar, R. 1999. Developing cultural sensitivity when your client is Filipino American. *Journal of Practical Nursing,* December, 16–20.
16. Hart, D.V. 1981. Bisayan Filipino and Malayan folk medicine. In G. Henderson & M. Primeaux (Eds.), *Transcultural Health Care.* Menlo Park, CA: Addison-Wesley.
17. Orque, M.S. 1983. Nursing care of Filipino American patients. In M.S. Orque, B. Bloch, & L.S.A. Monrroy (Eds.), *Ethnic Nursing Care: A Multicultural Approach.* St. Louis: Mosby.
18. Nowack, T.T. 2003. People of Vietnamese heritage. In *Transcultural Health Care,* 2nd ed. L.D. Purnell & B.J. Paulanka (Eds.). Philadelphia: FA Davis Company.
19. Orque, M.S. 1983. Nursing care of South Vietnamese patients. In M.S. Orque, B. Bloch, & L.S.A. Monrroy (Eds.), *Ethnic Nursing Care: A Multicultural Approach.* St. Louis: Mosby.
20. Stephenson, P.H. 1995. Vietnamese refugees in Victoria, BC: An overview of immigrant and refugee health care in a medium-sized Canadian urban centre. *Social Science and Medicine, 40,* 1631–1642.
21. Her, C., & Culhane-Pera, K. 2004. Culturally responsive care for Hmong adults. *Postgraduate Medicine, 116,* 39–46.
22. Capps, L.L. 1994. Change and continuity in the medical culture of the Hmong in Kansas City. *Medical Anthropology Quarterly, 8,* 161–177.
23. Cheyne, J.A. 2001. The ominous numinous: Sensed presence and "other" hallucinations. *Journal of Consciousness Studies, 8,* 133–150.
24. Hinton, D.E., Pich, V., Chhean, D., & Pollack, M.H. 2005. 'The ghost pushes you down': Sleep paralysis-type panic attacks in a Khmer refugee population. *Transcultural Psychiatry, 42,* 46–77.
25. Plotnikoff, G.A., Numrich, C., Wu, C., Yang, D., & Xiong, P. 2002. Hmong shamanism: Animist spiritual healing in Minnesota. Minnesota Medicine, online at http://mnmed.org/publications/MNMed2002/June/Plotnikoff.html
26. Gilman, S.C., Justice, J., Saepharn, K., & Charles, G. 1992. Use of traditional and modern health services by Laotian refugees. *Western Journal of Medicine, 157,* 310–315.
27. Hoang, G.N., & Erickson, R.V. 1982. Guidelines for providing medical care to Southeast Asian refugees. *Journal of the American Medical Association, 248,* 710–714.
28. Buchwald, D., Panwala, S., & Hooten, T.M. 1992. Use of traditional health practices by Southeast Asian refugees in a primary care clinic. Western *Journal of Medicine, 156,* 507–512.

29. Steinberg, R. 1970. *Pacific and Southeast Asian cooking.* New York: Time-Life Books.
30. Zibart, E. 2001. *The ethnic food lover's companion: Understanding the cuisines of the world.* Birmingham, AL: Menasha Ridge Press.
31. Gelle, G.G. 1997. *Filipino cuisine: Recipes from the islands.* Santa Fe: Red Crane.
32. Routhier, N. 1999. *The foods of Vietnam.* New York: Stewart, Tabori & Chang.
33. O'Neill, M. 2001, July 23. Letter from Cambodia: Home for dinner. *The New Yorker,* 55–63.
34. Foster, D. 2000. *The global etiquette guide to Asia.* New York: John Wiley & Sons.
35. Claudio, V.S. 1994. *Filipino American food practices, customs, and holidays.* Chicago: American Dietetic Association/American Diabetic Association.
36. Philippine Herbal Medicine. 2006. Online at http://herbal-medicine.philsite.net/index.htm
37. Bodo, K., & Gibson, N. 1999. Childbirth customs in Vietnamese traditions. *Canadian Family Physician, 45,* 695–697.
38. Muecke, M.A. 1983b. In search of healers—Southeast Asian refugees in the American health care system. *Western Journal of Medicine, 139,* 835–840.
39. Nguyen, M.D. 1985. Culture shock: A review of Vietnamese culture and its concepts of health and disease. *Western Journal of Medicine, 142,* 409–412.
40. Vu, H.H. 1996. Cultural barriers between obstetrician-gynecologists and Vietnamese/Chinese immigrant women. *Texas Medicine, 92,* 47–52.
41. Ikeda, J.P., Ceja, D.R., Glass, R.S., Harwood, J.O., Lucke, K.A., & Sutherlin, J.M. 1991. Food habits of the Hmong living in central California. *Journal of Nutrition Education, 23,* 168–175.
42. Rairdan, B., & Higgs, Z.R. 1992. When your patient is a Hmong refugee. *American Journal of Nursing,* (March), 52–55.
43. Lewis, J.S., & Glaspy, M.F. 1975. Food habits and nutrient intakes of Filipino women in Los Angeles. *Journal of the American Dietetic Association, 67,* 122–125.
44. Felicitas A. dela Cruz.Food Intake And Dietary Acculturative Changes Of First Generation Filipino Americans Western Institute of Nursing 2010 meeting abstract
45. Dirige, O.V. 1995, February 28. Filipino-American diet and foods. *Asian American Business Journal,* 11–17.
46. Tong, A. 1986. Food habits of Vietnamese immigrants. *Family Economic Review, 2,* 28–30.
47. Crane, N.T., & Green, N.R. 1980. Food habits and food preferences of Vietnamese refugees living in Northern Florida. *Journal of the American Dietetic Association, 76,* 591–593.
48. Story, M., & Harris, L.J. 1988. Food preferences, beliefs, and practices of Southeast Asian refugee adolescents. *Journal of School Health, 58,* 273–276.
49. Wiecha, J.M., Fink, A.K., Wiecha, J., & Hebert, J. 2001. Differences in dietary patterns of Vietnamese, white, African-American, and Hispanic adolescents in Worcester, Mass. *Journal of the American Dietetic Association, 101,* 248–251.
50. Richard C. Yang, Paul K. Mills. Dietary and Lifestyle Practices of Hmong in California. *Journal of Health Care for the Poor and Underserved*—Volume 19, Number 4, November 2008, pp. 1258–1269.
51. Story, M., & Harris, L.J. 1989. Food habits and dietary change of Southeast Asian refugee families living in the United States. *Journal of the American Dietetic Association, 89,* 800–803.
52. Sechena, R., Liao, S., Lorenzana, R., Nakano, C., Polissar, N., & Fenske, R. 2003. Asian American and Pacific Islander seafood consumption—a community-based study in King County, Washington. *Journal of Exposure Analysis and Environmental Epidemiology, 13,* 256–266.
53. L. Franzen, C. Smith Differences in stature, BMI, and dietary practices between US born and newly immigrated Hmong children. *Social Science & Medicine,* Volume 69, Issue 3, Pages 442–450 (2009).
54. Stang J, Kong A, Story M, Eisenberg M E, Neumark-Sztainer D. Food and weight-related patterns and behaviors of hmong adolescents. *Journal of the American Dietetic Assn. 2007,* vol. 107:6, pp. 936–941.
55. Tamir, A., & Cachola, S. 1994. Hypertension and other cardiovascular risk factors. In N.W.S. Zane, D.T. Takeuchi, and K.N.J. Young (Eds.), *Confronting Critical Health Issues of Asian and Pacific Islander Americans.* Thousand Oaks, CA: Sage.
56. Berg, J.A., Rodriguez, D., de Guzman, C.P., & Kading, V.M. 2001. Health status, health and risk behaviors of Filipino Americans. *Journal of Multicultural Nursing, 7,* 29–36.
57. Stavig, G.R., Igra, A., & Leonard, A.R. 1988. Hypertension and related health issues among Asians and Pacific Islanders in California. *Public Health Reports, 103,* 28–37.
58. Hoyert, D.L., & Kung, H.C. 1997. Asian or Pacific Islander mortality, selected states, 1992. *Monthly Vital Statistics Report, 46* (suppl.), 1–63.
59. Barnes PM, Adams PF, Powell-Griner E. Health characteristics of the Asian adult population: United States, 2004–2006. *Advance data from vital and health statistics; no 394.* Hyattsville, MD: National Center for Health Statistics. 2008.
60. Gardner, W.E. 1994. Mortality. In N.W. S. Zane, D.T. Takeuchi, & K.N.J. Young (Eds.), *Confronting Critical Health Issues of Asian and Pacific Islander Americans.* Thousand Oaks, CA: Sage.
61. Rossing, M.A., Schwartz, S.M., & Weiss, N.S. 1995. Thyroid cancer incidence in Asian migrants to the United States and their descendants. *Cancer Causes and Control, 6,* 439–444.
62. Cooper, G.S., Yuan, Z., & Rimm, A.A. 1997. Racial disparity in the incidence and case-fatality of colorectal cancer: Analysis of 329 United States counties. *Cancer Epidemiology Biomarkers & Prevention, 6,* 283–285.
63. Jenkins, C.N.H., & Kagawa-Singer, M. 1994. Cancer. In N.W.S. Zane, D.T. Takeuchi, & K.N.J. Young (Eds.), *Confronting Critical Health Issues of Asian and Pacific Islander Americans.* Thousand Oaks, CA: Sage.
64. National Center for Health Statistics. 2005. *Health, United States, 2005 with Chartbook on Trends in the Health of Americans.* Hyattsville, MD: Author.

65. Fuentes-Afflick, E., & Hessol, N.A. 1997. Impact of Asian ethnicity and national origin on infant birth weight. *American Journal of Epidemiology, 145,* 148–155.
66. Kieffer, E.C., Martin, J.A., & Herman, W.H. 1999. Impact of maternal nativity on the prevalence of diabetes during pregnancy among U.S. ethnic groups. *Diabetes Care, 22,* 729–735.
67. Rao, A.K., Cheng, Y.W., & Caughey, A.B. 2006. Perinatal complications among different Asian-American subgroups. *American Journal of Obstetrics and Gynecology, 194,* e39–41.
68. Cheng, Q., & Gould, J.B. 2006. The Asian birth outcome gap. *Paediatric and Perinatal Epidemiology, 20,* 279–289.
69. Lauderdale, D.S., & Rathouz, P.J. 2000. Body mass index in a US national sample of Asian Americans: The effects of nativity, years since immigration and socioeconomic status. *International Journal of Obesity, 24,* 1188–1194.
70. Popkin, B.M., & Udry, J.R. 1998. Adolescent obesity increases significantly in second and third generation U.S. immigrants: The National Longitudinal Study of Adolescent Health. *Journal of Nutrition, 128,* 701–706.
71. Araneta, M.R.G., Wingard, D.L., & Barrett-Conner, E. 2002. Type 2 diabetes and metabolic syndrome in Filipina American women: A high-risk nonobese population. *Diabetes Care, 25,* 494–499.
72. Cuasay, L.C., Lee, E.S., Orlander, P.P., Steffen-Batey, L., & Harris, C.L. 2001. Prevalence and determinants of type 2 diabetes among Filipino-Americans in Houston, Texas metropolitan statistical area. *Diabetes Care, 24,* 2054–2058.
73. Hedderson MM, Darbinian JA, Ferrara A. Disparities in the risk of gestational diabetes by race-ethnicity and country of birth. *Paediatr Perinat Epidemiol.* 2010 Sep; 24(5): 441–8.
74. Araneta, M.R.G., & Barrett-Conner, E. 2005. Ethnic differences in visceral adipose tissue and type 2 diabetes: Filipino, African-American, and white women. *Obesity Research, 13,* 1458–1465.
75. Palaniappan LP, Wong E, Shin JJ, Fortmann SP, Lauderdale DS. Asian Americans have a greater prevalence of metabolic syndrome despite lower body mass index. *Circulation.* 2009; 119: e363. Abstract.
76. Houa Vue, Marla Reicks. Individual and Environmental Influences on Intake of Calcium-rich Food and Beverages by Young Hmong Adolescent Girls. *Journal of Nutrition Education and Behavior.* 39:5, Pp 264–272, 2007.
77. Rosanhoff, A., & Calloway, D.H. 1982. Calcium source in Indochinese immigrants. *New England Journal of Medicine, 306,* 239–240.
78. Newman, V., Norcross, W., & McDonald, R. 1991. Nutrient intake of low-income Southeast Asian pregnant women. *Journal of the American Dietetic Association, 91,* 793–799.
79. Sargent, J.D., Stukel, T.A., Dalton, M.A., Freeman, J., & Brown, M.J. 1996. Iron deficiency in Massachusetts: Socioeconomic and demographic risk factors among children. *American Journal of Public Health, 86,* 544–550.
80. Ackerman, L.K. 1997. Health problems of refugees. *Journal of the American Board of Family Practice, 10,* 337–348.
81. Brown, J.E., Serdula, M., Cairns, K., Godes, J.R., Jacobs, D.R., Elmer, P., & Trowbridge, F.L. 1986. Ethnic group differences in nutritional status of young children from low-income areas of an urban county. *American Journal of Clinical Nutrition, 44,* 938–944.
82. Chow, R.T.P., & Krumholtz, S. 1989. Health screening of a RI Cambodian refugee population. *Rhode Island Medical Journal, 72,* 273–277.
83. Hann, R.S. 1994. Tuberculosis. In N.W.S. Zane, D.T. Takeuchi, & K.N.J. Young (Eds.), *Confronting Critical Health Issues of Asian and Pacific Islander Americans.* Thousand Oaks, CA: Sage.
84. Ziegler, V.S., Sucher, K.P., & Downes, N.J. 1989. Southeast Asian renal exchange list. *Journal of the American Dietetic Association, 89,* 85–92.
85. Elizabeth E. Dawson-Hahn MDa, Sarah L.M. Greenberg MDb, Joseph B. Domachowske MDc and Brad G. Olson Eosinophilia and the Seroprevalence of Schistosomiasis and Strongyloidiasis in Newly Arrived Pediatric Refugees: An Examination of Centers for Disease Control and Prevention Screening Guidelines. *J Pediatr.* 2010 Jun; 156(6): 1016–8.
86. Higgs DR, Weatherall DJ. The alpha thalassaemias. *Cell Mol Life Sci 2009; 66:* 1154–1162.
87. Molina, C.D., Molina, M.M., & Molina, J.M. 1988. Intestinal parasites in Southeast Asian refugees two years after immigration. *Western Journal of Medicine, 149,* 422–425.
88. National Cancer Institute, SEER Program. 1992. Five Most Common Cancers in Each Racial/Ethnic Group. Online at http://seer.cancer.gov/publications/ethnicity/topfive.pdf
89. Swallen, K.C., Glaser, S.L., Stewart, S.L., West, D.W., Jenkins, C.N., & McPhee, S.J. 1998. Accuracy of racial classification of Vietnamese patients in a population-based cancer registry. *Ethnicity & Disease, 8,* 218–227.
90. Bermingham, M., Brock, K., Nuguen, D., & Tran-Dinh, H. 1996. Body mass index and body fat distribution in newly-arrived Vietnamese refugees in Sydney, Australia. *European Journal of Clinical Nutrition, 50,* 698–700.
91. Van Minh, T., Thanh, L.C., Thi, B.N., Do, T.T., Tho, T.D., & Valensi, P. 1997. Insulinaemia and slight overweight: The case of Vietnamese hypertensives. *International Journal of Obesity Related Disorders, 21,* 897–902.
92. Duong, D.A., Bohannon, A.S., & Ross, M.C. 2001. A descriptive study of hypertension in Vietnamese Americans. *Journal of Community Health Nursing, 18,* 1–11.
93. Ross, M.C., Duong, D., Wiggins, S.D., & Daniels, S. 2000. Descriptive study of hypercholesteremia in a Vietnamese population of the Gulf Coast. *Journal of Cultural Diversity, 7,* 65–71.
94. California Department of Health Services. 2004. *Sentinel Health Indicators for California's Multicultural Populations 1999–2001.* Author.

95. Her, C., & Mundt, M. 2005. Risk prevalence for type 2 diabetes mellitus in adult Hmong in Wisconsin: A pilot study. *WMJ, 104,* 70–77.
96. Mills, P.K., Yang, R.C., & Riordan, D. 2005. Cancer incidence in the Hmong in California, 1988–2000. *Cancer, 104,* 2969–2974.
97. Yang, R.C., Mills, P.K., & Riordon, D.G. 2000. Cervical cancer among Hmong women in California, 1988 to 2000. *American Journal of Preventive Medicine, 27,* 132–138.
98. Hee Yun Lee and Suzanne Vang. (2010). "Barriers to Cancer Screening in Hmong Americans: The Influence of Health Care Accessibility, Culture and Cancer Literacy." *Journal of Community Health 35:* 302-314.
99. Davis, J.M., Goldenring, J., McChesney, M., & Medina, A. 1982. Pregnancy outcomes of Indochinese refugees, Santa Clara County, California. *American Journal of Public Health, 72,* 742–743.
100. Li, D.K., Ni, H., Schwartz, S.M., & Daling, J.R. 1990. Secular change in birthweight among Southeast Asian immigrants to the United States. *American Journal of Public Health, 80,* 685–688.
101. Ghaemi-Ahmadi, S. 1992. Attitudes toward breast feeding and infant feeding among Iranian, Afghan, and Southeast Asian immigrant women in the United States: Implications for health and nutrition education. *Journal of the American Dietetic Association, 92,* 354–355.
102. Romero-Gwynn, E. 1989. Breast feeding pattern among Indochinese immigrants in northern California. *American Journal of Diseases of Children, 143,* 804–808.
103. Serdula, M.K., Cairns, K.A., Williamson, D.F., Fuller, M., & Brown, J.E. 1991. Correlates of breast-feeding in low-income population of whites, blacks, and Southeast Asians. *Journal of the American Dietetic Association, 91,* 41–45.
104. Tuttle, C.R., & Dewey, K.G. 1994. Determinants of infant feeding choices among Southeast Asian immigrants in northern California. *Journal of the American Dietetic Association, 94,* 282–286.
105. Axtell, R.E. 1991. Gestures: *The do's and taboos of body language around the world.* New York: Wiley.
106. Melissa L. Finucane,Carmit K. McMullen. Making Diabetes Self-management education Culturally Relevant for Filipino Americans in Hawaii. *The Diabetes Educator* September/October 2008 vol. 34 no. 5 841–853.
107. Brown, E.R., Ojeda, V.D., Wyn, R., & Levan, R. 2000. *Racial and Ethnic Disparities in Access to Health Insurance and Health Care.* Los Angeles: UCLA Center for Health Policy Research and Kaiser Family Foundation.
108. Jenkins, C.N.H., Le, T., McPhee, S.J., Stewart, S., & Ha, N.T. 1996. Health care access and preventive care among Vietnamese immigrants: Do traditional beliefs and practices pose barriers? *Social Science & Medicine, 43,* 1049–1056.
109. Ngo-Metzger, Q., Massagli, M.P., Carridge, B.R., Manocchia, M., Davis, R.B., Iezzoni, L.I., & Phillips, R.S. 2003. *Journal of General Internal Medicine, 18,* 44–52.
110. Nguyen, D.N., & O'Connell, M.B. 2002. Asian and Asian-American college students' awareness of osteoporosis. *Pharmacotherapy, 22,* 1047–1054.
111. Miller, J.A. 1995. Caring for Cambodian refugees in the emergency department. *Journal of Emergency Nursing, 21,* 498–502.
112. Spring, M.A., Ross, P.J., Etkin, N.L., & Deinhard, A.S. 1995. Sociocultural factors in the use of prenatal care by Hmong women, Minneapolis. *American Journal of Public Health, 85,* 1015–1017.
113. Helsel, D., Mochel, M., & Bauer, R. 2005. Chronic illness and Hmong shamans. *Journal of Transcultural Nursing, 16,* 150–154.
114. Barrett, B., Shadick, K., Schilling, R., Spencer, L., Del Rosario, S., Moua, K., & Vang, M. 1998. Hmong/medicine interactions: Improving cross-cultural health care. *Family Medicine, 30,* 179–184.
115. Westermeyer, J., Neider, J., & Vang, T.F. 1984. Acculturation and mental health: A study of Hmong refugees at 1.5 and 3.5 years postmigration. *Social Science and Medicine, 18,* 87–93.
116. Gold, S.J. 1992. Mental health and illness in Vietnamese refugees. *Western Journal of Medicine, 157,* 290–294.
117. Katz, J., Prescott, K., & Woolf, A.D. 1996. Strychnine poisoning from a Cambodian traditional remedy. *American Journal of Emergency Medicine, 14,* 475–477.
118. Winters, E., & Swartz, M. 2000. Hawaiians. In *Gale Encyclopedia of Multicultural America,* R.V. Dassanowsky & J. Lehman (Eds.). Farmington Hills, MI: Gale Group.
119. *Selected Population Profile in the United States Native Hawaiian alone.* 2005 American Community Survey.
120. *Selected Population Profile in the United States Samoan.* 2005 American Community Survey.
121. Cox, P. 2000. Samoan Americans. In *Gale Encyclopedia of Multicultural America,* R.V. Dassanowsky & J. Lehman (Eds.). Farmington Hills, MI: Gale Group.
122. Spear, J.E. 2000. Guamanian Americans. In *Gale Encyclopedia of Multicultural America,* R.V. Dassanowsky & J. Lehman (Eds.). Farmington Hills, MI: Gale Group.
123. *Selected Population Profile in the United States Guamanian or Chamorro alone.* 2005 American Community Survey.
124. Swain, L. 2000. Pacific Islander Americans. In *Gale Encyclopedia of Multicultural America,* R.V. Dassanowsky & J. Lehman (Eds.). Farmington Hills, MI: Gale Group.
125. Harris, P.M., & Jones, N.A. 2005. *We the People: Pacific Islanders in the United States.* Washington, DC: U.S. Census Bureau.
126. Cooper, A. 2000. Tongan Americans. In *Gale Encyclopedia of Multicultural America,* R.V. Dassanowsky & J. Lehman (Eds.). Farmington Hills, MI: Gale Group.
127. Snyder, P. 1984. Health service implications of folk healing among older Asian Americans and Hawaiians in Honolulu. *Gerontologist, 24,* 471–476.
128. Chang, H.K. 2001. Hawaiian health practitioners in contemporary society. *Pacific Health Dialog, 8,* 260–273.

129. Hilgenkamp, K., & Pescaia, C. 2003. Traditional Hawaiian healing and western influence. *Californian Journal of Health Promotion, 1,* 34–39.
130. Judd, N. 1998. Laau lapaau: Herbal healing among contemporary Hawaiian healers. *Pacific Health Dialog, 5,* 239–245.
131. Norton, S.A. 1998. Herbal medicines in Hawaii from tradition to convention. *Hawaii Medical Journal, 57,* 382–386.
132. Howard, A. 1986. Samoan coping behavior. In P.T. Baker, J.M. Hanna, & T.S. Baker (Eds.), *The Changing Samoans: Behavior and Health in Transition.* New York: Oxford University Press.
133. Kinloch, P. 1985. *Talking health but doing sickness: Studies in Samoan health.* Wellington, New Zealand: Victoria University Press.
134. Janes, C.R. 1990. *Migration, social change, and health: A Samoan community in California.* Stanford: Stanford University Press.
135. Laudan, R. 1996. *The food of paradise: Exploring Hawaii's culinary heritage.* Honolulu: University of Hawai'i Press.
136. Dresser, N. 2005. *Multicultural manners.* New York: John Wiley & Sons.
137. Natalie N. Young, Kathryn L. Braun, La'au lapa'au and Western Medicine in Hawai'i: Experiences and Perspectives of Patients Who Use Both. *J Health Serv Res Policy 2010;* 15: 54–61.
138. Shovic, A.C. 1994. Development of a Samoan nutrition exchange list using culturally accepted foods. *Journal of the American Dietetic Association, 94,* 541–543.
139. Hanna, J.M., Pelletier, D.L., & Brown, V.J. 1986. The diet and nutrition of contemporary Samoans. In P.T. Baker, J.M. Hanna, & T.S. Baker (Eds.), *The Changing Samoans: Behavior and Health in Transition.* New York: Oxford University Press.
140. Pobocik, R.S., & Richer, J.J. 2002. Estimated intake and food sources of vitamin A, folate, vitamin C, vitamin E, calcium, iron, and zinc for Gaumanian children aged 9 to 12. *Pacific Health Dialog, 9,* 193–202.
141. Fitzpatrick-Nietschmann, J. 1983. Pacific Islanders—Migration and health. *Western Journal of Medicine, 139,* 848–853.
142. Galanis, D.J., McGarvey, S.T., Quested, C., Sio, B., & Afele-Fam'amuli, S. 1999. Dietary intake of modernizing Samoans: Implications for risk of cardiovascular disease. *Journal of the American Dietetic Association, 99,* 184–190.
143. LeonGuerrero, R.T., & Workman, R.L. 2002. Physical activity and nutritional status of adolescents on Guam. *Pacific Health Dialog, 9,* 177–185.
144. Karen L. Moy, James F. Sallis, Katrine J. David. Health Indicators of Native Hawaiian and Pacific Islanders in the United States. *J Community Health* (2010) 35: 81–92.
145. Mokuau, N., Hughes, C.K., & Tsark, J.U. 1995. Heart disease and associated risk factors among Hawaiians: Culturally responsive strategies. *Health & Social Work, 20,* 46–51.
146. Wang, C-Y., Abbott, L., Goodbody, A.K., & Hui, W-T. 2002. Ideal body image and health status in low-income Pacific Islanders. *Journal of Cultural Diversity, 9,* 12–22.
147. Haddock, R.L., Talon, R.J., & Whippy, H.J. 2006. Ethnic disparities in cancer mortality among residents of Guam. *Asian Pacific Journal of Cancer Prevention, 7,* 411–414.
148. Goldberg, D.L., Novotny, R., Kieffer, E., Mor, J., & Thiele, M. 1995. Complementary feeding and ethnicity of infants in Hawaii. *Journal of the American Dietetic Association, 95,* 1029–1031.
149. Aluli, N.E. 1991. Prevalence of obesity in a Native Hawaiian population. *American Journal of Clinical Nutrition, 53,* 1556S–1560S.
150. Crews, D.E. 1994. Obesity and diabetes. In N.W.S. Zane, D.T. Takeuchi, & K.N.J. Young (Eds.), *Confronting Critical Health Issues of Asian and Pacific Islander Americans.* Thousand Oaks, CA: Sage.
151. Juarez DT, Samoa RA, Chung RS, Seto TB. Disparities in health, obesity and access to care among an insured population of Asian and Pacific Islander Americans in Hawai'i. *Hawaii Med J.* 2010 Feb; 69(2): 42–6.
152. McNeely, M.J., & Boyko, E.J. 2004. Type 2 diabetes prevalence in Asian Americans. *Diabetes Care, 27,* 66–69.
153. Baruffi, G., Hardy, C.J., Waslien, C.I., Uyehara, S.J., & Krupitsky, D. 2004. Ethnic differences in the prevalence of overweight among young children in Hawaii. *Journal of the American Dietetic Association, 104,* 1701–1707.
154. Lenna L. Liu, Joyce P. Yi, Jennifer Beyer, Elizabeth J. Mayer-Davis, Lawrence M. Dolan, Dana M. Dabelea, Jean M. Lawrence, Beatriz L. Rodriguez, Santica M. Marcovina, Beth E. Waitzfelder, Wilfred Y. Fujimoto, and SEARCH for Diabetes in Youth Study Group. Type 1 and Type 2 Diabetes in Asian and Pacific Islander U.S. Youth. *Diabetes Care March 2009 vol. 32* no. Supplement 2 S133–S140.
155. Acanthosis Nigricans In Hawaiian Children Vs Non-Hawaiian Children. Chloe Edinger, Jessica Song, Karen Robbins. *Ethnicity & Disease,* Volume 18, Spring 2008 Page 25–27.
156. Maskarinec, G., Takata, Y., Pagano, I., Carlin, L., Goodman, M.T., Le Marchand, L., Nomura, A.M., Wilkens, L.R., & Kolonel, L.N. 2006. Trends and dietary determinants of overweight and obesity in a multiethnic population. *Obesity, 14,* 717–726.
157. Shintani, T.T., Hughes, C.K., Beckham, S., & O'Conner, H.K. 1991. Obesity and cardiovascular risk intervention through the ad libitum feeding of traditional Hawaiian diet. *American Journal of Clinical Nutrition, 53,* 1647S–1651S.
158. Centers for Disease Prevention. 2004. Physical activity among Asians and native Hawaiian or other Pacific Islanders—50 states and the District of Columbia, 2001–2003. *Morbidity and Mortality Weekly Report, 53,* 756–760.
159. Mamphilly, C.M., Yore, M.M., Maddock, J.E., Nigg, C.R., Buchner, D., & Heath, G.W. 2005. Prevalence of physical activity levels by ethnicity among adults in Hawaii, BRFSS 2001. *Hawaii Medical Journal, 64,* 272–273.

160. Rush, E., Plank, L., Chandu, V., Laulu, M., Simmons, D., Swinburn, B., & Yajnik, C. 2004. Body size, body composition, and fat distribution: A comparison of young New Zealand men of European, Pacific Island, and Asian Indian ethnicities. *New Zealand Medical Journal, 117,* U1203.
161. Brewis, A.A., McGarvey, S.T., Jones, J., & Swinburn, B.A. 1998. Perceptions of body size in Pacific Islanders. *International Journal of Obesity and Related Metabolic Disorders, 22,* 185–189.
162. Bruss, M.B., Morris, J., & Dannison, L. 2003. Prevention of childhood obesity: Sociocultural and familial factors. *Journal of the American Dietetic Association, 103,* 1042–1045.
163. Grandinetti, A., Chang, H.K., Mau, M.K., Curb, J.D., Kinney, E.K., Sagum, R., & Arakaki, R.F. 1998. Prevalence of glucose intolerance among Native Hawaiians in two rural communities. Native Hawaiian Health Research (NHHR) project. *Diabetes Care, 21,* 549–554.
164. Simmons, D., Thompson, C.F., & Voklander, D. 2001. Polynesians: Prone to obesity and Type 2 diabetes mellitus but not hypertension. *Diabetic Medicine, 18,* 193–198.
165. Rush, E.C., Plank, L.D., Mitchelson, E., & Laulu, M.S. 2002. Central obesity and risk for type 2 diabetes in Maori, Pacific, and European young men in New Zealand. *Food and Nutrition Bulletin, 23* (3 Suppl.), 82–86.
166. Yehieli, M., & Grey, M.A. 2005. *Health matters: A pocket guide for working with diverse cultures and underserved populations.* Boston, MA: Intercultural Press.
167. Mayeno, L., & Hirota, S.M. 1994. Access to health care. In N.W.S. Zane, D.T. Takeuchi, & K.N.J. Young (Eds.), *Confronting Critical Health Issues of Asian and Pacific Islander Americans.* Thousand Oaks, CA: Sage.
168. Jamie K Boyd, PhD, APRN and Kathryn L Braun, DrPH. Supports for and Barriers to Healthy Living for Native Hawaiian Young Adults Enrolled in Community Colleges. Prev Chronic Dis 4(4) pp. 1–12(2007) http://www.cdc.gov/pcd/issues/2007/oct/07_0012.htm. Accessed [3/14/20111].
169. Fujita, R., Braun, K.L., & Hughes, C.K. 2004. The traditional Hawaiian diet: A review of the literature. *Pacific Health Dialog, 11,* 250–259.
170. Ratner, M. 2000. Thai Americans. In *Gale Encyclopedia of Multicultural America,* R.V. Dassanowsky & J. Lehman (Eds.). Farmington Hills, MI: Gale Group.
171. Uhl, S. 2001. Exploring the culinary mysteries of Malaysia. Restaurants USA, January/February. Online at http://www.restaurant.org/business/magarticle.cfm?ArticleID=13
172. Owen, J. 2005. New rodent discovered at Asian food market. *National Geographic News,* May 16. Available online: http://news.nationalgeographic.com/news/2005/05/0516_050516_new_rodent.html
173. Gampell, J. 2006. Letter from Ubon: In northeast Thailand, a cuisine based on bugs. *New York Times,* June 22. Available online: http://travel2.nytimes.com/2006/06/22/travel/22webletter.html
174. Sukalakamala, S., & Brittin, H.C. 2006. Food practices, changes, preferences, and acculturation of Thais in the United States. *Journal of the American Dietetic Association, 106,* 103–108.
175. Pongstaporn, W., & Bunyaratavej, A. 1999. Hematological parameters, ferritin and vitamin B12 in vegetarians. *Journal of the Medical Association of Thailand,* 82, 304–311.
176. Storz, D. 1998. *Filipino culture.* News and Views: The Newsletter of the San Jose Peninsula Dietetic Association, 29, 7.
177. Daida, Y., Novotny, R., Grove, J.S., Acharya, S., & Vogt, T.M. 2006. Ethnicity and nutrition of adolescent girls in Hawaii. *Journal of the American Dietetic Association, 106,* 221–226.
178. Seakins, J.M., Elliott, R.B., Quested, C.M., & Mataumua, A. 1987. Lactose malabsorption in Polynesian and white children in the southwest Pacific studied by breath hydrogen technique. *British Medical Journal, 295,* 876–878.
179. Steele, J.C. 2005. Parkinsonism-dementia complex of Guam. *Movement Disorders, 12,* S99–S107.

People of the Balkans and the Middle East

CHAPTER 13

The southeast European nations of the Balkan Peninsula and the countries of the Middle East are in close proximity to central Europe, Africa, and Asia. The region has traditionally been a cultural crossroads of ideas, values, and material goods and is often considered the cradle of Western civilization. Many immigrants from the Balkans and the Middle East have come to the United States in search of economic opportunity and political stability. They frequently retain a strong ethnic identity, exhibited in their strong religious faith and in their maintenance of traditional food culture. This chapter examines the cuisine of the Balkans and the Middle East, its role in the U.S. diet, and the changes that have occurred in the United States.

Cultural Perspective

The Balkan nations include Greece, Albania, Bosnia-Herzegovina, Montenegro, Serbia (including the two autonomous provinces Kosovo and Vojvodina), the Republic of Macedonia, Croatia, Slovenia, Bulgaria, and Romania. Countries of the Middle East include Bahrain, Egypt, Iran, Iraq, Israel, Jordan, Kuwait, Lebanon, Oman, Saudi Arabia, Syria, Turkey, United Arab Emirates, and Yemen (see Figure 13.1). In addition, there are several notable groups who do not have a homeland in the region, including the Palestinians (an Arab ethnic group), the Kurds (an Indo-European ethnic group), and the Chaldeans and Assyrians (Semitic ethnic groups).

Geographically, much of the Balkans is considered temperate in climate and is suited to agriculture. In Greece, however, and in most of the Middle East, aridity limits cultivation. Even in the desert regions, distinct areas of arable land exist along the seacoasts and in some plains and valleys, such as the Fertile Crescent (a plain in Iraq fed by the Euphrates and Tigris rivers) and the Nile River valley of Egypt.

Greece dominated or greatly influenced its Balkan and Middle Eastern neighbors in ancient times, and, in turn, it was conquered and ruled by the Turkish Ottoman Empire for four centuries in the modern era. These hundreds of years of Greek and Turkish hegemony facilitated the spread of products, especially foods, throughout the region and stretched cultural influence to both the southern states of the former Soviet Union (see Chapter 7, "Central Europeans, People of the Former Soviet Union, and Scandinavians") and to North Africa. Despite such commonalities, however, populations within the nations of the Balkans and the Middle East are very diverse in religious affiliation. Judaism, Christianity (particularly Eastern Orthodox), and Islam all have substantial numbers of followers in the region. Faith is fundamental to daily life in both the Balkans and the Middle East.

An Arab is commonly defined as a person who speaks Arabic—the term does not refer to a particular religious belief. Iranians speak Farsi and call themselves Persians. The Turks, whose nation is geographically divided between Europe and Asia, speak modern Turkish, a language that evolved from Arabic but is written with the Latin alphabet.

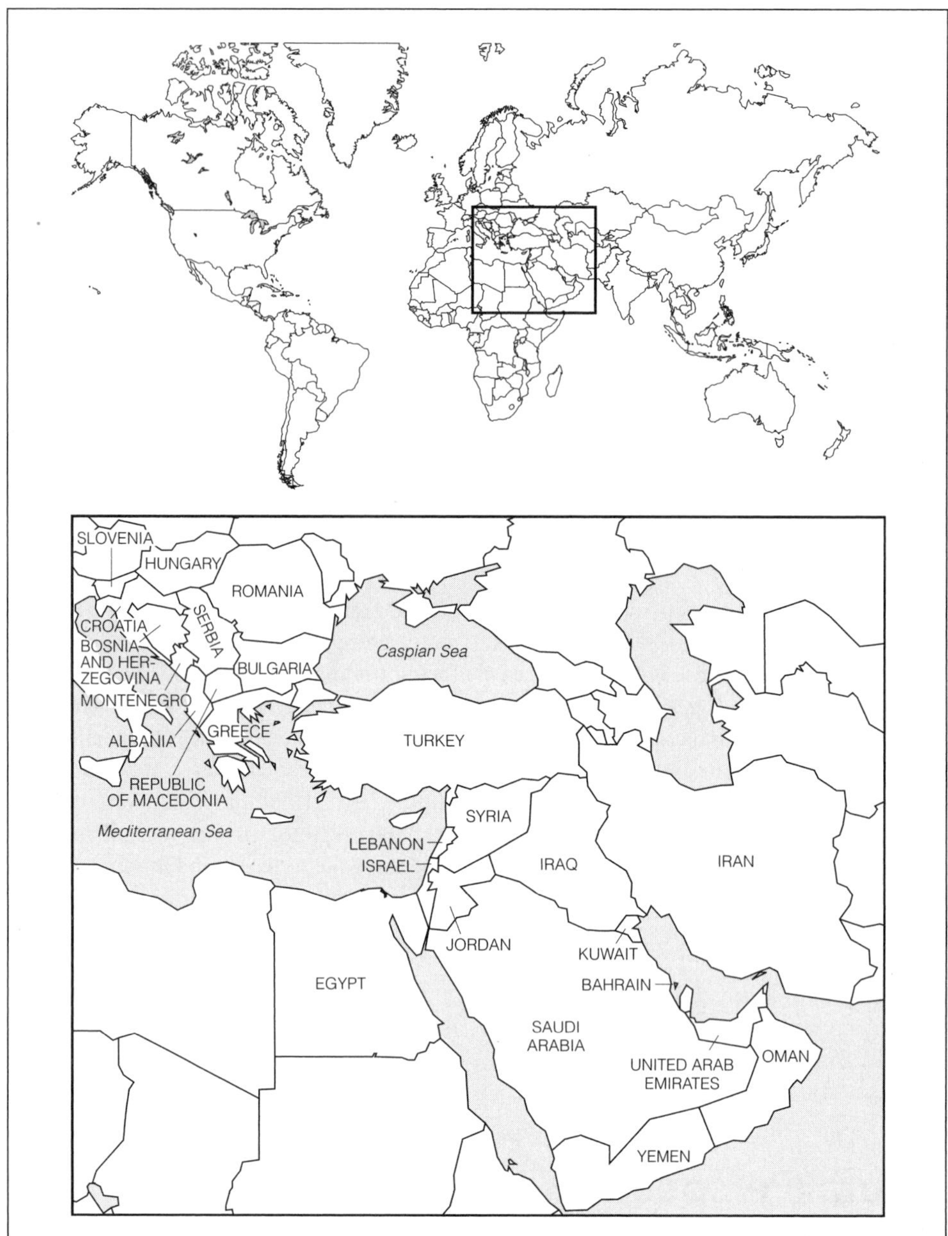

Figure 13.1
The Balkans and the Middle East.

History of People of the Balkans and Middle East in the United States

Balkans

Until 1992, the nations of Serbia, Croatia, Slovenia, Bosnia-Herzegovina, Montenegro, and Macedonia were the states of a single country known as Yugoslavia.

Immigration from Greece has occurred primarily in two waves. The first lasted from the late 1800s to the 1920s, when the restrictive Immigration Act of 1924 was imposed; the second wave started after World War II and has not yet ended. The early Greek immigrants were mostly young men from rural agricultural areas who came to America primarily for economic opportunities. Many came to make their fortune and go back to their homeland—approximately 30 percent of early Greek immigrants returned to Greece. A bitter civil war from 1946 to 1949 and a military coup in 1967 resulted in numerous Greek refugees who sought asylum in the United States. While most settled in New York,

◀ *A large Greek community developed in Tarpon Springs, Florida, based on diving for sponges.*

© Van Bucher/Photo Researchers, Inc

Detroit, Chicago, and other cities of the Midwest, a large community developed in Tarpon Springs, Florida, supported by diving for sponges. Some Greeks were attracted to mining and railroad work in the West.

During the 1850s and 1860s, numerous Croatian immigrants from the area of Dalmatia arrived in the United States. Most migrated to the southern and western regions where they had a substantial impact on the oyster fisheries of Mississippi and the fig, apple, grape, and plum horticulture of northern California. The many Serbs who immigrated to the United States in this period were unskilled laborers who obtained industrial jobs in the Northeast. Croatian and Serbian immigration increased following World War II, including professionals, such as engineers and physicians, seeking better job opportunities.

The largest number of Slovenian immigrants arrived between 1880 and World War I, though exact figures are not available because many were listed as Austrians. Most were farmers seeking economic opportunity. They settled initially in the rural Midwest, forming self-sustaining ethnic communities with Slovenian churches, schools, businesses, and social organizations.

Middle East

Statistics on immigration from the Middle East are inexact. Until 1900, all immigrants from the area were called "Egyptians." Later arrivals were typically termed "Syrians" or "Turks from Asia." More specific nationalities have been recognized since the 1930s, yet it has only been in recent years that Palestinian has been defined as an immigration category, and it is standard practice to combine Chaldeans with Assyrians.

Early Arab immigrants arrived at the turn of the twentieth century seeking economic opportunity. Most were Christians from the area that is today Lebanon and Syria, although small numbers of Turkish Muslims also came during this period. Most settled in New England and the Midwest and were employed in the textile, shoe, and wire factories.

The major influx of Middle Easterners began in the years following World War II. A high percentage of immigrants from Jordan, Egypt, Lebanon, Iraq, and Syria were Christian Palestinians who initially fled Israel when it was declared a state. After first settling in refugee camps, many later immigrated to the United States when Israel won the 1967 war against neighboring Arab countries. Political unrest and Islamic revolution in Iran led to a large exodus of Persians beginning in 1976. Many were members of the wealthy Iranian elite; others were Jewish and Christian minorities, including Chaldeans and Assyrians. In addition, many Turks and Arabs who came to the United States for their college education elected to stay in the country to establish their careers. A similar situation is reported for Israelis,

Though large numbers of Kurds have been displaced due to conflicts in the Middle East, most refugees—over 500,000—have gone to Europe. Much smaller numbers have come to the United States, particularly Nashville, Salt Lake City, and San Diego.

Around the turn of the twentieth century, an elite group of Arab artists, writers, and poets settled in New York City. They called their literary circle the Pen League; the best-known member was Kahlil Gibran, author of *The Prophet*.

who immigrated in small but steady numbers to America for advanced education and professional, managerial, or technical jobs.

Current Demographics and Socioeconomic Status

People of the Balkans

The 2001 Canadian Census lists over 215,000 residents of Greek ancestry, 97,000 Croatians, and much smaller populations of Serbians and Slovenians.

Nearly 1.4 million people claimed Greek ancestry in the 2007 U.S. Census estimates. New York City has the largest concentration of Greek Americans, followed by California, Illinois, Massachusetts, and Florida. In recent years, Greek Americans have moved progressively south and west in the United States. Early Greek immigrants preferred to live in urban areas. The Greeks sought economic independence by opening small businesses, especially in candy production and restaurants. The rate of self-employment among Greek Americans is still very high; unemployment rates are low, and the percentage of families living below poverty is also below the U.S. average. Greek Americans can be found in all occupations, particularly managerial, professional, technical, and service jobs.[1]

Over 420,000 Croatian Americans and 173,000 Americans of Serbian descent lived in the United States in 2007. Most originally settled in the mining regions of Pennsylvania, Ohio, and other Midwestern states; mining also attracted sizable populations to Colorado, Nevada, Arizona, and California.[2,3] Over 173,000 Slovenian Americans were identified in the 2007 U.S. Census data. Although many Slovenian immigrants also became involved in mining, others moved to industrial urban areas in search of jobs. Forty percent live in Ohio; Cleveland has the largest Slovenian community in the nation. Other states with notable Slovenian populations include Pennsylvania, Minnesota, Wisconsin, California, Colorado, Michigan, Florida, and New York. Many Croatians, Serbians, and Slovenians who first arrived in the United States were unskilled laborers in agriculture or industry.[4] Education was valued and second-generation immigrants obtained the training needed to secure white-collar employment. Over time, most Americans of Croatian, Serbian, and Slovenian descent have assimilated into the U.S. mainstream, working in all sectors of the economy. All three groups have incomes that are less than the Greeks', but approaching the U.S. average.

Beginning in the early 1990s, nearly 200,000 Bosnian refugees seeking refuge from ethnic cleansing in that region were resettled in the United States. Many left their homeland suddenly, with little time to pack or prepare for life in a new country. Nearly all are Muslim, and most were not comfortable in the small number of Bosnian-Serbian or Bosnian-Croatian communities already scattered throughout the United States. A majority have chosen to live in homogeneous Bosnian Muslim neighborhoods in New York, St. Louis, Chicago, Salt Lake City, St. Petersburg, Florida, and Waynesboro, Pennsylvania.[5] Significant populations are found in California as well.[6] Though approximately half of Bosnian refugees arrived unskilled in any profession, they are known for a strong work ethic and find employment wherever it is available, frequently as bakers, factory workers, or in the service industries, especially in hotels.[5] Those who do have professional training are often limited by a lack of English skills or by credentialing difficulties. Average income is believed to be low, and one small survey in California found that over half of Bosnian refugee families interviewed relied on some form of government assistance.[6]

Middle Easterners

Demographic figures for Middle Easterners are more problematic. Some older immigrants may deny their Arab ancestry due to a history of discrimination in the West, and more recent arrivals may refuse cooperation with government-sponsored surveys due to negative experiences in their homelands and fear of mistreatment following the September 11, 2001 attacks on the United States. Immigration figures suggest, however, that Arabs are one of the fastest growing ethnic groups in the nation, increasing from approximately 200,000 in 1970 to over 1.6 million in 2007. This growth is even more dramatic if a poll on Arab ancestry is correct in estimating that the Arab population is actually closer to 3.5 million.[7] In addition, there are over three quarter million Americans of Middle-Eastern descent from the non-Arab populations of Israel, Iran, and Turkey. One noteworthy change in recent immigration from the Middle East is in religious affiliation. In 1970 only 15 percent of immigrants from the region were Muslim; in 2000, that figure increased to 73 percent.[8]

In 2007, the census estimated 490,000 self-declared Americans from Lebanon, 150,000 from Syria, 194,000 from Egypt, 91,000 of Palestinian heritage, 76,000 from Morocco, 64,000 from Jordan, and 62,000 from Iraq. A movement toward pan-Arab Americanism is seen in the number of U.S.-born Arabs who are selecting "Arab" or "Arabic" as their background instead of a national affiliation.[7] Approximately 20 percent of Arabs are now listed in this category, 266,000 in 2007. Unofficial estimates of these populations are often two to three times larger.

Two-thirds of Arabs have settled in California, New York, and Michigan. The largest concentration of Muslims, numbering over 300,000 (including those from areas other than the Middle East) is in the Detroit metropolitan area.[9] Smaller Arab communities are developing in New Jersey (which has the largest population of Egyptians), Virginia, North Carolina, Connecticut, Alabama, and New Mexico.[7] In addition, Syrians often settle in Rhode Island, Moroccans often choose Washington, D.C., and Illinois is notable for hosting the largest number of Palestinians in the United States, as well as substantial populations of Iraqi, Assyrian, and Chaldean heritage.[7] Large numbers of Palestinians are also found in in California.[9,10] The Chaldean population, which is near 900,000, is also present in Detroit and San Diego.[9] Total numbers of Assyrians and Kurds are significantly smaller. Most Arabs choose urban areas of residence, and ethnic neighborhoods have developed in cities with substantial Arab populations. These communities help preserve Arab culture within what is often viewed as an alien American society. Many Arabs suffer discrimination and accept marginalization as a necessary part of living in the United States.[11] Despite these experiences by some, foreign-born Arabs become U.S. citizens at rates 25 percent higher than the average for all foreign-born residents.[12]

Early Arab immigrants found it easy to assimilate into American society, in part because a majority of them were Christian.[11] Many survived as peddlers and then later opened family businesses, usually dry goods or grocery stores. A tradition of entrepreneurship among Arabs continues today, and nearly one in five owns his or her own business. Even families that started with unskilled employment or small businesses have made substantial economic progress through schooling. Over four in every ten Arab Americans worked in management or professional jobs, and approximately 30 percent held sales and office positions in 2000.[12] Data also show smaller numbers than the U.S. average worked in the service industries, with the exception of Moroccans, who were employed in this sector at a rate more than double that of U.S. workers in general. Nearly every Arab American group in 2000 reported high school graduation rates exceeding those of the total U.S. population, and the percentage of Arabs with college degrees is far higher than the national norm.[12] Median family income was above average, yet differences were marked between different Arab groups. Some, such as Lebanese and Syrian Americans, fall below the U.S. family income norm. Poverty rates slightly exceeded the U.S. average for Egyptians and Jordanians and were approximately 50 percent higher for Palestinians and Moroccans. Poverty levels approximately double the U.S. rate were found in those groups identified as "Arabs" alone and Iraqis (about 25 and 26 percent, respectively). These figures suggest that while Arab Americans as a whole are doing well economically, there are significant subgroups that have not prospered.[8]

Over 400,000 Americans of Iranian heritage were living in the United States according to 2007 census figures; however, some experts estimate as many as one million Iranians live in the nation, many of whom are non-immigrants, including students and visitors.[13] Nearly two-thirds are thought to reside in Southern California, and another large population is found in New York City. Many who immigrated following the 1979 Iranian revolution were among the well-educated elite who spoke English fluently and were skilled in the professions, such as engineering, medicine, pharmacy, dentistry, and law. Others arrived in the United States with substantial savings from their livelihood in Iran and have significant earnings from rental properties and other investments. This group has been economically successful in the United States. It is also reported that many Iranian immigrants are self-employed.[13] Data on foreign-born Iranians show that over half hold college degrees, and median household income far exceeds that of the average U.S. family.[14] These statistics suggest that total Iranian American income is high, even though it is believed that estimates are skewed by large numbers of students who are employed only part-time, lowering average earnings.

McDonalds and Walmart sell halal foods, such as chicken McNuggets, at certain U.S. locations.

U.S. Census data and immigration figures suggest that nearly 10 percent of Middle Easterners in the country are here as undocumented residents.[8]

Persia was the conventional European name for Iran, used until the early twentieth century; many Iranian immigrants prefer the designation Persian American.

Turkish Americans numbered 164,000 in the census estimates of 2005. Immigration prior to 1965 was severely limited by quotas, yet included a disproportionate number of engineers and physicians. Since that time, the numbers of Turkish immigrants have increased, most seeking educational and occupational opportunity. Homogeneous neighborhoods, supported by Turkish American organizations, have been established in many cities, including New York City, Boston, Chicago, Detroit, Los Angeles, San Francisco, and Rochester. Smaller populations are found in New England, the Midwest, and parts of the South, including Maryland, Virginia, Texas, and Georgia.[15] Census statistics on foreign-born Turkish Americans indicate higher rates of high school and college graduation than the U.S. norm, and median household income slightly above the U.S. family average.[14] Turkish Americans are found in most employment sectors and are solidly middle class in socioeconomic status, such as Detroit and Chicago.[16]

Most Israelis today come to the United States for educational and professional opportunities; most are of European heritage, typically middle or upper class, and a significant number are single men and women.[17] Some arrivals are also seeking security and escape from the political unrest in the region. Most possess the job skills that assure an easy transition in the United States, and over half hold professional, managerial, or technical positions. They are prominent in the fields of medicine, technology, architecture, entertainment, and education. Approximately one-third are self-employed in the garment industry, electronic industy or in small businesses catering to the Israeli community, including restaurants, nightclubs, and retail shops.[16] Average earnings are believed to be higher than the U.S. norm.

Worldview

Religion

Religion is often the defining factor in Balkan and Middle Eastern life. Though affiliation varies, strong devotion is common. Many congregations remain insular in the United States, serving the needs of a specific ethnic group, and there is little interest in proselytizing to outsiders.

People of the Balkans. The ethnicity of Greek immigrants was affirmed mostly by religion; it was said that a person was not Greek by birth but through the active affiliation with the Greek Eastern Orthodox Church. The first Greek Orthodox Church in America was founded in New Orleans in 1864. Most Greek Americans today still belong to the church, which continues to be the center of Greek community life. The word *orthodox* comes from the Greek for "correct" (*orthos*) and "worship" (*doxa*). A fundamental belief of the Greek Orthodox faith is that an individual attains complete identification with God through participation in the numerous religious services and activities sponsored by the church. Although the Greek Orthodox Church is conservative and traditionally resistant to change, some accommodations have been made in the United States; for instance, the service typically is conducted both in Greek and English, and modern organ music accompanies the liturgy (see Chapter 4 for more information about Greek Orthodoxy).

Most Serbs also belong to the Eastern Orthodox faith, as members of the Serbian Orthodox Autonomous Church. A majority of Croatians and Slovenians are devout Roman Catholics, worshipping primarily in multiethnic congregations. A small number of Slovenians are Protestants known as Windish. Windish churches have helped to maintain Slovenian ethnic identity through services conducted in a Slovenian dialect. Small populations of Croatian Muslims are found in Cleveland and Chicago.

Early Bosnian-Croat immigrants were mostly members of the Catholic Church, and Bosnian-Serb immigrants followed the Eastern Orthodox religion. An estimated sixty-eight Serbian Orthodox churches exist in the United States, serving both the worship and social needs of each community. The majority of recent Bosnian refugees are Sunni Muslims. Regardless of faith, many Bosnians are more secular than observant, attending services on major holidays only, or not at all. Intermarriage between Bosnians of different religious beliefs is not uncommon.[5,6]

Middle Easterners. Early Arab immigrants were primarily Christians belonging to the Eastern Orthodox (particularly the Egyptian Coptic Church) or the Latin rite Maronite, Melkite, or Chaldean churches. Although Christian Arabs are still a majority in America, more recent immigrants follow Islam, and the number of Arab Muslims in the United States is growing rapidly. Most belong to the Sunni sect. Many Arab Muslims in the United States have made several adaptations to accommodate their religious practices to American society. Most significantly, the Friday Sabbath prayer has been moved to Sunday, and many Muslims cannot fulfill their obligation to pray five times daily due to work or school limitations (see Chapter 4 for more information about Islam).

Though the vast majority of Iranians in the United States are members of the Muslim Shiite sect, small numbers of Iranian religious minorities have come to the United States seeking freedom from persecution, including Jews, Catholics, Anglicans, Protestants, and members of the Baha'i faith (a Muslim offshoot that has renounced its ties to Islam and preaches gender equality, world brotherhood, and pacifism). Most Turkish Americans are Sunni Muslims, and many worship at Arab or Pakistani mosques.[15]

Presumably, most Israeli Americans are followers of Judaism. However, unlike other Middle Eastern groups, Israelis who immigrate may be among the least orthodox of Jews in Israel. Many, upon arrival, choose to join Reform congregations or are unaffiliated with a synagogue.

Family

People of the Balkans. The traditional Greek home is strongly patriarchal. The head of the household is the unquestioned authority, with responsibility for supporting the immediate family and elder parents. In addition, he is accountable for the family reputation within the community. The Greek term for this pride and obligation to family is *philotomo*, meaning "love of honor." Each family member is expected to behave in ways that maintain family dignity and status.

Obligation to family and community has lessened somewhat in the Greek-American community. Extended families have become less common due to assimilation pressures and secondary migration to other areas of the country. Children are doted on, and parents often put their welfare first. Yet obedience and respect for elders are expected, and intergenerational conflict is less common among Greek Americans than in many other immigrant groups.

Greek women have traditionally focused on family, home, and church. Even after coming to the United States, many Greek American women continue in this role. Some even cut themselves off from non-Greeks in an attempt to retain their ethnicity and pass on Greek culture to their children.[1] However, growing numbers are trying to balance their duty to family and community with personal interests in education or pursuing a career. Education is valued for Greek men but often considered unnecessary for women. Greek men earn college degrees at nearly twice the rate of Greek women. In addition, though many women are encouraged by their parents to marry within the Greek community, increased exposure at school and on the job has led to a rise in interethnic and interreligious marriages.

Croatian and Serbian families are also traditionally patriarchal. Extended families are the norm, often including friends as well as relatives. Among Croatians, communal living may involve taking in boarders. Both Croatians and Serbians have become well acculturated in the United States. The tradition of older generations caring for children has allowed many Croatian and Serbian women to take advantage of educational and career opportunities, and the authority of the father has lessened. Slovenian Americans are also assimilated. The extended family structure typical in Slovenia is rarely found among Slovenians in the United States, who prefer nuclear families. American women of Slovenian descent are active in the home, church, and Slovenian schools and are increasingly involved in politics.

Bosnians traditionally maintained extended family homes, but conflict in the region and migration to urban areas has resulted in more nuclear families living apart from relatives.[5] In the United States the extended family structure is important to newer arrivals but difficult for some refugees who had no family members already established in the country. While strong family bonds support new immigrants, Bosnian women (who are not employed outside the home) and elders are often dependent on extended family and may become isolated within the larger community.[6] Traditionally, Bosnian husbands and wives both work, but responsibility for the home remains with the woman. Bosnian children assimilate quickly to American culture, and some parents are frustrated by their inability to instill Bosnian cultural values and to maintain the Bosnian language within the home. Some refugees believe kinship ties and the integrity of the family are weakened with continuing exposure to American society.[6]

The tall white hats worn by professional chefs are thought to have originated with the tall black hats worn by Eastern Orthodox priests.

Of the 90,000 Jews living in Iran in 1987, 55,000 have since left, including 35,000 who have immigrated to the United States.

Middle Easterners. Traditionally, Arab cultures center on a strong patriarchal family whose honor must be maintained. The family demands conformity and subordination of individual will and interest, but in return the members of the family are protected and can identify with the family's status. Families often live with extended members in a single home or, for well-to-do Arabs, in a family compound. An exception is Egyptians, who traditionally live in nuclear family groups, often far from other relatives.[18]

The teachings of the Qur'an state that men and women are considered equal but with different roles and responsibilities. Children are valued in Arab families, and sacrifice for the good of the children is common. Men are obligated to provide economic security for children, while women are expected to socialize them, including the preservation of religious and cultural values. Due to the strict patriarchy, it is the role of women to provide the love and comfort in a home as well. The relationship between mothers and daughters is very intimate. There is also a strong bond between mothers and sons, especially the first-born son.[19]

Marriage contracts in the Middle East are often arranged between children to establish political and business alliances through kinship. Also among Arabs there is a preference for marriage between cousins, with the exception of Egyptians and some Arab Christians.[11] Only after an engagement announcement is made are young men and women allowed to date, and then only when chaperoned. A family's honor is related to the modesty and chastity of the women in the home: A woman is chaste before her wedding and faithful after. Actual or alleged violations of moral codes by a young woman are considered evidence that her mother has failed in her responsibilities; inappropriate sexual conduct also brings shame on her male relatives. Interethnic marriages are strongly discouraged by both Muslims and Christians, especially for women. It is preferable to marry someone of a different ethnic group and the same religious affiliation than to marry outside the religion. However, intermarriage between Egyptian men and American women is not uncommon and may occur in other Arab groups as well.[18] Children of these couples are usually raised in the Muslim faith.

Most Iranians and Turks establish homes similar to those of Arabs, headed by the father or eldest sons, and centered on an extended family network for support. Traditionally, Iranian families worked to place members in all significant sectors of the society (through marriage and employment) in order to ensure family status and security.[13] Immigration to the United States often results in difficulties maintaining an extended family household for many Middle Easterners. Family ties remain strong among Arabs, Iranians, and Turks; however, nuclear families are the norm for most immigrants. It is still common for grandparents to live with their children and grandchildren, yet in some more acculturated families (such as in many Lebanese homes) daughters are no longer solely responsible for the care of elders, a duty they now share with their brothers.[20]

The role of women of Middle-Eastern heritage in the United States is changing for those who are engaged in American society, and the customarily strict separation of private and public spheres has blurred with increased numbers of women seeking college degrees and professional careers.[13] For example, dating is becoming more accepted as segregation of the sexes cannot be maintained in most schools, nor in the U.S. workplace. Yet, more traditional Middle-Eastern women are sometimes isolated in their home life. Some Iranian women, for instance, feel that their limited authority has decreased since arriving in the United States, that they spend more time on household chores, and that they are less involved in family decisions and in mosque activities.[13] More acculturated Arab women may also experience serious setbacks in acceptance of their Americanized conduct every time a large influx of new and conservatively minded Middle Eastern women settles in their area.

Many second- and third-generation Middle Easterners in the United States think that their parents are old-fashioned. Adoption of American practices, such as personal autonomy and self-determination, contradicts the Middle Eastern emphasis on doing what is best for the good of the family, increasing intergenerational conflict between parents and children. Acculturation also may result in the reduction of a father's authority in some Middle Eastern homes, though filial respect is usually retained. Pressures for children to adhere to family expectations—from participation in arranged marriages to denial of college attendance—can be substantial, however, and not all children reject the values of their parents.

Researchers note that among Iranian Americans, the stresses of living in the United States lead some children of immigrants to embrace their heritage.[13] Although some Arab groups, such as Egyptians and Syrians, are especially well acculturated, and some subgroups within each population are fully integrated within the U.S. society, resistance to American culture can be strong among a small number of Middle Eastern immigrants. The attempt to retain ethnic identity takes many forms, and it cannot be assumed that assimilation is automatic, especially for new immigrants who may arrive with the intention of returning to their homeland when conflicts subside, or are fearful of discrimination in the United States. For example, some Arab immigrants seek to insulate their families from the influences of American society by isolating them within homogeneous enclaves, and ethnic identity is so important to most Turkish Americans that unlike most Arabs and Iranians (who have very high rates of citizenship), Turks who seek naturalization are often ostracized within their communities.[11,15]

A majority of Israelis live in nuclear families. A unique family structure is seen in the rural settlement cooperatives of Israel called *kibbutzim*. Families live and work communally in a kibbutz, and though many are agriculturally based, in recent years manufacturing or tourism activities have provided income for some. At one time children were raised by age level and all members participated in community meals. Today, most families live independently and may choose whether to eat with the group. Children are given community tasks, and teens are often assigned workdays once a week.

After arrival in the United States, the nuclear family structure continues with support from Jewish organizations and a network of Israeli American groups. Though Israeli homes are traditionally patriarchal, Israeli American women are typically well educated and over half are employed, often in professional positions. Israeli immigrants are often concerned that their cultural values will be lost if their children become acculturated, and that their children may become too competitive or too materialistic.[16] Many Israeli parents, like American Jewish parents, strive to preserve ethnic identity by enrolling their children in religious training, Jewish summer camps, and by sending them to Israel to learn about their heritage.

Traditional Health Beliefs and Practices

People of the Balkans. In many parts of the Balkans, physical fitness is thought essential to good health and is also considered necessary in the development of good character. Team sports, water activities, gymnastics, skiing, hiking, and bicycling are common forms of exercise.[4,21] Among some Greeks, eating a good diet, relaxation, adequate sleep, and keeping a positive mental attitude are equally important to maintaining health.[22,23]

Traditionally, most health care was provided by grandmothers or mothers at home. Many people kept an herbal pharmacy available for preparation of therapeutic teas. In Greece, castor oil is taken to clean the bowels, quinine is used to relieve pain, chamomile tea is sipped to get rid of cramps, and a sore throat is treated with honey and lemon. Examples of Bosnian cures also include chamomile, as well as elder, rose hips, and mint.[4,21,23] Cupping (see Chapter 2, "Traditional Health Beliefs and Practices," for more information) in a manner similar to the Asian method, except that the skin is cut with a razor first to allow the blood out, is done by a few Greeks to treat colds and chest ailments. More severe conditions and serious injuries were treated by neighborhood experts, for instance, the midwives and bone-setters used in Croatia.[2,23] Some people in the Balkans believe that the evil eye of one who envies a person can cause accidents or illness. Greeks may use ritual prayer, the sign of the cross, or wear blue amulets with an eye in the center or garlic as a precaution against a jealous gaze. When receiving a compliment (a form of envy), it is also customary to spit two or three times to keep harm away.[1,23]

Muslims often have serious concerns about raising their children in America. Permissive attitudes about dating and drinking and pressure to practice independent thought prompt some immigrants to send their children, especially their daughters, to the Middle East for schooling.

Middle Easterners. Cleanliness, diet, and keeping warm and dry are all Middle Eastern factors important to maintaining health. Some Middle Easterners believe that illness is due to wind or air in certain situations. Lebanese Muslims believe that following childbirth, a woman is especially susceptible to wind; showers and baths are avoided for ten days to prevent wind from entering the veins and causing sickness. An infant is vulnerable to wind through the umbilicus, so the baby's stomach is wrapped at all times with a band called a *zunaad*.[19]

Traditional humoral medicine is important in the health practices of Iranians. Though traditional humoral theories identify four bodily humors, in practice Iranians are only concerned primarily with hot and cold. Each person is born with a physiological temperament dependent on the ratio of humors, which varies by gender, age, and race—women, for instance, are considered colder than men, and younger persons are hotter than older people—and can be influenced by diet, climate, geographical location, and certain conditions, such as childbirth.[24] Sickness can be caused by the consumption of too many hot items or too many cold foods (see Therapeutic Uses of Foods section), but individual conditions and symptoms are not classified as being hot or cold. Iranians are also concerned with the amount of blood they have, which in turn is associated with numerous ailments, including thinness (due to lack of proper nourishment), weakness, irritability, lethargy, and headaches. *Kam kuhn*, blood deficiency due to excessive bleeding from injuries, menstruation, or a poor diet that prevents the making of blood, is the source of these symptoms.

The Sufis, members of an ascetic and mystical Islamic sect, define health as an existential state of abstinence, patience, and self-examination, resulting in harmony with the universe.

***Zamzam* water from the Mecca Valley is collected by Muslims pilgrims who complete the *Hadj* to be shared with family and friends at home—it is thought to have curative powers.**

Iranians use the term *narahati* for undifferentiated feelings of physical and emotional discomfort. Most often it is expressed privately and in a nonverbal form through sullenness, anorexia, and, among women, bouts of crying. Expressions of anger are considered the public expressions of the condition and are usually discouraged because anger is a lack of control that can upset the social order and cause others to become *narahat*. Sorrow or grief is another public expression, but in contrast to anger, sadness is considered the poetic manifestation of fully experiencing the tragedy of the human condition. *Naharati qalb* (heart distress) is a folk condition typified by fluttering of the heart due to the strong expression of anger or sadness.[25,24]

A long-standing tradition of home health care exists among Middle Easterners. Folk remedies are common, such as rubbing *ko'hl* (a dark powder made mainly from the chemical element antimony and used mostly as a cosmetic) on the umbilical cord of an infant to help it dry. Herbal therapies are especially prevalent, and it is believed that approximately 200 plant species are used in Arab traditional medicine today.[26] Examples include yarrow for diabetes and *khella* (a member of the parsley family) for kidney disorders. Palestinians use most traditional remedies as both food and medicine; for instance, caraway is used for digestive disorders and to increase milk flow in nursing mothers, mallow is used as a laxative, and olive oil to treat urinary tract infections, prostate conditions, and cancer.[27] Other cures include snakeroot as a diuretic, lavender for kidney stones, and rue as an analgesic and sedative. In Iran foxglove blossoms are used especially for nervous conditions, some digestive problems, to strengthen blood, relieve fear, and for pains of unknown etiology; arugula seeds are taken to clean dirty blood and for fever, constipation, and nausea; and mint tea with coriander seeds is used to promote sleep.[28,24] In Turkey, nettle, oleander, and thyme are used therapeutically for cancer. Other commonly used herbal remedies include St. John's wort, rosemary, sage, and hawthorne.[29]

Cupping (see previous Chapter 2 reference) is used by some Middle Easterners to cure chronic leg pain, paralysis, headaches, and obesity. Another therapy that is sometimes applied is called *wasm*, or cauterization. A heated iron rod is used to place symbolic burn marks on the patient, for example, below the anus to treat diarrhea and under the ear lope to cure a toothache. The burns are then treated with special herbal poultices.[30] Though supersition is discouraged in Islam, many Arabs retain some beliefs in magic and supernatureal causes of illness. As in the Balkans, the evil eye is feared by some Arabs, who may place blue beads on infants to protect them, or wear amulets.[19] In Iran, some believe the evil eye is the cause of *cheshm-i-bad*, the occurrence of a sudden or unexplained illness.[25] Some Arabs attribute mental illness to possession by the devil, or by *jinn* (spirits who can be good or evil).[31]

Iranians often put their health into the hands of God. *Tagdir*, meaning "God's will," is thought to determine all aspects of life and, ultimately, death. Throughout the Middle East, illness is sometimes seen as a punishment from God. However, biomedical practice is well established in the Middle East, and for the most part, Western therapies are considered strong and effective.[31]

Traditional Food Habits

The origins of most Balkan and Middle Eastern dishes may never be known because the geographical, political, and economic history of the region has resulted in similar food cultures. Many ingredients in the Balkans and the Middle East, including wheat, olives, and dates, are indigenous. Sheep were first domesticated in the region over 10,000 years ago. Other foods, such as rice, chickpeas, and lemons, gained widespread acceptance after introduction. Yet most countries in the area claim one dish or another prepared by all ethnic groups to be their own invention.

Culinary commonalities extend beyond the arbitrary designation of the Balkan and Middle Eastern nations. As previously mentioned, North Africa is sometimes considered part of the Middle East and shares numerous dishes of the region. In addition, the southern nations of the former Soviet Union, the southern regions of central Europe, and parts of South Asia exhibit influences as well.

The most significant differences in Balkan and Middle Eastern foods are due to various religious dietary restrictions and proximity to other global cuisines. For example, the Christian populations of Croatia, Serbia, and Slovenia frequently consume pork, a favorite of neighboring central Europeans. Yet Christian Greeks, who have no pork prohibitions, eat it only occasionally, preferring lamb and goat, similar to adjacent Middle Easterners. Alcoholic beverages are banned for Muslims and are avoided in most Middle Eastern nations, though widely consumed in Turkey, perhaps due to historical associations with nearby Russia (beet soup is also popular). But despite these distinctions, foods are far more similar than different.

Ingredients and Common Foods

Staples

The common ingredients used in Balkan and Middle Eastern cooking are listed in Table 13.1. Wheat, thought to have been cultivated first in this region, is consumed at every meal as bread. Leavened loaves are typical in Greece and the other Balkan nations, and leavened flat breads are more common in the Middle Eastern countries. However, both loaves and flat breads are found throughout the region. *Pita* or *pida,* a thin, round Arabic bread with a hollow center (known as pocket bread in the United States), is a common type, as is *lavash*, a larger, crisp flatbread (also called cracker bread).

Photo by Laurie Macfee

▲ ***Traditional foods of the Balkans and the Middle East. Some foods include almonds, chickpeas (garbanzo beans), couscous, cracked wheat, dates, eggplant, feta cheese, figs, filo dough, garlic, lamb, lemon, olives, pita bread, and yogurt.***

Besides bread, wheat doughs are also used to make pies and turnovers prepared in a variety of sizes and shapes. Bread dough, short crust, or paper-thin pastry sheets called *filo* or *phyllo* are all used. Savory pies may contain meat, cheese, eggs, or vegetables. Desserts are usually filled with nuts or dried fruits. An example of a fried meat- or cheese-filled pastry that can be served hot or cold is sanbusak. Traditionally half-moon shaped, it is popular in Syria, Lebanon, and Egypt. A similar turnover is called burek in Slovenia and boereg in Bulgaria and Romania. *Fatayeh* is another specialty served as a snack, featuring bread dough topped with cheese, meat, or spinach and baked like a pizza. Tiropetas are Greek flaky turnovers stuffed with cheese. In Serbia a cheese and egg pie called gibanjica is popular. A dessert made with *filo* dough, baklava or paklava, can be purchased at every bakery and café throughout the Balkans and Middle East. The sheets of dough are layered with a walnut, almond, or pistachio filling and then soaked in a syrup flavored with honey, brandy, rose water, or orange-blossom water. It is often cut into diamond shapes.

At ancient Greek weddings the bride and groom were showered with wheat kernels.

Raw kernels of cracked whole wheat are used in a number of Balkan and Middle Eastern dishes. When the kernels are first steamed and then dried and crushed to different degrees of fineness, the cracked wheat is called *burghul* or bulgur. Unripened dried but not cracked wheat kernels are known in Arabic markets as *fireek.* All varieties of wheat kernels are typically cooked as side dishes or made into tabouli, a popular salad containing

Table 13.1 Cultural Food Groups: Balkan and Middle Eastern

Group	Comments	Common Foods	Adaptations in the United States
Protein Foods			
Milk/milk products	Most dairy products are consumed in fermented form (yogurt, cheese). Whole milk is used in desserts, especially puddings. Sour cream and whipped cream are common in northern Balkan nations. High incidence of lactose intolerance is reported.	Cheese (goat's, sheep's, cow's, and camel's), milk (goat's, sheep's, camel's, and cow's), yogurt; buttermilk, cream	More cow's milk and less sheep's, camel's, and goat's milk are drunk. Ice cream is popular. Feta is the most common Middle Eastern cheese available in the United States.
Meat/poultry/fish/ eggs/legumes	Lamb is the most popular meat. Pork is eaten only by Christians, not by Muslims or Jews. Jews do not eat shellfish. In Egypt, fish is generally not eaten with dairy products. Legumes are commonly consumed.	*Meat:* beef, kid, lamb, pork, rabbit, veal, many variety cuts *Poultry:* chicken, duck, pigeon, turkey *Fish and shellfish:* anchovies, bass, bream, clams, cod, crab, crawfish, eels, flounder, frog legs, halibut, lobster, mackerel, mullet, mussels, oysters, redfish, salmon, sardines, shrimp *Eggs:* poultry, fish *Legumes:* black beans, chickpeas (garbanzo beans), fava (broad) beans, lentils, navy beans, red beans; peanuts	Lamb is still very popular. More beef and fewer legumes are eaten.
Cereals/Grains	Some form of wheat or rice usually accompanies the meal in the Balkans and Middle East.	Bread (wheat, barley, corn, millet), barley, buckwheat, corn, farina, millet, oatmeal, pasta, rice (long-grain and basmati), wheat (bulgur, couscous)	Bread and grains are eaten at most meals. Pita bread is commonly available.
Fruits/Vegetables	Fruits are eaten for dessert or as snacks. Fresh fruit and vegetables are preferred, but if they are not available, fruits are served as jams and compotes and vegetables as pickles. Eggplant is very popular. Vegetables are consumed often; sometimes stuffed with rice or a meat mixture.	*Fruits:* apples, apricots, avocado, barberries, bergamots, cherries, currants, dates, figs, grapes, lemons, limes, melons (most varieties), oranges, peaches, pears, plums, pomegranates, quinces, raisins, strawberries, tangerines *Vegetables:* artichokes, asparagus, beets, broccoli, Brussel sprouts, cabbage, carrots, cauliflower, celeriac, celery, corn, cucumbers, eggplant, grape leaves, green beans, green peppers, greens, Jerusalem artichokes, leeks, lettuce, mushrooms, okra, olives, onions, peas, pimientos, potatoes, spinach, squashes, tomatoes, turnips, zucchini	Fewer fruits and vegetables are consumed. Olives are still popular.
Additional Foods			
Seasonings	Numerous spices and herbs are used. Lemons are often used for flavoring.	*Ajowan,* allspice, anise, basil, bay leaf, caraway seed, cardamom, cayenne, chervil, chives, chocolate, cinnamon, cloves, coriander, cumin, dill, fennel, fenugreek seeds, garlic, ginger, gum arabic and mastic, lavender, linden blossoms, mace, *mahleh,* marjoram, mint, mustard, nasturtium flowers, nutmeg, orange blossoms or water, oregano, paprika, parsley, pepper (red and black), rose petals and water, rosemary, saffron, sage, savory, sorrel, *sumac,* tamarind, tarragon, thyme, turmeric, *verjuice,* verbena, vinegar	
Nuts/seeds	Ground nuts are often used to thicken soups and stews.	Almonds, cashews, hazelnuts, peanuts, pine nuts, pistachios, walnuts; poppy, pumpkin, sunflower, sesame seeds	
Beverages	Coffee and tea are often flavored with cardamom or mint, respectively. Aperitifs are often anise flavored. Alcoholic beverages are prohibited for Muslims, but consumed in most of the Balkans, Turkey, and Israel.	Coffee, date palm juice, fruit juices, tea and herbal infusions, yogurt drinks; beer, wine, brandy	
Fats/oils	Olive oil is generally used in dishes that are to be eaten cold. For most deep-frying, corn or nut oil is used; olive oil is preferred for deep-frying fish. Clarified butter (*samana*) is also popular. Sheep's tail fat is a delicacy.	Butter, olive oil, sesame oil, various nut and vegetable oils, rendered lamb fat	Olive oil is still popular.
Sweeteners	Coffee and tea are heavily sweetened. Dessert syrups are flavored with honey, rose water, or orange-flower water.	Honey, sugar	

onions, parsley, mint, and various fresh vegetables. Another Arab specialty is kish'ka, made by blending bulgur with yogurt, drying the mixture in the sun, then grinding it into a powder that can later be reconstituted with water to make a filling for pita or thinned enough for soup. In Serbia wheat kernels are cooked with sugar, dried fruits, and ground nuts to make koljivo.

Dumplings filled with meat are called cmoki in Slovenia and are stuffed with fruit for dessert. In Turkey, cheese or meat dumplings are known as manti. A few pasta dishes are found in the Balkans, such as the Greek pasta with baked lamb or goat and tomatoes called yiouvetsi and macaroni baked with cheese, ground meat, tomato sauce, and bechamel sauce called pastitsio. A dish similar to Italian pasta with beans is prepared. In the Middle East vermicelli is often added to rice, and noodles are found in a few dishes, such as Syrian chicken with macaroni.

In addition to wheat, rice is also a staple item in Middle Eastern cuisine and found in some Balkan regions. The long-grain variety is used to make pilaf or pilav, a dish that commonly accompanies meat or poultry. The rice is first sautéed in butter or oil in which chopped onions have been browned. It is then steamed in chicken or beef broth. Saffron or turmeric may be added to give the dish a deep yellow color.

Polo is the Iranian version of pilaf, but a final step in its preparation produces rice with a crunchy brown crust known as the tah dig. A more fragrant variety of rice, basmati, is used in Iran for khoresh, rice topped with stewed meat, poultry, or legumes. Rice is frequently added to soups and stuffings for poultry and vegetables in both the Balkans and Middle East.

A large variety of legumes are another important ingredient in both Balkan and Middle Eastern cooking. Cooked, pureed chickpeas are the base for hummus, often served as an appetizer or as a dip. Ground chickpeas or fava beans are sometimes formed into small balls and then fried and served as a main course (ta'amia) or in pita bread with raw vegetables (falafel). A common breakfast food is foul, slowly simmered fava or black beans topped with chopped tomato, garlic, lemon juice, olive oil, and cilantro (fresh coriander leaves). Lentils are especially popular in the soups of some areas.

Many vegetables are used, although eggplant is the most popular in the Middle East, Greece, and southern Balkan regions. A common cooking method for Middle Eastern vegetables (the Greek term is *yiachni*; the Arabic word is *yakhini*) is to combine them with tomatoes or tomato paste and sautéed onions together with a small amount of water, then cook until the vegetables are soft and very little liquid remains. Vegetable salads (freshly sliced tomatoes or cucumbers are common) and cold, cooked vegetable salads are eaten. Vegetables are frequently stuffed with a meat or rice mixture. Moussaka or musaka is a Balkan specialty made with minced lamb, eggplant, onions, and tomato sauce baked in a dish lined with eggplant slices. The Turks prefer *imam bayildi* (meaning "the priest fainted"), eggplant filled with tomatoes, onions, and garlic, stewed in olive oil and served cold.[32] Grape or cabbage leaves are stuffed to make the specialties known as *dolma* or *sarma*. Potatoes, particularly enjoyed in the northern Balkan nations, are also found in some Middle Eastern stews and side dishes. Vegetables are frequently enjoyed raw, mixed together in a salad, or preserved as pickles. Sauerkraut is eaten in Croatia, Serbia, and Slovenia.

The olive tree contributes in many ways, particularly to Greek and Middle Eastern cooking. Olives prepared in the Middle Eastern way have a much stronger flavor than European or American olives; they often accompany the meal or are served as an appetizer (see Table 13.2). The olive is also a source of oil, which is frequently used in food preparation, although butter, clarified butter (*samana*), and most vegetable oils, as well as rendered lamb fat and margarine, are

© Dave Bartruff/Documentary/Corbis

▲ ***Assorted Egyptian pastries made with filo dough, couscous, and nuts.***

In Bosnia-Herzegovina, Croatia, Serbia, and Slovenia, *pida* usually refers to *filo* dough, not pocket bread.

Eggplant "caviar," which is a cooked salad, and *baba ganouj*, an unusual smoked eggplant puree, are very popular appetizer dishes served throughout the Middle East.

Table 13.2 A Glossary of Selected Olives

Olives originated in the Middle East and spread throughout the Mediterranean. They are picked unripe (green, with dry, firm flesh and a bitter taste) or when fully ripened (black, oilier, soft textured, and milder in flavor). Raw olives are inedible and must be cured in salt (also called dry-cured) or in a brine, oil, wine, or lye solution before they can be consumed. Both the stage at which they are harvested and the type of curing process affect the flavor of the final product.

Aleppo (Middle East)	Small, dry-cured black olive (with wrinkled, chewy texture) named after a Syrian city
Amphissa (Greek)	Dark purple olives with nutty flavor
Gaeta (Italy)	Medium black olives, dry-cured (with wrinkled texture) or brine-cured (with smooth purplish flesh)
Kalamata (Greece/Middle East)	Large, deep-purple with crunchy texture—salt cured, packed in vinegar (sometimes with preserved lemon in Morocco)
Kura (Middle East)	Large green olive with hard flesh cracked to allow penetration of brine, bitter flavored, also called Middle Eastern cracked green
Manzanilla (Spain)	Small to large green olive usually pitted and stuffed with other ingredients (e.g., pimento, garlic, almonds)
Middle Eastern Green	Small, brine-cured olives packed with olive oil, herbs, and often chili peppers
Moroccan Dry-Cured	Medium black, dry-cured with wrinkled flesh and bitter flavor, used mostly in cooking
Nabali (Middle East)	Dark green olive with soft texture, brine-cured, and often packed with lemon, garlic, and vinegar—grown in Israel (including the Palestinian territories) and Jordan; also called *mushhan, baladi,* or Roman olives
Naphlion (Greece)	Dark green, brine-cured, packed in olive oil
Niçoise (France)	Small, sour, salt-cured purplish black olives
Picholine (France)	Small, mild, salt-cured light green olives
Sicilian (Italy)	Very large, green olive, brine-cured, somewhat sour
Thassos (Greece)	Small, dry-cured black olives (with wrinkled texture), intense tart flavor

The term yogurt is Turkish. In Syria and Lebanon the fermented milk product is called *laban*; in Egypt, *laban zabadi*; and in Iran, *mast*.

Camel's milk, unlike other commonly consumed animal milks, is high in vitamin C.

found in the Balkans and Middle East. Olive oil is generally used in dishes that are to be eaten cold. For most deep-frying, corn or nut oil is used, but olive oil is preferred for deep-frying fish.

Fruits are preferred fresh, eaten for dessert or as snacks. Apricots, cherries, figs, dates, grapes, melons, pomegranates, and quince are favorites. Pears, plums, and pumpkins are well liked in the more temperate climates of the northern Balkan nations. A distinctive characteristic of Middle Eastern fare is the addition of fruits to savory dishes; apricot sauce tops meatballs in Egypt and chicken in Syria. Fruits are also often served dried or as jams and compotes. Slatko is a Balkan specialty featuring fruits simmered in thick syrup. Fruit juices (especially lemon) and syrups flavor many foods.

Fresh milk is not widely consumed in the Balkans or Middle East, though it is used in puddings and custards. Dairy products are usually fermented into yogurt or processed into cheese. Yogurt is eaten as a side dish and served plain (unsweetened) or mixed with cucumbers or other vegetables. It is diluted to make a refreshing drink. Cheese is usually made from goat's, sheep's, or (in the Middle East) camel's milk. The most widely used cheese is *feta*, a salty, white, moist cheese that crumbles easily. *Myzithra*, a soft pot cheese, is a by-product of the feta process. *Lebneh* or *labni* is a fresh cheese made by draining the whey from salted yogurt overnight. *Haloumi* is a springy, semisoft cheese that is sometimes flavored with mint. It holds its shape when cooked, and pieces can be grilled quickly on both sides for a hot treat. *Kaseri* is a firm, white, aged cheese, mild in flavor and similar to Italian provolone. *Kashkaval* is a hard, tangy ewe's milk cheese, sometimes called the cheddar of the Balkans.

Almost all meats and seafood are eaten in the Balkans and Middle East, with the exception of pork in Muslim countries and pork and shellfish among observant Jews in Israel. Lamb is the most widely used meat, though pork is very popular in the northern Balkan areas. Grilling, frying, grinding, and stewing are the common ways of preparing meat in the region. A popular dish is *kabobs*, marinated pieces of meat threaded onto skewers then grilled over a fire. Vegetables, such as onions or tomatoes, are sometimes added to the skewers. *Souvlaki* or *shawarma* is very thin slices of lamb (or chicken) layered onto a rotisserie with slices of fat (resulting in a single roast), grilled, then carved and served. In Greece, thin slices of *souvlaki* are folded into pita bread with tomatoes, cucumber, and yogurt to make the sandwich-like treat gyros. Meatballs, called kofta, are favorites, eaten as snacks or served with stewed vegetables. A whole roasted lamb or sheep is a festive dish prepared for parties, festivals, and family gatherings.

Numerous spices and herbs are used in the Balkans and Middle Eastern seasoning as a result of a once-thriving spice trade with India, Africa, and Asia. Common spices and herbs are dill, garlic, mint, cardamom, cinnamon, oregano, parsley, and pepper. Sumac, ground red berries from a nontoxic variety of the plant, is sprinkled over salads to give a slightly astringent flavor; it is mixed with thyme to make the Arabic seasoning mix called *za'atar*. Other typical Middle Eastern spices include *mahleb*, made from the ground pits of a cherry-like fruit, and *ajowan*, small, black carom seeds with a thyme-like flavor. The juice of unripe lemons, verjuice, is used to provide a sour taste to dishes. Ground nuts are often used to thicken soups and stews. Sesame seeds are crushed to make a thick sauce, tahini, which is used as an ingredient in Arabic cooking and in a sweet dessert paste known as halvah.

Fruit juice is popular as a beverage throughout the Balkans and Middle East, and sometimes fruit syrups or flower extracts are mixed with ice (or in the past, snow) to make the refreshing beverage known as *sharbat* in Arabic or *şerbet* in Turkish (origin of the English word "sherbet"). Coffee is a favorite (see Is Coffee Beneficial for Health? later in this chapter) across the entire region, consumed throughout the day at home and in cafés. It is frequently flavored with cardamom and copious amounts of sugar. Traditionally, the drink is made in a long-handled metal *briki*, producing a strong, very thick, and often sweet brew served in small cylindrical cups. It is called "Turkish coffee" in Turkey and "Serbian coffee" in Serbia. Tea is equally popular in many nations and is served sweetened and flavored with mint, or fruit such as dates or raisins.

The Balkan countries are well known for their wines and distilled spirits. *Civek* is a rosé served throughout Slovenia, and a high-proof brandy made from plums called *sljivovica* is available in both Serbia and Slovenia. Best known in the United States are the Greek specialties *retsina*

© R. & S. Michaud/Woodfin Camp & Associates

▲ *Preparing Middle Eastern coffee. The coffee of the Balkans and the Middle East is made in a long-handled briki. It is preferred strong, thick, and sweet and is often flavored with cardamom.*

(white wine with a resinous flavor), *ouzo* and *arak* (anise-flavored aperitifs), and *metaxia* (orange-flavored brandy). Although observant Muslims do not drink alcohol, several Middle Eastern nations (e.g., Iraq, Israel, and Turkey) produce wines and spirits. *Raki*, a Turkish version of *arak*, is traditionally consumed with appetizers.

Regional Variations

Balkan. All Balkan nations combine both European and Middle Eastern elements in their cooking. The noteworthy division is between the more European-influenced fare of Bosnia-Herzegovina, Croatia, Romania (see Romanian Fare), Serbia, Slovenia, and the other northern nations and the foods of the southern countries, including Albania (see Albanian Fare) and Greece, with a decidedly more Middle Eastern flavor (Greek cooking is considered in the discussion of Turkish fare in the following section).

The use of pork and veal, the selection of fruits and vegetables, and the popularity of fresh dairy products are all characteristics of central European cooking in the northern Balkans. German-style sausages, pork roasts, and hams are frequently consumed. Veal is popular for stew and is sometimes seasoned with paprika. One popular dish found in most of the Balkan countries is ćevapi (or ćevapčići), grilled elongated kebobs of spicy minced meat that are often eaten on somun (a thick *pita* bread) or lepinja (a small, flat roll). In Bosnia they are usually made with beef, or a beef and lamb mixture, served with chopped

In the Middle East coffee is consumed highly sweetened at happy occasions; it is drunk black and bitter at funerals.

The Arabs were the first to mix gum arabic with sugar to produce chewing gum. The Greeks prefer to chew on the licorice-flavored resin *mastic* (source of the verb masticate).

CULTURAL CONTROVERSY—Is Coffee Beneficial for Health?

Despite the importance of wheat in the diet of Western nations, it can be argued that the most important Middle Eastern product consumed worldwide is coffee. Over 500 billion cups are consumed annually. Coffee is currently grown in over fifty countries and is second only to petroleum in global trade activity and value. The International Coffee Organization reports that exports increased 17 percent in 2006, and it is becoming increasingly common in tea-drinking nations such as Japan, China, and India.

Coffee is indigenous to Ethiopia, but it was the Arabs who first brewed and popularized the beverage after it was introduced to the region sometime around the tenth century.[89] It is said that a Sufi sheikh was first to note the ability of coffee to promote wakefulness, and it became widely used by worshippers to increase stamina and to produce a mystical euphoria.[38] The Sufis called it *qahwah* (thought to be the origin of the term "coffee"), a word originally used for wine. By the early 1500s it had become a secular beverage consumed in Middle Eastern social settings, especially coffeehouses, where men could drink and discuss the matters of the day. The coffeehouses attracted philosophers and poets, and in Istanbul they were known in jest as schools of knowledge.[38] Some Islamic leaders became concerned that men were spending more time drinking coffee than attending mosque services, and that politically subversive activities were being planned in the coffeehouses. This led to their closure in some areas, but efforts to enforce a permanent ban among Muslims failed due to coffee's broad popularity.

During this period, coffee was successfully grown in the Arabian Peninsula nations, and the product was improved by dark-roasting the beans. The beverage spread with the expansion of the Turkish Ottoman Empire, especially in southern Europe, where the social tradition of coffeehouses was well accepted. The Middle East became the first major exporter of beans through the Yemenese port of Moccha. However, some sixteenth-century Catholics believed that coffee was the beverage of Satan, due to its association with infidels. Popular legend is that plans to prohibit coffee were foiled when Pope Clement VIII asked to taste the brew and immediately claimed it so good that Christians should make it their own.

Historical controversies aside, the most significant issue regarding coffee in modern times has been its impact on health. Coffee contains numerous active ingredients, most notably caffeine and chlorogenic acid. Caffeine is an alkaloid that is classified with cocaine and amphetamines as a central nervous system stimulant. Chlorogenic acid is a phenolic compound that works as an antioxidant. Over the years, coffee has developed a bad reputation, related to studies on the development of ulcers, heart disease, cancer, and birth defects. Beginning in the 1980s, many health-conscious people began to cut consumption or switch to decaffeinated coffee.

Recent research, however, is leading to redemption of the beverage. Coffee has been exonerated as a causative agent in many gastrointestinal disorders, in most cardiovascular conditions, and in almost all cancers.[90,91,92,93] Even the association with birth defects has been disproved.[94] Instead, research suggests that moderate consumption of three to four cups each day may actually offer health benefits, reducing the risk of metabolic syndrome,[95] coronary heart disease,[96,97] type 2 diabetes,[98,99] several cancers,[100,101,102] rheumatoid arthritis,[103] and possibly Alzheimer's disease.[104] Studies on the effect of coffee on hypertension are contradictory.[91,105] Coffee is still suspect in bladder cancer, spontaneous abortion, and impaired fetal growth, and may aggravate some gastrointestinal disorders.[106,91] Researchers caution that coffee is not for everyone and that people with hypertension, children, elders, and pregnant women are most susceptible to adverse effects.

onions, cottage cheese, and an extra-rich sour cream called kajmak. Large, thin meat patties made from lamb and beef, known as pljeskavica, are considered the national dish of Serbia but are also a favorite with Bosnians and Croatians. Middle Eastern–style grilled meats are also found in some areas, especially in Bosnia-Herzegovina.

Potatoes, cabbage, and cucumbers are typical vegetables, and many families gather wild mushrooms. Ajyar, made with roasted red bell peppers and eggplant, seasoned with garlic and vinegar, is popular throughout the Balkans. It comes in many versions, from sweet to hot (flavored with chile peppers), and is served as a condiment with grilled meats, as a salad on a mezze plate (typically with a selection of sliced sausages or smoked meats, cheeses, hard-boiled eggs, and sliced tomatoes) or spread on bread. Vegetables are sometimes stuffed with meat and rice mixtures similar to those in the Middle East, but with a Balkan flavor due to the use of bell peppers, onions, potatoes, or cabbage leaves. In Bosnia-Herzegovina these stuffed items are called sarma or dolma and may be served in the tureen in which they have been heated.[5] Cooler weather fruits such as apples, berries, peaches, pears, and plums are more common. They are found in desserts, such as sweet dumplings and strudels, and preserved as compotes. Fruit juices are favorites and an important industry in the region.

Buttermilk is frequently consumed and fresh cheeses are well liked, often combined with herbs for mezze or mixed with eggs and honey or sugar for cheese-filled dessert pastries. Cream enriches soups, stews, casseroles, and sauces. Sour cream or whipped cream tops many dishes. A specialty dairy product found throughout the region is a heavy, crème fraiche–like product called *smetana* (in Slovenia), *vrhnje* (in Croatia), or *pavlaka* (in Bosnia-Herzegovina). In Croatia it is added to cottage cheese and seasoned with onions, garlic, radish or horseradish, and paprika, then eaten with cornbread.

The definitive northern Balkan treat is a sweet yeast bread rolled with a rich walnut, butter, cream, and egg filling. It is widely known as potica and in some areas as povitica or kolachki. Some versions are more savory, flavored with tarragon—others are sweeter, with dried fruits. Variations include cream cheese, poppy seed, and pumpkin fillings. Whipped cream tops many sweet versions.

Middle Eastern. There are two schools of thought about the number of regional cooking areas in the Middle East. One identifies three culinary areas: Greek/Turkish, Iranian, and Arabic, and the other defines five divisions: Greek/Turkish, Arabic, Iranian, Israeli, and North African (see Moroccan Cooking). Certainly every region has some unique recipes and cooking methods, but the similarity in fare throughout the region is striking.

The cooking of Greece and Turkey has evolved through an extensive exchange of ingredients and preparation techniques. Both cuisines feature more meat (especially grilled), fish and seafood, cheese, butter, and olive oil than in the fare of neighboring Middle Eastern countries. Both the Greeks and Turks prefer using flatware to fingers when eating. Similar distinguishing dishes include filo dough layered with spinach and feta filling (spanakopita in Greece, ispanakli borek in Turkey); fish roe (caviar) dip made with olive oil and bread (taramasalata in Greece and tarama in Turkey); salads with fresh greens, tomato, cucumbers, olives, and lemon juice–olive oil vinaigrette (the Greek version adds feta cheese; the Turkish recipe includes more vegetables, such as green peppers); the lemony, egg-enriched sauce used to thicken soups and top meat and vegetable dishes (avgolemono in Greece, tebiye in Turkey); yogurt cake; and anise-flavored alcoholic beverages. Yet many differences exist. Greeks prefer small pastries, such as the specialty butter cookie, kourabiedes, for snacking and dessert, while the Turkish sweet tooth is more often satisfied with fruit compotes, rich custards, and candy, including lokum, also known as Turkish delight. More significantly, consumption due to religious affiliation varies. Feasting and fasting rules for the Eastern Orthodox of Greece and the Muslims of Turkey differ tremendously (see Chapter 4 on religious food culture).

Arab fare, based originally on the cooking of nomadic tribes and later influenced by the foods of surrounding nations, features more grains, legumes, and vegetables than the Greek, Turkish, Israeli, or Iranian diet. In Syria and Lebanon, the national dish is kibbeh, a mixture of fine cracked wheat, grated onion, and ground lamb pounded into a paste. This mixture can be eaten raw or grilled, and with a great deal of dexterity it can be made into a hollow shell, then filled with a meat mixture, and deep-fried. In Jordan, a specialty is mansaf—flat breads layered with yogurt are placed on a communal platter and then topped with a mound of rice pilaf and shredded lamb or chicken. A broth mixed with whey or yogurt is poured over the top, then the dish is garnished with nuts. The national dish of Egypt is ful medames, cooked fava beans seasoned with oil, lemon, and garlic, sprinkled with parsley, and served with *baladi*, a whole-wheat type of pita bread. Many soups and stews include legumes, and some salads include grains, such as *tabouli*. Pieces of pita bread are added to many dishes as well. Tharid is a casserole of layered flatbread with meat stew found in many Arab nations. Fatout is a popular preparation in Yemen, combining toasted bread with honey, with scrambled eggs, or any other food; fattoush is a Lebanese favorite with greens, tomatoes, radishes, cucumbers, onions, and pieces of pita bread. Another feature of Arab cooking is the use of variety meats. Lamb, goat, and beef are costly; thus, all parts of the animal are used, with brains, chitterlings, heads, and feet considered delicacies. Pacha is an Iraqi soup

One specialty of North African cooking is *mirqaz* (or *merguez*) sausage—which is also popular in France—made from lamb or other meats, seasoned with chiles, cinnamon, dried rose petals, and other seasonings, then typically sun-dried and packed in olive oil.

Syrian food is often spicier than that of other Arab nations. They are known for their small baked lamb pies seasoned with cayenne called *sfeehas*.

***Tharid* was reputedly Mohammed's favorite dish.**

The Iranians call their breads *nan* and bake many of them, such as *nan-e lavash* and *nan-e barbari*, in a clay bread oven known as a *tanoor*. Both the name and oven were brought by the Persians to northern India, where *naan* flatbreads are baked in *tandoor* ovens.

Israelis with prickly personalities are sometimes called by the nickname *sabra* after the cactus of the same name. A U.S. native known as the prickly pear cactus, it was first exported to the Middle East in the nineteenth century; Israel is one of the largest *sabra* growers worldwide, selling the fruit to Europe, Africa, and Asia.

of sheep heads, stomach, and trotters served with ample bread and pickled vegetables.

Iran is the most eastern of Middle Eastern nations. It spans a region between the warm Persian Gulf and the cold Caspian Sea, encompassing several agricultural climates suited to a wide variety of fruits and vegetables. Its dominance of the Middle East, parts of the Balkans, and areas of India during the Persian Empire dispersed indigenous products such as spinach, pomegranates, and saffron throughout the region.

Later trade routes between China and Syria (the Silk Road) and between India and Africa crossed through Iran, introducing rice, tea, eggplants, citrus fruits, tamarind, and garam masala (the spice blend used in curried dishes) from these eastern cuisines. Though the cooking of Iran, usually called Persian cuisine, is very similar to other Middle Eastern foods, it is famous for its sophisticated rice dishes and its use of fruits for flavoring. The national dish is chelo kebab, which is thinly sliced pieces of marinated, charcoal-grilled lamb served over rice seasoned with butter, egg yolks, saffron, and sumac. Soups and sauces are given a sweet and sour taste by combining different ripe and unripe fruits, such as oranges, barberries, cherries, dates, grapes, plums, pomegranates, quinces, and raisins with astringent seasonings, including lemon juice, vinegar, tamarind, and sumac.

Israel probably has the most varied foods and food culture because its cuisine blends indigenous Middle Eastern cooking with that of the many Jewish immigrant groups who have settled in the area since nationhood. Hummus and pita bread may appear at the same meal as German schnitzel, Hungarian-style goulash, or Italian pasta.[33] American immigrants introduced bagels; Russians brought kasha and borscht. Chocolate mousse cake, linzertorte, and coconut macaroons are as popular as baklava. Furthermore, many of Israel's citizens adhere to the kosher laws of the Jewish religion (see Chapter 4 for more information on Jewish dietary practices).

Meal Composition and Cycle

Daily Patterns

The Balkans. People in Balkan countries eat three meals a day. The main meal is at midday, and in the hotter climates a short nap follows. Dinner is lighter and is served in the cooler evening hours. Snacking is prevalent.

In the northern regions a light breakfast of bread with preserves or honey and tea or coffee is most common. Lunch usually includes soup, a casserole of meats and vegetables, or a fish dish, bread and cheese, and a fruit compote or pastry for dessert. Dinner is often leftovers or another soup or stew; sweet dumplings may also be served. Wine is the typical beverage for both lunch and dinner,

EXPLORING GLOBAL CUISINE—Romanian Fare

Romania is a nation poised between the West and the East. Some describe Romanian food as "pastoral" with Turkish and Hungarian overtones. However, there are also many Italian and central European influences. Beef, veal, mutton, lamb, pork, chicken, geese, and duck are popular. Freshwater fish such as pike and catfish are harvested from the Danube and other rivers. Cabbage, red and green peppers, leeks, tomatoes, onions, radishes, and lettuce are common vegetables. Temperate fruits, particularly grapes, plums, and berries, are eaten. Other common foods include walnuts, filberts, olives, sour cream, and sheep and goat cheeses. The national bread is *mamaliga*, which is like the Italian *polenta*. It is sliced and spread with butter or topped with cheese, meats, or fruit for dessert. Another specialty is *pastrama* (from the Turkish meaning "to keep"), which is lamb, beef, pork, or goose cured (spicing varies, from garlic and black pepper to allspice, nutmeg, and hot red pepper) and then smoked. Ground meats are also popular, made into patties, stuffed into cabbage leaves, or as sausages. One-dish meals such as stews and soups are eaten with whole-grain bread; one example is *ciorba*, a soup made with vegetables (e.g., peppers, onions, sauerkraut, tomatoes) and meat (usually ground) or fish and then flavored with sour ingredients (e.g., sauerkraut juice, pickle juice, or vinegar). Cake is a traditional dessert, but custards (including one similar to Italian *zabaglione*) and soufflés are also eaten. Romanian beverages include wine (red, white, sweet, dry) and *tuica*, a plum and wheat brandy. Most Romanians belong to the Eastern Orthodox Church and adhere to the numerous feasting and fasting days of the church calendar.

EXPLORING GLOBAL CUISINE—Albanian Fare

Albania was not of interest to most Americans until the 1998 civil war in the Serbian province of Kosova focused attention on the plight of the Kosovar Albanians. Albanians, living in a country bordered by Greece, Macedonia, Serbia, and Montenegro, have often been involved in regional discord and shifting national boundaries.[107] Years of foreign rule have left their mark on Albanian cuisine: pastitsio and feta cheese from Greece, versions of *imam bayaldi* and *halvah* from Turkey, omelets and tomato sauces from Italy, *boereg* from Armenia, and *borscht* from Russia. Other dishes including *dolma, kofta, shish kebabs, moussaka,* and *baklava* are popular throughout the region.

In the poorest rural regions of Albania, farmers and shepherds are often limited to a diet of cornmeal bread, cheese, and yogurt, with added lamb or mutton when affordable. In wealthier areas, three meals a day are typical with a mid-afternoon snack of thick Turkish-style coffee or tea consumed with pastries, nuts, or fresh fruit, called *sille*. A complete lunch or dinner begins with mezze (appetizers), such as salads, pickles, fish and seafood, omelets, spit-roasted lamb or entrails, and baked variety meats. Examples include liptao, a feta cheese salad garnished with bell pepper, deli meats, sardines, and hard-boiled egg; and soup-like tarator, yogurt flavored with garlic and olive oil and mixed with vegetables, such as cucumber. These are usually consumed with a glass of the distilled Turkish specialty, *raki*, or a beverage made from fermented cabbage called *orme*. The meal follows with soups, meat, or cheese-stuffed vegetables or casseroles; pilaf-like dishes or pies filled with vegetables, cheese, and/or ground meats called byrek; and an assortment of vegetable side dishes and pickles. Dessert may include pastries but is typically a fruit compote. Few legumes are consumed, but nuts (especially walnuts) are added to numerous sweet and savory dishes.

One of the most distinctive characteristics of Albanian fare is the differentiation made between vegetables and fruit. Only vegetables are pickled and served as side dishes, and fruits are only eaten fresh, as desserts or as preserves. There are no crossover items, such as a fruit pickle or a vegetable-sweetened filling for a pie. Vegetables and fruits are also prepared separately and not mixed together in dishes.[108] In addition, though regional specialties were once common, years of communist rule during the first half of the twentieth century encouraged conformity, and many culinary differences have diminished.[109]

though buttermilk, fruit juices, and soft drinks are consumed in some areas. In urban regions street vendors ply pastries and ice cream throughout the day. Late evening visits to cafés or coffee houses often include small kebabs, meatballs, vegetable salads and pickles, and other tidbits to accompany coffee, wine, or plum brandy.

In Greece and the southern Balkan nations, the traditional breakfast typically consists of bread with cheese, olives, or jam plus coffee or tea. The main meal, eaten in early afternoon, usually begins with mezze or appetizers, such as hummus, baba ganouj, tiropetes, and dolmas, often consumed with a small glass of ouzo: the actual selection of included items varies by the inclination of the homemaker and affordability. Next a meat stew, meatballs, kebabs, vegetables stuffed with chopped meat, or a bean dish is served with a salad of raw seasonal vegetables, yogurt or cheese, and fruit for dessert. Roasted or baked whole meats are served on weekends, accompanied by cooked vegetables, salad, and dessert. Late afternoon and early evening are times when neighbors and friends may drop by for some sweets and a cup of coffee or glass of *ouzo*. A light supper is served in the late evening. Throughout the day, *mezze* are widely available from street vendors and cafes for snacking.

Middle East. In most Arabic countries coffee or tea is often served first for breakfast around 7:00 or 8:00 a.m., followed by a light meal that might include bread, cheeses, beans, eggs, olives, jam and bread, and plain yogurt. Lunch is the main meal of the day, eaten in the early afternoon. It is customarily bread, rice or bulgur, and a vegetable or legume casserole, a meat or poultry stew, or, where available, a fish or seafood dish. Fresh or cooked vegetable salad or onions and olives are common side dishes. Additional items, depending on region and affordability, may include a selection of mezze (such as hummus, tabouli, vegetables in yogurt, and bowls of nuts), a soup, and cheeses. Dessert is usually included, typically a piece of fresh fruit, a pastry, or a custard or pudding. Diluted yogurt drinks or water is served while eating, followed by sweetened tea or coffee. Dinner in the early evening is light, consisting of foods similar to those eaten at breakfast, soup, or leftovers from lunch. All the dishes in a meal are customarily served at once in Egypt, Iraq, and Yeman, and in courses in Jordan, Lebanon, and Syria.[34]

EXPLORING GLOBAL CUISINE—Moroccan Cooking

Morocco is one of five nations that make up the Maghreb, a region of North Africa differentiated from the Middle East by its substantial populations of nomadic Berbers, also including the countries of Algeria, Tunisia, Mauritania, and Libya. Although there are very few immigrants to the United States from the Maghreb, many Moroccan restaurants have opened, advertising an exotic dining experience at low tables amid pillowed opulence, often with belly-dancing entertainment.

The cooking of Morocco is predominantly Berber in origin, strongly influenced by neighboring Arabic fare and, to a much lesser degree, through interchange with sub-Saharan Africa and the southern European countries of the Mediterranean. It is noteworthy for its exquisite seasonings. Spices, such as allspice, anise, cardamom, cayenne, cumin, cinnamon, cloves, mace, malagueta pepper (see Chapter 8, "Africans"), nutmeg, turmeric, and saffron, are combined with herbs, including basil, fresh coriander, lavender, marjoram, mint, verbena, and *za'atar*. One mixture, *ras el hanout*, includes between ten and twenty-five ingredients, depending on the chef and its intended purpose; medicinal herbs such as belladonna or reputed aphrodisiacs (such as the pulverized beetle known as "Spanish fly") may also be added.[110] Garlic, onions, lemons (some preserved through brining), almonds, and sweet peppers also flavor many dishes, and some are heated with the chile pepper and garlic paste condiment called *harissa* (from Tunisia, where foods are preferred very spicy). Rose water and orange-blossom water are also commonly used. Foods are also flavored by the preferred cooking fats of the region, *zebeda* (a sour fresh butter) and *smen* (a preserved clarified butter often seasoned with herbs; it is traditionally stored for months underground until cheese-like).

Couscous is a staple eaten throughout the Mahgreb, where it is known by many names. It is made from a dough of hulled, crushed (not ground) grains of semolina wheat (other grains prepared the same way, such as barley and millet, are also called couscous) mixed with water and processed into very small pellets and dried. It is traditionally cooked in a specialized steamer known as a *couscousière*. The word couscous is also used to describe the finished dish: the steamed grain topped with a mixture of lamb with chickpeas and vegetables, fish with fennel, dates with cinnamon (for dessert), or other popular versions. Moroccan stews, tagines, are slow-cooked in ceramic pots and feature any combination of meats, poultry, fish, organ meats, vegetables, and fruits. Mechoui is spit-roasted lamb or kid. The meat is first rubbed with cumin and garlic, and then cooked until it can be pulled off with the fingers. Bastilla or b'stila (from the Spanish word for pastry or pie, *pastel*) is the quintessential Moroccan dish: sheets of *warqa* (a dough similar to filo, though thinner) enclose layers of ground almonds mixed with sugar and cinnamon alternating with pigeon or chicken meat. The layers are bound with a lemony egg sauce, and the pie is baked until golden. The crispy crust is often sprinkled with sugar and cinnamon before serving. Cooked or marinated vegetable salads usually start a meal, and fresh fruit and nuts add the finishing touch. All foods are eaten with the first three fingers of the right hand, and bread is used to sop up sauces.

Turkish meals vary slightly from the Middle Eastern pattern. Breakfast (often served a little later than in Arabic nations) varies regionally, but is often substantial, with leavened bread or *simit* (a chewy or crunchy ring-shaped roll resembling a bagel in shape, but richer in flavor—also found in Greece where they are called *koulouri*), cheese, butter, tomatoes, olives, and jam served with sweetened tea. Eggs, soups or sausages are common additions in some areas. Lunch, eaten around noon, and dinner, served between 6:00 and 8:00 p.m., are also plentiful meals, especially dinner, which is the main meal of the day. It begins with a selection of mezze (*mezeler* in Turkish) served with raki. Items may include lamb meatballs, dolma, stuffed mussels, fried squids, baba ganouj, hummus, and vegetable salads. These appetizers are followed by the olive oil course, featuring vegetables such as eggplant, tomatoes, or leeks stewed in olive oil. Kebabs, casseroles, or stews are the centerpiece of the meal, served with *pilaf and* bread. Fresh fruit such as melon, baklava, or rice pudding follows, and the sweets are consumed with coffee. In some regions, tripe soup with vinegar and garlic may be eaten after dinner, served with alcoholic beverages. Turkish meals are typically served in courses.

In Iran, breakfast is usually a selection of flat breads served with feta or other cheeses, sweetened whipped cream, and jam. In some regions, offal soup or halim (a savory wheat porridge with meats and vegetables) is preferred.[35] Lunch and dinner are similar, usually with rice, a meat or poultry dish (roasted, or as kebabs or ground meat), often a vegetable salad, flat breads, some feta cheese or yogurt, and a selection of chopped herbs (such as mint, basil, and dill). A meat or vegetable stew is frequently substituted for the meat or poultry course, served over the rice. Fruit, especially melon or grapes, is a typical dessert, and tea or a yogurt drink accompanies the meal. Traditionally, the dishes are all served at one time and eaten communally.

Weekday breakfasts in Israel are customarily light: coffee with some pita and olive oil and *za'atar*, European-style breads with jam or other spread, or a selection of cheeses, yogurt, and chopped vegetables and fruit. Sabbath breakfasts, however, are somewhat heartier. European Jews may choose coffeecakes or pancakes, and Middle Eastern Jews may select bureks, kataif (a sweet, stuffed pancake), or sabikh (an Iraqi dish of pita bread topped with fried eggplant, hard-boiled egg, tahini, and a mango pickle). The traditional Israeli breakfast buffet associated with kibbutz life is offered at some restaurants, featuring a more typically Middle Eastern selection of flat breads, cheeses, vegetables, and olives, often with added eggs, baked goods, and other selections. The midday meal is the largest in most homes, beginning with hummus or tahini served with pita bread, then a salad—often cucumbers, tomatoes, and onions—followed by items appropriate to a meat or dairy meal (see Chapter 4 for more information on Kosher rules). The evening meal, typically eaten around 8:00 or 9:00 p.m., is usually light with cheeses, yogurt, salads, and eggs. Some families serve all the dishes of the meal at once, while others serve them in courses, often depending on heritage. Street stands offer falafel, kebabs, shawarma, and other snacks, and fast-food restaurants, especially those serving hamburgers or pizza, are popular with many Israelis. Fruit juices, soft drinks, and beer are common meal beverages.

Etiquette

Throughout the Balkans and the Middle East, hospitality is a duty and a family's status is measured by how guests are treated. Guests, even uninvited ones, are made to feel welcome and are automatically offered food and drink. In the Balkans it is likely to be fruit compotes, candies, and buttermilk, coffee, plum brandy, or, in Greece, *ouzo* or *arak*. In the Middle East it may be a few dates and water or an extensive choice of mezze served with coffee, tea, or raki. Even if food is initially refused, it will be offered again, and a guest must accept because refusal is considered an insult. Invited guests bring a gift, often candy or other sweets, which the host must open immediately and serve. Hospitality is even offered to clients in the office setting,

[SAMPLE MENU]

A Greek Mezze

Olives and Cheeses (such as *Kaseri* or *Myzithra*)

***Taramosalata* (Caviar Dip)[a,b] or *Hummus* (Chickpea Dip)**

***Tzatziki* (Cucumber Yogurt Dip)[a,b]**

Pita Bread

***Spanakopita* (Spinach and Cheese Triangles)[a,b]**

***Dolmas* (Stuffed Grape Leaves)[a,b] with *Avgolemono* (Egg and Lemon Sauce)[a,b]**

***Ouzo* or Wine**

[a]Harris, A. 2002. *Modern Greek: 170 contemporary recipes from the Mediterrnean.* San Francisco: Chronicle Books.

[b]*Eat Greek Tonight* at http://www.greek-recipe.com

[SAMPLE MENU]

An Arab Sampler

***Baba Ganoush*[a,b] with Pita Bread**

***Kofta* (Kebabs) in Yogurt Sauce[a,b]**

***Tabouli* (Bulgur Salad)[a,b]**

Olive and Orange Salad[a,b]

***Baklava*[a,b] or Stuffed Dates[a] (or Apricots)[b]**

Arabic Coffee or Mint Tea

[a]Salloum, H., & Peters, J. 2001. From the land of figs and olives. New York: Interlink Books.

[b]*Middle Eastern Recipes* at http://www.sudairy.com

[SAMPLE MENU]

A Persian Lunch

Olives and Pistachios

***Khoresh-e Fesenjan* (Chicken Stew with Walnuts and omegranate over Rice)[a,b]**

Cucumber, Tomato, and Onion Salad[a,b]

Feta Cheese and *Lavash*

Fresh fruit or Sholeh Zard (saffron pudding)[a,b]

Tea

[a]Batmanglij, N.K. 2000. *New food of life: Ancient Persian and modern Iranian cooking and ceremonies.* Washington, DC: Mage Publishers.

[b]*Iranian/Persian Recipes* at http://www.iranchamber.com/recipes/recipes.php

"Spoon sweets" (seasonal fruits, vegetables, nuts, or rose petals preserved in a heavy, sweet syrup) are a Greek specialty specifically reserved for guests, offered by the spoonful on arrival as a sweet welcome.

and failure to make guests or clients comfortable may create extreme embarrassment for all parties.[24]

In the Middle East, a guest's status is indicated by which pieces of food are offered, and the order in which the items are served. Status is based on sex, age, family, and social rank. For example, a dignitary or head of the family is served the best portion first. In Saudi Arabia and other nations of the Arabian peninsula, the honored seat at the table is in the middle of the table, whereas in Egypt, it is at the head of the table.[36] In some areas, such as some parts of Yemen, it is customary for women to eat separately from men.[34] Guests are traditionally entertained in a separate room before the meal, at which time scented water is provided so they may wash their hands. The dining table might be a large, round metal tray, resting on a low stool or platform, and the diners sit around it on cushions. Western-style dining is found occasionally, especially in Middle Eastern restaurants. In Iran, food is traditionally served on a rug. The meal is set out in several bowls placed on the table or rug and then shared by the diners. After the meal, the guests leave the table, wash their hands, and then have coffee or tea.

Several rules of etiquette apply to eating in the Middle East. One should always wash one's hands before eating. In Muslim regions the guests thank Allah before and after the meal. Other Middle Easterners may say *"Sahain!"* ("good appetite!") to start the meal, and *"Daimah"* ("may there always be plenty") to end it.[36] Three fingers of the right hand are used if forks or spoons are not offered. The left hand should not be used in any food-related manner (including passing food), and women should not touch any food that is to be eaten by a Muslim man who is not her immediate family member. Rice should be taken from the communal bowl and rolled into a small ball with the fingers before dipping it into stews or sauces. Licking the fingers after eating is expected, and appreciation is shown in some areas by making eating noises.[37] It is rude to fill one's own cup, and it is expected that a diner will refresh his or her neighbor's cup as soon as it is half empty. It is also considered polite to continue eating until everyone else is finished because if one person stops the others feel compelled to stop, too. One should leave a little food on one's plate to indicate satisfaction with the abundance of the meal. Most conversation takes place before and after the meal, and limited discussion of pleasant and joyful things takes place while dining. It is important to compliment the host and hostess on their hospitality.

Special Occasions

In Balkan and Middle Eastern countries, food plays an important role in the celebration of religious occasions and in the observance of certain events such as weddings and births. In the Eastern Orthodox Church, there are numerous feast and fast days (see Chapter 4). The most important religious holiday for the Greeks is Easter. Immediately after midnight Mass on Holy Saturday, the family shares the first post-Lenten meal. It traditionally begins with red-dyed Easter eggs and continues with mayeritsa, a soup made of the lamb's internal organs, sometimes flavored with *avgolemono*, a tart egg, and lemon. The Easter Sunday meal usually consists of roast lamb, rice pilaf, accompanying vegetables, cheese, yogurt, and a special Easter bread called lambropsomo that is decorated with whole dyed eggs. Dessert usually includes sweet pastries made with filo dough and koulourakia, a traditional Greek sweet bread cookie, sometimes shaped into a hairpin twist or wreath or coiled in the shape of a snake (a creature that the pagan Greeks worshipped for its healing powers). Easter is preceded by the pre-Lenten holiday of Apokreas, which is similar to Carnival or Mardi Gras and features costumed events and parties with ample merrymaking, food, and music. In addition to religious holidays, Greek Americans typically celebrate Greek Independence Day on March 25. It is commemorated with parades in traditional dress, folk dancing, songs, and poetry readings.

The Easter meal in Croatia is typically lamb or ham and pogaca, an Easter bread with painted eggs on top that is similar to the Greek lambropsomo. Christmas Eve features a meal of cod, and a stuffed cabbage and sauerkraut dish is customary on Christmas. Among Serbians, the most auspicious day of the year is *Krsna Slava,* Patron Saint's Day. This holiday dates back to the worship of protective spirits in pagan times; today each family honors its self-chosen patron saint with a sumptuous feast and dancing that may last for two to three days. The family customarily announces the annual open house with a small advertisement in the local newspaper. Krsni kolac is a ritual bread prepared for the occasion, decorated with the religious Serbian emblem *"Samo sloga Srbina spašava"* ("Only unity will save the Serbs") as well as grapes, wheat, birds, flowers, barrels of wine, or other representations made in dough. Slovenians celebrate St. Nick's Feast. Gifts are distributed to children by St. Nick, dressed as a bishop, who admonishes the youngsters to be good. The grape harvest and winemaking are traditionally commemorated with numerous festivals and St. Martin's Feast.

There are also feasts and fasts connected with Islamic religious observances. Traditional festive foods vary from country to country and may also vary seasonally since the Muslim calendar is lunar, and holidays fall at different times each year. *Iftar* is the meal that breaks the fast during Ramadan, the month in which Muslims fast from sunrise to sunset; it is common to dine with relatives and neighbors. The meal usually starts with a beverage, preferably water, followed by an odd number of dates and coffee or tea. A large meal, served after prayers, includes moist and hearty dishes. Regular items eaten during Ramadan include soups, fruit juices, cheeses, and fresh or dried fruit. Traditional sweets include kataif, which refers to a pancake or a shredded wheat dough dessert and, in Turkey and Iran, rose-flavored rice puddings. In some Muslim homes, the post-fast meal is considered a feast with elaborate dishes that emphasizes the Muslim virtues of hospitality and community, while in other homes a more moderate meal is thought to be in keeping with the purposes of the fast.[38] The dawn meal is usually light, and salty foods are avoided because water is not allowed during the fast.

The holiday *Eid al-Fitr* follows the end of Ramadan and is described as a cross between the feasting of Thanksgiving and the festivity of Christmas. Typically family, friends, and neighbors gather to celebrate; in areas with large Muslim populations, *Eid al-Fitr* may be held at the local fairgrounds with games, rides, and many food vendors. The other major holiday observed by Arab Muslims is *Eid al-Adha,* the Feast of Sacrifice held in conjunction with the annual pilgrimage (*Hadj*) to Mecca (see Chapter 4).

In Turkey, *Eid al-Fitr* is known as *Seker Bayram,* meaning "sugar festival." It is traditional to exchange small gifts with friends, and for four days children are given sweet treats, such as *lokums* or chocolates. On the tenth day of the first lunar month of the Islamic calendar, Turks celebrate the martyrdom of Mohammed's grandson and the day Noah was able to leave the ark. They prepare *asure,* or Noah's pudding, made from the ingredients remaining after the receding of the flood waters: fresh and dried fruits, nuts, and legumes. *Kurban Bayram* is a day of remembrance for when the prophet Abraham nearly sacrificed his son Ishmael. Families customarily sacrifice a sheep or a goat and distribute it to family, friends, and community charities. Another special occasion in Turkey is National Sovereignty and Children's Day on April 23. It commemorates the establishment of the Grand National Assembly in 1923 and specifically honors all children. The following day has become Turkish American Day in the United States, featuring parades in traditional dress and other festivities.

In Iran the most significant holiday of the year is *Muharram,* which commemorates the martyrdom of the grandson of Mohammed in the seventh century. It is a time of communal mourning and penitence for Shiites, and often features *sholeh zard,* a sweet rice pudding flavored with saffron. Another celebration marking the spring

Yemen qat (the herb with amphetamine-like properties called _khat_ in Ethiopia—see Chapter 8) is frequently chewed in social and business settings.

In Greek Orthodox tradition, the egg represents life, and red is the color of the blood Christ shed. The breaking of the red-dyed egg symbolizes the resurrection.

For New Year's Day the Greeks prepare a sweet spicy bread called _vasilopitta_ with a coin baked into it—the person who gets the piece with the money has good luck in the upcoming year (the Serbs have the same tradition for Christmas Day).

Kahk is a sweet Egyptian bread made with ample butter and nuts that is served at all special occasions.

Some non-Christian Arab Americans celebrate the birth of Jesus on Christmas; Jesus is considered a prophet in Islam.

In Greece and the Middle East, sugared almonds (Jordan almonds) are served at weddings to ensure sweetness in married life.

The name of the Jewish Sabbath stew cholent may come from the French for "warm," *chaud*, and "slow," *lent*,[33] or from *cholent* (or *sholen*) from the Hebrew *she'lan*, which means "that rested [overnight]."[111]

equinox is *Nau Roz,* which features a meal, called *haft-sinn,* including a ceremonial table setting where the Qur'an, a mirror to reflect life, sweets, bread, cheese, and seven items starting with the letter s representing rebirth, good fortune, love, happiness, health, and other wishes for the new year are displayed. Foods and other goods starting with s may include *serke* (vinegar), *seeb* (apple), *sanjed* (dried fruit or olives), *sumagh* (sumac), *samanu* (a sweet sprouted wheat kernel pudding), *seer* (garlic), sonbul (hyacinth), *sabzi* (sprouted seeds), or *sekeh* (coins). Readings from the Qur'an are followed with a traditional meal of herbed rice (*sabzu polo*) and an herbed omelet (*kuku-ye sabzi*) served with fish. The number seven probably relates to the seven days of the week or the seven planets of the ancient solar system. On the thirteenth day of *Nau Roz,* it is customary to have a picnic.

The traditional holidays of the Jewish calendar are observed in Israel (see Chapter 4 for Jewish food practices). The Shabbat, or Sabbath, occurs from sunset Friday to sunset on Saturday evening. All businesses close for the day and work activities are prohibited. The Friday meal is served on a table set with white linen and includes the symbolic *Kiddush* cup of wine that is shared by all diners. In Ashkenazi homes a representative menu would include gefilte fish, a leavened, braided loaf of egg-rich *challah* bread, a roast chicken, a noodle pudding called kugel, and fruit or cake and tea for dessert. In Sephardic households a more Middle Eastern meal would be typical, such as pilaf, roast lamb, cooked eggplant or stuffed dolma, pita bread, and honey-soaked filo pastries with coffee for dessert. Jews from other regions have favorite Sabbath menus as well. For example, an Ethiopian family might serve chicken doro wat and caraway-flavored dabo bread. Because all work, including cooking, is banned on the Sabbath, only foods prepared during the day on Friday or left cooking overnight can be consumed on Saturday. Stews that can simmer overnight on the stovetop are popular dishes for the midday meal following morning services, such as cholent, also known as hamim in Israel. Every family has its own version, though most include beans and potatoes. The Lebanese and Syrians use white beans, the Brazilians use black beans, and the Moroccans add rice. Most Ashkenazi recipes use beef brisket, and some Sephardic versions include sausages. In most homes whole eggs cooked in their shells are buried in the stew. Other religious holidays offer a similar variety of food traditions based on nationality of origin. Israelis also observe the secular *Yom Ha-Atzma'ut*, Independence Day, on the fifth day of *Iyar* (a spring month in the Jewish lunar calendar). Celebrants view parades, hold barbecues, and watch fireworks. Street vendors sell falafel, ears of corn, and numerous sweets, including candied fruits and nuts, sesame seed candy, and European cakes and tortes with whipped cream.

Therapeutic Uses of Foods

Fresh foods are considered best, and canned or frozen foods are often avoided by Middle Easterners to preserve health. The amount of food eaten is of special concern in the diet. Ample meals are needed to prevent illness, and poor appetite is regarded as a disease in itself or as a generalized complaint signifying that one's life is not as it should be. Food deprivation is believed to cause illness.[39,40,24]

Some Middle Easterners also believe that illness can be triggered by hot-cold shifts in food, especially in people with weak or susceptible constitutions. In Iran, eating too many hot foods may result in headaches, sweating, itching, and rashes. Excessive amounts of cold foods can cause dizziness, nausea, and vomiting.[25,28,24] Foods and drinks of the opposite category can ameliorate these conditions. For example, citrus fruits or a sour lemonade called *ablimu* is used for headaches and acne. Nausea is treated with tea or a sweet similar to rock candy. Classifications can vary, but examples of hot foods include lamb, eggs, onions, garlic, carrots, bell peppers, apples, dates, quinces, chickpeas, wheat, almonds, walnuts, pistachios, honey, and tea. Cold foods can include beef, cucumbers, tomatoes, eggplant, grape leaves, grapes, lemons, sour cherries, apricots, mulberries, pomegranates, rice, yogurt, coffee, and beer. The temperature (not spiciness) can cause a shift in the body from hot to cold and vice versa, and it is believed the digestive system must have time to adjust to one extreme before a food of the opposite temperature can be introduced. In addition, though illness may be related to hot-cold imbalances, Iranians do not consider certain conditions as being hot or cold. Thus, a symptom such as coughing requires specific treatment unrelated to classification: consuming cold turnips is considered beneficial, whereas cold pickles are deemed harmful.[24]

Some Middle Easterners also believe certain combinations of incompatible foods are damaging to health. For example, Egyptians do not consume fish at the same time as dairy products. Other Middle Easterners avoid eating sour foods with milk and legumes with cheese. Iranians believe consumption of melon with yogurt causes wind in the stomach and gastrointestinal disorders.

Many special foods are associated with childbirth. Eggs cooked in garlic and chicken soup are frequently consumed by Lebanese women after childbirth. When a woman gives birth to a girl in Iran, coldness is neutralized with a diet high in hot foods to ensure a male child in the next pregnancy. Saffron custard garnished with nuts is thought to help Iranian women regain strength post-partum, while Palestinian women consume oats, coriander, or fennel.

The division between food and medicine is somewhat blurred in the Middle East, especially in Arabic nations.[27] Turnips are considered good for the kidneys and urinary tract, whereas cauliflower is beneficial for the respiratory system. Red onion bulbs and their leaves (which are added to salads) are consumed to help with diabetes and cancer. They are also eaten to ease liver disease, which is treated with asparagus and artichokes, too.[26] Many foods have a multiplicity of therapeutic uses. Some Palestinians, for example, consider garlic to be good for colic, nausea, kidney infections, intestinal worms, ulcers, genitourinary infections, prostate conditions, and tumors, and as an aphrodisiac.[27]

Contemporary Food Habits in the United States

Adaptations of Food Habits

There is scant information on the adaptation of Balkan or Middle Eastern diets in the United States. It is assumed that, as in other immigrant groups, increasing length of stay is correlated with Americanization of the diet, with traditional dishes prepared and eaten only for the main meal or for special occasions. It is less likely that religious dietary practices, such as adherence to halal or kosher law, change significantly after arrival in the United States.

Courtesy of World-Health Organization/P. Merchez

▲ *Middle Eastern market.*

Ingredients and Common Foods

Greek Americans still use olive oil extensively, although they use less of it than their immigrant relatives.[41] Salads still accompany the meal, and fruit is often served for dessert. Vegetables are prepared in the traditional manner. Lamb is still very popular; for special occasions, roasted leg of lamb is substituted for the whole animal. Consumption of beef and pork has increased, whereas consumption of legumes has decreased. Cereal and grain consumption among Greek Americans remains high, and bread, rice, or cereal products are usually included in every meal. Greek Americans consume more milk than their immigrant parents, and ice cream is very popular.

One small study of first-generation Egyptians found that traditional wheat bread remained commonly consumed, though intake of legumes, especially fava beans, was somewhat lower than when the immigrants had lived in Egypt. Snacking and eating out had become much more prevalent, and soft drinks were more popular.[42]

The demand for properly slaughtered (*halal*) meat among Muslims in the United States has led to increased numbers of Arab *halal* markets.

Meal Composition and Cycle

Greek Americans maintain traditional meal patterns, but the main meal of the day is now dinner.[41] They prefer an American-type breakfast and lunch, but dinner is more traditional.

Greek weddings in the United States offer a blend of Greek and American foods; for example, the wedding cake is served along with *baklava*.

However, they have adapted Greek recipes to make them less time-consuming to prepare and to include fewer fats and spices. It is assumed that the meal pattern for most Americans of Croatian, Serbian, and Slovenian heritage is much acculturated.

Reportedly many Arab Americans still eat their main meal at midday.[43] Members of the extended family may dine together daily, with the women who stay at home cooking for employed female relatives.[40] However, a recent study of Egyptian immigrants found that though the midday meal remained substantial, the main meal of the day had become dinner.[42]

Nutritional Status

Nutritional Intake

Life expectancy for adult Albanians is very high despite very low socioeconomic status, and some researchers hypothesize that this is due to a diet high in monounsaturated fats from olive oil and low intake of animal products due to privation.[109,112]

Very little has been reported on the nutritional composition of the Balkan-American or Middle Eastern–American diet. However, research on the diets of nations bordering the Mediterranean (particularly Greece) often report that the traditional diet there, one that is low in saturated fats and high in monounsaturated fats and omega-3 fatty acids (due to a low intake of meats combined with high consumption of olive oil, fruits, and vegetables), lowers the risk of cardiovascular disease and cancer.[44,45,46] Studies on the impact of the Mediterranean diet on the development of metabolic syndrome conditions (including obesity, hypertension, and type 2 diabetes) have been contradictory but supports improvement in risk factors associated with heart disease.[47,48,49,50] In addition, the role of alcohol consumption, particularly wine, in the Mediterranean diet is not yet fully understood.[51] What is known is that the number of people consuming this traditional diet is declining with the Westernization of the region. Since the 1960s, the Greeks have been consuming significantly less olive oil and more alcohol.[52] Rates of overweight and obesity in Greece are nearly 50 percent for women and almost 75 percent for men.[53] A recent study in Lebanon found younger adults were eating fewer fruits, vegetables, and legumes, while consuming more meat and sugar and drinking more soft drinks and alcoholic beverages, than older adults.[54] Similar results were found among male colleges students in Saudi Arabia.[55]

Sparse data on Bosnian immigrants has shown that 8 percent of first-time patsients seeking care at a refugee medical clinic were diagnosed with hypertensive disease, and another 4 percent presented with diabetes. Thirty-six percent of those in a smaller surveyed subset desired care for chronic disease management. Providers report a need for diet and exercise counseling due to diets high in sugar, fat, and meat, and low in salads, fruits, and grains. Some refugees have stated they have little time for exercise beyond work-related physical activity. Dental problems were significant, and alcohol abuse may be seen in some refugees.[6,21]

Research suggests that Arab men living in the Arabian Peninsula region may be as susceptible to the clustering of risk factors in metabolic syndrome as are some other ethnic groups, such as Asian Indians (see Chapter 14, "South Asians," for more information). High prevalence of undiagnosed type 2 diabetes and hypertension, and high rates of insulin resistance, low levels of high density lipoprotein (HDL) cholesterol, and a tendency toward abdominal obesity were found. Coronary heart disease, diabetes, hypertension, and cancer are the primary health concerns in Arab countries.[56] Studies of men and women in Turkey, where cardiovascular disease is the most common cause of death, have also reported strikingly low serum levels of HDL cholesterol unrelated to diet, obesity, or lifestyle.[57,58,59] Worldwide, one of the greatest relative increases in type 2 diabetes is expected to occur in the Eastern Mediterranean Region.[60]

Limited research on Arabs living in the United States has reported similar health trends. One study of Arab American adults found that 29 percent of men and 37 percent of women were obese (body mass index [BMI] ≥ 30), with central obesity (as measured by waist-to-hip ratio) found in over 50 percent of both men and women. High blood pressures were noted in about 20 percent of the adults.[61] Rates of type 2 diabetes in Arab Americans in Michigan aged twenty to seventy-five years were found to be very high: over 15 percent for women and 20 percent for men, with rates of impaired fasting glucose and glucose intolerance above 32 percent in women, and almost 50 percent in men.[62] When analyzed further, the data suggest that lack of acculturation was associated with diabetes risk, as was consumption of Arabic foods, being an older age at time of immigration, unemployment, and reduced activity in Arab organizations.[63] The age-adjusted prevalence of metabolic syndrome was found to be between 23 and 28 percent of Arab American adults in one study, and the most common component was low levels of HDL cholesterol.[64] Low HDL cholesterol levels in Arab American women were accompanied by high triglyceride levels in another study.[65]

Data on cancer incidence in Arab/Chaldean adults indicate that, when compared to the non-Arab white population, the men have disproportionately high rates of leukemia, multiple myeloma, and liver, kidney, and urinary bladder cancers, breast, while women have greater proportions of leukemia and thyroid and brain cancers.[66] However, the leading causes of cancer-related deaths in Arab Americans are lung, colorectal, and breast cancers.[67,68]

The effects of Ramadan fasting have been explored among Muslims. Although hunger increases in some fasters, there were no significant changes in body weight noted in one study.[69] Increases in uric acid blood levels have been noted when weight loss occurs in nonobese men, however, which may be related to high rates of kidney stones and angina pectoris reported in some epidemiological surveys conducted during the month-long fast.[70,71] A majority of pregnant women go through Ramadan, and one study suggested that with certain precautions, such as excluding women with medical risk factors (including diabetes and history of preterm deliveries or renal stones), increased prenatal visits, avoiding strenuous exercise, staying cool, and consumption of extra fluids before dawn, fasting was safe for many of the women. It was also noted that immigrant Muslim women fasted on average more days than did women born in the United States.[26] Individuals with type 2 diabetes who fasted during Ramandan were able to maintain normal blood sugar levels but fasting is not recommend for those with type 1 diabetes.[72]

Cross-cultural research on breast feeding reported that 82 percent of Iranian American mothers in the study exclusively breast fed their infants. This high rate was attributed to a strong social network of support for the practice. In Turkey, mothers sometimes nurse their sons longer than daughters because breast milk is believed to increase strength.[73,74]

Though rates of celiac disease in the Middle East are estimated to be below those in northern Europeans (see Chapter 6, "Northern and Southern Europeans"), it is considered the primary cause of chronic diarrhea in Iran and may contribute to iron deficiency, malnutrition, rickets, and short stature in children.[75,76,77,78] Thalassemia syndromes may also be prevalent in Middle Easterners.[79]

Counseling

Considerable discomfort and irritation have been noted between health care practitioners and their Middle Eastern–American clients, much of it due to cultural differences.[43] Most difficulties evolve from misunderstandings in the provider-patient relationship.[80] Though there are considerable similarities in culture between people of the Balkans and the Middle East, even among those immigrants who share Arabic as a language there can be notable differences in dialect, ethnic background, and socioeconomic status.[81] Religious affiliation and degree of orthodoxy are equally important. Furthermore, English language skills vary greatly. Among Arabs, for example, fewer than 14 percent of Lebanese Americans spoke English less than "very well," compared to 42 percent of Iraqi Americans in the 2000 census.[12] Language and communication problems are significant health care access issues according to some recent Arab immigrants.[31] Many Bosnian refugees, especially women and elders who are homebound, have poor English skills and lack of interpreter services is one barrier to health care services.[6]

Interactions are highly contextual throughout the Balkans and the Middle East, meaning that body language and general atmosphere are as important to communication as words, if not more so (see Chapter 3, "Intercultural Communication"). People in the Middle East spend time getting to know one another before any business is discussed. Offering coffee or tea at the beginning of any interaction helps to establish a warm and hospitable atmosphere with many Balkan or Middle Eastern clients.[36,43]

Direct eye contact is expected and necessary to interpret meaning, so Middle Easterners usually sit or stand quite close when conversing with intimates but may retain a larger distance with strangers.[82] Greeks may smile when angry. Nodding one's head up and down or back and forth can be very confusing. Traditionally, moving the chin up and down meant "no" and back and forth meant "yes" or "I don't understand." But many

Some researchers note that Muslim women who, for religious reasons, completely cover their bodies and heads with clothing when outdoors may be at risk for vitamin D deficiency.[40]

Kurdish and Iraqi refugees to the United States may have been exposed to chemical weapons. They have also had high rates of malnutrition, parasitic infection, hepatitis B, and tuberculosis; and the prevalence of glucose-6-phosphate dehydrogenase deficiency is also significant in this population.[113,114,115]

DISCUSSION STARTERS: DIET AND CULTURE OF BALKAN AND MIDDLE EASTERN AMERICANS

In small groups of three to four, compare and contrast the diet and culture of Balkan Americans and Middle Eastern Americans, with each group focusing on a different aspect of the diet and culture of these groups:

Group A: The food habits and the typical eating etiquette and meal composition of these two immigrant groups

Group B: Issues involved in counseling these immigrant groups on diet and health

Group C: Attitudes within each immigrant group toward diet, health, and medical treatment, notably attitudes toward traditional home culture medical treatment and U.S. biomedicine

Group D: Amount of obesity, diabetes, hypertension, and other diseases within each immigrant group

Given the scarcity of data in some cases, you may have to draw hypotheses about the diet, food habits, and so on, of these groups from what we know about diet, food habits, and the like, of non-immigrants.

Within your group, try to come to a consensus on what findings to report to the rest of the class. Before breaking up, assign a number to each group member: A1, A2, A3, A4; B1, B2, and so forth. Form new groups with all the 1's in a group, all the 2's in another group, all the 3's another group, and so on. In your new group, report the findings of your previous group, and as a group, discuss the relationship of traditional attitudes toward diet and health and changes in diet and health due to immigration to the United States.

Greeks and Middle Easterners use the American protocol, so it is difficult to know whether the gesture is affirmative or negative.

Touching between members of the same sex is frequent, including handshaking, patting, shoulder slapping, hugging, and kissing. Contact between members of the opposite sex is prohibited in some Middle Eastern–Muslim cultures and avoided in most. Extended eye contact between men and women can be considered a sexual overture (staring between members of the same sex is acceptable). The left hand is not used for any social purposes, nor to pass documents or administer medications.[36,83] In general, it is best to wait for Middle Easterners to extend their hand in greeting before making any unwanted contact. Proper posture is a sign of respect, and crossing one's legs, pointing with the foot, or showing the sole of the shoe is considered impolite. In Turkey, one should stand when an elder enters a room.

Due to the significance of how the message is communicated to Americans of Balkan or Middle Eastern background, providers may find that some of their clients are more receptive to verbal than to written information. A few minutes for general questions about the well-being of other family members or personal interests of the client should be allowed at the beginning of the interaction.[22] It is important to speak kindly, softly, and respectfully.[24,82] Individual clients may be inexperienced with making independent decisions, thus options should be kept to a minimum to avoid overwhelming the client with too many choices.[84] Family members, especially an elder male relative, may insist on participating in all conversations, even those that customarily take place in the office between only the practitioner and the patient. Because these family members may make the final decisions regarding care, their presence should be valued and their opinions fully solicited. However, if the client is a woman, she may ask to have her husband or father sign medical forms, presenting liability issues regarding consent.[82]

Balkan and Middle Eastern Americans value biomedicine and have considerable respect for authority figures.[82] They may be hesitant to ask questions when confused, however, or they may provide answers that are designed to please the provider. The health provider may have to assess and give advice about a medical or dietary problem without the client explaining his or her needs. If the provider does not repeat the offer to help, the client may believe that the provider is indifferent. Privacy is strongly protected, and clients may resist disclosing information about themselves and their families to strangers until a trusting relationship is established. Concerns regarding confidentiality may also inhibit discussion of some medical concerns, especially for Bosnian refugees and many Arab Americans. There may be suspicion regarding questions

NEW AMERICAN PERSPECTIVES—Egyptian

MIRAL MAAMOUN, MD

I am an Egyptian citizen and lived in Cairo for twenty-six years. Before coming to the United States, I graduated from medical school as a general practitioner and worked as a GP for two years before I got married to my husband who is a PhD graduate student in the United States.

In Egypt I used to eat three meals a day and snack on fruits once a day. Breakfast usually included fava beans, baladi bread, cheese, and a cup of milk. Sometimes falafel is added to the menu and occasionally a hot wheat cereal called Beleela. Lunch is the main meal of the day and is eaten around three o'clock. It contains rice or pasta together with vegetables and meat. Chicken was my favorite meat, but I also ate fish twice a month.

Dinner was served around 8:00 p.m., and it included a choice of three of the following: cheese, baladi bread, yogurt, eggs, falafel, fava beans, tomatoes, cucumbers, fried eggplant, and/or mashed potatoes. Water was the typical beverage with meals.

Once I came to U.S., the first thing that changed was the timing of my meals. Meals became two hours earlier than what I was used to. I tried to keep the quantity and quality of the food at lunch the same as in Egypt, but my husband felt too full to continue his workday with comfort after that, and he liked a lighter lunch of just soup, fruits, or one-third the quantity of the traditional lunch. Thus, dinner became the usual Egyptian lunch. In addition, we added one more meal before bedtime, which is usually a cake with milk.

The first time I invited friends over for lunch on Saturday, I was surprised that they knocked on my door at twelve o'clock. Of course, I was in the middle of cooking, and I was not even ready to host them. My husband was not even home. It was kind of embarrassing. I called my husband, who came right away, and the women ended up with me in the kitchen helping me with the cooking. They told me that it was too much food for lunch, and I should have mentioned that I was serving dinner. I was thinking dinner when I invited them, but I said lunch as it used to be the main meal we invite people over to share.

When I first came to the United States, I was impressed and tempted by the variety of food available and by how easy it was to prepare—all those ready-made sauces, dressings, appetizers that you didn't have to prepare; frozen food already peeled, cleaned, and ready to cook. I used to spend a lot of time cleaning and peeling in Egypt, so now cooking is easier for me. I tried not to change the use of fava beans and baladi bread in my diet. I can't resist the taste as these dishes made with fava beans can be done in many different delicious ways. I did have to replace the baladi bread with whole-wheat toasted bread as it is the closest to it, and I can't find the traditional baladi bread here.

My dietary advice to an Egyptian who has not yet acculturated is to have a good breakfast of traditional foods in reasonable quantity, eat lunch with more vegetables and less meat, and snack on fruits around three o'clock. Fruits can be the desserts, and only occasionally eat American-style desserts. Dinner can remain the same as the traditional Egyptian dinner of yogurt, low-fat cheese, milk, beans, salad, and vegetables. Before-bedtime meals can still be fruits. If they are acculturated, I would recommend that they watch the portion size of their diet and have extra lean meat if they have to eat meat more than once a day—and perhaps try to go back to the traditional meals on weekends for a change.

about religious affiliation or socioeconomic data. Fear of racial profiling may occur.[81] Further, shame about certain conditions may cause noncompliance when a client is in public situations. Diabetes, for example, may be associated with male impotence and female infertility. Strategies for adhering to diet modifications without disclosure of the cause can increase efficacy.[84]

A study of Middle Eastern immigrants from several countries representing five ethnic groups found that, in general, immigrants who perceived themselves as more traditionally ethnic experienced more physical complaints and had lower morale. Immigrants who were more acculturated reported better health.[85] Culturally, Arab immigrants tend to view good health as a state of balance and poor health as a state of imbalance.[86] Complaints by clients of Balkan or Middle Eastern heritage are frequently generalized or nonspecific, sometimes indicating anxiety or depression in a patient who does not distinguish culturally between physical and mental health. This may be particularly true with Bosnian refugees, who sometimes present with ill-defined symptoms and may suffer from anxiety, depression, and post-traumatic stress disorder.[6,82]

Middle Eastern Americans may expect the health care provider to make decisions for them and be responsible for the consequences. They may also demand the services of the top expert or the department head because the expectation is that they will receive the best care from the most senior, most powerful person in the system.[43] Female health care providers may face added difficulties in gaining the trust and respect of Balkan and Middle Eastern clients due to cultural norms regarding gender. Clients may believe that the more intrusive the medical procedure, the more effective the treatment.[39] A poor prognosis should be discussed with the family first and revealed in stages, preferably over several appointments. Among Iranians, the disclosure of bad news too quickly can cause a patient to become *narahati*.[24] It is considered sacrilegious to presume death because only God can make that final decision, and hope must always be maintained.

Muslim clients usually feel most comfortable with providers of the same gender.[83,82] One study of Middle Eastern–American parents indicated that they did not choose health care providers based on ethnicity.[87] Access, referrals, and effectiveness were factors in their choice. In contrast, another study found that less acculturated Greek Americans preferred seeing counselors (for psychological services) of the same ethnic background.[88] Consistent care from a single provider is most successful and may eliminate many communication difficulties.

Practitioners report that many Bosnian clients did not understand the importance of taking prescribed medications, and some did not follow instructions on how they should be taken or discontinued their use when side effects occurred.[21] Though few studies have been conducted to assess the continuing use of traditional health practices in the United States, it is believed that Bosnians frequently take herbal cures and alcohol-based tinctures simultaneously with biomedical therapies.[82] There are little data about the therapeutic ingredients in Balkan and Middle Eastern home remedies. *Ko'hl,* for example, used on the umbilical cords of newborns, is high in lead content and may present a danger, and the active ingredient in foxglove (used by some Iranians) is digitalis. Support for use of traditional remedies encourages clients to report their use, allowing the provider to prevent possible adverse interactions.

The need to pray during a medical care visit is a problem for some Muslims. Providers who have Muslim clients should consider setting up private areas with prayer rugs as a sign of respect. No talking can occur during prayers, and it is disrespectful to walk in front of someone who is praying.[83]

Many pregnant Muslim-American women in one study reported that they did not discuss Ramadan fasting with their prenatal care provider for fear of being treated disrespectfully or being told to stop outright.[26]

Traditional healers in Saudi Arabia sometimes recommend drinking sheep bile to treat diabetes, a practice that can result in acute toxicity.

Practitioners who work with Balkan and Middle Eastern clients should recognize that a high-context relationship is often intensive and time consuming.[43] The in-depth interview should be used to determine country of origin, degree of acculturation, and religious faith. Information on traditional health care beliefs still being practiced should be elicited.

REVIEW QUESTIONS

1. What food flavors and food ingredients are associated with the Balkan and Middle Eastern countries? Why might they be similar? Describe two recipes, one from the Balkans and one from the Middle East, that contain filo (phyllo) dough.
2. What is meant by the "evil eye"? How might you protect yourself from it?
3. What countries make up the Balkans and the Middle East? Pick either the Balkans or Middle East and map the religions found in that region. Pick one religion, describe a recipe eaten for a holiday of that religion, and explain how the recipe reflects the ingredients of the region.
4. What are common health problems associated with peoples from the Balkans or the Middle East? Select one group and a health disorder, and describe that group's cultural beliefs regarding the cause and appropriate treatment of that disorder.
5. In several countries from these two regions, food and illness may be classified as "hot" or "cold." What does this mean? Provide examples.

REFERENCES

1. Jurgens, J. 2000. Greek Americans. In *Gale Encyclopedia of Multicultural America,* R.V. Dassanowsky & J. Lehman (Eds.). Farmington Hills, MI: Gale Group.
2. Ifkovic, E. 2000. Croatian Americans. In *Gale Encyclopedia of Multicultural America,* R.V. Dassanowsky & J. Lehman (Eds.). Farmington Hills, MI: Gale Group.
3. Stevanovic, B. 2000. Serbian Americans. In *Gale Encyclopedia of Multicultural America,* R.V. Dassanowsky & J. Lehman (Eds.). Farmington Hills, MI: Gale Group.
4. Gobetz, E. 2000. Slovenian Americans. In *Gale Encyclopedia of Multicultural America,* R.V. Dassanowsky & J. Lehman (Eds.). Farmington Hills, MI: Gale Group.
5. Miller, O. 2000. Bosnian Americans. In *Gale Encyclopedia of Multicultural America,* R.V. Dassanowsky & J. Lehman (Eds.). Farmington Hills, MI: Gale Group.
6. Erwin, P., Leung, Y.Y., & Boban, D.I. 2001. *Bosnian refugees in San Francisco: A community assessment.* San Francisco: San Francisco Department of Public Health/International Institute of San Francisco.
7. Samhan, H. H. 2005. *By the numbers.* Arab American Business Magazine. Online at http://

www.allied-media.com/ArabAmerican/Arab_demographics.htm

8. Camarota, S.A. 2002. *Immigrants from the Middle East: A profile of the foreign-born population from Pakistan to Morocco.* Washington, DC: Center for Immigration Studies.
9. Arab American Institute http://www.aaiusa.org/pages/demographics/ Accessed 3/19/2011.
10. Kurson, K. 2000. Palestinian Americans. In *Gale Encyclopedia of Multicultural America,* R.V. Dassanowsky & J. Lehman (Eds.). Farmington Hills, MI: Gale Group.
11. Abraham, N. 2000. Arab Americans. In *Gale Encyclopedia of Multicultural America,* R.V. Dassanowsky & J. Lehman (Eds.). Farmington Hills, MI: Gale Group.
12. Brittingham, A., & de la Cruz, G.P. 2005. *We the People of Arab Ancestry in the United States.* Washington, DC: U.S. Census Bureau.
13. Gillis, M. 2000. Iranian Americans. *In Gale Encyclopedia Of Multicultural America,* R.V. Dassanowsky & J. Lehman (Eds.). Farmington Hills, MI: Gale Group.
14. U.S. Census Bureau. Immigration Statistics Staff. 2004. Foreign-Born Profiles (STP-159). Online at http://www.census.gov/population/www/socdemo/foreign/STP-159-2000tl.html
15. Altschiller, D. 2000. Turkish Americans. In *Gale Encyclopedia of Multicultural America,* R.V. Dassanowsky & J. Lehman (Eds.). Farmington Hills, MI: Gale Group.
16. Rudolph, L.C. 2000. Israeli Americans. In *Gale Encyclopedia of Multicultural America,* R.V. Dassanowsky & J. Lehman (Eds.). Farmington Hills, MI: Gale Group.
17. Gold, S.J. 2001. Gender, class, and network: Social structure and migration patterns among transnational Israelis. *Global Networks, 1,* 57–78.
18. Mikhail, M. 2000. Egyptian Americans. In *Gale Encyclopedia of Multicultural America,* R.V. Dassanowsky & J. Lehman (Eds.). Farmington Hills, MI: Gale Group.
19. Luna, L. 1994. Care and cultural context of Lebanese Muslim immigrants: Using Leininger's theory. *Journal of Transcultural Nursing, 5,* 12–20.
20. Hajar, P., & Jones, J.S. 2000. Lebanese Americans. In *Gale Encyclopedia of Multicultural America,* R.V. Dassanowsky & J. Lehman (Eds.). Farmington Hills, MI: Gale Group.
21. Lipson, J.G., Weinstein, H.M., Gladstone, E.A., & Sarnoff, R.H. 2003. Bosnian and Soviet refugees' experiences with health care. *Western Journal of Nursing Research, 25,* 854–871.
22. Papadopoulos, I. 2000. An exploration of health beliefs, lifestyle behaviors, and health needs of the London-based Greek-Cypriot community. *Journal of Transcultural Nursing, 11,* 182–190.
23. Rosenbaum, J.N. 1991. The health meanings and practices of older Greek-Canadian widows. *Journal of Advanced Nursing, 16,* 1320–1327.
24. Pliskin, K.L. 1992. Dysphoria and somatization in Iranian culture. *Western Journal of Medicine, 157,* 295–300.
25. Hafizi, H., & Lipson, J.G. 2003. People of Iranian heritage. In *Transcultural Health Care,* 2nd ed. L.D. Purnell & B.J. Paulanka (Eds.). Philadelphia: FA Davis Company.
26. Saad, B., Azaizeh, H., & Said, O. 2005. Tradition and perspectives of Arab herbal medicine: A review. *eCAM, 2,* 475–479.
27. Abu-Rabia, A. 2005. Herbs as food and medicine source in Palestine. *Asian Pacific Journal of Cancer Prevention, 6,* 404–407.
28. Lipson, J.G. 1992. The health and adjustments of Iranian immigrants. *Western Journal of Nursing Research, 14,* 10–29.
29. Ogur, R., Korkmaz, A., & Bakir, B. 2006. Herbal treatment usage frequency, types and preferences in Turkey. *Middle East Journal of Family Medicine, 4,* 38–44.
30. Ghazanfar, S.A. 1995. Wasm: A traditional method of healing by cauterization. *Journal of Ethnopharmacology, 47,* 125–128.
31. Kulwicki, A.D. 2003. People of Arab heritage. In *Transcultural Health Care,* 2nd ed. L.D. Purnell & B.J. Paulanka (Eds.). Philadelphia: FA Davis Company.
32. Nickles, H.G. 1969. *Middle Eastern cooking.* New York: Time-Life Books, Inc.
33. Nathan, J. 2001. *The foods of Israel today.* New York: Alfred A. Knopf.
34. Zibart, E. 2001. *The ethnic food lover's companion: Understanding the cuisines of the world.* Birmingham, AL: Menasha Ridge Press.
35. Batmanglij, N.K. 2000. *New food of life: Ancient Persian and modern Iranian cooking and ceremonies.* Washington, DC: Mage.
36. Foster, D. 2002. *The global etiquette guide to Africa and the Middle East.* New York: John Wiley & Sons.
37. Dresser, N. 2005. *Multicultural manners: Essential rules of etiquette for the 21st century.* New York: Wiley.
38. Seidal, K. 2001. *Serving the Guest: A Sufi Cookbook and Art Gallery.* Online at http://www.superluminal.com/cookbook/index_gallery.html Accessed on 3/18/2011.
39. Meleis, A.I. 1981. The Arab American in the health care system. *American Journal of Nursing, 81,* 1180–1183.
40. Packard, D.P., & McWilliams, M. 1993. Cultural foods heritage of Middle Eastern immigrants. *Nutrition Today,* (May–June), 6–12.
41. Valassi, K. 1962. Food habits of Greek Americans. *American Journal of Clinical Nutrition, 11,* 240–248.
42. Maamoun, M., Sucher, K.P., & Hollenbeck, C. 2006. *Food Habits and Acculturation among First Generation Egyptians Living in the San Francisco Bay Area (SFBA).* Unpublished Master's thesis, San Jose State University.
43. Lipson, J.G., & Meleis, A.I. 1983. Issues in health care of Middle Eastern patients. *Western Journal of Medicine, 139,* 854–861.
44. Estruch, R., Martinez-Gonzalez, M.A., Corelia, D., Salas-Salvado, J., Ruiz-Gutierrez, V., Covas, M.I., Fiol, M., Gomez-Garcia, E., Lopez-Sabater, M.C., Vinyoles, E., Aros, F., Conde, M., Lahoz, C., Lapetra, J., Saez, G., &

Ros, E. 2006. Effects of Mediterranean-style diet on cardiovascular risk factors: A randomized trial. *Annals of Internal Medicine, 145,* 1–11.

45. Knoops, K.T., de Groot, L.C., Kromhout, D., Perrin, A.E., Moreiras-Varela, O., Menotti, A., & van Staveren, W.A. 2004. Mediterranean diet, lifestyle factors, and 10-year mortality in elderly European men and women: The HALE project. *Journal of the American Medical Association, 292,* 1433–1439.
46. Manios, Y., Detopoulou, V., Visioli, F., & Galli, C. 2006. Mediterranean diet as a nutrition education and dietary guide: Misconceptions and the neglected role of locally consumed foods and wild green plants. *Forum of Nutrition, 59,* 154–170.
47. Esposito, K., Marfelia, R., Ciotola, M., Di Palo, C., Giugliano, F., Giugliano, G., D'Armiento, M., D'Andrea, F., & Giugliano, D. 2004. Effect of Mediterranean-style diet on endothelial dysfunction and markers of vascular inflammation in metabolic syndrome: A randomized trial. Journal of the *American Medical Association, 292,* 1490–1492.
48. Michalsen, A., Lehmann, N., Pithan, C., Knoblauch, N.T., Moebus, S., Kannenberg, F., Binder, L., Budde, T., & Dobos, G.J. 2006. Mediterranean diet has no effect on markers of inflammation and metabolic risk factors in patients with coronary artery disease. *European Journal of Clinical Nutrition, 60,* 478–485.
49. Esposito, K., Di Palo, C., Maiorino, M.I., Petrizzo, M., Bellastella, G., Siniscalchi, I. & Giugliano, D., (2011). Long-Term Effect of Mediterranean-Style Diet and Calorie Restriction on Biomarkers of Longevity and Oxidative Stress in Overweight Men. *Cardiology Research and Practice, 2011,* Article ID: 293916, 5 pages. doi:10.4061/2011/293916
50. Fung, T.T., Rexrode, K.M., Mantzoros, C.S., Manson, J.E., Willett, W.C., & Hu, F.B., (2009). Mediterranean diet and incidence and mortality of coronary heart disease and stroke in women. *Circulation, 119,* 1093–1100.
51. Rimm, E.B., & Ellison, R.C. 1995. Alcohol in the Mediterranean diet. *American Journal of Clinical Nutrition, 61* (Suppl.), 1378S–1382S.
52. Kromhout, D., Keys, A., Aravanis, C., Buzina, R., Fidanza, F., Giampaoli, S., Jansen, A., Menotti, A., Nedeljkovic, S., Pekkarinen, M., Simic, B.S., & Toshima, H. 1989. Food consumption patterns in the 1960s in seven countries. *American Journal of Clinical Nutrition, 49,* 889–894.
53. International Obesity Task Force. 2006. *Overweight and Obesity among Adults in the European Union.* Online at http://www.iotf.org/media/europrev.htm
54. Nasreddine, L., Hwalla, N., Sibal, A., Hamze, M., & Parent-Massin, D. 2006. Food consumption patterns in an adult urban population in Beirut, Lebanon. *Public Health Nutrition, 9,* 194–203.
55. Al-Rethaiaa, A.S., Fahmy, A.E., Al-Shwaiyat, N.M. 2010. Obesity and eating habits among college students in Saudi Arabia: a cross sectional study. *Nutrition Journal, 9,* 10pp. 10pp. doi:10.1186/1475-2891-9-39.
56. Musaiger, A.O. 2002. Diet and prevention of coronary heart disease in the Arab Middle East countries. *Medical Principles and Practices, 11,* 9–16.
57. Mahley, R.W., Pepin, J., Palaoglu, K.E., Malloy, M.J., Kane, J.P., & Bersot, T.P. 2000. Low levels of high density lipoproteins in Turks, a population with elevated hepatic lipase. High density lipoprotein characterization and gender-specific effects of apolipoprotein e genotype. *Journal of Lipid Research, 8,* 1290–1301.
58. Robinson, T., & Raisler, J. 2005. "Each one is a doctor for herself": Ramadan fasting among pregnant Muslim women in the United States. *Ethnicity & Disease, 15,* S1-99–S1-103.
59. Ucar, B., Kilic, Z., Colak, O., Oner, S., & Kalyoncu, C. 2000. Coronary risk factors in Turkish schoolchildren: Randomized cross-sectional study. *Pediatrics International, 42,* 259–267.
60. Draft nutrition strategy and plan of action for countries of the Eastern Mediterranean Region 2010–2019. World Health Organization, Regional Office for the Eastern Mediterranean. Dec 12, 2009. http://www.emro.who.int/nutrition/pdf/nutrition_strategy_2010_2019.pdf Accessed 2/20/2011.
61. Hammad, A., Herman, W.H., & Jaber, L.A. 2005. Cardiovascular risk factors in an Arab American population in southeastern Michigan. *Ethnicity & Disease, 15,* S1–29.
62. Jaber, L.A., Brown, M.B., Hammad, A., Nowak, S.N., Zhu, Q., Ghafoor, A., & Herman, W.H. 2003. Epidemiology of diabetes in Arab *Americans. Diabetes Care, 26,* 308–313.
63. Jaber, L.A., Brown, M.B., Hammad, A., Zhu, Q., & Herman, W.H. 2003. Lack of acculturation is a risk factor for diabetes in Arab immigrants in the U.S. *Diabetes Care, 26,* 2010–2014.
64. Jaber, L.A., Brown, M.B., Hammad, A., Zhu, Q., & Herman, W.H. 2004. The prevalence of the metabolic syndrome among Arab Americans. *Diabetes Care, 27,* 234–238.
65. Hatahet, W., & Fungwe, T.V. 2005. Obesity and cardiovascular disease risk factors are ethnicity based: A study of women of different ethnic backgrounds in southeastern Michigan. *Ethnicity & Disease, 15,* S1-23–S1-25.
66. Schwartz, K.L., Kulwicki, A., Weiss, L.K., Fakhouri, H., Sakr, W., Kau, G., & Severson, R.K. 2004. Cancer among Arab Americans in the metropolitan Detroit area. *Ethnicity & Disease, 14,* 141–146.
67. Darwish-Yassine, M., & Wing, D. 2005. Cancer epidemiology in Arab Americans and Arabs outside the Middle East. *Ethnicity & Disease, 15,* S1-5–S1-8.
68. Nasseri,,K., Mills, P.K., & Allan, M. 2007. Cancer Incidence in the Middle Eastern Population of California, 1988–2004. *Asian Pacific Journal of Cancer Prevention, 8,* 405-411.
69. Finch, G.M., Day, J.E., Razak, W.D.A., Welch, D.A., & Rogers, P.J. 1998. Appetite changes under free-living conditions during Ramadan fasting. *Appetite, 31,* 159–170.
70. Gumaa, K.A., Mustafa, K.Y., Mahmoud, N.A., & Gader, A.M. 1978. The effect of fasting in Ramadan: Serum uric acid and lipid concentration. *British Journal of Nutrition, 40,* 573–581.
71. Nomani, M.Z.A., Hallak, M.H., & Siddiqui, I.P. 1990. Effects of Ramadan fasting on plasma uric acid and

body weight in healthy men. Journal of the American Dietetic Association, 90, 1435–1436.

72. Azizi, F. 2010. Islamic fasting and health. *Annals of Nutrition and Metabolism, 56,* 273–282. doi: 10.1159/000295848
73. Geissler, E.M. 1998. Pocket guide to cultural assessment. St. Louis: Mosby.
74. Ghaemi-Ahmadi, S. 1992. Attitudes toward breastfeeding and infant feeding among Iranian, Afghan, and Southeast-Asian immigrant women in the United States: Implications for health and nutrition education. *Journal of the American Dietetic Association, 92,* 354–355.
75. Ertekin, V., Selimoglu, M.A., Kardas, F., & Aktas, E. 2005. Prevalence of celiac disease in Turkish children. *Journal of Clinical Gastroenterology, 39,* 689–691.
76. Rawashdeh, M.O., Khalil, B., & Rawily, E. 1996. Celiac disease in Arabs. *Journal of Pediatric Gastroenterology and Nutrition, 23,* 415–418.
77. Shahbazkhani, B., Mohamadnejad, M., Malekzadeh, R., Akbari, M.R., Esfahani, M.M., Nasseri-Moghaddam, S., Sotoudeh, M., & Elahyfar, A. 2004. Coeliac disease is the most common cause of chronic diarrhoea in Iran. *European Journal of Gastroenterology & Hepatology, 16,* 665–668.
78. Cataldo, F., Montalto, G. 2007. Celiac disease in the developing countries: A new and challenging public health problem. *World Journal of Gastroenterology, 13,* 2153–2159.
79. Vichinsky, E.P., MacKlin, E.A., Waye, J.S., Lorey, F., & Olivieri, N.F. 2005. Changes in the epidemiology of thalassemia in North America: A new minority disease. *Pediatrics, 116,* 818–825.
80. Aboul-Enein, B.H., Aboul-Enein, F.H. 2010. The cultural gap delivering health care services to Arab American populations in the United States. *Journal of Cultural Diversity, 17,* 20–23.
81. Jaber, L.A. 2003. Barriers and strategies for research in Arab Americans. *Diabetes Care, 26,* 514–515.
82. Yehieli, M., & Grey, M.A. 2005. *Health matters: A pocket guide for working with diverse cultures and underserved populations.* Boston, MA: Intercultural Press.
83. Pennachio, D.L. 2005. Caring for your Muslim patients. *Medical Economics, 82,* 46–50.
84. Khoury, S. 2001. Translating medical nutrition therapy approaches of diabetes into the Middle Eastern culture. *Diabetes Care & Education, 22,* 30–32.
85. Meleis, A.I., Lipson, J.G., & Paul, S.M. 1992. Ethnicity and health among five Middle Eastern immigrant groups. *Nursing Research, 42,* 98–103.
86. Abdulrahim, S., Ajrouch, K. 2010. Social and cultural meanings of self-rated health: Arab immigrants in the United States. Qualitative Health Research, 20, 1229–1240.
87. May, K.M. 1992. Middle-Eastern immigrant parents' social networks and help-seeking for child health care. *Journal of Advanced Nursing, 17,* 905–912.
88. Ponterotto, J.G., Rao, V., Zweig, J., Rieger, B.P., Schaefer, K., Michelakou, S., Armenia, C., & Goldstein, H. 2001. The relationship of acculturation and gender to attitudes toward counseling in Italian and Greek American college students. *Cultural Diversity & Ethnic Minority Psychology, 7,* 362–375.
89. Davidson, A. 1999. The Oxford companion to food. New York: Oxford University Press, Inc.
90. Botelho, F., Lunet, N., & Barros, H. 2006. Coffee and gastric cancer: Systematic review and meta-analysis, *Cadernos de Saúde Pública, 22,* 889–900.
91. Higdon, J.V., & Frei, B. 2006. Coffee and health: A review of recent human research. *Critical Reviews in Food Science and Nutrition, 46,* 101–123.
92. Johnsen, R., Forde, O.H., Straume, B., & Burhol, P.G. 1994. Aetiology of peptic ulcer: A prospective population study in Norway. *Journal of Epidemiology and Community Health, 48,* 156–160.
93. Kaltenbach, T., Crockett, S., & Gerson, L.B. 2006. Are lifestyle measures effective in patients with gastroesophageal reflux disease? An evidence-based approach. *Archives of Internal Medicine, 166,* 965–971.
94. Browne, M.L. 2006. Maternal exposure to caffeine and risk of congenital anomalies: A systemic review. *Epidemiology, 17,* 324–331.
95. Hino, A., Adachi, H., Enomoto, M., Furuki, K., Shigetoh, Y., Ohtsuka, M., Kumagae, S.I., Hirai, Y., Jalaldin, A., Satoh, A., & Imaizumi, T. 2007. Habitual coffee but not green tea consumption is inversely associated with metabolic syndrome: An epidemiological study in a general Japanese population. *Diabetes Research and Clinical Practice, 76,* 383–389.
96. Anderson, L.F., Jacobs, D.R., Jr., Carlsen, M.H., & Blomhoff, R. 2006. Consumption of coffee is associated with reduced risk of death attributed to inflammatory and cardiovascular diseases in the Iowa Women's Health Study. *American Journal of Clinical Nutrition, 83,* 1039–1046.
97. Lopez-Garcia, E., van Dam, R.M., Willett, W.C., Rimm, E.B., Manson, J.E., Stampfer, M.J., Rexrode, K.M., & Hu, F.B. 2006. Coffee consumption and coronary heart disease in men and women: A prospective cohort study. *Circulation, 113,* 2045–2053.
98. Salazar-Martinez, E., Willett, W.C., Ascherio, A., Manson, J.E., Leitzmann, M.F., Stampfer, M.J., & Hu, F.B. 2004. Coffee consumption and risk for type 2 diabetes. *Annals of Internal Medicine, 140,* 1–8.
99. Van Dam, R.M., Willett, W.C., Manson, J.E., & Hu, F.B. 2006. Coffee, caffeine, and risk of type 2 diabetes: A prospective cohort study in younger and middle-aged U.S. women. *Diabetes Care, 29,* 398–403.
100. Baker, J.A., Beehler, G.P., Sawant, A.C., Jayaprakash, V., McCann, S.E., & Moysich, K.B. 2006. Consumption of coffee, but not black tea, is associated with decreased risk of premenopausal breast cancer. *Journal of Nutrition, 136,* 166–171.
101. Dasanayake, A.P. 2005. Moderate coffee intake may reduce the risk of oral, pharyngeal, and esophageal cancer. *Journal of Evidence-Based Dental Practice, 5,* 31–32.
102. Gelatti, U., Covolo, L., Franceschini, M., Pirali, F., Tagger, A., Ribero, M.L., Trevisi, P., Martelli, C., Nardi, G., & Donato, F. 2005. Coffee consumption reduces the risk of hepatocellular carcinoma independently of its aetiology: A case-control study. *Journal of Hepatology, 42,* 526–534.

103. Karlson, E.W., Mandi, L.A., Aweh, G.N., & Grodstein, F. 2003. Coffee consumption and risk of rheumatoid arthritis. *Arthritis and Rheumatism, 48*, 3055–3060.
104. Arendash, G.W., Schleif, W., Rezai-Zadeh, H., Jackson, E.K., Zacharia, L.C., Cracchiolo, J.R., Shippy, D., & Tan, J. 2006. Caffeine protects Alzheimer's mice against cognitive impairment and reduces brain beta-amyloid production. *Neuroscience, 142*, 941–952.
105. Winkelmayer, W.C., Stampfer, M.J., Willett, W.C., & Curhan, G.C. 2005. Habitual caffeine intake and the risk of hypertension in women. *Journal of the American Medical Association, 294*, 2330–2335.
106. Cibickova, E., Cibicek, N., Zd'ansky, P., & Kohout, P. 2004. The impairment of gastroduodenal mucosal barrier by coffee. *Acta Medica, 47*, 273–275.
107. Jurgens, J. 2000. Albanian Americans. In *Gale Encyclopedia of Multicultural America*, R.V. Dassanowsky & J. Lehman (Eds.). Farmington Hills, MI: Gale Group.
108. Hysa, K., & Hysa, R.J. 1998. *The best of Albanian cooking: Favorite family recipes*. New York: Hippocrene.
109. Dosti, R. 1999. *After communism, a cuisine of survival*. Online at www.khao.org/albrecipe.htm
110. Wulfert, P. 1973. *Couscous and other good food from Morocco*. New York: Harper & Row.
111. Cholent http://en.wikipedia.org/wiki/Cholent# Etymology Accessed 3/18/2011.
112. Gjonca, A., & Bobak, M. 1997. Albanian paradox, another example of protective effect of Mediterranean lifestyle? *Lancet, 350*, 1815–1817.
113. Ackerman, L.K. 1997. Health problems of refugees. *Journal of the American Board of Family Practice, 10*, 337–348.
114. Hampl, J.S., Holland, K.A., Marple, J.T., Hutchins, M.R., & Brockman, K.K. 1997. Acute hemolysis related to consumption of fava beans: A case study and medical nutrition therapy approach. *Journal of the American Dietetic Association, 97*, 182–183.
115. Kobel, P.S. 2000. Iraqi Americans. In *Gale Encyclopedia of Multicultural America*, R.V. Dassanowsky & J. Lehman (Eds.). Farmington Hills, MI: Gale Group.

CHAPTER 14

South Asians

South Asia is the geographic region comprising the nations of India, Pakistan, Bangladesh, Sri Lanka, Nepal, and Bhutan (see Figure 14.1). Immigrants from South Asia, mostly India and Pakistan comprise one of the fastest-growing populations in the United States.

India is a culturally complex country with a population of more than 1.19 billion—nearly four times that of the United States. The sophisticated civilization began approximately 4,000 years ago and is the source of some of the most influential religions, art, architecture, and foods in the world. The South Asian subcontinent contains the fertile Indus and Ganges river basins, as well as parts of the Himalayan mountain range; it varies in climate from extensive desert regions to jungle forests to the world's largest mountain glaciers. The people of India are as diverse as its geography and climate. People from virtually every racial and religious group have migrated to or invaded India at some time in history, and each group has brought its own language and customs. As the different races and religions intermingled, other cultures were created. One result is that there are currently fifteen separate languages recognized by the Indian government. Nearly 300 languages are actually spoken in India, and there are approximately 700 dialects.

The Islamic Republic of Pakistan, located to the northwest of India, encompasses some of the most rugged territory in the world. The Himalayan Mountains stretch across the North, including the second-highest peak in the world, K2. The Hindukush range defines the northwestern region. From these mountains spills the Indus River, supplying the plains of the South before emptying into the Arabian Sea. Pakistan received its independence from India in 1947 to provide a homeland for the Muslim minority of that nation. It was a bitter split. Two wars between the countries have been fought since independence, and tensions continue over the province of Kashmir. Though Pakistan is an Islamic state, it is divided into four regions, each with its own cultural groups and languages: Punjab, Sindh, Baluchistan, and the North-West Frontier Province (NWFP).

Although India and Pakistan share a past, Asian Indians and Pakistanis each bring distinctive contributions when they move to the United States, particularly in their traditional foods and food habits. This chapter examines the customary diets of India and Pakistan and the changes that occur when immigrants from these countries move to America.

Fewer than 50,000 Sri Lankans, Nepalese, and Bhutanese combined have immigrated to the United States. The 2000 census reported that 57,000 Bangladeshis are estimated to live in the nation (over half of them in New York City), most of whom have arrived since 1990.[109,110]

Cultural Perspective

History of Asian Indians and Pakistanis in the United States

Given the complexity of South Asian culture, it is not surprising that the immigrants to the United States from India and Pakistan differ from other immigrant groups in several ways. Most significantly,

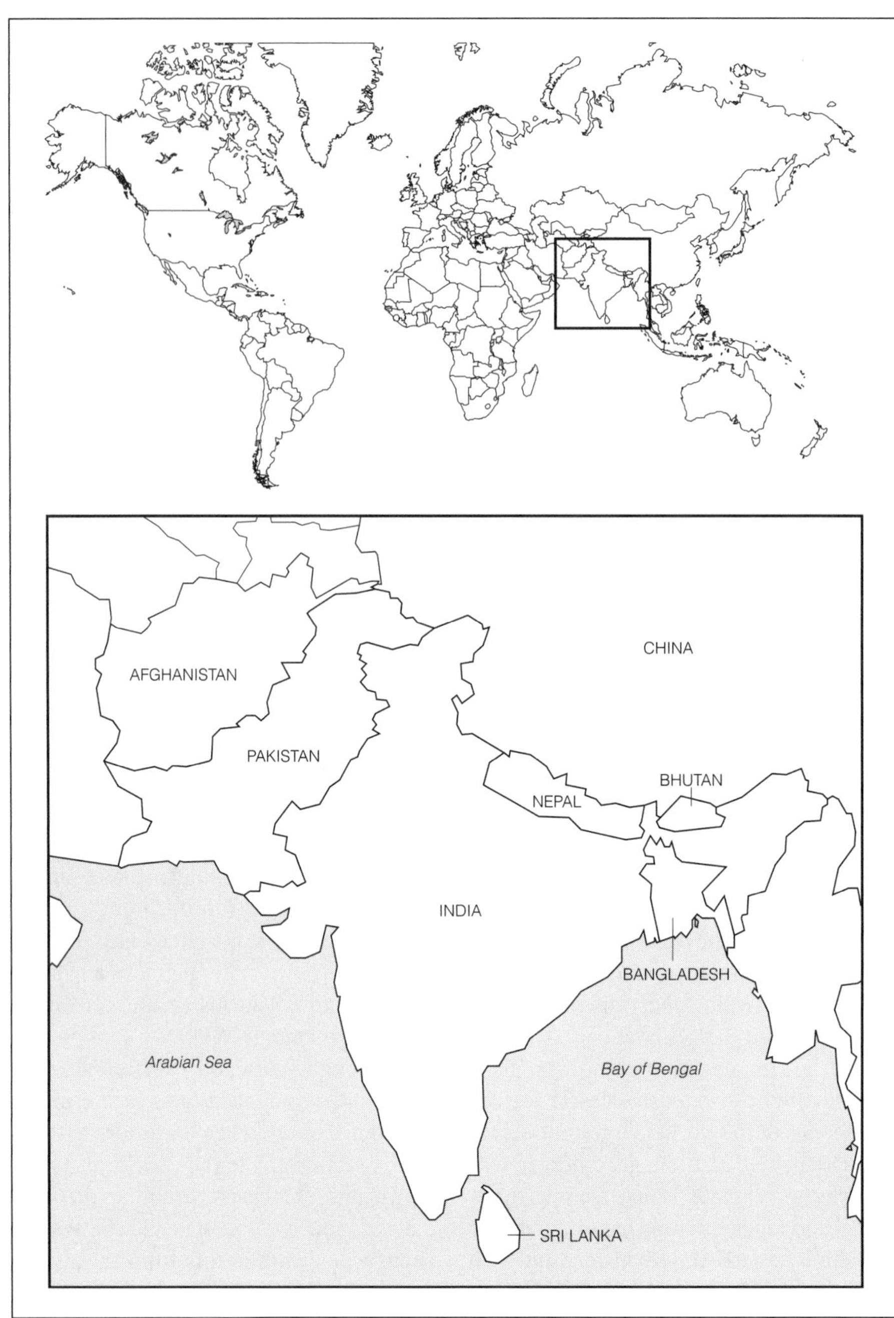

Figure 14.1
India and Pakistan.

Confusion over the term Indian (does it mean a Native American or an Asian Indian?) has made it difficult to find consensus on a designation for Asian Indians living in the United States. South Asian or Indian American has emerged as terms used by many Asian Indians.

the majority arriving in the United States are not escaping political or economic pressures in their homelands. Since 1965, when the national quota system was temporarily dropped from U.S. immigration laws, the majority of South Asian immigrants have been from the upper socioeconomic classes. They were somewhat acculturated at the time of arrival, often fluent in English, and acquainted with many American customs.

Immigration Patterns

Asian Indians. The first immigrants to the United States from India were members of the Sikh religion, who arrived on the West Coast in the early twentieth century. Many were employed by the railroads, and others established large farms. They faced overt discrimination and persecution. Newspapers warned of the "Hindoo invasion";

anti-Asian Indian feelings brought about the expulsion of Asian Indians from Washington logging communities and in 1907 sparked violent riots in California. Although such extreme bigotry lessened in time, the Asian-Indian immigrant population remained small until after World War II.

Relaxed immigration laws encouraged Asian Indians, especially well-educated urban professionals, to come to the United States in the 1960s and 1970s. Economic and social adjustment was a priority for this group, although many Asian-Indian traditions continue within the privacy of the home. These immigrants formed a self-reliant community and discouraged comparison or identification with other ethnic groups.[1]

Pakistanis. Though Muslims from northern India certainly came to the United States prior to the founding of Pakistan, immigration from the nation technically began in 1947. However, prior to 1965, only 2,500 Pakistanis moved to the United States. Beginning in 1965, when certain immigration quotas were lifted, thousands have been arriving from Pakistan each year.

Current Demographics and Socioeconomic Status

Asian Indians. According to U.S. Census estimates for 2005, the Asian-Indian population has increased more than 500 percent during the previous twenty-five years, from 400,000 to an estimated 2.3 million. Indian Americans are now the second largest Asian population in the nation. Over 75 percent of Indian Americans are foreign-born, and most of them have arrived since 1970.[2] Asian Indians have settled throughout the United States, but especially in the metropolitan areas of California, New York, and New Jersey, as well as in Illinois, Maryland, Massachusetts, Michigan, Ohio, Pennsylvania, and Texas. There are also several Asian-Indian settlements in the agricultural regions of California.[1]

Small numbers of Asian Indians coming to the United States today are from rural towns and villages. They are less educated and have experienced less exposure to Western society than previous Indian immigrants. There are also fewer professionals among them. Many recently arrived Indians are self-employed in jobs that serve their Asian-Indian immigrant community, such as restaurateurs, importers, and travel agents. These newcomers often identify more with their immediate ethnic subgroup community, such as the Gujaratis, Bengalis, Marathis, or Tamils, than with the wider, pan-Indian community.[1] The religion, language, and social class of each cultural subgroup retains greater importance for recent immigrants than for previous immigrants from India, and their adjustment to American culture is often more difficult.

Some Asian Indians coming to the United States in recent years are exiles from regions where Indian immigration in the past has been substantial, such as East Africa, Fiji, and Guyana. These immigrants often feel unwelcome in Indian American communities, and form separate enclaves.[3]

The relative affluence of Americans of Asian-Indian heritage is due mostly to a well-educated population: nearly 64 percent held a college or graduate degree in 2000.[4] Many were employed in professional or white-collar occupations in India, such as college professors, engineers, physicians, and scientists, and most continue their careers in the United States. Nearly 60 percent of Indian Americans hold management or professional jobs.[4] Newer immigrants have found success in small business and franchise ownership involving many members of their extended family. Motel and hotel ownership is especially common, and it is estimated that over 37 percent of these businesses nationwide are run by Indian Americans.[1] Median family income is 30 percent above the national norm, and family poverty rates are well below average.[4]

Many Asian Indians come to the United States to complete their college or postgraduate education. They are often unmarried or have left their spouses and children in India. It is not uncommon for the families to join the student in America after he or she has become financially established.

Pakistanis. As of 2007, census figures suggest the number of Pakistani Americans is over 200,000, fifty percent of whom are foreign-born immigrants, nearly all of whom have arrived since 1980.[5] A majority have come from large cities and have settled in the urban areas of New York City, Chicago, Philadelphia, and Los Angeles. Though many associate primarily with others from their ethnic groups, particularly Sindhis, Punjabis, and Baluchis, the Pakistani community is united at the national level.[6]

In the early 1900s it was not unusual for single Sikh men living in the United States to marry Mexican American women and start families.

Over 700,000 Canadians of Asian-Indian descent and nearly 75,000 of Pakistani descent were listed in the 2001 census.

U.S. Census data show median household income for foreign-born Bangladeshis is almost 25 percent below the U.S. norm, and nearly 18 percent of families live below the poverty level.[110]

The immigrants who first arrived after 1965 were typically well-educated professionals seeking employment in professions such as law, medicine, computer technology, and teaching. Many students obtaining advanced degrees also chose to stay in the United States. Over 54 percent of Pakistani Americans hold a college degree.[5] However, Pakistanis are often grouped with Asian Indians and Arabs in data collection, so little is known specifically regarding their socioeconomic status. It is believed that most Pakistani Americans are solidly middle class or upper middle class. Home ownership is valued and may be higher than average compared to other recent immigrant groups.

Though family income for all Pakistani Americans is similar to the U.S. median.[5] In recent years, less educated and less acculturated Pakistani Americans have immigrated, sometimes sponsored by relatives and sometimes seeking employment opportunities. Pakistani Americans may suffer employment and housing discrimination due to prejudice against Muslims.

Worldview

The Caste System in India

The untouchables were considered impure because of their contact with cattle carcasses and their consumption of beef.

A reverence for all life, called *ahimsa*, is fundamental to Asian-Indian ideology. It is reflected in the religions native to India, as well as in the vegetarian diet that many Indians follow.

The Moghul culture combined Persian and Indian influences, as seen in the architectural masterpiece of the period, the Taj Mahal.

The traditional Indian caste system, which influences the social structure of many Asian-Indian groups, is the Hindu method of ordering an individual's role in society. A more encompassing term is *jati*, which is the organization of all aspects of Hindu life, including actions, places, things, and symbols, not just people. Caste categories are hereditary. There are four main castes associated with certain professions (although members are not necessarily employed in these jobs): the Brahmans (priests), Kshatriyas (soldiers), Vaisyas (merchants or farmers), and the Sudras (serfs). These castes are divided into more than 1,000 subcastes, usually according to occupation. Existing outside the caste system are individuals considered so impure that they are called "untouchables." Although the laws discriminating against untouchables were repealed in 1949, this group of the desperately poor continues to occupy the lowest stratum in Indian society.

The caste system has permeated Indian society despite the fact that it is an exclusively Hindu classification. Americans of Asian-Indian descent often continue to identify proudly with their caste. Most come from the upper castes of Brahmans and Kshatriyas. As with all cultural practices, it is important to remember that even within a group there is great diversity of individual beliefs and customs.

Religion

Asian Indians. The influence of religion on Indian culture is ubiquitous. Every aspect of life and death is affected not only by individual religious affiliation, but also by the Hindu ideology that pervades Indian society.

Hinduism. Nearly 85 percent of Indians are Hindus. Hinduism is an ancient faith, believed to have developed in India between 2000 and 1500 b.c.e. from the Aryan hymns and prayers known as the Vedas mixed with elements from traditional Dravidian religion (see Chapter 4, "Food and Religion," for more information about Hinduism and other major Indian religions).

The Hindu Society of India established community organizations to serve the religious needs of early Asian-Indian immigrants to the United States. Many Hindu temples now exist in regions where Asian Indians have settled, with services and religious ceremonies conducted by Brahman priests (who are often employed part-time in other occupations). However, temple attendance may be limited to significant religious events. Small shrines are often created in Asian-Indian apartments or houses so that prayer and meditation may take place at home.

Islam. Today the Islamic religion in India is second only to Hinduism in number of followers; one in every nine Asian Indians is a Muslim. Islam was brought to India by traders from Persia, and it expanded with the Muslim invasions of the northern regions beginning about 1000. The Islamic Moghul Empire dominated the country for nearly 800 years. The influence of Islam is seen today mostly in northern India.

Buddhism. Buddhism developed as a protestant revolt against Hinduism. Its founder, known as Gautama Buddha, lived in India during the fifth century B.C.E. Although it is a popular religion in other parts of Asia, Buddhism is followed by less than 1 percent of Asian Indians today.

Jainism. This branch of Hinduism developed at about the time Buddhism emerged. The Jains believed that all living things have souls. Some wear masks to prevent breathing in insects and sweep a path in front of them to prevent stepping on any creatures. Orthodox Jains are strict vegetarians. Approximately 2 percent of Indians are Jains; in the United States, Jains have established their own temples for worship.

Sikhism. The Sikh religion differs from Hinduism in its belief in a single God. It is best known for its military fraternity, although most Sikhs in India are farmers. Male Sikhs wear a turban and follow the "five Ks": uncut hair (*kes*), a comb worn in the hair (*kanga*), short pants (*kaccha*), a steel bracelet worn on the right wrist (*kada*), and use of a special saber (*kirpan*). Each has a spiritual meaning; for example, the short pants symbolize self-restraint; the bracelet, obedience; and the comb, purity of mind. In the United States many Sikhs continue these traditions, though some men forgo uncut hair to better fit into American society.[1] Although Sikhs are only 2 percent of the population in India, it is reported that they make up nearly one-third of Asian Indians living in California.

Christians. It is estimated that Christians make up 3 percent of the Indian population. One form of Christianity emerged when the Syrians, who migrated to the Malabar Coast of southwest India in 345 C.E., intermarried with native people. Syrian Christians do not observe Hindu dietary laws, but they do participate in the caste system. Their agricultural community is operated with farm laborers who can be described as serfs. Another Christian community developed at the former Portuguese colony of Goa, farther north on the southwest coast. Approximately half of citizens are Catholic, known as Goan Christians, and the city is dedicated to St. Catherine.

Zoroastrianism. More than 1,200 years ago the Parsis fled from religious persecution in Persia to northern India. The religion they brought is known as Zoroastrianism, an ancient faith that venerates Ahur Mazda, the wise god of fire. The sacred fires of Zoroastrianism are tended in temples protected from the sun and from the eyes of unbelievers. The Parsis have adapted many of their practices to blend into Indian society but have maintained their faith through private schooling of their children. Parsis are considered the most Westernized of all Asian Indians, and significant Parsi communities are found in New York and Los Angeles.

Judaism. Four small Jewish communities were established in India when Jews fled persecution in Greece, Palestine (under Roman domination), Iraq, and Germany. The largest populations are found in Bombay and Calcutta.

Animism. The oldest religions in India are those practiced by the small tribal populations that live in isolated regions of the Himalayas. They worship spirits associated with natural phenomena, a religious practice known as animism. In the past they have practiced such varied social customs as polyandry (having more than one husband) and head hunting.

Pakistanis. Nearly 98 percent of Pakistanis are Muslim (approximately three-quarters are Sunni and one-quarter are Shiite). Small numbers of Pakistanis are Hindus, Christians, Sikhs, and Zoroastrians. In the United States, families attend a local mosque at least once a week, and men often attend daily. Religious education for children is expected, often occurring on weekends, and may include instruction in Arabic (so that the *Qur'an* can be read in its original language). There are few strictly Pakistani congregations, and some followers may only attend on special holidays if there is no local mosque and they must travel long distances for services.

The *kirpan* has become a civil rights issue in some U.S. public schools, pitting religious freedom against provisions restricting weapons on campus.

Family

Asian Indians. The husband is the head of the household in the traditional Indian family. The wife usually does not work outside the home and is expected to perform all duties related to housekeeping and childcare. She obtains help in these responsibilities from the extended family and, in some homes, from servants. If the wife does hold a job, she can depend on the help of relatives. Children are expected to show respect for their elders; parents may choose what career a child should pursue. Dating is uncommon in India, and many marriages are arranged by families based on similarities in caste, education, religion, and upbringing between prospective husbands and wives.

In the United States most Indian Americans live in nuclear families, and strains in the traditional structure often occur. Asian-Indian women are more likely to work in America than in their homeland, yet they lack the support system of an extended family. Elders may also find themselves cut off from their traditional role of advisors and

In India marriage is considered the beginning of a relationship from which love develops over time.

may not have opportunities for involvement in certain religious activities that would fill their lives in India. Some Asian Indians, especially women raised in India, find it difficult to adjust to these changes. More men than women choose to become U.S. citizens.

Children who grow up in the United States usually insist on making their own career choices. Dating has become more acceptable, but most parents strongly discourage relationships with persons of other ethnic or religious backgrounds.[1] Though most parents do not choose their child's spouse, many Asian-Indian children still defer to their parents' opinions; young male students in the United States sometimes ask their families in India to find suitable wifes for them.

Elders are well respected in Indian culture, and it is considered auspicious to have a senior at any social function.[7] Older women are considered experts in family matters. Traditionally, the oldest sons are expected to care for their parents, who in turn often help out with caring for children in the family. Some Indian Americans continue the practice of having elder parents in the home, and others (with parents who still live in India) host their mothers or fathers for months at a time.

The family is seen as the way to preserve Indian values and beliefs while living in the United States. They consider themselves as bicultural—Indians at home but Americans at work.[8] There is also strong interest in sponsoring the immigration of relatives to the United States. Most Asian Indians have found successful adjustment in the United States through educational and economic achievement in American public life, while maintaining an emphasis on Asian-Indian culture within the privacy of their home life.

Deepak Chopra, an Indian-born physician, has popularized Ayurvedic medicine in the United States through his best-selling books and videos.

Pakistanis. The traditional Pakistani home is strongly patriarchal. The husband is often the only wage earner in the family, and the wife is expected to stay inside the house to raise the children.[6] She is allowed out to do essential chores such as shopping, but other activities require that she be accompanied by her husband. Faith is the centerpiece of family life for Muslims, and modesty for women is prized.

Women are not allowed to have contact with unrelated men after puberty, and inappropriate touching could bring shame on the entire family and make a young girl unsuitable for marriage. In the United States most households include the immediate members, though close relatives such as grandparents or aunts and uncles may live in the home for extended periods. Some women prefer a traditional role. They may remain at home throughout the day and may never acquire English language skills. Others straddle a middle ground, working during the day, interacting with non-Pakistanis, and returning at night to don traditional garb and perform customary religious and family chores.

Many young Pakistani Americans are not allowed to date, and marriages are still frequently arranged. In themore conservative homes, girls may be withdrawn from public schools in the seventh grade to prevent mixing with boys. If private segregated schools are unavailable or unacceptable, girls may be schooled at home or not at all. Other Pakistani teens are well integrated into American life and education, continuing on to college and often attaining advanced degrees.

Traditional Health Beliefs and Practices

Asian Indians. Traditional medicine in India has a long and distinguished history. Several systems have developed over several thousand years, the most important of which is Ayurvedic medicine, which established the humoral concepts of the body that were later adopted in Greece and eventually evolved into biomedicine as it is practiced today.

Ayurvedic medicine developed into its current form between 500 B.C.E. and 500 C.E.; it is based on Sanskrit texts and the writings of practitioners. *Ayur* means "longevity" and *veda* means "science or knowledge." The purpose of the Ayurvedic system is to ensure a long and active life so that the wisdom of elders may be passed down to future generations. Ayurvedic physicians, called vaidyas, are trained at government-supported schools that grant degrees based on an established curriculum. Their diagnosis focuses on who the person is that has the illness: their tastes, their work habits, their character, and their life history. Evaluation of the pulse, the face, the eyes, and the nails provides further data. A person's constitution, including temperament and preferences in food, is believed to be determined at birth.

Ayurvedic therapy uses diet, herbal remedies, and meditation to reestablish equilibrium between the sick person and the universe, including the social, natural, and spiritual worlds. Diet is considered most significant.

Foods are classified as hot or cold depending on their effect on the body and must be balanced for each condition (see Therapeutic Uses of Food section later in this chapter). In addition, more than 700 plants and animal substances are listed in the Ayurvedic texts for prescriptive use. Because the mind, body, and soul are all considered to be interconnected parts of the whole system, meditation is used to address imbalance in the spirit of a person.

Ayurvedic medicine has declined somewhat in popularity in India as Westernized medicine has become more established and is frequently perceived as a paraprofessional practice. Folk beliefs about health and illness are found in some regions. For example, Yunani or Unani-Tibb medicine is common in the northern areas of India. It is an Arabic system that has been modified by Indian practitioners known as hakim. It is a humoral system that identifies four humors—yellow bile, black bile, phlegm, and blood—and four qualities—heat, cold, moisture, and dryness. Health is sustained through balance of these humors and qualities. Illness is treated by complementary remedies; for example, disease due to too much cold is cured with a hot therapy. Diet is an important therapeutic tool, and advanced conditions are often treated first with a fast, or limitation of intake, to allow the digestive system to rest.[9] Siddha medicine, another humoral system, is developed within Tamil culture and is found mostly in southern India. Older practices, such as the use of shamans, bonesetters, and snakebite healers, are found in some rural regions.

Home remedies such as herbal infusions and poultices are prevalent in India, often derived from Ayurvedic prescriptions or other traditional practices, but administered by home diagnosis. Many are known to have pharmacological activity, and several are contraindicated in certain medical conditions or toxic in some preparations. Examples include aloe vera for obesity, liver problems, and both high and low blood sugar levels. Licorice root is used for indigestion and stomach aches, urinary tract problems, constipation, colds, and coughs. Black nightshade is considered helpful in heart disease and liver problems. Diabetes is treated with numerous cures, including pellitory, neem, gu-dmar, and puncture vine.10 Recent surveys of more remote areas of the country have identified numerous previously unknown plants used by local inhabitants, many with demonstrated therapeutic properties.[11,12,13] Medications available only through prescription in the United States can be purchased over the counter in India. Widespread use of antibiotics and mixing of therapeutics have been reported.[14]

Pakistanis. Little has been reported on the traditional health beliefs and practices of Pakistanis, though it has been noted that complementary care may be sought concurrently with biomedicine.[15] A recent study found 23 percent of a sample of young, well-educated residents of Karachi sought the help of healers, known as hakims.[16] Similar to the same-named practitioners in India, hakims use Islami-Tibb, a humoral form of medicine adapted from traditional Arab systems and related to Indian Unani-Tibb. Therapeutic herbs or botanicals are used to maintain balance in the body and to cure a variety of ailments, such as common colds, coughs, cancer, leprosy, and reproductive disorders. Respondents who used such services reported that they believed hakims were reliable and inexpensive; those who did not visit hakims questioned their effectiveness and safety. Ayurvedic medicine is also available. Prophetic healing, prayer, and home remedies such as honey are often used to treat minor conditions or to seek protection from malign influence.[16,17]

Meditation is the quiet consideration of religious teachings to understand faith and to achieve spiritual enlightenment. It is practiced by Hindus and Buddhists, and by some Christians and Muslims as well. Transcendental meditation (TM) is associated with yoga and has no affiliation with any specific religion.

Homeopathy is well accepted throughout India and Pakistan.

Pakistani hakims sometimes use exotic preparations, such as those made from opium poppies or monitor lizard oil, in their treatment programs.

Traditional Food Habits

It is difficult to generalize about Indian cuisine because of the diverse geography and heterogeneous population of the country. Foods vary north to south, east to west, region to region, and among religious and caste groups. The cooking of Pakistan is considered similar to aromatic northern Indian fare, though with Persian and Afghani influences, including a greater emphasis on meat dishes and a preference for onions, ginger, and garlic.

Ingredients and Common Foods

Staples

India. Few foods are eaten throughout all of India. Grains and legumes predominate in the frequently vegetarian cooking, with added vegetables and fruits. Dairy items often supplement the diet. The types of ingredients and preparation

Photo by Laurie Macfee

▲ ***Traditional foods of India.*** *Some foods typical of the traditional Indian diet include* **amchoor** *(mango powder), basmati rice, broccoli, coconut, cucumber, eggplant, ghee, herbs and spices (black pepper, cardamom, chiles, fresh coriander, cloves, coriander seeds, cumin, garlic, ginger root, mint, mustard seeds, nutmeg, tamarind, and turmeric), lentils, peas, plantains, and yogurt.*

methods vary by locality and, often, according to religious practices.

Rice is the grain most commonly consumed, and the average Indian eats half a pound of it each day. This amount, however, varies considerably by region, and it is most popular in the southern and eastern areas of the nation. Wherever it is consumed, long-grained rice is preferred. Wheat, used primarily in breads, is another staple. Legumes are consumed daily by nearly all Asian Indians. *Dal* (or *dhal*) is the Hindi term for dried beans, peas, and lentils, which come hulled, skinned, whole, and split (in some literature the English word *pulse* is used instead). Dal is also the name for the dish made when they are boiled and seasoned. They are also commonly added to rice or soups, prepared as seasoned purees, or ground into flour to make distinctive breads (see Table 14.1).

Dairy foods are significant in most regions. Fermented milk products such as yogurt are found throughout most of the country, as is the cooking fat *ghee*, which is pure, clarified butter (this butter, known as *usli ghee*, is too expensive for daily use in many homes, so vegetable shortening, also called *ghee*, is often used instead). Seasonings are distinctive. Masalas are mixtures of spices and herbs that can be either fresh and "wet" or dried and powdered. Coriander, cumin, fenugreek, turmeric, black and cayenne pepper, cloves, cardamom, cinnamon, and chili peppers are a common blend that is called curry in Western countries. Other typical spices and herbs include *ajwain* (carom or loveage seeds), *amchoor* (unripe mango powder), asafetida (a pungent powdered resin), coconut, fresh coriander, garlic, mint, saffron, and tamarind (the sour pulp of a bean pod).[18] Beyond these generalities, the staples of the Indian diet are best classified by region.

The greatest division in diet is seen between northern and southern India. Northern cuisine is characterized by the use of wheat, tea, a large number of eggs, garlic, dried or pickled fruits and

Table 14.1 Selected South Asian *Dals*

English	Hindi	Common Preparations
Black lentils (black gram)	*Urad dal*	Black skins with creamy insides—boiled, added to rice or vegetables, seasoned with mustard oil (Bengal); often ground for flatbreads (e.g., pappadams) or fermented and combined with rice flour to make flatbreads (e.g., idli, dosas).
Black-eyed peas	*Lobhia*	Boiled, seasoned with onions, ginger, garlic (in the North), ginger, asafetida, mustard oil (in the West), or coconut (in the South).
Chickpeas (Bengal gram)	*Channa dal*	Most commonly used *dal* in India—boiled, added to curries, chutneys, rice (pulao); pureed (sambar); roasted whole for snacks; ground into flour (besan) and added to curries, used for deep-fried fritters; made into thick, sweet puddings for dessert.
Green peas	*Mutter dal*	Boiled, added to curries, rice (pulao) or pureed.
Hyacinth beans	*Valor*	Boiled, often seasoned with coconut, ginger, and jaggery; sprouted in soups, salads.
Lima beans	*Pavti*	Boiled, mixed with vegetables (especially potatoes, eggplant), added to curries; made into fritters or patties.
Madras beans (horse gram)	*Kulith*	Assertive earthy flavor—boiled, added to curries; powdered for soup.
Mung beans (green gram)	*Moong dal*	Brownish green—boiled with spices, added to rice (khichri), made into dumplings; sprouted for salads.
Red lentils	*Masur dal*	Salmon colored—boiled, often mashed and added to meat for kebabs or curries (most common in the North).
Yellow lentils (yellow split peas)	*Toor (Arhar) dal*	Pale yellow—boiled, often pureed with seasonings, or added to rice (khichri); mashed with other dals or rice to make pancakes (adai).

EXPLORING GLOBAL CUISINE—Vegetarianism in India

The ancient Indian diet featured a variety of meats, such as cows, bulls, buffalos, horses, rams, goats, and pigs, in addition to wild game including deer, alligator, and tortoise. The vegetarian ethic entered India slowly, probably beginning with the bulls and barren cows used for sacrifice by the Aryans. Later, prohibitions were extended to the milk cow and the draft bull, as well as the village pig (a useful scavenger) and the village cock. Over time, a more general concern for animal life developed, though meat eating (especially in the upper classes) was difficult for many Indians to forgo. Buddhist and, later, Jain doctrines reinforced the concept of *ahimsa*, and vegetarianism became more widely practiced. It is often suggested that only in India, with its enormous variety of available fruits, vegetables, and grains, could such a broad acceptance of a vegetarian diet prevail.[22]

Yet the definition of vegetarianism in India is elusive. It is usually considered a symbol of piety in the Brahman castes and may be a necessity among the poor. Abstinence from meat and poultry is most common; however, nearly all Indian vegetarians consume milk products, and some eat eggs. Fish is problematic because it is an inexpensive food where available. Except in the state of Gujerat (where the influence of Jainism has been especially strong), a large percentage of people living in coastal regions eat fish, sometimes justifying it as fruit of the sea. Other Indians practice vegetarianism only on days of religious observance or as they age and become more devout. Some sources suggest that vegetarianism is most prevalent in southern India due to the Muslim influence found in the North. Indian census data show this is not the case. Overall, it is believed that 30 percent of Indians are strict vegetarians, abstaining from all meat, poultry, fish, and eggs but consuming milk, yogurt, and other dairy products.

vegetables, and use of dry masalas that are aromatic rather than hot. These foods are typical of a cooler climate, where wheat grows better than rice and where fruits, vegetables, herbs, and spices are available only seasonally. Boiling, stewing, and frying are the most common forms of cooking. In the south steaming is the preferred method of food preparation. Rice, coffee, fresh pickles (some known as chutney), pachadi (seasoned yogurt side dishes called raytas in northern India), "wet," spicy-hot masalas, and fresh fruits, vegetables, herbs, and spices are fundamental to the cuisine. Again, these foods reflect the regional agricultural conditions.

Many Asian Indians are vegetarians, and most use some milk products but avoid eggs (see Exploring Global Cuisine: Vegetarianism in India). Pork is eaten in some communities in the west, lamb and beef are eaten in many areas of the north, and fish and poultry are eaten in several coastal regions. The cultural food groups list is found in Table 14.2.

Pakistan. Pakistani fare combines the spices of India, such as cumin, turmeric, and chili peppers, with the more typically Arab flavors of cinnamon, cloves, and cardamom. It is distinctive for its ample use of garlic, ginger, and onions in many savory and even some sweet dishes. Wheat is the staple of Pakistan, and flatbreads, similar to those in northern India, accompany every meal. *Dalia*, the Pakistani version of Middle-Eastern bulgur, is cooked with water or milk to make a porridge. Other commonly consumed grains include rice, usually the nutty-flavored basmati rice cooked as pulao (cooked in a manner similar to Turkish pilau) and khichri (mildly spiced mixture of rice and legumes). Corn is popular in some areas, typically ground into meal and made into bread flavored with mustard greens and served with butter. Barley, sorghum, and millet are available but consumed less frequently. Legumes, especially chickpeas and lentils, are served daily, usually as one of several side dishes. One favorite is cholay, chickpeas or whole dried peas cooked with ginger, garlic, onions, tomatoes, chile peppers, cumin, and turmeric. Besan (chickpea flour) is used for breads and batters for fried foods.

Dairy foods from cows and water buffaloes are another staple in the diet. Whole-milk yogurt (*dahi*) is used to prepare raytas (yogurt and vegetable side dishes) that are eaten with every meal. Lassi (the diluted yogurt beverage found in India), paneer (Indian-style pot cheese), fresh milk, cream, and ice cream are other common dairy foods, consumed regularly or added to other dishes.

Lamb, mutton, goat, beef, and chicken are Pakistani favorites. Pork is rarely eaten due to the Muslim majority, and nearly all meats are processed according to Islamic *halal* guidelines (see Chapter 4). Meat or poultry is served at lunch and dinner if affordable. Beef stew, called nihari, is an example of pot roasting (*dum*), a popular preparation technique. Braising (*korma* or *qorma*) is also common, as is the Indian charcoal *tandoori* style

The word curry is believed to be the English adaptation of a southern Indian term for "sauce," *kari*. Curry powder is not a single spice but a complex blend of seasonings that varies according to the cook and the dish. The closest Indian equivalent is *garam masala*.

In Bombay, it is legal for a vegetarian to refuse to sell his or her property to a non-vegetarian.

Table 14.2 Cultural Food Groups: South Asian

Group	Comments	Common Foods	Adaptations in the United States
Protein Foods			
Milk/milk products	In general, milk is considered a beverage for children in India; consumed by some adults in Pakistan. Fermented dairy products are popular.	Fresh cow's, buffalo's, ass's milk; evaporated milk; cream used in Pakistan; fermented milk products (yogurt, *lassi*); fresh curds very popular; fresh cheese (paneer); milk-based desserts, such as kheer, khir, *kulfi*, *barfi*, and Pakistani puddings.	Much cheese is consumed by Asian Indians; ice cream is popular with Asian Indians and Pakistanis.
Meat/poultry/fish/ eggs/legumes	Assophisticated vegetarian cuisine exist in India; legumes are a primary protein source; meat and poultry are very popular in Pakistan. Hulled, split legumes, grains, and seeds, such as lentils, are known as *dals*. Legumes are typically prepared whole or pureed, or used as flour to prepare baked, steamed, or fried breads and pastries. Beef avoided by Hindus; pork prohibited for Muslims.	Meats: beef, goat, mutton, pork Poultry: chicken, duck Fish and seafood; Bombay duck, carp, clams, crab, herring, lobster, mackerel, mullet, pomfret, sardines, shrimp, sole, turtle Eggs: chicken Legumes: beans (kidney, mung, etc.), chickpeas, lentils (many varieties and colors), peas (black-eyed, green)	Consumption of legumes decreases; meat intake increases. Meat may be added to traditional Indian vegetarian dishes. Fast foods are popular.
Cereals/Grains	More than 1,000 varieties of Indian rice are cultivated. Basmati is preferred in Pakistan. Wheat used mostly in northern India and Pakistan; rice in southern India. Most breads (*roti*) are unleavened.	Rice (steamed, boiled, fried, puffed), wheat, buckwheat, corn, millet, sorghum	Use of American-style breads occurs in place of *roti*; breakfast cereals are popular
Fruits/Vegetables	More than 100 types of fruit and 200 types of vegetables are commonly used in India. Fruits and vegetables may be used in fresh or preserved pickles, called *rayta* (northern India/Pakistan), *pachadi* (southern), or chutney. Fruits often costly in Pakistan.	Fruit: apples, apricots, avocados, bananas (several types), coconut, dates, figs, grapes, guava, jackfruit, limes, litchis, loquats, mangoes, melon, nongus, oranges, papaya, peaches, pears, persimmons (*chicos*), pineapple, plums, pomegranate, pomelos, raisins, starfruit, strawberries, sugar cane, tangerines, watermelon Vegetables: agathi flowers, amaranth, artichokes, bamboo shoots, banana flower, beets (leaves and root), bitter melon, Brussels sprouts, cabbage, carrots, cauliflower, collard greens (*haak*), corn, cucumbers, drumstick plant, eggplant, lettuce, lotus root, manioc (tapioca), mushrooms, mustard greens, okra, onions, pandanus, parsnips, plantain flowers, potatoes, pumpkin, radishes (four types, leaves and roots), rhubarb, sago palm, scallions, spinach, squash, sweet potatoes (leaves and roots), tomatoes, turnips, yams, water chestnuts, water convolvulus, water lilies	Decreased variety of fruits and vegetables is available; decreased vegetable intake results for Asian Indians. More fruit juice is consumed by Asian Indians. Salad is well accepted by Asian Indians. Use of canned and frozen produce increases.
Additional Foods			
Seasonings	Aromatic (northern) and hot (southern) combinations of fresh or dried spices and herbs accentuate or complement food flavors. Pakistani fare similar to northern Indian but with ample use of ginger, garlic, and onions.	*Ajwain*, *amchoor*, asafetida, bay leaf, cardamom (two types), chiles, cinnamon, cloves, coconut, fresh coriander, coriander seeds, cumin, dill, fennel, fenugreek, garlic, *kewra*, lemon, limes, mace, mint, mustard, nutmeg, pepper (black and red), poppy seeds, rose water, saffron, tamarind, turmeric	Spice use depends on availability.
Nuts/seeds	Nuts and seeds of all types are popular; used to thicken korma sauces in India and garnish desserts in Pakistan.	Almonds, betel nuts and leaves, cashews, peanuts, pistachios, sunflower seeds, walnuts	*Paan* tray may be limited to betel nuts and spices.
Beverages	Tea is common in northern India/ Pakistan, coffee in southern India. Coffeehouses are favored meeting places.	Coffee, tea, water flavored with fruit syrups, sugar cane, spices, or herbs; alcoholic beverages such as fermented fruit syrups, rice wines, beer	Increased consumption of soft drinks and coffee is noted for Asian Indians. Alcoholic beverages are widely accepted by Asian Indians (women may abstain); consumed by very few Pakistanis.
Fats/oils		Coconut oil, *ghee* (clarified butter), mustard oil, peanut oil, sesame seed oil, sunflower oil	Purchased *ghee* is often made from vegetable oil instead of butter.
Sweeteners		Sugar cane, *jaggery* (unrefined palm sugar), molasses	Candy and sweets are enjoyed by Asian Indians but not overconsumed; cookies may replace flatbreads as snacks for Pakistanis.

of cooking. *Bhuna* is a method of slowly frying wet seasonings (such as onions, ginger, and garlic), then adding dry spices to make a thick paste, then vegetables, and finally bits of meat to make a curried dish. Biryani rice is another specialty, a highly seasoned pilau (including saffron) with added meat. Yogurt or *amchoor* is used to marinate both meats and poultry. Minced and ground meat dishes are especially popular, and meats are sometimes extended with ground legumes. Kababs can be grilled or pan-fried patties; koftay are fried meatballs (sometimes dipped in besan batter first) served with a curry sauce. *Ghee* is the preferred cooking fat, although some Pakistanis must use less costly vegetable oils.

Both tropical and temperate vegetables and fruits are available, though not consumed in large amounts. Apples, apricots, cabbage, carrots, cauliflower, cucumbers, dates, grapes, guavas, mangos, onions, oranges, papayas, peas, plums, pomegranates, potatoes, pumpkin, spinach, tamarind, and watermelon are common. Vegetables are typically added to raytas, chutneys, curried dishes, and stews. Fruits are expensive in many regions.

Desserts are popular, especially ice cream and puddings made from rice, besan, carrots, bread, or vermicelli noodles. Cardamom, cloves, ginger, poppy seeds, aniseed, saffron, almonds, or pistachios flavor many sweets. One unique pudding dating from the Moghul period includes both ginger and garlic. Fried fritters are consumed, as are ladoos, balls made from sweetened besan and garnished with nuts. A Pakistani dessert specialty is ras malai, which may be best described as a rich cheesecake without a crust. Special-occasion desserts may be garnished with silver leaf. Tea is consumed throughout the day. It is usually heavily sweetened and boiled with milk, flavored with cinnamon or cardamom. Other popular beverages include lassi, sharbat (fruit juice), and sugarcane juice. Carbonated drinks are less common, and alcohol is prohibited for Muslims.

Regional Variations

Northern India. The influences of the Moghul period are still found in the cooking of northern India, where Muslim influence was most prominent. The royal court fare of that time featured lavish meat and rice dishes flavored with expensive aromatic seasonings, nuts, dried fruits such as raisins, and yogurt or cream. Ample use of *ghee* and sugar was also characteristic. Many similarities between the foods of this region, Pakistan, and modern-day Iran are still evident due to this shared history.

Basmati rice is commonly served as a pilaf in northern India, and biriyani rice with seasoned chicken, lamb, or beef is popular. Meatballs (kofta) made with ground meats or with meat and dal mixtures are a specialty, as are skewered pieces of broiled or grilled meats (kababs). Northern specialties include korma—a curried lamb dish with a nut- and yogurt-thickened sauce—and masala chicken. Peanut and sesame oils are used in many preparations.[9] The dishes of the North, particularly in Kashmir (which boarders Pakistan and has a significant Muslim population), are often seasoned with saffron.[3,19]

Bread, which is called *roti* in northern India, is eaten daily. Examples include whole-wheat flatbreads, such as chapatis, which are cooked on a griddle without oil until they puff up, and puris, which are deep-fried, usually in *ghee*. Paratha, a griddle-fried roti, is used as a wrapping for spiced vegetable fillings. A rich, leavened bread of the region, called *sheermal*, is flavored with rose water. Fresh cheese made from buffalo milk (similar to cottage cheese), called *paneer,* is added to many dishes, or skewered and grilled. Milk desserts are favored, such as carrot pudding (gajar halva) and rice pudding with cardamom (kheer).

[SAMPLE MENU]

Pakistani Midday Meal

Lamb *Korma (Qorma)*[a,b] or Beef *Kofta*[a]

***Sambals: Imli* (Tamarind) Chutney,[a]**

***Rayta* (Yogurt Condiment),[a] *Cholay*[a]**

Naan*

Tea

*Store bought

[a]*Desi Cookbook* at http://www.desicookbook.com/

[b]*PakiRecipes* at http://www.pakirecipes.com

Green tea is made in a samovar in the Indian state of Kashmir, a method that may have been introduced to the region from Russia via central Asia.[9]

Samosas—angular, deep-fried turnovers with spicy potato, vegetable, cheese or meat stuffings and served with chutney—are thought to be variations of Middle Eastern sanbusak.[111]

A comfort food in many parts of India is khichri, a combination of vegetables sautéed with rice and *dal* in an ample amount of *ghee*. Khichri is usually served with *kadhi*, a curry made with yogurt and besan.

A French Christian colony on the east coast of India at Pondicherry introduced baguettes, croissants, pâté, and French-style desserts into the regional fare.

In northern and northwestern India a special cylindrical clay oven heated with charcoal and called a tandoor is used. Tandoori cooking is identified particularly with lamb and chicken dishes (the meat is often marinated in a spicy yogurt sauce before cooking), although the leavened bread known as naan is also typically baked in a tandoor. This method of cooking is associated with the state of Punjab, and though few homes have tandoor ovens, it has been popularized throughout the nation (and with many visitors) by specialty restaurants.

The northwestern region is characterized by the large percentage of Hindus and Jains and high numbers of vegetarians. In the state of Punjab, where the national capital Delhi is situated, many cooler weather vegetables associated more with temperate climates than tropical ones, such as cabbage, carrots, cauliflower, potatoes, tomatoes, and turnips, are used. Onions and garlic, infrequently used in most of India, are common seasonings. Dairy products, including milk and buttermilk in addition to yogurt and paneer, are consumed more often in this area than in any other Indian region.[9] In the state of Rajasthan, barley, millet, and, later, corn were the primary grains grown in the region and are featured in many breads. Today, wheat is becoming much more prominent. Aromatic spices such as cumin and cardamom are found in many dishes, and red chiles add zing. A little farther south is the state of Gujarat, which specializes in vegetarian dishes flavored with green chiles and ginger.[9]

▼ ***Samosas, spicy deep-fried turnovers.***

Courtesy Raga Restaurant, New York, NY, Taj Group of Hotels

Sweet-and-sour dishes are also featured, usually achieved by pairing sugar with a sour fruit indigenous to the region called kokum (related to mangosteen and tamarind). The combination is found in savory dishes as well as in desserts, and especially in the drink *kokum sharbat*.

Coastal India. The coastal region offers a number of seafood specialties and fish prepared in a variety of ways, including fried, steamed, boiled, curried, and stuffed with herbs. For example, in the northeastern state of Bengal (which includes the city of Kolkata—formerly known as Calcutta), prawns are a specialty even in the more inland areas. Freshwater fish, mostly those from numerous rivers and estuaries, are consumed by most Bengalis every day. A favorite along the coast is bhapa, steamed packets of fish (or vegetables) seasoned with mustard seed and spices, such as cumin, asafetida, and nigella (a small black seed with subtle bitterness called *kalonji* in Hindi). A dessert version features sweetened yogurt. The inland dishes of Bengal are noteworthy for their use of poppy seeds. Mumbai (formerly known as Bombay), located on the west coast in the state of Maharashtra, boasts a dried, salted fish—which is thin, bony, and strongly flavored—known as Bombay duck. It is usually prepared fried. Other coastal foods eaten by the people of the Marathis region include numerous curried fish dishes as well as shrimp, crab, and lobster. In inland areas, Marathis are known for adding peanuts to their dishes, and for bhakris, a crispy, traditional flatbread made from rice flour (sorghum flour is used in some rural areas) and cooked on an ungreased griddle. The tiny state of Goa, which is south of Mumbai, is home to many Christians. Fish is eaten daily, but pork is also popular. The most famous dish of the region is vindaloo, a hot-and-sour pork curry seasoned with coconut, vinegar, tomatoes, and ample chili peppers.

Southern India. The menus of the south feature numerous steamed and fried rice dishes. A coarse red rice with a smoky flavor called rosematta is favored in some parts of southeastern India, including the state of Tamil Nadu (home of the Tamils), and may be mixed with other grains.[20] Rice is even served puffed, as in a snack called bhelpuri. Other grains, such as semolina wheat, are also popular cooked as a cereal known as uppama, which may include vegetables. *Dals*, particularly chickpeas and lentils, accompany nearly every meal in the form of a spiced purée known as sambar or as a thin, crisply fried roti called pappadams. Fermented black lentil flour mixed with rice flour is used at breakfast for steamed cakes called idli and for spicy, fried pancakes called dosas. A mixture of different *dals* (and sometimes rice) is cooked, seasoned with chili peppers, and mashed into a thick, unfermented puree that is fried for the savory pancakes known as adai, traditionally served with jaggery or coconut chutney. Fresh milk curds are also served for breakfast.

Highly spiced vegetable curries, such as aviyal, include such southern ingredients as bananas, banana flowers, bittermelons, coconut, drumstick plant, green mango, and jackfruit seeds, in addition to potatoes, cauliflower, and eggplant. Pandanus leaves, with a flavor reminiscent of mown hay, are used to season some dishes. Coconut milk is often used in curries and sauces, and coconut oil is commonly used for frying.[9] Refreshing yogurt-based pachadi and spicy, pickled fruits or vegetables, such as chutney, accompany the main course. In the state of Kerala (in the southwest), where there is the smallest percentage of vegetarians in the nation, fish or seafood is eaten often, as is chicken. Black pepper is a favorite seasoning, often combined with coconut, green chiles, and karhi (curry leaves), an herb with a citrus-like, tangerine flavor). A large Muslim population prepares traditional biryanis and cooks lamb with garlic, anise, and ground chili peppers. In the large state of Andhra Pradesh, dishes are typically seasoned with tamarind (which is also used for beverages), gongura (the leaves of roselle, a type of hibiscus also used in African cooking), and red chili peppers. Andhran fare is reputedly the hottest in all of India. Throughout the South, deep-fried salty foods and sweets are favored snacks, such as the syrup-soaked, orange-colored pretzels called jalebis.

Pakistan. Pakistani fare consists of many regional variations. In Punjab, the royal cooking style of the Moghul period still influences a preference for elaborate, rich dishes. Tandoori fare is popular, and the *karahi*—a deep, cast-iron pot shaped something like a wok—is used to deep-fry foods. Fish is a common food in Sindh, which has a lengthy coastline, and is prepared as fritters, kababs, steamed, or curried. Spit-roasted meats are a specialty in Baluchistan. Called sajji, the whole lamb or chicken is skewered on a small pole, then the poles are inserted into the dirt around a large fire. The poles are rotated by hand as the meat cooks, assuring even roasting. The North-West Frontier Province, which is populated by eight different tribes, has a simple cuisine that emphasizes rice, *dal*, and lamb. More locally, in the valley of Hunza, a distinctive fare developed due in part to the limitations of its high altitude. Wheat predominates, traditionally baked as a flatbread (phitta) in hot ashes. *Maltash*, a strongly flavored aged butter, is prized, and *kurutz*, a salty dry cheese, flavors soups. Wild thyme and turmeric are common seasonings. Apricots and apricot kernels are eaten as snacks, while oil extracted from the kernels is used in cooking.[21]

The English words pepper, sugar, and orange are all derived from Asian-Indian terms for those foods.

Balti cooking, from the Kashmir region (claimed by both Pakistan and India), uses a wok-like karahi pan to stir-fry aromatic curries seasoned with fresh coriander, mint, and fenugreek served with flatbreads instead of rice.

[SAMPLE MENU]

A Southern-Indian Vegetarian Dinner

***Aviyal* (Spicy South Indian Vegetable Curry)[a,b]**

***Sambar* (Seasoned *Dal*)[a,b,c]**

Steamed Rice

Pineapple *Pachadi*[a,c]

***Pappadams** (Spicy Fried Flatbread)**

Water (with the meal)

Mango *Lassi*[b,c] or *Chai*[a,c] (following the meal)

*Store-purchased, fried before serving

[a]Kaimal, M. 2000. *Savoring the Spice Coast of India: Fresh Flavors from Kerala*. New York: HarperCollins.

[b]*Indian and Pakistani Recipes* at http://www.recipesource.com

[c]*South Indian Recipes* at http://www.south-indian-recipes.com

Chai, a north Indian specialty, was created as a way of using lower-grade tea leaves by boiling them with spices, sugar, and milk.[9]

Religious Variations

In addition to region, religious affiliation may greatly influence food habits, especially in India. Religious groups have varying dietary practices, yet their cooking is Indian in flavor.

The relationship between food and spirituality is very complex in Hinduism. Eating is an integral part of each person's spiritual journey and defines one's role within society (see Special Occasions and Role of Food in Indian Society). For example, each caste traditionally was associated with different food habits.[9] Brahmans were generally vegetarians. Kashatriyas consumed meat, and vaishyas consumed meat depending on their locale and whether it was available—farmers often had more access than merchants. Sudras ate meat, but typically only when it was provided for them as leftovers.

Muslims avoid all pork and pork products but are not vegetarians. Orthodox Jains may only eat innocent foods that avoid injury to any life and are therefore strict vegetarians. In addition, there are twenty-two prohibited foods (e.g., fruit with small seeds or tender new greens) and thirty-two other items that may have the potential for life to exist, including root vegetables, because insects might be killed when the tubers are harvested and honey, because bees might be killed when it is gathered from the hive. They also refuse to eat any foods made with eggs, or blood-colored foods such as tomatoes and watermelon. Water must be boiled (and re-boiled after six hours)—if boiled water is unavailable, distilled water may be permitted.[22,9]

Sikh cuisine is noted for its use of wheat, corn, and sugar and the complete abstinence from alcohol and beef (pork is permitted). Sikhs are also prohibited from consuming *halal* meat. Some Sikhs are vegetarians and may avoid eggs. Many Sikh dishes are prepared in pure usli ghee, which gives them a richness not found in some other religious fare.[23,9] The Syrian Christians are renowned for their beef (tenderized by mincing or marinating), duck, and wild boar dishes. Goan Christians are unique in Indian cooking for their use of pork. They make western-style sausages and have such specialties as a vinegar-basted hog's head stuffed with vegetables and herbs.[24] Most Jews in India keep kosher. The Parsis blend Indian and Persian elements in their cuisine, exemplified by dishes such as dhansak, an entrée combining lamb, tripe, lentils, and vegetables. Eggs, such as ekuri—spicy scrambled eggs—are especially popular.

Dietary variations due to religious practice are limited in Pakistan because of a large Muslim majority. The small number of Hindu, Christian, Sikh, and Zoroastrian Pakistanis are assumed to adapt their food habits in ways appropriate to their faith (see Chapter 4).

Meal Composition and Cycle

Daily Patterns

Asian Indians. Meal patterns in India, though not consistent across regions and classes, vary less than the foods served. Two full meals with substantial snacks are typical. Early risers enjoy a rich coffee or tea boiled with milk and sugar.

Breakfast, usually eaten between 9:00 and 11:00 a.m., consists of rice or roti, a pickled fruit or vegetable, and a *sambar* or other *dal* dish, which may be left over from the previous evening. At 4:00 or 5:00 p.m., similar foods or snack items are eaten with coffee or tea. The main meal of the day follows between 7:00 and 9:00 p.m.[24] Texture, color, and balance of seasoning are all important factors in an Indian meal. A menu customarily includes at least one rice dish; a curried vegetable, legume, or meat dish; a vegetable legume side dish; a baked or fried roti; a fruit or vegetable pickle; and a yogurt rayta or pachadi. Sometimes a dessert is served, usually fruit.

Water is the most common drink consumed with meals, though milk and buttermilk are also prevalent, especially in the North and West. Sugarcane juice, fruit juice, and sodas are popular in urban areas. Alcoholic beverages are not widely consumed, though rice beer, home-brewed rum made from molasses, toddy (a brandy-like drink made from palm sap), and melon wine are a few traditional beverages still popular in some rural regions. Many Westernized Asian Indians, particularly men, drink beer or scotch.

Courses are not presented sequentially in an Indian meal. They are placed on the table all at once, with savory dishes eaten at the same time as sweets. Typically, an individual serving of rice or breads is served surrounded by a selection of other foods, such as curried dishes, dals, rayta, or pachadi, and pickled fruits or vegetables. Diners may combine tastes and textures according to personal preference. The meal concludes with the passing of the paan tray. Paan is a combination of betel (areca) nuts and spices, such as anise seed, cardamom, and fennel, wrapped in large, heart-shaped betel leaves secured with a clove. It is chewed to freshen the breath and to aid digestion.

Snacking is very popular in India. In cities and small towns snacks are sold in numerous small shops and by street vendors. In villages they are prepared at home. A clear distinction is made between meals and snacks. Many Indian languages have specific words to define each form of eating. In southern India the word *tiffin* is used to distinguish a snack from a meal. The coffee or tea drunk before breakfast or in the late afternoon is considered *tiffin*. A meal is not a meal unless the traditional staple prepared in the traditional manner, such as boiled rice in southern India (or roti in northern India), is served. This means that no matter how substantial the snack—and some include more food than a meal—it is still called *tiffin*.

Spicy snacks served with chutney often consist of batter-fried vegetables, pancakes with or without a filling, or fried seasoned dough made from wheat or lentils. Savory salad-like mixtures of diced fruit and vegetables (sometimes with added meat or shrimp) and flavored with *amchoor* or tamarind, called chaat, are popular. Snacks sweetened with sugarcane, molasses, or jaggery are usually milk-based, as are the saffron-spiced khir, the Indian ice cream called kulfi, and the candy barfi, although nuts, coconut, sesame seeds, or lentils are also used. Bengalis are noted for their sweetshops, which prepare numerous specialties, such as sandesh, a delicate curd candy, and singhara (a sweet version of the samosa filled with coconut and jaggery).[19] A snack may also include a cooling beverage, such as the sweetened, diluted yogurt drink called *lassi* or the fruit juice known as *shurbut*.

Restaurants are becoming increasingly popular in India, and Western fast-food franchises are found in many regions. Chinese and Thai establishments are also common, particularly in urban areas.

Pakistanis. Breakfast, if consumed, is a light meal in Pakistan, consisting of fried flatbreads such as puris, a sweetened porridge, or a legume dish. Traditionally, however, only two meals a day are eaten. Lunch and dinner are large meals and if affordable, include a meat, poultry, or fish dish, and sambals: side dishes such as curries, cholay, raytas and other fresh vegetable or fruit salad-like mixtures, chutneys, and pickles selected for a balance of flavors and textures. Flat breads and tea are served with the meal.

© Staffan Windstrand/Encyclopedia/Corbis

▲ ***Fast-food street vendor sells his food in Karachi, Pakistan.***

Dessert often follows, and *paan* may be chewed afterward. *Khat*, a plant with mild amphetamine-like properties, is often added to the betel-leaf roll (see Chapter 8, "Africans"). Snacking is common and hearty, including fried items such as meat, poultry, or fish fritters and patties, stuffed pastries and flatbreads, kababs, sandwiches, spicy salad-like mixtures, roast beef or chicken, and, in urban areas, Western fast food.

Traditionally, meals were served on large trays and eaten with the hands while sitting on the floor. Many Pakistanis today have been influenced by European customs and consume their meals at tables using flatware and cutlery.[25]

Special Occasions

Asian Indians. Another aspect of Indian culture affecting daily diet is the concept of feasting and fasting. As with other Indian food habits, feasting and fasting activities are complex and vary greatly from person to person and group to group. No occasion passes in India without some special food observance: regional holidays, community celebrations, and personal events such as births, weddings, funerals, and illness. A devout Hindu may feast or fast nearly every day of the year (see Chapter 4).

Feasting. Feasts serve as a method of food distribution throughout the community. They are

It is illegal to bring betel leaves into the United States. There is a black market for paan obtained in Canada and other countries.

Traditionally, six tastes (sweet, sour, salty, bitter, pungent, and astringent) and five textures (foods that need to be chewed, those that need no chewing, those that are licked, those that are sucked, and those that are drunk) were balanced in an Indian meal.

McDonald's restaurants in India do not serve beef hamburgers but offer instead selections such as *paneer* wraps and spicy potato burgers—all items are made with egg-free mayonnaise.[112]

The decorative red or yellow dot that Hindu-Asian Indian women apply to their foreheads represents joy or prosperity. It is omitted during fast days.

generally observed by presenting large amounts of everyday foods and sweets of all kinds to the appropriate holy figure; all members of the community then eat the food. Feasts may be the only time that the poor get enough to eat.

Some foods are associated with certain concepts. Rice and bananas both symbolize fertility, for example. Betel leaves represent auspiciousness; *ghee*, purity; salt, hospitality and pleasantness; mango, hospitality and auspiciousness; and betel nuts and coconuts, hospitality, sacredness, and auspiciousness.

Most festivals are Hindu in origin, and although many are observed nationwide, each is celebrated differently according to the region.

Holi is a spectacular holiday in the North, featuring reenactments of Krishna's life, fireworks, and colored powders tossed everywhere. Celebrants snack at bazaar booths. *Dussehra* is a ten-day holiday observed in both the North and the South. A special dish is prepared each day, and every day that dish is added to those prepared the previous days, culminating in an enormous feast on the last evening after a torchlight parade of ornamented elephants. *Divali*, the festival of lights, is the New Year's holiday celebrated everywhere with gifts of sweets. Another holiday, *Janmashtami*, commemorates the birth of Krishna. As a boy, Krishna and his friends would steal butter or curds hung high in earthen containers. This story is recreated during the celebration as young boys attempt to break elevated clay pots full of curds.

Sweets represent prosperity because they often include costly ingredients and may be decorated with silver or gold leaf.

Non-Hindu harvest festivals also feature feasts. They are dedicated to wheat in the North and rice in the South. At the three-day rice festival in Pongal, dishes made from the newly harvested rice are ceremonially fed to the local cows. The ten-day festival of Onam in Kerali culminates with a feast served by the local women, including thirty to forty dishes ranging from fiery curries to foods sweetened with a combination of molasses, milk, and sugar.

Asian-Indian Muslims may dine with friends on *Eid al-Fitr* at the end of Ramadan and *Eid al-Azha* (see Chapter 13, "People of the Balkans and the Middle East"). Christians celebrate Christmas and Easter in India.

Fasting. Fasting is also associated with special occasions in India. It accompanies both religious and personal events. An orthodox Hindu may fast more days a week than not. However, the term fast includes many different food restrictions in India, from avoidance of a single food item to complete abstinence from all food. A person might adopt a completely vegetarian diet for the day or eat foods believed to be spiritually purer, such as those cooked in milk (see the section Purity and Pollution in this chapter). Individuals rarely suffer from hunger because of fasting in India. In fact, more food may be consumed on fast days than on a non-fast day.

Muslims in India also fast, notably during the month of Ramadan. No food or drink is consumed between sunrise and sunset (see Chapter 4 for more information). Sikhs may fast on the days of the full moon.[23]

Pakistanis. Most Pakistanis follow the Islamic calendar, fasting for the month of Ramadan and celebrating the feast days of *Eid al-Fitr* and *Eid al-Azha* (see Chapter 4). Several secular holidays are also observed, including Pakistan Day (March 23), Independence Day (August 14), and the birthday of the national founder, Jinnah (December 25). Special occasions are marked by dishes that use costly ingredients, such as silver leaf and nuts, and feature numerous sweets.

Role of Food in Indian Society and Etiquette

The importance of food in Indian culture goes far beyond mere sustenance. Sanskrit texts describe how:

> "From earth sprang herbs, from herbs food, from food seed, from seed man. Man thus consists of the essence of food. . . . From food are all creatures produced, by food do they grow. . . . The self consists of food, of breath, of mind, of understanding, of bliss." (Achaya, 1994, p. 61)

In a society that traditionally experienced frequent famines and chronic malnutrition, food is venerated. Complex traditions have developed around when, how, and why foods are prepared, served, and eaten.

Purity and Pollution

Many Hindu dietary customs are meant to lead to purity of mind and spirit. Pollution is the opposite of purity, and polluted foods should be avoided or ameliorated. To be pure is to be free of pollution.

The Hindu classification system of *jati* is used to evaluate the relative spiritual purity of all foods. Purity is determined by the ingredients, how they are prepared, who prepares them, and how they are served. Some foods, such as milk, are inherently pure. Raw foods that are naturally protected by a husk or a peel are less susceptible to pollution. *Pakka* (meaning "cooked") foods are those that are fried or fat-basted during preparation, preferably in *ghee*. *Pakka* foods are relatively unrestricted due to their high degree of purity and often include fried breads and many sweets. *Pakka* foods are considered appropriate for serving at temples and at community feasts because they are pure enough for anyone to consume.[9] *Kaccha* (meaning "undercooked") foods are those that are boiled in water, baked, or roasted. *Kaccha* foods are more susceptible to pollution than *pakka* foods and must therefore be treated carefully during serving and consumption. They include many of the foods that are central to the daily diet, such as rice and *dal*, and are typically only served within the home.

Some foods, such as alcohol and meat, are *jhuta*, meaning that they are innately polluted. *Jhuta* foods are those that are by their very nature impure. All leftovers, unless completely untouched by the consumer or by other foods that have been eaten, are *jhuta*. Those foods that are identified as *jhuta* vary by religious sect. The term *jhuta* is also used for garbage and offal.

Asian-Indian Women and Food

The role of women in food preparation is extremely important throughout Indian culture. Feeding the family is an Indian woman's primary household duty. She is responsible for overseeing the procurement, storage, preparation, and serving of all meals. Because arranged marriages are common, training in kitchen management is considered essential for a Hindu woman in obtaining good marriage offers. It is generally believed that a woman cannot be completely substituted for in the kitchen, for she imparts a special sweetness to food. If the wife is unable to perform food-related duties, a daughter or daughter-in-law may substitute, and in multi-family homes, the mother-in-law assumes control of the kitchen. If servants help in meal preparation, it is still important for the woman of the house to serve the food directly from the *chula* (stove) to the table. This often requires many trips.

Etiquette

Traditionally, only foods cooked and served by a member of an equal or superior caste could be consumed by any Hindu. The customs were such that a Brahman would not eat food if the shadow from a member of another caste fell on it.[9] Only members of the same caste ate together. Untouched leftovers can be given to a member of a lower caste, such as servants, but polluted leftovers are eaten only by scavengers of the lowest caste. Today, Asian Indians adhere to these commensal rules to various degrees, however. An orthodox Hindu attempts to follow them at all times. A more modern Hindu might adhere to them only during holy services and holidays. Most Westernized Indians eat in restaurants and use convenience products in cooking, ignoring how and by whom the food was prepared.

The consumption of *jhuta* foods also varies among Hindus. Historically, laborers and warriors were allowed to eat meat to help keep up their strength. Some Brahman subcastes, though primarily vegetarian, permit consumption of impure foods that are plentiful in their region, such as fish in the coastal areas and lamb in the north. Other sects are so rigid that even inadvertent intake of polluted food results in spiritual disaster. Members of the International Society for Krishna Consciousness (see Chapter 4) believe that if they accidentally eat a prohibited animal food, they will lose human form in their next life and assume the form of an animal that is the prey of the animal they ate.

Hospitality is highly valued in Hindu homes, where serving a guest is considered equivalent to serving God.[9] Traditionally, the male head of the household was responsible for seeing that any guests, pregnant women, or elderly persons were well fed before he could sit down to eat. The order of serving today is more likely to be guests, oldest men, remaining male diners, children, and then women. In some situations men and women may be separated while eating.[26] However, more Westernized Asian Indians are often relaxed about these customs, and in some homes each diner goes his or her own way at meals.[9]

Food is served in small individual bowls from serving trays called *thalis*. The *thalis* may be silver or brass, with matching bowls. Originally, the *thalis* were simply banana leaves and the bowls

Food served in brass dishes is less vulnerable to pollution than is food in clay dishes.

A Hindu woman is traditionally considered impure during her menses and is prohibited from cooking or touching any food that is to be consumed by others.

Many Brahmans were historically employed as cooks, because everyone could eat food prepared by this caste.

Courtesy Raga Restaurant, New York, NY Taj Group of Hotels

▲ *Traditional Indian thali (individual silver serving tray), featuring a selection of roti (flat breads), fruit and vegetable pickles, such as chutneys, and yogurt-based pachadi or rayta.*

Cow's milk is thought to increase intelligence; buffalo's milk is believed to strengthen the body.

Garlic is found in many Ayurvedic remedies, and an old Indian proverb states: "Garlic is as good as ten mothers."

Curcumin, a chemical found in turmeric (used in many curry spice blends), has anti-inflammatory and immunomodulatory properties, and has been reported to be beneficial in arthritis, asthma, cardiovascular disease, diabetes, Alzheimer's disease, and some cancers.[113]

were earthenware, and these are still used today in some rural areas or when disposable trays and bowls are desired.

Only the right hand is used in dining, which may be done with spoons (the most common utensil), forks, and knives, or with just the fingers. When eating with the hand in the North, only the fingertips are used to delicately scoop up food with bits of bread, whereas in the South, the bread and food are dexterously rolled into the palm, then popped into the mouth.[27] Food being served to others should never be directly touched with the hand, nor should a diner refill his or her glass, waiting instead for neighboring diners to do so, and carefully tending to their drinks whenever their glass is half empty. If alcohol is served, a guest is expected to make a toast to the health of the host (after the host toasts the guests). However, any verbal thanks for the meal are considered very rude, and only a quick nod with the head while holding the hands palm to palm and saying "namaste" (see Counseling) is appropriate.[26]

Therapeutic Uses of Food

Asian Indians

Ayurvedic medicine is based on the premise that each human is a microcosm of the universe. As such, the body experiences the three inevitable laws of nature (also called universal tendencies) of creation (*sattwa*), maintenance (*rajus*), and dissolution (*tamas*).[10] The fundamental elements of fire, water, and wind also have their counterparts in the humors of the body—bile (*pitta*), phlegm (*kapha*), and wind (*vata*). *Pitta* regulates metabolic activities and resulting heat. *Kapha* provides structure and support through bone and flesh, and vata represents movement of muscle and semen. Health is maintained through a careful balance of humors and substances in the body according to each person's internal constitution and external experiences.

When *pitta* is in balance, digestion is comfortable and a person is content; balanced *kapha* produces physical and emotional stability, strength, and stamina; vitality and creativity are the results of balanced *vata*.[28] Good digestion is critical because food is transformed into the body humors and substances when it is cooked by the digestive *agnis* ("fires"), producing food juices and wastes. Food that is indigestible is harmful because it is believed to accumulate in the intestines and decompose, sending toxins into the bloodstream; excessive waste or too little waste is an imbalance that causes illness.[14]

Foods are classified according to which humors they enhance or inhibit. For example, pomegranate increases *vata* and reduces *pitta* and *kapha*. Molasses does the opposite: it increases *pitta* and *kapha* and reduces *vata*. Some foods are also grouped according to their universal tendencies. Mung beans, for instance, are considered *sattawic*, chili peppers are *rajasic*, and nutmeg is *tamasic*. Furthermore, the hot-cold classification system is used for foods, depending on how they affect the body. The specific identification of an item as hot or cold varies regionally; for example, lentils and peas are considered hot in western India, but cold in northern India. Generally wheat, spices, and seasonings (except mustard and sesame seeds), chicken, and oils are classified as hot; rice, leafy vegetables, fruits (except mango, papaya, and jackfruit), dairy products, honey, sugar, pickles, and condiments are considered cold.[22] The hot or cold nature of a food can be altered through the method of preparation. The use of hot spices or roasting may make a cold food hot; conversely, soaking a hot food in water or blending it with yogurt can change it into a cold food.[14] Many foods are considered incompatible in Ayurvedic medicine, such as honey with *ghee*, rice with vinegar, and honeydew melon with yogurt, because of conflicting properties which overwhelm the *agnis* and diminish digestion.

Although a balance of foods according to humoral effect, universal tendencies, and hot-cold is essential to health, the exact proportions of each change with age, gender, physical condition, and the weather. Traditionally, six seasons are recognized, each with certain dietary recommendations. During winter, when digestion is thought to be strongest, roasted or sour and salty dishes are preferred as well as sweets; in summer and during the monsoons, when digestion is thought to be weak, salty, sour, and fatty foods are avoided.[22] The way foods are eaten is as important as which foods are consumed. To maximize digestion of foods, a person should eat in a quiet atmosphere, sip warm water throughout the meal, and sit for a short while after dining.[28]

Pregnancy is considered to be a normal and healthy condition; however, certain food taboos are sometimes followed. Women especially avoid extremes in foods that are too hot or too cold. Lime juice with honey is a general tonic, believed to prevent excessive bleeding at birth, while cow's milk (particularly with almonds and saffron) and rice porridge are thought to ensure proper development of the fetus. Fenugreek seeds in buttermilk are given for nausea, and butter or *ghee* is believed to make the body supple and ease delivery of the baby.

Food taboos for infants and young children may also be practiced. A survey of Indian mothers found that many believed spicy foods and mangoes were too hot and caused diarrhea, bananas caused colds, and fried foods were considered difficult to digest and the source of coughs. A small percentage also reported that sweets and salty foods were avoided to prevent diabetes and hypertension, respectively.[29]

Numerous dietary remedies are listed for minor illness. Barley water is consumed for a fever; vomiting is treated with milk. Coconut water, buttermilk, anise seed oil, and pomegranate flowers are all considered helpful for diarrhea. A powder called ashtachooran (a mixture of asafetida, salt, ginger, pepper, cumin, and ajwain) is added to honey for indigestion. Ginger tea or garlic soup is used to treat colds. Gooseberries and hibiscus flower tea are considered general tonics.[22,30] Bittermelon and fenugreek seeds—used to treat diabetes—have been found to work clinically as hypoglycemic agents.[31]

It is widely believed that a highly spiced diet is necessary in the tropical Indian climate to stimulate the liver. One food habit that may result in ill health is the common practice of disguising otherwise undrinkable water with flavorful herbs and spices. Cholera, dysentery, and typhoid are endemic in many regions.

Pakistanis

Limited data suggest that a hot-cold system of classification is used by some Pakistanis. Items considered hot and therefore avoided during summer include beef and potatoes. Cold foods avoided during winter include chicken, fish, and fruit.[32] Folk remedies are very common in Pakistan, for everything from colds and flu to asthma and jaundice.[16] Eggs, curds, ginger, honey, and poppy seeds are just a few of the foods used therapeutically. One study of infant feeding practices during periods of diarrhea found that most mothers continued to breast-feed, and those who also provided solid foods considered khichri and bananas good for curing diarrhea.[33]

Contemporary Food Habits in the United States

Adaptations of Food Habits

Asian Indians

Americans of Asian-Indian descent have usually been exposed to American or European lifestyles in India and may be familiar with a Westernized diet before immigration to the United States. Yet even the most acculturated Indian Americans continue some traditional food habits. Most accept American foods when eating out, but many prefer Asian-Indian foods when at home.[34,35]

Ingredients and Common Foods. Very little research has been conducted on the food habits of Asian Indians in the United States. Two study of Asian Indians college students in Pennsylvania suggested that acculturation takes place in two phases.[36] Typically,[35] the first lasts for two to three years, often while the immigrant is a student. Interaction with mainstream American society may be limited during this period. The recent immigrant prefers to associate with members of the same caste, regional, or linguistic group; experience with American foods often

In 2006 a shortage of *dal* (lentils) in the United States caused wholesale prices to rise over 500 percent and caused home cooks to dilute favorites, such as sambar, with water.[31]

includes only fast foods. Male Asian-Indian students are often unable to cook and may rely heavily on purchased meals. Many Asian-Indian immigrants will eat hamburgers because of their availability and low cost. Sometime during the next ten years, Asian Indians who stay in the United States longer than four years enter the second phase of acculturation. They are usually employed by American businesses and are raising families. They keep their social interactions with Americans separate from those with other Indian Americans. They might serve meat and alcohol to American guests, for example, and vegetarian dishes to Indian guests.

Early research on Asian-Indian immigrants reported some vegetarians become meat eaters when living in America. In one early study, one-third of those who were vegetarians in India became non-vegetarians in the United States.[37] Eating eggs began soon after arrival, progressing to chicken, and finally to beef.[36] It is estimated that non-vegetarian Asian Indians take between two months and one year to accept beef. One survey on acculturation patterns revealed that nearly 75 percent of Indian Americans believe it is acceptable for non-vegetarians to eat beef; 44 percent indicated that it is acceptable for vegetarians to eat beef.[34] A recent study of software engineers living in Northern California also found that acculturation resulted in increased acculturation and consumption of meat.[38]

Asian Indians who are practicing Muslims rarely begin eating pork in the United States. They may drive long distances to purchase *halal* or kosher meats to fulfill traditional Muslim dietary laws.

As yet, the reasons vegetarians become non-vegetarians have not been stated conclusively. It has been suggested that vegetarianism may lose its social and cultural significance in the United States. Data regarding the influence of factors such as gender, income, region of origin in India, and length of stay in America have been contradictory. Variables that affect acculturation include gender (men tend to change their food habits more readily than women because women are the traditional food preparers in Indian society), age (children raised in the United States prefer American foods), marital status (single unmarried men are the most acculturated, married men with families in India next, and married men with families in the United States are least acculturated), caste (depending on whether caste members used meat or alcohol in India), and region (Asian Indians from rural areas are often stricter vegetarians than those from the cities).

A study of Asian Indians living in Cincinnati found several changes in the types of foods they ate.[37] Foods that subjects used frequently in India but that were in only low to moderate use in the United States were *ghee*, yogurt, *dal*, roti, rice dishes, and tea. Foods that were in low to moderate use in India but in frequent use in the United States were fruit juice, canned or frozen vegetables, American bread, dry cereals, cheese and cheese dishes, and soft drinks. Coffee consumption also increased.

A small study of Asian Indians living in New York City and Washington, DC, compared the diets of residents who lived in the United States for more than ten years to those who had lived in the United States for less than ten years. Few significant differences were identified, though longtime residents reported a greater preference for traditional meals at dinner and on weekends than did more recent immigrants. Regardless of length of residence, consumption of roots/tubers, vegetables oils, legumes, white bread, and tea remained about the same as in India. *Ghee* intake decreased for both groups; cheese, fruit juice, and whole-wheat bread intake increased. Cola beverages, low-fat milk, pizza, mayonnaise, and cookies were popular American items. Nontraditional foods that were never or rarely consumed included egg substitutes, nondairy creamer, nonfat milk, peanut butter, hot dogs, and hamburgers. The authors note that many American processed foods and baked goods are now available in India and that recent immigrants have incorporated these items into their diets before coming to the United States.[39]

Research that compared dietary intake of Asian-Indian immigrants originally from different regions of India found some differences in food use. Those from the North were more likely to use fat spreads, such as butter or margarine, while those from the southern areas of India were more likely to eat starchy foods and fried chicken. Immigrants who came from western India ate significantly less fruit and eggs than those from the North and South.[40]

A majority of Americans of Asian-Indian descent make an effort to obtain traditional food products. Many markets in the United States specialize in Indian canned and packaged food products, including spices, and many stores provide mail orders. Fresh foods are more difficult to find. Some fruits and vegetables can be bought at Asian specialty markets, and Indian bakeries featuring sweets and *tiffin* items have opened in some areas. Even in an older study done in the Midwest, 100 percent of Indians interviewed reported that their traditional foods were available.[37]

Meal Composition and Cycle. Asian-Indian eating patterns may become more irregular in the United States, possibly because of the pressures of a faster-paced lifestyle. Breakfast is the meal most commonly omitted; snacking occurs between one and three times per day and may be more common in women than in men.[39,41] Many Americans of Asian-Indian descent eat American foods for breakfast and lunch. Traditional Indian evening meals are preferred if native foods and spices are available. Yet dinners at home may also be influenced by U.S. food habits in that more meat, poultry, or fish may be eaten, and American breads may be served in place of roti.

Recent research on Bengali Americans illustrates many of these changes. Dinner is now the main meal of the day, and breakfast is a little larger than is traditional, usually consisting of toast or cereal and milk, with tea.[42] Lunch, unless brought from home, was typically pizza, a salad or a sandwich. Rice remains the core of the evening meal, and 60 percent of households reported serving it daily. Fish consumption, which is closely associated with Bengali ethnic identity, actually increases in American homes compared to those in India. Fish is served with rice and *dal*, seasoned with cumin, fennel, fenugreek, nigella seed, and mustard seed. The portion size of fish has doubled to about eight ounces, and this meal is eaten at approximately half of all dinners. Rice with other items, for example, roast chicken, is consumed at other main meals. Dishes are usually prepared Bengali-style by sautéing, stewing, or braising.

Pakistanis

Many Pakistani Americans are believed to consume at least one traditional meal each day, usually dinner, when the family can gather and discuss the day's events.[6] American-style convenience foods are popular for breakfast and lunch; cereals, pizza, hamburgers, sandwiches, fried fish, and cookies replace the flatbreads, stews, and curries typically consumed for these meals. Research on Pakistani immigrants in Norway found that dinner had become the main meal of the day and that meals on weekends included more traditional foods than did meals during the week.[43] Restaurant meals may be avoided by some Pakistani Americans if *halal* meats are unavailable.[44]

Nutritional Status

Nutritional Intake

Research on Asian Indians is noteworthy for the dramatic health changes that have occurred among urban Indians in India and immigrants to Western nations, suggesting adverse effects of dietary differences and a sedentary lifestyle. In comparison, there is scant data on Pakistani and other South-Asian immigrants, especially those in the United States. However, the limited studies show some similarities in trends, which may be expected due to a shared genetic heritage with Asian Indians.

Asian Indians. Limited data on the nutritional status of Indian Americans suggest that in general, many meet recommended intakes for grains and vegetables but do not meet those for fruits, dairy products, or meats, poultry, and fish.[45] Intake of dietary fat approximates that of the U.S. population but is often higher than fat intake in India. Energy, carbohydrate, and protein intakes may increase with length of stay.[46] Figures on fiber intake are less conclusive. Although one study found older immigrants consumed ample amounts, others suggest fiber intake is low, even among vegetarians.[46,40,47]

Micronutrient intake is less studied. A study on older Gujarti Americans found that intakes of vitamins D and B_6 were low and that intakes of calcium, magnesium, potassium, manganese, copper, zinc, and selenium were marginal. The researchers questioned dietary sufficiency in this group given the possible reduced bioavailability of some micronutrients due to a relatively high intake of fiber, especially among vegetarians.[46]

Obesity rates in Indian Americans are lower than among African Americans and whites in the United States, and average body mass index (BMI) is less than that found in blacks, Mexican Americans, and whites.[48,49] However, BMI tends to increase with urbanization and migration, and Asian Indians have a higher percentage of body fat in relation to BMI than other groups.[50,51] In

***Chaat* houses, specializing in the small, usually cold dishes of mixed fruits, vegetables, legumes, and meats topped with tangy dressing, are trendy gathering spots for Asian-Indian Americans.**

Rates of vegetarianism vary among Indian American groups. A study of immigrants from different regions of India found that the highest rates (61%) were from the western states, compared to rates of only 21 percent in immigrants from the South and 14 percent in those from the North.[40]

Iron intake may be low among some Indian American vegetarians; however, substantial amounts of iron are obtained through the use of traditional iron cookware.[114]

addition, Asian Indians have increased amounts of visceral fat, even in nonobese persons.[52,53,54] Since a higher percentage of body fat, fat patterning, and abdominal adiposity are associated with increased rates of insulin resistance and dyslipidemia, this suggests that some Asian Indians who are not overweight by national standards are, nonetheless, metabolically obese.[55,56,57]

High rates of insulin resistance and dylipidemia, especially high triglyceride levels and low high-density lipoprotein (HDL) cholesterol levels, are associated with increased risk for type 2 diabetes and cardiovascular disease in Asian Indians.[55,58,59,60] Data show that Asian Indians develop diabetes at an earlier age and at higher prevalence rates than whites and most other ethnic groups.[61,54,60,62,63] Overall prevalence of diabetes in Indian Americans has been estimated to average 18 percent in adults,[64,54] compared to 6 percent for the general U.S. population, and is higher than rates for other Asians, blacks, Hispanics, whites, and many Native American groups. However, research suggests the rates are skewed toward younger persons, and after the age of 70 years, the prevalence is similar to that found in similarly aged blacks and Hispanics. Rates of cardiovascular disease for Indian Americans are estimated to be three times higher than for whites, and data show that it has earlier onset and higher mortality rates when compared to deaths from all other causes.[65,66,67] Cardiovascular risk was found to increase in Asian-Indian immigrants with length of stay in one study.[68] Hypertension rates in Indian Americans are variable—slightly below the prevalence in whites in some studies and above the average in others.[69,53,70,71]

This clustering of conditions—a high waist-to-hip ratio indicative of abdominal fat, insulin resistance or glucose intolerance, high triglyceride levels, low HDL cholesterol levels, and hypertension—is considered the hallmark of metabolic syndrome It is especially associated with Asian Indians, and they may have the highest rates of all ethnic groups.[49,72,73] Researchers are studying numerous other factors related to these issues in an attempt to fully understand the role of genetic predisposition and environmental influence (such as diet, inactivity, and stress) in the condition.[74,75,76,61]

Bangladeshi men living in Britain had substantially greater risk for cardiovascular disease based on smoking, dyslipidemia, and diabetes rate (26 percent of the sample) than did Asian Indians and Pakistanis, suggesting variability in South Asian subgroups.[116]

It should be noted that a vegetarian diet does not necessarily provide protection from these health problems seen in many Asian Indians. High carbohydrate intake may increase triglyceride levels, and low protein consumption was associated with increased visceral fat in one study.[77,41] Asian Indians on a vegetarian diet were less insulin sensitive than white vegetarians and consumed less fiber.[78] Asian-Indian vegetarians have also been found to have higher BMIs, more body fat, and more abdominal fat than do white vegetarians.[79,47] In addition, Asian Indians have significantly higher plasma homocysteine levels (a risk factor for cardiovascular disease), thought to be due in part to low levels of vitamin B_{12} (from low animal protein consumption).[80]

The most prevalent cancers in Asian-Indian men are of the lung and prostate. Colorectal cancers, leukemias, and liver cancers are also common. Asian-Indian women suffer from breast cancer most often, followed by lung and colorectal cancer. Women also have high rates of ovarian and uterine cancers, as well as cancers of the pancreas.[81] Oral cancers are also common in South Asians, due in some subgroups to chewing of *paan* with betel nuts.[82]

Despite high socioeconomic status, few births to teen mothers, and good prenatal care, Asian-Indian women in one study were more likely to have adverse birth outcomes than blacks, Hispanics, and whites.[83,84] These included high levels of low birth weight, mental retardation, and fetal mortality. Other research comparing perinatal outcomes among Asians and Pacific Islanders reported Asian-Indian/Pakistani women had the highest risk for preterm delivery, gestational diabetes, and low birth weight at term.[85] Studies in Canada and Britain have found that South Asian women were more likely than whites to become insulin resistant during late pregnancy, and nearly half of all Asian-Indian women who developed gestational diabetes had metabolic syndrome following birth.[86,87,88] Scientists are uncertain as to why this paradox of poor birth outcomes in a population with few environmental risk factors exists.

A study comparing neonatal feeding practices of whites and Indian Americans in the United States and Asian Indians in India found that whites relied on health professionals for advice, while Indian Americans sought information from family and friends during the first six months and from health professionals thereafter.[89] It is common practice in India to avoid breast-feeding colostrum to newborns due to various concerns such as causing indigestion or diarrhea, or that it is bad for the infant's health. The researchers report that all white, 76 percent of Indian American, and 32 percent of Indian mothers initiated breast-feeding within an hour of birth. Twenty-four percent of Indian American women and 68 percent of Indian mothers delayed breast-feeding for several days, in part to find an auspicious time to start based on astrological conditions.[90] Foods such as *ghee* and fenugreek are believed to increase milk production.[91] Prelacteal feeds were prepared for the infants during this period, including honey (which is thought by some Asian Indians to help rid the infant of meconium) and water with sugar, glucose, or jaggery and seasonings such as asafetida, cumin, and aniseed. Another study noted that the longer an Indian American woman has lived in the United States, the less likely she is to use traditional feeding practices.[90] These same researchers found Asian-Indian American women breast-fed for shorter durations and introduced formula and solids earlier than did white women.[92]

Pakistanis. Little has been reported on the consumption patterns of Pakistani Americans. Malnutrition has been noted in some regions of Pakistan where droughts limit the food supply, and some studies estimate stunting occurs in 40 to 50 percent of all Pakistani children, with stunting in female children three times more likely than in male children.[93] Rural children have been found at potential risk for underconsumption of micronutrients even when ample food is available. Vitamin D and iron deficiencies have been found in pregnant Pakistani women living in Norway, in part due to low intake of supplements; low levels of plasma vitamin D are also a concern in South Asian children living in England. Low bone mineral density and a risk for early osteoporotic fractures in both men and women from South Asia have been noted.[94,95,96,97,98]

In contrast, overconsumption is more of a concern for Pakistanis living in urban areas and in Westernized nations. Children living in Pakistan cities were found to have a high intake of calories, sugar, total fats, and cholesterol.[96] Overweight and obesity in adults is estimated at 25 percent in the nation, associated with increased age, being female, urban residence, being literate, higher socioeconomic status, and higher intake of meat.[99] Adults living in wealthier neighborhoods of Karachi reported higher rates of type 2 diabetes and cardiovascular disease when compared to those living in impoverished Karachi households.[100] Of particular concern is a fat distribution pattern that favors abdominal fat, even at normal weight levels.[101]

Pakistanis in Britain had a higher average BMI and higher intake of calories from fat than Europeans or Afro-Caribbean immigrants in one study.[102] When compared to other South Asians, Pakistani men were found to have higher rates of diabetes than Asian-Indian men (22 percent and 15 percent, respectively). Dyslipidemia in Pakistani Americans, including high serum triglyceride levels and low HDL cholesterol levels, was identified in a Chicago study.[103] Although these data on Pakistanis are limited, this clustering of risk factors is similar to that found in Asian Indians and may be indicative of metabolic syndrome.

Counseling

Asian Indians. The majority of Indian Americans are very familiar with biomedicine, and health care access is considered good. Though information on health care insurance rates of Asian Indians is limited, aggregate data on South Asians show most insurance is obtained through employment, and it is assumed that high employment rates in the Indian American community translate into high rates of coverage.[104] English fluency is prevalent, and only 23 percent of Indian Americans report they speak English "less than well"—a small number given the high percentage of first-generation immigrants in the population. The largest dialects spoken in the United States are Hindi, Gujerati, and Bengali. Many Asian Indians whose native tongue is another dialect will also speak Hindi.[1]

Traditional expectations regarding social interactions may present problems in the provider–client relationship.[14] A client may not disagree with or correct a provider in situations where the client believes that the provider is superior in status, and compliance may be assumed because the client does not indicate otherwise. If the client does not feel that treatment is satisfactory, he or she will typically change providers rather than confront the problem. Conversely, if a client believes that she or he is an equal with the provider, the client may desire a social rather than professional relationship with the provider. The client may also feel entitled to special privileges, such as immediate access to the provider and longer appointments. Some Americans

Iodine deficiency and goiter are endemic in the Kashmir region.

Asian Indians may have difficulty digesting fresh milk; estimates also indicate that 60 percent of adults in Pakistan are lactose intolerant.[117]

Conditions that may be of concern in some Asian Indians and Pakistanis include beta-thalassemia, which is a commonly inherited disorder, and a high prevalence of glucose-6-phosphate dehydrogenase deficiency in some tribal groups from the North-West Frontier Province of Pakistan.

One Ayurvedic remedy for diabetes uses ginger and rock candy.

Mental illness is highly stigmatized in India. Complaints of headache, leg tingling, or burning on the soles of the feet may be related to psychological distress in recent immigrants.

of Asian-Indian descent may feel cheated if exhaustive testing procedures or invasive therapies are not provided for their condition.[14]

A polite, respectful, direct yet leisurely style of communication is preferred by most South Asians.[7] Loudness is discouraged, whereas self-control and occasional periods of silence are valued. Small talk is significant, and rushing is considered rude. A direct "no" is considered impolite, so an evasive negative response is the norm. Feelings and faith may be more important than objective facts in making decisions.[26]

Though India is a relatively low-context culture,[26] several nonverbal communication customs among Asian Indians are significant. Most Indian Hindu, Sikh, and Christian men and women avoid contact with the opposite sex in public, and contact between men and women in public is completely prohibited by Muslims. Direct eye contact between men and women who are not related may be interpreted by some less acculturated South Asians as suggestive.[7] The traditional greeting is to nod with the head while holding the hands palm to palm beneath the chin and saying "namaste" ("I honor God within you"), but Westernized Indians will use a handshake (it is best to wait and follow the client's lead). Some Asian Indians (especially those from South India) indicate agreement by head wobbling, which resembles the way Americans shake their heads back and forth to indicate the negative.[26] Informal smiles are used between equals only—superiors do not generally smile at subordinates or vice versa.

In both India and Pakistan the left hand is never used for any social purposes, including handshakes, giving an item to another person, or pointing. The head is considered the seat of the soul; patting or touching another person's head should always be avoided. Conversely, the feet and shoe soles are considered the dirtiest parts of the body, and it is impolite to point with the foot or show the bottom of one's shoe.[26] All that said, it should be noted that most Indian Americans are highly acculturated and can converse in a style comfortable for Americans.

Asian-Indian women are often less vocal than men and may be very uncomfortable with a male health care provider. Men do not touch women in either informal or formal situations; physical exams of women are unusual. Female gynecologists and obstetricians may be preferred. Family members may accompany a patient, especially a woman, as a chaperone during an appointment. The chaperone expects to participate in any discussion of the client and to oversee the exam (privacy among spouses and siblings is limited). Family members may also assume responsibility for all but the most technical care of a patient during hospitalization; it is their obligation to feed, bathe, and support a relative. If cost is an issue in obtaining health care, the male head of the household may receive priority over women, girls, or elders in a traditional Asian-Indian home.

Americans of Asian-Indian descent may provide more information regarding their condition than the practitioner may feel is warranted. Ayurvedic medicine focuses on a person's role within the cosmos, so details about what a client is eating, sleep patterns, and changes in the weather may be important to a client. Most clients expect lifestyle advice, particularly regarding diet, as part of their therapy.[28,14] Some clients may expect a complete history and physical exam to be undertaken at each appointment, while others assume that the practitioner can diagnose simply by taking a pulse and without prying into what may be perceived as personal issues. The practitioner should proceed cautiously with questioning until the client's expectations have been determined. Indian Americans have strong opinions about the therapeutic value of medications. Injections are believed most potent, and in India a patient might receive between one and four shots each visit. Capsules are thought to be more effective than tablets, and colorful tablets more potent than white tablets. Furthermore, the medication becomes more effective through the skill of the provider; thus, it is imperative that the provider personally hand the prescription to the client.

Little information has been reported regarding the use of Ayurvedic medicine or folk remedies by Asian Indians living in the United States. Research suggests that home remedies, Ayurvedic medicines, and homeopathic prescriptions are used frequently, most often for minor ailments or when biomedical therapies were ineffective. Traditional cures are often used concurrently with biomedicine.[105] Although most herbal preparations are compatible with biomedical therapies, some have potent therapeutic effects (see Traditional Health Beliefs and Practices discussed previously), and thus it is important to ascertain which remedies are in use. Further, a study of herbal medicines purchased in South Asian grocery stores found that one in every five contained potentially harmful levels of lead, mercury, or arsenic.[106]

An in-depth interview should be used to establish the client's religious affiliation and degree of adherence, length of residency in the United States, and degree of acculturation, as well as vegetarian or non-vegetarian preferences. Clients should also be asked about Ayurvedic practices, particularly those regarding diet and home remedies.

Pakistanis. There has been little reported specifically on counseling Pakistani American clients. Most are familiar with biomedicine. Few data are available on health insurance coverage for Pakistanis. A study of South Asians (combining Asian Indians, Pakistanis, and Bangladeshis) found that one in every five was uninsured and that lack of insurance prevented 31 percent of adults and 29 percent of children from obtaining medical care at least once during a two-year period.[104] Higher-than-average rates of poverty may contribute to health care access problems. Though fewer than 10 percent of Pakistanis speak only English at home, another 60 percent reported speaking it well.[4] The remaining 32 percent speak English less than "very well," which presents potential language difficulties in some clinical situations. It should be noted that although Urdu is the official language of Pakistan, only 8 percent of Pakistanis speak it as their native tongue. Nearly half of Pakistanis speak Punjabi, and the remainder are divided between six or seven other languages or dialects. In addition, the literacy rate in the country is only 38 percent.[6]

In contrast to Asian Indians, Pakistanis tend to prefer indirect, restrained communication and avoid confrontation. They are more high-context than Asian Indians, and meaning is interpreted through the specifics of the social situation and the use of symbolic language.[26] In general, loud expression and body motions are avoided, and many gestures, including winking and whistling, are considered vulgar. Pakistan is essentially a non-touching culture, but hugs and other expressions of warmth are found between intimates of the same gender. A very soft shake of the right hand, then touching the heart, is known as the salaam and is the traditional greeting between men. Handshaking may be used with Westerners, though it tends to be an adaptation of the salaam and is light. Men and women do not use the salaam or shake hands in the American style unless the woman is westernized. Many Pakistanis maintain an expression of serious attention in interactions with acquaintances.

Pakistanis are similar to Asian Indians regarding certain nonverbal communication styles.[26] No eye contact or only peripheral glances may be preferred between men and women, for example. A Pakistani should never be touched on the head without requesting and receiving permission to do so, the foot should never be used to point, and the soles of shoes should not be shown when seated. Only the right hand should be used to touch people and pass objects, including papers and money. Slouching and leaning are inappropriate in most circumstances. Pakistanis may stand closer than is comfortable for Americans, and it is considered rude to back away.

Islamic rules regarding modesty may require women to seek care only from other women. Exposure of clients, even when men are examining men, should be limited.[15] In Pakistan, families may be responsible for providing food or medicine for ill members who are hospitalized. Pakistanis in Great Britain report that family care often includes prayer and massage.[107] Women may be discouraged from decision-making, which is traditionally the responsibility of a husband or older male relative.

Many South Asians in Scotland (Pakistanis and Asian Indians) believe American medications are superior to those from their homeland. A study on use of oral hypoglycemic agents found that some patients assumed they should only take the medication if they had symptoms, while others worried about long-term use or adverse effects if taken with other drugs or certain traditional foods.[108] Clients sometimes reduce intake without supervision.

An in-depth interview, including assessment of education level and acculturation status, is necessary to determine appropriate therapeutic approaches for Pakistani American clients. Accommodation of Muslim dietary laws and individual preferences in nutritional counseling is most effective.

DISCUSSION STARTERS: COMPARING NATIVE DIET AND CULTURE TO THAT OF IMMIGRANTS IN THE UNITED STATES

In small groups of three to four, compare and contrast the diet and culture of Asian Indian and Pakistani immigrants to the United States with the diet and culture of immigrants to the United States from the Balkans and the Middle East (see Discussion Starter from Chapter 13) and from Vietnam, Cambodia, and Laos (see Discussion Starter from Chapter 12). Again, each group of students is to focus on a different aspect of the diet and culture of these groups:

Group A: The food habits and the typical eating etiquette and meal composition of these three immigrant groups

Group B: Issues involved in counseling these immigrant groups on diet and health

Group C: Attitudes within each immigrant group toward diet, health, and medical treatment, notably attitudes toward traditional home culture medical treatment and U.S. biomedicine

Group D: Amount of obesity, diabetes, hypertension, and other diseases within each immigrant group

Within your group, decide on what findings to report to the rest of the class. Before breaking up, assign a number to each group member: A1, A2, A3, A4; B1, B2, and so on. Form new groups with all the 1's in a group; all the 2's in another group; all the 3's another group; and so on. In your new group, report the findings of your previous group, and as a group, discuss the how traditional attitudes toward diet and health in these immigrant groups relate to the changes in diet and health due to immigration to the United States.

PRACTITIONER PERSECPTIVE—Asian Indian

GITA PATEL, MS, RD, CDE, LD

I am from India and have been a practitioner in the United States for close to thirty years, and many of my clients are South Asians. I would describe the typical South Asian meal as containing a lot of variety on the plate—grains (rice and bread), beans (lentils), salad ingredients in the raytas, along with vegetables. The meal also contains several condiments, such as pappadams and pickled mangos—all these ingredients help to balance the meal, which is important because many South Asians from India are vegetarians. For South Asians living in the United States, the typical fast foods that have crept into the diet tend to be those that can be vegetarian, such as pizza. Inexpensive convenience foods are bagels, pasta, and ramen noodles. Many South Asians will go out to eat in Mexican and most Asian restaurants.

Preparation of South Asian food is labor intensive, and back home there are often servants to help in the kitchen. Here, especially for younger South Asian men who are not married, the ingredients may be hard to find, and they will have to prepare their own food. So it is not uncommon that they will go out to eat and, even though they are vegetarian, will sometimes even eat hamburgers.

South Asians have a very high rate of type 2 diabetes and heart disease, so my advice to many of my clients is to eat smaller portions of rice and bread, increase their intake of vegetables and lentils, and be aware of the sodium content in food, especially pickles. I always ask about desserts and sweets in their diet since they are common in the diet. I never tell my clients to eliminate bread and rice in their diet because I know they won't do it. Instead, I tell them how much they can eat. Same with fruit, but I tell them to eat whole fruit and not consume it as juice.

REVIEW QUESTIONS

1. List the countries that comprise South Asia. List and briefly describe at least four religions practiced in this region. What are the similarities and differences in regards to religion between Pakistan and India?
2. Describe the vegetarian diet of Hindus. Which animal foods are allowed and which are not consumed? What are the staples of the diet? How would the Hindu diet differ from that of the Sikhs and Muslims? What would make a food pure or polluted? Are there regional differences in staples used in India and Pakistan? Describe at least three types of bread consumed in India.
3. What are masalas, and when are they used in South Asian cooking? Describe regional variations in South Asian cuisine. What is curry?
4. Briefly explain Ayurvedic medicine. How does food fit into the Ayurvedic system?
5. What is metabolic syndrome? How does it affect Asian Indians, and how may their diet contribute to its development?

REFERENCES

1. Pavri, T. 2000. Asian Indian Americans. In *Gale Encyclopedia of Multicultural America,* R.V. Dassanowsky & J. Lehman (Eds.). Farmington Hills, MI: Gale Group.
2. Yearbook of Immigration Statistics: Fiscal Years 1820 to 2006. Accessed March 28, 2011.
3. Berger, J. 2004. Indian, twice-removed. *New York Times,* December 17.
4. Reeves, T.J., & Bennett, C.E. 2004. *We the People: Asians in the United States.* Washington, DC: U.S. Census Bureau.
5. U.S. Census Bureau, 2007 American Community Survey.
6. Pavri T. 2000. Pakistani Americans. In *Gale Encyclopedia of Multicultural America,* R.V. Dassanowsky & J. Lehman (Eds.). Farmington Hills, MI: Gale Group.
7. Yehieli, M., & Grey, M.A. 2005. *Health matters: A pocket guide for working with diverse cultures and underserved populations.* Boston: Intercultural Press.
8. Arpana G. Inman, Erin E. Howard, Robin L. Beaumont, Jessica A. Walker. Cultural Transmission: Influence of Contextual Factors in Asian Indian Immigrant Parents' Experiences. *Journal of Counseling Psychology 2007, 54*(1), 93–100.
9. Sen, C.T. 2004. *Food culture in India.* Westport, CT: Greenwood Press.
10. Tirtha, S.S. 1998. *The Ayurvedic encyclopedia: Natural secrets to healing, prevention and longevity.* New York: Ayurveda Holistic Center Press.
11. Chhetri, D.R., Parajuli, P., & Subba, G.C. 2005. Antidiabetic plants used by Sikkim and Darjeeling Himalayan tribes, India. *Journal of Ethnopharmacology, 99,* 199–202.
12. Katewa, S.S., Chaudhary, B.L., & Jain, A. 2004. Folk medicines from tribal areas of Rajasthan, India. *Journal of Ethnopharmacology, 92,* 41–46.
13. Mahishi, P., Srinivasa, B.H., & Shivanna, M.B. 2005. Medicinal plant wealth of local communities in some villages in Shimoga District of Karnataka, India. *Journal of Ethnopharmacology, 98,* 307–312.
14. Ramakrishna, J., & Weiss, M.G. 1992. Health, illness, and immigration. East Indians in the United States. *Western Journal of Medicine, 157,* 265–270.
15. Geissler, E.M. 1998. *Pocket guide to cultural assessment.* St. Louis: Mosby.
16. Qidwai, W., Alim, S.R., Dhanani, R.H., Jehangir, S., Nasrullah, A., & Raza, A. 2003. Use of folk remedies among patients in Karachi Pakistan. *Journal of Ayub Medical College, 15,* 31–33.
17. Shaikh, B.T., & Hatcher, J. 2005. Complementary and alternative medicine in Pakistan: Prospects and limitations. *Evidence Based Complementary and Alternative Medicine, 2,* 139–142.
18. Bharadwaj, M. 2000. *The Indian Spice Kitchen.* New York: Hippocrene Books, Inc.
19. Bladholm, L. 2000. *The Indian Grocery Store Demystified.* Los Angeles: Renaissance Books.
20. Zibart, E. 2001. *The ethnic food lover's companion: Understanding the cuisines of the world.* Birmingham, AL: Menasha Ridge Press.
21. Flowerday, J. 2006. *Cooking in Hunza.* Saudi Aramco World, May/June. Available online: http://www.saudiaramcoworld.com/issue/200603/cooking.in.hunza.htm accessed March 29, 2011.
22. Achaya, K.T. 1994. *Indian food: A Historical Companion.* Delhi: Oxford University Press.
23. Kirkwood, N.A. 2005. *The Hospital Handbook on Multiculturalism and Religion.* Harrisburg, PA: Morehouse.
24. Rau, S.R. 1969. *The Cooking of India.* New York: Time-Life Books.
25. Albyn, C.L., & Webb, L.S. 1993. *The Multicultural Cookbook for Students.* Phoenix: Oryx Press.
26. Foster, D. 2000. *The Global Etiquette Guide to Asia.* New York: John Wiley & Sons.
27. Kaimal, M. 2000. *Savoring the Spice Coast of India: Fresh Flavors from Kerala.* New York: HarperCollins.
28. Larson-Presswalla, J. 1994. Insights into Eastern health care: Some transcultural nursing perspectives. *Journal of Transcultural Nursing, 5,* 21–24.
29. Sivaramakrishnan, M., & Patel, V.L. 1993. Role of traditional knowledge in the explanation of childhood nutritional deficiency by Indian mothers. *Journal of Nutrition Education, 25,* 212–129.
30. Khanna, G. 1986. *Herbal remedies: A handbook of folk medicine* (4th ed.). New Delhi: Tarang Paperbacks.
31. Jung, M., Park, M., Lee, H.C., Kang, Y.H., Kang, E.S., & Kim, S.K. 2006. Antidiabetic agents from medicinal plants. *Current Medicinal Chemistry, 13,* 1203–1218.
32. Galanti, G.A. 2004. *Caring for Patients from Different Cultures,* 3rd ed. Philadelphia: University of Pennsylvania Press.
33. Malik, I.A., Azim, S., Good, M.J., Iqbal, M., Nawaz, M., Ashraf, L., & Bukhtiari, N. 1991. Feeding practices for young Pakistani children: Unusual diet and diet during diarrhoea. *Journal of Diarrhoeal Disease Research, 9,* 213–218.
34. Sodowsky, G.R., & Carey, J.C. 1988. Relationships between acculturation-related demographics and cultural attitudes of an Asian-Indian immigrant group. *Journal of Multicultural Counseling and Development, 16,* 117–136.
35. Mahadevan, Meena and Blair, Dorothy (2009) 'Changes in Food Habits of South Indian Hindu Brahmin Immigrants in State College, PA', *Ecology of Food and Nutrition, 48*(5), 404–432.
36. Gupta, S.P. 1976. Changes in food habits of Asian Indians in the United States: A case study. *Sociology and Social Research, 60,* 87–99.
37. Karim, N., Bloch, D.S., Falciglia, G., & Murthy, L. 1986. Modifications of food consumption patterns reported by people from India, living in Cincinnati, Ohio. *Ecology of Food and Nutrition, 19,* 11–18.
38. Shruti Maheshwary and Ashwini Wagle. *Acculturation, Food Habits, And Physical Activity In South Asian Software Engineers Living In The United States,* MS project, San Jose State University 2010.

39. Raj, S., Gangnna, P., & Bowering, J. 1999. Dietary habits of Asian Indians in relation to the length of residence in the United States. *Journal of the American Dietetic Association, 99*, 1106–1108.
40. Jonnalgadda, S.S., & Diwan, S. 2002. Regional variations in dietary intake and body mass index of first-generation Asian-Indian immigrants in the United States. *Journal of the American Dietetic Association, 102*, 1286–1289.
41. Yagalla, M.V., Hoerr, S.L., Song, W.O., Enas, E., & Garg, A. 1996. Relationship of diet, abdominal obesity, and physical activity to plasma lipoprotein levels in Asian Indian physicians residing in the United States. *Journal of the American Dietetic Association, 96*, 257–261.
42. Ray, K. 2004. *The migrant's table: Meals and memories among Bengali American households*. Philadelphia: Temple University Press.
43. Mellin-Olsen, T., & Wandel, M. 2005. Changes in food habits among Pakistani immigrant women in Oslo, Norway. *Ethnicity & Health, 10*, 311–339.
44. Balagopal, P., Ganganna, P., Karmally, W., Kulkarni, K., Raj, S., Ramasubramanian, N., & Siddiwui-Mufti, M. 2000. *Indian and Pakistani food practices, customs, and holidays* (2nd ed.). Ethnic and regional food practices. Chicago: The American Dietetic Association/The American Diabetes Association.
45. Jonnalgadda, S.S., Diwan, S., & Cohen, D.L. 2005. U.S. Food Guide Pyramid food group intake by Asian Indian immigrants in the U.S. *Journal of Nutrition, Health, and Aging, 9*, 226–231.
46. Jonnalgadda, S.S., & Diwan, S. 2002. Nutrient intake of first generation Gujarati Asian Indian immigrants in the U.S. *Journal of the American College of Nutrition, 21*, 372–380.
47. Reddy, S., & Sanders, T.A. 1992. Lipoprotein risk factors in vegetarian women of Indian descent are unrelated to dietary intake. *Atherosclerosis, 95*, 223–229.
48. Lauderdale, D.S., & Rathouz, P.J. 2000. Body mass index in a US national sample of Asian Americans: The effects of nativity, years since immigration and socioeconomic status. *International Journal of Obesity, 24*, 1188–1194.
49. Misra, A., & Vikram, N.K. 2004. Insulin resistance syndrome (metabolic syndrome) and obesity in Asian Indians: Evidence and implications. *Nutrition, 20*, 482–491.
50. Dudeja, V., Misra, A., Pandey, R.M., Devina, G., Kumar, G., & Vikram, N.K. 2001. BMI does not accurately predict overweight in Asian Indians in northern India. *British Journal of Nutrition, 86*, 105–112.
51. Lubree, H.G., Rege, S.S., Bhat, D.S., Raut, K.N., Panchnadikar, A., Joglekar, C.V., Yajnik, C.S., Shetty, P., & Yudkin, J. 2002. Body fat and cardiovascular risk factors in Indian men in three geographical locations. *Food and Nutrition Bulletin, 23*, 146–149.
52. Bajaj, M., & Banerji, M.A. 2004. Type 2 diabetes in South Asians: A pathophysiologic focus on the Asian-Indian epidemic. *Current Diabetes Report, 4*, 213–218.
53. Ivey, S.L., Mehta, K.M., Fyr, C.L., & Kanaya, A.M. 2006. Prevalence and correlations of cardiovascular risk factors in South Asians: Population-based data from two California surveys. *Ethnicity & Disease, 16*, 886–893.
54. Venkataranman, R., Nanda, N.C., Baweja, G., Parikh, N., & Bhatia, V. 2004. Prevalence of diabetes and related conditions in Asian Indians living in the United States. *American Journal of Cardiology, 94*, 977–980.
55. Anand, S.S., Yusuf, S., Vuksan, V., Devanesan, S., Teo, K.K., Montague, P.A., Kelemen, L., Yi, C., Lonn, E., Gerstein, H., Hegele, R.A., & McQueen, M. 2000. Differences in risk factors, artherosclerosis, and cardiovascular disease between ethnic groups in Canada: The Study of Health Assessment and Risk in Ethnic Groups. *Lancet, 356*, 279–284.
56. Raji, A., Seely, E.W., Arky, R.A., & Simonson, D.C. 2001. Body fat distribution and insulin resistance in health Asian Indians and Caucasians. *Journal of Clinical Endocrinology and Metabolism, 86*, 5366–5371.
57. Ruderman, N., Chisholm, D., Pi-Sunyer, X., & Schneider, S. 1998. The metabolically obese, normal-weight individual revisited. *Diabetes, 47*, 699–713.
58. Misra, A., Reddy, R.B., Reddy, K.S., Mohan, A., & Bajaj, J.S. 1999. Clustering of impaired glucose tolerance, hyperinsulinemia and dyslipidemia in young north Indian patients with coronary heart disease: A preliminary case-control study. *Indian Heart Journal, 51*, 275–280.
59. Zoratti, R., Godsland, I.F., Chaturvedi, N., Crook, D., Stevenson, J.C., & McKeigue, P.M. 2000. Relation of plasma lipids to insulin resistance, nonesterfied fatty acid levels, and body fat in men from three ethnic groups: Relevance to variation in risk of diabetes and coronary disease. *Metabolism, 49*, 245–252.
60. Enas, E.A., Mohan, V., Deepa, M., Farooq, S., Pazhoor, S., Chennikkara, H. The metabolic syndrome and dyslipidemia among Asian Indians: a population with high rates of diabetes and premature coronary artery disease. *J Cardiometab Syndr*. 2007; 2: 267–275.
61. Retnakaran, R., Hanley, A.J., & Zinman, B. 2006. Does hypoadiponectinemia explain the increased risk of diabetes and cardiovascular disease in South Asians? *Diabetes Care, 29*, 1950–1954.
62. Oza-Frank, R., Ali, M.K., Vaccarino, V., Narayan, K.M. Asian Americans: diabetes prevalence across U.S. and World Health Organization weight classifications. *Diabetes Care. Sep*. 2009, *32*, 1644–1646.
63. Kanaya, A.M., Wassel, C.L., Mathur, D., Stewart, A., Herrington, D., Budoff, M.J., Ranpura, V., Liu, K. Prevalence and correlates of diabetes in South Asian Indians in the United States: findings from the metabolic syndrome and atherosclerosis in South Asians living in America study and the Multi-Ethnic Study of Atherosclerosis. *Metab Syndr Relat Disord*. 2010, *8*(2), 157–164.
64. Jonnalagadda, S.S., & Diwan, S. 2005. Health behaviors, chronic disease prevalence and self-rated health of older Asian Indian immigrants in the U.S. Journal of Immigrant Health, 7, 75–83.

65. Enas, E.A., Garg, A., Davidson, M.A., Nair, V.M., Huet, B.A., & Yusuf, S. 1996. Coronary heart disease and its risk factors in first-generation immigrant Asian Indians to the United States of America. *Indian Heart Journal, 48*, 343–353.
66. McKeigue, P.M., Ferrie, J.E., Pierpoint, T., & Marmot, M.G. 1993. Association of early-onset coronary heart disease in South Asian men with glucose intolerance and hyperinsulinemia. *Circulation, 87*, 152–161.
67. Palaniappan, L., Wang, Y., & Fortmann, S.P. 2004. Coronary heart disease mortality for six ethnic groups in California, 1990–2000. *Annals of Epidemiology, 14*, 499–506.
68. Sunderam, B., Holley, D.C., Cornelissen, G., Naik, D., Hanumansetty, R., Singh, R.B., Otsuka, K., & Halberg, F. 2005. Circadian and circaseptan (about-weekly) aspects of immigrant Indians' blood pressure and heart rate in California, USA. *Biomedical Pharmacotherapy, 59* (Suppl. 1), S76–S85.
69. Agyemang, C., & Bhopal, R.S. 2002. Is the blood pressure of South Asian adults in the UK higher or lower than that in European white adults? A review of cross-sectional data. *Journal of Human Hypertension, 16*, 739–751.
70. Mohanty, S.A., Woolhandler, S., Himmelstein, D.U., & Bor, D.M. 2005. Diabetes and cardiovascular disease among Asian Indians in the United States. *Journal of General Internal Medicine, 20*, 474–478.
71. Leenen FH, Dumais J, McInnis NH, Turton P, Stratychuk L, Nemeth K, Lum-Kwong MM, Fodor G Results of the Ontario survey on the prevalence and control of hypertension. *Canadian Med Assn J. 2008* May 20, *178*(11), 1441–9.
72. Merchant, A.T., Anand, S.S., Vuksan, V., Jacobs, R., Davis, B., Teo, K., & Yusuf, S. 2005. Protein intake is inversely associated with abdominal obesity in a multi-ethnic population. *Journal of Nutrition, 135*, 1196–1201.
73. Ramachandran, A., Snehalatha, C., Satyavani, K., Sivasankari, S., & Vijay, V. 2003. Metabolic syndrome in urban Asian Indian adults—a population study using modified ATPIII criteria. *Diabetes Research and Clinical Practice, 60*, 199–204.
74. Brunner, E.J., Hemingway, H., Walker, B.R., Page, M., Clarke, P., Juneja, M., Shipley, M.J., Kumari, M., Andrew, R., Seckl, J.R., Papadopoulos, A., Checkley, S., Rumley, A., Lowe, G.D., Stansfeld, S.A., & Marmot, M.G. 2002. Adrenocortical, autonomic, and inflammatory causes of the metabolic syndrome: Nested cases-control study. *Circulation, 106*, 2659–2665.
75. Chandalia, M., Cabo-Chan, A.V., Devaraj, S., Jialal, I., Grundy, S.M., & Abate, N. 2003. Elevated plasma high-sensitivity C-reactive protein concentrations in Asian Indians living in the United States. *Journal of Clinical Endocrinology and Metabolism, 88*, 3773–3776.
76. Dhawan, J., & Bray, C.L. 1997. Asian Indians, coronary artery disease, and physical exercise. *Heart, 78*, 550–554.
77. Pais, P., Pogue, J., Gerstein, H., Zachariah, E., Savitha, D., Jayprakash, S., Nayak, P.R., & Yusuf, S. 1996. Risk factors for acute myocardial infarction in Indians: A case-control study. *Lancet, 348*, 358–363.
78. Scholfield, D.J., Behall, K.M., Bhathena, S.J., Kelsay, J., Reiser, S., & Revett, K.R. 1987. A study on Asian Indian and American vegetarians: Indications of racial predisposition to glucose intolerance. *American Journal of Clinical Nutrition, 46*, 955–961.
79. Dhurandhar, N.V., & Kulkarni, P.R. 1992. Prevalence of obesity in Bombay. International *Journal of Obesity and Related Metabolic Disorders, 16*, 367–375.
80. Chandalia, M., Abate, N., Cabo-Chan, A.V., Devaraj, S., Jilal, I., & Grundy, S.M. 2003. Hyperhomocysteinemia in Asian Indians living in the United States. *Journal of Clinical Endocrinology and Metabolism, 88*, 1089–1095.
81. Chu, K.C., & Chu, K.T. 2005. 1999–2001 Cancer mortality rates for Asian and Pacific Islander ethnic groups with comparisons to their 1988–1992 rates. *Cancer, 104*, 2989–2998.
82. Ahluwalia, K.P. 2005. Assessing the oral cancer risk of South-Asian immigrants in New York City. *Cancer, 104*, 2959–2961.
83. Gould, J.B., Madan, A., Qin, C., & Chavez, G. 2003. Perinatal outcomes in two dissimilar immigrant populations in the United States: A dual epidemiologic paradox. *Pediatrics, 111*, e678–e682.
84. Alexander GR, Wingate MS, Mor J, Boulet S. Birth outcomes of Asian-Indian Americans.2007 Apr 3. *Int J Gynaecol Obstet. 2007 Jun, 97*(3), 215–20.
85. Rao, A.K., Daniels, K., El-Sayed, Y.Y., Moshesh, M.K., & Caughey, A.B. 2006. Perinatal outcomes among Asian American and Pacific Islander women. *American Journal of Obstetrics and Gynecology, 195*, 834–838.
86. Kousta, E., Efstathiadou, Z., Lawrence, N.J., Jeffs, J.A., Godsland, I.F., Barrett, S.C., Dore, C.J., Penny, A., Anuaoku, V., Millauer, B.A., Cela, E., Robinson, S., McCarthy, M.I., & Johnston, D.G. 2006. The impact of ethnicity on glucose regulation and the metabolic syndrome following gestational diabetes. *Diabetologia, 49*, 36–40.
87. Retnakaran, R., Hanley, A.J., Connelly, P.W., Sermer, M., & Zinman, B. 2006. Ethnicity modifies the effect of obesity on insulin resistance in pregnancy: A comparison of Asian, South Asian, and Caucasian women. *Journal of Clinical Endocrinology and Metabolism, 91*, 93–97.
88. Hedderson MM, Darbinian JA, Ferrara A. Disparities in the risk of gestational diabetes by race-ethnicity and country of birth. *Paediatric and Perinatal Epidemiology 2010.*
89. Kannan, S., Carruth, B.R., & Skinner, J. 2004. Neonatal feeding practices of Anglo American mothers and Asian Indian mothers living in the United States and India. *Journal of Nutrition Education and Behavior, 36*, 315–319.
90. Kannan, S., Carruth, B.R., & Skinner, J. 1999. Cultural influences on infant feeding beliefs of mothers. *Journal of the American Dietetic Association, 99*, 88–90.

91. S. Varma, A. Wagle, & K. Sucher. *Dietary Behaviors and Practices of Pregnant and Lactating South Asian Women Living in the United States*. San Jose State University, San Jose, CA. Masters project. 2010.
92. Kannan, S., Carruth, B.R., & Skinner, J. 1999. Infant feeding practices of Anglo American and Asian Indian American mothers. *Journal of the American College of Nutrition, 18*, 279–286.
93. Baig-Ansari, N., Rahbar, M.H., Bhutta, Z.A., & Badruddin, S.H. 2006. Child's gender and household food insecurity are associated with stunting among young Pakistani children residing in urban squatter settlements. *Food and Nutrition Bulletin, 27*, 114–127.
94. Alekel, D.L., Mortiallaro, E., Hussain, E.A., West, B., Ahmed, N., Peterson, C.T., Werner, R.K., Arjmandi, B.H., & Kukreja, S.C. 1999. Lifestyle and biologic contributors to proximal femur bone mineral density and hip axis length in two distinct ethnic groups of premenopausal women. *Osteoporosis International, 9*, 327–338.
95. Brunvand, L., Henriksen, C., Larsson, M., & Sanberg, A.S. 1995. Iron deficiency among pregnant Pakistanis in Norway and the content of phytic acid in their diet. *Acta Obstetricia et Gynecologica Scandinavica, 74*, 520–525.
96. Hakeem, R., Thomas, J., & Badruddin, S.H. 1999. Rural-urban differences in food and nutrient intake of Pakistani children. *Journal of the Pakistani Medical Association, 49*, 288–294.
97. Henricksen, C., Brunvand, L., Stoltenberg, C., Trygg, K., Huaga, E., & Pederson, J. 1995. Diet and vitamin D status among pregnant Pakistani women in Oslo. *European Journal of Clinical Nutrition, 49*, 211–218.
98. Lawson, M., Thomas, M., & Hardiman, A. 1999. Dietary and lifestyle factors affecting plasma vitamin D levels in Asian children living in England. *European Journal of Clinical Nutrition, 53*, 268–272.
99. Jafar, J.H., Chaturvedi, N., & Pappas, G. 2006. Prevalence of overweight and obesity and their association with hypertension and diabetes in an Indian-Asian population. *Canadian Medical Association Journal, 175*, 1071–1077.
100. Hameed, K., Kadir, M., Gibson, T., Sultana, S., Fatima, Z., & Syed, A. 1995. The frequency of known diabetes, hypertension, and ischaemic heart disease in affluent and poor urban populations of Karachi, Pakistan. *Diabetic Medicine, 12*, 500–503.
101. Bose, K. 1995. A comparative study of generalized obesity and anatomical distribution of subcutaneous fat in adult white and Pakistani migrant males in Peterborough. *Journal of the Royal Society of Health, 115*, 90–95.
102. Vyas, A., Greenhaigh, A., Cade, J., Sanghera, B., Riste, L., Sharma, S., & Cruickshank, K. 2003. Nutrient intakes of an adult Pakistani, European and African-Caribbean community in inner city Britain. *Journal of Human Nutrition and Dietetics, 16*, 327–337.
103. Kamath, S.K., Hussain, E.A., Amin, D., Mortillaro, E., West, B., Peterson, C.T., Aryee, F., Murillo, G., & Alekel, D.L. 1999. Cardiovascular disease risk factors in 2 distinct ethnic groups: Indian and Pakistani compared with American premenopausal women. *American Journal of Clinical Nutrition, 69*, 621–631.
104. Brown, E.R., Ojeda, V.D., Wyn, R., & Levan, R. 2000. *Racial and ethnic disparities in access to health insurance and health care*. Los Angeles: UCLA Center for Health Policy Research and Kaiser Family Foundation.
105. Rao, D. 2006. Choice of medicine and hierarchy of resort to different health alternatives among Asian Indian immigrants in a metropolitan city in the USA. *Ethnicity & Health, 11*, 153–167.
106. Saper, R.B., Kales, S.N., Paquin, J., Burns, M.J., Eisenberg, D.M., Davis, R.B., & Phillips, R.S. 2004. Heavy metal content of ayurvedic herbal medicine products. *Journal of the American Medical Association, 292*, 2868–2873.
107. Cortis, J.D. 2000. Perceptions and experiences with nursing care: A study in Pakistani (Urdu) communities in the United Kingdom. *Journal of Transcultural Nursing, 11*, 111–118.
108. Lawton, J., Ahmad, N., Hallowell, N., Hanna, L., & Douglas, M. 2005. Perceptions and experiences of taking oral hypoglycaemic agents among people of Pakistani and Indian origin: Qualitative study. *BMJ, 330*, 1247.
109. Jessica S. Barnes; Claudette E. Bennett (February 2002). "The Asian Population: 2000." *U.S. Census Bureau*. U.S. Department of Commerce. "http://www.census.gov/prod/2002pubs/c2kbr01-16.pdf". Retrieved 30 September 2009.
110. U.S. Census. Immigration Statistics Staff. 2004. *Foreign-Born Profiles (STP–159)*. Online at http://www.census.gov/population/www/socdemo/foreign/STP-159-2000tl.html.
111. Davidson, A. 1999. *The Oxford companion to food*. New York: Oxford University Press.
112. Rai, S. 2003. *Tastes of India in U.S. wrappers. New York* Times, April 29.
113. Jagetia, G.C., & Aggarwal, B.B. 2007. "Spicing up" of the immune system by curcumin. *Journal of Clinical Immunology, 27*, 19–35.
114. Kollipara, U.K., & Brittin, H.C. 1996. Increased iron content of some Indian foods due to cookware. *Journal of the American Dietetic Association, 96*, 508–510.
115. Yajnik, C.S. 2001. The insulin resistance epidemic in India: Fetal origins, later lifestyle, or both? *Nutrition Review, 59*, 1–9.
116. Bhopal, R., Unwin, N., White, M., Yallop, J., Walker, L., Alberti, K.G., Harland, J., Patel, S., Ahmad, N., Turner, C., Watson, B., Kaur, D., Dulkarni, A., Laker, M., & Tavridou, A. 1999. Heterogeneity of coronary heart disease risk factors in Indian, Pakistani, Bangladeshi, and European origin populations: Cross sectional study. *BMJ, 319*, 215–220.
117. Ahmed, M., & Flatz, G. 1984. Prevalence of primary adult lactose malabsorption in Pakistan. *Human Genetics, 34*, 69–75.

CHAPTER 15

Regional Americans

In Boston, they eat beans. In Philadelphia, they eat cheesesteak. In Kansas City, they eat barbecue. And in Seattle, they drink café lattes. A person from Montana is no more likely to eat grits (ground hominy) than a person from Mississippi is likely to eat Rocky Mountain oysters (deep-fried beef testicles). Just as certain fare is associated with ethnicity and religious affiliation, local food preferences are central to American regional identity.

Although most Americans expect uniformity in some food products, such as the breakfast cereal they buy in the grocery or the burger they purchase at a fast-food franchise, U.S. cuisine is anything but homogeneous. Consumption data illustrate many differences. People in the Northeast spend the most money annually on frozen meats, bread and cracker products, certain meats and fish (especially lamb, hot dogs, lunch meats, ham, and fresh fish), butter, tea, and iced tea mixes. Those in the South spend more on pork (particularly sausage), hamburger and round steak, cornmeal, shortening, canned vegetables, dried beans and peas, and refrigerator biscuits. Beef roasts, frozen foods, salad dressings, canned pie filling, baking chocolate, brown sugar, potato chips, and carbonated beverages are above-average sellers in the Midwest; fresh and dried fruits, fresh vegetables, cereals and grains (especially flour, rice, pasta, and non-white bread), chicken and canned ham, cheese, spices and seasonings, Mexican foods, pies and tarts, and coffee are purchased more often in the West.[1,2] Beer is preferred in the Midwest and South, while wine dominates in the West and Northeast.[3]

These figures reflect regional tastes and specialties. Tea in the Northeast dates back to colonial times, and the strong German influence in some areas accounts for the popularity of hot dogs and lunch meats. Pork and corn have long been a foundation of southern fare. The popularity of spices, seasonings, and Mexican food in the West reflects a history of ethnic diversity in the Southwest and California; the significant purchasing of coffee drinks began in the Pacific Northwest. This chapter profiles the Northeast, South, Midwest, and West regions and examines traditional fare, noting significant culinary variations and trends in U.S. regional food habits.

American Regional Food Habits

What Is Regional Fare?

Regional fare has traditionally been home-style food prepared with local ingredients, dependent on agricultural conditions and seasonal availability. Most families made do with what they could grow, gather, or barter. Local foods are the most significant of several factors that influence the development of a particular regional cuisine. The spicy cooking of the Southwest with its emphasis on corn, beans, and chiles could not have been

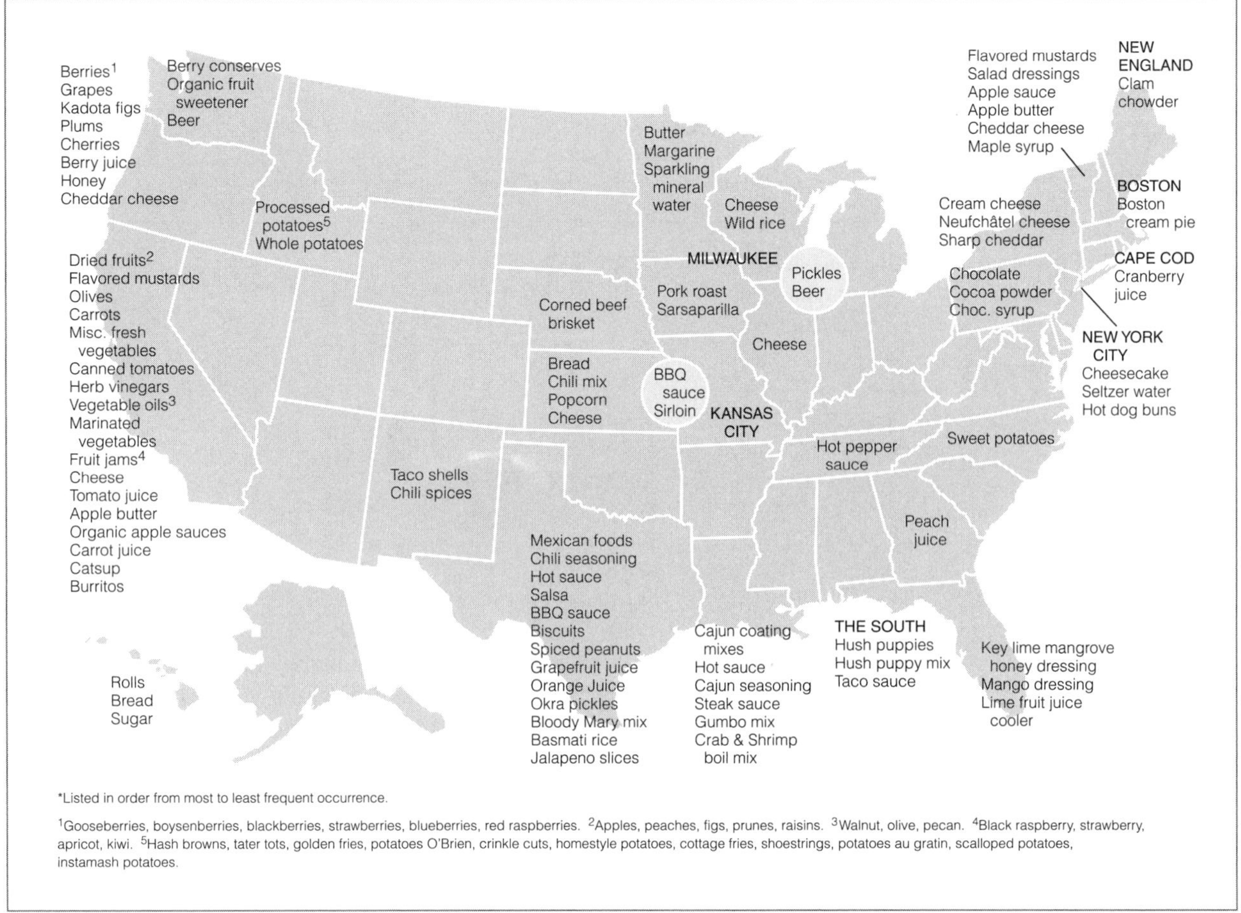

Figure 15.1
Foods associated with places.
Source: de Wit, C. W. Food-place associations on American product labels. *Geographical Review, 83,* 323–330. Copyright 1992. Reprinted with permission by American Geographical Society.

created in the upper Great Lakes area, which produces wheat, fish, and dairy foods. Before the advent of food preservation and refrigerated shipping, local items were not only freshest and tastiest, but also most economical. Strong associations with place and food suggest the importance placed on the superior quality of local items. Even today, with a global assortment available, there is a certain cachet to Maine lobsters, Vermont maple syrup, Georgia peaches, Florida oranges, Idaho potatoes, and Washington apples (see Figure 15.1).

Ethnic and religious practices also affect the development of regional fare, particularly specialty foods. Jewish bagels in New York, German doughnuts called *fastnachts* in Pennsylvania, Cajun French–style sausages in Louisiana, West African–influenced hoppin' John in South Carolina, southern Italian–flavored pizza in Chicago, Cornish pasties in Michigan, and Mexican-inspired chili con carne in Texas are just a few examples. Most regional cuisine is a blend of several ethnic elements, such as the British–Native American dishes of New England, the African-French-Spanish-British–Native American mélange that is southern fare, and the northern Italian-Mexican-Asian mix found in California cuisine. Religious food habits are a factor in areas where large numbers of a specific faith have congregated. For instance, the majority of Mormons live in Utah, and Mormons typically do not drink alcohol, tea, coffee, or other stimulating beverages. Alcohol purchases in Utah are strictly controlled through limited availability at state-run outlets. Sweets are allowed, however, and are well integrated into family activities. The people of Utah eat twice as much candy per capita as the U.S. average.[4]

A third factor in regional foods is local history, which is often associated with certain dishes. A good example is the Kentucky stew called burgoo. Legend has it that the mixture of poultry and red meat with vegetables dates back to the Civil War when a chef for the Confederate cavalry at Lexington created the stew from native blackbirds, game, and greens. There is no single recipe for the dish, but today it typically includes chicken, pork, beef, or lamb; and cabbage, potatoes, tomatoes, lima beans, corn, and okra. It is seasoned with cayenne. Burgoo is traditionally served at picnics, political rallies, and sporting events, including Derby Day. Current trends can be an influence as well. Some dishes sweep through one region but never gain national acceptance, such as caviar pie (layered hard-boiled eggs, scallions, caviar, and sour cream) in the Southeast, or loco moco (a bowl with rice topped with a ground beef patty, then an egg over easy, and gravy) in Hawaii. Other trends start out regionally and then catch on countrywide, such as the salsas of the Southwest.

Economics contribute to the popularity of certain foods. One study found that some of the best markets for beer in the country are in the poorest areas. Wine is popular in upper-income regions, which are often the worst beer markets despite the growing popularity of local brewpubs.[5] Households with incomes below $20,000 a year spend about $1,400 per capita on food, buying more white bread, bacon, pork chops, luncheon meats, and eggs per person than the U.S. average. On the other end of the socioeconomic spectrum, households with incomes more than $100,000 annually spend over $4,725 on food, with above-average purchases in almost every category, especially expensive cuts of beef, lamb, fresh fish and seafood, dairy products, biscuits/rolls, fruits, prepared salads, nuts, and snack foods.[2] Upscale consumers are more likely to be exposed to unique culinary ideas when dining out in trendy full-service restaurants and to be willing to pay for new or unusual food items.

This blending of physical, cultural, historic, current, and economic conditions in a region produces what researchers call a "taste of place," from the French *gout de terroir*.[6] Elementally, the taste of place is the identification of certain ingredients or dishes with an area. At a deeper level, it is the emotional connection between people and a local heritage, an appreciation for the regional characteristics that create flavors unlike those found anywhere else in the nation, or the world.

Regional Divisions

The United States has been divided numerous ways. Sometimes regions are delineated by terrain, as in the Great Plains, or marked by major rivers or mountain ranges, as in the Mississippi River Valley or the Appalachians. Sometimes areas are defined by similarities in climate, as in the Sun Belt; by economic affiliation, as in the Steel Belt and the Silicon Valley; or by the characteristics of the population, as in Indian Lands and the Bible Belt. Historical divisions, such as the Mason-Dixon Line, or political divisions, including state boundaries, can also be used. Geographers suggest that traditional regions contain elements of all these variables, characterized as a synergistic relationship between a people and the land that develops over time and is specific to the locality. Such regional identity is dynamic, more of a process than a delineation, subject to changes in population, economy, ecology, and other factors.[7,8] Ideally, the people who live in an area, through their voluntary affiliation and their mutual agreement on group boundaries, define each region.[9] Most often, however, geographic considerations are used to set arbitrary regional divisions independent of cultural relationships.

The U.S. Census Bureau and the U.S. Department of Commerce list four regions with nine subdivisions for data collection purposes: Northeast, Midwest, South, and West. Although these categories group states with distinctively different cuisines together, such as Florida and Texas, or Alaska and Hawaii, most demographic and food consumption data are presented in this four-region format (see Figure 15.2). It is useful for detecting broad trends, as long as results are not overgeneralized to smaller populations who may observe alternate regional boundaries.

The Northeast

Regional Profile

The states of the U.S. Northeast include those of New England (Connecticut, Maine, Massachusetts, New Hampshire, Rhode Island, and Vermont) and those of the Mid-Atlantic region (New Jersey, New York, and Pennsylvania). The New England area features a rugged, irregular Atlantic coastline with many protected bays. Rolling hills and

Table 15.1 Estimates of Ethnicity in the Northeast Region

	United States		Northeast Region	
	Estimate	Margin of Error	Estimate	Margin of Error
Total:	301,461,533	*****	54,906,297	*****
White alone	224,469,780	+/–45,816	41,742,418	+/–18,923
Black or African American alone	37,264,679	+/–22,810	6,322,332	+/–8,686
American Indian and Alaska Native alone	2,423,294	+/–12,407	138,061	+/–3,355
Asian alone	13,201,056	+/–16,395	2,761,224	+/–4,362
Native Hawaiian and Other Pacific Islander alone	447,591	+/–4,938	15,672	+/–1,169
Some other race alone	16,986,453	+/–46,059	2,984,872	+/–24,097
Two or more races:	6,668,680	+/–65,623	941,718	+/–13,538
Two races including some other race	1,351,590	+/–25,511	219,042	+/–5,950
Two races excluding some other race, and three or more races	5,317,090	+/–44,32 5	722,676	+/–10,318

Source: U.S. Census Bureau, 2005–2009 American Community Survey

valleys that gradually become densely forested mountains extend west. The region is noted for its spectacular autumn weather and colorful fall foliage, followed by harsh winters. The Mid-Atlantic states are farther south and more temperate in climate. Sandy beaches and estuaries line the long coast. The ridges, river valleys, and fertile plateaus of the Adirondack, Appalachian, Blue, Catskill, and other mountain ranges crisscross much of the three states. Freshwater lakes dot the

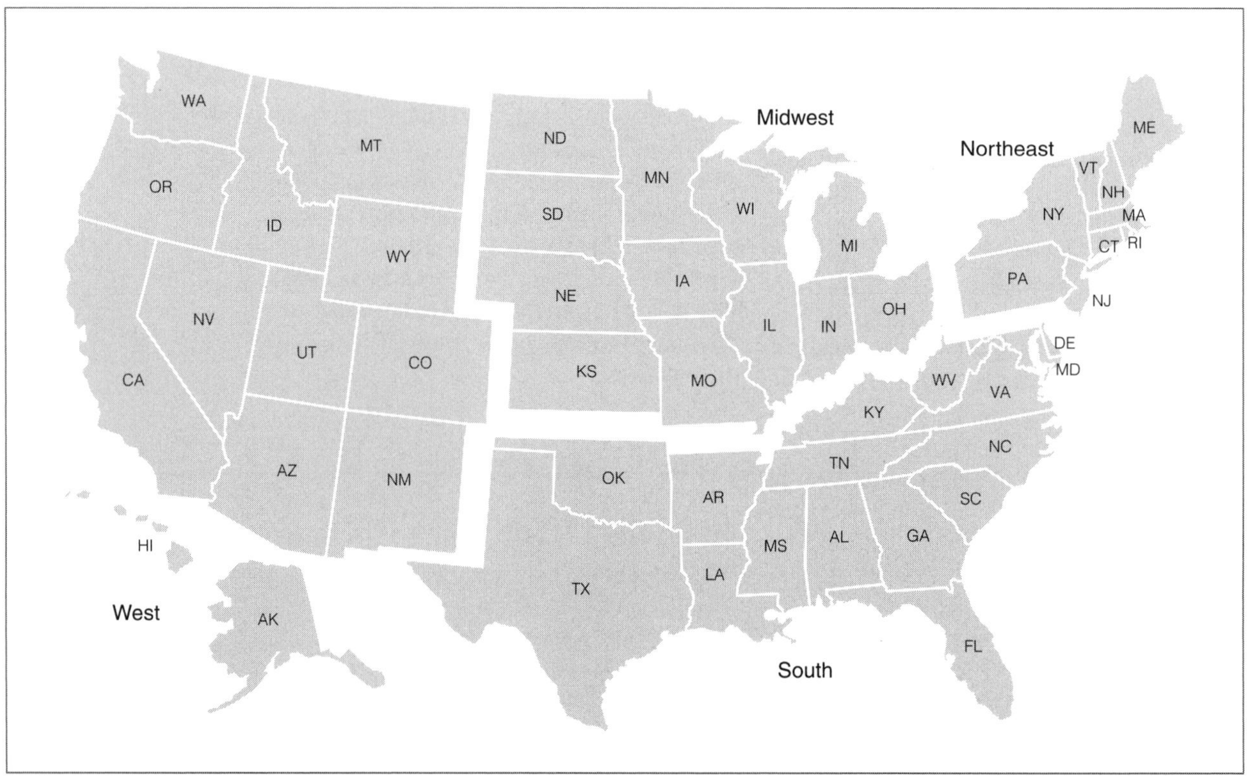

Figure 15.2
Regional divisions in the United States.

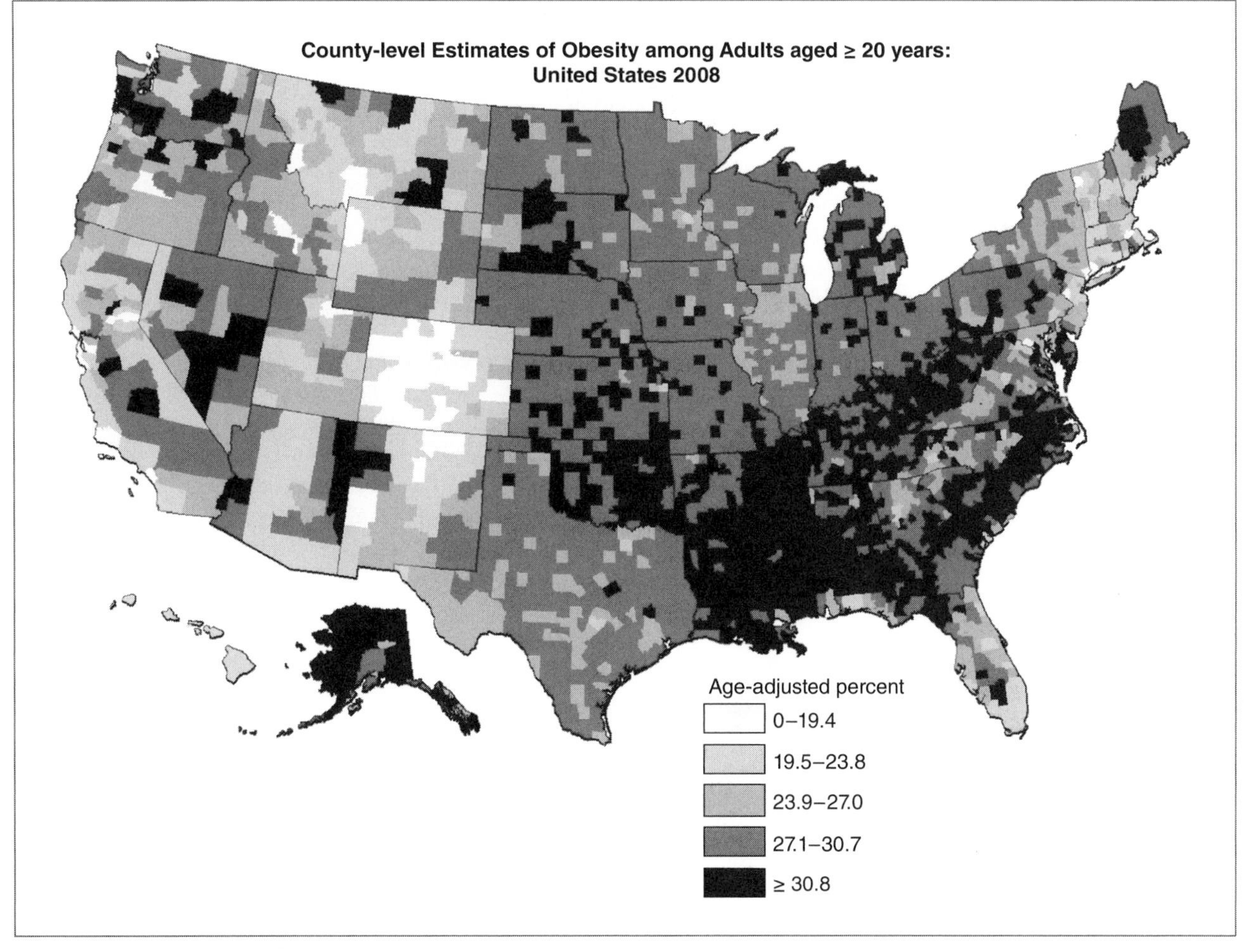

Figure 15.3
Estimates of Obesity in the United States, 2008 by region.
Source: CDC: www.cdc.gov/diabetes

region and provide the northern boundary along Lake Ontario and Lake Erie.

Despite differences in climate and geography, the entire Northeast shares a common early history of sophisticated Native American societies supplanted by European settlements. The colonial immigrants from England, Germany, the Netherlands, and France were followed by newcomers from Ireland, Italy, Portugal, Poland, and other central and eastern European countries, particularly in the Mid-Atlantic states. African Americans from the South and more recent immigrants from the Caribbean, Central America, Africa, and Asia have added to the diversity of some parts of the Northeast (see each chapter on ethnic food habits for more immigration details and Table 15.1).

Eighteen percent of the total U.S. population is found in the Northeast,[10] Over eight million individuals (15%) living in the northeast region are born in another country.[11] Over two-thirds of Puerto Ricans living on the mainland reside there, and more than one-third of all Asian Indians in the United States have settled in the region. Disproportionately large populations of blacks and Latinos reside in New York and New Jersey. Yet the rest of the Northeast is still predominantly white, including large percentages of Italians, Ukrainians, Portuguese, French Canadians, Russians, Lithuanians, Slovaks, and Poles. Nearly half of all American Jews also live in the Northeast, primarily in New York, New Jersey, and Massachusetts.

Compared to national figures, the Northeast has a lower percentage of persons under the age of fifteen, and a higher proportion of persons over the age of sixty-five. The Northeast Region had a total population of 54.9 million in 2009. The median age was 38.6 years with 23%

Fudge was originally a maple sugar candy popular in New England in the 1800s. When cocoa became widely available, the chocolate version was developed.

Shad migrate from the ocean up freshwater rivers to spawn in early summer. Shadberries (also known as Juneberries) are a popular treat that ripen at the same time the fish arrive each year.

Salmon served with fresh peas is a Fourth of July tradition for many New Englanders.

of the population being under 18 years and 14 percent 65 years and older. Data from the 2009 American Family Survey also show that the Northeast has the highest population density of all regions and the highest number of people living in metropolitan areas (nearly 90%). Average household income was $57,208 per year. An estimated 12 percent of the Northeast population live in poverty. Sixteen percent of related children under 18 were below the poverty level, compared with 10 percent of people 65 years old and over.[11]

Traditional Fare

The cooking of the Northeast features the abundance of the Atlantic, the plenty of native and introduced produce, and the freshwater wealth of the many rivers and lakes (see Table 15.2). In New England, seafood, such as clams, lobster, scallops, and fish, especially cod, has been prominent. Indigenous game, including wild turkey and venison, supplemented the poultry, pork, and beef brought by early immigrants. The foundation of the diet was traditionally corn, and many dishes of the region reflect its importance. Beans have also made a substantial contribution. Root vegetables (potatoes, onions, beets, turnips, rutabagas, carrots, etc.) quickly added to the vegetable selection. Wild berries, including blueberries, cranberries, gooseberries, and cloudberries (which look like bleached blackberries), grapes, and beach plums were the main fruits consumed until the apple orchards planted by immigrants became productive. Maple sugar sweetened many foods in New England. Even

Table 15.2 Northeastern Specialties

Group	Foods	Preparations
Protein Foods		
Milk/milk products	Cream; cheddar and cream cheese	Cream soups, sauces, puddings; ice cream
Meats/poultry/fish/eggs	Native game, particularly venison and turkey	New England boiled dinner; scrapple; red-flannel hash
	Preserved meats, such as corned beef, pastrami, salt pork, ham, bacon, sausages	
	Seafood prevalent, especially clams, lobster, oysters, scallops	Fish stews, soups; clam chowder; clam bakes; oyster or lobster loafs
	Salt and freshwater fish, such as salt cod and shad	Cod cakes; shad bakes; *gefilte* fish
	Numerous beans (e.g., cranberry, kidney, lima)	Baked beans; succotash
Cereals/Grains	Corn, wheat, rye	Cornmeal porridges, puddings, and breads Dumplings Baked goods—savory and sweet pies, cakes, doughnuts, waffles, pretzels, bagels
Fruits/Vegetables	Apples, blueberries, cranberries, grapes Cabbage, fiddlehead fern fronds, potatoes	Applesauce, apple butter; fruit puddings, pies Coleslaw; sauerkraut; ferns on toast; potatoes—mashed, fried, creamed, baked, scalloped, hashed, as croquettes, salad
Additional Foods		
Seasonings	Salt, pepper, onions, saffron	
Nuts/seeds	Black walnuts, butternuts	
Beverages	Apple cider, hard cider, applejack; ale, beer; rum, whiskey; wine	New York State white, red, sparkling wines; sherry, port
Fats/oils	Lard, butter	
Sweeteners	Maple sugar; molasses	Maple sugar candies, maple syrup pie; chocolates

when molasses and cane sugar became widely available, maple syrup was preferred for many dishes.

The warmer weather and fertile lands of the Mid-Atlantic states have provided more native foods than New England. The coastal waters offer clams, oysters, mussels, scallops, and crabs, while the estuaries shelter ducks, geese, and turtles. Passenger pigeons once darkened the skies with their massive numbers, and bison roamed the area around Lake Erie. Freshwater fish such as catfish, eels, pickerel, salmon, shad, smelt, trout, and whitefish were plentiful; at one time, shad, a flavorful relative of the herring, was the most numerous of all freshwater fish in the United States. Both the flesh and roe were commonly eaten.

Introduced foods also flourished in the region. Cabbage, potatoes, yams, carrots, peas, apples, pears, cherries, peaches, and strawberries were easily grown. Tomatoes thrived in the hot summers. Although New Jersey is now one of the most industrialized states in the nation, it is still known as the Garden State due to the success of these early agricultural efforts. Wheat, which was difficult to grow in New England, did well in the Mid-Atlantic. At one time, New York provided all the wheat consumed in the Northeast and much of the South.

New England

The cuisine of New England has been shaped predominantly by Native American preparation techniques combined with British homestyle cooking. Roasting, boiling, and stewing are preferred. Dishes are often made with cream, and strong seasonings are avoided. People of the region take pride in simple fare.

The immigrants of the early seventeenth century were mostly tradespeople, inexperienced in farming and husbandry. History abounds with tales of how the first settlers were dependent on the skills and generosity of local Indians in preventing starvation (see Chapter 5, "Native Americans"). Corn dishes were especially significant. Cornmeal porridge cooked into a mush-like consistency was a Native American food called *samp* by the early colonists. It was often prepared with another cornerstone of the Indian diet, beans. New Englanders used cornmeal to make an adaptation of the traditional English dish known as hasty pudding. The settlers would pour cornmeal porridge into a loaf pan to firm up overnight, then slice it and serve it topped with cream. This new dish, often flavored with maple syrup or molasses, was named Indian pudding.

Steamed, baked, and boiled puddings were eaten daily in New England homes. They were known as grunts (steamed dough and berries), slumps (baked puddings), and flummeries (a British molded oatmeal or custard pudding). As in England, a pudding could be savory or sweet and was generally served at the beginning of the meal. Breads were also a mainstay. Many were dense, baked without any leavening (reliable leavenings such as baking powder and baking soda were not available until the mid-1900s). Homegrown yeast from potatoes, hops, or the dregs of beer barrels was used in some recipes. Cornbread cooked in a skillet over the fire was most common. Rye, which grows well in cooler climates, was often combined with cornmeal to make a popular bread called ryaninjun (from "Rye 'n' Indian"). Stewed pumpkin was sometimes added for a moister loaf. Boston brown bread is a traditional recipe of the region—a steamed loaf made with whole wheat and rye flours (sometimes with cornmeal as well) and flavored with molasses.

Pork, cod, or beef flavored most main dishes. Long winters required that most meats and fish be preserved, and few recipes called for fresh cuts. Salt pork, bacon, smoked pork, dried salt cod, corned beef, and dried beef were common, usually braised or stewed with vegetables. The New England boiled dinner is typical. This one-pot meal is still popular throughout the region and usually includes corned beef brisket simmered for hours with potatoes, onions, carrots, turnips, and, traditionally, beets. Cabbage is added toward the end of the cooking time. Seasoning is mild, often just a little black pepper. Leftovers are often chopped up the next day and heated in a skillet with a little cream and sometimes some bacon and more onions to make another New England specialty known as red-flannel hash, so-called because the cooked beets would bleed into the other ingredients during frying. A dried salt cod and potato version of the New England boiled dinner is prepared in Massachusetts, called Cape Cod turkey. Plymouth succotash is another example. Although this Native American dish (see Chapter 5) is often associated with southern fare today, it was popular in the Northeast during colonial times. It combined corned beef, turkey or chicken, beans, corn, potatoes, and turnips.

The oldest continually operating cheese factory in the United States was founded in 1822 in Healdville, Vermont. Manufacturers have expanded into the European-style cheese (e.g., Brie, Camembert) market in recent years; the largest producer of feta cheese in the nation is found in Vermont.

The Maine bean pot is based on the Indian method of placing the ingredients in a pot that is then buried in a pit over embers, a so-called "bean hole."

During the colonial period, dried salt cod was exchanged for fruit in the Mediterranean and for molasses in the Caribbean (the molasses was then used to make rum). Cod traders in Massachusetts became wealthy and were nicknamed the "codfish aristocracy."

Other variations featured just vegetables or combinations of meats, poultry, and fish.

Other popular dishes included dried beef rehydrated in boiling water and served with cream sauce over bread or potatoes (precursor to what is called chipped beef today), and fried salt pork topped with cream gravy. New England states without access to the coast were more dependent on meat, poultry, and dairy products. Today, New Hampshire is acknowledged for the quality of its butter, and Vermont is famous for its cheeses, such as cheddar and the similar but milder Colby.

Beans were eaten regularly in early New England. Best known are baked beans flavored with molasses or maple syrup and salt pork, a recipe adapted from the Native American dish. Traditionally the Puritans prepared a large pot of beans on Saturday morning, simmered them over the fire all day, and then ate them with Boston brown bread for dinner to start the Sabbath (observed from sundown to sundown). Leftovers were kept warm on the hearth for Sunday breakfast with codfish cakes and for Sunday lunch.[12] Boston is still known as Bean Town due to its long association with baked beans. In Maine, a version of baked beans called bean pot is made with indigenous yellow-eye, cranberry, or kidney beans.

Jeff Greenberg/ Visuals Unlimited, Inc

▶ *Lobster is a specialty of Maine, though it is also trapped in other New England coastal states.*

Pies made with suet pastry were served at most meals. Savory kinds included an American version of the British steak and kidney pie, chicken pot pie (later topped with biscuits instead of pie crust), a ground pork and onion pie seasoned with allspice called tourtière introduced by French Canadians in the region (served traditionally at Christmas or on New Year's Day), clam pie, lobster and oyster pie, and a salt-cod pie covered with mashed potatoes. Sweet pies were also popular, especially apple pies, made with fresh apple slices, dried rings, or even applesauce. Mincemeat, a traditional English treat combining savory and sweet ingredients, was featured at many meals because the filling of meat, dried fruits, nuts, and rum or other alcoholic preservative aged well, becoming tastier over time. Today, Vermont fried pies (fried applesauce turnovers flavored with cinnamon) and apple pie topped with sliced cheddar cheese are Vermont favorites, while in New Hampshire, apple pie is sometimes drenched in maple syrup. Blueberry pie is a specialty in Maine.

No discussion of New England fare is complete without further detailing the use of fish and shellfish in coastal areas. In Massachusetts, for example, cod helped sustain the earliest populations, and Cape Cod, the peninsula that curls out into the Atlantic, was so named in 1602 for an abundance of the fish. The Puritans used it in boiled and baked dishes, soups, stews, hash, and, most notably, codfish cakes. These cakes, which are also called codfish balls, are still a sign of regional affiliation for some residents of the Boston area. In Connecticut shad was enjoyed by the Native Americans of the region but disdained by the earliest Europeans in the area due to its multitude of tiny, difficult-to-remove bones. By the mid-eighteenth century Connecticut residents had changed their opinion of shad, especially the roe, which they fried quickly in butter. Traditionally, American Indians would plank the fish and slowly cook it at the edge of hot coals, a method still practiced today at shad bakes where the fillets are placed on an oak board with strips of bacon, then grilled or smoked.

Lobster is a specialty, especially in Maine. The Indians of the area consumed the meat, used the discarded shells for fertilizer, and formed the claws into pipes. British settlers mostly added the meat to mixed fish dishes, and later colonialists added it to salads, sauces, soups, and fried croquettes.[12] Commercial trapping began in the late 1800s with the advent of shipping by train and the development of the canning industry. The lobster supply diminished rapidly, increasing its prestige and popularity—today trapping regulations are strict. In addition to steamed or grilled lobster tail, a specialty in Maine and other coastal areas is lobster rolls, which take two forms (both served on toasted, fluffy white bread buns): plain meat drenched in butter, or meat mixed with mayonnaise, celery, onions, and lemon juice.

Clams, oysters, and scallops are also New England favorites. The clambake, in which clams, corn, and other items such as onions, potatoes, or lobster are steamed in a pit on the beach, shares some similarities with American Indian seafood feasts, and was enjoyed as a way of connecting with what New Englanders in the early nineteenth century believed was their Puritan past.[3] Clams are also featured in the cream-based soup known as clam chowder. One version, the cream-based Boston clam chowder, is known nationally. It is typically garnished with Boston crackers or oyster crackers, the slightly sweet, small, dry biscuits invented by Massachusetts sea captains for use on long journeys aboard ship. In Rhode Island tomato-based red clam chowder, a soup inaccurately attributed to Manhattan, is popular. Clams called steamers are just that—steamed and served with the broth and melted butter (a bucket of steamers is often the first course of a lobster dinner). Oysters were typically prepared with cream and breadcrumbs in a dish called scalloped oysters, or served in oyster stew. In Rhode Island they were especially popular among the nineteenth-century elite, who served them in pies (raw oysters in cream sauce topped with biscuit dough), as patties, creamed, curried, and, for New Year's Eve, pickled, with eggnog to wash them down. Today they are commonly broiled with bacon or breaded and deep-fried. Bay, sea, and Digby scallops are prepared similarly to oysters. In Maine two less common shellfish specialties are found. The first is mussels (the state provides nearly two-thirds of those shellfish consumed nationally), and the second is sea urchin roe (*uni*), served at local restaurants and sushi bars, or exported to Japan.[4]

There are two fruits particularly associated with the New England area. The first is cranberries, known as *sassamanesh* and *ibimi* (meaning bitter or sour) by some Native Americans, who ate them fresh with maple syrup, or dried and added to pemmican. It was the Dutch who introduced the term *Kranbeere*, meaning "crane berry," because the flower resembles the head of a crane. Cranberries grow exceptionally well in the sandy peat bogs of eastern Massachusetts, where they were first cultivated in the early 1800s. They are used primarily in juices and sauces, though in recent years, dried cranberries have become popular as snacks or added to baked goods. The second fruit is wild or low-bush blueberries, which are used mostly in baked goods. Maine grows nearly 100 percent of this variety in the United States.

New England desserts are mostly fruit based. In addition to the puddings and pies already discussed, pandowdies (baked fruit layered with bread), shortbreads (fruit preserves, biscuits, and cream), and roly-polys (fruit rolled up in biscuit dough, then baked) were other favorites. Pound cakes and fruitcakes were enjoyed but were difficult to make before commercial leavening and reliable ovens were available.

No sweet is as associated with New England, particularly Vermont, as is maple syrup. The sweet sap of the sugar maple tree had long been used by

[SAMPLE MENU]

A New England Supper

New England Boiled Dinner[a,b,c]

Boston Brown Bread[a,c]

Blueberry Pie[b,c] or Apple Pandowdy with Maple Syrup[a]

[a]Jamison, C.A., & Jamison, B. 1999. *American home cooking.* New York: Broadway Books.

[b]Oliver, S.L. 1995. *Saltwater foodways.* Mystic, CT: Mystic Seaport Museum.

[c]*New England Recipes* at www.newenglandrecipes.com

One New England dessert popular throughout the nation is chocolate chip cookies, which were created by Ruth Wakefield in 1930 at the Toll House Inn in Whitman, Massachusetts.

Boston cream pie, a favorite in New England, is not a pie but a custard-filled white cake covered with chocolate icing. It probably derives from a popular colonial dessert called "pudding cake" that included cake, custard, and usually fruit or jam.

Indians of the Northeast to cook beans and meats and to flavor other items (see Chapter 5). The syrup was an everyday sweetener in colonial kitchens throughout the region until cane sugar became more affordable. Maple syrup production peaked in the 1880s, and the sweet has since become a costly item. Vermont specialties include sugar-on-snow (hot syrup poured over fresh snow to make a chewy taffy eaten with pickles or doughnuts), maple syrup pie (with a filling of cream, eggs, and syrup), and maple-sugar candies.

Tea and apple cider were consumed daily in colonial times. Hard cider, an alcoholic beverage caused by the fermentation of sugars in apple cider, was also favored. Many New Englanders would start their day with a pint of beer or ale made from barley, corn, pumpkins, persimmons, or spruce bark. Rum, as well as whiskey made from rye, was available. Wine from dandelions or gooseberries was a specialty, and an American version of the English drink called syllabub, containing apple cider, sherry or wine, and whipped cream, was served at special occasions. Today, apple cider remains a regional specialty, particularly in New Hampshire.

Mid-Atlantic

Many of the influences on New England fare are seen in the foods of the Mid-Atlantic states as well. Native American fare was combined with immigrant preferences to produce a new regional cuisine. Unlike New England, where most of the colonists were from England, many settlers in New Jersey, New York, and Pennsylvania came from the Netherlands and Germany. They provided a distinctively different flavor to foods, including a greater use of pork (especially sausages) and dairy products, more baked goods, and stronger seasonings. Later immigrants from southern and eastern Europe contributed many specialties. Further, the warmer climate and fertile farmlands offered a greater variety of ingredients to the cooks of New York, New Jersey, and Pennsylvania.

The Dutch in the mid-1600s brought wheat to the New York area, which at the time was known as New Netherland. They also grew barley, buckwheat, and rye. Although these were preferred grains, the Dutch used what they called "turkey wheat" (corn) to make a boiled milk and cornmeal porridge known as Suppawn that was eaten daily at breakfast. This same porridge was topped with meats and vegetables for lunch, then baked to make the hearty dish called Hutspot, an American adaptation of a stew common in the Netherlands.

Dairy cattle provided ample milk, butter, and cheese. The Dutch were among the first settlers wealthy enough to import sugar, brandy, chocolate, and numerous spices, including pepper, cloves, cinnamon, and saffron. Many Dutch specialties of the region have made their way into American cooking, including pickled cabbage; Kool sla (from the Dutch word for "cabbage"), now known as coleslaw; and headcheese, a ball-shaped sausage made from the head and feet of the hog. Doughnuts, crullers, pancakes, and waffles were also introduced by the Dutch.

During the same period, German immigrants arrived in the United States. Many were religious outcasts (mostly Mennonites, with smaller numbers of Amish, Schwenkfelders, and other sects) who made their home in the tolerant colony of Pennsylvania (see Chapter 7, "Central Europeans, people of the Former Soviet Union, and Scandinavians"). They became known as the Pennsylvania Dutch, a corruption of the German word *Deutsch*, which means "German." Although some German religious communities remained isolated (and are even to this day), many German immigrants gradually became integrated into the broader populations of Pennsylvania and surrounding states. Likewise, many German foods of the region have become an indistinguishable part of U.S. cuisine.

Pork was the foundation of the German diet, and immigrants brought ham, pork chops, pork Schnitzel (pounded into thin slices), bacon, salt pork, pickled pig's knuckles, Souse (jellied pig's feet loaf), maw (stomach stuffed with meat and vegetables), and a German version of headcheese. Every part of the hog was used, and leftovers would be stretched with lima beans to make a Pennsylvania version of baked beans or with dried green beans and potatoes (*Bohne mit Schinken un'Grumberra*). The best-known leftovers dish is scrapple, still popular throughout the state. Scrapple is a combination of ground pork or sausage, cornmeal porridge, and spices formed into a loaf, sliced into thick slabs when firm, and fried in butter. It is typically served with fried eggs, applesauce, and maple syrup. In addition, smoked and fresh sausages were consumed daily. Chicken stews and soups, made substantial with homemade noodles or dumplings, were also popular with the Pennsylvania Dutch. Beef was used in the braised roast known as sauerbraten and in the smoked, cured dried beef called Bündnerfleisch.

Asparagus, green peas, sugar peas (called Mennonite pod peas), and rhubarb are a few of the vegetables favored by the Pennsylvania Dutch. Potatoes are eaten mashed, fried, creamed, baked, scalloped, hashed, as croquettes, as dumplings, in stews and soups, and as potato salad. Cabbage is also ubiquitous, mostly as sauerkraut and slaw. Apples are particularly popular—fresh, as applesauce, in pastries, as cider, and in preserves such as the thick, sweet spread known as apple butter. Many fruits and vegetables are pickled or preserved. Examples include spiced pears, pickled watermelon rind, sweet pickles, and corn relish. Dark rye breads, cornbreads, yeast rolls, potato rolls, cinnamon rolls and sticky buns, Streuselkuchen (coffeecakes with a sugar-crumb topping), doughnuts (called Fastnachts), and buckwheat pancakes are just a few of the baked goods found in the region. The Pennsylvania Dutch also make numerous desserts, especially pies (see Chapter 7).

Though generally considered a rural cuisine, Pennsylvania Dutch fare was well accepted in the early urban centers of the state, such as Pittsburgh, Allentown, Bethlehem, and Reading. Even Philadelphia, which was founded by English Quakers, favored German foods. Scrapple has become so associated with the city that it is often called Philadelphia scrapple, despite its country beginnings. It is eaten for breakfast, often drizzled with catsup, and is used to make deep-fried croquettes or to stuff vegetables like green peppers and cabbage for dinner. Lebanon bologna is a Pennsylvania Dutch smoked beef sausage that has become a state specialty. It is traditionally sliced, battered, or dipped in bread crumbs, fried, and served with sauerkraut and mashed potatoes. Although the origins are lost to history, Philadelphia pepper pot, a soup made with tripe, onions, potatoes, and black peppercorns, is most likely a Pennsylvania Dutch recipe and is sometimes served with dumplings. Cheesesteak (grilled strips of beef topped with American cheese and grilled onions in a toasted Italian roll), the quintessential Philadelphia sandwich, was supposedly invented during the 1930s when a frankfurter pushcart vendor was accidentally sent beef instead of his standard order of hot dogs.

The hearty fare of the Dutch and Germans combined with many traditional items also found in New England, such as puddings, savory pies, and seafood soups and stews, to produce Mid-Atlantic cuisine. Later immigrants to the Mid-Atlantic region introduced foods that have become associated with certain cities and states. Notably, southern Italians in New York and New Jersey brought pizza, spaghetti with tomato–meat sauce, calzone, cannoli, gelato, and espresso. Eastern European Jews introduced pastrami, smoked salmon and whitefish, chopped liver, and other deli items. Particularly in New York, other Eastern European, Russian, Greek, Chinese, Caribbean Island, and Middle Eastern cuisines became popular, due in part to numerous ethnic eateries (see chapters on each group for more information). New York is noteworthy for the influence of its restaurant fare. Taverns, boarding houses, oyster houses, and coffeehouses served the needs of those eating out in the late eighteenth and early nineteenth centuries. The first European-style bakery was opened in 1825, and delicatessens serving the Jewish community were established in the 1880s. Full-service continental-style restaurants became popular in the mid-1800s. By the turn of the century, New York City had become the gastronomic

Buffalo wings, deep-fried chicken wings drenched in spicy (often using Tabasco sauce) seasoned butter and served with celery and blue cheese dressing, evolved in Buffalo-area bars during the 1960s. No one has established the exact origins of the appetizer.

The word "cookie" is derived from the Dutch word for a small cake, *Koeckje*.

© Mia Foster/PhotoEdit, Inc

◀ *Bagels, introduced by Polish immigrants, were paired with an 1872 New York invention, cream cheese; the chewy doughnut-shaped roll was popularized nationwide during the 1990s.*

center of the nation. Many dishes created for elite diners are now American specialties, such as Waldorf salad (originally a mixture of apples and celery in mayonnaise served at the Waldorf-Astoria), vichyssoise (chilled leek and potato soup from the Ritz-Carlton), Lobster Newburg (lobster tail topped with a Madeira-flavored cream sauce), and the dessert baked Alaska (from Delmonico's).

New Jersey, often called the garden basket of New York due to its numerous commercial crops, is also known for its contributions in food technology. Scientific work on hybridization has yielded new, improved varieties of peaches, tomatoes, and sweet potatoes. Food-processing techniques developed in New Jersey include condensed, canned soups (the Campbell Soup Company); inspected, bottled milk (the Borden Company); and the first application of pasteurization to milk at a small farm outside Princeton. Black tea, in convenient individually sized bags, was introduced in Hoboken in 1880 (the Thomas J. Lipton Company).

Coffee milk, similar to chocolate milk but made with coffee syrup, is the official state beverage of Rhode Island.

[SAMPLE MENU]

A Mid-Atlantic Brunch

Philadelphia Scrapple[a,c] with Eggs

Pecan Sticky Buns[b,c]

Toast with Apple Butter[a,b,c]

Bagels and Cream Cheese

[a]Cunningham, M. 1996. *Fannie Farmer cookbook: Anniversary.* New York: Knopf.

[b]Jamison, C.A., & Jamison, B. 1999. *American home cooking.* New York: Broadway Books.

[c]*Recipe Source* at www.recipesource.com

In addition to the Dutch and Pennsylvania Dutch cookies, doughnuts, pies, pancakes, and waffles of the region, other sweets have gained nationwide acceptance, especially those from Pennsylvania. Philadelphia was one of the first cities to enjoy ice cream, perhaps as early as 1782. An ice cream parlor with frozen treats, cakes, syrups, and cordials was opened in 1800, and the following years saw the first wholesale distributor of ice cream and the first ice cream soda. The city gained a reputation for a high-quality product, and ice cream molded into flowers, fruits, animals, or holiday icons is still a specialty. Another confectionery contribution was affordable chocolate. Commercial production of chocolate for beverages and bonbons began in Pennsylvania during the late 1700s, though it was so costly it was considered a luxury item. Milton Hershey of Derry Church was the first manufacturer of chocolate for the mass market beginning in 1905 when he reduced his expenses by making uniform bars instead of fancy novelties. Two years later, he introduced the chocolate candy, Hershey's Kisses.

Several beverages are associated with the Mid-Atlantic states. American beer, a heavy, top-fermented beverage similar to English ale, was first commercially produced during the late seventeenth century in Pennsylvania. Two hundred years later, a German immigrant to Philadelphia founded the first brewery that made a bottom fermented beverage. The new, lighter beer known as lager, or pilsner, soon became synonymous with beer in the United States (see Chapter 7). New York is second only to California in wine production, including white (Chardonnay, White Riesling, and Seyval-Villard varietals), red (small amounts of Cabernet Sauvignon and Merlot), and sparkling wines as well as some fortified wines, such as sherry and port. New Jersey is the state where hard apple cider was first distilled to produce the apple brandy known as applejack, sometimes called New Jersey lightning.

Health Concerns

State-specific data suggest that people living in the New England states are often healthier than the U.S. average, while those in the Mid-Atlantic states are closer to national norms (see chapters on each ethnic group for population-specific

Table 15.3 Northeast State-Specific Health Data Compared to National Averages, 2001–2002

	CT	ME	MA	NH	NJ	NY	PA	RI	VT
Overweight[a]	↓	AVG	↓	↓	AVG	AVG	AVG	AVG	↓↓
No Leisure-time Exercise[b]	↓	↓↓	↓↓	↓↓↓	AVG	↑↑	AVG	AVG	↓↓↓
Diabetes[c]	AVG	AVG	↓↓	↓↓	↑	AVG	AVG	AVG	↓↓↓
Hypertension[d]	↓	AVG	↓	↓↓	AVG	AVG	↑	AVG	↓↓
High Blood Cholesterol[e]	AVG	AVG	AVG	AVG	AVG	AVG	↑	↑	AVG
Don't Consume 5 Fruits/Vegs.[f]	↓	↓	↓	↓	AVG	AVG	AVG	↓	↓↓
Heavy Drinking[g]	AVG	↑	↑↑↑	↑↑↑	↓↓↓	AVG	AVG	↑↑↑	↑↑↑
Low-Birth weight[h]	↑	↓↓↓	↓↓	↓↓↓	AVG	AVG	AVG	↓	↓↓↓
Deaths from Heart Disease[i]	AVG	↑	AVG	↓	↑	↑↑↑	↑↑↑	↑↑	AVG
Death from Stroke[j]	AVG	↑↑	AVG	↓↓	↓↓	↓↓↓	↑↑↑	AVG	↓
Death from Cancer[k]	↑	↑↑↑	↑↑	AVG	↑↑	AVG	↑↑↑	↑↑	AVG

[a]U.S. prevalence = 58.4 percent (overweight defined as body mass index [BMI] >25.0)

[b]U.S. prevalence = 25.8 percent (persons who did no leisure-time physical activity in past month)

[c]U.S. prevalence = 6.6 percent (self-reported data based on number of persons who were told they had condition by a health professional)

[d]U.S. prevalence = 25.7 percent (self-reported data based on number of persons who were told they had condition by a health professional)

[e]U.S. prevalence = 30.4 percent (self-reported data based on number of persons who were told they had condition by a health professional)

[f]U.S. prevalence = 77.6 percent (adults who do not consume at least 5 fruits/vegetables per day)

[g]U.S. prevalence = 5.1 percent (>2 drinks/day in the past month for men, >1 drink per day in the past month for women)

[h]U.S. prevalence = 7.3 percent (live births of infants weighing <2,500 grams)

[i]U.S. age-adjusted death rate per 100,000 = 245.8

[j]U.S. age-adjusted death rate per 100,000 = 57.9

[k]U.S. age-adjusted death rate per 100,000 = 194.4

AVG – similar to national average

↑ – slightly above national average

↓ – slightly below national average

↑↑ – significantly above national average

↓↓ – significantly below national average

↑↑↑ – exceptionally above national average

↓↓↓ – exceptionally below national average

Sources: Ahluwalia, I.B., et al. 2003. State-specific prevalence of selected chronic disease-related characteristics—Behavioral Risk Factor Surveillance System, 2001. National Center for Health Statistics. 2004. *Chartbook on trends in health of Americans*. Hyattsville, MD: U.S. Government Printing Office. *MMWR Surveillance Summaries*, 52, (SS08); 1–80. Centers for Disease Control. 2003. *Profiling the leading causes of death in the United States: Heart disease, stroke, and cancer.* Available at http://www.cdc.gov/nccdphp/publications/factsheets/ChronicDisease/

data).[13] As in all of the regions of the United States, obesity and rates of diabetes have consistently increased in the Northeast and Mid-Atlantic states (see Figures 15.3 and 15.4). Obesity rates in the Northeast increased from 10–14% in 2000 to an estimated 20–29% in 2009.[14] A noteworthy risk behavior is heavy drinking; the prevalence is 20 percent higher than the national average in Massachusetts, New Hampshire, Rhode Island, and Vermont. Pennsylvania is notable for its exceptionally high death rates (more than 20% above the U.S. average) from heart disease, stroke, and cancer (see Table 15.3).

Connecticut, Maine, Massachusetts, New Hampshire, and Vermont repeatedly rank in the top ten healthiest states in various annual assessments using factors such as mortality rates, risk behaviors, crime rate, motor vehicle accidents, per-capita spending on health care, and health insurance coverage.

The Midwest

Regional Profile

The Midwest is known as the Great Plains region of the United States. The earliest American settlers and European immigrants in the area found a vast, flat terrain covered by tall prairie grasses. Oak-wooded hills and low mountain ridges ringed the territory. The rich soil irrigated by the extensive Mississippi and Missouri river systems proved ideal for wheat, corn, and numerous fruits. The region is still renowned for its agricultural productivity, which is why it is nicknamed "America's breadbasket."

The Midwest encompasses twelve states with 21 percent of the total land area and just over 22 percent of the total U.S. population.[11] It is divided into the east north central region (Illinois, Indiana, Michigan, Ohio, and Wisconsin), and the west north central region (Iowa, Kansas, Minnesota, Missouri, Nebraska, North Dakota, and South Dakota).

The states of the east north central (ENC) area are bounded by the Great Lakes, which temper the climate, ensuring milder weather than that experienced by other areas of the Midwest. Although the French were the first Europeans to explore the region, it was Americans from the Northeast states, as well as Virginia and Delaware, who were the first pioneers. Later immigrants from Germany, Switzerland, Scandinavia, Central Europe, and the Cornwall area of England were attracted by the fishing, dairy, mining, lumber, and meat-packing industries. The west north central (WNC) states are geographically near the center of North America, exposed to long winters, short summers, and extreme temperatures. Most Americans who settled the territory were homesteaders, interested in the inexpensive land and farming opportunities. They came from New England and the Mid-Atlantic states, followed by new immigrants from Germany, Scandinavia, and Central Europe, particularly Poland.

As suggested by the history of immigration to the area, the Midwest has the largest percentage of whites in the nation. Over half of all U.S.

Figure 15.4
Estimates of Diabetes Mellitus in the United States, 2008 by region.
Source: CDC: www.cdc.gov/diabetes

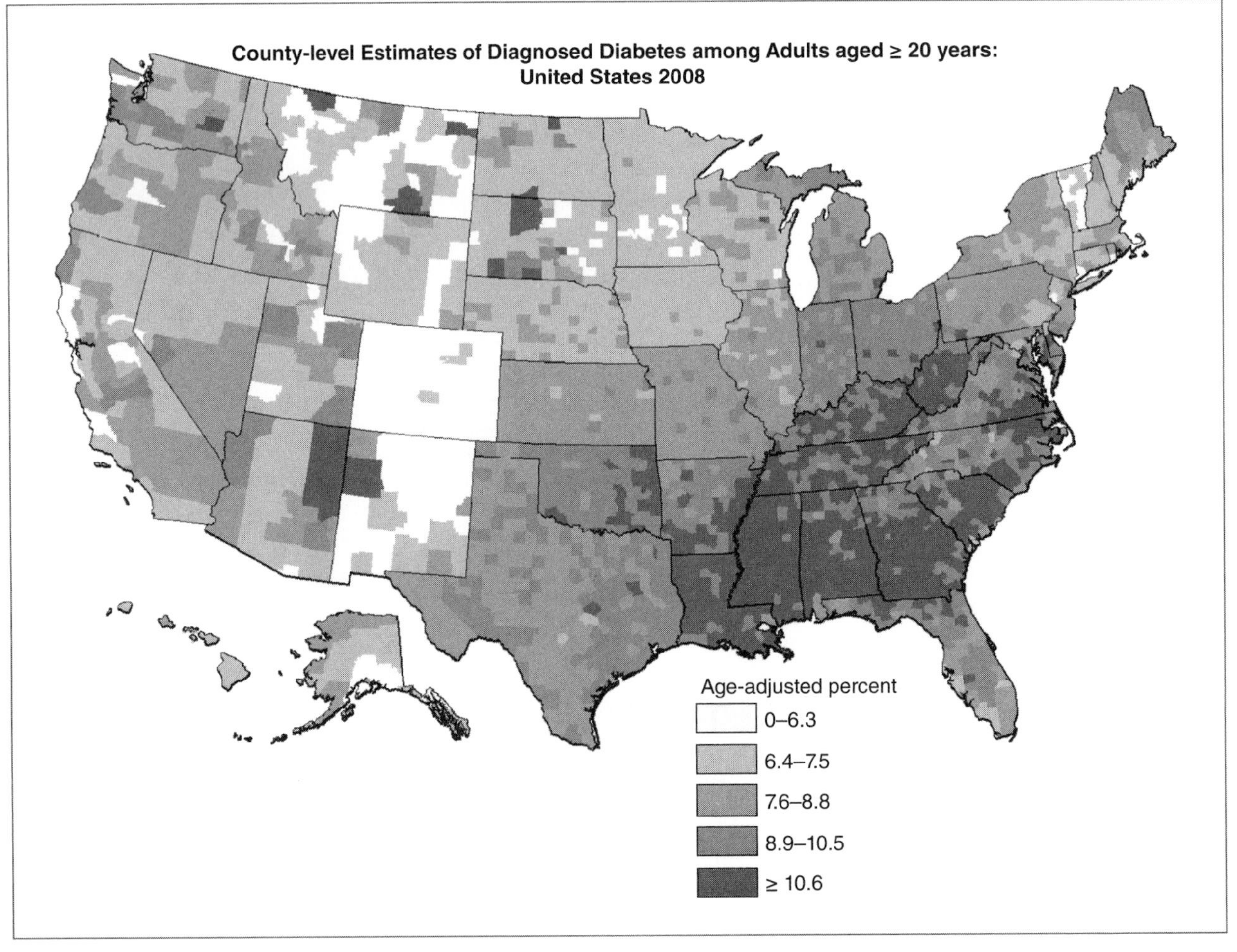

citizens of Czech and Norwegian ancestry live in the Midwest, as well as large numbers of people of Finnish, Croatian, Swedish, German, and Polish heritage. There are below-average numbers of blacks (approximately 10% of the total population) throughout the Midwest; the exceptions are in Ohio, Michigan, and Illinois, which have slightly above-average African American populations. Native Americans, Latinos, and Asians/ Pacific Islanders are also underrepresented, although there is a large population of Latinos in Illinois, approaching the national average, and above average numbers of Native Americans in the Dakotas. Overall, Illinois has the most diverse population in the Midwest; nearly one in every three residents is nonwhite (Table 15.4).

One immigrant group that has made the Midwest home is Laotians, including Hmong, who have arrived since the 1970s (see Chapter 12, "Southeast Asians and Pacific Islanders," for immigration history and food habits). Over 45 percent of all Hmong counted in the 2000 U.S. Census lived in the region, with large populations in Minnesota and Wisconsin. Other Southeast Asians have not tended to settle in the area, however. Additional recent immigrant populations of note include Asian Indians in Illinois and Michigan, Koreans in Illinois, Middle Easterners in Michigan, and Russians in Illinois and Ohio.

The Midwest has the lowest percentage of people living in metropolitan areas in the United States, approximately 74 percent. Average household income is slightly above the national average, at $$49,932 according to 2009 American Community Survey data. Approximately 13 percent of individuals living in the midwest are considered to be living in poverty .[11]

The number of farms in the United States has dropped from 5.7 million in 1900 to 2.1 million in 2000; during this period the average size of each remaining farm has more than tripled.

Traditional Fare

Midwestern fare is usually described as no-frills homestead and farm food, exemplifying what is called typical American cuisine (see Table 15.5). Prime meat or poultry is prepared simply, served with vegetables, potatoes, and fresh bread. Hearty breakfasts start the day, robust soups and stews replenish body and spirit, and homespun desserts round out the meal.

The Midwest is a region of food traditions maintained over generations. Beef and pork are preferred meats, although better cuts are affordable today, and variety cuts may be consumed less often than in settlement days. Canning and freezing to preserve summer's bounty is still a common activity. Bread is sometimes baked at home, and pies make use of seasonal fruits. Midwestern hospitality, which began with festive sorghum pulls, group apple peelings, and canning parties, is continued through buffets, box socials, pitch-in or potluck suppers, strawberry socials, corn roasts, and fish boils popular throughout the region.

East North Central

The earliest American settlement in the East North Central (ENC) region was in Marietta,

Table 15.4 Estimates of Ethnicity in Midwest

	United States		Midwest Region	
	Estimate	Margin of Error	Estimate	Margin of Error
Total:	301,461,533	*****	66,336,038	*****
White alone	224,469,780	+/−45,816	54,701,944	+/−14,985
Black or African American alone	37,264,679	+/−22,810	6,711,525	+/−9,126
American Indian and Alaska Native alone	2,423,294	+/−12,407	385,976	+/−3,976
Asian alone	13,201,056	+/−16,395	1,568,234	+/−4,091
Native Hawaiian and Other Pacific Islander alone	447,591	+/−4,938	23,827	+/−1,585
Some other race alone	16,986,453	+/−46,059	1,779,275	+/−16,076
Two or more races:	6,668,680	+/−65,623	1,165,257	+/−11,872
Two races including some other race	1,351,590	+/−25,511	172,334	+/−4,017
Two races excluding some other race, and three or more races	5,317,090	+/−44,325	992,923	+/−9,890

Source: U.S. Census Bureau, 2005–2009 American Community Survey.

Ohio, in 1788. The people who came to the area from the original colonies were mostly farmers who survived in their new homes on hogs, corn, beans, squash and pumpkins, cabbage, and potatoes. Corn was eaten at every meal as porridge or as baked or fried breads. Sun-dried or smoked meat strips called jerky were adopted from the Indians of the region, used first for game such as venison, then later for beef. Other wild meats, poultry, and fish, such as rabbit, squirrel, woodchuck, opossum, raccoon, skunks, duck, quail, sturgeon, and trout, were widely available. Even bear meat was consumed.[15] Native fruits included persimmons, blueberries, bush cranberries, gooseberries, ground cherries, grapes, and many types of nuts. Later settlers brought wheat and oats, as well as apples, cherries, peaches, and berries. Fishing provided salmon, smelt, trout, and other freshwater fish; dairying, particularly cheese making, offered further food variety.

The origins of meatloaf, the quintessential Midwestern beef dish, are unknown. It may have come with German immigrants who sometimes added rye bread as an extender, or sauerkraut for moisture, to ground meat dishes.[25]

Fish boils combining fish, potatoes, onions and salt were begun by Scandinavians as an efficient way to feed the workers at lumber camps. Today, fish boils at the edge of Lake Michigan are annual tourist events in Wisconsin.

Today, agricultural products are still significant in the region. In addition to wheat and corn, soybeans are a primary crop in Illinois and Ohio, grown for oil (used in products such as margarine, mayonnaise, salad dressing, and for industrial purposes), meat substitutes, and animal feed. Apples are a major crop in Michigan—local preparations include apple salad, apple meat loaf, and apple bread. The French introduced sour European cooking cherries to Michigan, where nearly all of these nationally used fruits are produced. One unusual Illinois specialty is horseradish. German immigrants brought the eastern European food to the state, and the pungent, gnarly root thrived. Nearly 85 percent of the global supply is grown in the Illinois.

▼ Wisconsin is known for its dairy foods, especially cheeses such as Colby and brick.

© 2002 Wisconsin Department of Tourism

Dairying remains important in some regions as well. Wisconsin is the leading U.S. producer of milk, sweetened condensed milk, butter, and cheese. Dairying was sparked by the arrival of Swiss farmers to the state in the 1840s. They brought their expertise in breeding livestock and making cheese. Colby, a hard cheese similar to cheddar, is an original Wisconsin cheese that was created in 1885. Another variety developed in Wisconsin is brick, a semisoft cheese with holes and a flavor described as sweet, nutty, and spicy. Italian cheeses, including ricotta, mozzarella, provolone, Romano, and Parmesan, are specialties of northern Wisconsin, while blue cheese is made in the caves near Milwaukee.

Contributions in food processing from ENC states have extended beyond regional importance to influence the development of American cuisine. Many of these changes took place in Illinois during the late 1800s and early 1900s. Historical accounts include stories about Philip Armour, who made millions in pork sales when he founded the Chicago meatpacking industry; Gustavus Swift, who made his fortune in hams and sausages; Oscar Mayer, a German immigrant who got his start in the hot dog business as a butcher in Chicago; Louis Rich, a Russian immigrant who became involved in poultry processing and founded a turkey luncheon meat empire; and James Lewis Kraft, a grocery clerk who came up with the idea that home-delivered, uniform pieces of cheese would be more popular than freshly cut wedges from a large wheel. He later introduced processed and prepackaged cheeses, including Velveeta.

Developments were not limited to Illinois. In Ohio, an Austrian immigrant, Charles Fleischmann, created the first standardized yeast cakes for baking (he later formed a distillery that produced the first American gin). Michigan is probably best known for its role in the creation of the U.S. cereal industry. The city of Battle Creek was home to two health sanitariums during the late nineteenth century. The first was founded by Seventh-day Adventist leader Ellen Harmon White, who advocated vegetarianism. Her medical director was Dr. John Kellogg, inventor of cornflakes (see Chapter 4, "Food and Religion"). C. W. Post, a dissatisfied Kellogg patient, started his own health institute in Battle

Table 15.5 Midwestern Specialties

Group	Foods	Preparations
Protein Foods		
Milk/milk products	Milk, buttermilk, butter, cream, cheeses	Cream gravy, fondue, *rømmegrøt, skyr*
Meat/poultry/fish/eggs/legumes	Native game, including buffalo, venison, beaver, raccoon, opossum, turkey, prairie chickens (grouse), pheasant	Jerky, *booyaw, Hasenpfeffer*
	Pork in all forms, especially salt pork, hams (country ham and Westphalian ham), and sausages (bratwurst, weinerwurst, kielbasa); beef	Ham with gravy, pork chops, barbecued pork, hot dogs, *Bubbat* Beef pot pie, stew, barbecued brisket, bierocks, pasties, Cincinnati chili
	Oysters shipped from the East Coast	
	Freshwater fish, especially smelt, sturgeon, trout, and whitefish	Fish boils, fried trout or smelt
	Dried beans	Baked beans with salt pork or bacon
Cereals/Grains	Corn, wheat, rye, oats, wild rice	Cornbreads, porridges; oatmeal; *bannocks*; rye breads, pumpernickel; biscuits, dumplings (including stuffed, such as *pierogi* and *verenikas*) Baked goods, especially fresh fruit pies (apple, cherry, persimmon, rhubarb), iced cakes, strawberry shortcakes, strudel, *kolaches*, butter cookies, pancakes, *aebelskivers*, Danish pastries
Fruits/Vegetables	Apples, berries (blueberries, elderberries, strawberries), cherries, grapes, peaches, persimmons, rhubarb	Applesauce; apple butter, fritters, bread, salad; fried apples, candied apples; fruit jams and jellies
	Cabbage, onions, peas, potatoes, rutabagas, turnips, wild mushrooms	Sauerkraut, sauerkraut balls; coleslaw; potatoes—boiled, fried, baked, as dumplings, salad; onion pie
Additional Foods		
Seasonings	Salt, pepper; parsley, dill; cinnamon, ginger, nutmeg, saffron; molasses	Most foods are preferred mildly spiced
Nuts/seeds	Almonds, black walnuts, hickory nuts, pecans; poppy seeds	Nut pies, almond paste; nut candies; poppy seed cakes and pastries
Beverages	Apple juice, beer, wine, apple brandy	Lager-style (American) beer
Fats/oils	Butter, lard	
Sweeteners	Sugar, honey, molasses, sorghum	

Creek. He created a coffee substitute, Postum (a blend of wheat berries, bran, and molasses), and a cereal based on his own recipe for digestive problems, called Grape-Nuts.

Each group of pioneers in the region brought favorite dishes. Baked beans, meat pot pies with biscuit topping, and succotash were preferred by settlers from New England. In Ohio these settlers stuffed meats with breadcrumbs, a practice still popular in the state. The people from New York and Pennsylvania favored sausages, sauerkraut, pickles, and relishes when they moved westward. In Indiana, where the earliest pioneers came from the South, pork is especially popular, including roasts and chops, and sometime the whole roasted pig. Sausage patties and ham are common for breakfast, typically served with pancakes or biscuits, cream gravy, and fried apples. The southern influence is also seen in batter-fried chicken served with fried biscuits (made with a yeast dough that puffs up into spheres when dropped into hot oil, then slathered with butter while still warm).

In areas where European immigrants congregated in numbers, regional ethnic fare developed. For example, the Michigan Dutch (actually from the Netherlands, not Germans like the Pennsylvania Dutch) brought ham croquettes, pea soup, saucijzenbroodjes (now known in English as pigs-in-a-blanket), and double-salted licorice. In Ohio the fare of European immigrants was more broadly integrated into the regional cuisine: Germans popularized sausage, ham, potato, and cabbage dishes, such as the unusual Ohio specialty called sauerkraut balls (deep-fried sauerkraut and ham fritters served with mustard sauce); Polish immigrants introduced pierogis (boiled dumplings traditionally stuffed with potatoes, cabbage, onion, and/or meat, or fruit), kielbasa sausage, and strudel (flaky pastry rolls filled with sweetened fruit, nuts, poppy seeds, or cheese). Fish boils, pickled fish, and meatballs are just a few of the items adopted from Scandinavian immigrants in Wisconsin. Some European foods are so well accepted that their ethnic associations have been forgotten. Eastern Europeans brought to Wisconsin their pork or veal sausages, which are now considered state specialties—Sheboygan is the self-proclaimed "Bratwurst Capital of the World." In other cases, European influence has been more limited, seen mostly in one or two dishes, such as Swiss cheese fondue in Indiana, and in Michigan, a dish with French roots known as booyaw or boolyaw (perhaps from the term bouillon), a game stew featuring venison or whatever else was available (including rabbit, woodchuck, squirrel, muskrat, or duck), salt pork, carrots, potatoes, and onions.

Other dishes with ethnic origins have been adapted to Midwestern tastes, losing much of their heritage along the way. For example, Ohio is probably best known nationally for Cincinnati chili. It is an all-beef version created by Greek and Macedonian immigrants in the 1920s, flavored with a balanced blend of sweet spices (e.g., cinnamon, allspice, cloves, and nutmeg) and hot spices (garlic, cumin, black pepper, and dried chiles). Some researchers note the similarities between the seasoning of Cincinnati chili and dishes such as pastitsio or moussaka (see Chapter 13, "People of the Balkans and the Middle East").[17] Chili parlors found throughout Ohio (and parts of nearby Kentucky) serve the mild chili "one-way" (just the meaty stew alone), "two-way" (over spaghetti), "three-way" (spaghetti, chili, grated-cheese topping), "four-way" (spaghetti, chili, cheese, and diced onions), or "five-way" (spaghetti, chili, cheese, onions, and kidney beans).

Wisconsin fare is sometimes called "white cooking." Whitefish (from the Great Lakes) and white meat (pork, veal, or chicken) combined with white dairy products (such as farmer's cheese, cottage cheese, cream cheese, fresh cream, or sour cream) are favored.[16]

Deep-dish Chicago-style pizza is baked in a skillet. It is an American adaptation of the pizza brought to the region by Italian immigrants from Naples.

The origins of some dishes reflect the succession of immigration to an area. Cornish pasties are an example. Miners from Cornwall arrived in Michigan's Upper Peninsula to excavate iron and copper in the 1840s, bringing their traditional lunch specialty called pasties (see Chapter 6, "Northern and Southern Europeans"). This complete meal-in-a-turnover often featured venison in the Michigan versions, with potatoes and turnips the common vegetable filling. Apples were the most popular fruit used for the dessert end of the pastry. When immigrants from Finland came in the following years, they adopted the dish, which was similar to piiraat and kukko, Finnish pastries filled with meat or fish, rice, and vegetables. The origins of the dish are claimed by many Finns in the region, though those of Cornish descent point out that the Finnish turnovers are not pasties because the filling is mixed instead of layered.[17] Today's pasty shops, often featuring untraditional fillings (e.g., pizza ingredients), are common throughout the Upper Peninsula (U.P.) and in cities where "U.P.ers" ("Yoopers") have settled, such as Detroit.

Sweets, especially baked goods, hold a special place in the cooking of the East North Central states. Traditional items such as hickory nut cookies and pies are found in many areas. In Indiana dessert favorites include steamed or baked persimmon pudding; pork cake, a moist dessert made with sausage or salt pork, molasses, brown sugar, flour, dried fruits, and spices that is a Christmas tradition in some Indiana homes; and sweet cream pie, pastry filled with a heavily sweetened custard that is popular all year. In Michigan apple fritters, caramel-covered apples, candied apples, and Dutch apple kock (cake) are common. Elderberry-flower fritters dipped in powdered sugar are also a specialty. In Wisconsin many popular desserts have retained their foreign names, including German kuchen (yeasted coffee cake, often with a fruit and cream filling and crunchy, sugary streusel topping) and schaum torte (meringue topped with ice cream and/or whipped cream and fresh fruit); and Danish kringle (pretzel- or ring-shaped flaky pastry with fruit, nut, cheese, or butterscotch filling).

Beer is especially associated with the Midwest, particularly Wisconsin. The first breweries were located in the southwestern section of the state and produced the ales and stouts favored by English settlers. By the middle of the nineteenth century, however, ten breweries producing German-style lagers and pilsners had been founded in Milwaukee, including plants owned by Frederich Miller, Frederich Pabst, and Joseph Schlitz. At the beginning of the 1900s, there were over 300 breweries statewide, but Prohibition and consolidation in recent years have reduced that number to eight.

West North Central

In the West North Central (WNC) states, the settlers of the mid–nineteenth century came to farm the fertile land of Iowa, Kansas, Minnesota, and Missouri. Harsh winters and a scarcity of provisions limited variety in many early pioneer homes. Homemakers of the period describe burying melons in sand, which with luck would stay fresh until Christmas. Other cooks would prepare up to a hundred fresh fruit pies at a time, covering the extras with snow for use throughout the winter months. Parched corn, herbs, bark, or root brews replaced coffee.[18] Prestige foods were often unavailable, so ample, even excessive, amounts of common foods became symbolic of hospitality in the Midwestern frontier. The more western areas of the WNC states, which are drier and less suitable for crops, provided limited opportunities for agriculture. The region attracted trappers, traders, and prospectors. Wild game, such as bear, buffalo, elk, deer, and small mammals, as well as turkeys, prairie chickens (grouse), quail, doves, and frogs, was hunted; the meat or oil was often sold in settlement towns. These regional foods were first documented in an American cookbook in 1902.[19]

As in the ENC region, pioneers from New England and the Mid-Atlantic states contributed dishes eaten frequently in their eastern homes: Baked beans and pies of all sorts became as common as sausages and sauerkraut. One northeastern specialty that became surprisingly popular throughout the Midwest was oysters. By the mid-1800s the live shellfish were shipped regularly to the region packed in barrels filled with wet straw. One English visitor commented that the rich "consumed oysters and Champagne and the poor [ate] oysters and lager bier."[20] One 1859 recipe for small birds, such as magpies, suggested stuffing a breaded oyster into each bird before roasting over a hot fire.[17]

[SAMPLE MENU]

A Great Lakes Sampler

Cheese *Pierogi*[a,c]

Bratwurst or Kielbasa

Apple Sauerkraut[b,c]

Danish Kringle[b,c] or Sour Cherry Pie[a,c]

[a]Fertig, J.M. 1999. *Prairie home cooking*. Boston, MA: Harvard Common Press.

[b]Fussell, B. 1997. *I hear America cooking*. New York: Penguin Books.

[c]*Cooks.com* at www.cooks.com

Irrigation improved crop production throughout the region, and today corn, wheat, soybeans, and sugar beets are widely cultivated. Barley, oats, sunflowers, rutabagas, and rye are other crops in some areas. Wild rice, called a grain but actually the triangular-shaped seed of an aquatic grass found in shallow rivers and lakes, is a Minnesota specialty (see Chapter 5). Though Missouri is too hilly for grain crops, the terrain is well suited for nut trees. Eastern black walnuts are native to the area. The nuts are strongly flavored with a slightly bitter aftertaste and are the primary ingredient in black walnut pie. Pecans, too, are indigenous to the state. They are popular in pies, candies, cookies, and cakes. Further, over 7,000 beekeepers take advantage of the woodlands in the state to provide another specialty of the region—honey.

Iowa is noteworthy for its commercial hog farms, which produce 25 percent of all pork consumed in the nation. Pork is also a favorite in Missouri due to the large number of southerners who settled the state. Missouri is well known for its country hams, which are cured with salt, then smoked and hung to age in the cool winter months. The resulting meat is red, salty, and dry in texture.

It is traditionally served with biscuits and red-eyed gravy made from ham drippings, coffee, and flour.

Beef is more significant in Kansas, Nebraska, and the Dakotas. Before the introduction of cattle to the region, over 30 million bison roamed the Great Plains, providing sustenance, clothing, shelter, and fuel to the local Native Americans. American settlers and European immigrants also ate what they called hump-backed beef, at least until other meats became widely available. However, the huge bison herds interfered with expanding settlements, the railroads, and Texas cattlemen, who drove their longhorns through the prairie states to the slaughterhouses of the North and East. The bison were systematically eradicated, providing unimpeded access to the grazing lands of the plains. Today Kansas is famous for its corn-fed beef, and cattle are the most important agricultural commodity in the state. Steaks, beef stews, barbecued beef, and hamburgers are all Kansas favorites. In South Dakota, early settlers introduced longhorn cattle from Texas; they were soon joined by Scottish cattle, including Aberdeen, Angus, and Herefords. Immigrants from Scotland were soon exporting beef to their homeland. Cattle ranching is a major industry in the state, though sheep and hogs are also important commodities. South Dakota is one of the only states in the WNC that features lamb dish specialties.

Dairying is common in several WNC states. Minnesota is one of the top butter and cheese producers in the nation, and Iowa is known for the development of American blue cheese, introduced in the 1920s by Maytag Dairy Farms.

Pierogi **are known as** ***pelmeni*** **by some Russians;** ***varenyky*** **by Ukrainians; and** ***verenikas*** **by German Russian Mennonites.**

The Danish community of Askov, Minnesota, has an annual festival commemorating the rutabaga. Although the tuber is commonly called a "Swede" or "Swedish turnip," Danes are believed to have introduced the rutabaga to the region.

Despite a preference for pork or beef, poultry is well represented in WNC state fare. Chicken with dumplings or noodles, pan-fried chicken with cream gravy, and chicken or turkey pot pies (topped with pastry or biscuits) are classic Midwestern dishes.

Several notable religious communities were founded in WNC states. In Iowa the Amish who settled around Kalona grew all of their own food and butchered all of their own meat, traditions still practiced today. Cornbread with tomato juice gravy, stews or hashes with potatoes and peas, fried meats and eggs, and fresh fruit pies are common dishes. A group of German Lutherans, known as True Inspirationists, settled in seven Iowa villages in 1859 to form what is known as the Amana Colonies. They lived communally, with everyone eating three enormous meals and two coffee breaks each day in a large dining hall. The weekly menu was set and included Mehlspeisen (literally "flour desserts," such as simple puddings) on Tuesdays and boiled beef every Wednesday. The Colonies now serve German specialties to visiting tourists in several large restaurants. German-Russian Mennonites (who had first migrated from Germany to southern Russia) came to Kansas in the 1870s. They brought German-style foods familiar in Pennsylvania Dutch areas, such as chicken noodle soup, pancakes, sausages, and buttermilk pie. They also introduced verenikas (their term for pierogis) served with cream gravy. Beef rolls stuffed with bacon, onions, and pickles similar to German Rouladen, and sausage-filled buns called Bubbat are other Kansas dishes brought by the German-Russian Mennonites.

European influence was seen in secular settlements as well. German-Russian yeast dough turnovers (typically filled with beef, cabbage, and onions) are common throughout the Great Plains. They were derived from the Russian pirozhki. The turnovers are called bierocks in Kansas and the Midwestern regions east and south of that state, and they are known as runsas in Nebraska and the northern Midwest areas.

In Minnesota, German immigrants brought hogs and dairy cattle and introduced their dark rye breads, including pumpernickel. Specialties such as Hasenpfeffer (stewed rabbit), Spätzle (tiny dumplings), and Maultaschen (a sort of German ravioli filled with ground ham, eggs, onions, and sometimes spinach) were other common German dishes. Preserved fish were a mainstay for the Scandinavians. Pickled fish, smoked fish, and salt-cured fish were popular, particularly the Norwegian dish known as lutefisk (see Chapter 7), served with butter and potatoes. Ham, bacon, Swedish meatballs, and Danish frikadeller (fried, breaded ground beef and veal patties) were consumed. Dark breads and the thin Norwegian potato pancake called lefser are still common, as are butter cookies (especially at Christmas) and Danish aebleskivers, traditionally served with chokecherry or blueberry syrup or jam. The Scandinavian concept of the smörgåsbord was introduced to the nation in Minnesota (see Chapter 7). Several ethnic communities in Minnesota maintain their culinary heritage at holidays and festivals, including the German Catholic city New Ulm and the Danish town Askov.

In Nebraska, the Swiss introduced plum tarts and a specialty called Thuna, breadsticks topped with creamed greens thickened with flour. Czech settlers brought jaternice (pork sausage), jelita (blood sausage), and houska (a sweet, braided bread). Swedish yeasted waffles and Hungarian chicken paprika are other examples of European contributions in the state. In Missouri, the French introduced crêpes and brioche to the region. They also made hard cider from apples, wine from native grapes, and brandy from peaches. In North Dakota the Norwegians brought spekejøtt (smoked, dried lamb), rullepølse (cold, spicy rolled beef), and søtsuppe (fruit soup) and baked goods, including the large pyramid of almond paste and meringue rings called kransekake. A large population of settlers from Iceland smoked mutton, made skyr (a sweet, cultured milk product similar to yogurt), fried kleinur (doughnuts), and baked vinarterta (a multilayered cardamom-flavored cake with fruit fillings) for dessert. The Scotch Irish introduced colcannon (mashed potatoes, onions, and cabbage), and the French Canadians came with croissants and cassoulet (see Chapter 6).

A unique cuisine of the Midwest is found in the Ozark Mountains of Missouri. Contrary to immigration trends in urban areas, the people who came to the Ozarks gradually arrived from other states in small groups and were scattered throughout the region. They were known as backwoodsmen, and they subsisted on hunting, fishing, gathering, and cultivation of corn, beans, squash, and various tubers.[21] Hogs were let loose to forage until butchering time in December or January. The people of the Ozarks were known for their stews made from opossum, raccoon, or squirrel. Sorghum was used to sweeten foods, ginger root was brewed for beer, and sassafras was steeped for tea. Today, the Ozarks are best known as a vacation and retirement destination.

Popular desserts in the WNC states include fruit pies and frosted cakes. Czech kolaches are a specialty found throughout the region. These yeasted buns are baked with an indentation on top that is filled with sweetened cheese, poppy seeds, or fruit (apple, apricot, cherry, and prune are traditional) and sprinkled with sugar or streusel before they are baked. The Scandinavian dessert kransekake is also found in many communities.

[SAMPLE MENU]

A Hearty Plains Lunch

Chicken Noodle Soup[a,b,c]

Meatloaf[a,b,c]

Mashed Potatoes

Pickled Cucumbers (Cucumbers in Vinegar)[a,c]

Apricot Kolaches[a,c]

[a]Fertig, J.M. 1999. *Prairie home cooking*. Boston, MA: Harvard Common Press.

[b]Stern, J., & Stern, M. 2001. *Square meals: America's favorite comfort food cookbook*. New York: Lebhar-Friedman Books.

[c]*Babi's Czech Recipes from the Dumpling Newsletter* at www.dumplingnews.com/recipes

Health Concerns

Measures of health in the Midwest approach the national average in many ways, with a few notable state differences (see chapters on each ethnic group for population-specific data).[13] Several states have lower-than-average rates of low birth weight, including Iowa, Minnesota, Nebraska, North Dakota, South Dakota, and Wisconsin. Significantly higher-than-average rates of heavy drinking are found in Michigan and Minnesota, with the highest rate in the nation found in Wisconsin (more than 40% above average). Deaths from coronary heart disease are exceptionally high in Iowa and North Dakota (see Table 15.6).

The South

Regional Profile

Most southerners say the South is more an attitude than a location. This perhaps explains why there are so many definitions of the region, such as those states below the historic Mason-Dixon Line or those south of the culinary grits line (the divide between where grits are eaten and where they aren't). While no one questions that

Table 15.6 Midwest State-Specific Health Data Compared to National Averages, 2001–2002

	IA	IL	IN	KS	MI	MN	MO	NE	ND	OH	SD	WI
Overweight[a]	AVG	AVG	AVG	AVG	AVG	AVG	AVG	AVG	↑	AVG	AVG	AVG
No Leisure-time Exercise[b]	AVG	AVG	AVG	AVG	↓	↓↓↓	↑	↑↑↑	↓↓	AVG	AVG	↓↓↓
Diabetes[c]	↓↓	AVG	AVG	↓	↑	↓↓↓	AVG	↓	↓↓↓	↑	↓	↓↓
Hypertension[d]	AVG	AVG	AVG	↓	↑	↓↓	AVG	↓↓	↓	AVG	↓	↓
High Blood Cholesterol[e]	AVG	AVG	AVG	AVG	↑↑	AVG	AVG	↓	AVG	↑	AVG	AVG
Don't Consume 5 Fruits/Vegs.[f]	↑	AVG	AVG	↑	AVG	AVG	AVG	↑	AVG	AVG	↑	AVG
Heavy Drinking[g]	↓	↓	↑	↓	↑↑	↑↑	↓	↓↓	↓	↑	↓↓↓	↑↑↑
Low Birth Weight[h]	↓↓	↑	AVG	↓	↑	↓↓↓	AVG	↓↓	↓↓↓	AVG	↓↓↓	↓↓
Deaths from Heart Disease[i]	↑↑	AVG	AVG	AVG	↑	↓↓↓	↑↑	AVG	↑	↑↑	↑	AVG
Death from Stroke[j]	↑↑↑	AVG	↑↑	↑↑	AVG	↓↓	↑↑	↑↑	↑↑↑	↑	↑↑	↑↑
Death from Cancer[k]	↑↑	AVG	↑	AVG	AVG	↓	↑↑	AVG	↑↑	↑↑	↑	AVG

[a]U.S. prevalence = 58.4 percent (overweight defined as BMI >25.0)

[b]U.S. prevalence = 25.8 percent (persons who did no leisure-time physical activity in past month)

[c]U.S. prevalence = 6.6 percent (self-reported data based on number of persons who were told they had condition by a health professional)

[d]U.S. prevalence = 25.7 percent (self-reported data based on number of persons who were told they had condition by a health professional)

[e]U.S. prevalence = 30.4 percent (self-reported data based on number of persons who were told they had condition by a health professional)

[f]U.S. prevalence = 77.6 percent (adults who do not consume at least 5 fruits/vegetables per day)

[g]U.S. prevalence = 5.1 percent (>2 drinks/day in the past month for men, >1 drink per day in the past month for women)

[h]U.S. prevalence = 7.3 percent (live births of infants weighing <2500 grams)

[i]U.S. age-adjusted death rate per 100,000 = 245.8

[j]U.S. age-adjusted death rate per 100,000 = 57.9

[k]U.S. age-adjusted death rate per 100,000 = 194.4

AVG – similar to national average

↑ – slightly above national average

↓ – slightly below national average

↑↑ – significantly above national average

↓↓ – significantly below national average

↑↑↑ – exceptionally above national average

↓↓↓ – exceptionally below national average

Sources: Ahluwalia, I.B., et al. 2003. State-specific prevalence of selected chronic disease-related characteristics—Behavioral Risk Factor Surveillance System, 2001. National Center for Health Statistics. 2004. *Chartbook on trends in health of Americans.* Hyattsville, MD: U.S. Government Printing Office. *MMWR Surveillance Summaries,* 52, (SS08); 1–80. Centers for Disease Control. 2003. *Profiling the leading causes of death in the United States: Heart disease, stroke, and cancer.* Available at http://www.cdc.gov/nccdphp/publications/factsheets/ChronicDisease/

Alabama, Arkansas, Florida, Georgia, Louisiana, Kentucky, Maryland, Mississippi, the Carolinas, Tennessee, Virginia, and West Virginia are clearly part of the South, the borderline states of Delaware, Missouri, Oklahoma, and Texas can be argued for inclusion either way. Using the U.S. government definition, Missouri is considered part of the Midwest, whereas Delaware, Oklahoma, and Texas are part of the South.

The lands of the South are varied. They include the fertile coastal plains along the Atlantic and Gulf coasts, the rolling hills leading up to the mountains (called the Piedmont in most states), the rugged Appalachian and Ozark mountain territories, the lowlands of the Mississippi Delta, and the high desert plains of the western reaches. The climate also ranges from the warm, moderate Atlantic states and the hot, humid Gulf Coast states to the hot, dry weather in parts of Texas and Oklahoma.

The development of the South was in many ways independent from that of the northern United States. During colonial times, southern states were predominantly agricultural, growing tobacco, wheat, corn, rice, and indigo (a blue dye). The plantation system that emerged in the coastal regions was characterized by commercial farms owned by aristocratic English or French immigrants and worked by African slaves. Each plantation was a self-sufficient, independent operation providing cash crops and food products for use by each household. It was a comfortable, leisurely lifestyle for the upper classes, enlivened by occasional visits to the cultural centers of Atlanta, Charleston, or New Orleans. For the lower classes, which included the slaves and the poor farmers of the inland hill and mountain regions, it was a hand-to-mouth existence.

During the period when the northern areas of the nation became more urbanized and industrialized, the South remained rural and agricultural, adding cotton as a major crop. Differences of opinion regarding the role of the federal government in state issues, particularly slavery, led to the Civil War in the mid-1800s. After losing the war, the South regrouped in the late nineteenth century. The traditions and practices that give the South its character became more important than ever. The South continues to preserve its identity, in part, through its cuisine.

Over one-third (36 %) of Americans make their home in the South, the highest percentage of the U.S. population in any region.11 It is divided into the South Atlantic states of Delaware, Florida, Georgia, Maryland, North Carolina, South Carolina, Virginia, West Virginia, and the District of Columbia; the East South Central (ESC) states of Alabama, Kentucky, Mississippi, and Tennessee; and the West South Central (WSC) states of Arkansas, Louisiana, Oklahoma, and Texas. Over-all, the South has below-average numbers of Asians and Pacific Islanders, Latinos, and Native Americans, but above-average numbers of African Americans: 56 percent of all U.S. blacks live in the South. However, the very size of the southern population means that significant numbers of most ethnic groups reside in the region. For example, 35 percent of all Native Americans and 33 percent of Latinos live in the South; both Florida and Texas host above-average populations of Latinos. While only 20 percent of U.S. Asians are found in the South, larger numbers of some groups, such as Vietnamese, Pakistanis, and Asian Indians, reside there. In addition to African Americans, groups with disproportionately large representation in the South include persons of British ancestry, Scotch-Irish ancestry, Cuban ancestry, and Cajun ancestry (see Table 15.7).

Table 15.7 Estimates of Ethnicity in the Southern Region

	United States		South Region	
	Estimate	Margin of Error	Estimate	Margin of Error
Total:	301,461,533	*****	110,450,832	*****
White alone	224,469,780	+/−45,816	79,199,532	+/−31,425
Black or African American alone	37,264,679	+/−22,810	20,946,185	+/−12,397
American Indian and Alaska Native alone	2,423,294	+/−12,407	704,356	+/−6,145
Asian alone	13,201,056	+/−16,395	2,776,230	+/−6,311
Native Hawaiian and Other Pacific Islander alone	447,591	+/−4,938	66,961	+/−2,251
Some other race alone	16,986,453	+/−46,059	4,743,190	+/−26,319
Two or more races:	6,668,680	+/−65,623	2,014,378	+/−24,163
Two races including some other race	1,351,590	+/−25,511	377,783	+/−10,079
Two races excluding some other race, and three or more races	5,317,090	+/−44,325	1,636,595	+/−16,991

Source: U.S. Census Bureau, 2005–2009 American Community Survey.

The population of the South is notable for its high numbers of Protestant Christians and low numbers of people without religious affiliation. Baptist and evangelical faiths are especially popular. The highest percentages of Protestants in the nation are found in the East South Central states and the West South Central states: approximately 70 percent of persons in Alabama, Mississippi, and Louisiana.

Three-quarters of the population in the South live in metropolitan areas. Census data from 2000 show that average household income in the South is near the national average in the South Atlantic approximately $47,000. Poverty rates are at approximately 15% for all individuals.

Traditional Fare

The foods most associated with the South reflect both the bounty of the plantation and the scarcity of the slave diet. Corn dishes, pork, sweet potatoes, and greens began as the foundation of southern fare and remain characteristic components today. (See Chapter 8, "Africans.")

The southern lifestyle has fostered a culture of graciousness and cordiality. The isolation of the plantations meant socialization was limited in frequency but lengthy in duration. Hours of travel to nearby homes typically resulted in overnight visits or extended stays. Parties, balls, picnics, barbecues, and seafood feasts were all occasions for get-togethers. For the slaves, Sunday meals with extended kin were the primary way to maintain family connections. In the hills and mountains of the South, the difficulties of subsistence farming necessitated friendly relationships between neighbors. Poor families often survived through regular sharing of food. As a result of these conditions, the South has become synonymous with hospitality.

The first European explorers in the South Atlantic states were the Spanish, who arrived in Florida in 1513 and founded St. Augustine in 1565. They were soon followed by the English, who started in Virginia and spread north and south along the Atlantic coastline during the seventeenth and eighteenth centuries into Delaware, Maryland, the Carolinas, Georgia, and eventually into Florida. The Native American population at the time numbered in the hundreds of thousands, including the members of the Powhatan, Cherokee, Chickasaw, Choctaw, Creek, and Seminole tribes (see Chapter 5).

The white settlers discovered a region with plentiful fruits, nuts, game, fish, and seafood. Native strawberries, blackberries, blueberries, huckleberries, ground cherries, persimmons, muscadine grapes, beechnuts, hickory nuts, and pecans covered the land. Bream, catfish, perch, pike, and trout filled the rivers, while oysters, clams, and crab were abundant along the coast. In Florida pompano, red snapper, shrimp, spiny lobster, and conch were widely available. Diamondback terrapin, sea turtles, and alligators were found in many waterways; and bear, deer, opossum, rabbits, raccoons, squirrels, turkey, grouse, ducks, and quail were prevalent in woodland areas. Indians of the region grew corn, beans, pumpkin, squash, sweet potatoes, and sunflowers.

Most of the first white settlers in the region were farmers who established plantations. They brought wheat, hogs, cattle, poultry, cabbage, potatoes, and fruit trees, including apples. Africans were imported as laborers. They introduced southern staples, such as peanuts, okra, watermelon, and sesame seeds, and taught the farmers of the lowland coastal areas how to successfully grow and harvest rice.[22] It was these traditional foods of the Native Americans, the European settlers, and the African slaves that combined to create the foundation of southern fare (see Table 15.8).

South Atlantic

Plantation hospitality was famous in the South Atlantic region. A description of a meal served to guests in Georgia from the early 1800s listed turtle soup, trout, ham with sweet potatoes, turkey with a cornmeal and walnut stuffing, rice, asparagus, and green beans, followed by orange sherbet to cleanse the diners' palates before continuing with cold venison, cheese, corn fritters with syrup, and sweet potato pie.[4] Traditional southern fare, such as Georgia squirrel stew, ham, hoecakes, okra with tomatoes, and biscuits served with preserves, was served at family meals as well.

Hot breads are the cornerstone of every meal in the states of the South Atlantic, primarily cornbreads (see Table 15.9) or biscuits. In Virginia, spoon bread is a specialty; it is a cornbread enriched with eggs and milk then cooked until it forms a crust on the top but remains custardy underneath. In Delaware the biscuits are made

Table 15.8 Southern Specialties

Group	Foods	Preparations
Protein Foods		
Milk/milk products	Buttermilk, milk	Cream gravy
Meat/poultry/fish/ eggs/legumes	Native game, including buffalo, venison, raccoon, opossum, badger, squirrel, turkey, ducks, alligator, diamond back terrapin	Brunswick stew, squirrel stew, possum 'n' taters, turtle soup
	Pork in all forms, especially country-cured and Smithfield hams; beef, mutton, kid	Ham on beaten biscuits, sliced ham and red-eye gravy; barbecued pork; *souse* (head cheese); chitterlings; Texas-style barbecued beef, chili con carne, son-of-a-bitch stew; *cabrito*
	Chicken	Fried chicken with cream gravy, chicken and dumplings
	Crab (blue, stone), crawfish, conch, oysters, shrimp, spiny lobster; ocean fish, such as mullet, pompano, shad; freshwater fish, particularly catfish	Crab, shrimp, or crawfish boils; crab cakes; she-crab soup; conch chowder; oyster stew; shrimp pilau; shrimp Creole; jambalaya; gumbo; *étouffée*; fish muddle; fried catfish
	Chicken eggs	Scrambled eggs and brains, scrambled eggs and ramps
	Dried beans; peanuts	Baked beans, butter bean custard; peanut soup, peanut brittle
Cereals/Grains	Corn, rice, wheat, buckwheat	Hominy, grits, corn pone, hush puppies, cornbread, spoon bread; rice pilaus; beaten biscuits; buttermilk or sour milk biscuits; buckwheat pancakes
Fruits/Vegetables	Apples, huckleberries, key limes, oranges, mayhaw, peaches, watermelon	Preserves and pickles; fried pies; key lime pie; ambrosia; peach pie
	Wild greens (cochan, creases, dandelion, dock, lamb's quarters, poke, sorrel, and ramp), domesticated greens (e.g., mustard, turnip), black-eyed peas, cabbage, okra, ramps, sweet potatoes	Greens simmered with fat back or salt pork, consumed with pot likker; poke salad (sallet); fried ramps; hoppin' John; coleslaw, fried okra, okra stews; sweet potato pie
Additional Foods		
Seasonings	Chile peppers (especially bird's eye); *filé*; celery, garlic, onions, green peppers; bourbon, sherry, whiskey	Pepper sherry, chili powder, hot sauce; High Holy mayonnaise; barbecue sauce
Nuts/seeds	Black walnuts, hickory nuts, pecans; sesame (benne) seeds	Nut cakes, brittles, glazed pecans, pecan pie, pralines; sesame seed candies and cookies
Beverages	Buttermilk; bourbon, corn whiskey, Sherry, Tennessee whiskey	Whiskey and bourbon are added to barbecue sauces, baked goods, candies
Fats/oils	Lard	
Sweeteners	Sorghum syrup	Used over pancakes, grits, cornbread, in coffee

with sour milk. Beaten biscuits prepared by hitting the dough repeatedly with a rolling pin to produce pockets of air for leavening, are a favorite in many areas.

Country hams, ribs, fatback, cracklings, and chitterlings were traditionally produced from hogs, and remain important today. In Maryland, one recipe that is popular, particularly in the southern sections of the state, is stuffed ham. It calls for inserting greens (e.g., cabbage, kale, and/or watercress) flavored with onions, mustard seeds, and cayenne into deep slits of the ham. The ham is served cold and is often the centerpiece of the Easter meal. In Virginia, Smithfield ham is a specialty, adapted from the process used by the local Powhatan Indians to salt-cure and smoke venison. A Smithfield ham is similar to a country ham (see the West North Central states of the Midwest), but it is made with the shank end of the leg and with the bone in. It is first rubbed with salt, sugar, and pepper for curing, then smoked over hickory, and then hung to age. The meat differs from a country ham in that it is saltier, darker in color, and leaner. The flavor is very strong, and it is traditionally eaten in very thin slices on biscuits or fried with red-eye gravy (made with the

Table 15.9 Southern Cornbreads

Cornbread	Made with white cornmeal, eggs, and water. No sugar is added. Baked in a pan, sliced into squares, served with butter, honey, or sorghum syrup.
Cracklin' bread	Usually yellow cornmeal bread with added pork cracklings for flavor, traditionally cooked in a frying pan on the stove.
Spoon bread	Yellow cornmeal bread made with eggs and milk. Baked slowly in a pan until golden crest forms on top and center remains custard-like.
Corn pone	Yellow cornmeal and water (lard added if available) mixed into a stiff dough, formed into sticks (sometimes called "corn sticks") or patties (sometimes called "hoecakes") and cooked in a skillet.
Hush puppies	Yellow cornmeal and water dough, with added egg and buttermilk if available, formed into balls and deep-fried.

ham drippings and coffee) and served with fried apples. Another Native American game dish adopted by the southern settlers, Brunswick stew, became a mainstay throughout the region. There are many variations, but most contain chicken, ham or salt pork, corn, beans, potatoes, onions, tomatoes, and lots of black pepper.

Chicken dredged in cornmeal or flour and fried in lard, traditionally served with cream gravy, is the quintessential dish of the region. Though popular throughout the entire South, poultry is especially associated with Delaware. The first broilers in the nation were marketed in the state during the 1920s when an excess of chicks prompted an enterprising egg producer to sell the birds when they reached about two pounds at sixteen weeks of age. This was far younger than most chickens were sold at the time and yielded a tender bird that could be roasted or broiled instead of stewed or fried. It was the beginning of a national industry, and today broiled chicken is the state dish of Delaware. Roasters are also popular in Maryland where other chicken specialties include pot pies and chicken seafood stews.

© Matthew Klein/Royalty-Free CORBIS

Fried chicken is the quintessential dish of the South. It is often served with cream gravy and biscuits.

Seafood is especially important in the South Atlantic coastal areas. Maryland, for example, is famous for its shellfish. The state is indented by the largest estuary in the nation, Chesapeake Bay, which teems with oysters, clams, scallops, and crabs. Oysters were so common that many settlers in the region ate them three times a day: raw, fried, baked, fricasseed, in seafood stews, in chowder, in oyster stuffing for turkeys, and over steaks. Crabs were equally versatile. A regional specialty is blue crabs, a swimming crab so named because the underside of the large claws is blue. They are traditionally steamed over water flavored with vinegar and seasoned with salt, pepper, ground ginger, celery and mustard seeds, and paprika. Because they are small, half a dozen or more are served to each diner, with plenty of beer to wash them down. The meat is used to make one of Maryland's most esteemed dishes, crab cakes. The crab is mixed with a little mayonnaise, cracker crumbs, and a spicy seasoning of cayenne, dry mustard, and hot sauce, and then formed into small patties and fried. They are served with lemon wedges and tarter sauce. Crab soup (with beef stock and bacon) and deviled crabs (baked in the shell and topped with bread crumbs) are other common preparations. Another noteworthy shellfish of the region is soft-shell crab—a blue crab that has shed its hard shell during a molt. The new papery shell is completely edible, but it begins to harden after only a day. Blue crabs are often kept in tanks until they shed their hard shells to time harvesting of the soft-shell crabs. The whole crab is served deep-fried.

The Florida waterways and coastline also offer a profusion of seafood. Red snapper, pompano (a very large, meaty fish), mullet, and tarpon are a few of the fish commonly available; shellfish includes shrimp (several varieties), spiny lobster (similar to those of New England, but without claws), conch (a large mollusk), and stone crabs (only the very large claw is eaten—the claw is removed when the crab is caught, and then the crab is thrown back in the water to grow a new one). Many of Florida's specialties have developed out of this unique ocean larder. Red snapper fillets are baked with orange juice. Pompano is stuffed with shrimp, seasoned with Sherry, and baked, or prepared en papillote (with a nod to the French influence of the Gulf Coast states). Spiny lobster tails are stuffed with fish and grilled, while stone crab claws are traditionally boiled and served with garlic butter or mustard sauce. Rock shrimp, a hard-shelled, white shrimp that tastes like a cross between lobster and shrimp, has become a trendy restaurant item throughout the country. Conch fritters and conch chowder (made with onions and tomatoes, seasoned with Worcestershire sauce, oregano, and bay leaves) are popular.

Elsewhere in the region, oyster roasts (similar to a New England clam bake) are favorites in South Carolina, served with hoppin' John, biscuits, and small sandwiches, such as a crab omelet on slices of bread. Oyster suppers, informal feasts featuring oysters cooked over a fire in the moonlight, then served with melted butter, are popular in Georgia. Shrimp are common in the Carolinas, including shrimp pâté or butter-sautéed shrimp with grits for breakfast, and deep-fried shrimp and rice croquettes. In Delaware, one specialty called muddle (a stew of miscellaneous fish with potatoes and onions) capitalizes on coastal resources. A variation unique to South Carolina is pinebark stew, a muddle flavored with bacon, named for the tiny roots of pine trees that seasoned it traditionally, or because it was cooked over a pine bark fire. Also common in the state are frogmore stew, a spicy seafood, sausage, and corn combination similar to gumbo, and she-crab soup similar to that of Maryland but garnished with a spoonful of Sherry and a dollop of unsweetened whipped cream.

Long-grain rice is common in many parts of the South Atlantic region. A variety of rice native to Madagascar was found suitable for the coastal plain climate of South Carolina, and thousands of acres of tidal lands were diked and flooded to support the crop. By 1700, it was well established, thanks in part to the skills of slaves from the rice-growing regions of Africa. It became known as "Carolina Gold" (due to its amber color when ripe). The rice was traditionally boiled instead of steamed to produce individual fluffy grains that did not stick together. French Huguenots who settled in South Carolina during the seventeenth century are thought to have introduced pilau (also spelled purlow, or pullow), which has become a specialty of the region. It is characterized by combining a single additional ingredient with rice, which is first simmered in an aromatic broth (reserved from cooking the secondary ingredient) until dry, then mixed with the other food. Shrimp pilau and okra pilau are examples. African-influenced hoppin' John, made with black-eyed peas and rice, is also a pilau. Molded rice dishes that are baked until they form a golden crust are called rice pies or rice casseroles. Some include layers of meat or fish. One unusual rice dish found in Georgia is Country Captain Chicken, invented by a sea captain from Savannah who used Indian spices to liven up his routine fare aboard ship. It is a curried chicken that includes tomatoes and green peppers and is served over rice. In the Carolinas, rice breads, such as philpy (cooked rice added to cornbread), and desserts, such as rice pudding, are also found.

Certain crops historically associated with the South Atlantic states have been in the region so long they are occasionally mistaken as native foods. Some accounts state that Native Americans of the region cultivated melons. Melons are not native to the New World, although it is possible they were brought to Florida by the Spanish explorers of the sixteenth century, in which case the Indians may have been growing them for perhaps one hundred years by the time white settlers arrived from the North. Tomatoes are a food that the Spanish may have brought to the region

Hoppin' John is served on New Year's Eve in South Carolina and other parts of the South because eating the rice and black-eyed peas is thought to bring good luck in the upcoming year.

Unlike most states of the South Atlantic, neither corn nor rice grows well in the cool, damp climate of West Virginia. Buckwheat, however, thrives. Buckwheat pancakes served with whole-hog sausage and applesauce are a specialty.

from elsewhere in the Americas. They also introduced peaches to the Carolinas, which at one time were so plentiful they were used as hog feed. Today South Carolina is the largest producer of peaches in the South, second only to California in the nation.

Oranges, the foundation of the Florida citrus industry, were another early introduction by the Spanish. Later, grapefruit were hybridized from pummelos that had been brought from the Caribbean, and other citrus fruits, such as tangerines, tangelos, and Persian (also known as Tahiti) limes, were introduced. Key limes, small, thin-skinned yellow limes with juicy, green flesh, were discovered in the Florida Keys. It is not known where the limes came from, but it is assumed that they drifted to the islands from the Caribbean. They are grown mostly in home gardens and are renowned for their tangy flavor. Today, 70 percent of the U.S. citrus crop is grown in Florida (nearly all the oranges are processed into juice). Florida is also known for other subtropical crops, such as avocados, guavas, kumquats, mangoes, papaya, and pineapples, as well as early-ripening crops, such as tomatoes and strawberries. Sugarcane is grown in the south of the state, and sabal palmetto palms grow like weeds, providing the delicacy known as hearts of palm. In Georgia, pecans, peanuts, and watermelon are commonly cultivated. Vidalia onions, thought to be exceptionally sweet due to the mild Georgia weather and the low-sulfur soil around Vidalia in Toombs County, Georgia, are a specialty crop sold throughout the nation. Mayhaw jelly, made from the cranberry-like fruit of the native mayhaw tree, is a particular favorite in Georgia and other states along the Gulf Coast. Other native fruits found in the South Atlantic are muscadine grapes and scuppernongs (the bronzy white version of muscadines), which are used to produce jams, jellies, pies, and wine.

Thomas Jefferson brought many French specialties to his home in Monticello, such as *boeuf à la daube* (jellied beef) and crêpes. He also brought Italian foods, including pasta, to the United States. His daughter, Mary Randolph, is credited with introducing macaroni and Parmesan cheese, which evolved into the American dish, macaroni with cheddar cheese.

Immigrants from Europe have contributed only limited ingredients and dishes to the foods of the South Atlantic. In Virginia, English settlers favored roasted beef dishes, mutton, and Yorkshire pudding. In Georgia, a French nuance can be seen in the popularity of dishes such as crab soufflé; common German-style dishes include sauerkraut and pepper pot soup; and the Scots brought scones and haggis (hog's stomach stuffed with oatmeal—see Chapter 6). In South Carolina, a French influence was seen in many dishes, particularly elaborate desserts like Huguenot torte (a sponge cake with pecans and apples) and charlotte russe (a special cylindrical mold lined with ladyfingers, then filled with Bavarian cream and garnished with strawberries and whipped cream). In Florida, Greek immigrants who came to Tarpon Springs for sponge-fishing jobs at the beginning of the nineteenth century (see Chapter 13) introduced traditional dishes such as moussaka (stuffed eggplant), spanakopita (spinach- or cheese-filled phyllo dough pastries), and gyros (pita bread sandwiches). In West Virginia lasagna, fagiole (pasta with beans), minestra (vegetable soup), and cannoli are popular in the area around Clarksburg where more than half the population is of Italian descent.[23]

One notable ethnic group in North Carolina is the German Moravians, persecuted German Protestants who had immigrated to Pennsylvania originally but moved south in the early 1700s when they discovered that much of that land was already claimed. The Moravians established an insular German community near the Winston-Salem area, founding a wholesale produce business that sold local fruits and vegetables in markets extending to Philadelphia. They were best known for their baked goods, such as sugar cakes (a yeasted, potato bread dough covered with brown sugar and cinnamon before baking) and citron tarts (tarts with lemon curd filling). Moravians commemorate special occasions, including November 17 (the founding of North Carolina), with Love Feasts featuring wine, creamy coffee, and cakes topped with a nut frosting. At Christmas, paper-thin ginger spice cookies and a sweet bread studded with raisins and candied citron, sprinkled with sliced almonds, are specialties.

In recent years a more significant culinary influence in Florida has been the contributions of Cuban immigrants to the Miami area (see Chapter 9, "Mexicans and Central Americans"). Arroz con pollo is made with chicken and rice, flavored with the Cuban combination of tomatoes, olives, capers, raisins, and chile peppers. Black beans, traditionally prepared with rice and salt pork or ham, are common. So-called Cuban sandwiches, with roast pork, ham, sausage, cheese, and dill pickle filling mounded on Cuban bread, are fast-food favorites. Flan, a baked custard with caramel topping (sometimes flavored with orange), has become a popular dessert.

Cuban cuisine is not the only spicy food found in the South. Many settlers, especially in South Carolina and Georgia, had lived first in Barbados and other Caribbean Islands. They brought a taste for tropical flavors and spicy seasonings. Fruit and vegetable pickles were common, for example, mango chutney from India, which was also made with other local fruits and called "Indian pickle." Today, Jerusalem artichoke, okra, green tomato, squash, and watermelon rind pickles are still popular condiments in the region. In Georgia, very small (one-fourth to one-half inch) scorching-hot bird's eye peppers (also known as tepin chiles) are sometimes crushed and placed at the bottom of a bowl before adding soup or stew. Pepper sherry, made by infusing incendiary Scotch Bonnet chiles in sherry, is a popular condiment added to dishes for zing in South Carolina.

Key lime pie, which has a lime custard filling and is traditionally covered with meringue, can also be made as a chiffon pie (folding the meringue into the custard to lighten it and then topping the pie with whipped cream).

© Michael Lamotte, Cole Group/PhotoDisc/Getty Images

Desserts have always had a place on the South Atlantic table. Tea breads and cakes (such as Sally Lunn cake, best described as a sponge-cake–like bread, which is popular throughout the region), fruitcakes, and pies are common. Peach pie is the consummate Georgia dessert, although recipes vary. Some are custard pies topped with sliced peaches, others are two-crust pies, and some are individual deep-fried pies. Pecan pie is popular as well. Key lime pie, a specialty from Florida now found throughout the South, traditionally includes a lime custard filling covered with a meringue topping but can also be made as a chiffon pie (folding the meringue into the custard to lighten it and then topping the pie with whipped cream). Ambrosia, made with sliced oranges and grated coconut, is another Florida dessert common in other states of the region. Puddings and custards were an everyday treat in the early days of settlement, made with leftover cornmeal, rice, or bread; chocolate was a favorite but costly, so it was used only at special occasions. Today, bread puddings are still favorites. Candies, such as divinity with nuts and nut brittles, are specialties.

The cooking of the more rural inland areas of the South Atlantic states differed from the more populated coastal areas. During the early 1800s, Scotch-Irish immigrants searching for religious freedom began making their homes in the Blue Ridge, Cumberland, and Great Smoky mountains of the Appalachians. They also spread west to the Kentucky and Tennessee frontier. English and some Welsh settlers moved from the coastal South Atlantic states inland to the hilly Piedmont areas. Germans from Pennsylvania traveled south along the Shenandoah Valley into Virginia and North Carolina. Hogs 'n' hominy (pork and corn) kept the pioneers going until they established small farms. Frontier meals were robust. For example, the noon meal might consist of ham, bacon or sausage, chicken or grouse, game meat, dumplings or biscuits, cornbread or grits, gravy, sweet potatoes, and boiled greens served with coffee, milk, or corn whiskey.

Coca-Cola was invented by an Atlanta pharmacist in 1886 as a headache remedy. "Dope" is a slang term for cola drinks in parts of the South.

Traditionally every bit of the pig was consumed on Appalachian farms, including the snout, or rooter (which was roasted), the tail (which was added to stews), and the brains (which were usually boiled, mashed, and scrambled with eggs). Bacon and cabbage, and ham with cream gravy were typical entrees, while barbecued pork with spicy hot sauce on the side was prepared for special occasions. Most families kept a dairy cow and a breeding cow for a few calves each year. Fresh beef was preferred. When a cow was slaughtered, it would be shared with neighbors, who would later return the favor.

Game supplemented the diet, especially squirrel, rabbit, raccoon, opossum, turtle, and frogs. Badger, considered by some a dish of last resort, was known as bombo in North Carolina hill country. Brunswick stew is a favorite. Wild greens were well loved by adults in the Appalachians, but not so popular with children.[24] Poke, creases (similar to watercress), dandelion, lamb's quarters, dock, sorrel, and ramp (a particularly assertive wild onion) were added to soups, stews, potatoes, or eggs, or cooked as a side dish. Poke salad is representative: a cooked salad (from the English tradition) in which the greens are parboiled, then fried in bacon or fatback grease until tender. They are seasoned with salt, pepper, and hot sauce or vinegar. Domesticated greens such as mustard and turnip greens were also common.

Other than greens, green beans, hominy, sweet potatoes, potatoes, okra, and beets were the most frequently consumed vegetables. Cornbread (sometimes with cracklings added), biscuits, dumplings, and/or grits were served at every meal. Pinto beans, called soupbeans, were common, served with cornbread crumbled on top or a dollop of pickled vegetable relish (i.e., cabbage, bell peppers, green tomatoes, onions, chili peppers). Watermelon was a favorite fruit, eaten fresh or preserved as pickles or jam. Applesauce, apple pies, and fried apple slices were popular. Honey was the most common sweetener, even consumed alone as a dessert. Thick, caramel-colored sorghum syrup was also used, poured over cornbreads or used to sweeten coffee.

Many of these foods are still favored in the Appalachians today, though research suggests changes in preparation techniques and the use of convenience food items are increasing as dependence on hunting and farming decreases.[25] A survey of senior adult Appalachians found that they were interested in lower-fat diets and had switched to baking and broiling instead of frying. Shortening has replaced lard in many dishes. Many also use items such as cornbread mixes instead of preparing foods from scratch.[26] However, even those who have relocated to other areas may maintain their heritage by regularly consuming fried chicken with gravy, soupbeans, skillet cornbread, biscuits and gravy, fried potatoes, green beans cooked in lard, and other foods typical of Appalachian fare.[25]

[SAMPLE MENU]

A Southeastern Luncheon

Fried Oysters[a,b,c]

Shrimp Pilau[b,c]

Okra with Tomatoes[a,c]

Biscuits or Cornbread

Ambrosia[a]

[a]Lewis, E., & Peacock, S. 2003. *The gift of Southern cooking*. New York: Alfred A. Knopf.

[b]Jamison, C.A., & Jamison, B. 1999. *American home cooking*. New York: Broadway Books.

[c]*Southern U.S. Cuisine* at www.southernfood.about.com

East South Central and West South Central

The early fare of the East South Central and West South Central states was similar to that of the Atlantic states but with more significant French overtones. Immigrants from France settled in the Gulf Coast region during the seventeenth century, and at the end of the eighteenth century, French Acadians from Canada relocated to Louisiana (see Chapter 6). They were joined by white American and English settlers arriving from the North. Plantation life in the region was similar to that of the South Atlantic, except that it was more dependent on cotton than on tobacco. The tradition of the big southern breakfast and dinner may have originated in the region and was the norm for plantation owners and their city associates. Coffee and mint juleps were available for early risers. Late morning repasts included eggs, grits, biscuits, cornbreads or muffins, waffles, and several meats, such as ham, sausage, or fried chicken. A large dinner with soups, stews, and dishes similar to those at breakfast was consumed in the early afternoon; supper was a lighter version of dinner.

Pork and corn remained key to the cooking of the ESC and WSC. The cornbread in this area is made from white cornmeal without the addition of sugar. The French added their recipes for soups, stews, fricassees, and baked goods to the southern mix, as well as their appreciation for good eating. The resulting cuisine is found in some form throughout the Gulf Coast, from Mobile, Alabama to Beaumont, Texas. The French factor accounts for such adopted and adapted specialties as bouillabaisse (a French fish stew); fish cooked *en papillote* (in paper packets with a *velouté* sauce); and *sauce mahonaise* (homemade mayonnaise), particularly High Holy mayonnais" (a fanciful, Anglicized term for aïoli) made with fresh garlic and served with shrimp or cucumbers. Creole cuisine, a blending of French, Spanish, African, English, and Native American cooking, is unique to New Orleans. It is a complex fare with many refined dishes; celery, tomatoes, bell peppers, onions, and garlic are the hallmark flavorings.

Cajun fare, created by the French Acadians, is mostly limited to the bayou country of Louisiana, though its gumbos, jambalayas, and *étouffées* have become popular throughout the region (see Chapter 6). While Oklahoma and Texas are both southern in attitude and enjoy many specialties of the South, such as grits, greens, Gulf Coast seafood, and Brunswick stews, their dishes are also influenced by Native American, central European, and Latino cooking. Beef is the dominant meat; barbecue is prevalent; and hot, spicy seasoning emboldens their dishes.

The fare of the ESC states is more homogeneous than that of the WSC region. Alabama, Kentucky, Mississippi, and Tennessee share many culinary traditions. The French influence is limited to the coastal areas, where dishes feature seafood as the main ingredient. In Alabama, shrimp are especially prevalent, prepared fried, boiled in seasoned water, with rémoulade sauce, and stuffed into *mirleton* (chayote squash), avocados, and other vegetables. Plump, local oysters, called Bon Secour oysters, are plentiful and popular throughout the Gulf Coast. In Mississippi, rock shrimp and blue crabs are typically boiled, served with an assortment of seasonings, such as vinegar, lemon juice, bird's eye chiles, and cloves. Outdoor oyster bakes and fish muddle served with corn dumplings are other Mississippi coastal favorites.

It is the inland foods of the ESC states that are most associated with the region, however. Sumptuous breakfasts are still common in some areas. In western Tennessee, for instance, the meal may feature eggs, tomatoes, potatoes, and cornmeal biscuits with sorghum syrup. During the winter, thick slices of Tennessee country ham with grits and red-eye gravy are often served with the meal; in the summer, fried chicken is more common. ESC dinners and suppers also include many traditional items. Fried chicken is found throughout the region. In Alabama chicken and dumplings, ham balls (fried fritters), and Brunswick stew (made with a whole hog's head) are specialties. In many areas biscuits and cornbreads such as sweet potato biscuits, crackling bread, hoecake bread (cornmeal and water cooked in a frying pan), and beaten biscuits are eaten daily. Tennessee pork sausages are a specialty, as is spiced beef (marinated in vinegar, brown sugar, and seasonings, then simmered and sliced thinly). In the eastern region of the state, barbecued ribs prepared with a tomato-whiskey sauce are a favorite. Hominy, greens, okra, green beans, black-eyed peas, peas, butter beans (similar to lima beans, but slightly smaller), rutabagas, and turnips are typical side dishes of the region. Many of these foods are cooked in lard or flavored with pork. In Kentucky, for example, green beans are simmered with bacon throughout the day to make a smoky, mushy stew.

Game meats are prevalent in some areas. Squirrels and frogs are featured in certain dishes from Alabama. Early settlers in Kentucky depended on game. Bear meat was popular because it could be smoked like pork and was fatty enough to provide bacon. Burgoo, a stew traditionally made with wild birds and game meats such as squirrel, is the signature dish of Kentucky. It is still made this way in some areas, though most current versions use chicken, pork, beef, or lamb; cabbage, potatoes, tomatoes, lima beans, corn, okra, and cayenne—and some variations add filé powder, curry powder, or bourbon. In eastern Tennessee the diet was historically closer to Appalachian fare than the plantation style of the western half of the state. Deer, raccoon, opossum, squirrel, and wild turkey were primary meats for the settlers of the area and are still consumed occasionally today.

In Mississippi, catfish up to one hundred pounds can still be caught in the rivers and lakes of the state, but most are now farmed in ponds. Although the first catfish farms were started in Arkansas, Mississippi is the leading producer in the nation, with over 250 million pounds harvested annually. Traditionally, catfish is deep-fried in a cornmeal crust and served with hush puppies and coleslaw. Newer recipes include

Moon pies are a Chattanooga, Tennessee, confection that have become an obsession in the South. They are graham cracker sandwiches with a marshmallow filling covered in chocolate, vanilla, banana, or caramel icing. During the Great Depression, a moon pie and an RC Cola were called a "working-class dessert" because both could be had for a dime.

fried strips served with barbecue sauce or mustard, and catfish pâté.[23]

Sweets in the ESC states are favorites. In Alabama seasonal pies were popular, especially dewberry (the first ripe fruit of the summer season) and peach. Fried pies, a southern specialty, are thought to have originated in the state. Small circles of pie crust are filled with fruit (typically peaches or peach preserves in Alabama), then folded into a half-moon shape, crimped, deep-fried, and sprinkled with powdered sugar. Rich, chocolaty Mississippi mud pie has become popular nationwide, while butter bean custard pie is a local specialty of the region, made with mashed butter beans cooked as a sweet pudding flavored with cinnamon, cloves, and nutmeg. Banana pudding is another favorite. Pecans are native to Mississippi and added to breads, sugar glazed, orange glazed, and baked in the syrupy sweet pecan pie. Many farms in eastern Tennessee had at least one apple tree, providing fruit for apple butter and pies. Funnel cakes, undoubtedly introduced by German immigrants, are topped with sorghum syrup. Fried pies were also popular.

Traditional beverages consumed in the ESC states include buttermilk and coffee, though iced tea and soda (most often called pop) are more popular now. Sassafras tea is common in eastern Tennessee. Perhaps the best known food products of the region are alcoholic beverages. Bourbon was developed in Kentucky. Many of the early Scotch-Irish settlers in the state discovered farming corn for corn whiskey was more profitable than farming it for cornmeal. It is thought that the first corn whiskey aged in oak barrels, creating the characteristic flavor of bourbon, was produced in Bourbon County, Kentucky, in the late eighteenth century. In 1860, a further refinement occurred when it was accidentally discovered that charred oak barrels added not only a touch of color, but also a favorable smoky taste. The favorite bourbon drink of Kentucky is the mint julep (bourbon sweetened with a touch of sugar or syrup and a hint of fresh mint), traditionally served in a silver cup. Bourbon also flavors stews, hams, pound cakes, fruitcakes, and bourbon balls (a candy made with chocolate, crushed vanilla wafers, pecans, corn syrup, and bourbon). Whiskey is associated with Tennessee. In 1866 Jack Daniel purchased a corn whiskey still and added an extra refinement to the distillation process, using maple wood charcoal to filter the whiskey before aging it in charred oak barrels. This produced a flavor distinct from bourbon, and the liquor became known as Tennessee whiskey.

The foods of the WSC states (Arkansas, Louisiana, Oklahoma, Texas) share some similarities due to geographical proximity but also vary due to historical influences. Arkansas exemplifies the region. It is at the crossroads of the South, the Southwest, and the Midwest. The diverse terrain in the state includes the fertile alluvial plains of the Mississippi River in the southeast of the state, the dry pasturelands of the Southwest, the orchards and wheat fields of the Northwest, and the rocky hills and mountains of the Ozarks in the Northeast. Settlers were mostly of English or Scotch-Irish heritage, and they brought the foods they prepared in their home states, such as cured hams, sausages, baking-soda biscuits, and molasses pies from the North, and fried chicken, buttermilk biscuits, sweet potatoes, and peach cobblers from the South. Barbecued beef and pinto beans are found in the areas of the state adjacent to Texas,[25] and in the Ozarks the fare is similar to that found in the Missouri section of the mountains (see previous section on the cooking of the Midwest), with pork, game meats (especially baked opossum and raccoon), corn, beans, and greens the foundation of the diet.

Arkansas specialties include pork chops with cream gravy (sometimes made with bits of sausage in it) and pan-fried chicken that is then baked with a Creole sauce. Arkansas is also the leading producer of rice in the nation. Ducks, which are attracted to the rice paddies, are a specialty in the region, roasted over a fire, baked with bacon and basted with wine or port, and prepared as gumbo. Catfish have long been an Arkansas favorite, dredged in cornmeal and fried, or in catfish stew. Catfish is traditionally served with hush puppies (deep-fried cornmeal biscuits) and coleslaw.

Fare in the other WSC states overlaps with that of Arkansas. The hilly north areas of Louisiana feature dishes with pork and cornmeal. The southern portions of Oklahoma are called Little Dixie, and a study of foods in the eastern portion of the state found that pork, fried chicken, catfish, biscuits and cream gravy, cornbread, fried okra, and black-eyed peas were frequent items in local eateries. However, grits and buttermilk were rarely offered.[27] In the affluent eastern region of Texas, southern-style dishes frequently feature costly ingredients and tend to be richer (with extra butter, eggs, and cream) than versions from other southern states. Cornbreads, biscuits, hominy and grits, black-eyed peas, okra, sweet potato pie, bread pudding, and pralines are a few common items. Rice is an important crop, and southern-style rice dishes are popular.

Other similarities in the WSC states are found. Cooking in parts of Oklahoma is similar to food in southwestern Arkansas, as seen in the greater use of flour instead of cornmeal. Although the Oklahoma territory was not officially opened up to settlement by whites until 1889, land-hungry homesteaders invaded the state before then (they were called Sooners). African Americans, many of whom had been held as slaves by the Indians, purchased land in the region after abolition. Most settlers established small family farms. The plains regions in the state are arid, and droughts occurred regularly. The fare in this region of Oklahoma derived more from scarcity than from ethnic and regional preferences of the settlers. Rabbit and turnip stew was flavored with flour-thickened gravy, while beef and wheat berries were the primary ingredients of Oklahoma Stew. Baking-soda biscuits were common, and black blizzard cake (a pound cake whose name refers to the frequent dust storms in the region) was a specialty.

In the Northeast and panhandle areas of Texas, settlers also scraped out a living on small family farms, surviving on corn, beans, and native game and fish. When wheat proved a successful crop in the region, cornbreads were replaced with biscuits. In the western areas of the state, beef has always been popular. It is served traditionally as stews and steaks. Chicken-fried steak was one specialty created to treat tough cuts—the steak is cut thinly, then pounded with a mallet, coated in flour, and fried. It is served with a ladleful of gravy made with coffee. Bread or tortillas and pinto beans often round out the meal. Today, Texas pasturelands are the leading producer of cattle, sheep, and lambs in the nation.

It is the differences in the cuisines of the WSC states that are most noteworthy. Unlike Arkansas, Louisiana was colonized by the French, who established several fortified settlements along the Gulf Coast, including Nouvelle-Orléans (New Orleans) in the 1700s. African slaves were brought in to work the plantations, and thousands of French Acadians from Canada and French Creoles from Haiti seeking refuge arrived.

Fish and seafood are more important than pork in the southern regions of Louisiana. The famous stews of the area, bouillabaisse, gumbo, and jambalaya, are examples of dishes made from coastal plenty. Shrimp is the primary seafood industry in Louisiana, marketed throughout the nation fresh and frozen. It is commonly served boiled with lemon butter or with *sauce piquante* (tomatoes, green peppers, onions, bay, vinegar, and hot sauce) over rice, a dish often called shrimp Creole. Shrimp is also added to stews and to stuffings for vegetables. Oysters are commonly served raw, on the half shell, and by the dozen in the many oyster bars of New Orleans. They are traditionally slurped with a squeeze of lemon juice and a dash of hot sauce or a sauce mixed to taste by each diner with catsup, vinegar, and horseradish. Oysters, too, are added to soups and stews.

Crawfish, which look like miniature lobsters, are found in the all the fresh waterways of the state. They have become the ethnic emblem of Cajuns and the regional symbol of southern Louisiana. Over 100 million pounds are produced annually. Some are harvested from the wild, but most are cultivated in approximately 300 crawfish farms. They are typically served at a crawfish boil, where they are cooked in water seasoned with cayenne, salt, and herbs. Potatoes or corn are often added for side dishes. The crawfish are placed in a gigantic mound in the center of the table, and each person takes and peels as many as desired. Only the meat in the tail and the claws is edible, along with the fat found in the head, which is extracted with a finger or sucked out appreciatively. Crawfish are also prepared fried, stuffed, as fritters, in soups and stews, in pies, and *étouffée* (meaning "smothered") in a spicy tomato sauce.

Other regional specialties include rice dishes, such as the fried cakes called *calas*, red beans and rice, and dirty rice (cooked with gizzards). Rice is also the foundation of dishes like gumbo and jambalaya. Baked goods and sweets are specialties, including French petits fours, crêpes, *beignets* (deep-fried squares similar to doughnuts), and pralines (pecan candies). Café au lait, a favorite beverage in New Orleans, is a dark-roasted coffee (sometimes flavored with chicory root) prepared with equal amounts of hot

Chuck wagon fare was a cooking style all its own, dependent on the skills and whims of the cowboy chefs called "Cookie" or "Miss Sally." Beans, cornbread, sourdough biscuits, and coffee were the staples, but some specialties were created on the trail, including "son-of-a-bitch stew" (known as "son-of-a-gun stew" in more genteel circles) made with beef organs, including tongue, brain, liver, heart, and kidneys.

The term "Creole" is often used to describe Europeans born outside Europe and is applied, for example, to the descendants of the original French and Spanish immigrants to New Orleans.

Louisiana Tourism

▲ *Crawfish is a specialty in southern Louisiana that has become popular in other areas of the South.*

milk. Café brûlot is a sweetened dessert coffee flavored with brandy and *curaçao* (orange liqueur).

Restaurant fare in New Orleans is renowned. Among the nationally recognized dishes created by local chefs are oysters Rockefeller (baked on a bed of salt with a rich spinach sauce), oysters Bienville (baked with a béchamel sauce and green pepper, onions, pimento, and cheese), Bananas Foster (sliced bananas cooked in butter, brown sugar, rum, and banana liqueur served over vanilla ice cream—it started out as a breakfast specialty), and Ramos gin fizz (a shaken or blended cocktail with cream, gin, lemon juice, orange flower water, and egg whites). Street food is equally tasty in the city. Fried oysters, sliced tomatoes, and onions with tarter sauce on a french bread roll are especially popular. They are called peacemakers, from the nineteenth century, when men would bring one home as a surprise for dinner after a fight with their wives. Po' boy (for "poor boy") is another name for the sandwich, although a po' boy may also refer to a sandwich with deli meats, sausages, and cheeses with or without gravy or tomatoes. A muffeletta sandwich is yet another version, usually including a chopped olive salad with the meats and cheeses on a whole round loaf of seeded Sicilian-style bread.

Sandwiches with deli meats and cheese on a french bread roll are found throughout the country. In addition to being called "po' boys" in New Orleans, they are also known as "bombers" (upstate New York), Cuban sandwiches (made with roast pork in Miami), "grinders" (New England), heros (New York City), "hoagies" (Philadelphia), Italian meat sandwiches (Chicago), and submarine sandwiches (from a World War II naval base in Connecticut).

Oklahoma started its U.S. history as Indian Territory, lands set aside in the 1820s for the Native American tribes that had been dispossessed of their homes in the Gulf Coast areas. Five major Indian groups lived in the region: Cherokee, Chickasaw, Choctaw, Creek, and Seminole. They were primarily agrarian, growing corn, beans, and squash. They gathered indigenous foods (such as acorns, chestnuts, creases, grapes, Jerusalem artichokes, hickory nuts, persimmons, ramp, and sorghum) and hunted small game (see Chapter 5). Today, Native Americans make up approximately 8 percent of the total population in the state, over eight times the national average. Traditional foods, such as a Cherokee soup made with hickory nut cream, called kanuche, and game dishes, are served mostly at ceremonial occasions. Fry bread and adapted dishes, such as scrambled eggs with spring onions, are more common but have not been accepted into the broader Oklahoma cuisine.[27] Other ethnic fare is available in Oklahoma but not widely consumed. Some Italian American foods are consumed, such as spaghetti and meatballs, particularly around Krebs in the southeastern region of the state. Sauerkraut, potato soup, and dark breads are evidence of German-Russian influence, and central European traditions are maintained at heritage festivals. A few Tex-Mex items such as chili con carne have become very popular.

Ethnic cuisine is much more evident in Texas. The state is the size of New England, the Mid-Atlantic states, Ohio, and Indiana combined. It was occupied by Native Americans, claimed by the Spanish and French, ruled by Mexico, and existed as an independent nation before it became part of the United States in 1846. Germans, Czechs, and Poles emigrated from Europe to central Texas, attracted by land grants. Sausages, ham, sauerbraten, sauerkraut, pumpernickel bread, potato salad, potato dumplings, bierocks, and strudel are popular in areas where the Germans and other central Europeans settled.

The most distinctive Texan fare evolved in the south of the state, where Mexican and Spanish influence added their flavors to dishes. Some authentic Mexican foods, such as tortillas, tamales, chalupas, salsas, guacamole, and buñuelos (see Chapter 9) were accepted by white settlers in the region. However, most foods in the area are adapted dishes with Mexican overtones, often referred to as "Tex-Mex cuisine." Examples include tamale pie, nachos, and most tacos and enchiladas, which usually feature nontraditional fillings. One regional specialty is chili con carne, known in Texas as "a bowl of red," which began as beans, progressed to beans with beef, and is now typically an all-beef stew flavored primarily with hot chili powder. Barbecue is also favored. Unlike barbecue in other regions of the country (e.g., Kansas City), there are two sauces involved in Texas barbecue. The first, called the mop, or sop, is used to marinate the meat before cooking and for basting the meat on the spit or grill. (The term *mop* for the sauce basted on barbecued meat may have come from the use of a clean mop to slap the sauce on whole carcasses.) The second sauce is served on the side with the cooked meat. Although barbecued beef is most associated with the state, barbecued goat kid (*cabrito*) is almost as popular in the southern sections. The unifying element in most of these foods is that they are preferred hot and spicy. In addition to chili powder, chili peppers are used in many dishes. Numerous varieties are added, but worth mention is the tepín chile, the indigenous precursor to domesticated pequin chiles. They are among the hottest of all chiles (also called bird's eye peppers, described previously). Chiles often flavor foods in Texas not normally associated with the spice, such as cornbread and jelly. In addition to chili peppers, other fresh fruits and vegetables are now prevalent in southern Texas due to irrigation. Cantaloupe, pink grapefruit, peaches, sugarcane, and tomatoes are a few examples of specialty crops.

Health Concerns

Health risk indicators in the South tend to be higher than in the rest of the nation (see chapters on each ethnic group for population-specific data).[13] Florida and West Virginia have average or above average rates in every health risk and mortality category—Alabama, Arkansas, Kentucky, Mississippi, and Oklahoma show a similar profile, with the exception of heavy drinking, which is below average. Rates of obesity are the highest in the nation with nine states having >30% prevalence of obesity.[14] (See Figure 15.3.) Lack of leisure time exercise, diabetes, low birth weight, and mortality rates are of concern in many states. Only the percentage of heavy drinking is substantially lower in some areas. Death rates in West Virginia are particularly high, greatly exceeding the national averages for heart disease (30 percent higher), stroke (20 percent higher), and cancer (25 percent higher). (See Table 15.10.)

© J. Griffis Smith/TxDOT

▲ ***Barbecue is a traditional Tex-Mex method for preparing food.***

The West

Regional Profile

The western United States is the largest region in the nation, encompassing an enormous diversity of lands, from the icy tundra of Alaska to the tropical volcanic islands of Hawaii. The

[SAMPLE MENU]

A Gulf Coast Supper

Deviled Crab[a,b]

Chicken and Sausage Gumbo[a,b,c]

Pecan Pralines[a,b,c] or Mississippi Mud Pie[a,b]

[a]Barker, A. 2003. *The best of Cajun and Creole cooking*. New York: Gramercy Books.

[b]Claiborne, C. 1987. *Craig Claiborne's Southern cooking*. New York: Times Books.

[c]*The Creole and Cajun Recipe* Page at www.gumbopages.com

Table 15.10 South State–Specific Health Data Compared to National Averages, 2001–2002

	AL	AR	DE	FL	GA	KY	LA	MD	MS	NC	OK	SC	TN	TX	VA	WV
Overweight[a]	↑	AVG	AVG	AVG	AVG	↑	AVG	AVG	↑	AVG	↑	AVG	AVG	AVG	AVG	↑
No Leisure-Time Exercise[b]	↑↑↑	↑↑↑	AVG	↑	↑	↑↑↑	↑↑↑	↓	↑↑↑	AVG	↑↑↑	AVG	↑↑↑	↑	↓	↑↑↑
Diabetes[c]	↑↑↑	↑↑	↑	↑↑↑	AVG	AVG	↑↑	AVG	↑↑↑	AVG	↑↑	↑↑↑	↑↑	↑	↓	↑↑↑
Hypertension[d]	↑↑↑	↑↑	↑	↑	↑	↑↑	↑	AVG	↑↑↑	↑	↑↑	↑↑	↑↑	AVG	AVG	↑↑↑
High Blood Cholesterol[e]	↑	AVG	AVG	AVG	↑	AVG	↓	AVG	AVG	AVG	AVG	↓	↓↓	AVG	AVG	↑↑↑
Don't Consume 5 Fruits/Vegs.[f]	AVG	AVG	AVG	AVG	AVG	↑	↑	↓	↑	AVG	↑	AVG	AVG	AVG	AVG	↑
Heavy Drinking[g]	↓	↓	↑↑↑	↑	↓↓↓	↓↓↓	↓↓	AVG	↓	↓↓	↓↓↓	↑	↓↓↓	↑	AVG	↓↓↓
Low Birth Weight[h]	↑↑↑	↑↑	↑↑	↑	↑↑↑	↑	↑↑↑	↑↑	↑↑↑	↑↑↑	AVG	↑↑↑	↑↑↑	AVG	↑	↑
Deaths from Heart Disease[i]	↑↑	↑↑↑	↑	↑↑↑	↓↓	↑↑	↑	↓	↑↑↑	↓	↑↑↑	AVG	↓↓	↓↓	↓↓	↑↑↑
Death from Stroke[j]	↑↑↑	↑↑↑	↓↓↓	↑↑	↓↓	↑↑↑	↑	↓	↑↑↑	↑↑↑	↑↑	↑↑↑	↓↓↓	↓↓	AVG	↑↑↑
Death from Cancer[k]	↑↑	↑↑	↑↑	↑↑↑	↓↓	↑↑	↑↑	AVG	↑↑	AVG	↑↑↑	AVG	↓↓	↓↓	AVG	↑↑↑

[a]U.S. prevalence = 58.4 percent (overweight defined as BMI >25.0)

[b]U.S. prevalence = 25.8 percent (persons who did no leisure-time physical activity in past month)

[c]U.S. prevalence = 6.6 percent (self-reported data based on number of persons who were told they had condition by a health professional)

[d]U.S. prevalence = 25.7 percent (self-reported data based on number of persons who were told they had condition by a health professional)

[e]U.S. prevalence = 30.4 percent (self-reported data based on number of persons who were told they had condition by a health professional)

[f]U.S. prevalence = 77.6 percent (adults who do not consume at least 5 fruits/vegetables per day)

[g]U.S. prevalence = 5.1 percent (>2 drinks/day in the past month for men, >1 drink per day in the past month for women)

[h]U.S. prevalence = 7.3 percent (live births of infants weighing <2,500 grams)

[i]U.S. age-adjusted death rate per 100,000 = 245.8

[j]U.S. age-adjusted death rate per 100,000 = 57.9

[k]U.S. age-adjusted death rate per 100,000 = 194.4

AVG – similar to national average

↑ – slightly above national average

↓ – slightly below national average

↑↑ – significantly above national average

↓↓ – significantly below national average

↑↑↑ – exceptionally above national average

↓↓↓ – exceptionally below national average

Sources: Ahluwalia, I.B., et al. 2003. State-specific prevalence of selected chronic disease-related characteristics—Behavioral Risk Factor Surveillance System, 2001. National Center for Health Statistics. 2004. *Chartbook on trends in health of Americans.* Hyattsville, MD: U.S. Government Printing Office. *MMWR Surveillance Summaries*, 52, (SS08); 1–80. Centers for Disease Control. 2003. *Profiling the leading causes of death in the United States: Heart disease, stroke, and cancer.* Available at http://www.cdc.gov/nccdphp/publications/factsheets/ChronicDisease/

tallest mountains in the country, vast fertile valleys and coastal plains, stretches of scenic desert, and temperate rainforest add to the variety. It is the history of the open wilderness that links this region. Indigenous peoples adapted their lifestyles to fit each climate and terrain. Pueblo Indians made their homes in the cliffs and cultivated corn, beans, chiles, and squash; the Inuits of Alaska lived in ice igloos and hunted sea mammals and fish for food; and the native Hawaiians enjoyed such fresh abundance that they cooked few dishes (see Chapter 5 and Chapter 12). The first whites in the West were explorers, trappers, miners, and traders—hardy individuals (mostly men) seeking their fortune. Emigrants came from every direction: the Spanish and Mexicans from Mexico in the South, Russians from the North, Chinese and Japanese from the West, and the numerous pioneers of northern and southern European descent (mostly English, Scottish, Welsh, Danes, Swedes, Slavs, Italians, and Greeks) from the Midwest, looking for new farming, ranching, and fishing opportunities. The West is the most diverse region not only in climate and terrain but also in population.

The West is divided into the Mountain states of Arizona, Colorado, Idaho, Montana, Nevada, New Mexico, Utah, and Wyoming and the Pacific states of Alaska, California, Hawaii, Oregon, and Washington. Approximately 23 percent of all Americans reside in the Western region,[11] and, of these, over half live in California. Large numbers of many ethnic groups reside in the West. Compared to total U.S. figures, five times as many Pacific Islanders and nearly twice the Asians, Latinos, and Native Americans make up approximately 23% of the western population. This includes disproportionate numbers of total Inuit, Japanese, Aleut, Filipinos, Salvadorans, Chinese, Vietnamese, Mexicans, American Indians, and Koreans. African Americans and whites fall below the U.S. average in the region. Among whites residing in western states, there are large numbers with Danish and Spanish ancestry and Yugoslavian heritage (including those from what are currently Croatia and Serbia). In 2009, > 32 percent of immigrants to the United States settled in the West, with approximately 26 percent in California (see Table 15.11).[28]

Furthermore, a few individual western states report notable ethnic population figures. Asians and Pacific Islanders account for over half the population in Hawaii, for example. Latinos are a large percentage of both the New Mexico and California populations. Arizona and New Mexico host large numbers of American Indians; Alaska Natives, including Inuits and Aleuts, who make up 15 percent of the population in Alaska. Though only 40 percent of persons living in the West adhere to a Christian faith, the highest percentage of Christians in the nation reside in Utah (80 percent—nearly all are Mormons) and large numbers are also found in New Mexico (58 percent—many are Roman Catholics).

Table 15.11 Estimates of Ethnicity in the Western Region

	United States		West Region	
	Estimate	Margin of Error	Estimate	Margin of Error
Total:	301,461,533	*****	69,768,366	*****
White alone	224,469,780	+/−45,816	48,825,886	+/−32,520
Black or African American alone	37,264,679	+/−22,810	3,284,637	+/−8,877
American Indian and Alaska Native alone	2,423,294	+/−12,407	1,194,901	+/−8,713
Asian alone	13,201,056	+/−16,395	6,095,368	+/−9,926
Native Hawaiian and Other Pacific Islander alone	447,591	+/−4,938	341,131	+/−3,729
Some other race alone	16,986,453	+/−46,059	7,479,116	+/−29,482
Two or more races:	6,668,680	+/−65,623	2,547,327	+/−29,105
Two races including some other race	1,351,590	+/−25,511	582,431	+/−12,310
Two races excluding some other race, and three or more races	5,317,090	+/−44,325	1,964,896	+/−20,983

The West has a higher than national average proportion of young people under age fifteen and a lower than average percentage of persons over age sixty-five.[11] Despite the vast open space of the region, over 86 percent of the population lives in metropolitan areas. Average household income according to American Community Survey data is approximately $51,187. Persons living in poverty are just about the national average at 13 percent.[11]

Traditional Fare

The West was largely unknown to whites before the nineteenth century. Adventurous trappers and traders made their way into the territory from the Great Plains, often surviving on dried bison meat. Miners who followed the gold and silver strikes in the California Mother Lode, Pike's Peak in Colorado, Montana's Grasshopper Creek, and the Alaska Klondike prepared their own meals, usually pork, beans, and hardtack (tough, dry, unleavened bread or biscuits) three times a day. Some were dependent on the way stations, hotels, and boarding houses that opened to support the rush.[29] Neighborly hospitality, so common in the Midwest and South, disappeared in the name of profit; miners were charged the maximum for supplies (e.g., eggs for a half dollar each, potatoes for a dollar a pound, and a box of apples for $500), and a meal would cost about three dollars in a local establishment. The farmers and ranchers who later made their way westward frequently consumed game with cornbreads and potatoes to complete the meal. Sourdough breads and biscuits were common with settlers in the Mountain states, California, and Alaska. Mashed potato or a milk and flour starter was left out to catch wild yeast and begin fermentation. Once going, the starter was kept indefinitely, replenished each time a little was used as leavening (see Table 15.12).

The growth of towns and the success of irrigation increased the food supply. Expensive goods such as wines and chocolates became available, and eastern specialties, including Long Island duck and Smithfield ham, were offered at restaurants. Depending on the region, potatoes, corn, apples, wheat, and hops prospered; cattle, dairy cows, and sheep became plentiful. In Alaska and Hawaii, white settlers faced different challenges. With the arrival of experienced fishermen, more of the Pacific coast seafood was utilized. Salmon, crab, oysters, and clams were especially popular.

Immigrants from other countries came to the West in search of mining and railroad jobs, including the Chinese and Mexicans. Both of these groups enjoyed highly seasoned foods and promoted the use of chile peppers. Other groups, such as the Italians, Japanese, and some Greeks, became involved in fishing and introduced specialties such as seafood cioppino (a seafood stew using local fish and shellfish) and teriyaki. Many immigrants opened restaurants and markets to serve the needs of the booming towns. German sausages, Italian cannoli, and Chinese stir-fried dishes were all available. Still other immigrants arrived looking for farmland; they planted the fertile Pacific coastal regions and the California Central Valley with temperate fruits and vegetables such as apples, pears, dates, grapes, plums, prunes, cherries, artichokes, avocados, broccoli, brussels sprouts, lemons, grapefruit, and oranges.

The Mountain States

Cuisine in the Mountain states varies considerably between the North and South. Cooking in Idaho, Montana, Utah, and Wyoming was influenced by the American and European settlers and features the foods available in the cooler climates of the northern ranges and plains. Meats are a specialty. The fare in the southwestern states of Arizona and New Mexico is shaped by the limitations of the desert and the significant Native American, Spanish, and Mexican presence in the area. The foods of Colorado and Nevada are mostly northern states in their cooking, with some southern state influences.

Bighorn sheep, deer, pronghorn antelope, elk, moose, javelina (wild pig), bear, and bison were prevalent in many parts of the North. Recreational hunting is popular in the region, and game meats are favorites. Venison with huckleberry sauce is a specialty in Idaho, and in Montana it is prepared roasted, or as chili con carne, or into meatballs in a spicy tomato sauce. Both venison and antelope are favorites in Wyoming. Tenderloin, sirloin, and T-bone steaks are cut; the ribs and sirloin tips make roasts (sometimes marinated and braised in wine, vinegar, and spices); the brisket, flank, and plate are used for stews or hamburgers; the hams are smoked; and miscellaneous meat is used to make Polish sausage or

Table 15.12 Western Specialties

Group	Foods	Preparations
Protein Foods		
Milk/milk products	Milk, cheese (Cheddars such as Cougar Gold, Monterey Jack, Tillamook); Basque sheep's milk cheeses	
Meat/poultry/fish/eggs/legumes	Native game, including buffalo, deer, elk, moose, antelope, mountain sheep, mountain goats, bear, javelina (wild pig), beaver, rabbit	Game meat steaks, roasts, stews (such as chili con carne), hamburger, sausages; beaver tail
	Beef, mutton and lamb, pork	Steaks; beef *enchiladas, tamales, chimichangas, pirozhki; teriyaki*; Indian tacos; *Pueblo pozole*; lamb spit-roasted or roasted with chiles; *chorizo*; luau (pit-cooked) pork
	Clams (e.g., geoducks), crab (Dungeness, king, snow), oysters, shrimp, squid Salmon, tuna, halibut, mackerel, sardines, anchovies, mahi mahi, bonito, marlin, snapper; freshwater fish, particularly trout	Clam chowder, Seattle clam hash; *cioppino*, steamed crab, crab cocktails, fried *calamari*; grilled or poached salmon, *lobimuhenno's* (salmon chowder); *sushi, sashimi, teriyaki*; trout grilled with bacon
	Chicken eggs	Hangtown fry
	Dried beans	Chickpeas with lamb, chickpea pudding; Basque beans; lentil soup with lamb, lentil and sausage casserole, white beans cooked with pimento and cheese; split-pea soup
Cereals/Grains	Wheat, corn	Sourdough breads, biscuits, pancakes; *sopapillas*; fry bread; *panocha; capriotada*; whole-wheat Mormon bread; *bara brith; malasadas*; Hawaiian bread; tortillas (corn or wheat); *piki*; Asian noodle dishes (e.g., *saimin*) and dough-wrapped foods (egg/spring rolls, *lumpia*, wonton); fortune cookies
Fruits/Vegetables	Apples, apricots, wild and cultivated berries, cactus fruit, cherries, dates, figs, grapes, kiwifruit, lemons, oranges, peaches, pears, pineapple, plums, prunes, sugarcane	Fresh fruit desserts; fruit added to roasts or poultry stuffings; preserves, jellies, wines; cold fruit soups
	Artichokes, avocados, asparagus, broccoli, breadfruit, cauliflower, chile peppers, eggplant, *jicama, nopales*, olives, onions, specialty lettuces (arugala, radicchio, rocket), tomatoes, *tomatillos*, potatoes, taro root, zucchini	Fresh vegetable side dishes; mesclun salads; guacamole; Basque potatoes; squash patties; *poi*
Additional Foods		
Seasonings	Chile peppers (especially New Mexico/Anaheim, Jalapeño/Chipotle, Serrano); cinnamon, cilantro, epazote, cumin, garlic, oregano, mint, safflowers (dried petals), *yerba buena*; chocolate; vanilla	Fresh chiles, dried chile powders, smoked chiles, pickled chiles; salsas; red or green chile sauces; mole sauce; fresh, dried, powdered, roasted, pickled garlic
Nuts/seeds	Almonds, hazelnuts, macadamia nuts, pine nuts (piñon seeds), pumpkin seeds	
Beverages	Varietal wines; coffee; tea (chamomile, Brigham Young); hot chocolate	Coffee drinks (lattes, etc.); *picón* punch
Fats/oils	Olive oil	
Sweeteners	Sugar from beets, cane	Sugarcane is eaten fresh in Hawaii

salami. In Nevada deer are ranched for consumption. Numerous game birds are found as well, including geese, ducks, pheasant, partridge, grouse, and wild turkeys. Pheasant roasted with apples or in a pie is a favorite in Wyoming. Fish, such as sockeye salmon, bass, and catfish, is available in thousands of freshwater lakes and streams. Mountain trout is a regional specialty.

Cattle and sheep ranching are the dominant agricultural activities of the Mountain states.

Colorado is the leading producer in the nation of lamb, known as Rocky Mountain lamb. In addition, bison is raised and processed as a specialty meat in the region. Pork is farmed in Montana, and poultry, especially turkey and eggs, is produced in Utah.

Colorado is the highest state in the nation, with an average elevation of nearly 6,800 feet. Pioneer cooks had difficulties making baked goods at high altitudes. Extensive experimentation found that leavening must be reduced and oven temperatures raised to produce satisfactory breads and cakes.

Legend is that the famous Denver (or Western) sandwich, made with an omelet containing ham, onions, and green peppers between slices of bread, was invented by a Chinese chef making eggs *foo yung* with available ingredients.[23]

Basques from Nevada may have introduced sourdough bread to San Francisco, where it has become a signature item.

Forage is grown to support meat production in the region, though wheat, oats, barley, sugar beets, hops, lentils, beans, cherries, and apples are cash crops in some areas. Potatoes are synonymous with Idaho. They are grown primarily in the volcanic soils of the Snake River plain, where ten billion pounds are harvested annually, approximately one-third of national production. Sixty percent are frozen, dehydrated, or milled into flour. Peppermint and spearmint (grown for their oils used in flavorings) are specialty crops in the state. Native berries are a regional favorite, especially in Montana, including huckleberries, which are made into breads and pies, and sour-tasting chokecherries, which are used in pies, cakes, preserves, jellies, and wine.

Settlers of the region often brought their favorite foods. For example, Wyoming attracted a diversity of immigrants, many of whom opened bakeries, confectionery stores, and restaurants. French croissants, Middle Eastern halvah (sesame seed candy), German schnitzel, and Chinese wonton soup were reportedly available in southeastern Wyoming as early as 1900.[4] In Montana Scandinavians who arrived from Minnesota to work in lumbering brought yellow split-pea soup, cold fruit soups, Swedish meatballs, and ham with cherry sauce. Borscht, cheese-filled pastry shells (vatroushki), and cherry desserts were favored by the Russians. The Scots made oatmeal porridge and Mulligatawny stew with mutton, and central Europeans brought stuffed cabbage, dumplings filled with fruit or cheese, and pancakes rolled around cherries. In Idaho lobimuhenno's (a salmon chowder) was brought by the Finns; and bara brith, a bread studded with currants, was favored by the Welsh. Many dishes in Idaho today feature local ingredients with European nuances, such as split-pea soup, lentil soup with lamb, white beans cooked with pimento and cheese, and ham with apple casserole. Apple jelly is added to mayonnaise to make salad dressing. Prunes are used in preserves and desserts, such as prune-whip pie and prune pudding.[4] In Colorado, chili peppers and other spices came with the Mexicans, and dishes such as chicken or turkey cooked in mole sauce (a rich blend of spices, nuts, and unsweetened chocolate) became popular throughout the state.

One of the most notable ethnic groups in the northern Mountain states is the Basques, who settled in Idaho and Nevada, working first as shepherds and later as land-owning sheep ranchers. Initially only men came, later bringing their families. Women did most of the cooking, providing meals for the ranchers such as biscuits with sheep's milk cheese and coffee for breakfast, and Basque beans (pinto beans with lamb or pork), lamb stews, or Spanish-style potato omelets for the main meal. They introduced sourdough bread and a pencil-thin version of the spicy Spanish sausage chorizo. Basque potatoes" still a favorite in the region, is a sliced potato, onion, and bacon casserole. Basques often established hotels, which served as meeting places for Basques doing business in the area and for new immigrants. The hotels became famous for their four- or five-course meals served family-style, and non-Basque visitors often came for the food. Chickpea and meat stews, spit-roasted lamb, and even traditional seafood dishes such as bacalao al pil-pil (dried salt cod cooked with garlic and olive oil) were offered when the ingredients were available. Red wine, chamomile tea, and picón punch (a beverage no longer common in Spain but still available where Basques live in the United States, made from bitter orange picón liqueur, brandy, grenadine, and soda water) were popular drinks.

In Utah, another group that has maintained many of their food traditions is the Roman Catholic Italians who originally came from Calabria for mining and railroad jobs in the late nineteenth century. They often grow Mediterranean vegetables and seasonings, such as eggplants, tomatoes, endive, fava beans, fennel, zucchini, garlic, parsley, and basil, in home gardens. Everyday fare includes bread sticks, pastas, minestras (see Chapter 6), salads, and fresh fruits for dessert. Specialties such as goat meats and goat cheeses, variety meats cooked up with eggs in a frittata or in a spicy stew, boiled chicken's feet, and deep-fried squash patties made with chopped squash and squash blossoms are traditional favorites. Outdoor baking ovens used by the first settlers are still found in the region.[29,30]

Utah is also home to a large population of Mormons (nearly 80 percent of all residents in the state), who settled there in the early 1800s in order to escape the persecution suffered in Ohio, Illinois, Missouri, and other areas where the members of the Church of Jesus Christ of Latter-Day Saints had lived. Many were of northern European descent, particularly British and Scandinavians, and they brought a preference for hearty foods. Ham, pot roast, roast beef, stews, and fried chicken remain favorite entrees, often served with homemade whole-wheat bread or buttermilk biscuits. Spicing is usually mild. Hamburger bean goulash, a Utah specialty, is thought to be a denatured version of chili con carne.[4] Milk gravy, made with browned flour, pork drippings, milk, and seasoned with black pepper, was served with so many foods that cowboys riding through the area dubbed it "Mormon dip." Potatoes, red cabbage, green beans, and peas are still common side dishes. The Mormons are well known for their love of sweets, and desserts are prominent in the diet. Layer cakes, fruit pies, strawberry shortcake, fruit candies, chocolates, and ice cream are still commonly consumed. Sour cream raisin pie is a popular Utah dessert that recalls the sweetened milk custards and dried fruits of early pioneer days. Another notable favorite is pepparkakor, a Scandinavian ginger cookie often Anglicized as "pepper cookies." Mormons do not drink alcohol or stimulating beverages such as coffee or tea; lemonade and Brigham Young tea (sweetened hot water with milk) are traditional beverages.

In the southwestern regions of the Mountain states, a warm climate conducive to agriculture is combined with insufficient rain to grow most crops without irrigation. Some of the most scenic terrain in the nation, from the majestic Grand Canyon to the broad plains of the Sonoran Desert, is found in Arizona. New Mexico is a mix of high desert plateaus, portions of the Great Plains, and sections of the Rocky Mountains. Southwestern fare reflects the arid conditions and the preferences of the people who settled the region, especially the Native Americans, the Spanish, and the Mexicans.

Native Americans who lived in the region prior to European contact, particularly Pueblo peoples, cultivated small amounts of corn, beans, chili peppers, squash, and pumpkins. Pine nuts (piñon seeds) were indigenous foods. Juicy fruits (called tunas, or Indian fig") and the young pads (nopales) of the saguro and prickly pear cactus were other native items. Small game, such as rabbit, provided meat in the diet. They prepared stews flavored with chiles and very thin, blue corn tortillas called piki. Other tribes, such as the Navajo and Apache, were initially hunters

© Stockbyte/Picture Quest/Jupiter Images

Meat stews flavored with chile peppers are popular in New Mexico.

[SAMPLE MENU]

A Southwestern Supper

Posole/Pozole[a,b,c]

Fresh Flour Tortillas

Jicama Salad[b] or Nopalitos Salad[a]

Biscochitos[a,b,c] or Flan[c]

[a]Cox, B., & Jacobs, M. 1996. *Spirit of the West: Cooking from ranch house to range*. New York: Artisan.

[b]Fussell, B. 1997. *I hear America cooking*. New York: Penguin Books.

[c]*Cocinas de New Mexico* at www.vivanewmexico.com

who roamed the region in search of big game and wild plants (see Chapter 5).

When the Spanish arrived in the sixteenth century, they introduced many of the foods that became elemental in the cooking of the Indians in both the United States and Mexico, including wheat, hogs, sheep, and cattle, as well as chocolate and other items obtained through trade in the Americas. The Spanish explorers were followed by Mexican-born settlers who established small farms called haciendas, ranchos, and estancias beginning in the late 1500s along the banks of the Rio Grande in New Mexico and in southern Arizona. Anglo Americans (those who spoke English) followed, first to New Mexico and then into Arizona, adding to the population of Spanish Americans (mostly of Mexican and Native American ancestry who spoke Spanish) and Native Americans. The ethnic heritage of the region is evident in the many Native American celebrations, adobe villages, and numerous examples of Spanish architecture that still exist.

Nowhere is the blending of Native American, Spanish American, and Anglo American culture more evident than in southwestern fare. Wheat tortillas became a staple in parts of the American Southwest and in northern Mexico. Hearty beef or lamb stews replaced the mostly vegetarian versions, and fry bread, made with wheat flour and cooked in lard, became common. Popular dishes in Arizona and New Mexico with Native American and Spanish or Mexican roots include menudo (spicy tripe soup flavored with mint), pozole (hominy flavored with pork, chili peppers, and often epazote or oregano), chimichangas (wheat tortillas wrapped around a beef or chicken and vegetable filling, then deep-fried), green chili stew (beef, tomatoes, onions, and a variety of green chili peppers), chickpeas with lamb, and Indian tacos (fry bread folded and filled with meats, cheeses, vegetables, and salsa). Indian roasted lamb with red and green chili peppers, Spanish arroz con pollo (updated with chiles), and Mexican chilaquiles, enchiladas, tamales, quesadillas, and flautas are other common dishes in the region. Chili peppers, onions, garlic, oregano, *yerba buena*, *epazote*, safflower blossoms (dried and powdered, reminiscent of saffron), and abundant amounts of mint flavor savory dishes. Toasted pumpkin seeds or pine nuts from the native piñon tree were added to spice blends for sauces, providing a nutty flavor to roasted meats and poultry. Today almonds are used in the same way.

Sweets depend on chocolate, vanilla, cinnamon, and other spices. Popular desserts include anise cookies called biscochitos (a Christmas specialty), the light, deep-fried wheat flour puffs called sopaipillas served with honey or cinnamon syrup, and flan. Puddings are also common, including panocha (similar to Indian pudding from the Northeast, but made with wheat kernels and flavored with cinnamon), and capriotada (bread pudding with pine nuts, raisins, and mild cheese, traditionally served with a caramel sauce).

Agriculture in Arizona has expanded with irrigation. Grapefruit, lemons, melons, and figs flourish in the region. Two vegetables popular in Mexico have found success in the state as well. Tomatillos, a relative of the ground cherry that looks a little like a green tomato in a papery husk, and jicama, a crispy tuber related to the morning glory, have both been transplanted from Mexico to Arizona. Tomatillos are added to salsa verde and other green sauces, and the sweet pea–flavored jicama frequently provides a crunch in local salads. New Mexico is the leading producer of chili peppers in the United States. There is no consensus on the common names for the hundreds of varieties, nor is there a single heat classification system. Mild New Mexican (also called Anaheim) chiles, medium hot ancho chile (the name for dried, ground poblano chiles), hotter jalapeños (sliced, pickled jalapeños are sprinkled on nachos; smoked jalapeños are called chipotles), and very hot serrano chiles are the most popular in southwestern cuisine.

The Pacific States

No region in the United States is as diverse as the Pacific states. Significant variations in climate, terrain, and settlement history have led to the development of very different cuisines in the three coastal states of California, Oregon, and Washington and that of the two states separated from the continental nation: Alaska and Hawaii.

The climate of the Pacific coast states ranges from cool and moist in the northern areas near the Canadian border to hot and dry in the southern deserts abutting Mexico. Parallel mountain ranges run north-south through the region, dividing the states into different agricultural zones. The foods of the Native Americans depended on their location. Those near the Pacific survived on clams, mussels, and fish, with local berries and greens. In inland areas of California, acorns were the foundation of the diet, leached of tannins and processed as a meal or flour. Western Oregon and Washington are endowed with a wealth of native foods, including deer, elk, antelope, rabbits, beaver, muskrats, ducks, geese, greens, wild mushrooms, and a multitude of berries. In the eastern sections of the states, however, high prairie-like plateaus and near-desert conditions exist (see Chapter 5).

The first whites to settle in California were the Spanish in the eighteenth century. They built several presidios (forts), pueblos (small farms), and a series of missions (each one day apart in travel time), with the purpose of protecting their claim to the territory. The Spanish cultivated numerous crops, including wheat, olives, grapes, and oranges in the lands surrounding the missions, and planted native foods including corn, beans, and tomatoes as well. Cattle and hogs were raised. Many of the local Indians were forced into the missions as laborers. When Mexico took control of the state in the early nineteenth century, much of the mission territory was redistributed as grants to resident families who founded wealthy rancheros producing mostly beef. Despite Spanish claims to the contrary, many other European powers sought access to California riches. Russia, in particular, established colonies in northern California to support its fur trade along the Pacific coast and in Alaska. Trappers were the first whites to explore Oregon and Washington. Significant settlement from other states did not occur in the region until the discovery of gold in California during the 1840s and the opening of the Oregon Trail to pioneers from New England, the Midwest, and the South, who came to make their fortunes in the new frontier of the West. Regardless of ancestry, most early settlers adapted their cuisines to the local natural pantry.

Coastal seafood was a mainstay for many settlers. In Oregon items familiar in Atlantic fare, particularly clams, were favored. Over two dozen varieties (e.g., butter, horse, Japanese littleneck, Manila, razor) live in the sands of the Pacific Northwest. Clam chowder, adapted with whatever local settlers could rustle up, including rice, tomatoes, and cabbage, was common.[31] Geoduck (pronounced "gooey-duck") clams are a large, bivalve native to Oregon and Washington with necks (siphons) that can stretch several feet. A geoduck may weigh up to fifteen pounds, although most are in the two- to three-pound range. The body is sliced and pounded into thin steaks (similar to abalone), and the neck is usually minced or ground for soups or stews. Washington also has an abundance of seafood. Beginning in the 1800s the five species of salmon in the region (king or chinook, sockeye, dog, humpback, and silver) were sent fresh to West Coast markets or pickled and shipped in barrels to Hawaii, South America, the East Coast, and Europe. Iced fish was also sent to Europe, where it was smoked, then returned to the United States. Washington still leads the nation in total salmon catch and products. Oysters are another specialty. The small native Olympia oyster is a favorite, but overharvesting has greatly reduced availability. Pacific oysters, which were imported from Japan in the early 1900s, are the species consumed most often on the West Coast, though European flat oysters are also harvested. Oysters are typically served on the half shell in the region, although bacon-wrapped broiled oysters, oyster fritters, barbecued oysters, oyster loaf, and oyster stew are common cooking preparations. Dungeness crab, named for the town of Dungeness on the Olympic Peninsula, are caught during the winter months. In California, sardines were the leading catch during the early twentieth century. As their numbers diminished due to overfishing, other fish, such as tuna, salmon, halibut, mackerel, and anchovies, became the predominant catch. Dungeness crab, traditionally eaten steamed with melted butter and lemon or in crab cocktails, and squid, served deep-fried or over pasta, are popular. Abalone, a large, flat-shelled mollusk that clings to rocks off the California coast, is a specialty. The tough muscle must be pounded into a thin steak to tenderize it before cooking. It has a delicate, sweet flavor and is typically lightly floured and sautéed in butter.

Many early settlers in the Pacific coastal states started small farms, which over the years have grown into a significant industry. California produces more than half the fruits and vegetables consumed nationally and accounts for nearly the entire U.S. production of avocados, artichokes, garlic (some of which are processed as granules and powder), walnuts, almonds, apricots, nectarines, olives, dates, figs, pomegranates, prunes, and persimmons. It leads the nation in growing numerous other crops as well, such as lettuce, broccoli, grapes, lemons, strawberries, and melons. Furthermore, 90 percent of all raisins consumed in the world are produced in California. Specialty fruits, such as kiwifruit and *feijoa* (a small, green, egg-shaped fruit native to Brazil that tastes a little like pineapple and eucalyptus with mint nuances), are now being grown. In Oregon, fruits such as pears (mostly Bartlett, with some Anjou, Bosc, Comice, Seckel, and Winter), apples, prunes, plums, cherries, and domestic berries, have proved very successful. An Oregon specialty is hazelnuts (also called filberts). The trees were introduced from France in the 1800s, and today nearly all hazelnuts consumed in the United States are grown in the state. Washington leads the nation in the production of both apples and cherries. Apples are especially associated with the state: Grown on the eastern slopes of the Cascade mountains, they require extensive hand labor to thin each cluster of blossoms to a single king blossom and to pick any fruit that appears after the initial crop sets. This process produces exceptionally large, well-formed apples. Nearly the entire crop is devoted to Red Delicious and Golden Delicious varieties, although several other types are grown in small amounts, such as Fuji, Gala, Granny Smith, and Winesap. Hops, Walla Walla onions (a cool weather sweet onion that is a different variety from the sweet onions grown in Georgia and Hawaii), mint, and spearmint are specialty crops.

▶ *California produces more than half of the fruits and vegetables consumed in the nation and takes pride in a cuisine based on fresh, local ingredients.*

© Paul Barton/CORBIS

Dairying is significant in the Pacific states, with California first in milk production nationwide. Best known are the cheeses of the region, including several Cheddar styles such as Cougar Gold from Washington, Tillamook from Oregon, and Monterey Jack (a mild, white cheese) from California. More recently, French-style *chèvre* goat cheeses and blue cheeses have become specialties.

The abundance of seafood, dairy products, and fruits and vegetables in the region has led to the creation of cuisines that emphasize what is fresh and local. In California, celebrity chefs, such as Wolfgang Puck and Alice Waters, have popularized dishes such as pasta with chanterelles and grilled duck breasts, goat-cheese salad with arugula and radicchio, poached salmon with fresh basil and olive butters, and pears and figs poached in Zinfandel with Cassis cream (by Alice Waters, founder of the restaurant Chez Panisse in Berkeley)—all examples of the California approach to cooking. Other trendy items started in California include mesclun salads (made with a mixture of baby lettuces) and roasted garlic, which can be added to salads and stews or spread on bread like butter. Grilled fish, especially salmon and halibut, is a specialty in all three states. In Oregon, abundant use of fruit is seen in both savory and sweet dishes, such as fruit soups, poultry stuffed with prunes or apricots, and salads with fresh berries or dried fruit, as well as various fruit pies, soups, preserves, jams, and jellies.

Few distinctively ethnic flavors are found in the cooking of Washington, and limited influence from the settlers of Oregon is seen in German items (schnitzels, sauerbraten, sage sausages, stuffed cabbage, sauerkraut, and strudels) and southern-style fried chicken served with biscuits and hominy. In contrast, thousands of Chinese, Italian, and Japanese immigrants came to California in the early twentieth century, lured by jobs in farming, ranching, fishing, fish processing, meat packing, and the canning of fruits and vegetables. Agriculture continues to draw immigrants from Mexico and Central America looking for migrant farm work. Other recent immigrants adding to the diversity of the state include Vietnamese, Cambodians, Laotians (particularly Hmong), Koreans, Asian Indians, Ethiopians, Filipinos, and Samoans. Many California specialties are attributable to the ethnic preferences of the population. The Mexicans brought corn tortillas, refried beans, guacamole, and popular filled dishes like tacos and enchiladas. Italians introduced northern Italian favorites, such as polenta and pesto, as well as seafood dishes, such as the tomato-based fish stew made with local fish and Dungeness crab called cioppino, as well as fried calamari (squid). The Chinese offered authentic stir-fried dishes, wontons, eg g rolls, and adapted dishes such as chop suey and fortune cookies. The Japanese added sukiyaki, teriyaki, tempura, sushi, and other favorites to the mix. Armenians in the Central Valley brought flavorful lamb skewers, dolma (rice- and lamb-stuffed eggplant or grape leaves), and specialty desserts such as baklava. Newer immigrants have popularized hot Thai dishes, Vietnamese pho, Filipino lumpia, and Indian tandoori cooking and flat breads in the state (see the chapter specific to each ethnic group for more details).

California, Oregon, and Washington are famous for their wines. The European settlers of the mid-nineteenth century first introduced the superior varietal grapes used in French and German wine making to California. Today there are many premier wine regions in the state, including Napa Valley, Sonoma, Santa Cruz, Monterey, San Luis Obispo, and Santa Barbara. The Central Valley accounts for most of the grapes cultivated for the bulk wine market. Successful varietals include Cabernet Sauvignon, Merlot, Syrah, and Zinfandel among red wines and Chardonnay and Sauvignon Blanc among whites. Sparkling wines, which may be made from red or white grapes, and dessert wines are also specialties of the state. Oregon wines have gained a reputation for quality, especially for cool weather varietals such as Pinot Gris (a white wine) and Pinot Noir (a red grape used in French-style burgundies). Washington is best known for its white wines, such as Chenin Blanc, Gewürtztraminer, and White Riesling. In more recent years, fruit wines, such as those made from blackberries, currants, cranberries, or peaches, have become common, especially in Oregon.

The cooking of Alaska has been hampered in its development by the limited variety of foods. The climate and terrain are by no means uniform throughout the state, with warmer regions found along the coastal panhandle extending south into Canada, around the Yukon River Delta, and the protected Matanuska Valley in the south-central region, but there is little land suited for agriculture. The Indians, Inuits, and Aleuts of the region lived primarily on seafood and game. Wild berries and roots were harvested in the short summer months (see Chapter 5). The first permanent white settlement in Alaska was on Kodiak Island. The Russians arrived in the 1700s to hunt fur seals. They brought kasha, a cooked buckwheat porridge, buckwheat blini, and soups of fish or cabbage. Pirogs, large filled pastries, were made with fish, game, or cabbage. At Easter, they prepared traditional dishes such as the rich fruitcake called kulich (see Chapter 7). A rush of prospectors searching for gold arrived in the Klondike area in the 1800s. Supplies were severely limited, and the new settlers lived on little more than flour, bacon, salt pork, lard, and a bit of coffee or tea. Most kept a sourdough culture going to make breads and biscuits. Sourdough

E. & J. Gallo, headquartered in Modesto, California, is the largest wine producer in the world. It controls approximately 30 percent of the national market and over 50 percent of exported U.S. wines.

Coffee drinks such as lattes have become a specialty item associated with Seattle. The city is home to the Starbucks Corporation, which started the trend of U.S. coffee bars (modeled after Italian espresso bars) in 1984. There were almost 8,000 Starbucks coffee bars nationwide in 2006.

▼ ***Salmon and other seafood are favorites in Alaska and the Pacific Northwest.***

© Buddy Mays/CORBIS

specialties included poppy seed potato bread, rye bread with caraway seeds, and whole-wheat bread. Kelp was collected at the coast, and wild flowers were boiled to make syrup. Some miners hunted to supplement their diets, and game, such as deer, caribou, moose, Dall sheep, rabbits, and ptarmigan, was available. Game meats are still popular today, including steaks, roasts, and hamburgers made from moose meat, and caribou meat sausages and Swiss steaks.

Sheep are now raised on the Aleutian Islands, providing lamb for the mainland. Cattle ranches are found on Kodiak Island and in the Delta and Matanuska. Reindeer herds were imported to Alaska in the late 1800s from Siberia and Scandinavia in the hopes they would become a profitable meat source. Many starved to death when the ranges were overgrazed, and others are thought to have become part of caribou herds. Today, some reindeer in the Seward Peninsula are raised for meat and for supplying antlers to Asia. More important are the dairy operations providing fresh milk, butter, and cream. Potatoes are the most successful crop, but vegetables such as cabbage, cauliflower, and rhubarb also grow in the region. Many of the vegetables attain gigantic proportions in the long daylight hours of summer; for example, cabbages may reach seventy pounds, and rhubarb sometimes grows four feet tall.

Seafood is the main commodity in Alaska, ranking first nationally in quantity and value of the yearly catch. Salmon, herring, and halibut are the most prevalent fish. Shrimp and crab, including Dungeness, Snow (also known as Tanner), and limited King crabs, are trapped during the winter months. Most are frozen or canned for export to the rest of the United States and Japan.

In Hawaii, a mild tropical climate and abundant natural food resources greeted the earliest inhabitants. The volcanic islands are believed to have been first settled by Polynesians from the Marquesa Islands and Tahiti in the fifth century. The foundation of their diet was starchy vegetables such as taro root (traditionally made into poi—boiled, pounded into a paste, and slightly fermented), breadfruit, plantains, cassava, and yams. Seafood and possibly pork and chicken were also eaten (see Chapter 12). British explorers discovered the islands in the late 1700s; the area became a major American port for whaling ships in the nineteenth century, and Japanese, Chinese, Korean, Filipino, and Asian Indian agricultural workers came to support the developing pineapple and sugarcane industries.

Traditional native dishes and foods introduced to the islands by the many immigrants coalesced into Hawaiian fare. Unlike some areas of the nation, where various foreign contributions have melded into a single cuisine with occasional European, Latino, or Asian overtones, many dishes in Hawaii maintain their ethnic integrity. Foods from different cultures are commonly served at the same meal, however, representing the state's diverse heritage. Popular Hawaiian foods include those with Japanese origins, such as teriyaki-grilled meats and fish, sashimi (raw, thinly sliced fish), and noodle dishes such as saimin, an island adaptation of ramen noodles, topped with pork and frequently eaten for lunch. The Chinese brought wok cooking, dim sum, long-grain rice, soybeans, bok choy, lotus root, kumquats, litchi, and ginger to the region. Scottish scones and shortbreads are available, and Portuguese sweet bread is so common it is often called Hawaiian bread. Another Portuguese specialty, malassadas (fried doughnuts without a hole), is especially popular, and local variations made with poi or macadamia nuts are novelty items. Filipino fish sauces and lumpia (Filipino-style egg rolls), Korean kimchi (hot cabbage relish) and spicy beef dishes, and Indian curries are other contributions.

Historically sugarcane is the most important crop in Hawaii, accounting for 20 percent of the raw sugar produced nationally. But soaring costs and environmental concerns have put most sugarcane plantations out of business. Pineapples have also been a significant commodity, though pressures from Asia are reducing their profitability. Hawaii grows most of the world's supply of macadamia nuts and also exports famous kona coffee, grown on the western slopes of the island of Hawaii. Cattle ranches are found on the islands of Maui, Hawaii, and the privately owned Niihau; most provide beef for local consumption. Seafood, another Hawaiian specialty, is also mostly fished for Hawaiian markets, although some tuna is canned and exported. In addition to tuna, common food fish include mahi mahi (also called dorado), bonito, mackerel, and snapper.

Health Concerns

In general, people living in the West are healthier than the national average (see chapters on each ethnic group for population-specific data).[13] Idaho, New Mexico, Utah, and Washington all boast average or below U.S. average figures in every health risk and mortality category. Certain specific state data are noteworthy. High rates of heavy drinking are found in Arizona, California, and Nevada. Death from stroke is nearly 25 percent above U.S. rates in Oregon. (See Table 15.13.) Rates of obesity range from the lowest in Colorado at 18.6 percent to the highest among the western region in New Mexico, California, and Alaska.[32] (See Figure 15.3).

Table 15.13 Western State–Specific Health Data Compared to National Averages, 2001–2002

	AK	AZ	CA	CO	HI	ID	MT	NV	NM	OR	UT	WA	WY
Overweight[a]	↓	AVG	AVG	↓↓	↓↓	AVG	AVG	AVG	AVG	AVG	↓	AVG	AVG
No Leisure-Time Exercise[b]	↓↓	↓↓	AVG	↓↓↓	↓↓↓	↓↓	↓↓	↓↓	AVG	↓↓	↓↓↓	↓↓↓	↓↓
Diabetes[c]	↓↓↓	↓	AVG	↓↓↓	↓	↓↓	↓↓	↓	↓	↓↓	↓↓↓	↓↓	↓↓↓
Hypertension[d]	↓↓	↓	↓	↓↓	↓	AVG	AVG	AVG	↓↓↓	AVG	↓↓	↓	↓↓
High Blood Cholesterol[e]	↓	AVG	AVG	AVG	↓↓	AVG	AVG	↑↑↑	↓↓	↑	AVG	AVG	AVG
Don't Consume 5 Fruits/Vegs.[f]	AVG	AVG	AVG	AVG	↓	AVG	AVG	AVG	AVG	AVG	AVG	AVG	AVG
Heavy Drinking[g]	↑↑	↑↑↑	↑↑↑	↑	AVG	↓↓	↓	↑↑↑	AVG	↑↑	↓↓↓	AVG	AVG
Low Birth Weight[h]	↓↓↓	↓	↓↓	↑↑	↑	↓↓↓	↓↓	AVG	AVG	↓↓↓	↓↓	↓↓↓	↑↑
Deaths from Heart Disease[i]	↓↓↓	↓↓	↓↓	↓↓↓	↓↓↓	↓↓	↓↓	↓↓	↓↓↓	↓↓	↓↓↓	↓↓↓	↓↓
Death from Stroke[j]	↓↓↓	↓↓	↓	↓↓↓	↑	AVG	↑↑	↓↓↓	↓↓↓	↑↑↑	↓↓↓	↓	↓
Death from Cancer[k]	↓↓↓	↓↓	↓↓↓	↓↓↓	↓↓	↓↓	↑↑	↓	↓↓	AVG	↓↓↓	↓	↓↓

[a] U.S. prevalence = 58.4 percent (overweight defined as BMI >25.0)

[b] U.S. prevalence = 25.8 percent (adults who did no leisure-time physical activity in past month)

[c] U.S. prevalence = 6.6 percent (self-reported data based on number of adults who were told they had condition by a health professional)

[d] U.S. prevalence = 25.7 percent (self-reported data based on number of adults who were told they had condition by a health professional)

[e] U.S. prevalence = 30.4 percent (self-reported data based on number of adults who were told they had condition by a health professional)

[f] U.S. prevalence = 77.6 percent (adults who do not consume at least 5 fruits/vegetables per day)

[g] U.S. prevalence = 5.1 percent (>2 drinks/day in the past month for men, >1 drink per day in the past month for women)

[h] U.S. prevalence = 7.3 percent (live births of infants weighing <2500 grams)

[i] U.S. age-adjusted death rate per 100,000 = 245.8

[j] U.S. age-adjusted death rate per 100,000 = 57.9

[k] U.S. age-adjusted death rate per 100,000 = 194.4

AVG – similar to national average

↑ – slightly above national average

↓ – slightly below national average

↑↑ – significantly above national average

↓↓ – significantly below national average

↑↑↑ – exceptionally above national average

↓↓↓ – exceptionally below national average

Sources: Ahluwalia, I.B., et al. 2003. State-specific prevalence of selected chronic disease-related characteristics—Behavioral Risk Factor Surveillance System, 2001. National Center for Health Statistics. 2004. *Chartbook on trends in health of Americans*. Hyattsville, MD: U.S. Government Printing Office. *MMWR Surveillance Summaries*, 52, (SS08); 1–80. Centers for Disease Control. 2003. *Profiling the leading causes of death in the United States: Heart disease, stroke, and cancer.* Available at http://www.cdc.gov/nccdphp/publications/factsheets/ChronicDisease/

DISCUSSION STARTERS: EXAMINATION OF ETHNICITY, CULTURE, DIET, AND HEALTH

For Chapter 1, you described your cultural identity (your race and ethnicity) and the foods that you typically eat (what you like to eat, what foods are eaten in your home). Now, identify the regional area (Northeast, Midwest, South, or West) that your diet most reflects—and any more specific area(s) within that larger region (New England, Mid-Atlantic, East North Central, West North Central, South Atlantic, East South Central, West South Central, Mountain States, or Pacific States). Next, identify any even more specifically localized area and/or culture within that specific area that your diet seems to reflect.

Are the region, area, and culture that your diet seems to reflect the same as the region, area, and culture that apply to you? Do other factors affect your diet, such as ethnicity?

Look at the appropriate table of health concerns for the area that your diet reflects and think about your family members' individual health histories, as well as your racial and ethnic background(s). Next, identify health concerns that you might need to address in the future. For example, if your diet is southern and in particular West Virginian, you would definitely identify high blood cholesterol, hypertension, and diabetes as possible future health concerns. On the other hand, if your diet is Western and in particular Californian but also heavily influenced by a Greek heritage, you will want to analyze whether that Greek influence on your diet is strong enough that you should identify overweight and obesity as possible future health concerns, even though studies suggest that Californians generally are similar to the national average in those two categories.

In small groups, share your findings. Together, brainstorm about what each person in your group could begin doing now to address possible future health concerns.

REVIEW QUESTIONS

1. List and describe three factors influencing regional cuisine. Pick one region and summarize its influences.
2. How did Native American foods/cooking methods influence regional U.S. cuisine?
3. Compare and contrast the preparation of beans, corn, and apples in different regions of the United States.
4. Describe one unique recipe associated with a particular region of the United States that you were not familiar with prior to reading Chapter 15. Would you try it? Why or why not?
5. Based on health statistics, which region of the country would you choose to live in to stay healthy? What dietary factors may be influencing these health statistics? If you wanted to eat "unhealthy" one day, which regional cuisine would you try? Why?
6. You have decided to eat local. What does this mean, and what foods would be available for you to purchase? List some of the arguments for eating local.

REFERENCES

1. Jekowski, M.D., & Binkley, J.K. 2000. Food spending varies across the United States. *FoodReview, 23*, 38–43.
2. New Strategist. 2005. *Who's buying groceries* (3rd ed.). Ithaca, NY: Author.
3. Neustadt, K. 1992. *Clambake: A history and celebration of an American tradition*. Amherst, MA: University of Massachusetts Press.
4. Lee, H.G. 1992. *Taste of the states: A food history of America*. Charlottesville, VA: Howell.
5. Edmondson, B. 1998. The line between beer and wine. *American Demographics, 1*, 8–19.
6. Trubek, A.B. 2005. Place matters. In *The Taste Culture Reader*, C. Korsmeyer (Ed.), New York: Berg.
7. Pitzer, G.R. 2004. *Encyclopedia of Human Geography.* Westport, CN: Greenwood Publishing.
8. Lipard, L.R. 1997. *The lure of the local: Senses of place in a multicentered society*. New York: The New Press.
9. Brown, L.K., & Mussell, K. 1984. *Introduction. In Ethnic and regional foodways in the United States.* Knoxville: University of Tennessee Press.
10. US Census. US Population Projections. Available from: http://www.census.gov/population/www/projections/projectionsagesex.html. Accessed: March 19, 2011.
11. US Census. 2005-2009 American Community Survey. Available from: http://factfinder.census.gov/servlet/DatasetMainPageServlet?_program=ACS&_submenuld=&_lang=en&_ds_name=ACS_2009_5YR_G00_&ts=. Accessed: March 19, 2011.
12. Oliver, S.L. 1995. *Saltwater foodways*. Mystic, CT: Mystic Seaport Museum.
13. Ahluwalia, I.B., Mack, K.A., Murphy, W., Mokdad, A.H., & Bales, V.S. 2003. *State-specific prevalence of selected chronic disease-related characteristics—Behavioral Risk Factor Surveillance System*, 2001.
14. Sherry B, Blanck HM, Galuska DA, Dietz W. Vital Signs: State-Specific Obesity Prevalence Among Adults—United States, 2009. *Morbidity and Mortality Reports Weekly. 59*, 1–5. 2010. Available from: Accessed: March 20, 2011.

15. Wilcox, E.W., Buckeye Publishing Company. 2002. *Buckeye Cookery and Practical Housekeeping, 1877 (facsimile edition)*. Bedford, MA: Applewood Books.
16. Lloyd, T.C. 1981. The Cincinnati chili culinary complex. *Western Folklore, 40*, 28–40.
17. Lockwood, Y.R., & Lockwood, W.G. 1991. Pasties in Michigan's Upper Peninsula. In S. Stern & J.A. Cicala (Eds.), *Creative Ethnicity: Symbols and Strategies of Contemporary Ethnic Life*. Logan: Utah State University Press.
18. Luchetti, C. 1993. *Home on the range*. New York: Villard.
19. Anderson, J. 1997. *The American century cookbook*. New York: Clarkson Potter.
20. Mandel, A. 1996. *Celebrating the Midwestern table*. New York: Doubleday.
21. Matson, M. 1994. *Food in Missouri*. Columbia: University of Missouri Press.
22. Hess, K. 1992. *The Carolina rice kitchen*. Columbia: University of South Carolina Press.
23. Carlson, B. 1997. *Food festivals*. Detroit: Visible Ink.
24. Page, L.G., & Wigginton, E. 1984. *The foxfire book of Appalachian cookery*. New York: Dutton.
25. Smith, C. 2003. Food and culture in Appalachian Kentucky: An ethnography. *Journal for the Study of Food and Society, 6*, 64–71.
26. Tribe, D.L., & Oliveri, C.S. 2000. *Changing Appalachian Foodways: Perceived Changes and Rationale for Food Habits of Appalachian Ohioans (A Preliminary Study)*. At http://food.oregonstate.edu/ref/culture/tribe.html.
27. Milbauer, J.A. 1990. The geography of food in Eastern Oklahoma: A small restaurant study. *North American Culture, 6*, 37–52.
28. Grieco EM, Trevelyan EN. *Place of Birth of the Foreign-Born Population: 2009. American Community Survey.* Available at: http://www.census.gov/prod/2010pubs/acsbr09-15.pdf. Accessed: March 21, 2011.
29. Notarianni, P.F. 1994. Italians in Utah. In *Utah History Encyclopedia*, A.K. Powell (Ed.), Salt Lake City: University of Utah Press.
30. Raspa, R. 1984. Exotic foods among Italian-Americans in Mormon Utah: Food as nostalgic enactment of identity. In *Ethnic and regional foodways in the United States*. Knoxville: University of Tennessee Press.
31. Jones, E. 1981. *American food: The gastronomic story* (2nd ed.). New York: Vintage.
32. Centers for Disease Control. *2009 State Obesity Rates*. Available at: http://www.cdc.gov/obesity/data/trends.html#State. Accessed: March 21, 2011.

GLOSSARY OF ETHNIC INGREDIENTS

Abalone (paua): Large, flat mollusk with finely textured, sweet flesh in the broad muscular foot that holds it to rocks (must be pounded before use). It is common in the waters off Asia, California, Mexico, and New Zealand. Available fresh, frozen, canned, and dried.

Abiu (caimito): Yellow egg-shaped or round fruit native to the Amazon; popular throughout Brazil and Peru. Translucent white flesh with caramel-like flavor.

Acerola cherries (Barbados cherries): Exceptionally sour Caribbean berries resembling small, bright red cherries with orange flesh.

Achiote: See *Annatto.*

Adzuki bean (aduki, azuki; red bean): Small, dark red bean used primarily in Japanese cooking, often as a sweetened paste.

Ahipa: See *Jicama.*

Ajowan (ajwain; carom; omum or lovage seeds): Similar to celery seeds in appearance and to thyme in flavor. Used in Asian-Indian and Middle Eastern cooking.

Ajwain: See *Ajowan.*

Akee (ackee, ache; seso vegetal; pera roja): Red fruit with three segments containing large inedible seeds and flesh resembling scrambled eggs. Nearly all parts toxic, causing fatal hypoglycemia. Fresh, dried, frozen akee banned in United States; some canned types permitted.

Alligator: Reptile native to rivers and swamps throughout the southern Gulf Coast region, from Florida to Texas. Mild white meat, with texture similar to veal. Tail and other parts eaten.

Almond paste: Arab confection of ground almonds kneaded with sugar or cooked sugar syrup (some brands also contain egg white) used in many European and Middle Eastern desserts. Marzipan is a type of almond paste made with finely ground, blanched almonds.

Amaranth (tampala; yien choy; Chinese spinach): Leafy, dark green vegetable similar to spinach; red and purple leaf varieties, also. The high-protein seeds can be ground into flour and used in baked products, or boiled and eaten as cereal. Popular throughout Asia and Latin America.

Ambarella (hog or Jew plum; kadondong; otaheite or golden apple; vi-apple): Small, oval-shaped fruit with very strong flavor native to Polynesia but also found in Southeast Asia and Caribbean. Used unripe for preserves and ripe in desserts.

Amchoor (amchur; khati powder): Dried, unripe mango slices or powder, with a sour, raisin-like flavor.

Angelica root: Herb with a licorice-flavored root common in European dishes. Usually available candied. Used medicinally in China.

Annatto (achiote, atchuete): Seeds of the annatto tree used to color foods red or golden yellow. Used in Latin America, India, Spain, and the Philippines. In the United States annatto is added to some baked goods, Cheddar-style cheeses, ice creams, margarines, and butter for color. May be cooked whole in oil or lard to produce the right hue or used as a ground spice.

Apio: See *Arracacha.*

Apios: See *Groundnuts.*

Apon seeds (agonbono): Seeds of the wild mango commonly used in West Africa. Basis of the soup known as agonbono.

Areca nuts: See *Betel.*

Arracacha (apio; Peruvian carrot): Starchy white root of the carrot family with flavor similar to chestnuts and parsnips used in South America, especially Colombia, Peru, and Venezuela.

Arrowroot (chee koo): Many varieties of a bland, mealy tuber found in Asia and the Caribbean. When made into a powder, it is used to thicken sauces and stews.

Artichoke (carciofo): Globelike vegetable member of the thistle family, with multiple edible bracts (leaves) crowning the undeveloped edible flower (the heart). The flavor is slightly sweet. Popular in Middle Eastern and southern European dishes.

Arugula (rocket): Small member of the cabbage family native to the Mediterranean; the peppery leaves are popular in salads throughout Europe.

Asafetida (devil's dung): Dried resin with a pungent odor reminiscent of burnt rubber, which nonetheless imparts a delicate onion-like flavor. It is available as a lump or powder and is commonly used in Asian-Indian dishes.

Asian pear (apple pear): Round, yellow fruit from Asia with the crispness of an apple and the flavor of a pear.

Atemoya: Hybrid of the cherimoya and sweetsop. See *Cherimoya; Sweetsop.*

Aubergine: See *Eggplant.*

Avocado (aguacate; alligator pear; coyo): Pear-shaped to round fruit with leathery skin (green to black) and light green, buttery flesh. Native to Central America. Numerous varieties; eaten mostly as a vegetable, though considered a fruit in some cuisines.

Bacalao (bacalhau, baccala): Cod preserved by drying and salting, popular in northern and southern European cooking (especially Portuguese). Must be soaked, drained, and boiled before use.

Bagoong: See *Fish paste.*

Bagoong-alamang: See *Shrimp paste.*

Bambara groundnut (Congo goober; kaffir pea): Legume very similar to peanuts, native to Africa.

Bamboo shoot (juk suhn): Crisp, cream colored, conical shoot of the bamboo plant. Used fresh (stored in water) or available canned in brine (whole or sliced).

Banana flower (plantain flower): Native to Indonesia and Malaysia, bananas are now found in most tropical regions. Male inflorescence of the plant (female inflorescence that develops into fruit not eaten) is sheathed in inedible red-purple petals. Starchy interior must be boiled repeatedly to remove bitterness; used fresh in salads, cooked in curries, soups, or as side dish in palm oil or coconut milk.

Bangus: See *Milkfish.*

Baobab (monkey bread, lalu powder): Slightly sweet seeds from the large fruit of the native African baobab tree. Used roasted or ground. Pulp of the fruit is also consumed.

Basmati rice: See *Rice.*

Bean curd (cheong-po, tempeh, tofu, tobu): Custard-like, slightly rubbery white curd with a bland flavor made from soybean milk. Japanese bean curd (tofu) tends to be softer than Chinese, which is preferred for stir-fried dishes. A chewier version common in Southeast Asia is called tempeh. Cheong-po, a Korean bean curd, is made from mung beans.

Beans: See specific bean type.

Bean sprouts (nga choy): The young sprouts of mung beans or soybeans popular in Asian cooking (sprouts may also be grown from the tiny seeds of alfalfa or peas, also from legumes). The crisp 1- to 2-inch sprouts are eaten fresh or added to stir-fried dishes.

Belgian endive: See *Chicory.*

Berbere: Ethiopian spice mix (typically very hot) used to season many foods, usually including allspice, cardamom, cayenne, cinnamon, cloves, coriander, cumin, fenugreek, ginger, nutmeg, and black pepper.

Bergamot orange: Pear-shaped orange with exceptionally tart flesh. Rind used to flavor dishes in the Mediterranean and North Africa; oil extracted from rind flavors Earl Grey tea.

Betel (areca nuts; catechu): The heart-shaped leaves of the betel vine (related to black pepper) are used to wrap areca nuts (from the Areca palm; the nuts are usually called betel nuts because of their use with betel leaves) and spices for paan in India. Betel nuts and leaves are chewed together in many Southeast Asian countries and in India to promote digestion. May stain teeth red.

Bindi: See *Okra.*

Bird's nest: Swallows' nests from the cliffs of the South China Sea made from predigested seaweed; added to Chinese soups or sweetened for dessert. Must be soaked before use.

Bitter almond: An almond variety with an especially strong almond flavor, often used to make extracts, syrups, and liqueurs. Grown in the Mediterranean region, bitter almonds are used in European dishes. They contain prussic acid and are toxic when raw (they become edible when cooked) and are unavailable in the United States.

Bitter melon (balsam pear; bitter gourd, foo gwa): Bumpy-skinned Asian fruit similar in shape to a cucumber; pale green when ripe. The flesh has melon-like seeds and an acrid taste due to high quinine content (flavor and odor become stronger the longer it ripens).

Bitter orange: See *Seville orange.*

Black bean (frijol negro; turtle bean): Small (less than 1.2 inches) black bean used extensively in Central American, South American, and Caribbean cooking.

Black beans, fermented: Black soybeans salted and fermented to produce a piquant condiment. Used in Chinese cooking as a seasoning or combined with garlic, ginger, rice wine, and other ingredients to make black bean sauce.

Black-eyed peas (cow peas; crowder peas): Small legume (technically neither a pea nor a bean), white with a black spot, native to Africa and southern Asia.

Black mushrooms: See *Mushrooms.*

Blood orange: Old variety of orange with deep maroon–colored flesh, sometimes streaked with white. Intense sweet-tart flavor. Common in Spain and North Africa.

Blowfish (bok; fugu; globefish; puffer): A popular Japanese specialty, blowfish contain a deadly neurotoxin in the liver and sex organs. Must be carefully prepared by expert; flesh has a slight tingle when eaten.

Bok choy (Chinese chard; pak choi; white cabbage): Vegetable of the cabbage family with long, white leaf stalks and smooth, dark green leaves used in Chinese cooking.

Boonchi: See *Long bean.*

Bottlegourd: See *Calabash.*

Boxthorn: See *Matrimony vine.*

Breadfruit: Large, round, tropical fruit with warty green skin and starchy white flesh popular in nearly all tropical regions. It must be cooked. Unripe, green fruits are generally prepared as a vegetable, boiled, fried, or even pickled. In South Pacific may be fermented to make poi-like starchy dish. Ripe, yellow-fleshed fruit usually sweetened and served as dessert. Available canned; frozen.

Breadroot (Indian breadroot; prairie turnips; *timpsila; tipsin*): Hairy perennial plant *(Psoralea esculenta)* with large brown root eaten by Native Americans of the Plains and adopted by European immigrants who knew it as *pomme de prairie.*

Brinjal: See *Eggplant.*

Buckwheat (kasha): Nutty-flavored cereal native to Russia (where it is called kasha), sold as whole seeds (groats) and ground seeds (grits if coarsely ground, flour if finely ground). It is common in Russian and eastern European cooking.

Buffalo berry: Scarlet berry of the *Sheperdia* genus, so called because it was usually eaten with buffalo meat by Native Americans of the plains.

Bulgur (bulghur, burghul): Nutty-flavored cracked grains of whole wheat that have been precooked with steam. Available in coarse, medium, and fine grades.

Burdock root (gobo): Long thin root with thin brown skin and crisp white flesh and an earthy, sweet flavor. Popular in Asian cooking.

Cactus fruit (cactus pears, cholla, Indian figs, pitaya, sabra, strawberry pear, thang long): Succulent fruit of various cacti popular in numerous nations. Red prickly pear cactus fruit—cactus pears, cholla, Indian figs, sabra, tuna—common in Mexico, U.S. Southwest, Central America, Israel and some other Middle Eastern countries, Australia, South Africa, and Italy. Fruit of the organ pipe cactus sold in the United States as strawberry pear or pitaya. Fruit of saguaro cactus, nopales cactus, and apple cactus eaten in desert areas of Mexico and U.S. Southwest. Climbing epiphytic cacti common in South America, Australia, Israel, and Vietnam; one variety called thang long red pitaya or dragon fruit.

Cactus pads (nopales, nopalitos): Paddles of the prickly pear cactus or nopales cactus commonly eaten in Mexico and parts of the U.S. Southwest, fresh, cooked, or pickled. Available canned.

Cactus pears: See *Cactus fruit.*

Caimito: See *Star apple.*

Cajú: See *Cashew apple.*

Calabash (bottlegourd; calabaza; West Indian pumpkin): Gourd-like fruit of a tropical tree native to the New World.

Calabaza: See *Calabash; Cushaw*

Calamansi (calamondin, Chinese or Panama orange, golden or scarlet lime, musk lime): Small sour lime native to China but widely distributed in Indonesia and the Philippines, also available in Southeast Asia, Malaysia, and India. Prized for its sour flavor in Filipino cooking.

Callaloo (cocoyam): Edible leaves of root vegetables, especially amaranth, malanga, and taro. Callaloo is sometimes the name of a dish made from these leaves.

Camass root: Sweet bulb of the camass lily common in the U.S. Pacific Northwest.

Candlenut (kemini; kukui nut): Oily tropical nut sold only in roasted form (toxic when raw). Popular in Malaysia, Polynesia, and Southeast Asia.

Càng cua: See *Peperomia.*

Cannellini: See *Kidney bean.*

Capers: Small gray-green flower buds from a bush native to the Mediterranean; commonly pickled.

Carambola: See *Star fruit.*

Cardoon: Member of the artichoke family resembling a spiny celery plant, popular in Italian cooking.

Cashew apple *(cajú):* The fleshy false fruit attached to the cashew nut. Native to Brazil, it is also eaten in the Caribbean and India.

Casimiroa (white sapote, zapote blanco): Dark green to yellow fruit native to Central America; resembles an Asian pear. Soft, white flesh is eaten fresh or prepared as jellies, ices, milkshakes, and fruit leather.

Cassarep: Caribbean sauce made from the juice of the bitter variety of cassava cooked with raw sugar.

Cassava (cocoyam; fufu; manioc; yuca): Tropical Latin American tuber (now eaten in most tropical areas of the world) with rough brown skin and mild white flesh. Two types exist: bitter (poisonous unless leached and cooked) and sweet. Flour used in Africa (gari), the Caribbean, and Brazil (farinha). Cassava starch (fufu) is used to make the thickening agent tapioca. Leaves also consumed.

Caviar (red caviar, ikura, tarama, tobikko): Fish roe from a variety of fish eaten worldwide, including sturgeon (technically the only roe that is called caviar), salmon (red caviar, ikura in Japan), flying fish (tobikko), carp (tarama, most often made into a paste with lemon juice and other ingredients, in Greece called taramasalata), herring, and mullet. Sturgeon caviar graded according to size and quality.

Celeriac (celery root): Gnarled, bulbous root of one type of celery, with brown skin, tan flesh, and nutty flavor.

Cèpes: See *Mushrooms.*

Chanterelles: See *Mushrooms.*

Chayote (christophine, chocho, huisquil, mirliton, vegetable pear): Thin-skinned, green (light or dark), pear-shaped gourd. Native to Mexico, it is now common in Central America, the Caribbean, the southern United States, and parts of Asia.

Cheong-po: See *Bean curd.*

Cherimoya (anona, custard apple, graviola): Large, dimpled, light green fruit native to South America. White, creamy, flesh has a flavor reminiscent of strawberries, cherries, and pineapple. See also *Custard apple.*

Chicharrónes (pork cracklings): Deep-fried pork skin, fried twice to produce puffy strips.

Chickpeas (Bengal gram dal, chana dal, garbanzo bean): Pale yellow, spherical legume popular in Middle Eastern, Spanish, Portuguese, and Latin American cooking. Can be purchased canned or dried.

Chico: See *Zapote.*

Chicory (Belgian endive, witloof): European chicory plant. Leaves used as salad green; bitter root roasted to prepare a coffee substitute. Often added to dark coffee in Creole cooking.

Chile pepper: Although chile peppers, or chiles, are often called hot peppers, the fruits are not related to Asian pepper (such as black pepper) but are pods of capsicum plants, native to Central and South America. The alkaloid capsaicin, found mostly in the ribs of the pods, is what makes chile peppers hot. In general, the smaller the chile, the hotter it is. More than 100 varieties are available, from less than one-quarter inch in length to over eight inches long. Used fresh or dried. Common types include mild pods (see *Peppers*), slightly hot peppers such as Anaheim (also called *California* or *New Mexico chile*) and Cayenne (used mostly dried and powdered as the spice cayenne); dark green, medium hot Jalapeño (often available canned—when smoked are known as Chipotle); spicy, rich green Poblano (used fresh, or ripened and dried, called Ancho); hot Serrano (small, bright green or red); and very hot Chile de Arbol, Japones, Péquin (tiny berry-like pepper, exceptionally hot, also known as bird or bird's eye peppers), Piri-piri (favored in West Africa for sauces and marinades; also name of dishes that include some form of the pepper) and Tabasco (small, red chiles, often used dried and for sauce of same name). Those with extreme heat include Habanero and Scotch Bonnet; similar varieties native to the Caribbean.

Chile pepper sauce/paste (harissa, kochujang, pili-pili, Tabasco): Fiery condiments based on hot chile peppers. Sauce typically made from fermented chile peppers, vinegar, and salt (Tabasco sauce is the best-known U.S. brand). Pastes often include other ingredients, such as garlic and oil (Chinese-style and North African harissa). Pili-pili used in West Africa made with the piri-piri chile (see *Chile peppers*) and other ingredients such as tomatoes, onions, or horseradish. Korean kochujang includes soybeans and is fermented.

Chili powder: Ground, dried chile peppers, often with added spices such as oregano, cumin, and salt.

Chinese date (dae-chu; jujube): Small Asian fruit (not actually belonging to the date family) usually sold dried. Red dates are the most popular, but black and white are also available.

Chinese parsley: See *Coriander.*

Chitterlings (chitlins): Pork small intestines, prepared by boiling or frying.

Chokecherry: Tart, reddish black cherry *(Prunus virginiana)* native to the Americas.

Cholla: See *Cactus fruit.*

Chrysanthemum greens (chop suey greens, crowndaisy greens, sook-gat): Spicy leaves of a variety of chrysanthemum (not the American garden flower), popular in Asian stir-fried dishes, especially in Korea.

Cilantro: See *Coriander.*

Citron: Yellow-green, apple-size citrus fruit. Valued primarily for its fragrant peel that is used raw to flavor Indonesian foods, and candied in European baked goods. Available crystallized and as preserves.

Citronella: See *Lemon grass.*

Clotted cream (Cornish cream, Devonshire cream): Very thick cream made by allowing cream to separate from milk, then heating it and cooling it so that it ferments slightly. Finally, the cream is skimmed from the milk (although Cornish cream is skimmed before heating and cooling). Popular in southwest England, where it is spread on bread or used as a topping for desserts.

Cloud (wood) ears: See *Mushrooms.*

Coconut cream: High-fat cream pressed from fresh grated coconut.

Coconut milk: Liquid extracted with water from fresh grated coconut.

Cocoplum: Bland plum with white flesh native to Central America, found in the Caribbean, Central America, and Florida. Eaten fresh or dried.

Cocoyam: See *Callaloo; Cassava.*

Conch: Large, univalve mollusk found in waters off Florida and Caribbean (where it is sometimes called lambi). Chewy meat valued for its smoky flavor; can be bitter. Used especially in soups and stews.

Copra: Dried coconut kernels used in the extraction of coconut oil.

Coriander (cilantro, Chinese parsley, dhanyaka, yuen sai): Fresh leaves of the coriander plant with a distinctive "soapy" flavor, common in Asian, Middle Eastern, Indian, and Latin American cooking. Seeds used as spice; root used in Thai cooking.

Corn smut (huitlacoche): Fungus *(Ustilaginales)* that grows on corn ears. Prized in Chinese, Mexican, and Native American cooking.

Couscous (cuscus, cuzcuz): Small granules of semolina flour used as a grain in African, Italian, Brazilian, and Middle Eastern dishes.

Cow pea: See *Black-eyed pea.*

Cracked wheat: Cracked raw kernels of whole wheat used in Middle Eastern cooking.

Crawfish (crawdad, crayfish, mudbug): Small freshwater crustacean, 4 to 6 inches long, that looks and tastes something like lobster. Found in Europe and the United States (California, Louisiana, Michigan, and the Pacific Northwest). The names *crawfish* and *crayfish* are also applied to the langostino, a saltwater crustacean that lacks large front claws.

Crème fraîche: Slightly thickened, slightly fermented cream popular in France.

Culantro (bhandhani, ngo gai, recao, siny coriander): Herb *(Eryngium foetidum)* that is close relative of cilantro (see *Coriander*); however, looks more like a dandelion with a pungent flavor reminiscent of crushed beetles. Used interchangeably with cilantro in the Caribbean and Central America, especially associated with Puerto Rican sofrito. Seasons Thai curries, Malaysian rice dishes, Indian chutneys and snacks; larger leaves used as a wrap for foods in Vietnam. Reportedly high in riboflavin, carotene, calcium, and iron.

Curry leaves (kari): Herb with tangerine overtones used throughout India, Sri Lanka, and in parts of Malaysia. Fresh leaves are briefly fried in ghee, then added to dishes before other seasoning. Not usually a component of curry powder.

Curry powder: The western version of the fresh Asian-Indian spice mixture (garam masala) used to flavor curried dishes. Up to twenty spices are ground, then roasted, usually including black pepper, cayenne, cinnamon, coriander, cumin, fenugreek, ginger, cardamom, and turmeric for color.

Cushaw (calabaza, green pumpkin): Round or oblong winter squash with yellow flesh and a flavor similar to pumpkin.

Custard apple (anona roja, bullock's heart, mamon): Green-skinned, irregular (heart-, spherical-, or ovoid-shaped) fruit about 3 to 6 inches in diameter, with granular, custardy flesh. Flavor sweet but considered inferior to related fruits such as cherimoya and sweetsop. See also *Cherimoya.*

Cuttlefish (inkfish): A mollusk similar to squid, but smaller. Available fresh or dried.

Daikon (icicle radish, white radish, mooli): Relatively mild white radish common in Asian cooking. The Japanese variety is the largest, often 12 inches long, and is shaped like an icicle. The Chinese variety tends to be smaller.

Dals: Indian term for hulled and split grains, legumes, or seeds. Many types are available, such as lentils and split peas.

Dashi: Japanese stock made from kelp and dried fish (bonita). *Dashi-no-moto* is the dried, powdered, instant mix.

Dilis (daing): Small fish related to anchovies, dried and salted. Used in Filipino dishes.

Dragon's eyes: See *Longan.*

Drumstick plant (horseradish tree, malunggay, reseda, sili leaves): Small, deciduous tree native to India, now popular in India, Southeast Asia, the Philippines, and West Africa. Fern-like leaves (very spicy flavor), flowers, seeds (resembling bean pods but not a legume), and roots (indistinguishable from horseradish) consumed.

Duhat: See *Jambolan.*

Durian: Football-size spiked fruit with a strong odor reminiscent of gasoline or rotten onions and sweet, creamy flesh prized in Malaysia, Southeast Asia, and parts of China.

Edamame: See *Soybean.*

Eddo: See *Taro.*

Eggplant (ai gwa, aubergine, brinjal, melananza, nasu): Large, pear-shape to round member of the nightshade family with smooth, thin skin (white or deep purple in color) and spongy, off-white flesh. Native to India, where it is called brinjal, it has a mildly bitter flavor. Especially popular in Mediterranean and Asian cuisine. Asian varieties known as Japanese (nasu) and Chinese (ai gwa) eggplant are widely available; the Thai type is small, round, and white with green stripes and is less common.

Egusi: See *Watermelon seeds.*

Elderberries: Small shrubs up to 20 feet. Numerous species found throughout northern hemisphere. In the United States the small, dark purple berries used fresh and in preserves, pies, and wine. Blossoms fried as fritters.

Enoki: See *Mushrooms.*

Epazote (Mexican tea; pigweed, wormseed): Pungent herb related to pigweed or goosefoot (and sometimes called by these names). Found in Mexico and parts of the United States. Often added to bean dishes to reduce gas.

Farinha: See *Cassava.*

Fava bean (broad bean, brown bean, horse bean, Windsor bean): Large, green, meaty bean sold fresh in the pod. Smaller white or tan fava beans are dried or canned and cannot be used interchangeably with the fresh beans. Common in Italian and Middle Eastern cooking.

Feijoa (pineapple guava): Small (up to 3 inches), ovoid fruit with greenish skin and white flesh. Flavor is similar to strawberries and pineapple with minty overtones. Shrub native to central regions of South America, but now also found in California, Australia, and New Zealand.

Fennel (finnochio, sweet anise): Light green plant with slightly bulbous end and stalks with feathery, dark green leaves, a little like celery. Used as a root vegetable, especially in Italy (known as finnochio). Delicate licorice or anise flavor.

Fenugreek (methi): Tan seeds of the fenugreek plant, with a flavor similar to artificial maple flavoring. Essential in the preparation of Asian-Indian spice mixtures. Leaves, called methi, also commonly eaten.

Fiddlehead ferns: Young unfurled fronds a specialty dish of the U.S. Northeast and southeastern Canada. Roots were eaten by Native Americans.

Filé powder: See *Sassafras.*

Fish paste (bagoong, kapi, pa dek, prahoc): Thick fermented paste made from fish, used as a condiment and seasoning in the Philippines and Southeast Asia.

Fish sauce (nam pla, nam prik, nuoc mam, patis, tuk-trey): Thin, salty, brown sauce made from fish fermented for several days. Asian fish sauces vary in taste from mild to very strong, depending on the country and the grade of sauce. Filipino patis is the mildest; Vietnamese nuoc mam is among the most flavorful. Nuoc cham is a sauce made from nuoc mam by the addition of garlic and chile peppers.

Five-spice powder: A pungent Chinese spice mixture of anise, cinnamon, cloves, fennel seeds, and Szechuan pepper.

Fufu: See *Cassava; Yam.*

Fugu: See *Blowfish.*

Fuzzy melon (hairy melon, mo gwa): Asian squash similar to zucchini with peach fuzz–like skin covering. Called fuzzy.

Gai choy: See *Mustard.*

Gai lan (Chinese broccoli, Chinese kale): Thick, broccoli-like stems and large, dark or blue-green leaves, with slightly bitter flavor. Used especially in stir-frying.

Garbanzo bean: See *Chickpea.*

Gari: See *Cassava.*

Geoduck: Large (up to 15 pounds) clam native to U.S. Pacific Northwest, with neck or siphon as long as 3 feet. Neck used in soups, stews; body sliced for steaks.

Ghee: Clarified butter *(usli ghee)* from cow's or buffalo milk used in India. The term *ghee* is also used for shortening made from palm or vegetable oil.

Ginger root: Knobby brown-skinned rhizome with fibrous yellow-white pulp and a tangy flavor. Used sliced or grated in Asian dishes. Immature root with milder flavor used in some preparations, particularly pickled ginger popular in Japanese cuisine and candied ginger. Dried, ground ginger provides ginger flavor without the bite of fresh.

Ginkgo nut: Small pit of the fruit of the ginkgo tree (ancient species related to the pine tree), dried or preserved in brine, common in Japan.

Ginseng: Aromatic forked root with bitter, yellowish flesh, used in some Asian dishes and beverages; best known for therapeutic uses.

Glutinous rice: See *Rice.*

Granadilla: See *Passion fruit.*

Grape leaves: Large leaves of grape vines preserved in brine, common in Middle Eastern cooking.

Graviola: See *Cherimoya.*

Gravlax: See *Salmon, cured.*

Greens: Any of numerous cultivated or wild leaves, such as chard, collard greens, creases, cochan (coneflower), dandelion greens, dock, kale, milkweed, mustard greens, pokeweed, purslane, and spinach.

Grits: Coarsely ground grain, especially hominy, which is typically boiled into a thick porridge or fried as a side dish. Served often in the U.S. South.

Ground-cherries (Cape gooseberries, poha, golden berries): Yellow fruit that looks similar to a tiny husked tomato, from a bush native to Peru or Chile. Now popular throughout Central and South America, Central and South Africa, and the South Pacific. Also available in Australia, China, India, Malaysia, and the Philippines.

Groundnuts (apios, Indian potatoes): South American tuber *Apios americana* eaten by Native Americans, adopted by European settlers. Different from Africa groundnuts (referring to either peanuts or Bambara groundnuts).

Guanabana: See *Soursop.*

Guapuru: See *Jaboticaba.*

Guarana (Brazilian cocoa): Shrub, *Paullinia cupana* indigenous to the Amazon. Dried leaves and seeds of the fruit are used to make a stimulating tea (containing caffeine) or mixed with cassava flour to form sun-dried sticks.

Guava (araca de praia, cattley guava, waiwai): Small sweet fruit with an intense floral aroma, native to Brazil. Skin is yellow-green or yellow, and the grainy flesh ranges from white or yellow to pink and red. Many varieties are available, including strawberry guava (also known as cattley guava, araca de praia, and waiwai) and pineapple guava. Guava is popular as jelly, juice, or paste.

Guayo: See *Mamoncilla.*

Guineps: See *Mamoncilla.*

Headcheese: Loaf of seasoned meat made from the hog's head and sometimes also feet and organs.

Heart of palm (palmetto cabbage, palmito): White or light green interior of the palm tree, especially popular in the Philippines. Available canned.

Hickory nuts: Tree indigenous to North America, in same family as pecans. Eaten fresh, roasted, or ground into meal or pressed for a cream-like fluid by Native Americans; used in confections in the U.S. South.

Hog peanut: A high-protein underground fruit that grows on the root of the vine *Falcata comosa* in the central and southern United States. The peanut has a leathery shell that can be removed by boiling or soaking. The nut meat can be eaten raw or cooked.

Hoisin sauce: Popular Chinese paste or sauce, reddish brown in color, with a spicy sweet flavor. It is made from fermented soybeans, rice, sugar, garlic, ginger, and other spices.

Hominy (posole, pozole): Lime-soaked hulled corn kernels (yellow or white) with the bran and germ removed. Traditionally prepared by some Native Americans with culinary ash, which increases potassium, calcium, iron, phosphorus, and other mineral values. Ground, commonly called grits (see *Grits*).

Hot pepper: See *Chile pepper.*

Huisquil: See *Chayote.*

Icicle radish: See *Daikon.*

Ikura: See *Caviar.*

Imli: See *Tamarind.*

Indian breadroot: See *Breadroot.*

Indian fig: See *Cactus fruit.*

Indian potato: See *Groundnuts.*

Irish moss (carrageen): Gelatinous seaweed extract added to milk or rum as a beverage in the Caribbean.

Jaboticaba (guapuru, sabara): Brazilian shrub or small tree with 0.5- to 1.5-inch fruit clustered like grapes. Gelatinous pulp is mild and sweet.

Jackfruit: Large (up to 100 pounds) fruit related to breadfruit and figs, native to India, now cultivated in Asia, Malaysia, and Southeast Asia. Two varieties are widely eaten, one with a crisp texture and bland flavor, the other softer and sweeter. Immature fruit is usually prepared like other starchy vegetables such as breadfruit and plantains, or pickled. Sweeter types are popular as dessert. Available dried or canned.

Jaggery: Unrefined sugar from the palmyra or sugar palm common in India.

Jagua: See *Mamoncilla.*

Jambolan (duhat, Indian blackberry, jaman, Java plum, rose apple, voi rung): Small sour fruit grown in India and Southeast Asia, especially the Philippines. Used primarily in preserves, juices, and sherbets.

Jerusalem artichoke (sunchoke, sunroot): Small nubby-skinned tuber that is the root of a native American sunflower. It is neither from Jerusalem nor related to the artichoke, though the flavor when cooked is similar. It is used raw and cooked.

Jicama (ahipa, sa got, singkamas, yambean): Legume with medium to large tuber with light brown skin and crisp white flesh, indigenous to Brazil. Used raw in Latin American cuisine, it has a sweet, bland flavor, similar to peas or water chestnuts. Also found in Asia, where it is typically stir-fried or added to other cooked dishes.

Jujube: See *Chinese date.*

Juneberries (saskatoons, serviceberries; shadbush): Red to deep purple berries on large bush native to the Great Plains region of the United States and Canada. White blooms in June associated with shad migratory run on East Coast; favorite of Native Americans.

Juniper berry: Distinctively flavored dark blue berry of the juniper evergreen bush, native to Europe. Used to flavor gin.

Kadondong: *See Ambarella.*

Kaffir lime (ichang lime, makrut, wild lime): Aromatic citrus popular in Southeast Asia, especially in Thai cooking. Juice, rind, and leaves used to flavor curries, salad dressings, and sauces.

Kamis: Sour, cucumber-like vegetable native to the Philippines. Used to achieve a sour, cool flavor in Filipino cooking.

Kang kong: See *Water convolvulus.*

Kanpyo (kampyo): Ribbons of dried gourd used mostly for garnishing dishes in Japan.

Kaong: See *Palm nuts.*

Kapi: See *Fish paste.*

Kasha: See *Buckwheat.*

Kava: See *Pepper plant.*

Kemini: See *Candlenut.*

Kewra: See *Pandanus.*

Key lime (dayap, nimbu, West Indian or Mexican lime): Small, tart lime indigenous to the Caribbean, popular in Florida Keys; also used in east and north Africa, India, and Malaysia. Known best as primary ingredient in key lime pie.

Khati powder: See *Amchoor.*

Kidney bean (cannellini, red peas): Medium-size, kidney-shaped bean, light to dark red in color (a white variety is popular in Europe, especially Italy, where they are known as *cannellini*). The flavorful beans are common in Europe, Latin America, and the United States.

Kochujang: See *Chile pepper sauce/paste.*

Kohlrabi (tjin choi tow): Light green or purple bulbous vegetable that grows above the soil and produces stems bearing leaves on the upper part. A member of the cabbage family, it can be eaten raw or cooked.

Kola nut: Bitter nut of the African kola tree (extracts from this nut were used in the original recipe for Coca-Cola).

Kudzu (ge gen, Japanese arrowroot): Japanese vine valued for its tuberous root (up to 450 pounds) that is dried and powdered for a starch used in sauces and soups and to coat foods before frying. Now found in much of Asia and U.S. Southeast where it is best known for its growth rate of up to 1 foot per day. May alleviate hangovers or induce sobriety.

Kukui nut: See *Candlenut.*

Kumquat (kin kan): Small, bright orange, oval fruit with a spicy citrus flavor common in China and Japan. Also available in syrup and candied.

Laverbread: Thick purée of laver (see *Seaweed*) that is baked. Used in sauces and stuffings in Great Britain.

Lemon grass (citronella root): Large, dull green, stiff grass with lemony flavor common in Southeast Asian dishes. Available fresh, dried, or powdered.

Lily buds (golden needles, gum chum): The buds of lily flowers used both fresh and dehydrated in the cooking of China.

Lingonberry (low-bush cranberry): Small wild variety of the cranberry found in Canada and northern Europe. Usually available as preserves.

Litchi (lychee): Small Chinese fruit with translucent white flesh and a thin brown hull and single pit. The flavor is grape-like but less sweet. Available fresh and canned. Dried litchis, also called litchi nuts, have different flavor and texture.

Lobster: Ocean-dwelling crustacean valued for its sweet flesh. Two main species consumed in United States. American lobster *(Homarus americanus)* found from Labrador to North Carolina; meat from large claws and tail, premature eggs called *coral,* and liver eaten. Spiny lobster (*Panulirus argus* and other species) looks similar to American lobster but is a different animal. Found in warm waters from North Carolina to Brazil; small claws, only tail meat eaten.

Longan (dragon's eyes): Fruit of an Asian Indian tree related to litchis. Used fresh, canned, or dried.

Long bean (boonchi, dau gok, sitao, yardlong bean): Roundish Asian bean, 12 to 30 inches long. Similar in taste to string beans, long beans are softer, and chewier, less juicy, and less crunchy than string beans.

Long-grain rice: See *Rice.*

Loquat (nispero): Slightly fuzzy yellow Asian fruit about 2 inches across, easily peeled, with tart peach-flavored flesh. Cultivated worldwide; available fresh, dried, and in syrup.

Lotus root (lian, lin gau hasu, renkon, water lily root): Tubular vegetable (holes, as in Swiss cheese, run the length of the root, producing a flower-like pattern when the root is sliced) with brownish skin and crisp, sweet, white flesh. Becomes starchy when overcooked or canned.

Lox: See *Salmon, smoked.*

Luffa (cee gwa, Chinese okra, loofa, padwal, silk melon): Long, thin-skinned Asian vegetable, a member of the cucumber family, with spongy flesh. Immature luffas consumed fresh, stir-fried, and in curries; mature luffa becomes bitter. Also see *Sponge gourd.*

Lulo: See *Naranjillo.*

Lupine seeds (tremecos): Bitter seeds of a legume used primarily for fodder. Must be leached in water before eating.

Macadamia nut: Round, creamy nut native to Australia, now grown in Africa, South America, and Hawaii.

Mahi-mahi (dolphinfish, dorado): A saltwater finfish found in parts of the Pacific and the Gulf Coast (not the mammal also known as dolphin).

Mahleb: Middle Eastern spice made from ground black cherry kernels, which impart a fruity flavor to foods.

Makrut: See *Kaffir lime.*

Malagueta pepper (grains of paradise, guinea pepper): Small West African berries related to cardamom, with a hot, peppery flavor. In Brazil the term refers to a tiny Pequin chile pepper.

Malanga (cocoyam, tannier, yautia): Caribbean tuber with cream-colored, yellow, or pinkish flesh, dark brown skin, and nutty flavor. Name also applied to other tubers (see *Taro*).

Mamey (sapote): Medium-size egg-shaped fruit with brown skin and soft flesh ranging in color from orange to yellowish to reddish. It has a flavor similar to pumpkin. See also *Mammea.*

Mammea (mamey apple): South American fruit with reddish-brown skin and bright yellow flesh that tastes like peaches.

Mamoncilla (guayo, guineps, jaguar, macao, Spanish lime): Small 1- to 2-inch green fruit found in the Caribbean and South America that grow in clusters like grapes but have thicker skin and distinctive sweet, citrusy flesh around a large seed.

Mango (mangoro, mangue): Fruit native to India, now found throughout Africa, Asia, Latin America, and parts of the South Pacific. Yellow to red when ripe, averaging 1 pound in weight. The flesh is pale and sour when the fruit is unripe, bright orange and very sweet when it is ripe. Used unripe for pickles and chutneys, ripe as a fresh fruit.

Manioc: See *Cassava.*

Marzipan: See *Almond paste.*

Masa: Dough used to make tortillas and tamales. Made fresh from dried corn kernels soaked in a lime solution, or from one of two flours available: masa harina (tortilla mix made from dehydrated fresh masa) or masa trigo (wheat flour tortilla mix).

Mastic: Resin from the lentisk bush that has a slightly piney flavor, used to flavor Middle Eastern foods. Available in crystal form.

Matai: See *Waterchestnut.*

Mate: Plant in holly family native to South America. Dried, powdered leaves, called *yerba,* are brewed to make a stimulating tea (containing caffeine) that is popular in Argentina, Brazil, and Paraguay.

Matrimony vine (boxthorn, wolfberry): Asian vine with culinary and medicinal uses; both leaves and fruit are used in China.

Mayhaw: Type of hawthorn tree found in U.S. South. Its fruit looks like cranberries. Tart apple flavor. Used in preserves, syrups, and wines.

Methi: See *Fenugreek.*

Mikan: Japanese citrus related to tangerines and mandarin oranges. Eaten fresh, frozen, and canned in syrup.

Milkfish (awa, bangus): Silvery, bony fish with oily flesh especially popular in Filipino cooking.

Millet: Cereal native to Africa, known for its high-protein, low-gluten content and ability to grow in arid areas. The variety common in Ethiopia is called *teff.*

Mirin: Sweet rice wine used in Japanese dishes.

Miso: Fermented soybean-barley or soybean rice paste common in Japanese cooking. Light or white (shiro miso) is mild flavored; dark or red (aka miso) is strongly flavored. Also available sweetened and as powder.

Mizuna: See *Mustard.*

Morels: See *Mushrooms.*

Mullet (ama ama): Finfish of two families that can be black, gray, or red. The flesh is a mix of dark, oily meat and light, nutty-tasting meat. The texture is firm but tender.

Mung beans (green gram dal, mung dal): Yellow-fleshed bean with olive or tan skin used in cooking of China, India. See also *Bean curd; Bean sprouts.*

Mushrooms: Fresh or dried fungi used to flavor dishes throughout the world. Common Asian types include *enoki* (tiny yellow mushrooms with roundish caps), *oyster mushrooms* (large, delicately flavored gray-beige caps that grow on trees), *shiitake* (dark brown with wide flat caps, available dried as Chinese black mushrooms), *straw mushrooms* (creamy colored with bell-like caps), and *cloud ears* or *wood ears* (a large, flat fungus with ruffled edges, available dried). Popular mushrooms in Europe, available both fresh and dried, include *chanterelles* (a golden mushroom with an inverted cap), *morels* (a delicately flavored mushroom with a dark brown wrinkled cap), and *porcini* or *cèpes* (large brown mushrooms with caps that are spongy underneath; also called *boletus*).

Musk lime: See *Calamansi.*

Mustard (Chinese green mustard, gai choy, kyona, mizuna, potherb): Though best known for the condiment made from its seeds, greens of several varieties are popular in Asia, called gai choy in China (dark green-reddish leaves), mizuna (small yellowish, notched leaves) in Japan. Usually steamed, boiled, or stir-fried. Root also consumed.

Nam pla: See *Fish sauce.*

Nam prik: See *Fish sauce.*

Nance: Small, yellow tropical fruit native to Central America and northern South America. Similar to cherries with a slightly tart flavor. Two varieties are available.

Napa cabbage (celery cabbage, Chinese cabbage, Peking cabbage, wong bok): Bland, crunchy vegetable with broad white or light green stalks with ruffled leaves around the edges. Several types are available, similar in taste.

Naranjilla (lulo): Walnut-size, orange-skinned, green-fleshed fruit indigenous to the Americas, used mostly for its juice. Particularly popular in Central America.

Naseberry: See *Zapote.*

Nigella seed ("black cumin," "black onion," kalonji): Small, black seeds native to Europe, North Africa, and the Middle East. Sometimes used as a substitute for black pepper, the flavor of the seeds (which are related neither to cumin nor onions) is pungent, slightly bitter. Added to spice mixtures in India and the Middle East, sprinkled on savory breads and cakes in both regions, as well as in Eastern Europe.

Nispero: See *Loquat.*

Nku: See *Shea nut.*

Nongus (palmyra): Fruit of the palmyra palm, grown in India, Indonesia, and Malaysia primarily as a source of sugar. See also *Jaggery.*

Nopales, Nopalitos: See *Cactus pads.*

Nuoc cham: See *Fish sauce.*

Nuoc mam: See *Fish sauce.*

Oca: Tuber of Andean plant *(Oxalis tuberosa).* Resembles a pink potato. Tastes lemony when fresh, sweet after storage. Used in South America, prepared like potatoes or eaten fresh.

Okra (bindi, lady's fingers): Small, green, torpedo-shaped pod with angular sides. A tropical African plant valued for its carbohydrates that are sticky and mucilaginous. Used as a vegetable and to thicken soups and stews.

Olive: Fruit of a tree native to the Mediterranean. Green olives are preserved unripe. Large, soft Kalamata olives are a medium size, purplish Greek olive. Dark olives (such as Niçoise) are picked in autumn, often cured in salt, with a tannic flavor. Ripe, black olives are smooth-skinned and mild-flavored or wrinkled with a strong tannic flavor.

Olive oil: Extracted from the olive flesh, it is labeled according to percent acidity, from *extra virgin* to *virgin* (or *pure*). U.S. labeling laws restrict the use of the term *virgin* to only olive oil made from the first press; virgin olive oils mixed with refined olive oils to reduce acidity are labeled *pure.*

Ostiones: Oyster native to the Caribbean that grows on the roots of mangrove trees.

Otaheite apple: See *Ambarella.*

Oyster mushrooms: See *Mushrooms.*

Oyster sauce: Thick, brown Chinese sauce made with soy sauce, oysters, and cornstarch.

Pacaya bud: The bitter flower stalk of the pacaya palm found in Central America. The edible stalk is about 10 inches long and is encased in a tough green skin, which must be removed before cooking.

Pa dek: See *Fish sauce.*

Palillo: Peruvian herb, used dried and powdered to provide a yellowish-orange color to foods.

Palmetto cabbage: See *Heart of palm.*

Palm nuts (kaong): Seeds from palms; pounded into palm butter in West Africa. Also boiled and added to halo-halo mix in Philippines. Available canned, in syrup.

Palm oil (aceite de palma, dende oil): Oil from the African palm, unique for its red-orange color, used extensively in West African and Brazilian Bahian cuisine. Crude oil contains high levels of carotenoids and tocopherols; refined oil deodorized and decolorized, significantly reducing nutritional value. Oil from the seed of the palm fruit high in saturated fats; should be labeled palm kernel oil, but often mislabeled as palm oil.

Pandanus (flowers—kewra, screw pine; leaves—duan pandan, pandan, rampa, screw pine): Perfume essence of the male screw-pine flower *Pandanus fascicularis* used primarily in north Indian cooking. Screw-pine leaves *Pandanus amaryllifolius* reminiscent of mown hay, used to flavor the foods of Southeast Asia, Malaysia, South India, Bali, and New Guinea. Fresh withered leaves used in rice puddings and as wrappers for steaming foods in Thailand. Bright green screw-pine essence also available.

Papaya (kapaya, pawpaw, tree melon): Thin-skinned green (under-ripe), yellow, or orange fruit with sweet flesh colored gold to light orange to pink; native to Central America, now found throughout the tropics. Mexican (large and round) and Hawaiian (smaller and pear shaped) varieties are commonly available. The shiny round black seeds are edible. Unripe papaya is used in pickles; the ripe fruit is eaten fresh.

Paprika: Powdered red peppers especially popular in Hungarian cooking. Paprika is made from several types of pods related to bell and chile peppers. Paprika is usually designated sweet or hot. Spanish paprika, used in Spanish and Middle Eastern dishes, is more flavorful.

Passion fruit (granadilla, lilikoi): Small oval fruit with very sweet, gelatinous pulp. Its berries are used dried; leaves brewed to make herbal tea.

Patis: See *Fish sauce.*

Pawpaw (Hoosier banana, Poor Man's banana, tree melon): Light orange fruit that tastes like a cross between a banana and a melon. Native to the Americas, it is approximately 6 inches long. See also *Papaya.*

Peanuts (groundnuts, goobers, monkey nuts): Legume native to South America, introduced to Africa by the Portuguese, then brought to the United States in the 17th century by black slaves. Eaten raw, roasted, or pulverized into peanut butter. Popular in Africa and the United States; used in some Chinese, Southeast Asian, and Asian-Indian dishes.

Pejibaye (peach palm): Fruit of a Central American palm, especially popular in Costa Rica.

Peperomia *(càng cua):* Small plant with heart-shaped leaves *Peperomia pellucida* found throughout Central and South America, Africa, and Southeast Asia. Used as a culinary herb in Vietnam, and as a medicinal herb in the Philippines, Polynesia, and parts of Latin America.

Pepitas (cushaw seeds): Pumpkin or squash seeds, typically from cushaw, common in Latin-American cooking. May be hulled or unhulled, raw or roasted, salted or unsalted.

Pepper plant *(Piper methysticum):* Leaves of the South Pacific plant used to produce the intoxicating beverage called *kava* or *awa.*

Peppers: Misnamed pods of the capsicum plants native to South and Central America (not actually related to Asian pepper plants, which produce black pepper). Peppers are divided into sweet and hot types (see *Chile pepper*). Sweet peppers include bell peppers (green, red, yellow, and purple), pimentos, and peppers used to make paprika (see *Paprika*).

Perilla (shiso; beefsteak plant; quen-neep): Aromatic herb with distinctive minty flavor; green or red. Available fresh or pickled. Used mostly as a seasoning or garnish in many Japanese and Korean dishes; sometimes served as a side dish or to wrap rice and other items.

Pigeon pea: Small pea in a hairy pod (a member of the legume family, but not a true pea) common in the cooking of Africa, the Caribbean, and India. Yellow or tan when dried.

Pignoli: See *Pine nut.*

Pigweed: See *Amaranth; Epazote.*

Pili nut: Almond-like nut of a tropical tree found in the Philippines eaten raw and toasted. Popular also in Chinese desserts.

Pili-pili: See *Chile pepper sauce/paste.*

Pine nut (pignoli, piñon seed): Delicately flavored kernel from any of several species of pine tree. Pine nuts are found in Portugal (most expensive type), China (less costly, with a stronger taste), and the U.S. Southwest. Common in some Asian, European, Latin American, Middle Eastern, and Native American dishes.

Pink bean (rosada): Small oval meaty bean that is a light tannish pink in color.

Pinto bean: Mottled bean similar to kidney beans, especially popular in U.S. Southwest and Mexico.

Pitanga (Surinam cherry, Brazilian cherry): Small, bright red, ribbed fruit of shrub or small tree *Eugenia uniflora* native to northeastern South America; found also in the Caribbean and Florida. Thin skin with orange flesh that melts in the mouth. Sweet with a slightly bitter bite.

Pitaya, Pitahaya, Pitajaya: See *Cactus fruit.*

Plantain: Starchy type of banana with a thick skin, which can be green, red, yellow, or black. There are many varieties, ranging in size from 3 to 10 inches. The pulp is used as a vegetable and must be cooked. It is similar in taste to squash. Flower also consumed (see *Banana flower*).

Poha: See *Ground-cherries.*

Poi: See *Taro.*

Porcini: See *Mushrooms.*

Posole, Pozole: See *Hominy.*

Prahoc: See *Fish paste.*

Prairie turnips: See *Breadroot.*

Prickly pear: See *Cactus fruit; Cactus pads.*

Pulses: Term used especially in India for edible legume seeds, including peas, beans, lentils, and chickpeas.

Quinoa: Cereal native to the Andes, typically prepared like rice. Also available as flour and flakes (hojuelas).

Radicchio: Magenta-colored, slightly bitter member of the chicory family used throughout southern and northern Europe.

Rambutan: Bristly, juicy, orange or bright red fruit used in Southeast Asian cooking; related to the litchi.

Ramp: Strong-flavored indigenous American onion that tastes somewhat like a leek. Both leaves and bulbs are edible.

Recao: See *Culantro.*

Red bean: Small, dark red bean native to Mexico and the southwestern U.S.

Red caviar: See *Caviar.*

Red pea: See *Kidney bean.*

Rice: Grain native to India. More than 2,500 varieties are available worldwide, including basmati rice (small grain with a flavor similar to popcorn, very popular in India and the Middle East); brown rice (unmilled rice with the bran layer intact; can be short-, medium-, or long-grain); glutinous rice (also called *sweet* or *pearl rice;* very short grain and very sticky when cooked); long-grain rice (white, polished grains that flake when cooked, common in China and Vietnam); and short-grain rice (slightly sticky when cooked, popular in Japan and Korea). Rice flour is used to prepare rice noodle, rice paper, and baked products.

Roseapple (pomarrosa, kopo): Small, thin-skinned pink or red fruit native to Southeast Asia with somewhat spongy flesh that has slightly acidic flavor.

Roselle (Florida cranberry; karkadeh; red sorrel; sorrel): Pods of a hibiscus plant relative, common in Africa, the Caribbean, Southeast Asia, Australia, and Florida. Used to make a tart tea popular in Egypt and Senegal and a rum-laced punch in the Caribbean. Also used for chutneys, preserves, and candies. Young leaves are eaten raw as salad or cooked as greens.

Sabra: See *Cactus fruit.*

Saewujeot: See *Shrimp paste.*

Saffron: Dried stamens of the crocus flower. It has a delicate, slightly bitter flavor and bright red-orange color. Available as threads or powder.

Sa got: See *Jicama.*

Salal: Thick-skinned black berries of a native American plant in the heath family. Used fresh and dried, good for preserves. Leaves used for tea.

Salmon, cured: Salmon fillets cured in a mixture of salt, sugar, and dill weed, common in Sweden (where it is known as gravlax), Finland, and Norway.

Salmon, smoked: Raw, tender salmon slices lightly smoked and cured in salt produced in Norway, Nova Scotia, and Scotland. Smoked salmon soaked in a brine solution is called *lox,* a Jewish specialty.

Salmon roe: See *Caviar.*

Salt pork: White fat from the side of the hog, streaked with pork meat, cured in salt.

Saluyot (jute, okra leaves, rau day): Leaves from Southeast Asian jute bush with slippery texture when cooked (not related to okra). Added to soups and stews in Filipino cooking.

Samphire (beach asparagus, glasswort, sea pickle, pousse-pied): Several species of samphire thought to have originated in Brazil, but now found worldwide, especially in Australia and the South Pacific. Yellow- and purple-skinned varieties are available. Passion fruit is often made into juice.

Sapodilla: See *Zapote.*

Sapote: See *Zapote.*

Saskatoons: See *Juneberries.*

Sassafras (filé powder): Native American herb used to thicken soups and stews.

Screwpine: See *Pandanus.*

Sea cucumber (sea slug): Brown or black saltwater mollusk up to 1 foot in length. They lack a shell, but have a leathery skin and look something like smooth, dark cucumbers. Sold dried, they are rehydrated for Chinese dishes, becoming soft and jellylike, with a mild flavor.

Sea urchin roe (uni): Small, delicate eggs of the spiny sea urchin, popular in Japan.

Seaweed (kim): Many types of dried seaweed are used in Chinese, Korean, and Japanese dishes, including *aonoriko* (powdered green seaweed), *kombu* (kelp sheets), and *nori* (tissue-thin sheets of dark green seaweed, also known as laver). Also popular in the Pacific Islands. See also *Irish moss; Laverbread.*

Serviceberries: See *Juneberries.*

Sesame seeds (benne seeds): Seeds of a plant native to Indonesia. Two types are available: tan colored (white when hulled) and black (slightly bitter). Untoasted sesame paste popular in the Middle East (tahini); toasted sesame paste and powdered seeds common in Asia, especially Korea. Widely grown for their oil. Light sesame oil is pressed from raw seeds, dark oil from toasted seeds; the dark oil has a strong taste and is used as a flavoring.

Seville orange (bitter orange; naranja aria, sour orange): Orange with tough skin and dark flesh native to Mediterranean. Inedible raw; juice used in liqueurs (Grand Marnier, Cointreau, Curaçao) and in cooking of the Mediterranean, Caribbean, Central America, and Korea.

Shadbush: See *Juneberries.*

Shallot: Very small bulb covered with a reddish, papery skin, related to onions but with a milder, sweeter flavor.

Shea nut (bambuk butter, nku): Nut from the African shea tree, grown for its thick oil, called shea nut butter or shea nut oil.

Shiitake mushrooms: See *Mushrooms.*

Shiso: See *Perilla.*

Short-grain rice: See *Rice.*

Shoyu: See *Soy sauce.*

Shrimp paste: Strongly flavored fermented Asian sauce or paste made from small dried shrimp or similar crustaceans. Many types are available (bagoong-alamang is the Filipino variety; saewujeot is the Korean type).

Singkamas: See *Jicama.*

Snail (escargot): Small, edible land snail (a common variety of garden snail, cleansed with a commercial feed), popular in France. Giant, baseball-sized snails popular in parts of Africa and the South Pacific.

Snow pea (Chinese pea pod, ho lan dow, mange-tout, sugar pea): Flat, edible pod with small, immature peas.

Sorghum (guinea corn, kaffir corn): Cereal common to tropical regions of Africa with seeds produced on a stalk. In the Appalachians, Ozarks, and the U.S. South, sorghum is often processed to make sweet syrup.

Sorrel (dock, sour grass, wild rhubarb): Small, sour green popular in Europe and parts of United States. See also *Roselle.*

Sour orange: See *Seville orange.*

Soursop (guanabana): Large (often 12 inches long) rough-skinned fruit with cottony, fluffy flesh that can be white, pink, or light orange. Native to northern South America or the Caribbean, now found in many parts of the Americas, Africa, India, China, Southeast Asia, Malaysia, and South Pacific. Often made into juice or conserves.

Soybean: Small high-protein bean common in Asia. Many varieties of different colors, including black, green, red, and yellow, are available; immature beans in the pod (called edamame) popular in Japan. They are used fresh, dried, and sprouted, most often processed into sauces, condiments, and other products (see *Bean curd; Bean sprout; Hoisin sauce; Miso; Oyster sauce; Soy milk, Soy sauce*).

Soy milk: Soybeans that are boiled, pureed, then strained and boiled again to produce a white milk-like drink.

Soy sauce (shoyu, tamari): Thin, salty, brown sauce made from fermented soybeans. Several types are available. Chinese and Korean soy sauces tend to be lighter in flavor than the stronger, darker Japanese shoyu. Very dark soy sauces, such as Chinese black soy sauce and Japanese tamari may be thickened with caramel or molasses.

Spicebush: Shrub *(Lindera benzoin)* with spicy-smelling bark and leaves; red berries. Used to make Native American teas.

Spiny lobster: See *Lobster.*

Sponge gourd (luffa): Immature vegetable consumed in Asia fresh and in soups; tough fibrous skin used for sponges (loofah), filters, and stuffing.

Star anise: Eight-armed pods from a plant in the magnolia family, with an anise-like flavor. Native to China.

Star apple (caimito): Purple, apple-size fruit with mild, gelatinous, lavender-colored flesh native to the Caribbean. Seeds form a star around the center.

Star fruit (carambola): Small, deeply ribbed, oval fruit with thin skin shaped like a star when sliced. Green and sour when unripe, yellow and slightly sweet (though still tart) when ripe. Unripe fruit is used in Indian and Chinese dishes. Ripe it is eaten fresh.

Strawberry pear: See *Cactus fruit.*

Straw mushrooms: See *Mushrooms.*

Sumac: Sour, red Middle Eastern spice made from the ground berries of a nontoxic variety of the sumac plant.

Sunflowers: Native to the United States (genus *Helianthus*); over 60 varieties. Seeds eaten by Native Americans raw, dried, and powdered (in breads). Unopened flower head can be cooked and eaten like an artichoke. Petals are dried and used like saffron in Southwest.

Sweet peppers: See *Peppers.*

Sweetsop (annona blanca, ata, sugar apple): Sweet, white-fleshed fruit related to the cherimoya, custard apple, and soursop.

Szechwan pepper (fagara): Aromatic berries with a hot flavor popular in some Chinese and Japanese dishes.

Tabasco sauce: See *chile pepper sauce/paste.*

Tahini: See *Sesame seeds.*

Tamarind (imli, tamarindo): Tart pulp from the pod of the tamarind bean. Available in the pod, as a paste, in a brick, or as a liquid concentrate. Unripe pulp used extensively in flavoring numerous foods and beverages, especially Asian Indian and Latino dishes, as well as Worcestershire sauce and prepared salad dressings. Ripe pulp eaten fresh.

Tampala: See *Amaranth.*

Tannier (tannia): See *Taro; Malanga.*

Tapioca: See *Cassava.*

Taramasalata: See *Caviar.*

Taro (cocoyam, eddo, dasheen, tannier, malanga, yautia): Starchy underground vegetable similar to cassava with brown hairy skin and white to grayish flesh, common in the Caribbean and Polynesia. In Hawaii the boiled, pounded taro paste called *poi* is a staple in the traditional diet. The young shoots and large leaves are also eaten (see *Callaloo; Malanga*).

Tarpon: Large silver fish of the herring family found off the coasts of Mexico and Central America.

Teff: See *Millet.*

Tempeh: See *Bean curd.*

Tepary beans: Small, high-protein bean with wrinkled skin. Grows wild in the U.S. Southwest.

Ti: Tropical plant popular in Polynesia (not related to tea). Ti leaves are used to wrap food packets, and the root is eaten and brewed for a beverage.

Tilapia: Small freshwater fish with sweet, firm, white flesh.

Timpsila: See *Breadroot.*

Tipsin: See *Breadroot.*

Tobikko: See *Caviar.*

Tobu, Tofu: See *Bean curd.*

Tomatillo (husk tomatoes, miltomate): Small, light green, tomato-like fruit surrounded by a green or tan papery husk, common in Mexico. The flesh is slightly tart and is eaten cooked, usually in sauces and condiments. Available fresh or canned.

Tremecos: See *Lupine seeds.*

Truffle: Black (French) or white (Italian) fungus found underground. Truffles vary from the size of small marbles to as large as tennis balls and are distinctively flavored, similar to a wild mushroom. Available fresh or canned.

Tuk-trey: See *Fish paste.*

Tuna: See *Cactus fruit.*

Turtle: Popular in Caribbean, Central America, and U.S. South. Diamondback terrapin (*Malaclemys terrapin)* is the primary ingredient in turtle soups of the Atlantic states. Green turtle (*Chelonia mydas)* is a sea turtle, commonly eaten as steaks or stews. Other turtles eaten occasionally (including eggs) are alligator snapping turtle, common snapping turtle, and loggerhead turtle.

Ugli fruit: Citrus fruit that is a cross between a pommelo and a mandarin orange, with a very bumpy yellow-orange skin and a sweet orange-like flavor. Especially popular in Jamaica.

Uni: See *Sea urchin roe.*

Usli ghee: See *Ghee.*

Verjuice: Juice of unripe lemons used in Middle Eastern fare to give a tang to dishes.

Voi rang: See *Jambolan.*

Wasabi: Light green Japanese condiment from root of plant similar to horseradish with a powerful pungency. Available fresh or powdered; green-dyed horseradish often sold as wasabi.

Water chestnut (matai): Aquatic, walnut-size tuber with fibrous brown peel and crunchy, sweet, ivory-colored flesh. Available fresh or canned.

Water convolvulus (kang kong, ong choi, rau muong, water spinach) Plant related to sweet potato valued primarily for its sprouts and young leaves. Natives to China; significant crop in Southeast Asia, Malaysia, and South India.

Watermelon seeds: Seeds often eaten in Africa (called *egusi,* toasted and ground or pounded into meal or paste for thickening soups and stews) and in Asia (toasted as a snack; sometimes flavored or dyed red).

White bean: Three types of white bean are widely used: cannellini (see *Kidney bean*); Great Northern beans, which are large, soft, and mild tasting; and the smaller, firmer navy beans.

White radish: See *Daikon.*

Wild rice: Seeds of a native American grass.

Winged bean: Edible legume called the soybean of the tropics. All parts of the plant are consumed, including the shoots, leaves, flowers, pods and seeds, and tuberous root. The pods are large, from 12 to 24 inches long, and feature wing-like flanges.

Winter melon (dong gwa, petha, wax melon/gourd): Round green-skinned member of the squash family with a waxy white coating and translucent white green or pink flesh. Similar in taste to zucchini, it is used cooked in Chinese dishes. Called fuzzy melon when immature, winter melon when mature. See also *Fuzzy melon.*

Witloof: See *Chicory.*

Wolfberry: See *Matrimony vine.*

Wong bok: See *Napa cabbage.*

Worcestershire sauce: Sauce developed by the British firm of Lea and Perrins including anchovies, garlic, onions, molasses, sugar or corn sweetener, tamarind, and vinegar, among other ingredients.

Yacón (yakon, leafcup): Sweet-tasting root, *Polymnia sonchifolia,* with brown skin and white flesh native to Andes. Eaten throughout South America; in some regions confusingly called *jicama* (See *Jicama*).

Yam (ñame; yampi; cush-cush; mapuey): Tuber with rough brown skin and starchy white flesh (not related to the orange sweet potato called *yam* in the United States). Numerous varieties; may grow quite large, up to 100 pounds. Found in all tropical regions. Yam paste called *fufu* in West Africa.

Yambean: See *Jicama.*

Yard-long bean: See *Long bean.*

Yautia: See *Malanga; Taro.*

Yerba buena: A variety of mint used in some Native American teas.

Yuca: See *Cassava.*

Yucca (Navajo banana): Spiky-leaved desert plant *(Yucca baccata)* with large, pulpy fruit that ripens in summer. Eaten fresh, boiled, baked, or dried into fruit leather.

Zapote (chico, black sapote, naseberry, sapodilla): Drab-colored fruit of the sapodilla tree (which is the source of chicle used in chewing gum). It has granular, mildly sweet flesh, which can be yellow, red, or black. The zapote is a member of the persimmon family. Potato valued primarily for its sprouts and young leaves. Native to China; significant crop in Southeast Asia, Malaysia, and South India.

RESOURCES

In many ways a book like this poses more questions than it answers. Knowledge of cultural foods is neither balanced nor complete. Many interested readers are undoubtedly asking why there is so little research on adaptations of food habits in the United States or why there are such limited data available on certain cultural groups. As stated in the Preface, only the major American cultural groups are presented in this book. Although the authors reviewed many resources, the resulting text is undeniably inadequate in some areas.

Thus, the most urgent question is: "Where to go from here?" Classes, seminars, association memberships, research, client interaction, and community involvement are all useful ways to learn more about cultural foods. In the nutrition field, many departments of dietetics and home economics offer courses in food and culture. Culinary schools often offer similar classes. Nutrition, dietetic, and food service organizations occasionally sponsor seminars on general topics in food habits, as well as on culturally specific diets.

Academic research on foods habits and nutrition is published in a limited number of journals. *American Journal of Clinical Nutrition, American Journal of Public Health, Ethnicity & Disease, Ethnicity & Health, Journal of the American Dietetic Association, Journal of Nutrition Education & Behavior, Journal of Transcultural Nursing,* and *Social Science and Medicine* are a few that frequently feature articles on diet and culture. Food service, hospitality, and restaurant journals sometimes address cultural food issues. Anthropology, folklore, history, home economics, human resources, geography, management, medicine, nursing, psychology, and sociology publications are occasionally good resources as well.

There are numerous books that have contributed to research on foods and food habits. Cookbooks and other popular literature often include anecdotal information of interest. Online Web sites are an additional source of data. Some of the many resources available follow.

Observation and participation in the community, at markets, at festivals and fairs, and at public religious events are also good ways to learn about a population. Community leaders, traditional healers, restaurateurs, and grocers can all contribute to cultural knowledge about foods and food habits. Accumulated experience with a minority population can be an important adjunct to printed research.

The authors encourage food professionals to undertake and publish studies on cultural foods to expand the limited information on the topic. We also hope that all professionals share their diversity experiences through associations, journals, and online nutrition Web sites or list-servers. Such research and communication help us to become more culturally competent and thus become more effective health care providers and educators.

GENERAL BIBLIOGRAPHY

Abala, K. (Ed.) 2003–present. *Food culture around the world series* (14 vols.). Westport, CT: Greenwood Press.

American Dietetic Association/American Diabetes Association. 1998–2000. *Ethnic and regional food practices: A series* (11 booklets). Chicago: Author.

Anderson, E.N. 2005. *Everybody eats: Understanding food and culture.* New York: New York University Press.

Avakian, A.V., & Haber, B. 2005. *From Betty Crocker to feminist food studies: Critical perspectives on women and food.* Amherst, MA: University of Massachusetts Press.

Beardsworth, A., & Keil, T. 1997. *Sociology on the menu: An invitation to the study of food and society.* New York: Routledge.

Belasco, W. (Ed.) 2001. *Food nations: Selling taste in consumer societies.* New York: Routledge.

Bell, D., & Valentine, G. 1997. *Consuming geographies: We are where we eat.* New York: Routledge.

Bryant, C.A., Dewalt, K.M., Courtney, A., & Schwartz, J. 2003. *The cultural feast: An introduction to food and society* (2nd ed.). Belmont, CA: Brooks/Cole.

Caldwell, M., & Watson, J.L. 2005. *The cultural politics of food and eating.* Oxford: Blackwell Publishing Ltd.

Carlson, B. 1997. *Food festivals: Eating your way from coast to coast.* Detroit: Visible Ink.

Civetello, L. 2003. *Cuisine and culture: A history of food and people.* New York: John Wiley & Sons.

Counihan, C.M. (Ed.) 2002. *Food in the USA: A reader.* New York: Routledge.

Counihan, C.M., & Van Esterik, P. (Eds.) 1997. *Food and culture: A reader.* New York: Routledge.

Dalby, A. 2002. *Dangerous tastes: The story of spices.* Berkeley: University of California Press.

Dassanowsky, R.V., & Lehman, J. (Eds.) *Gale Encyclopedia of Multicultural America* (3 vols.). Farmington Hills, MI: Gale Group.

Davidson, A. 1999. *The Oxford companion to food.* Oxford: Oxford University Press.

Farb, P., & Armelagos, G. 1980. *Consuming passions: The anthropology of eating.* New York: Washington Square Press.

Fieldhouse, P. 1995. *Food and nutrition: Customs and culture* (2nd ed.). London: Chapman & Hall.

Gabaccia, D.R. 2000. *We are what we eat: Ethnic food and the making of Americans*. Cambridge, MA: Harvard University Press.

Germov, J., & Williams, L. 1999. *The social appetite: A sociology of food and nutrition*. South Melbourne, Australia: Oxford University Press.

Harris, M. 1998. *Good to eat: Riddles of food and culture*. Long Grove, IL: Waveland Press.

Heiser, C.B. 1990. *Seed to civilization: The story of food*. Cambridge, MA: Harvard University Press.

Hess, J.L., & Hess, K. 2000. *The taste of America*. Chicago: University of Illinois Press.

Hopkins, J. 1999. *Strange foods: Bush meat, bats, and butterflies: An epicurean adventure around the world*. North Claredon, VT: Periplus.

Inness, S.A. (Ed.) 2001. *Pilaf, pozole, and pad thai: American women and ethnic food*. Amherst, MA: University of Massachusetts Press.

Jacobs, J. 1995. *The eaten word: The language of food, food in our language*. New York: Birch Lane.

Katz, S., & Weaver, W.W. (Eds.) 2003. *Encyclopedia of food and culture* (3 vols). New York: Thomson-Gale.

Lee, H.G. 1992. *Taste of the states: A food history of America*. Charlottesville, VA: Howell.

Levenstein, H. 2003. *Paradox of plenty: A social history of eating in modern America* (Rev. ed.). Berkeley: University of California Press.

Livingston, A.D., & Livingston, H. 1993. *Edible plants and animals: Unusual foods from aardvark to zamia*. New York: Facts on File.

MacClancy, J. 1992. *Consuming culture: Why you eat what you eat*. New York: Holt.

McClelland, D.A. 1991. *Good as gold—Foods America gave the world*. Washington, DC: National Museum of History/Smithsonian Institution.

McIntosh, W.A. 2002. *Sociologies of food and nutrition*. New York: Springer Publishing.

Meiselman, H.L. (Ed.) 2000. *Dimensions of the meal: The science, culture, business, and art of eating*. Gaithersburg, MD: Aspen Publishers.

Meiselman, H.L., & MacFie, H.J.H. 1996. *Food choice, acceptance and consumption*. Gaithersburg, MD: Aspen Publishers.

Menzel, P., & D'Alusio, F. 1998. *Man eating bugs: The art and science of eating insects*. Berkeley, CA: Ten Speed Press.

Miller, W.I. 1997. *The anatomy of disgust*. Cambridge, MA: Harvard University Press.

Mintz, S. 1997. *Tasting food, tasting freedom: Excursions in eating, power, and the past*. Boston: Beacon Press.

Montanari, M. 2006. *Food is culture*. New York: Columbia University Press.

Morgan, L. 1997. *The ethnic market guide: An ingredient encyclopedia for cooks, travelers, and lovers of exotic food*. New York: Berkeley.

Newman, J.M. 1993. *Melting pot: An annotated bibliography and guide to food and nutrition information for ethnic groups in America* (2nd ed.). New York: Garland.

Pilcher, J.M. 2005. *Food in world history*. New York: Routledge.

Powers, J.M., & Stewart, A. (Eds.) 1995. *Our northern bounty: A celebration of Canadian cuisine*. Toronto: Random House of Canada.

Purnell, L.D., & Paulanka, B.J. 2003. *Transcultural health care: A culturally competent approach* (2nd ed.). Philadelphia: FA Davis Company.

Rappoport, L. 2003. *How we eat: Appetite, culture, and the psychology of food*. Toronto: ECW Press.

Roberts, C. 1992. *Cultural perspectives on food and nutrition*. Beltsville, MD: National Agricultural Library.

Rundle, A., Carvalho, M., & Robinson, M. 2002. *Cultural competence in health care*. New York: Jossey-Bass.

Satcher, D., & Pamies, R.J. 2005. *Multicultural medicine and health disparities*. New York: McGraw-Hill Professional.

Schivelbusch, W. 1992. *Taste of paradise: A social history of spices, stimulants and intoxicants*. New York: Pantheon.

Shepard, S. 2000. *Pickled, potted, and canned: How the art and science of food preserving changed the world*. New York: Simon & Schuster.

Shortridge, B.G., & Shortridge, J.R. (Eds.) 1998. *The taste of American place*. Lanham, MD: Rowman & Littlefield.

Simoons, F.J. 1994. *Eat not this flesh: Food avoidances from prehistory to the present* (2nd ed.). Madison: University of Wisconsin Press.

Smith, A. (Ed.) 2004. *Oxford encyclopedia of food and drink in America*. New York: Oxford University Press.

Sokolov, R. 1991. *Why we eat what we eat: How Columbus changed the way the world eats*. New York: Simon & Schuster.

Spector, R.E. 2003. *Cultural diversity in health and illness* (6th ed.). New York: Prentice-Hall.

Tannahill, R. 1995. *Food in history*. New York: Three Rivers Press.

Trager, J. 1995. *The food chronology*. New York: Holt.

Van Wyk, B.E. 2005. *Food plants of the world: An illustrated guide*. Portland: Timber Press.

Vissar, M. 1991. *The rituals of dinner: The origins, evolution, eccentricities and meaning of table manners*. New York: Grove Weidenfeld.

Whit, W.C. 1995. *Food and society: A sociological approach*. Dix Hills, NY: General Hall.

Wilson, D.S., & Gillespie, A.K. 1999. *Rooted in America: Foodlore of popular fruits and vegetables*. Knoxville, TN: University of Tennessee Press.

Wood, R.C. 1995. *The sociology of the meal*. Edinburgh: Edinburgh University Press.

Zibart, E. 2001. *The ethnic food lover's companion*. Birmingham, AL: Menasaha Ridge Press.

PERIODICALS

Flavor and Fortune: Dedicated to the Art and Science of Chinese Cuisine. Expert articles, reviews, and recipes on Chinese food and cooking. Published quarterly by the Institute for the Advancement of the Science and Art of Chinese Cuisine (PO Box 91, Kings Park, NY 11754; www.FlavorandFortune.com).

Food, Culture & Society. A multidisciplinary, international approach with an emphasis on the social aspects of food and food habits.

Published by the Association for the Study of Food and Society (ASFS). Journal available with membership (http://food-culture.org/).

Food and Foodways: Explorations in the History and Culture of Human Nourishment. Interdisciplinary research on the history and culture of human nourishment. Quarterly peer-reviewed journal (phone: 800-345-1420; fax: 215-625-8914; www.tandf.co.uk/journals/titles/07409710.html).

Food History News. Food historians tackle subjects in America's culinary past in this newsletter. Published quarterly by S.L. Oliver (19061 Main Road, Islesboro, ME 04848; http://foodhistorynews.com).

Gastronomica: The Journal of Food and Culture. This gorgeous quarterly offers provocative articles on every aspect of cultural foods, from history, sociology, anthropology, and geography to literature, poetry, art, and film. (University of California Press–Journals Division, 2000 Center St. #303, Berkeley, CA 94704-1223; www.Gastronomica.org).

ONLINE RESOURCES

Ethnic

Asian and Pacific Islander Health Forum (http://www.apiahf.org): Health fact sheets on Asian and Pacific Islander groups; links to other health sites.

Cultural and Ethnic Food and Nutrition Education Materials: A Resource List for Educators (http://www.nal.usda.gov/fnic/pubs/bibs/gen/ethnic.html): annotated site run by the Food and Nutrition Information Center (FNIC) of the United States Department of Agriculture.

EthnoMed (http://ethnomed.org/ethnomed/): An ethnic medicine guide from the Harborview Medical Center, University of Washington. Provides somewhat inconsistent information on the groups profiled, but everything presented is reviewed by members of the target community for accuracy.

Indian Health Service (http://www.ihs.gov): Information about health programs and resources for American Indians and Alaska Natives.

MedlinePlus (www.nlm.nih.gov/medlineplus/): Government site is a great resource for quick updates on African-American, Asian (including Pacific Islander), Hispanic, and Native American (including Alaska Native) health issues. Each listing (under Health Topics) includes latest news, prevention, research, diseases and conditions, organizations, statistics, and information specific to seniors, teens, and women.

Multicultural Health Clearinghouse (http://slic.njstatelib.org/Research_Guides/Health_Resources/Cultural_Health.php): An exceptional site run by the University of Illinois with information on U.S. ethnic group health needs and links to other cultural health sites.

National Alliance for Hispanic Health (www.hispanichealth.org): Health information of interest to Latinos. Catalog of resource materials offers several useful brochures.

New Mexico State University Transcultural and Multicultural Health Links (http://web.nmsu.edu/~ebosman/trannurs/): Listing of sites for numerous religious, ethnic, and special populations.

Refugee Health/Immigrant Health (https://bearspace.baylor.edu/Charles_Kemp/www/): This site is one of the best for exploring the traditional health beliefs and practices of recent immigrant populations.

World Food Habits Bibliography (http://lilt.ilstu.edu/rtdirks): An outstanding listing of research in the field of food and culture. Search by region or topic (such as eating attitudes, festivals and feasting, and taboos).

Religious

Buddhism and Medical Ethics (www.changesurfer.com/Bud/BudBioEth.html): An introduction to Buddhist perspective on abortion, death and dying, and euthanasia with numerous links.

Catholic Encyclopedia (http://newadvent.org/cathen): This comprehensive resource includes articles on feasting, abstinence, and fasting with detailed histories.

India Divine (www.indiadivine.com): Comprehensive listing of articles on philosophy, mysticism, meditation, alternative health practices, scriptures, and more.

Islamic Information Page (http://www.missionislam.com/knowledge/index.htm): A well-organized, comprehensive site providing links on all aspects of Islamic life, from faith and family to health, nutrition, and medical issues.

Judaism and Jewish Resources (http://shamash.org/trb/judaism.html): Well-organized resource listing of Jewish Web sites, with links including kosher organizations, kosher recipes, and kosher wines, as well as those on religion and holidays. For information on Jewish genetic diseases (with information on Crohn's, Gaucher's, ulcerative colitis, etc.) see the Mazornet Guide (www.mazornet.com/genetics/).

Orthodox Christian Information Center (http://www.orthodoxinfo.com/): Provides extensive guidance on fasting. Click on "Church (Old) Calendar" within the text to get more information on the Orthodox calendar.

Seventh-Day Adventist Church (http://www.adventist.org): The official site of the faith; the church manual provides information on health and temperance under the standards for Christian living.

Botanical/Alternative Health

A Mini-Course in Medical Botany (www.ars-grin.gov/duke/syllabus/): Excellent source of information on phytochemicals, ethnic plant uses, GRAS botanicals, and more, with a link to the search site to obtain data by plant name or active ingredient.

Herbal Remedies Index (www.pccnaturalmarkets.com/health/Index/Herb.htm): Lengthy listing of botanical remedies with description, traditional use, and helpful information on activity and contraindications. A commercial site.

HerbMed (www.amfoundation.org): Database maintained by the Alternative Medicine Foundation with information on evidence for activity, warnings, preparations, mixtures, and mechanisms of action. An enhanced version requires subscription, though daily use rates are available.

Jiva Ayurvedic (www.ayurvedic.org/ayurveda/): A site that offers extensive information on beliefs and practices. Includes brief research reviews and an online consultation system.

National Center for Complementary and Alternative Medicine (http://nccam.nih.gov): This center run by the National Institutes of Health conducts research and disseminates information on complementary and alternative medical practices. Their CAM on PubMed (www.nlm.nih.gov/nccam/camonpubmed.html) limits your search to their citation index.

Tropical Plant Database (www.rain-tree.com/plants.htm): Search by common or scientific name, condition, plant properties, or recorded ethnobotanical uses. A commercial site.

Foods and Cooking

Culinary History Timeline (www.foodtimeline.org/food1.html): This fun links site is set up chronologically—from articles on prehistoric diet to the slow food movement with numerous ethnic and cultural topics. A companion site (www.foodtimeline.org) provides links on specific foods.

Epicurus (www.epicurus.com): Sponsored by a leading food and hospitality industry consultant, this site features a monthly e-zine with short, informative articles on foods, beverages, herbs, and spices; interviews with chefs and hoteliers; and breaking food news, as well as tantalizing recipes. Information on some ethnic ingredients and cultural food events.

Food Composition Resource List for Professionals (http://riley.nal.usda.gov/nal_display/index.php?tax_level=1&info_center=4&tax_subject=279): These food composition resources are all available from the National Agricultural Library, including books, U.S. government publications, software databases, and journals. There is also a contact for assistance with specific requests and a bulletin board.

Food, Nation and Cultural Identity (http://www.bl.uk/learning/citizenship/foodstories/Accessible/foodnationidentity/foodnationidentity.html): A British website that explores the cultural diversity of food within the UK and be able to listen to actual stories of individuals about their experiences trying new foods.

Food Museum (www.foodmuseum.com): Described as a "one stop source for food exhibits, news/issues, resources, food history, answers to your food questions, book reviews and just plain fun." Good for history information.

Foodbooks.com (www.foodbooks.com): "Serious books for serious cooks" and an outstanding collection of food history volumes as well.

Fost—Social and Cultural Food Studies (http://www.vub.ac.be/FOST/fost_in_english/): Well organised website that organizes information on Cultural Food Studies over several academic disciplines.

Gernot Katzer's Spice Pages (http://www.uni-graz.at/~katzer/engl/): Well-maintained site is run by an Austrian chemist. Over 100 herbs and spices listed, indexed alphabetically, by region, and by part used in cooking; glossary of spice mixtures. Great links.

International Food Composition Directory (http://www.fao.org/infoods/directory_en.stm): Resources of the International Network of Food Data Systems (INFOODS) under the auspices of the Food and Agricultural Organization (FAO) of the United Nations, including listings for published and online food composition tables worldwide.

RecipeSource.com (www.recipesource.com): This site has a searchable online archive of recipes that has catalogued over 70,000 recipes from throughout the world. Search by ethnic group or type of dish. Excellent selection, though recipes in ethnic categories are often not traditional.

Sally's Place (http://www.sallys-place.com): An overview of several international cuisines can be found at this site. Recipes, restaurants, and ingredient sources are listed in some. Another useful page on this site lists professional food organizations, including descriptions and addresses.

PHOTO CREDITS

i [tl] Matka_Wariatka/Shutterstock.com,
[cl] Vicky German/Shutterstock.com,
[cr] Jiri Hera/Shutterstock.com,
[tr] ALNOOR/Shutterstock.com
2 © Tom McCarthy/PhotoEdit, Inc.
6 © Peter Menzel/Stock, Boston Inc.
11 © Robert Brenner/PhotoEdit, Inc.
16 © Kostenko Maxim/Shutterstock
25 © Becky Luigart-Stayner/Terra/Corbis
36 © Michael Newman/PhotoEdit, Inc.
42 © Mitch Hrdlicka/Photodisc/Getty Images
46 © Tony Freeman/PhotoEdit, Inc.
48 © Mitch Hrdlicka/Photodisc/Getty Images
49 © Wolfgang Spunbarg/PhotoEdit
57 © Tom McCarthy/PhotoEdit, Inc.
61 © Michael Newman/PhotoEdit
63 © Steve Kaufman/Documentary/Corbis
64 © Jeff Dunn/Index Stock/Photolibrary
70 © David Young-Wolff/PhotoEdit
85 © Michael Newman/PhotoEdit, Inc.
87 Courtesy of Grossich and Bond, Inc.
92 © Nigel Blythe/Cephas Picture Library/Alamy
97 © United Nations/J. Isaac
103 Photo by Laurie Macfee
106 © Lionel Delevingne/Stock, Boston, Inc.
111 From A Pictorial History of the American Indian Copyright © 1956 by Oliver La Farge Reprinted by permission of Frances Collin, Literary Agent
113 © Bettmann/CORBIS
115 © Bettmann/CORBIS
133 Photo by Laurie Macfee
137 © Photodisc/PhotoLink/Getty Images
138 Courtesy of Grossich and Partners
141 © K Sanchez/Cole Group/Getty Images
143 © Robert Brenner/PhotoEdit, Inc.
144 Courtesy of the Louisiana Office of Tourism
147 Photo by Laurie Macfee
152 © M Lamotte/Cole Group/Getty Images
153 © Photodisc/PhotoLink/Getty Images
155 © Cole Group/Getty Images
157 © PhotoLink/Photodisc/Getty Images
164 Photo by Laurie Macfee
170 AP/World Wide Photos
175 © Eising/Photodisc/Getty Images
177 © Lottie Davies/Getty Images
178 Courtesy of Florida State News Bureau
181 Courtesy Pennsylvania Dutch Visitors Bureau
188 Courtesy of Denmark Cheese Association
189 © Susanna Blavarg/Getty Images
199 Photo by Laurie Macfee
201 Photo courtesy of the World Health Organization/P. Almasy
204 © Bonnie Kamin/PhotoEdit, Inc.
207 © Merritt Vincent/PhotoEdit, Inc.
226 Photo by Laurie Macfee'
227 San Antonio Light Collection, UTSA's Institute of Texan Cultures, No. L-1535
229 © PhotoLink/Getty Images
245 Mexican School/Bridgeman Art Library/ Getty Images
249 Courtesy of the Florida Division of Tourism
263 Courtesy of the Metro Dade Department of Tourism
267 © Francoise de Mulder/Corbis
272 Courtesy of the Florida News Bureau
289 © Yoshio Tomii/SuperStock
306 Photo by Kathy Sucher
310 Photo by Laurie Macfee
320 © Wellcome Library, London, Museum No: A164587
321 Kevin Fleming/Encyclopedia/Corbis
324 Photo by Laurie Macfee
328 © Gary Conner/PhotoEdit, Inc.
337 Michael Freeman/Documentary/Corbis
338 © Gary Conner/Photo Edit, Inc.
355 © Michael Newman/PhotoEdit, Inc.
359 Photo by Laurie Macfee
361 Photo by Kathy Sucher
365 © PhotoLink/Photodisc/Getty Images
372 David Weintraub/Stock, Boston, Inc.
378 © Bettmann/CORBIS
381 © Isabelle Rozenbaum/PhotoAlto Agency RF Collections/ Getty Images
383 © Porterfield/Chickering/Photo Researchers, Inc.
397 Van Bucher/Photo Researchers, Inc
405 Photo by Laurie Macfee
407 © Dave Bartruff/Documentary/Corbis
409 © R. & S. Michaud/Woodfin Camp & Associates

419 Courtesy of World-Health Organization/
P. Merchez
436 Photo by Laurie Macfee
440 Courtesy Raga Restaurant, New York, NY, Taj Group of Hotels
443 © Staffan Windstrand/Encyclopedia/Corbis
446 Courtesy Raga Restaurant, New York, NY Taj Group of Hotels
466 Jeff Greenberg/Visuals Unlimited, Inc
469 © Mia Foster/PhotoEdit, Inc
474 © 2002 Wisconsin Department of Tourism
484 © Matthew Klein/Royalty-Free CORBIS
487 © Michael Lamotte, Cole Group/PhotoDisc/Getty Images
492 Louisiana Tourism
493 © J. Griffis Smith/TxDOT
499 © Stockbyte/Picture Quest/Jupiter Images
502 © Paul Barton/CORBIS
503 © Buddy Mays/CORBIS

Color Inserts

Starch Foods, page 1:
a) © Jupiter Images/comstock.com
b) © Becky Luigart-Stayner/CORBIS
c) © Danny Lehman/CORBIS
d) © Wolfgang Kaehler/CORBIS
e) © Andrew Unangst/Spirit/Corbis
f) © Siede Preis/Photodisc/Getty Images

Protein Foods, page 2:
a) © Chris Shorten/Cole Group/Photodisc/Getty Images
b) © 2002 Wisconsin Department of Tourism
c) Courtesy of United Soybean Board
d) © Pat O'Hara/CORBIS
e) © Comstock/Jupiter Images
f) © S. Meltzer/PhotoLink/Getty Images

Vegetables, page 3:
a) © Royalty-Free PhotoDisc/Getty Images
b) top, © Royalty-Free PhotoDisc/Getty Images; middle, © R.F. Images; bottom, © Cole Group/Getty Images
c) © Jack Star/PhotoLink/Getty Images
d) © Dan Lamont/CORBIS

Fruits, page 4:
a) © Dean Uhlinger/Royalty-Free CORBIS
b) © Andrew Unangst/Royalty-Free CORBIS
c) © L. Hobbs/PhotoLink/Getty Images
d) © Wolfgang Kaehler/CORBIS
e) © Kevin R. Morris/CORBIS
f) © E. Carey/Cole Group/Getty Images

INDEX

6313 004

CINCINNATI STATE LIBRARY